Birkhäuser

Science Networks. Historical Studies

Founding Editors

Erwin Hiebert
Hans Wußing

Volume 68

This series is devoted to historical studies in the fields of mathematics, physics, and astronomy, including their applications. It is mainly composed of monographs, although it may occasionally include critical editions of primary sources and biographies of important actors. It aims to attract a multiple audience including scientists, historians, philosophers, and graduate students throughout the academic world. Submitted projects are examined by an international and multidisciplinary committee of experts. The publication language is preferentially English, and the diffusion is international.

Markus Ehberger

Representing the Unobservable

The Formation of the Virtual Particle Concept in the Practice of Theory (1923–1949)

Birkhäuser

Markus Ehberger
Technische Universität Berlin
Institut für Philosophie, Literatur-,
Wissenschafts- und Technikgeschichte
Fachgebiet Wissenschaftsgeschichte
Berlin, Germany

ISSN 1421-6329 ISSN 2296-6080 (electronic)
Science Networks. Historical Studies
ISBN 978-3-032-09187-1 ISBN 978-3-032-09188-8 (eBook)
https://doi.org/10.1007/978-3-032-09188-8

Zugl.: Berlin, Technische Universität, Diss., 2024 u. d. T. From virtual oscillators to virtual transitions to virtual particles: Practices and representations in the formation of the virtual particle concept.

This work was supported by Technische Universität Berlin, RWTH Aachen and the Deutsche Forschungsgemeinschaft.

Mathematics Subject Classification: 01-XX, 01-02, 81-83

Cover illustration: From Waller Ms de-00215, August Beer: Über die Correction des Cosinusgesetzes bei der Anwendung des Nicol'schen Prismas in der Photometrie, after 1850. With friendly permission by the Waller Manuscript Collection (part of the Uppsala University Library Collections).

This book is published under the imprint Birkhäuser, www.birkhauser-science.com by the registered company Springer Nature Switzerland AG
The registered company address is: Gewerbestrasse 11, 6330 Cham, Switzerland

Acknowledgements The dissertation on which this book is based has been, to put it mildly, a long-term project. In all generality, I want to thank everyone involved, whether in academic circles or in my private life, for their incredible patience and long-lasting support!

Friedrich Steinle, my *Doktorvater*, and Adrian Wüthrich, who also supervised this dissertation, had to chew through a lot of (to quote Wolfgang Pauli) half-understandable, because only half-understood, sentences and arguments. Their criticism and feedback, well formulated, constructive, and to the point, has been very valuable to me. Without their constant input, this work would have never taken the form it has. I appreciate their intellectual guidance and their institutional support, their patience and encouragement to the highest degree. Olivier Darrigol, who took on the task to externally review my dissertation, sharply noticed shortcomings or inconsistencies of my arguments and contributed to clearing them. I want to thank him for his precise and constructive comments. I also valued the remarks by an anonymous referee during the review process of the manuscript very much.

It is also a pleasure to thank the members of the DFG-funded research group *The Epistemology of the LHC* (grant number: FOR 2063), especially the members of Project A1 (which my research was part of): Robert Harlander, Jean-Philippe Martinez, and Daniel Mitchell. Their comments on presentations and chapters of this dissertation, reading as well as discussion sessions shaped my understanding of the history and the concept of the virtual particle. I want to thank Alexander Blum for letting me join an excellent course on the history of quantum electrodynamics for physicists and for the possibility to present some of my ideas and findings at the seminar of his research group. I also want to thank anyone with whom I discussed my work, who has attended one of my talks, or encouraged my research, especially Arianna Borrelli (whose writing and thinking has shaped my understanding of conceptual development), Martin Jähnert, Christian Joas, and Marta Jordi Taltavull.

Although he can tragically no longer read these lines, Christian Forstner deserves special mentioning. I owe him a lot, for example, the possibility of financing myself during parts of the writing process of this dissertation or the publication of first results of my investigation in a collected volume he edited together with Mark Walker. I am thankful for the advices he gave and the opportunities he offered me. I will always and gratefully keep him in good memory.

I want to thank Nina Krampitz, who helped me in all the organizational details during my employment at the TU Berlin. I also want to thank my current employer and especially my direct superior, Ulf Hashagen, for allowing me to arrange my working hours as freely as organizational restrictions allowed in the last months of my writing process.

Rory Grennan and the team of the Special Collections and Archives at the Florida State University Library as well as Karl Grandin and the team of the Center for History of Science at the Royal Swedish Academy of Science were incredibly helpful by providing me with digitized versions of the archival material of the P. A. M. Dirac Papers and the Ivar Waller Archives, respectively. The same holds for

Antonio Augusto Passos Videira and the Guido Beck Archives at the University of Rio de Janeiro. Finally, I want to thank the staff of the library of the Max Planck Institute for the History of Science (Berlin) for providing me access to the microfiches of the AHQP and the BSC.

My family and friends as well as my partner's family have always been sympathetic and understanding during this (at times and also for them) frustratingly long endeavor. Their unconditional support has made this work possible. I especially want to thank my partner, Lara, who has never been the biggest fan of my idea to start a dissertation. Although not the least interested in the topic I was investigating, she is among the persons who heard me rambling on about it the most. But she respected my decision, carried the burdens with me, and always kept a watchful eye on me and the workload I could take. I feel deeply thankful and admire her endurance! Last but not least, I am grateful that we have our Dachshund mixed breed, Hanni, in our life: Having racked my brains for hours over abstract entities in the hands of historical actors, the mental and physical recreation of a long walk and the feeling of being in this exact moment and this exact place, the presence in the present a dog can make you feel, was invigorating. Thank you all for keeping me going!

Funding This work was supported by the research unit "The Epistemology of the LHC," funded by the Deutsche Forschungsgemeinschaft (DFG) under the grant number FOR 2063.

We acknowledge support by the Open Access Publication fund of TU Berlin.

Competing Interests The author has no competing interests to declare that are relevant to the content of this manuscript.

Contents

List of Figures

Chapter 1
Introduction

The aim of this book is to reconstruct the historical development of the concept of virtual particles, roughly from the early 1920s to the late 1940s. These particles, although (or maybe because) they form an integral part of the conceptual framework of quantum electrodynamics (QED) and quantum field theory (QFT), are rather strange things. At least this is what a first, superficial look at the concept, its role in modern QED and QFT, and the corresponding philosophical literature suggests. Virtual particles have strikingly counter-intuitive features: For instance, they are unobservable by definition and, as a popular narrative has it, violate energy conservation for the short time of their existence. Actually, the mainstream of philosophical research suggests that these particles do not exist. Yet, in the standard, perturbative approach to QED, without virtual particles no interaction would take place at all. One might even argue that the physical parameters of the theory, the electric charge and the electron mass, are defined by the effects of virtual particles.

The centrality of the concept for the theoretical framework is also reflected in the diagrammatic techniques applied by physicists. The best known representation of quantum electrodynamical effects, their calculation, or both, depending on who you ask, are Feynman diagrams. In Fig. 1.1, you see the first Feynman diagram published by its inventor, Richard Feynman. It describes the first approximation to the interaction between two electrons, which Feynman himself conceived of as the "fundamental interaction" (Feynman 1949a, 772) of QED. At its center, a virtual photon is represented, diagrammatically by the wavy line and mathematically by the function $\delta_+(s_{56}^2)$.

Through this mode of representation, the virtual particle made it into the realms of popular culture and art, and I would suspect that nearly every person with only a slight interest in particle physics has already seen a representation of a virtual particle. Feynman diagrams, and therefore virtual particles, figured regularly in the sitcom *The Big Bang Theory*, they are used in sculptures and prints, the best-known of which being created by Edward Tufte, and they were incorporated in the redesign of the

M. Ehberger, *Representing the Unobservable*, Science Networks. Historical Studies 68,
https://doi.org/10.1007/978-3-032-09188-8_1

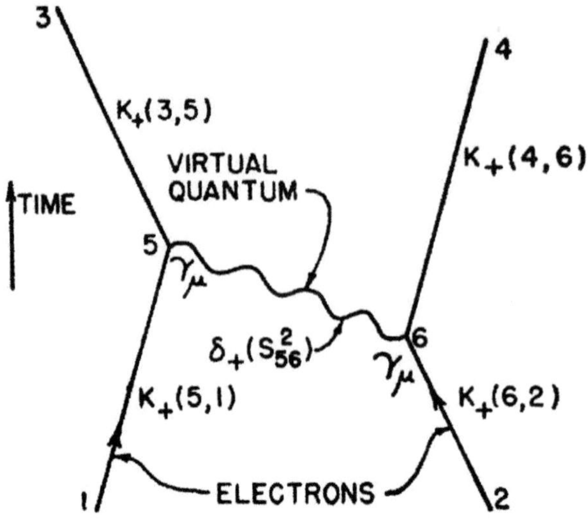

Fig. 1.1 The "fundamental interaction" of QED: the exchange of one photon between two electrons, i.e. the first order contribution to electron-electron scattering. Reprinted figure with permission from Feynman, Richard, "Space-time approach to quantum electrodynamics", *Physical Review* 76, 769–789, here 772, 1949, https://doi.org/10.1103/PhysRev.76.769 (cited as Feynman (1949a) in the bibliography). Copyright 2025 by the American Physical Society

windows of the St. Nicolai church in Kalkar (North-Rhine-Westphalia, Germany) in the early 2000s (Pawlak and Hartmann 2004). In Munich, there is even an unofficially named "Feynman Platz": The road layout resembles the Feynman diagram of the Compton effect (see Fig. 1.2). Put into physics jargon, the part of the Kanalstraße between the corners of Liebherrstraße and Mannhardtstraße is virtual.[1]

Consequently, from a first look at the concept there is quite some tension within it: Virtual particles seem to be conceptually central to the practice of physicists and the explanations given by them, yet philosophers tend to deny their reality. Within the physics community, the stances towards virtual particles and Feynman diagrams seem also to be diverging. Judgements stretch from virtual particles being "indeed real particles" (Kane 2006), "disturbance in a field that is *not* a particle" (Strassler 2011, emphasis in the original), or only "(intuitive imagery for) internal lines of Feynman diagrams" (Neumaier 2016).

Quite curiously, not only the virtual particle itself but also the polyphony of stances towards its ontological status has made it into popular culture: Recently, an article was featured on *Motherboard*, the technical branch of the life-style magazine *Vice*, that presented the various views of physicists and philosophers under the rather

[1] I do not know the history of the naming of the "Feynman Platz", but one should be aware that one can find such structures in road maps all over the place. The current status of the "Feynman Platz" seems still to be unofficial as it is missing from the official street map of the city of Munich: https://geoportal.muenchen.de/portal/stadtplan/ (visited on April 13. 2023).

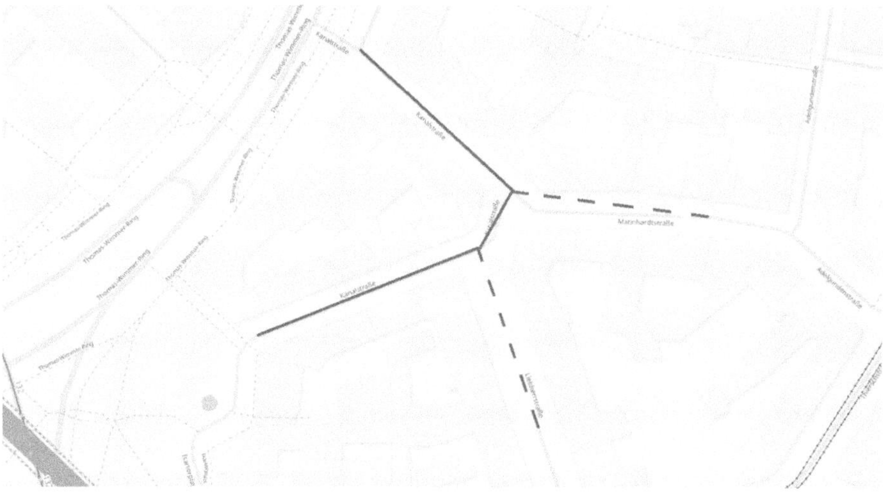

Fig. 1.2 Snapshot of the street-map near Isartor in Munich. The lines indicate how the Kanalstraße (electron), the Liebherrstraße (photon) and the Mannhardtstraße (photon) form the Feynman diagram of the Compton effect. Diagram added by the author. Map data/snapshot from OpenStreetMap (copyright information under openstreetmap.org/copyright)

colloquial title: “OK, WTF are ‘Virtual Particles’ and Do They Actually Exist?” (Rorvig 2020).

1.1 A First Look at Virtual Particles

To understand the concept of virtual particles we have to engage with the theoretical treatment of quantum electrodynamical effects.[2] QED aims at a description of the relativistic and quantum-theoretical interaction of the electromagnetic field and charged particles (described by the electron-positron field). The primary problem of the theory is that one cannot solve its basic equations exactly. To this day, we have no means of calculating what QED is designed to describe directly from the fundamental equations of the theory. But what to do then? How can one solve any problem with a theory that is in its basis to complicated to be solved?

The solution of this difficulty, as applied in QED and large parts of quantum theory, is perturbation theory. Originally invented to deal with analytically unsolvable problems in celestial mechanics it has become an integral part of modern physics. The basic idea behind this method can be stated as follows: You split the equations describing your system into a part which you know how to solve and a part which you treat as a perturbation. The perturbation normally depends on a small parameter, referred to here as ϵ. The procedure, which can take diverse manifestations, yields a

[2] For a more comprehensive evaluation the reader is referred to Sect. 2.1 and the references therein.

series expansion in ϵ (unsurprisingly called "perturbative series") as a representation of the full solution (S) of the problem:

$$S = S_0 + \epsilon S_1 + \epsilon^2 S_2 + \ldots = \sum_{n=0}^{\infty} \epsilon^n S_n. \tag{1.1}$$

In essence, if one calculated the full series to infinity one would arrive at an exact solution.[3] In practice, through the expansion in the small parameter the successive orders get smaller and smaller and the approximate procedure quite often gives a reasonable solution, i.e. sufficient in accuracy to be compared with experiment, after a small number of terms.

In QED, the freely developing quantized electron-positron and electromagnetic fields form the unperturbed parts of the system. The interaction between the two fields is conceived of as the perturbation. This means that in QED you have a theory that gives you a perturbative description of the interaction between the electron-positron field and the electromagnetic field: the higher the order, the more accurate the approximation. Which process one is calculating is modelled through the so-called in- and out-states. One imagines that freely moving particles come in from infinity and go out to infinity, where the interaction is no longer acting. These particles are referred to as real particles: They leave tracks in cloud chambers and are detected by particle detectors.[4]

Now let's bring in Feynman diagrams. Everyone can agree that Feynman diagrams are a representation of the above described perturbative expansion and that they are highly effective in singling out the mathematical terms to be calculated. That the diagram in Fig. 1.1 represents a term in the expansion for electron-electron scattering can be inferred from the external lines, the ones which reach out of the diagram. They represent the above mentioned in- and out-states. In general, electrons are drawn as straight lines and photons as wavy ones. The line in the middle of the diagram, connecting the two electrons and therefore establishing an interaction between them, is a virtual photon. All internal lines, the ones with no connections out of the diagrams, represent virtual particles. Higher order terms involve further virtual particles (see e.g. Fig. 2.3), which are unobservable (reflected by the lines having no connections out of the diagram) and "off shell", i.e. they do not have to obey the relativistic energy-momentum relation while energy and momentum are conserved at the interaction points (vertices).

The diagram in Fig. 1.1 can be read as, note the arrow of time on the left, two electrons coming in, exchanging a (virtual) photon, thereby being scattered, and then going off to infinity. And this is how Feynman actually intended his diagrams

[3] At least this is what one hopes. Things are more complicated for QED and it is likely that the perturbative expansion is actually divergent. See also Sect. 2.1 and the literature referred to therein.

[4] This is a standard conceptualization in QED. I refrain to go into any detailed philosophical discussion of the concept of measurement or detection in particle physics.

to be read.[5] But there are diverse problems with this interpretation. First of all, this is only the first approximation so that higher order terms might be added and they contribute to exactly the same measured process of electron-electron scattering. Even worse, the quantity one calculates after having translated the diagrams into a mathematical expression is a probability or quantum amplitude[6] and only the absolute square of this quantity has a connection to observation. It is not the individual terms of the perturbative series that are of physical importance but the square of a sum of those: Contributions of diagrams with different numbers of virtual particles are superimposed and interfere with each other. This is the basis of the most important argument against a realistic interpretation of virtual particles and Feynman diagrams, the so-called superposition argument.[7]

But virtual particles do not only figure as force carriers in scattering processes. They also turn up in the description of self-energies, i.e. the interaction of a particle with its own field. The given conceptualization is that an electrically charged particle emits and reabsorbs a virtual photon: This effect is actually divergent and posed a huge obstacle to the development of QED until the late 1940s. Similarly, vacuum fluctuation and polarization effects are modelled through the "popping-up" of virtual electron-positron pairs from the vacuum which then mutually annihilate. Again, the calculation of the corresponding diagrams gives a divergent result. The renormalization procedures of the late 1940s can be read as incorporating such effects into the definition of the physical parameters of the theory: the electron mass and its charge. In this sense, a perfectly isolated electron is a concept which no longer has any definite meaning but can only be interpreted by the inclusion of the "cloud of virtual particles" formed around it.

For the moment, this short outline should suffice to exemplify that the intuitive appeal of Feynman diagrams, their reading as representations of physical processes and therefore their parts representing constituents of the physical world is problematic. But it should also point to the fact, that without virtual particles no interaction would be conceivable in the standard approach to QED, that the definition of parameters of the theory depends on their "presence", and that the concept is deeply connected to the theoretical practice of perturbation theory. I will reserve a more detailed discussion of perturbation theory in QED and the concept of the virtual particle, with all its philosophical quandaries, for Chap. 2, where I will also elaborate on my understanding of a "concept". But, one might ask, why should I do so? Why devote the larger part of a chapter to the modern concept and the philosophical debate around it in a book that is primarily interested in the historical genesis of the concept?

---

[5] Compare Feynman's own reading of the diagram and the connected equation in Feynman (1949a, 772–773).

[6] I will use the terms "probability amplitude" and "quantum amplitude" synonymously throughout this book. Most often, I will use the term "quantum amplitude" to emphasize the quantum-theoretic nature of these amplitudes.

[7] See also Sect. 2.2.1.

1.2 The Necessity of Constructing a Historiographic Framework for the Study of Concepts

To answer this question[8] I want to refer the reader to a short but very insightful quote by philosopher of science Vasso Kindi:

> How we understand concepts affects the way we study them historically. (Kindi 2012, 37)

In the initial question guiding this historical research, namely "How did the concept of a virtual particle emerge?", the concept occurs twice: once in the concrete form of the virtual particle and once in the category "concept" itself. Similarly, we can understand the "understanding of concepts" mentioned in the quote in two different ways. On the one hand, we need to be clear about how we think of the investigated concept within the practice of modern-day science. Although the historian's job is to unravel the understanding and use of the studied concept by the historical actors as well as its trajectory, having an own understanding is inescapable. Instead of sweeping such personal conceptions under the rug, making them explicit helps the audience of the historian to understand his or her choices. In addition, we might actually use them as our understanding of the concept, its features and applications give indications on the relevant lines of the historical development: In the case of the virtual particle, the short outline above already suggests a closer look on perturbation theory and the description of the interaction of charged particles, self-energies and scattering phenomena.

On the other hand, the understanding of the category "concept" itself has high relevance for the story to be told. This is the sense which Vasso Kindi (2012, 37–42) intended. For example, if an author understood concepts as elastic things, formed through their use by historical actors, this author will most probably arrive at a different historical reconstruction than another author who is convinced that a concept needs some hard core, some sort of stable meaning throughout its development. Although the two authors might use the same historical resources, the emphasis put, in the example on either change or continuity, will always be implicit in the work growing out of the historical investigation.

I think there is little doubt about the fact that reference to the present is unavoidable when engaging with the history of science and that we should not confound presentism with Whiggism or anachronism.[9] In writing the history of concepts that are still in use, a presentist perspective becomes even more prominent since the

[8] In slightly modified and shortened form this section has already been published as Ehberger (2024, Sect. 1, pp. 281–283).

[9] For a very convincing argument about the necessity of presentism and its possible fruitfulness in the history of science, see Alvargonzález (2013) or Loison (2016, Especially Sects. 2 and 3). While Alvargonzález does not discern between a Whig history and presentism, Loison characterizes anachronism and Whiggism as an abuse of possibly fruitful applications of presentist ideas.

reference point of the question guiding the historical work (How did the (modern) concept of a virtual particle develop?) is the present. Instead of sweeping such forms of presentism under the rug and purporting to write an "objective" history, I think it is best to make my own stance as explicit and therefore the particular choices taken as transparent as possible.

Although the above may suggest a one-way street from the modern day understanding of the author to the historical reconstruction, it is important to note that this is far too simplistic. My own work on the concept of the virtual particle rather suggests to me a strong interrelation between the historian's understanding of

1. the modern version of the concept under study,
2. historical versions or what we construct as concepts connected to its historical development,
3. the historical development with all its intricacies and
4. the category "concept" itself.

As already described, I am convinced that our modern understanding of the concept under study and the category of a concept we initially have in mind will unavoidably inform our historical analysis. But also the historical versions of the concept under study, their genesis, their place in scientific discussions and practices unravelled in our historical investigation will reflect on our modern-day understanding of the (category of a) concept. I consider it necessary for a historian to pay great attention to finding some kind of balance between these aspects of the history to be written. In essence, this balance is constructed iteratively by the author in revisiting and questioning all of these aspects in the course of the investigation.[10]

Naturally, I am not the first one to explicate that the past (and our reconstruction of it) is structured by the present (state of science). I emphasize this fact at this point because I hold that it is particularly important for the historical development of concepts and an approach which is interested in scientific practice. Just to indicate one possible reference point where similar thoughts have been expressed, I want to mention Hans-Jörg Rheinberger's reflections on the contributions by George Canguilhem and Jacques Derrida to the project of historical epistemology. Rheinberger connected his insights strongly with his analytical categories of experimental systems and epistemic things but also drew general historiographic conclusions.[11] Rheinberger emphasized that the history of science will always have to be rewritten under the premises of the current state of knowledge. As he put it, the historian of science who is interested in the epistemic process has to take a "position of a reflected anachronicity [Position einer reflektierten Anachronizität]" (Rheinberger

[10] Here, I have something in mind like the relation between history and philosophy as it was proposed by Chang (2012, especially Sect. 8.4). When put to work several times and by the same person, this can be conceived of as coming close to the iteration process mentioned above.

[11] See Rheinberger (2013, Chap. 5) and Rheinberger (2002, Chaps. 4 and 10).

2002, 202) and we have to acknowledge that, since at least our conception of the past is (partly) structured by the present (and vice versa), history has to be rewritten from time to time in the "iterative business of the historian [iterativen Geschäft des Historikers]" (Rheinberger 2002, 194).[12]

In my attempt to provide what Rheinberger has called a "reflected anachronicity", I will argue in Sect. 2.3 that we are best of in conceiving of virtual particles and the concepts we will encounter in its development as physical-mathematical concepts, the historical development of which should be studied through their verbal and visual representations, the connected mathematical and theoretical practices and their role in the argumentation of the historical actors. But before going into this historio-philosophical discussion, let's see what we already know about the development of the virtual particle concept from the secondary historical literature.

1.3 What Do We Already Know About the Development of the Virtual Particle Concept?

The time span I will investigate in this book stretches from the early 1920s and the genesis of the virtual oscillator concept to the late 1940s, invariant perturbation theory and Feynman's representation of virtual particles in his diagrammatic technique. Such an extended study of the development from the first signs of virtuality in the old quantum theory[13] to a (nearly) modern representation of virtual particles in Feynman diagrams is as yet not available. While the concepts mentioned above figure prominently in the historical literature because of their centrality in the respective theoretical frameworks, a historical investigation of their conceptual (dis-)connections is only partly given and, at times, the literature is equivocal about the advent of these concepts and their relations (see below).

[12] I do not consider it a coincidence that the work of the historian-philosophers Rheinberger builds his insights on was concerned with aspects and branches of the history of science that are important for this investigation, i.e. conceptual development (Canguilhem) and the role of representational formats (the written word in the case of Derrida). All translations, if not otherwise indicated, by the author.

[13] This is simply meant as a chronological umbrella term for theoretical investigations using quantum ideas prior to the advent of matrix mechanics. It should not indicate that any kind of shared and general theoretical framework existed.

The standard references to the history of quantum mechanics and quantum field theory[14] surely form an invaluable basis for this project. Yet, a discussion of the development of the virtual particle is mostly absent therein: For example, Olivier Darrigol (1982) and Sam Schweber (1994) of course mentioned the concept and use it in their reconstruction but made next to no comment about its genesis. Even Hentschel (2017, Sect. 3.12), who included the virtual photon as one of the "semantic layers" in his study of the photon concept, discussed the virtual photon's historical origin only superficially and went straight to Feynman diagrams for its evaluation. Similarly, Tian Yu Cao (1997) referred to the virtual particle concept only in passing in his *Conceptual development of 20th century field theory*. The latter, in turn, led philosopher of physics Tobias Fox to state that an exclusion of the virtual particle from the historical literature "is quite adequate, since there was no conceptual development of 'virtuality' in the history of physics, and therefore it need not be subsumed under the scope of this [Cao's] book" (Fox 2008, 44). Unsurprisingly, I think that Fox's premise is mistaken. In the following, I will show how we can construct a broad picture of the development of concepts that (would at some point) carry the epithet "virtual"[15] from the currently available secondary literature.[16]

In essence, there are five interrelated concepts to be discussed: virtual oscillators and the virtual radiation field, virtual transitions and virtual states, and virtual particles. While the virtual oscillators and the virtual radiation field are clearly perceived of as (semi-)classical concepts in the literature, the same cannot unambiguously stated about virtual transitions or states. The virtual particle, on the other hand, is only connected to quantum electrodynamics or quantum field theory. As a matter of fact, in the literature and in my own reconstruction to follow, the virtual transitions and the connected concept of an intermediate (or virtual) state, will take the role of a bridging element in the conceptual development.

---

[14] For the history of quantum mechanics, see for example Hund (1967), Jammer (1966) and Darrigol (1992). Besides providing translations of important papers for the genesis of quantum mechanics, van der Waerden (1967) also gives an account of the developments in his introductory essay. The most comprehensive studies on the development of quantum electrodynamics can be found in Schweber (1994) (focusing on the development of the late 1940s), Darrigol (1982) (stretching over the period before renormalization) and Pais (1986, Chaps. 15–18). Dresden (1993) focuses on the history of renormalization. In Miller (1994a) translations of important papers for the development of quantum electrodynamics can be found. Miller also provides an overview of the developments throughout the late 1920s and 1930s. Reminiscences and accounts of the historical actors themselves are part of Fierz and Weisskopf (1960), Mehra (1973), and Brown and Hoddeson (1983). The most comprehensive account on the development of the theory of nuclear forces is probably given by Brown and Rechenberg (1996).

[15] I will focus on concepts that were used within quantum theory and exclude classical concepts such as "virtual work" or "virtual displacements".

[16] This study had already been finished when Aaron Sidney Wright's book *More than nothing. A history of the vacuum in theoretical physics, 1925–1980* was published (Wright 2024). Therefore, a critical review of Wright's points and its impact on the story to be told was no longer possible. Wright's and my analysis, of course, touch upon the same fields of inquiry and have different overlaps. But our narratives are often complementary, so that I did not rewrite the respective sections, but refer the reader to parts of Wright's work in the given context, most importantly Chap. 7 and Sect. 10.1.

1.3.1 From Virtual Oscillators and Radiation Field to Virtual Transitions and States (1923–1935)

The concepts of the virtual oscillators and the virtual radiation field have already found a lot of historical attention and there is a twofold reason for this.[17] On the one hand, virtual oscillators and the connected concept of *Ersatzoszillatoren* have been introduced in the theoretical discussion of dispersion, which has traditionally been framed as leading the path to Werner Heisenberg's so-called *Umdeutung paper* (Heisenberg 1925), which is generally accredited to have opened the field of matrix mechanics.[18] On the other hand, virtual oscillators and the virtual radiation field are the fundamental concepts of the infamous BKS theory, named after its creators Niels Bohr, Hendrik Anthony Kramers and John Clarke Slater (Bohr et al. 1924b). This theory attracted a lot of attention in the course of the above mentioned development and is today best-known for its renunciation of exact energy conservation.

The traditional framing in the historical literature held that virtual oscillators were a means for the explanation of and had their origin in dispersion phenomena, i.e. the dependence of the refractive index on the frequency of the impinging radiation.[19] The description of the phenomenon includes the scattering of light the frequency of which does not coincide with a frequency connected to a transition of the scattering material. This constituted an obstacle for a model that described the interaction of light and matter as exclusively governed by transitions between stationary states. It would eventually lead physicists to describe the interaction between light and matter by the incorporation of oscillators that were connected with, but not equivalent to, the possible transitions of the scattering material.[20] In more general terms, Blum and Jähnert (2022) as well as Blum and Jähnert (2024) have traced the virtual oscillators to the concept of an *Ersatzstrahler*, having its origin in the application of the correspondence principle in Kramers' 1919 dissertation. As such, it was a way of circumventing the breach between classical radiation theory and Bohr's atomic model for the light-matter interaction. Thus, the BKS theory and the concepts used within it served a more general purpose: They should provide for a possible (although not realistic) picture of the dynamical description of the light-matter interaction, which was inaccessible for a considerable amount of time in the old quantum theory.

---

[17] For the relevant and quite extensive literature on BKS I refer the reader to my discussion of the BKS theory in Chap. 4.

[18] See for example Jammer (1966, Chaps. 4 and 5), Hund (1967, Chap. 10) or the introductory chapter of van der Waerden (1967). This framing has been upheld for a considerable amount of time (see for example Duncan and Janssen (2007a)) and has only been questioned quite recently by Blum et al. (2017).

[19] The role of the virtual oscillator, as is noted in the literature, goes well beyond the phenomenon of dispersion, but for this shot review the focus on this phenomenon shall suffice.

[20] For a comprehensive history of dispersion in the transition from classical physics to quantum theory, see Jordi Taltavull (2017).

The concept of the virtual oscillator allowed for the conception that "something in the atom [was] vibrating at the right frequency", to paraphrase Heisenberg's recollections.[21] We also know that the *Ersatzoszillatoren* were envisioned in analogy to the classical model of dispersion while in its further development classical perturbation theory and the correspondence principle were used to incorporate the concept in quantum-theoretical explanations. Furthermore, it is known that the mathematical structures of the virtual oscillators actually survived the demise of the BKS theory and can still be found in quantum mechanics as matrix elements of the position operator.[22]

But the disagreement in the literature starts when we ask about the connection between virtual oscillators and virtual transitions/virtual states, especially when we try to reconstruct when the second concept was eventually developed and ask about the relevance of the BKS theory for its invention. Virtual states, just to make the corresponding idea accessible, are intermediaries in the description of multi-stage processes. Virtual states are unobservable and are reached by energy non-conserving transitions. These transitions are called virtual transitions. Dresden (1987, especially Part 2, Chap. 13, Sect. C) has argued that in a paper by Kramers and Heisenberg (1925) on dispersion, that can either be conceived of as the high-point of physical explanations envisioned in the BKS framework or as a prime example of rather abstract and mathematical reasoning, the authors applied the concept of virtual transitions in their derivation and connected it to energy non-conservation as implied in the BKS theory. MacKinnon (1977), on the other hand, framed Heisenberg's *Umdeutung* paper as the place where a shift in perception from virtual oscillators to virtual transitions occurred. Konno (1993, 161) suggested that "the real solution of these problems was given by P.A.M. Dirac [in 1927]", where "dispersion is regarded as one successive process; an atom absorbs radiation and simultaneously emits it via an intermediate [or virtual] state." Pais (1986, 337–338) seems to agree with Konno. Finally, Kragh (1990, 126) stated that "Dirac [1927] introduced a concept of great importance, later known as the concept of 'virtual states'."

At the latest, and this is again clearly formulated in the historical literature, with his application of time-dependent perturbation theory in his theory of a second quantized radiation field,[23] Paul Adrien Maurice Dirac "set the basic language and concepts of the modern conception of the elementary processes as sequences of absorption/emission of light quanta, passing through a sequence of intermediate states" (Lacki, Ruegg, and Telegdi 1999, 486). Dirac (1927b) unmistakably stated

---

21 As cited in van der Waerden (1967, 29).

22 Already MacKinnon (1977) pointed towards this fact, but it was most strikingly argued for by Duncan and Janssen (2007a). Also Victor Weisskopf remembered, while reflecting on his and Eugene Wigner's results on the breadth of spectral lines (published in 1929), that before the quantization of the radiation field "thinking about emission was so closely tied to the picture of the atom as an orchestra of oscillators [...]" (Weisskopf 1983, 62).

23 Here and in the following I take up the language of the historical actors and refer to the transversal part of the electromagnetic field as the "radiation field".

that the transitions to the intermediate states do not have to conserve energy, i.e. every possible, otherwise not forbidden[24] state of the system could act as such an intermediate state. Only the process in total, the transition from initial to final state, had to conserve energy. Although the concepts Dirac applied would later be called "virtual states" and the transitions to and from them "virtual transitions", Dirac did not use this terminology in his verbal description. Knowing that virtual oscillators, or better some representations of them, survived the demise of the BKS theory and that they can also be found in the early days of quantum mechanics, this necessarily raises the question if there is a deeper connection between Dirac's description of the light-matter interaction and the one given by the virtual oscillator model. Is there any sort of generative connection between the two concepts?

At the same time, we can ask when and why the terminology was connected to the concepts Dirac used and which meaning was connected to the word thereby.[25] The possible connotations of the term "virtual" in the BKS paper have found some attention in the literature,[26] but up until now only Jean-Philippe Martinez (2024) has commented on the connection between the concept of virtuality in BKS and the later concept of virtuality (firmly established for intermediate states and transitions in the mid 1930s) in quantum electrodynamics and quantum field theory.[27] By a slight modification of Charles Sanders Peirce's pragmatist definition of the virtual,[28] which can be captured by an "as if",[29] Martinez showed how a general connection between the usages of the term "virtual" during the 1930s and the BKS concept (and even older ones) can be established.[30] Although I will mostly concern myself with the physical characteristics and the functions of the virtual entities, I will provide my own take on a general definition of the virtual in quantum physics.[31]

Martinez' contribution points to another aspect of the term "virtual", which makes the above question of the introduction of the term into quantum field theory even

[24] There might be conservation principles that forbid such transitions, as for example the ones for charge or (angular) momentum.

[25] I already provided for a first tentative history of the connection between intermediate states and the term "virtual" in Ehberger (2020).

[26] I will review these attributions in Sect. 4.1.3.

[27] Martinez discussed the two (initially) distinct concepts of "virtual states" as explicitly introduced by Guido Beck in 1930 and "virtual transitions", which are connected to the development of the virtual particle. I will not discuss the "virtual states" of Guido Beck and I will also refer, following the actors' terminology, to intermediate states connected to "virtual transitions" as "virtual states".

[28] See also Steinle (2024) for discussion of Peirce's concept of the virtual.

[29] The definition of Peirce (1902, 763) reads: "A virtual X (where X is a common noun) is something, not an X, which has the efficiency (virtus) of an X." Martinez (2024, 363) modified this to: "a virtual X (where X is a common noun attributed to a physical entity or process) is a theoretical construct, similar to X but not an X, which, in its associated theoretical framework, has the efficiency (*virtus*) of X in describing physical phenomena."

[30] Also Blum and Jähnert (2024) have argued that Peirce's definition might fit the general usage of the term in the BKS theory.

[31] I gestured at it in Ehberger (2024).

more pressing. Part of his investigation was concerned with "virtual states" (terminology introduced by Guido Beck in 1930) in neutron scattering, which were (at least initially) unconnected to the virtual states and transitions of Dirac's description of scattering in QED. Also other concepts carried the epithet "virtual" although they were at first sight not directly related to the modern virtual particle concept, e.g. the "virtual electrons" as negative energy electrons in the Dirac sea (terminology introduced by Guido Beck in 1933) or "virtual photons" of the Weizsäcker-Williams method (terminology introduced by Williams in 1935).[32] Although these concepts will not be prominent in my discussion of the development of the virtual particle concept, the necessity of accounting for them will be reflected in my recapture of the terminology in the concluding chapters of this book. But up until the mid 1930s, a coherent use of the terminology of the "virtual" in connection to concepts we can connect with the virtual particle was seemingly not completely established.[33]

There is yet another aspect connected to the modern concept of the virtual particle that is strikingly absent in Dirac's original paper: the Heisenberg uncertainty relation for time and energy $\Delta E \Delta t \geq \hbar$. Modern descriptions often invoke it for the explanation of the energy non-conservation connected to the virtual particles.[34] Certainly, Heisenberg's uncertainty paper was published after Dirac's paper on dispersion, but an influence cannot be definitely excluded since Heisenberg and Dirac spent the prior months together in Copenhagen and kept corresponding afterwards. But if Dirac did not invoke the uncertainty relations in his published paper, how then, if at all, did he explain energy non-conservation in the model he envisioned and when, and for which reasons, was the concept of virtual transitions connected to the uncertainty relations?

1.3.2 *From Virtual Transitions and States to Virtual Particles (1927–1949)*

After Dirac's introduction of the (later to be called) virtual transitions and states we are left with very little clues when the concept of a virtual *particle* actually emerged. Following expositions such as Kaiser (2005, 29–37) or Schweber (1994, Chap. 2), we get the impression that the virtual particle was there nearly as soon as quantum electrodynamics emerged. Surely, the evaluation of QED and QFT was always perturbative, the predictions, neglecting renormalization for the moment, of these theories as practised in the 1930s and late 1940s were essentially identical, and we therein find mathematical structures we might identify as connected to the modern concept of virtual particles. We could even argue that particles in virtual states *are* virtual particles.

32 This method is well taken into account in the historical literature, see for example Cassidy (1981, 15–16), while it is to note, that Cassidy suppresses the term "virtual".

33 See also my short discussion of the development of the terminology in Ehberger (2020, 268–269).

34 See for example Weinberg (1977, 24) and Kaiser (2005, 29–30).

This reasoning is certainly not completely off bounds, but I think we should be more cautious here. Although we cannot find any argument in the secondary literature when the concept of a virtual particle emerged, a close look at scholarly work discussing effects, representations and methods connected to the virtual particle seems to indicate that the solution is not as simple as the above line of thought suggests. For example, it is well-known that QED and QFT were closely modelled after non-relativistic quantum mechanics in their infancy (i.e. the conception of scattering was based on successive transition processes[35]) and the perturbation theory used was not relativistically covariant. Simply reflecting on one of the defining features of virtual particles, the way how energy non-conservation enters the description, the above identification of virtual particles as particles in virtual states becomes problematic. It has been pointed out in QED textbooks after the advent of Feynman's and Freeman Dyson's methods[36] as well as by historians of science[37] that in a non-covariant description the energy is not conserved in the transitions while in a covariant treatment the particles are actually off shell, i.e. they do not obey the relativistic energy-momentum relation, while energy and momentum are conserved at the interaction points.

The above idea might be conceived of by some readers as an argument about definitions or words, but I will argue, that there is more to this difference than the question of what *we define* a virtual particle to be. When we conceive of the virtual particle and related concepts as *concepts of practice*, there are already hints in the literature that we should be prepared to find that the above mentioned difference between virtual particles and virtual transitions actually has further conceptual implications.

The cumbersomeness of the calculations during the 1930s, the extreme amount of matrix elements to be calculated in comparison to the late 1940s, already directs attention to such practical differences.[38] But the problem seems even deeper: Adrian Wüthrich has argued by comparing term scheme diagrams as used in the 1930s to represent light-by-light scattering with the modern representation in terms of Feynman diagrams that

> [t]he statement 'Imagine a process that can be carried out by electrons and photons' differs [...] in meaning, depending on whether one is working with Feynman diagrams or with the modified term schemes. (Wüthrich 2010, 20)

Pushing this line of thought a little further, the following question arises: Did the change in conception have implications on the conceivability of processes in the respective frameworks? Did the conceptual framework of the 1930s allow for the conception of different processes than the one in the late 1940s and vice versa?

---

[35] See for example Wüthrich (2010, Sect. 2.1).

[36] See for example, Jauch and Rohrlich (1959, 158).

[37] Lacki et al. (1999, 496) comment on the above described fact, when discussing a covariant perturbation theory introduced by Ernst Stueckelberg in 1934, which was largely ignored by the scientific community.

[38] See for example the discussion of Hans Euler's work on light-by-light scattering in Kaiser (2005, 34–35).

Certainly, with the application of covariant calculational schemes and the stringent application of renormalization techniques in the late 1940s there was a kind of radical change in the practice of QED.[39] But we might ask if already in the 1930s conceptual innovations took place, that are connected to the concepts of interest for this investigation. Completely new effects like vacuum polarization and fluctuation were proposed in the early to mid 1930s, but the role virtual transitions and virtual particles played in the derivations or how such newly found effects altered these concepts is only partially discussed in the literature.[40] The same statement can be made about self-energy[41] and the idea of renormalization in QED, which also has its roots in the 1930s.[42]

Another phenomenon, which is today described by virtual particles, the idea of forces being mediated by particle exchange, was traditionally framed as being introduced by Enrico Fermi and Hans Bethe (Bethe and Fermi 1932). Cathryn Carson (1996b) has questioned this view and argued that a full establishment of exchange forces in QED only arose with Feynman in the late 1940s, but had its origins in the conception of nuclear forces of the 1930s. This argument not only shifts attention to the quantum field theoretical description of nuclear mechanisms, but is also an example of the conceptual innovations during the 1930s I gestured at above.

Similarly, Arthur Miller (1984) indicated that shifts in the perceptions of processes and the concepts used to describe them might have occurred during the 1930s but were only fully established in the late 1940s.[43] Like Wüthrich, Miller invoked two diagrammatical representations, in his case of β-decay: One was closely modelled after a term scheme diagram (left of Fig. 1.3) while the other one seems to resemble a Feynman diagram (right of Fig. 1.3). Miller remarked that from Gregor Wentzel's diagrams to Feynman's there is still a huge step to be taken, but:

> By 1943 the mental imagery of most physicists had already undergone several transformations. For is it not the case that Fig. 4.7 [diagram resembling a Feynman diagram] is another way of "seeing" the energy levels in Fig. 4.6 [energy level diagram]? In turn, Fig. 4.6 is a descendant of the Copernican atom that classical physics had imposed on the Bohr theory. (Miller 1984, 166)

---

[39] Whether this was a purely technical innovation or if something of the kind of a Kuhnian revolution took place, is an often debated question. I will turn my attention to this problem and the relevant literature shortly.

[40] Meanwhile, Wright (2024, Chap. 3 and Sect. 4.2) has discussed the alterations to the concept of the vacuum during this time period in some detail, highlighting the role of perturbation theory and its application to the vacuum state of the theory, and I will comment on his work in the respective sections.

[41] The infinities during the 1930s are amongst the best studied phenomena of this period. See, for example, Darrigol (1982, Chap. 3), Pais (1986, Chap. 16), Rueger (1992), Brown (1993), Schweber (1994, Especially Sect. 2.5), Miller (1994b, Sects. 3 and 4), or Cao (1997, Especially Sect. 7.6). Yet, if the explicit derivation of the effects is discussed it is often combined with a modern terminology and conception. See also my discussion in Sect. 10.1.

[42] See, for example, Aramaki (1987), Dresden (1993), Blum (2015) and Blum and Joas (2016).

[43] Miller mainly focused his discussion on the idea of *Anschaulichkeit* in Heisenberg's work but also gave hints at some more general developments.

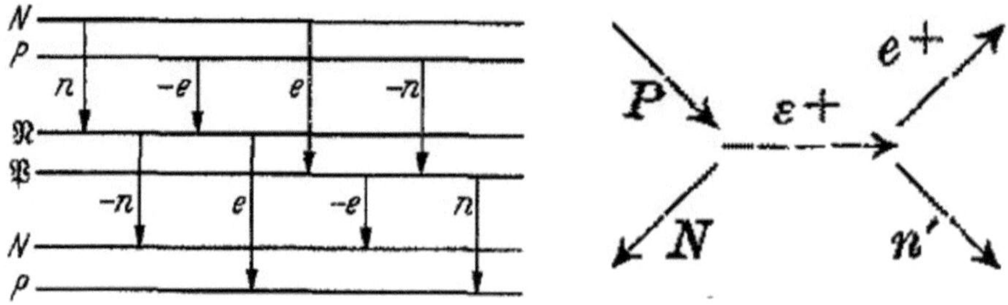

Fig. 1.3 The two types of diagrams discussed by Miller (1984, 165–166). Both diagrams describe β-decay as a stepwise process. The left one uses a term scheme diagram to do so. The right one (Miller used a later diagram of the same type, taken from Wentzel (1943, 104)) seems to come close to a Feynman diagram (although it had no calculational purpose). Left figure in Wentzel, Gregor. 1937b. "Zur Theorie der Beta-Umwandlung und der Kernkräfte. I." *Zeitschrift für Physik* 104, 34–47, here 37, https://doi.org/10.1007/BF01337892. Right figure in Wentzel, Gregor. 1938. "Schwere Elektronen und Theorien der Kernvorgänge". *Die Naturwissenschaften* 26, 273–279, here 276, https://doi.org/10.1007/BF01590308. Both reproduced with permission from Springer Nature

Although my reading of Wentzel's diagrams digresses from that of Miller (see Sect. 10.2.4), I share his conviction that conceptual innovation during the 1930s is a subject that deserves closer investigation, especially for the study of the virtual particle concept.

Where and how exactly such changes in conception took place is at times very hard to pinpoint and the consequences on and the role of the concepts like the virtual particle are hard to trace. I consider it a prime task of this investigation to shed some light on these interrelations by looking at the physical and mathematical analyses as well as the application of the connected theoretical concepts in detail: an evaluation that is as yet only partially available for the concepts of the virtual particles/transitions/states during the 1930s.

The last step in my reconstruction of the development of the virtual particle concept from the current literature is connected to the interpretation of the development of covariant techniques and renormalization during the late 1940s and how this reflected on the concept of a virtual particle. We might take up the traditional view that the developments of the late 1940s were foremost of a technical nature and no far-reaching conceptual reformulation took place.[44] In regards to the virtual particle, the covariance of the methods, that was only fully established in the late 1940s, still implies a different look upon energy conservation in the theoretical description of the processes. Furthermore, it is generally acknowledged that the manifest covariance of the methods was of prime importance for the construction of renormalization techniques of the late 1940s.[45] So even if we treat this development as a foremost technical innovation, simplifying calculations and making the reasons for and struc-

[44] Just consider Steven Weinberg's famous assessment that "things changed only just enough so that they could stay the same" (Weinberg 1977, 18).

[45] See e.g. Pais (1986, 457).

tures of the divergencies of QED more transparent, the difference between virtual transitions and virtual particles might have been decisive for this development.

In recent years the idea that actual conceptual shifts took place in the late 1940s has forcefully been argued for. In his *The Genesis of Feynman Diagrams* Adrian Wüthrich (2010) argued through a detailed study of archival material that Feynman developed his technique, with which the concept of the virtual particle is so closely connected today, in struggling with a suitable representation of the Dirac equation and the incorporation of interactions therein.[46] Wüthrich further argued that one of the driving forces behind Feynman's overall approach was the search for a solution to the divergences plaguing QED. In terms of these divergences, Wüthrich particularly stressed that through the systematization of Dyson the "free" electron became a concept that no longer made sense in QED but only the "dressed" one, with the inclusion of self-energy processes represented by virtual particles. Concerning the nature of the development this shift in perception is connected with, Wüthrich argued for a kind of "*creeping* revolution: a series of inconspicuous but fundamental revisions to the foundations of QED" (Wüthrich 2010, 21, emphasis in the original) which he studied through the advent of the diagrammatical approach of Feynman. The latter gave, according to Wüthrich, a suitable picture for avoiding the divergences of QED.

Most recently, Alexander Blum has presented penetrating studies on the history of quantum electrodynamics that engage with the conceptual differences between QED in the 1930s and the late 1940s. Blum (2015) provided a historical study of the infrared divergence, again a phenomenon that is today described by the invocation of virtual particles, and the connected failed calculations of radiative corrections to scattering by Sidney Dancoff in the late 1930s. Blum argued against the traditional framing of Dancoff's paper as simply being erroneous, but traced the failure in the calculation back to the different conceptualizations of the 1930s and the late 1940s.[47] He identified the distorted historical conception as due to a difference in the understanding of the term "renormalization" also used by Dancoff himself.[48] Concluding, Blum stressed "the divide, both in concepts and in methods, that separated it [Dancoff's calculation] from the renormalized QED of the 1940s" (Blum 2015, 92).

Finally, the conceptual transition that was studied by Wüthrich through the focus on the genesis of Feynman diagrams was investigated by Blum in more general terms: He followed the path from QED as a theory focusing on the characteristics of states to "a theory of scattering" (Blum 2017). Blum argued that the transformation from QED of the 1930s to the late 1940s can be considered as a genuine paradigmatic shift, but not in the conceptual basis of the theory but rather in its exemplar, its paradigmatic

---

[46] Nevertheless, it must be mentioned that there is still disagreement about the exact emergence of this diagrammatical technique. For different reconstructions see Schweber (1994, Sect. 8.8) or Darrigol (2019). See also Chap. 11.

[47] In this case, the use of the Coulomb gauge and the dropping of the electro-static self-energy during the 1930s.

[48] Olivier Darrigol (1982, 203, Second Footnote) had priorly pointed to this difference in the conception of the term "renormalization" and the distorted historical narrative.

problem (Kuhnian terminology). The transition was, according to Blum, from a theory that focuses on the properties of quantum states to a theory which has scattering as its exemplar.

In the prior subsection, we already saw that at least with the shift from a semi-classical conception of the radiation field to a second quantized theory the secondary literature agrees that the transformation (if it was such) from virtual oscillators to the (later to be called) virtual transitions and states was fully established. In the same vein, the development of the concepts traced in this book will have to take into account the conceptual shifts identified by Carson, Miller, Wüthrich, and Blum. It will have to place the development of the concepts in this wider conceptual framework as it will have to judge the relevance of these shifts for the conceptual development under study.

Above I have identified some blind spots in the secondary literature and disagreements within it. I posed some of the questions that will have to be answered in the following. But most importantly I wanted to show that, on the one hand, there was a historical development of concepts that are connected to the virtual particle and, on the other hand, we are not yet in possession of a coherent understanding of the same. It should have also become clear that the concepts connected to the virtual particle all lay at the heart of the theoretical description of the light-matter interaction.[49] This connects the concepts not only directly to the conception of light and matter in the respective theoretical frameworks, but also to the methods used for the evaluation of their interaction and the possible physical effects derived from it.

Consequently, the investigation to be carried out will have to comment on the (dis)connections between the concepts of semi-classical methods and second-quantized ones, non-covariant and covariant ones, between the concepts of a theory that is concerned with the characteristics of stationary states while resting on a model of successive transitions and the ones applied in a theoretical treatment that focuses on scattering phenomena. It will have to acknowledge the differences in the terminology, the representations used, the methods applied, and the theoretical description resulting thereby.

In short: The study of the historical development of the virtual particle will have to trace the (dis)connections in the genesis of a still debated concept, its representations, and its application incorporated in the evolving conceptual framework of quantum mechanics and quantum field theory. In the following section, I will delineate my plan of how to achieve this.

[49] Of course, this also holds for the interaction of different matter fields, as soon as QFT had been developed.

1.4 Goals and Outline of This Book

Since this work has a considerable scope I want to point out some short cuts and Chapters with specific attention to the different groups of potential readers, before delineating the contents below. For a general overview of the historical results the reader is referred to Sect. 12.3. Readers who are interested in general approaches towards concept formation processes, the philosophy of the virtual particle or historiographical tools and their application gain most from Chaps. 2 and 12, especially Sects. 12.1 and 12.2. For readers who are interested in the social environment and dynamics of the research discussed in this historical investigation (and those who want to brush up their knowledge on the respective schools of theory), Chap. 3 is advised.

For the main part of this work, the readership might have chronological preferences and might be interested in specific fields of research: For quantum-theoretical research before quantum mechanics see Chap. 4, for quantum electrodynamics see Chaps. 5, 7 through 10.1 (mainly 1930s) and Chap. 11 (late 1940s), for nuclear physics and meson theory, see Sect. 10.2, and for the early discussions on the Raman effect, see Sect. 6. Readers who are interested in the development of the conception of the vacuum are referred to Chap. 7 and Sect. 10.1.2, for those with specific focus on the divergences of quantum electrodynamics during the 1930s Sects. 10.1 and 9.3 (theory of radiation damping) are advised.

The historical investigation will focus on theoretical practices and representations of theoretical entities. Chapters 8 and 9 are particularly rich in in-depth studies of the theoretical practice of QED during the 1930s, while Sect. 9.4 provides for a condensed analysis. Through the analytical category of "false predictions" the difference in the practice of QED during the 1930s and the late 1940s is accessed in Sect. 12.2. If the development of diagrammatic representations is the major interest, the reader is referred to Sect. 12.1.2 for an overarching description, for the terminology see Sect. 12.1.1.

Coming to the delineation of the contents, my starting point is the short outline of the concept of the virtual particle (Sect. 1.1) and the necessity of constructing a suitable historiographic framework (Sect. 1.2) that also takes our modern understanding of the particular concept into account. Trying to meet my own standards, I will discuss my conception of the virtual particle and argue for a suitable way of treating its historical development in Chap. 2. First, I will treat the concept's role in theoretical explanations in some more detail. My selective review of the philosophical debate around the virtual particle concept will allow for a deeper understanding of the concept itself. Combining these insights with current trends in the historical study of concepts of theoretical physics, I will argue that we should conceive of the virtual particle as a concept of theoretical practice, which is best studied through its representations and their usage. I will introduce my key analytical concepts of representations, verbal models, narratives, and false predictions, and discuss the historiographical consequences of this approach. I will also discuss the usage of specific source material for the specific chapters in this connection. Finally, I will exemplify my understanding of the category of a concept and the virtual particle by the example of Feynman's *Space-Time Approach to Quantum Electrodynamics* (Feynman 1949a), the historical endpoint of my investigation.

Before going into the detailed discussion of the historical development, Chap. 3 will provide for an overview of the historical network of practitioners. This chapter will serve a twofold purpose: First of all, it will introduce some of the personal connections and delineate different stances towards the theoretical study of physics, which will become important in the development of the virtual particle concept. Secondly, it will serve the reader as a backdrop when the overall development is lost in the nitty-gritty details discussed in the following chapters.

The central part of this book, the historical reconstruction of the development of the virtual particle concept, will be divided into three parts. In the first part (Chaps. 4–6), I will discuss the historical trajectories leading from the first prominent occurrences of virtuality in the old quantum theory to the first explicit connection between the term "virtual" and the conception of scattering as a succession of emission and absorption processes in quantum electrodynamics: I will study the development from "virtual oscillators" to "virtual transitions". In order to do so, I will focus on the conception of dispersion and scattering, the effects in which the discussed concepts played the most prominent role. The second part (Chaps. 7–9) will be concerned with the practice of time-dependent perturbation theory in quantum electrodynamics during the 1930s: the field in which the application of the conception of successive emission, absorption, and transition subprocesses was most common. The guiding question will be: How did physicist apply the concept of virtual transitions (modern terminology) in theoretical evaluations and how did their concepts and the representations change thereby? I will discuss the diverse alterations of the practice and conception of physical processes and study the interrelation between the concepts under study and the overall conceptual framework. The third and final historical part (Chaps. 10–11) will examine the transition from "virtual states" and "virtual transitions" to "virtual particles", from the conceptual overlap that started to show in the mid to late 1930s to the full establishment of the virtual particle concept in the work of Richard Feynman and Freeman Dyson in the late 1940s.

Chapter 4 will focus on the period before quantum mechanics (roughly until mid 1925). On the one side, I will discuss the development and contents of the Bohr-Kramers-Slater theory of the "virtual oscillators" and the "virtual radiation field" and reconstruct the meaning of the term "virtual" used therein. Their relevance for the development of the Kramers-Heisenberg formula, the key formula for quantum theoretical accounts of dispersion, but also the technical basis from which this formula was derived will be evaluated in some detail. As mentioned, some historians have ascribed the concept of virtual transitions to this dispersion-theoretical account and I will revisit this judgement by evaluating the different representations the actors applied. In order to strengthen my argument that a story of successive transitions had not yet been (fully) established at the time Heisenberg published his *Umdeutung paper*, I will contrast the wave-theoretical accounts with light-quantum accounts of dispersion and scattering. In the latter, dispersion was conceived of as successive but real emission and absorption processes through "intermediate states" (lying both energetically and temporally in between initial and final state). This historical detour will exemplify the relevance of the conception of quantized radiation for the verbal model of successive emission, absorption, and transition subprocesses, i.e. the relevance of the conceptual framework for the conceptions I am investigating.

Of course, the route to follow was set by the development of matrix and wave mechanics. And Dirac's introduction of an explicit verbalization of energy non-conserving transitions through virtual states (modern terminology) has to be placed within this tradition. Yet, it took the development of second-quantization, and thus a light-quantum framework, to make these ideas explicit. In Chap. 5, I will revisit how Dirac came up with the verbal model that still resonates in the modern approach to QED. I will do so by discussing the technical, conceptual, and philosophical background of Dirac by a close investigation of the archival material in the P.A.M. Dirac papers. Through comparison with the concept of virtuality in the BKS theory and of the virtual oscillators, I will argue that the conceptual gap between "virtual oscillators" and "virtual transitions" is larger than normally assumed. A direct, generic connection is rather unlikely.

In Chap. 6 I will thus ask when and why the term "virtual" was eventually connected to genuine quantum concepts, i.e. states and transitions. It was Chandrasekhara Venkata Raman who first introduced a direct connection between the term "transition" and the term "virtual" in a speech before the Faraday Society in late 1929. Locating Raman's speech in the contemporary discussions around the effect named after him, the first empirical proof of the Kramers-Heisenberg formula, as well as his hopes of receiving a Noble Prize will make this first invocation understandable. At the same time, the considerable break with the prior connotation of the term "virtual" will be emphasized. Simultaneously, Yakov Il'ich Frenkel revived the conception of scattering as a succession of real transitions, which was immediately shown to be untenable by Georg Placzek. The idea of energy non-conserving transitions through intermediate states had been put forward, it was connected to the term virtual, and the conception of real/actual transitions was excluded. The development from "virtual oscillators" to "virtual transitions" will, thus, find a first conclusion at around 1929.

Turning in the following to the practice of time-dependent perturbation theory, I will first discuss the epistemic relevance of virtual transitions (modern terminology) for the inception and reception of the Dirac sea: I will study the impact of the concept on its conceptual surroundings. I will argue in Chap. 7, that the development of the Dirac sea was motivated by the occurrence of negative energy intermediate states. The most detailed historical investigation of this argument is so far due to Grandin (2008). I will judge its resilience by bringing together all the identifications of the above fact, their relevance for the modelling of the scattering phenomena, and the ontological connotations of the virtual processes (modern terminology) the actors respectively invoked. Following the development further, the relevance of the description of the scattering mechanism and the concepts of virtual transitions (modern terminology) to negative energy intermediate states for the refutation of the identification of a hole in the Dirac sea with a proton will be discussed. A connection, that has as far not found the required attention. Finally, the conception of scattering within the Dirac sea will lead to the inception of the first published occurrence of what I call a “false prediction”. Igor Tamm, closely associated with Dirac at the time, proposed that light could be scattered by a pair of non-interacting electrons to second order. This conception was allowed by the modelling of the effects at the time but was proven to be mistaken by the transformation of the verbal model into a mathematical representation.

Chapter 8 will focus on the initial evaluation of Dirac’s radiation theory and its representation in the most important didactic literature of the time. I will start out with the work of two Göttingen doctoral candidates: Maria Göppert-Mayer and Victor Weisskopf. Göppert-Mayer introduced the concept of the *Schüttelwirkung* to the perturbative evaluation of radiation theory which would, in the late 1930s, eventually contribute to Walter Heitler’s theory of radiation damping (see below). Victor Weisskopf’s work, before coming under the influence of Niels Bohr and Wolfgang Pauli, introduced the second pillar (Weisskopf-Wigner theory) of the theory of radiation damping and led up to the first general perturbation formula to all orders, at times referred to as “Weisskopf’s formula”. Weisskopf’s and Göppert-Mayer’s work will further elucidate the fuzzy boundaries between virtual and real processes and introduce new modes of representations, diagrammatically and verbally. Turning to the didactic literature, we will see how the specific kind of modelling applying the virtual transitions and states allowed, in contrast to prior judgements, for a general algorithm for the description of the light-matter interaction in QED during the 1930s. I will close by describing this algorithm and point towards the relevance of the different representational formats of the virtual transitions and states for its application.

Chapter 9, the final chapter on the application of time-dependent perturbation theory during the 1930s, will at first engage with the theoretical modelling of two interrelated effects: light-by-light scattering and the Delbrück effect, probably the most advanced (finite) calculations in radiation theory of the time. Those effects are not only relevant as they show how different representational formats were applied to investigate the respective effects within a specific culture of theory. But the term “virtual” was rather prominently connected to the intermediate states and transitions to and from them by Hans Euler, student of Werner Heisenberg in Leipzig. This invocation, as the simultaneous ones by Heisenberg, will not only lead to a widespread use of the terminology in the community. Through Heisenberg’s conception of quantum theory, and the measurement process in particular, a connection to the connotation of the term “virtual” of the BKS theory can be established. The historical study of the Delbrück effect will not only unravel the different conceptualizations and applications of the concepts under study, formed through the prior work at the respective institutes, but also reveal a false prediction regularly made, namely that this effect could occur to third order. This false prediction constituted one of the motivations for the evaluation of the perturbative series and eventually to the inception of Furry’s theorem. Finally, I will revisit the development of the theory of radiation damping, one of the attempts to circumvent the divergences of QED. It was conceptually based on the work of Göppert-Mayer, specifically the *Schüttelwirkung*, and of Weisskopf. In this theory, the concept of a (virtual) intermediate state and the perturbative series as a whole were altered. Its failure shows that concepts directly connected to the virtual particle concept could be manipulated, but not at will.

These three chapters each focus on specific effects, their evaluation through the application of perturbation theory and the role the virtual transitions and states played for the respective theoretical investigation. Through this focus, different modes of representation, terminological variants, little-known concepts directly connected to the virtual states and transitions, as well as their characteristics and rules for their application will be revealed. But even processes which cannot be thought of in terms of Feynman diagrams and a covariant perturbative evaluation will come to the fore. All the same, although physicists talked about "virtual particles", from time to time, the terminology of "virtual transitions" was far more common and actually in line with the mode of perturbation theory. Thus, the transformation from virtual states and transitions to virtual particles will be the main subject of the last historical part.

In Chap. 10, the theoretical description of the divergences during the 1930s will be discussed at first. I will argue that the applicatcha:Verbalizatioion of a verbal model of successive transitions only gradually was applied within the self-energy calculations. It found its way into the discussion when the effects were treated at higher orders, which eventually necessitated a complementary mode of representation to the mathematical formalism. In the mid 1930s, the concept of a charge cloud, the idea that a particle does not exist on its own but always has a cloud of virtual particles surrounding it, emerged. This idea was the pillar of the discussions around vacuum polarization and the so-called anomalous magnetic momenta of the nucleons. The emergence of the idea of exchange forces, i.e. the conception of forces being transmitted by virtual particle exchange, will be revisited. As I will argue, the invocation of a dynamical force-carrier picture and related ideas depended heavily on the respective conceptual and technical framework with which the historical actors worked. Nuclear theory will also provide the key for understanding the invocation of what I will call a "narrative": a way of arguing for the inclusion of a concept, in this case virtual transitions and particles, in theoretical evaluations. The most popular narrative for virtual particles and transitions, connecting these concepts to the energy-time uncertainty, was prominently invoked for qualitative conclusions when the formalism of quantum field theory was on particularly shaky grounds. Also another aspect of the virtual particle concept, its diagrammatical representation in terms of Feynman-like diagrams by Gregor Wentzel, will be discussed in the respective historical connection. Combining the insights gathered through the respective historical episodes, I will show that at the end of the 1930s a considerable overlap of (if not tension between) the concepts of virtual transitions and virtual particles had emerged. It manifested itself in the diverse modes of representations the historical actors used in connection with the discussed theoretical entities. All the same, it took the work of Richard Feynman to coherently establish the concept of a *virtual particle*.

As I will argue in Chap. 11, the change in practice of QED in the hands of Feynman and Dyson at the late 1940s had enormous impact both on the concept of the virtual particle and its status. Not only the representations and the respective perturbative practice changed considerably, but also the characteristics of particles in intermediate states, i.e. virtual particles, were subject to change. Connected to the paradigmatic shift of the practice of QED that Alexander Blum had identified, one of the prime concepts of the theoretical practice, the virtual particle, considerably changed in character, representation, and finally status.

My conclusion (Chap. 12) consists of three parts. According to my agenda of investigation, I will first revisit the development of specific representational formats: the terminology used to refer to the concepts under study and the diagrammatical representations used in the theoretical evaluations. Drawing these developments together and embedding them in the respective theoretical and conceptual environment, I will elucidate how conceptual continuities and breaks can be simultaneously accessed when conceiving of concepts as manifested through their representations and scientific practice. Secondly, I will discuss the occurrences of what I called a false prediction. As I will show, false predictions constitute a historiographical tool that allows for the diagnosis of subtle changes in the representation, practice, and conception of a specific theory. Through this analysis, I will also contribute to the question of which kind of historical process had taken place at the end of the 1940s.

Finally, I will revisit the whole historical development and turn to the question guiding this investigation: When, how and why did the virtual particle emerge? To take this upfront: I will not give a straightforward and unambiguous answer to any of these questions. But I will argue that by the taken perspective on the conceptual development a far more refined picture of the development from virtual oscillators over virtual transitions to virtual particles has finally been made available. All besides the diverse breaks and readjustments in the historical development, a specific function of all concepts lying in the historical line of the virtual particle could be established: In their own way and within their conceptual framework, all of these concepts served as bridging elements in theoretical description. They extended the set of *possible* processes and turned this extended set of possibilities into effective physical entities.

Chapter 2
How to Conceive of the Concept of Virtual Particles in a Historical Study of Its Development

In the previous chapter I announced the construction of a suitable philosophical and historiographic framework for the study of the development of the virtual particle concept. In order to do so, I will first and in more detail describe how the virtual particle shows up in the modern version of QED and discuss the role it plays in theoretical explanations (Sect. 2.1). In Sect. 2.2, I will discuss a few philosophical stances towards the virtual particle concept and its most prominent representation, Feynman diagrams. This will serve to increase the apprehension of the concept. But it will also motivate my own stance towards concept development and formation, which was mainly influenced by pragmatic accounts and will be discussed in Sect. 2.3. Therein, I will introduce the basic historiographical categories I will use for my analysis: the concept as tool, representations of concepts and models, what I call "false predictions" and my understanding of the term "narrative". With this background I will finally revisit the concept of the virtual particle, both to provide my own understanding of the concept and to make the historiographical part on concepts concrete, taking Feynman's 1949 presentation, the endpoint of my historical investigation, as an example.

2.1 Virtual Particles in Contemporary Physics

There are a few things we can safely say about virtual particles. First of all, they are an integral part of the everyday work of a large part of the practising quantum field theorists community: The concept is effectively used in the calculation and modelling of diverse different physical effects within this branch of physics.[1] Secondly, virtual particles have some strange features: They are off shell, i.e. they

[1] Although one surely can go to the current research literature in particle physics to acknowledge the role of virtual particles, one does not have to. A look at the didactic literature on particle physics such as Peskin and Schroeder (1995) should be sufficient.

M. Ehberger, *Representing the Unobservable*, Science Networks. Historical Studies 68,
https://doi.org/10.1007/978-3-032-09188-8_2

do not obey the relativistic energy-momentum relation, and they are unobservable by definition: Everything that leaves tracks in cloud chambers or causes signals in particle detectors are real particles. Thirdly, virtual particles are closely related to a specific calculational technique, namely perturbation theory and its representation in terms of Feynman diagrams. The principle idea of perturbation theory and its practice in modern-day QED has already been explained in Sect. 1.1 and it is now time to add some more details to this description.[2]

Although one normally starts out with a Lagrangian density for the dynamical description of the electron-positron and the electromagnetic field as well as their interaction, for the given purpose a highly abstracted version of the equations in Hamiltonian form is best fitting:

$$H\Psi = (H_0 + H_I)\Psi = i\hbar\frac{\partial}{\partial t}\Psi \tag{2.1}$$

Ψ describes the state of the two fields and the dynamical behaviour is governed by the explicit form of the Hamiltonian H. With hindsight, I have already divided the Hamiltonian in the second step into two parts: H_0 refers to the so-called free or unperturbed part of the Hamiltonian and H_I describes the interaction between the electromagnetic and the electron-positron field, i.e. the perturbation.[3]

As already mentioned, you cannot solve this equation exactly in QED and the prime workaround to this problem, at least for QED, is perturbation theory.[4] For modern-day QED, the most common treatment is the so-called S-matrix expansion ("S" stands for scattering) within the interaction picture.[5] In this representation, the temporal development of the (transformed) states of the system is governed by the (transformed) interaction Hamiltonian:

$$i\hbar\frac{d}{dt}\Psi = H_I\Psi \tag{2.2}$$

Without interaction the state vector of the system is constant. The S-matrix might now be introduced in the following way:

An initial "in"-state of the system $|i\rangle$, which is assumed to lie in the far past at $t \to -\infty$, and a final "out"-state $|f\rangle$, which is assumed to lie in the far future at $t \to +\infty$, are envisioned. These states are considered to be free states, i.e. solutions of the dynamical equations without interaction. The number of electrons, positrons, or photons (with given properties) in the in- and out-states is dictated by the problem

[2] I will follow Mandl and Shaw (1984, Chap. 6) in the next view paragraphs. Other than Mandl and Shaw I did not set $\hbar = 1$.

[3] I have subsumed the small parameter ϵ, as it figures in Eq. (1.1), in the interaction Hamiltonian.

[4] The importance of perturbation theory for QED has most strikingly been expressed by Abraham Pais (1986, 374): "From the first quantum electrodynamical calculation ever performed [...] until the present [1986], all we know about quantum electrodynamics is based on perturbation theory."

[5] We arrive at this representation by a canonical transformation of both the operators and the state of the system given by: $U_0 = e^{-iH_0(t-t_0)/\hbar}$; $\Psi_I = U_0^\dagger\Psi$; and $O_I(t) = U_0^\dagger O U_0$.

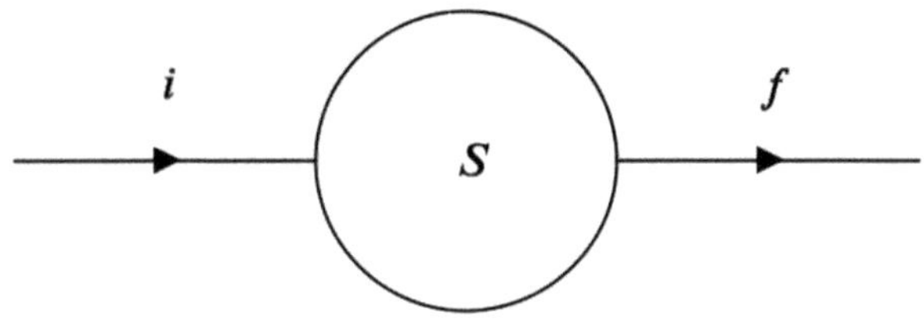

Fig. 2.1 Completely schematic representation of an interaction process. An initial state i changes through the action of S into some final state f. Diagram by the author

studied. The S-matrix is the operator which connects these in- and out-states. It transforms the initial into the final states:

$$S_{fi} = \langle f|S|i\rangle \tag{2.3}$$

A diagrammatic representation of this conception is given in Fig. 2.1. The circle in the middle is at times referred to as the "bubble of ignorance",[6] since, by construction, we do not have any empirical means of accessing it.

Basically, the operator S tells you which initial and final states couple, i.e. which transitions are actually allowed, and how strongly they couple.[7] The explicit dependence of S on the interaction Hamiltonian is given by:

$$S = \mathcal{T}\left(e^{\frac{-i}{\hbar}\int_{-\infty}^{\infty} d^4x\mathcal{H}_I}\right) \tag{2.4}$$

where $\mathcal{T}$ is the time-ordering operator[8] and $\mathcal{H}_I$, defined by $\int d^3x\mathcal{H}_I = H_I$, is the Hamiltonian density corresponding to H_I.

An expansion of the exponential in a power series results in:

$$S = \sum_{n=0}^{\infty} \frac{(-i)^n}{\hbar^n n!} \int_{-\infty}^{\infty} d^4x_1 ... \int_{-\infty}^{\infty} d^4x_n \mathcal{T}[\mathcal{H}_I(t_1)...\mathcal{H}_I(t_n)] \tag{2.5}$$

---

6 See, for example, Passon (2019, 7).

7 I choose the terminology of "coupling" on purpose, since I do not want to use any probability talk when we are actually not dealing with probabilities but their amplitudes. See further below.

8 For an intuitive introduction of the time-ordering operator the reader is referred to Sakurai (1967, Sect. 4.4). Sakurai also discussed how this operator is important for manifest covariance and how the idea of positrons being electrons travelling backwards in time can be intuitively constructed by its application. For any one interested in the technical relations between non-covariant methods, as practised in the 1930s, and the later covariant ones, Chaps. 2 and 4 of Sakurai's book are recommended.

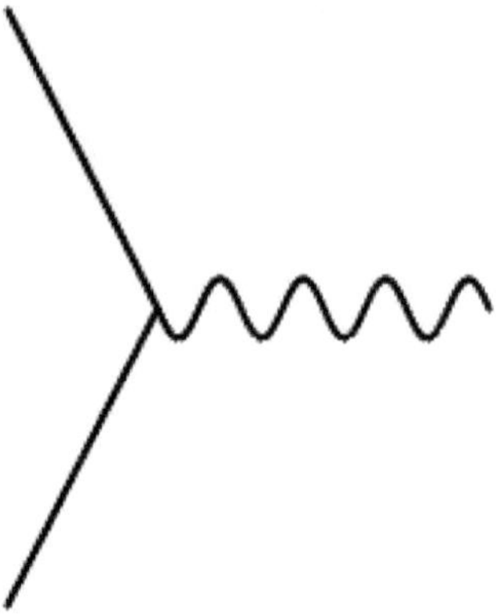

Fig. 2.2 The basic (unphysical) building block of every Feynman diagram. Corresponding to the time-direction, this might either be conceived of as an emission/absorption process of an electron/positron, or the creation/annihilation of a pair. Diagram by the author

which is the so-called Dyson series.[9] The summation over n can be stopped at each order and then gives the nth order approximation to the process under study. In essence this can also be written as:

$$S_{fi} = \sum_{n=0}^{\infty} S_{fi}^{(n)} \approx S_{if}^{(0)} + S_{if}^{(1)} + ... + S_{if}^{(k)} \tag{2.6}$$

where the superscript (n) refers to the order of the approximation and is equal to the number of expressions referring to the interaction occurring in the integrand of Eq. (2.5). For the 0th order approximation no interaction takes place at all.

So far, this was only a basic and very abstract outline of the perturbative approach towards QED. It forms the background of much of the calculations done within this theory. Yet, in essence, this expansion can be applied to any theory, as long as the Hamiltonian or Lagrangian allows for the splitting we started from and the time-ordering applied. For QED in particular, the mathematical structure of $\mathcal{H}_I \propto \bar{\psi}\gamma_\mu\psi A^\mu$ suggests that every interaction involves the annihilation of one material particle, the creation of another and one photon (see also Fig. 2.2). The actual starting point of calculations is then to find out which S-matrix elements to which order have to be considered in the derivation of some process. This is where Feynman diagrams enter.

Feynman diagrams,[10] this is certainly safe to say, are a diagrammatic representation of the expansion of the S-matrix and an indispensable part of the tool kit of a modern quantum field theorist. But why are Feynman diagrams so useful? Part of the answer lies in the fact that there are simple rules which diagrams can be drawn, how they are interpreted, and how they are translated into mathematical expressions:

[9] See for example Mandl and Shaw (1984, 101–102).

[10] For a textbook introduction to Feynman diagrams and rules see Mandl and Shaw (1984, Sect. 7).

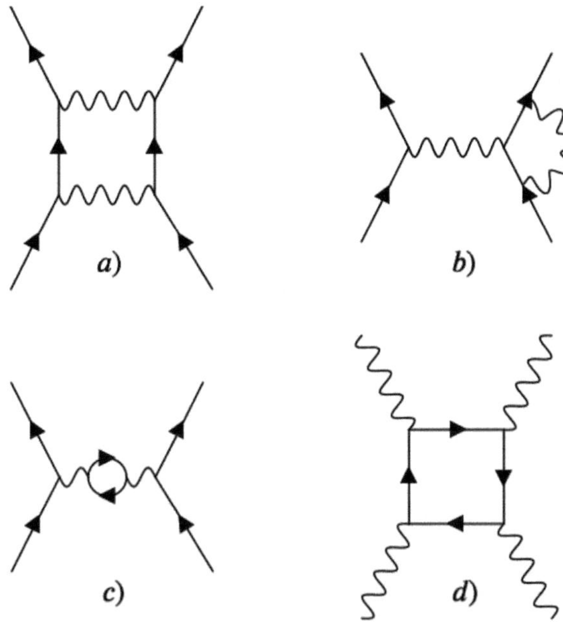

Fig. 2.3 Diagrams (**a**), (**b**), and (**c**) refer to higher order corrections to the scattering of two electrons. The diagram (**d**) refers to light-by-light scattering. The diagrams are read from bottom to top. Diagrams by the author

the so-called Feynman rules. These rules allow for the construction of all non-zero matrix elements involved in the theoretical description of the process.[11]

Depending on which theory is involved and which representation is chosen, the Feynman rules might differ, yet there is always a correspondence between terms to be calculated and diagrams to be drawn. The evaluation of the mathematical expressions is still a messy business and a lot of the work actually carried out by theorists is dedicated to the solution of these equations. While the perturbative expansion and the corresponding conceptualization of free fields and interactions as perturbations gives the background of a calculation in QED, Feynman diagrams stand at the beginning of the same. The basic building block out of which Feynman diagrams are constructed in QED is depicted in Fig. 2.2.[12] Wavy lines refer to photons and straight lines to electrons or positrons. Arrows on the solid lines (see the diagrams in Fig. 2.3) indicate, in reference to the time-direction chosen for the respective diagram, whether an electron (arrow pointing towards the future) or a positron (arrow pointing towards the past) is represented.

One Feynman diagram was already presented in Fig. 1.1 and shortly discussed in Sect. 1.1. It corresponds to the first order in which electron-electron scattering can occur. I also pointed out that the reading of Feynman diagrams in terms of

[11] For QED-Feynman rules in momentum space see e.g. Mandl and Shaw (1984, Sect. 7.3).

[12] This basic structure on its own does not refer to any observable process, since we cannot, at the same time, put all involved particles on the mass-shell and conserve energy and momentum at the vertex. For example, a free electron cannot absorb or emit a photon due to a necessary breach with energy or momentum conservation.

physical processes in which virtual particles, represented by the internal lines of these diagrams, pop in and out of existence or are exchanged is very suggestive, but, as will be outlined more fully below, also problematic. In Fig. 2.3 the reader can find further Feynman diagrams representing either higher order corrections to electron-electron scattering (a–c) or the first order of light-by-light scattering (d).

It is important to note, that a lot of abstraction goes into the drawing of these diagrams. One can relate every Feynman diagram, more specifically every component of it, to a mathematical term, but the constructed mathematical expressions involve integrations. In configuration space representation, an integration has to be carried out over all coordinates of vertices, i.e. interaction points. In momentum space representation, all momenta associated with internal lines, i.e. virtual particles, have to be integrated over, while a delta function is included for each vertex to ensure four-momentum conservation at the interaction points.[13] In essence, because of this integration a *single* Feynman diagram already represents a lot of different possible versions of the processes that are included in the mathematical representation.

This has two further effects: On the one hand, the coordinates associated to the virtual particles will not show up in the final results of the calculation, again an instantiation of them being unobservable. Secondly, as can easily be seen by a transformation into momentum space, energy and momentum are conserved at each interaction point but the virtual particles are not bound to fulfil the relativistic energy-momentum relation, i.e. $E^2 = (pc)^2 - (mc^2)^2$: they are also said to be "off shell". This is often explained by the short lifetime of virtual particles and by invoking the energy time-uncertainty: $\Delta E \Delta t \geq \hbar$.[14]

The diagrams shown (especially Fig. 1.1), the general structure of the approach to QED, and the occurrence of virtual particles therein already provide enough insight to understand one common idea the virtual particle is connected with: Since any interaction in QED takes place through virtual particles, they take the place of forces in classical physics. Therefore, they are often referred to as force carriers. To be sure, this is a very picturesque understanding of virtual particles, but an often invoked one. Yet, there is also another interpretation of their actions and characteristics to be explained.

In order to do so, I want to point the reader to the diagrams shown in Fig. 2.4. These are the so-called self-energy diagrams. In a physical interpretation, the left diagram represents an electron emitting a photon and then absorbing it again, i.e. the electron is interacting with its own field. The right diagram represents a photon creating or hitting–according to your favoured interpretation–a virtual electron-positron pair which in turn annihilates and creates a photon. The problem with these diagrams is that their mathematical counterparts diverge as they include an integration over an unbound virtual momentum. This actually happens every time when one calculates a next-to-leading order process. The issue was only resolved in the late 1940s by the renormalization techniques.

[13] See, for example, Mills (1993, Sect. 3.4).

[14] See for example Kane (2006) or Peskin and Schroeder (1995, 13). Further possibilities of explaining the violation of the relativistic energy-momentum relation can be found in Jaeger (2019, 13).

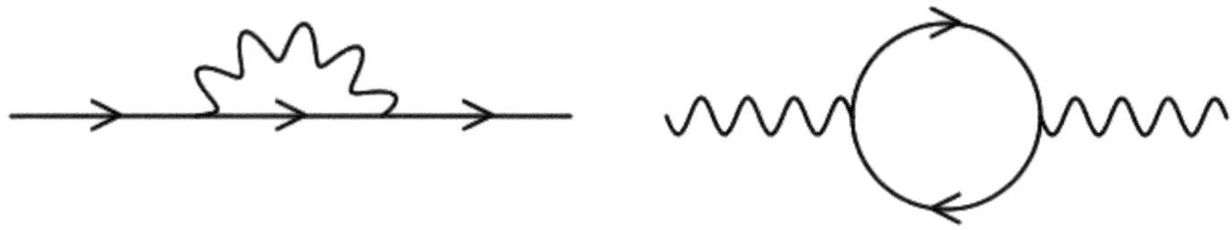

Fig. 2.4 To the left: self-energy diagram of the electron. To the right: self-energy diagram of the photon, i.e. vacuum polarization diagram. The diagrams are read from left to right. Diagrams by the author

As already pointed out in Sect. 1.1, the method of renormalization does not only allow for the handling of the theoretical structures represented in Fig. 2.4. It has consequences for QED's conceptual structure. Through the incorporation of the divergent parts of the diagrams into the parameters of the theory, the idea of an electron without a photon or electron-positron cloud, the so-called "bare" electron, is no longer a physical entity but only its "dressed" counterpart.[15] Furthermore, especially the right diagram in Fig. 2.4 suggests another interpretation of virtual particles, namely as vacuum fluctuations. The story connected to this line of conceptualization goes like this: The uncertainty relation for time and energy or, more technically, the non-commutativity of number operators and the Hamiltonian imply that the number of pairs (or photons) cannot be sharply defined. Therefore, vacuum fluctuations of the electromagnetic and the electron-positron field occur and give rise to the self-energies, but also to other effects.[16]

There are two further points to note that are highly important for the interpretation of Feynman diagrams and virtual processes. First of all, we do not actually know if the S-matrix expansion, even after renormalization, is convergent or not: Most probably it is not.[17] Therefore, it is far from clear how much credence can be given to the overall procedure, although it is so highly effective and accurate in quantitatively predicting empirical results. The divergence of the S-matrix plays an essential role in more recent approaches to the interpretation of Feynman diagrams.[18]

[15] For an easily accessible, only slightly technical introduction to renormalization, see Mills (1993).

[16] Naturally, the line between virtual particles as force carriers and virtual particles as vacuum fluctuations is not sharp. For example, the diagrams of Fig. 2.4 can be included in other diagrams and act as part of the force carrying machinery therein.

[17] The first one to point this out in print was Freeman Dyson (1952).

[18] See for example Bacelar Valente (2011a), Brown (2018), or Stöltzner (2018). To be sure, there are far worse conceptual problems for perturbative QED: The question whether QED can be coherently interpreted either in terms of particles or in fields is not settled; Haag's theorem which states that the interaction picture of a relativistic QFT does not exist in a mathematically well-defined way also questions the conceptual coherence of QED (see for example Kuhlmann (2018)). But the mentioned points rather indicate a deeper conceptual tension in the interpretation of QED and, in contrast to the divergence of the S-matrix, do not figure in the philosophical debate around the virtual particle.

Secondly, Feynman diagrams and the corresponding matrix elements do not refer directly to probabilities of an event to occur. Only the absolute square of S_{fi} refers to a probability.[19] This essentially means that

$$|S_{fi}|^2 \approx |S_{fi}^{(0)} + S_{fi}^{(1)} + ... + S_{fi}^{(k)}|^2 \neq |S_{fi}^{(0)}|^2 + |S_{fi}^{(1)}|^2 + ... + |S_{fi}^{(k)}|^2 \qquad (2.7)$$

i.e. the probability of the process $|i\rangle \rightarrow |f\rangle$ to occur is not proportional to the sum over probabilities but to the square over the sum of probability amplitudes. The different contributions of Feynman diagrams interfere with each other.

2.2 Some Philosophical Stances Towards Virtual Particles

With the above background we can now proceed to the philosophical debate around the virtual particle. Therein the concept is exclusively studied through its post-Feynman articulation, and–taking an overall perspective–it can be stated that the majority of philosophers tends to deny an onologically realistic reading of Feynman diagrams and virtual particles.[20] All the same, there are important reasons why a closer look at the philosophical literature is advisable in the context of this historical work. First of all, the philosophical analysis directly aims at problematic features, either epistemological, ontological, or representational, and therefore it will provide for a better understanding of the concept before delving into its history. Secondly, and following the methodological guideline I set myself in Sect. 1.2, the philosophical debate has certainly informed my own understanding of the concept and is reflected in the historiographical framework discussed below.

Yet, the following discussion of the philosophical literature will be highly selective and far from comprehensive.[21] I chose a small subset of philosophical arguments which–in my view–best enhance our understanding of the virtual particle concept and its intricacies.

[19] This is equivalent to the difference between a wave function ψ, i.e. the quantum amplitude, of a system, and the absolute square $|\psi|^2$, which refers to a probability.

[20] That a physical reading of Feynman diagrams, and therefore also of virtual processes, was not shared in the physics community right from the start of the spreading of this diagrammatic technique is evidenced by what David Kaiser has called the "Feynman-Dyson split" (Kaiser 2005, 263): "[the] tension between Feynman's and Dyson's position, with their varying emphases on 'intuition' versus derivation, physical pictures versus topological indicators.".

[21] For a concise and rather comprehensive evaluation of philosophical arguments until 2007, see Fox (2008). In 2018, a Special Issue of *Perspectives on Science* (Volume 26, Issue 4), edited by James Robert Brown, presented mostly philosophical articles on Feynman diagrams. An article by myself on the philosophical stances towards the virtual particle, providing a more comprehensive discussion, is in preparation.

2.2.1 *Why Virtual Particles Might Not "Really" Be There*

While the literature on the philosophy of virtual particles and Feynman diagrams has gathered a considerable scope, I picked three arguments against the reality of virtual particles to present here. Each of them argues against a realistic conception of virtual particles from a completely different point of view. They highlight specific features of the concept which will show up in the historical reconstruction to follow. Let me start out with the most prominent and most robust one, the super-position argument by Robert Weingard (1982).

The Super-Position Argument [SP]

In 1982, Robert Weingard[22] presented an argument which is accepted by the majority of philosophers. As yet, I have not seen a philosophical argument which has countered or circumvented Weingard's observation. Tobias Fox (2008, 35) considered [SP] as *the* argument to be "sufficient to deny the realistic interpretation of virtual particles".

Weingard first accepts the concept and then questions its reality due to the consequences drawn. Let's assume we have some initial state $|p, q\rangle$ and some final state $|p', q'\rangle$, and the scattering matrix S is expanded perturbatively, so that we get:

$$\langle p', q'|S|p, q\rangle = \sum_k \langle p', q'|S^{(k)}|p, q\rangle \tag{2.8}$$

As already pointed out, this sum still refers to *probability amplitudes* and not to *probabilities*. This means, when we want to calculate the actual probability of a scattering event to occur, we need to square Eq. (2.8):

$$|\langle p', q'|S|p, q\rangle|^2 \approx |\langle p', q'|S_1|p, q\rangle + \langle p', q'|S_2|p, q\rangle + ...|^2 \tag{2.9}$$

The probability of going from initial to final state is not the sum of probabilities but the square of the sum of probability amplitudes.

But this means that we cannot separate out the contributions of Feynman diagrams with different kinds and numbers of virtual particles in them. The number of virtual particles for one and the same process is not sharp. Different kinds and numbers of virtual particles will contribute (see Fig. 2.5 for the scattering of two electrons to second order) to the probability of one and the same observable process. Therefore, Weingard concluded:

---

[22] The following is a short recapture/reformulation of Weingard (1982) and Weingard (1988, Sect. I). Weingard extended the argument in Section II of his 1988 article to a path integral approach and showed in Section III that through the so-called Faddeev-Popov ghost fields the number and types of virtual particles included in the theory become gauge dependent. Yet, for the current purpose a review of those parts is unnecessary.

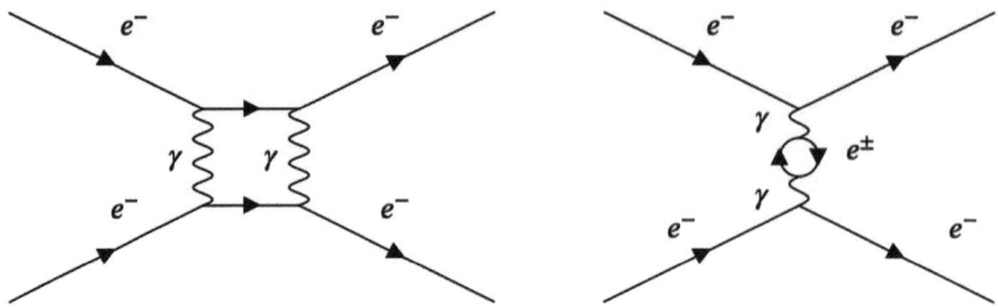

Fig. 2.5 Two Feynman diagrams contributing to the second order in the calculation of Møller scattering: In the left diagram an additional virtual photon and two virtual electrons, in the right diagram an additional electron-positron loop are included. The diagrams are read from left to right. Diagrams by the author

> According to the standard view [of superposition in quantum mechanics] *neither the number nor kinds* of 'virtual' particles will be sharp. Given this result, it is hard to see how the diagrams in any literal sense, can be thought of as picturing or representing the mechanism by means of which transitions such as electron-electron scattering take place. (Weingard 1982, 240, emphasis in the original)

The Argument of False Premises [FP] (Bunge 1970)

While [SP] constructs seemingly absurd consequences by first accepting the concept, Mario Bunge's argument, to my knowledge the first one by a philosopher of physics, argues against the reality of virtual particles by questioning the premises for a realistic interpretation. Bunge (1970) identified the breach with energy conservation by virtual particles and its explanation through the energy-time uncertainty as the premises for a realistic reading. He argued that these premises are false: Energy conservation, a law of nature, cannot be violated; rest mass and duration of a process cannot be considered mean standard deviations which enter the uncertainty relations; and the energy-time uncertainty is not a theorem of quantum mechanics. Thus, a realistic interpretation lacks grounding. He even argued that the concept is methodological unsound[23] and, because of its fictitious nature, "the weird metaphysical notion of a virtual object, as something real but not quite, must go" (Bunge 1970, 508).

Whether or not [FP] holds seems not to be settled. Fox (2008, 45–46), for example, remained undecided. The reference of Δt to the duration of a process is certainly spurious, but there is the question if the argument is still applicable to a modern formulation of particles being off shell, and there might be other possibilities to take account of the seeming violation of energy conservation.[24] Regardless of the philosophical and physical correctness of [FP], it points to two deep-seated problems:

[23] "[I]t consists in hanging one fiction from another in the style of Ptolemaic astronomy." (Bunge 1970, 508) The first fiction is most probably a particle ontology for QFT.

[24] For a discussion see Jaeger (2019, 13).

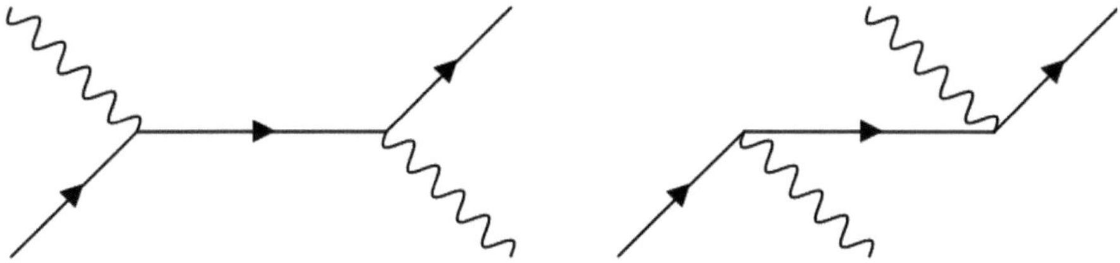

Fig. 2.6 Two Feynman diagrams contributing to the first order of Compton scattering. The diagrams are read from left to right. Diagrams by the author

(1) The basis for a realistic reading might be unsound, and, pushed to the limits already by Bunge himself, (2) this can entail that the concept should not be used in physics.[25] Both of these points, will show up during the historical reconstruction and motivated my definition of the analytical category of a "narrative".

The Argument of Topological Equivalence [TE]

Finally, I want to point the reader to a rather recent argument by Oliver Passon (2019). It is based on the interpretation of diagrams as representing a particular physical process: The Feynman diagrams depicted in Fig. 2.6 are the lowest order diagrams contributing to Compton scattering. In terms of processes, a verbal description would run as follows: In the left diagram first the incoming photon is absorbed and the final photon is emitted. The second diagram also contributes to the first approximation and the order of the two subprocesses is reversed. If we look at Fig. 2.7 it seems like there is another process in which the initial photon creates the final electron and a positron, which annihilates the initial electron under emission of the final photon.

Passon's point is that the diagram on the right in Fig. 2.6 and the one in Fig. 2.7 are topologically equivalent[26] and therefore only one of the diagrams has to be taken into account in the derivation of the scattering amplitude. Consequently, it seems as if two diagrams which can be interpreted as depicting two different sequences of events are one and the same. Consequently, Passon drew the conclusion that "Feynman graphs do not support a specific 'story' to be told" (Passon 2019, 12).

[25] It should be noted, that the refutation of the reality of virtual particles does most often not entail a call for the abandonment of the concept. Some of the few examples, where philosophers explicitly argued against the usage of virtual particles are Kristin Shrader-Frechette (1977, 421) and, quite recently, Richard Arthur (2012). Arthur is rather careful in his analysis: He does not argue against the use of fictions in science as such. He accepts the philosophical insight that scientists can use fictions in an epistemologically fruitful way. Building upon Alisa Bokulich's distinction (Bokulich (2009) and Bokulich (2012)) between "mere" or "non-explanatory fictions", Arthur still argues that virtual particles belong in the second category and therefore should not be used in physics. I believe that we can argue against Arthur's conclusion on several grounds, but this would definitely go beyond the scope of this overview.

[26] Roughly speaking, this means that you can deform one diagram into the other without breaking it up.

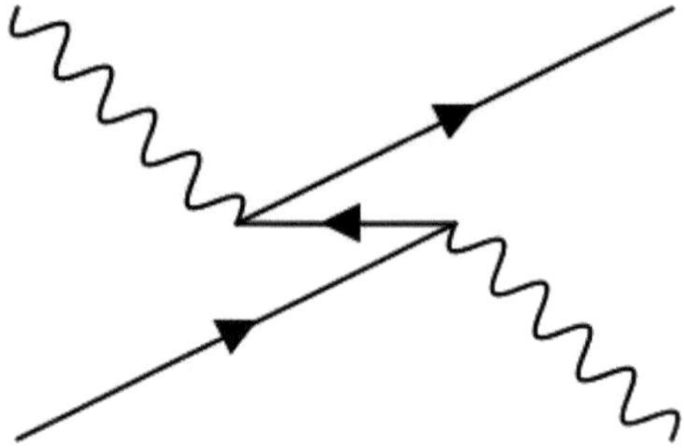

Fig. 2.7 Diagram topologically equivalent to the right diagram in Fig. 2.6. The diagram is read from left to right. Diagram by the author

Passon's argument against a realistic reading of virtual particles is aimed at the fact, that different "stories", i.e. different interpretations of Feynman diagrams in terms of processes, can boil down to one and the same process (even on a quantum-amplitude level). My historical reconstruction will highlight the "story to be told" as an epistemic tool for the protagonists when constructing mathematical representations of the investigated effects. And, as we will see, the equivalence of the two diagrams only holds for a covariant version of perturbation theory. In the practice of the physicists during the 1930s, the two processes depicted in the diagrams were truly different ones.[27]

An Interim Conclusion

The super-position argument is–to this day–the most robust argument against the reality of virtual particles.[28] It uses one of the prime features of the concept: Virtual particles play their role on a quantum-amplitude level of the theory. [FP], while on shaky grounds, points to the assumptions that ground a realistic reading of the concept, and it questions them. Such grounding pillars told around the concept, connecting the concept to the theoretical framework, will be called "narratives" in my analytical approach. Finally, [TE] highlights the fact that two diagrams, which we might read as different processes when we follow the direction of time through the diagram, eventually refer to the same mathematical structures. All the same–and possibly motivating Passon's argument–, textbooks actually

[27] Turning Passon's argument on the head and taking a pragmatist perspective, we might argue that it is a particular strength of the medium of Feynman diagrams that we have the freedom to envision processes differently and check their equivalence later on.

[28] Some arguments in the philosophical literature can actually be understood as variants of [SP] or [TE]. Paul Teller (1995), for example, pointed to the fact, that a Feynman diagram represents an integral over all space and time, and therefore is a representation of infinitely many processes itself. Michael Redhead (1988) highlighted the fact that we are unable to switch of the perturbation during the process–this would result in a different process–, and thus the state during the process cannot be "cashed out physically" (Redhead 1988, 20).

advise students to envision processes when starting out their calculations in QED.[29] Such (pseudo-)causal verbal statements, which find their graphical manifestations in Feynman diagrams, complement and enrich the mathematics, and facilitate the handling of the theoretical apparatus, partially motivated my introduction of the term "verbal model" (see below).

2.2.2 *For a (Mildly) Realistic or Representational Interpretation of Virtual Particles*

All the above points question not only a realistic reading of virtual particles, virtual processes or Feynman diagrams, but also the modelling or representational functions we can ascribe to them. Yet, the interpretational debate around the virtual particle and Feynman diagrams has not been settled, at least not finally. In the following, I will describe three arguments which point to the possibility of a realistic interpretation or, at least, a representational and epistemic function of Feynman diagrams and virtual particles. Again, I will not aim at comprehensiveness and my focus is not on the philosophical correctness of the arguments. Rather, these arguments will show how the characteristics of the concept on which some of the above arguments were built can be used in favour of a realistic or, at least, representational reading.

Dynamical Causal Role Arguments [DCR]

I will call the first line of reasoning dynamical causal role arguments [DCR]. The gist of these arguments is that virtual particles play a causal role in the description of the processes and, importantly, they carry with them the dynamical properties which are exchanged: momentum and energy. Such arguments do not defend any naive interpretation of Feynman diagrams in terms of space-time trajectories, and they are often combined with other forms of epistemological arguments.

Johannes Röhl (2005), for example, pointed to a category mistake if we assume that the energy and momentum exchange between real particles is ascribed to non-existent entities, i.e. virtual particles. At the same time, he provided for an epistemological argument why the concept should be taken seriously: "The whole [the scattering process] is not reducible to its parts, but the analysis of the components is reasonable in order to make properties of the whole intelligible."[30] The resolution of

[29] See, for example, the quote by Peskin and Schroeder in Sect. 2.4.

[30] "Das Ganze [der Streuprozess] ist nicht auf die Teile reduzibel, aber die Analyse der Komponenten ist sinnvoll, um bestimmte Eigenschaften des Ganzen verständlich zu machen." Röhl (2005, 475).

the real scattering process through an analytical decomposition using virtual particles can warrant genuine insights into its properties.[31]

Adrian Wüthrich (2012) gave a more context-sensitive version of this argument. He highlighted the virtual particles' role as the "common cause" (Wüthrich 2012, 178) for the exchange of momentum between the two scattered particles.[32] But instead of implying a classical causality, he showed how the dynamical relations between real and virtual particles, or rather their off shellness, can be used to gather information about the statistic of the represented events. Wüthrich combined this with a no-miracle argument: Since we can rely on physical intuition in constructing Feynman diagrams (Wüthrich 2012, 175–176), it seems highly unlikely that these construction methods do not "contain at least some correct statements about the state of affairs in question" (Wüthrich 2012, 176).

[DCR] with a Pragmatic Twist: [PDCR]

Mario Bacelar Valente (2011a) combined a [DCR] argument with a pragmatist attitude towards the divergence of the S-matrix expansion and a mathematically informed reading of Feynman diagrams. I shall call his argument [PDCR]: Pragmatist [DCR]. Bacelar Valente used the perturbative nature of QED and the divergence of the S-matrix to invest virtual particles with more than computational meaning. As he noted: "We do not 'stop the perturbation progression due to practical reasons'; we only really have few lower-order terms to count on." (Bacelar Valente 2011a, 45) Thereby, he mitigated [SP], since only a limited numbers of different possible processes overlap, but he could not circumvent it.

Importantly, Bacelar Valente rejected any "simplistic view of virtual quanta" (Bacelar Valente 2011a, 45) because of the overall space-time nature of Feynman diagrams. He refused the use of "temporal language in an overall description of the interaction in quantum electrodynamics" (Bacelar Valente 2011a, 47).[33] Conceiving of Feynman diagrams in such an informed way, he argued that "we can take the intricate description of one Feynman diagram [...] as the physical description of interactions in quantum electrodynamics, giving at the same time the only available physical-mathematical meaning to what we understand by the term 'exchange of quanta'." (Bacelar Valente 2011a, 49)

---

[31] It must be noted that Tobias Fox (2008, 47) has already encountered this version of [DCR]: Feynman diagrams do not describe causal processes but represent statistical mathematical objects. There is no classical causal connection.

[32] "The photon *mediates* the interaction between the electron and the muon; it *links* or *connects* the two particles." Wüthrich (2012, 178, emphasis in the original).

[33] As will become clear in my own approach to the language used to describe the interaction in perturbative QED, I consider the temporal description highly important in providing for an epistemic relevance of virtual particles.

Bacelar Valente did not argue for a realistic reading of virtual particles, but he used the above [DCR] point to argue for an epistemic role of virtual particles: "They convey a relevant physical description or explanation of the quantum interaction in terms of providing the link for a quantized exchange of energy and momentum between the real particles." (Bacelar Valente 2011a, 50)

Stöltzner's Models as Mediators Perspective

By explicitly referencing Bacelar Valente as a basis for his considerations, Michael Stöltzner (2017a, 2018) gave an interpretation of Feynman diagrams from a models as mediators perspective [MaM].[34] According to Stöltzner (1) a single diagram is best understood as representing a mathematical object, but (2) when used to denote the leading order term of a subseries, that is connected to a specific event or effect, it might be thought of as a model of a physical process. (3) The part of such a subseries that is actually calculated as it gives sufficient accuracy in connection with experiment could as well be interpreted physically in accordance with Bacelar Valente's pragmatic account. (4) The complete expansion of the S-matrix, which is at times accredited as actually being the theory, faces the problem of possibly not being well defined, as it may diverge. Consequently, the independence of the model of Feynman diagrams from the higher level theory, as it is important from a models as mediators perspective, is warranted.

The modelling of physical processes in terms of subseries of the S-matrix can surely not be taken as independent of the overall approach of describing scattering by a perturbative expansion of the same, rather "their representative features emerge within [...] the Feynman diagram analysis of an S-Matrix. Yet [they] are physically reasonable enough for experimental physicists to attempt to measure individual subseries, be it [...] subprocesses in the S-matrix, or effects being part of the Lamb shift" (Stöltzner 2018, 498). The representational feature of Feynman diagrams is therefore further supported by the actual practice of contemporary theoretical and experimental physicists.

Decoupling Representation and Understanding from Ontological Commitments

Letitia Meynell (Meynell (2008) and Meynell (2018)) argued for a representational function and epistemological relevance specifically of Feynman diagrams, while her argument can be extended to verbal descriptions of virtual processes as well. Her starting point is the distinction between denotation and representation. She argued

[34] For more details on this approach towards the analysis of modelling in the sciences see the influential edited volume by Morgan and Morrison (1999). Roughly speaking, in this approach models are conceived of as mediating between high-level theory and the world, while still having some independence of both of these realms.

that Feynman diagrams can have representational features, although not necessarily realistic ones, and can thereby contribute to an understanding of quantum physics. We might play a "game of make-believe" when discussing virtual processes and picturing Feynman diagrams, but these representations can nevertheless function as "props for our imagination" and "allow viewers to engage in coordinated imaginings of subatomic events" (Meynell 2018, 462) as they refer to shared "visual schemata" (Meynell (2008, 45); originally taken from Kaiser (2005, 372)).

Following the growing insight that scientists constructively use fictions in their reasoning,[35] Meynell (2018) reiterated and argued for a distinction of understanding and the category of knowledge, closely connected to factivity. In the consequent conception of scientific theories as a "carefully organized bricolage of fact and felicitous falsehood that points to the relevant features of nature" (Meynell 2018, 471)[36] scientists show the extent to which they understand the theory by the extent they are able to "fruitfully and creatively navigate" (Meynell 2018, 471) within it. Through the usage of Feynman diagrams physicists can engage in a more informed, even if not realistic, fashion with the quantum world: "they [Feynman diagrams] 'may prove to be entirely haywire' or 'absurd' [but] may nonetheless 'help'." (Meynell 2018, 475) Even though Feynman diagrams and virtual processes may be unrealistic, they provide understanding and are consequently epistemologically relevant.

2.2.3 Intermediate Conclusion

In my discussion of the arguments against the reality of virtual particles, I already pointed to some prominent features of the concept: (1) The virtual particle plays its role on the quantum-amplitude level of the theory [Basis of SP]. (2) There is a widespread, although doubtful argument connecting the virtual particle to the basic pillars of the theory, i.e. the energy-time uncertainty [FP]. (3) Reading Feynman diagrams in physical terms as describing a process in space and time is practised, but already for one diagram doubtful [TE].

Most of the arguments in favour of a realistic interpretation or representational function of Feynman diagrams and virtual particles acknowledge that naive classical concepts of particles or causality should not be applied. A mathematically well-informed understanding of the mediating function of virtual particles as an interlink is the key to the [(P)DCR] arguments. And even if we follow the majority of philosophers in denying a realistic reading of virtual particles, the epistemic function of

[35] Meynell specifically named Catherine Elgin as her source, while it has to be noted that Meynell went beyond Elgin's conception of understanding and knowledge. For further reading on *Fictions in Science*, the introduction to the volume of the same title (Suárez 2009) forms a good starting point.

[36] It should be noted, that Meynell argued that the concept of felicitous falsehood of Elgin is still too strong a claim for Feynman diagrams and virtual particles.

the virtual particle can be argued for from two distinct points of view: (1) from a sociological point of view as Feynman diagrams are often used as exemplaric representatives of physical processes (see [MaM]), or (2) from a practical point of view as the handling of those diagrams and the imagination it entails are means of navigating in and exploiting the theory.[37]

Whatever side we might take in the debate, we have to acknowledge that virtual particles are an important part of theoretical practice, that they are the mediating entities in a perturbative evaluation of QED, that a corresponding processual language might be beside the point but can be used to navigate this evaluation, and that there exists a (doubtful) argument connecting the concept with the basis of the theory. These points have informed my own pragmatist approach to the history of the virtual particle concept and two important analytical tools I will describe in the following section: the "verbal model" and my understanding of a "narrative".

2.3 Physical-Mathematical Concepts and the Practice Turn

Following my own stance towards concepts and their historical study discussed in Sect. 1.2, the modern concept of the virtual particle might render first hints at how to assess its formation and development. My own elaborations on such an approach have already been published (Ehberger 2024).[38] In this section, I want to shortly summarize its basic pillars and elaborate on some of its consequences. I will focus on two modes of verbal representations, the historiographical notion of a false prediction and, finally, spell out the historiographical consequences with specific attention to the relevant sources.

2.3.1 Studying the Virtual Particle Concept After the Practice Turn

My approach towards conceptual development has been informed by the study of the philosophical status of the virtual particle concept, my investigation of the historical episodes and the historio-philosophical literature on conceptual development, especially by scholars who take a pragmatically informed approach.

[37] For full closure: Arthur (2012) has argued that virtual particles are "non-explanatory" fictions and therefore should not be used. I will comment on Arthur's argument in a paper in preparation.

[38] Section 2.3.1 of this book is based on Sect. 2 ("Using Current Debates in Historical Reconstructions") and Sect. 3 ("Studying Concepts after the Practice Turn"), pp. 283–293, of Ehberger (2024); Sect. 2.3.4 of this book has in modified form been published as Sect. 4 ("Conclusion"), pp. 293–294 of Ehberger (2024). A preprint of this paper is available at: https://philsci-archive.pitt.edu/23037/1/02-Ehberger.pdf (accessed 22 April 2025).

From the above discussion we can already draw some inferences about the virtual particle concept which most scholars (and physicists)–all besides the ontological status of the virtual particle–can agree on. (1) The virtual particle is a *physical-mathematical concept*, it is a concept of the highly mathematiced branch of theoretical physics. (2) It is often *not considered as fundamental*, but is mostly discussed as *part of larger theoretical structures*, such as Feynman diagrams. It is connected to the respective *epistemological practices*. (3) The ontological commitment of a respective actor or philosopher does (mostly) not interfere with the recognition of the virtual particle's *tool character*[39] and its *fruitful application*.

Taking these three observations, the insights from the philosophical debate on the virtual particle and the literature which takes a pragmatic outlook on conceptual development as my starting points, I developed an approach which I deem suitable for the historical study of the virtual particle concept: I will focus on the usage of the concept, on the involved scientific practices, but not necessarily on the virtual particle's ontological status. I take the viewpoint[40] that concepts should be conceived of as tools which are embedded in historically situated epistemic constellations and practices. This in turn means that their application, fruitfulness, and epistemic power depends on the framework in which they are put to use.

Furthermore, concepts should not be understood as some disembodied (and therefore inaccessible) ideas located in the brains of the historical actors, but as manifested in concrete representations–which include not only diagrams, tables, etc., but also, and importantly, terminology and mathematical formalism. Historical actors use, shape, manipulate, etc. such concrete representations in their everyday practice. The representations guide the reasoning process. By conceiving of concepts as embodied by representations, the epistemic function of a concept as a tool becomes manifest and accessible to a historical reconstruction.

As we will see throughout the historical investigation, studying the different representations of a concept allows us to think about breaks and continuities as manifested in the representational formats and as guided by the conceptual environment. Since often more than one concept is represented in one representation (mostly we deal with representations of models), shifts in the conceptual framework can also be traced and will reflect back on the concept under study. But physicists also and often use different representations of one and the same concept (or model) in their reasoning. As such a concept becomes a multi-dimensional set of different representations, embedded in its environment and manipulated by scientists.

---

[39] Bacelar Valente (2011a), for example, titled his contribution to the philosophical debate by: "Are Virtual Quanta Nothing but Formal Tools?".

[40] For details and the consulted literature I refer the reader to Ehberger (2024). Most important for my approach were certainly the work by Friedrich Steinle and Arianna Borrelli.

2.3.2 Verbal Models and Narratives

As this book engages with the development of a concept of theoretical physics, a particular important representational format will be the mathematical formalism. Also the terminology used to refer to the concept under study will gather some attention, as there is, on first sight, a tension between the claim of science to engage with the actual world and the usage of the category of a virtual X "as something real but not quite" (to use Mario Bunge's wording, Bunge (1970, 508)), or as something which behaves as if it were an X but actually isn't.[41] A representational format that might not have immediately come to the mind of the reader but which will gather some prominence in my historical investigation is the verbal representation of concepts and models.

Physicists use words to describe what they do and the structures they use. And they do so in different contexts: They make comments in notes and drafts, they communicate with peers in private, they present their work to other physicists, they engage with students or the public. In all of these contexts, the things to verbalize and the things to make apparent through language differ appreciably. As I have a particular interest in theoretical practice, the main focus will lie on the communication amongst physicists and only on some occasions include popular texts. But even in this smaller social group, language can be used in different forms in the reasoning and communication process. As two specific forms of verbal statements, both in the repertoire of practising theoretical physicists, take prominence in the formation of the virtual particle concept, I want to explicate and discern them upfront.

The focus on these two kinds of (mostly) verbal statements has been informed by the literature on narratives in science and narrative explanations.[42] For clarity's sake, I will only refer to one of the verbal structures as a narrative. Whether or not both types of verbal explications can be understood as a narrative structure and, hence, the following is a typology of narratives, is for some one else to decide. I can only state that setting the two types of verbal statements apart helped me in thinking about the historical development (but also the philosophy) of the virtual particle concept. Following a pragmatic approach, this delineation will be one of the respective usage, of the place the respective arguments take in the reasoning of the historical actors.

I will call the first kind of these verbal statements "verbal model", as a shorthand for "verbal representation of a model". A verbal model, as I define it, is a specific form of representing the particular model physicists use to evaluate the investigated effects, namely through the combination of words. A verbal model is thus to some extent a verbal description of a model, possibly an interpretation of its mathematical representation. But, as I define it, a verbal model goes beyond a mere interpretation as it is used within the context of quantitative analysis. Its main purpose is to allow

---

[41] See for example Charles Sanders Peirce's definition of "virtual" in the *Dictionary of Philosophy and Psychology* (Peirce 1902, 763). See also Steinle (2024) for a discussion of Peirce's definition.

[42] The special issue of *Studies in History and Philosophy of Science. Part A* on *Narrative in Science*, and especially its introduction, forms a good starting point for the interested reader. See Morgan and Wise (2017).

for or, at least, facilitate the calculation of the effects the community is interested in. Knowledge about the usage of a verbal model is part of the methodological tool kit of a practising physicists. In Kuhnian terminology, a verbal model is part of the puzzle-solving techniques. It serves analytic purposes. It might even contradict accepted guidelines of the interpretation of the respective theory. The justification of the verbal model is not consistency with the interpretation of the theoretical framework but its fruitful application in the quantitative evaluation.

But scientists do not only argue for a specific model or a concept on the grounds of empirical adequacy and fruitfulness. They also try to motivate specific models and concepts by embedding them into the conceptual framework, by connecting them to general principles of the respective theory and its interpretation. Following Hartmann (Hartmann 1999), I will call this second kind of verbal statement a "narrative". I am well aware that the analytical category of a "narrative" has been developed further and was more strongly connected to scientific explanation and knowledge, also for the physical sciences,[43] but Hartmann's definition serves my purpose best. According to Hartmann, a narrative "is not a deductive consequence of the model nor of the underlying theory. It is, however, *inspired* by the underlying theory (if there is one). [It] takes advantage of the vocabulary of the theory [...] and refers to some of its features [...]. Using more general terms, the narrative fits the model in a larger framework (a 'world picture') in a non-deductive way." (Hartmann 1999, 344) The narrative, as Hartmann defined it, is "told *around* the formalism of the model" and is "an integral part" of the same, since "it complements" the model (Hartmann 1999, 344).

The narrative, in the sense I will use it, is an argument for the application of the model and its representations. It is not necessarily concerned with the quantitative evaluation of the theory. If any direct conclusions are drawn from its use, these are mostly qualitative. Its main purpose is to motivate the application of a particular model or concept and it does so by embedding the model or concept in the broader theoretical and conceptual framework, the "world picture" as Hartmann has framed it. The narrative thus aims at establishing a consistency between the concepts used within the model and the principles of interpretation of the theory.

The verbal model is exclusively a verbal representation, although it is certainly informed by other representational formats, in my case most often the mathematical apparatus and previously established interpretations. The narrative, in my understanding, although most often and primarily framed verbally, may also include other representations as, for example, mathematical ones.[44] But these representations are

[43] See, for example, Stöltzner (2017b) or Borrelli (2015, 2021).

[44] In this sense, I certainly follow Arianna Borrelli's idea of "a hybrid narrative combining words, formulas, numbers and analogies" (Borrelli 2015, 69). But I digress from Borrelli's understanding of a narrative in a twofold way. On the one hand, Borrelli focused on explanations while in my understanding the narrative is mostly motivational. Secondly, Borrelli convincingly showed how concepts such as "naturalness" (Borrelli 2015) or "spontaneous symmetry breaking" (Borrelli 2021) can be conceived of as embodied through narrative knowing, how the concepts themselves are only accessible through such narrative structures. I, on the other hand, conceive of a narrative as motivating the use of a model and the concepts figuring in it.

not used for quantitative evaluation of specific effects, but rather qualitatively to show how the concept fits into the larger conceptual framework and the general interpretation of the theory. If the narrative is followed in its consequences, as we will encounter it in the discussion of nuclear theory in Sect. 10.2.3, the conclusions drawn are qualitative, at best order of magnitude estimations.

Both verbal models and narratives use the language of the theory and bring the concepts into contact with the conceptual framework. But they do so in very different ways. The verbal model is used in quantitative evaluations and serves as part of the inferential techniques physicists learn when solving problems. The narrative connects the model and the concepts to general principles of the interpretation of the theory. And even though they might invoke the same storyline, the verbal model is a calculational tool, while the narrative is a motivational structure.

Narratives and verbal models do not only differ in their function and their mode of connecting the concept to the conceptual and theoretical framework. Physicists invoke them in different places in the literature. Verbal models, as part of quantitative analyses, are part of research articles and the didactic literature. We can also encounter them in popular accounts, but then often in a rather stylized or simplified form. Narratives as motivational structures, on the other hand, can be part of research articles but do not necessarily have to accompany the corresponding verbal model. They might be part of didactic texts but, again, do not necessarily have to be invoked.[45] In overview articles or popular accounts, physicists tend to explicate such narrative structures to motivate their reasoning in terms of the corresponding verbal model. At times, and this is where the line between narrative and verbal model partially gets blurred, physicists follow the narrative in its consequences, also in research articles, but mostly to draw qualitative conclusions, while the actual calculation and quantitative evaluation is performed through the invocation of the verbal model.

2.3.3 False Predictions

There is one last piece of my analytical tool kit. It is motivated by an observation of a historical protagonist who was working on quantum field theory during the 1930s: Wendell H. Furry. Furry spent his post-doctoral years from 1932 to 1934 with Robert Oppenheimer partly in Berkeley and Caltech. In order that Furry would acquaint himself with QED, Oppenheimer first assigned him to study the derivation of the Klein-Nishina formula by Ivar Waller (1930b). In his AIP oral history interview, Furry did not only point towards the extremely crude and often overly complicated calculational methods of the 1930s (at least they seem so in hindsight). After Furry had performed the calculation of the Klein-Nishina formula, Oppenheimer suggested to Furry to make his first own calculations in QED, and Furry remembered that the following happened:

---

[45] As the verbal model is not necessarily warranted by the general interpretation of the underlying theory, physicists might even argue against the inclusion of the same by questioning the narrative.

> So he [Oppenheimer] thought of a calculation for me to do, and I can't remember exactly what it was, but it was of course interaction of radiation with an atom and probably also giving off an electron, but *there was some strange arrangement here whereby one was trying to get an effect*, and the *final conclusion of our thoughts on the matter was that the effect didn't exist*. If [sic!] didn't exist because while you could find large matrix elements for it, if you wrote all the matrix elements out for it, they always cancelled. And this is one of the cancellation effects, one which I don't think anybody ever did write a paper on.[46] There was another cancellation effect later on, on the Spurries(?) theorem,[47] but this wasn't that. But this indicated then that there wasn't really any phenomenon here. That happened after I worked on that for a number of weeks.[48]

What Furry describes here can surely be captured by the term "hypothesis" or "assumption" of an effect, which eventually turned out to be non-existent. But, judging from Furry's reference to the cancellation of the large matrix elements and the historical instances I will discuss, this episode is an example of my slightly narrower definition of a "false prediction".

Most generally the analytical term of a false prediction refers to a prediction that is made by a historical actor on the basis of a particular theory or model and its representations while, in the course of the theoretical investigation and importantly by the transformation of one representation into another, the purported prediction is found not to be a prediction of the theory or model.

Physicists might think that some effect should be part of a particular theory, as Oppenheimer did when he suggested the problem to Furry. Oppenheimer and Furry obviously could conceive of the effect within the conceptual boundaries set by the theory, its interpretation and its representations. Obviously, Furry was even able to construct matrix elements and most probably had to apply the current verbal model for this purpose. But when Furry investigated the phenomenon systematically and combined the matrix elements, he found after "a number of weeks" that there was no phenomenon at all.

Let me elaborate on this historiographical concept of a "false prediction" a little, as I will use it at different places in this historical investigation. First of all, I want to stress that the falseness of the prediction is not an anachronistic one. It is not the case that a new theory had superseded the former and the prediction priorly made is now ruled out. On the contrary, in the examples I will identify, the theory and the method of its evaluation did not change during the period from prediction to

---

[46] *Comment by the author*: This judgement by Furry is definitely wrong. Physicists wrote papers on such effects. But it is most probably true that a lot more calculations on such effects were performed than were published.

[47] *Comment by the author*: Here Furry might either refer to Furry's theorem (Furry (1937); see also Sect. 9.2.4) or to a specific calculation method introduced by Hendrik Casimir (1933), today known as "Casimir's trick". It operators with the traces (German: Spur) of Dirac matrices. The reference to Furry's theorem is much more likely.

[48] Interview of Wendell Furry by Charles Weiner, 1971 August 9, Niels Bohr Library & Archives, American Institute of Physics, College Park, MD USA. https://doi.org/10.1063/nbla.jcro.dirn. P. 35 of the PDF; emphasis added.

refutation (mostly within days to months). As a matter of fact, in nearly all cases I identified, the same historical actors found that their prediction was a false one, i.e. the phenomenon they investigated could not occur although they initially believed that their theory predicted it.

Then, one might wonder, how can something like a false prediction occur (both the prediction and its refutation)? Importantly, it is not (necessarily) the case that the false prediction is due to an incompetent use of the theory or its representations by a historical actor. Quite in contrast, the point of this historiographical notion is that the predictions might be perfectly reasonable given the theoretical tool kit of the respective physicist. In the cases I identified, the effects were envisioned on the basis of the verbal model of QED during the 1930s, while the actors found that the effects would not occur by transforming the said phenomenon into a mathematical format. Some of these false predictions, but not all, are effectively ruled out by the representational format of Feynman diagrams. Although, as is regularly stressed, the conceptual basis of the theory did not change in the late 1940s, the conclusions to be drawn from the theoretical tool kit were different ones. Thus, the concept of a false prediction is directly related to two of the main themes of my understanding of concepts: Concepts are manifested in theoretical practice and in specific representational formats.

False predictions might actually spur the development of the theory, as it was for example the case for Furry's theorem (see Sect. 9.2.4). Whatever the respective physicist or the community which becomes aware of the false prediction does, for the historian the concept of a false prediction allows for making explicit the differences in theoretical practice, even when the basis of the theory purportedly did not change. Whether on purpose or by accident, new or altered representational formats or even theorems about the theoretical evaluation can limit or extend the range of (false) predictions. As such, the concept of a false prediction provides for a window into the development of the thinking (im-)possibilities connected to different representational formats and their respective refinements.

At the moment, I can neither judge whether false predictions are a phenomenon that occur generally across theoretical disciplines nor the frequency with which they occur. But as will be seen in the historical discussion of the practice of QED during the 1930s, within this theory a number of false predictions occurred. At times, their evaluation was conceived of important enough to be published. Taking Furry's judgement in the quote opening this subsection seriously, the instances I identified constitute only a small subset of the whole of false predictions encountered by the historical actors. Thus, at least for the history of QED, the analytical concept of a false prediction is applicable and may be used in further evaluations of the practice of theory.

2.3.4 *Summary, Historiographical Consequences and Sources*

To summarize: I will conceive of concepts as tools for historical practitioners, embedded in a specific historically situated epistemic constellation. To study concepts and their development, I will focus on the representations which the actors used, which structured their reasoning process and which shaped the practice of a theoretical community. The epistemic function of a concept as a tool thereby becomes manifested and accessible to a historical reconstruction.

Since we mostly deal with representations of models, shifts in the conceptual framework can also be traced and will reflect back on the concept under study. However, concepts are not only connected to their conceptual environment through these representations. Scientists tend to motivate the application of concepts and models by pointing towards the connection to the overall interpretation of the theory. They invoke motivational structures of argumentation, a "narrative", in the sense I defined it above.

But representations of models and concepts do not always lead physicists to the prediction of phenomena which can actually occur according to the theory they use. Even informed and competent usage of the representations can lead to what I called "false predictions". Within the conceptual framework and by the invocation of specific representational formats historical actors could conceive of effects which showed to be non-existing when transferred into another representational format, often the mathematical one during the quantitative evaluation.

As a first specifically historiographical consequence of this approach an important shift in the question guiding the historical study of a concept emerges. We have to ask: "What did the concept or a specific representation do for the historical actor?" and "What did the actor use it for?" From this point of view, the objects of the historical study are to some extent the reasoning processes and possibilities of the historical actors. This might be considered a detour, but when trying to historically engage with concepts in scientific practice, I think it is a fruitful and necessary one.

The study of concepts through their representations has two further important consequences for the historical description of concept formation and development processes:

1. Studying the different representations allows to think about breaks and continuities as manifested in the representations. Since not all representations of one and the same concept necessarily change at the same pace, we can think of and trace continuities and breaks in the historical development simultaneously.
2. Focusing on representations further allows us to trace the origins of a concept even before the actors themselves conceived of a set of representations as referring to one and the same theoretical object. There might be abrupt moments,[49] but it can take a considerable amount of time until a set of representations, or parts of

[49] Potentially motivated by what Friedrich Steinle has called an "epistemic challgenge" (Steinle 2009, 310).

> it, is understood by an actor or the scientific community to refer to one and the same theoretical object. It might even be the case, and I think this is true for the virtual particle, that we cannot pinpoint any clear-cut moment in the development at which we can say that the concept was introduced.[50]

The line between concept formation and development is blurred.

Following the proposed line of thought, we will have to engage with the theoretical and representational repertoire of the actors[51] and the historical community in detail. There are also representational traditions which we must be aware of. This theoretical and representational repertoire of a historical community forms the resources and the boundaries for concept formation processes. Alterations in representations and the introduction of new ones have to be judged in comparison to such traditions. A concept or a representation and its implementation in scientific practice has to be accepted by the historical community to become part of exemplaric solutions, to become a genuine building block in scientific explanations. Furthermore and thinking about the scientific community, we must keep in mind that a public representation is also a mode of communication,[52] and that messages can also be misunderstood, leading to unforeseen paths of conceptual development.

In total, this seems like a very demanding approach and one question obviously comes to mind: Which sources can we draw from to implement the above conception of concepts in a historical study? The answer obviously depends on the detail with which we need to be aware of the reasoning process in order to judge the development and changes in the concepts and representations. For example: Since I consider the introduction of a verbal model of successive energy non-conserving transitions, emission or absorption processes crucial for the further development, I will engage with calculational and draft notes of Paul Dirac in this connection (Chap. 5). Also in the case of the first direct impact of virtual transitions on their conceptual surroundings–the development and evaluation of the Dirac sea (Chap. 7)–I engaged with the correspondence of Paul Dirac with Ivar Waller, Niels Bohr, and Igor Tamm. At places, where I considered a deeper look into the actual derivation

[50] Before taking the above perspective towards concept formation and development, I already argued against a sharply defined moment of concept creation in the case of the virtual particle by criticizing the application of biographical concepts such as "birth" (see for example Arabatzis (2006)) for concept formation process. See Ehberger (2020).

[51] For the role of the historical actor in the concept formation process, best see Steinle (2009, Sects. 5, 6, and 7).

[52] The communication of scientists amongst peers has always been a major theme in the history of scientific concepts and I refrain from giving any detailed elaborations. Just to consider one prominent example which also covers a wide range of publication types, take the "Denkkollektiv" of Ludwik Fleck (1980) (originally published 1935): It is defined by the "Gemeinschaft der Menschen, die im Gedankenaustausch oder in gedanklicher Wechselwirkung stehen" (Fleck 1980, 54). And the exchange of ideas did not necessarily have to be direct. From the esoteric circles of the informed and specialized scientists to the interested layman, Fleck correspondingly considered forms of publication from the journal article over the handbook and textbook to popular scientific publications.

processes necessary, I had to face the fact that next to no archival material exists, as in the case of the work by Werner Heisenberg and Hans Kramers (Sect. 4.2) or the first calculations on light-by-light scattering and the Delbrück effect (Sect. 9.2). But for the complete work the materials of the Archive for the History of Quantum Physics (AHQP), the Werner Heisenberg papers at the Archive of the Max Planck Gesellschaft (Berlin) and the published correspondence of Wolfgang Pauli were used for further contextualization.

Publications, as a means of communication, can also be extremely explicit about the scientists'[53] methodology and thus a valuable source for the historian engaging with the practice of science. I consider it a reasonable assumption that, if a specific means of representation or a specific concept was important (necessary, guiding, helpful, etc.) for the conclusions drawn, the historical actors will try to explain and reproduce it in publication. Hence, the reproduction in a publication points towards the epistemic upshot of the representation or concept for the respective actor. Vice versa, the non-invocation of a concept or representation, when it could have been at the actor's disposal, deserves attention and needs explanation.[54]

I consider it suitable to mostly engage with published literature in Chaps. 8, 9 and 10, where a wider perspective on the development of the practices and (published) representations as well as their application is the focal point of my investigation, while the concept of virtual transitions (modern terminology) had already been established. Finally, didactic literature is especially important to judge in how far a specific concept or model has already been accepted as a resource for theorizing. The main intention of textbooks is to spread ideas, methods and representations, to teach students how to handle them. As such, they allow us to judge whether a concept or representation has become part of the tool kit of the scientific community. They allow us to gather information about the shared (representational) practices of a community and not only of one specific historical actor. Therefore, I will also engage with the most popular textbooks for the application of QED in Sect. 8.2.

But before going into the historical discussion itself, let me elaborate my approach with an example. In order to provide for a better apprehension of my understanding of concepts, I will apply the ideas spelled out in the previous sections on one central example: the 1949 paper on the *Space-Time Approach to Quantum Electrodynamics* by Richard Feynman (1949a).

---

[53] According to Ludwik Fleck (1980, 156–157), scientists surely have a tendency to make their personal stances invisible in publication. Yet, as Fleck (1980, 156) observed: "Die Zeitschriftwissenschaft trägt also das Gepräge des Vorläufigen und Persönlichen."

[54] I will use this kind of reasoning at different places. It is particularly important for the relevance of the picture of successive transitions in the Kramers-Heisenberg paper, see Sect. 4.2.

2.4 Representations of the Virtual Particle in Scientific Practice: The Example of Feynman's Space-Time Approach

The following analysis[55] will, on the one hand, allow the reader to get a better grasp of my analytical framework. On the other hand, I want to tentatively explicate a perspective on Feynman diagrams and virtual particles that is definitely practised in (the didactics of) physics and is implicit in the philosophical discussion, but, to the best of my knowledge, has not been made explicit therein. I will draw from Feynman's original presentation and application of his approach to quantum electrodynamics (Feynman 1949a) in order to show how different representations of the concepts and the model of the light-matter interaction were intertwined and each served its purpose in allowing for conclusions to be drawn.

Feynman's aim was to find a consistent way of describing the interactions taking place in quantum electrodynamics within his, at the time, idiosyncratic propagator approach.[56] When Feynman described his methods, he gave an explanation of the scattering of two electrons in a diagrammatic representation (see Fig. 1.1). The corresponding mathematical representation was given by Feynman in the following terms:

$$\begin{aligned} K^{(1)}(3,4;1,2) = -ie^2 \int\int K_{+a}(3,5)K_{+b}(4,6)\gamma_{a\mu}\gamma_{b\mu}\times \\ \times\delta_+(s_{56}^2)K_{+a}(5,1)K_{+b}(6,2)d\tau_5 d\tau_6 \end{aligned} \tag{2.10}$$

These two representations were connected by Feynman through the following verbal statement ("a" and "b" refer to the scattering electrons and the numbers to space-time points):

> It can be understood (see Fig. [1.1]) as saying that the amplitude for "a" to go from 1 to 3 and "b" to go from 2 to 4 is altered to first order because they can exchange a quantum. Thus "a" can go to 5 (amplitude $K_+(5,1)$) emit a quantum (longitudinal, transverse, or scalar $\gamma_{a\mu}$) and then proceed to 3 ($K_+(3,5)$). Meantime "b" goes to 6 ($K_+(6,2)$), absorbs the quantum ($\gamma_{b\mu}$) and proceeds to 4 ($K_+(4,6)$). The quantum meanwhile proceeds from 5 to 6, which it does with the amplitude $\delta_+(s_{56}^2)$. We must sum over all possible quantum polarizations μ and positions and times of emission 5, and of absorption 6. Actually if $t_5 > t_6$ it would be better to say that "a" absorbs and "b" emits but no attention need be paid to these matters, as all such alternatives are automatically contained in [2.10]. (Feynman 1949a, 772–773)

Feynman, thus, provided his reader with three different representations–mathematical, verbal, and diagrammatic–with which he worked and which all described the first order of the interaction between two electrons. Surely, there were a lot of things going on in the background and only with a backdrop to the overall development and a concretization of the functions playing a role in the mathematical

[55] It is underlying the introduction of Ehberger (2022, 532–535).

[56] I will not go into a full discussion of the development of the theoretical and conceptual framework here. The interested reader is referred to Chap. 11 and the literature cited therein.

equations the picture can be considered complete. Yet, these three representations already allow for the evaluation of their mutual dependency and their functions within Feynman's reasoning–at least, the function in the reasoning Feynman presented to his readers.

The function of the mathematical expression is quite clear: It was a means for calculating the cross section. Yet, this mathematical representation would be impotent without any physical interpretation in terms of what the functions and, therefore, the entire expression referred to. This points to the general theoretical and conceptual background that Feynman was working in. It was partially verbalized by Feynman in the quote above.

Yet, this was not the only function the explicit verbal statement in terms of a causal process served. To acknowledge this, I want to point the reader to the framing Feynman gave his verbal representation[57]:

> We shall now interpret Eq. [2.10] in a manner which will permit us to write down the higher order terms. [Then comes the interpretation as given in the quote above and Feynman closed his interpretation with the following:][...]
> The correct terms of higher order in e^2 or involving larger numbers of electrons (interacting with themselves or in pairs) can be written down by the same kind of reasoning. (Feynman 1949a, 772–773)

For Feynman, the storyline implicit in Eq. (2.10) allowed for the derivation of higher order terms when made explicit, in his case both verbally and diagrammatically.

Actually, as Feynman presented his reasoning, the interpretation in terms of processes was underlying the construction of the diagrams, but from historical investigations (Wüthrich 2010) we know that this was most probably not (entirely) the case. Nevertheless, the verbal representation had a tool character and was used for what a lot of everyday tools are used for: construction. This is an instance of what I called a verbal model: The verbal representation was applied by Feynman for the construction of diagrams which he could then translate into mathematical equations. As such, the verbal representation's most prominent function was its role in the quantitative analysis.

In the quote, Feynman also explicitly pointed to the abstractions involved in the verbal and diagrammatic representations (the summations involved and the alteration in the order of emission/absorption processes) but he advised the reader not to worry: The mathematical representation was of such a nature that these abstractions were actually allowed. The form and the features of one representation, reflected back on the usage of the others.

In Feynman's representations of the light-matter interaction, we also encounter representations of the concept I am trying to write the history of, the virtual particle: (1) the wavy line in the middle of the diagram which connects the two electrons, (2) the mathematical representation in terms of the function $\delta_+(s_{56}^2)$, (3) the term "virtual quantum", so the name Feynman used to refer to the concept, and, finally, (4) the description of the behaviour of the quantum in the verbal model as quoted above.

---

[57] Feynman made very similar statements on the verbal representation in momentum space, see Feynman (1949a, 774–776).

Feynman (1949a, 773) also described some of the features of the virtual particles in connection to his conception of the scattering process. In particular, he named the characteristics which were important for the quantitative evaluation: For particles in intermediate states, the Pauli exclusion principle could be completely neglected; energy and momentum were conserved at the interaction points; emission and absorption times of the virtual quanta were irrelevant; and, finally, longitudinal and transverse components were described by the same functions. All of these points digressed from the conception of the process during the 1930s. Feynman traced these alterations to a reinterpretation of the concept of negative energy particles, hence an alteration in the conceptual environment,[58] and to the "unrelativistic basis" of previous separations, so a methodological step.

In Feynman's description, the virtual particle served an important function: It was the connecting link in the (probably pseudo-)causal and physical interpretation Feynman provided. Without the concept, no interaction would take place at all. But Feynman did not give any definition of the concept. Reading the article, one gets the impression as if Feynman expected the reader to be familiar with the concept. The closest Feynman (1949a, 773, Footnote 8) came to a definition was that "they [virtual quanta] have a known source and are eventually absorbed".[59]

Narrative structures, either to motivate his overall approach or the inclusion of the virtual particle in the conceptual framework of QED, did not occur in Feynman's research article. He devoted a whole section to argue for his space-time approach, but his points were mostly technical. He also argued for a particular piece of his overall conception, namely that the interaction was understood as an interruption of the mostly freely moving particles (Feynman 1949a, Second Paragraph of Sect. 2). But he did so by referring to the "enormous advantage" (Feynman 1949a, 771) of the concept in calculations. Feynman provided purely pragmatic reasons. False predictions, as I defined them, did not occur in the publication. But, as I will argue, Feynman was already led to a false prediction of his diagrammatic technique in autumn of 1948.[60]

All the same, we have already encountered the most popular narrative for the virtual particle in the prior discussion and also arguments against it (e.g. Bunge's argument of False Premises as presented in Sect. 2.2.1).[61] According to this narrative,

[58] Feynman proposed this reinterpretation of a positron being an electron travelling backwards in time in Feynman (1949b).

[59] Feynman blurred the lines between real and virtual particles in this footnote as he stated: "One way to deduce the correct rules for real quanta from (4) is to note that in a closed system all quanta can be considered virtual (i.e., they have a known source and are eventually absorbed) so that in such a system the present description is complete and equivalent to the conventional one." Feynman (1949a, 773, Footnote 8).

[60] See Sect. 11.3.2. This instance is well-known, but the interpretation in terms of a false prediction is mine.

[61] This narrative is questioned in textbooks as well. David Griffiths (2004, 60), for example, stated that "[m]any authors interpret this [the off shellness of virtual particles] to mean that virtual processes violate conservation of energy [...]. Personally, I consider this misleading, at best. Energy is always conserved." Interestingly, although Griffiths rejected the narrative normally told around the virtual

virtual particles are only present for a short amount of time and, hence, for the short time of their existence they are allowed to violate energy conservation as implied by the energy-time uncertainty relation $\Delta t \Delta E \geq \hbar$. This argumentative structure connects the virtual particle and virtual processes directly to one of the foundational pillars of quantum theory, the Heisenberg uncertainty relations. It motivates the virtual particle's inclusion in the conceptual framework.[62] But the narrative has no calculational purpose. At best, the conclusions drawn are qualitative (see also the discussion in Sect. 10.2.3). Forms of this narrative would have been available to Feynman, but he chose not to present them. Similarly, when energy non-conserving transitions to intermediate states were introduced by Paul Dirac (see Chap. 5), Dirac did not provide any explicit justification.

All besides this analysis, the explicit physical reasoning with virtual particles, the use of the verbal model and, hence, the way of thinking presented above was idiosyncratic to Feynman. Freeman Dyson, for example, followed a different approach towards the diagrams, namely without invoking a procedural language and rather arguing on the basis of their topological features.[63] To be sure, there are definitely merits in thinking about Feynman diagrams in such a way and I do not want to argue that the merits of the diagrams are completely taken into account by the possibility of interpreting them as processes. Especially in modernday physics, where Feynman diagrams are no longer necessarily read by humans but created and evaluated by computer programs,[64] the topological merits of the diagrams as calculation techniques gather particular importance. But calculational merits do not exclude a physical interpretation or a modelling function of the diagrams.

In the beginning, actually Dyson himself acknowledged:

> In Feynman's theory the graph corresponding to a particular matrix element is regarded, not merely as an aid to calculation, but as a picture of the physical process which gives rise to that matrix element. [A short interpretation is given.] This interpretation of a graph is obviously consistent with the methods, and in Feynman's hand has been used as the basis for the derivation of most of the results of the present paper. (Dyson 1949b, 496)

Also in modern textbooks we find similar statements. Take for example Peskin and Schroeder's *An introduction to Quantum Field Theory*, one of the standard textbooks in the field. They opened their book with a heuristic introduction to scattering events and stated, in connection with the methodology of QED:

---

particle, he seemed to favour a realistic reading (see the comments on the difference between real and virtual particles in Griffiths (2004, 60)). Hence, the negation of the narrative does not actually have to come with the negation of the ontological status of the virtual particle. For Griffiths, the narrative was simply besides the point as virtual particles do not violate energy conservation but are off shell.

[62] Not to be misunderstood: I, personally, do not think that this is particularly convincing and we know that such an argument is, at the least, on shaky grounds. The above statement is simply an observation on what the users of the narrative want to imply.

[63] This difference in interpretation has been dubbed the "Feynman-Dyson split" by David Kaiser (2005, Especially the first section of Chap. 5).

[64] I want to thank Robert Harlander for pointing this out to me.

> *Imagine a process* that can be carried out by electrons and photons, draw a diagram, and then use the diagram to write down the mathematical form of the quantum amplitude for that process to occur. (Peskin and Schroeder 1995, 3, emphasis added)

Following Peskin and Schroeder, the verbal model and the envisioning of a causal process stands at the beginning of any calculation in QED. But also textbook authors who noted that "the reader must be warned not to take this pictorial description of the mathematics as a literal description of a process in space and time" (Mandl and Shaw 1984, 56) use a physical and processual description to introduce the idea of Feynman diagrams (Mandl and Shaw 1984, 54–55).

The basic concepts involved in the verbal model, through which we can build up higher and higher order terms, are rather simple: emission, absorption, pair creation and annihilation. Although these processes on their own, as they are displayed graphically in Fig. 2.2, are physically impossible, we can use these basic ingredients, either, more formally, through the Feynman rules connected to the diagrams or, more heuristically, by envisioning a process that can be carried out by electrons, positrons, and photons, to initialize a computation in QED. We might use reasoning that is connected to thinking in physical, space-time concepts for the construction, although we know that the story is based on unphysical building blocks and does not take place in space and time.

I want to connect the laid out interpretation to the question of the epistemic status of Feynman diagrams, virtual particles and processes, and to the question why these objects function so highly effectively. Both Letitia Meynell and David Kaiser (see Sect. 2.2.2) have specifically highlighted the "shared visual schemata" encoded in Feynman diagrams and have explained their heuristic power by focussing on their visual features. My attention, in contrast, lies on the verbal representation which can be used to construct the diagrams. The basic idea I argue for is that Feynman diagrams form such an highly effective tool in research and specifically education and the means how they unfold their epistemic power as "props for our imagination" lies partly in them providing us with the means for interpreting them as causal processes and for constructing such from very simple basic building blocks. These basic building blocks and the way to handle and apply them are of course shaped by the theoretical and conceptual environment. But through their interpretation they serve as tools for the quantum field theorist.

When conceiving in this way of the procedures of Feynman, we see the virtual particle concept emerging in the application of it. Within the verbal model, the virtual particle always takes the form as the causal link between the unperturbed, free solutions. In this manner, when consistently conceiving of QED as a perturbative theory of the relativistic interaction between quantized fields, the virtual particles form an integral part of genuine theoretical explanations.

This is certainly not an ontological argument. It aims at the explanation of the epistemological relevance of the concept of the virtual particle by pointing to its role in the practice of QED. It should also provide for an explanation why this kind of reasoning is so appealing: It is not only the creation of shared visual schemata that makes Feynman diagrams appealing. It is also the storyline that might be used in the construction of the diagrams that explains the attraction of the method, although it eventually leads us to a picture of processes that are no longer temporally but only causally ordered.

As a matter of fact, the concepts we will meet throughout my historical discussion will all take a similar position as the virtual particle in theoretical explanations given by the historical actors, namely as intermediaries in the description of interaction processes. I will also lay specific focus on what effects can be envisioned in the particular conceptual framework and how the representations served the actors in their investigation. The focus will lie on the diverse ways in which concepts and their representations allow for "surrogate reasoning", to borrow a term of Suárez (2004). I will show how concepts and their representations enabled the physicists to better understand the theoretical framework with which they were working and to provide for the prediction of new, as yet unforeseen empirically testable effects.

Chapter 3
The Community of Practitioners

The conception of concepts discussed in the previous chapter located the development of the abstract concepts in their concrete manifestations, i.e. their representations. This chapter aims at locating the conceptual trajectories in the real-world network of practitioners. In the following, I will discuss the broader social, political, economical, and epistemic constellations which play a huge role in forming scientific communities.

One might wonder why such a chapter is necessary in the historical investigation of a concept. First of all, it should serve the reader as a backdrop when the overall development of the community and networks is lost in the nitty-gritty study of the reasoning processes in the chapters to come. Secondly, it should further underline the fact that, when we are dealing with scientific practice and with representational formats, the communication (possibilities) amongst peers, their networks, and the emergent "cultures of theory"[1] form the necessary backdrop of a historical analysis. Concepts are neither discovered nor abstract ideas, they are formed, communicated and altered by people in a specific social and professional constellation.

The investigated period stretches from roughly 1920 to 1950 and encompasses four generations of theoretical physicists and major disruptions of theoretical research, both in its content and its social environment. Changes in the network of practitioners have to be understood as shaped through the changing social, economic and political conditions but also in reference to these deep-lying problems (as well as their solutions) in the content of the physical research.[2]

The nature of this chapter, essentially an overview of the historical development of the network of practitioners, will thus differ from the chapters to come, which

[1] See Galison and Warwick (1998). Joas and Hartz (2019) have taken up this theme under the term "quantum cultures". A collection of papers focusing on examples of such can be found in the corresponding special issue of *Berichte zur Wissenschaftsgeschichte*.

[2] This also means that the classical (and probably already overcome) dichotomy between internal and external factors is rather unimportant for the following. Also Galison and Warwick (1998, 288–289) refuted this distinction for the study of "cultures of theory".

M. Ehberger, *Representing the Unobservable*, Science Networks. Historical Studies 68,
https://doi.org/10.1007/978-3-032-09188-8_3

focus on the conceptual development in connection to theoretical practice. It will set the stage for the main part of this historical investigation. As such, it is based on two pillars: On the one hand, it is informed by the reconstruction in the following chapters. I will respectively refer to the corresponding sections. On the other hand, this reconstruction will draw from the rich literature on the social development of the physical sciences in the first half of the 20th century. My main sources from the secondary literature will thus be delineated upfront.

Helge Kragh (1999) provided a comprehensive study of the development of quantum physics in the 20th century, which forms an important basis for the following. All throughout his *QED and the man who made it*, Sam Schweber (1994) discussed the networks and biographies of practitioners in some detail. Schweber's general distinction between the particle and wave viewpoint is followed in this outline. For the discussion of Soviet science, I mainly consulted the work of Alexei Kojevnikov (2004). For Japanese physics on quantum field theory before WWII, the works of Laurie M. Brown, a lot of which is captured in Brown and Rechenberg (1996, Part B), are a very valuable resource. Also the dissertation of Kenji Ito (2002) has been consulted. For funding policies and political matters in relation to the US community of physicists, Daniel Kevles (1979, especially Chaps. 11–21) provided a comprehensive discussion. Katherine Sopka (1988) focused on US quantum physics and in particular quantum theory until 1935. Sam Schweber (1986) has further emphasised the pragmatic attitude of theoretical physics in the US. Aramaki (1987) and Rueger (1992) formed my most important sources for the reactions of quantum field theorists on the divergences during the 1930s. Finally, but very importantly, the *Physics History Network* of the American Institute of Physics is an incredibly valuable (although a bit US/Western-centred) tool comprising, inter alia, a huge collection of biographical data of prominent physicists.[3]

In the first section of this chapter I will describe the centres of research of the so-called old quantum theory, the respective schools of theoretical research, and how both the inception of the virtual oscillators and virtual radiation field as well as first models that described scattering as a two step process of emission and absorption subprocesses have to be understood in reference to the methodological and conceptual preferences at the respective institutes. The second section will deal with the spread of quantum theory and the development of quantum electrodynamics in Central Europe. Based on the distinction between the "field and particle viewpoint", I will shortly introduce their main proponents and describe how these viewpoints interrelate with the development of non-mathematical representations of concepts connected to the virtual particle, foremost the verbal model introduced by Dirac. My description of the dissemination of these ideas, most importantly to the Soviet Union and the US, will focus on the pragmatist attitudes of the theoretical communities. I will discuss how the reactions of different (groups of) physicists to the divergences and the epistemological constellation of the field partially hindered, but also contributed to the further development of the virtual particle concept. Finally, I will shortly

[3] Accessible under https://history.aip.org/phn/ (visited on August 28. 2022).

portray the development of the European research centres after the rise to power of the National Socialists and shortly turn to the pragmatic attitude of the post WWII generation of American scientists.

3.1 The Centres of the Old Quantum Theory and Their Approaches to Theoretical Physics

Chapter 4 will deal with the so-called old quantum theory, i.e. with quantum-theoretical reasoning prior to the development of matrix and wave mechanics (roughly mid 1925). From the "patchwork of problems",[4] dispersion and scattering will be central for the development of concepts closely connected to the virtual particle. The main focus will be on the period from roughly 1923 to 1925, for which the development of networks and institutional settings is extremely well researched.[5] In the beginning of the 1920s, the physical institutes in Copenhagen, Göttingen, and Munich, were gathering or had already gathered their status as major centres of quantum-theoretical research.[6] Since the output of physicists associated with specific institutes was still largely controlled by the residing professors,[7] the development of concepts historically related to the virtual particle can only be understood in connection to the particular epistemic values favoured by these key figures.

In Munich, Arnold Sommerfeld, mathematician trained in Göttingen and since 1906 Professor at the LMU, established a "nursery of theoretical physics", as Sommerfeld himself has called his institute.[8] By the end of WWI, the Munich school had made important contributions to the evaluation of Bohr's atomic model. The

---

[4] In reference to the framing of the period of the old quantum theory as a "patchwork of problems", Martin Jähnert (2019, 3) has further introduced the analytical term "state transition model". The main point is that theoretical research applying quantum ideas had not yet found a coherent theoretical underpinning and Bohr's atomic model was not used by all practitioners. But the conception of transitions between quantized states was generally shared. Although I find this a valuable outlook on the respective field of research, I will still use the term old quantum theory as a chronological umbrella term for quantum-theoretical research prior to wave and matrix mechanics.

[5] Quite recently, a series of four books have engaged with the social and institutional settings of the four centres of quantum-theoretical research in this period: Berlin (Castagnetti and Goenner 2020), Munich (Eckert 2020), Copenhagen (Kojevnikov 2020) and Göttingen (Schirrmacher 2019).

[6] I will not go into much detail about the development of their status or the initial spread of quantum ideas. For a very concise description of the developments of the institutes in Munich, Copenhagen, and Göttingen, see Eckert (2001).

[7] At the time, as Kojevnikov (2020) stressed, the influence and the control of residing professors on their assistants and their scientific publications was still enormous: "Professors typically suggested to their 'scientific helpers' (*wissenschaftliche Hilfsarbeiter*) topics for research and calculations and also, by the accepted social norm, controlled all publications coming from the institutes they directed. Prior to sending a paper to a journal, even such qualified researchers as Pauli and Heisenberg required the manuscript to be approved by the professor they worked for at the moment." Kojevnikov (2020, 70).

[8] See e.g. Eckert (2020, 16).

application of advanced mathematical techniques on concrete problem situations, "the physics of problems", as Suman Seth (2010) has called it, and the modelling of increasingly complicated systems in terms of simpler, solvable ones, are generally framed as the major characteristics of the Munich school. Although a broader conceptual and theoretical framework was certainly also sought by Munich physicists, the conclusions drawn from specific solutions were (at least so for the application of the correspondence principle; see Jähnert (2019, 262)) headed towards the evaluation of the specific problem-solving techniques.

At Niels Bohr's Copenhagen institute for theoretical physics, officially opened in 1921, emphasis was more thoroughly put on the general principles of the theory. From its establishment, Bohr intended his institute to be a place to foster the careers of young academics.[9] Bohr's reputation due to his atomic model (1913) and its further development, Bohr's charismatic nature, the deep influence he could establish on his co-workers, and the informal as well as stimulating atmosphere Bohr created at the Copenhagen institute made it the "Meccah of physics", at least until the beginning of WWII. But to fully comprehend the institute's status, Denmark's political neutrality during WWI is key. It soon allowed for international cooperation, also with German scientists,[10] and the acquisition of funding for young physicists. Bohr's scientific diplomacy and his possibility of ensuring appropriate funding eventually paved the way for the institute to becoming the major centre of quantum-theoretical research in the interwar period.

In its theoretical and conceptual approach the Copenhagen institute differed from Sommerfeld's school in Munich. Quantitative analysis was often left to Bohr's assistants, a position that was held from 1919 to 1925 by Hendrik Anthony "Hans" Kramers. The solutions of concrete problems was certainly as well central in Copenhagen, but the conclusions drawn (often by Bohr) aimed more directly at the establishment of a new conceptual framework. The BKS theory (see Sect. 4.1), for example, can be understood as the attempt to find a theoretical description, or possibly a world-picture, which could (under certain conceptual premises) incorporate the known experimental data in its entirety. Non-Copenhagen scientists at times considered Bohr's methodological tool kit, a central piece of which was the "magic wand" of the correspondence principle, as rather qualitative and vague.[11]

The third centre of the old quantum theory was Göttingen. Already before WWI, quantum-theoretical research started in Göttingen under the mathematician David

---

[9] For the economic and conceptual design of the institute, see Kojevnikov (2020, Chap. 1).

[10] Bohr's close contact with German physicists was actually criticized by scientists from the allied countries of WWI. The political constraints on German scientists included an exclusion from newly formed international organizations and their conferences until 1926. See e.g. Kojevnikov (2020, Sects. 2.3 and 3.2).

[11] This should not be misunderstood. As Martin Jähnert has shown, the basic idea behind the correspondence principle, the possibility of connecting the movement of the atom to the radiation phenomena, has been understood by the majority of physicists and it could be applied without direct interaction with the Copenhagen institute. Only in its further development and not in every case, the application of the correspondence arguments rested on closer contact with Copenhagen (Jähnert 2019, 256–258).

Hilbert, was fostered by the professorship of Peter Debye (1914–1920)[12] and eventually took a leap with the appointment of Max Born (theory) and James Franck (experiment). In Göttingen, the interrelation between experiment and theory was particularly close, as Born, well versed in higher mathematics, also had an experimental background and cooperated closely with his personal friend James Franck. James Franck, vice versa, was in close contact with theoretical physicists at the other major institutes and mentored theoretical research as well. Born's strategy to solve quantum riddles further strongly incorporated perturbative approaches, which will receive particular importance for the development of concepts connected to the virtual particle. From 1923 onwards, Max Born and James Franck's conceptual outlook on quantum theory rather sided with Niels Bohr than with Arnold Sommerfeld.

The interrelation between these research centres was close. Either through direct correspondence between the residing professors or through communication with post-doctoral fellows an information flow between the institutes was established. At conferences, such as the famous Bohr Festspiele (1922) in Göttingen, idea exchange between the different institutes was fostered. Particularly promising doctoral and post-doctoral students, as Werner Heisenberg and Wolfgang Pauli, had the opportunity to study and work at each of the institutes. Direct contact to the places which are traditionally framed as peripheral to the development of quantum mechanics, such as Breslau, Vienna, Cambridge, Leiden, or Harvard, was often established through these research centres and their central figures.

All besides the connections between the institutes, the local modes of theoretical practice, commitments to general principles of theory and the (often well-founded) conceptual (dis-)likings of the residing professors were central for the development I am about to investigate. The demarcation line between the different approaches I will follow can be pinpointed to the conceptualization of the radiation field, namely either as classical waves or as light quanta. The watershed was the year 1923 and the discovery of the Compton effect.[13]

The central figure of the first approach leading up to the BKS theory (Sect. 4.1) was Niels Bohr. Through his position in the quantum-theoretical network at around 1923, he acquired first hand knowledge about diverse attempts to solve specific quantum-theoretical riddles, especially the dispersion problem. The infamous BKS theory, with its virtual oscillators and virtual radiation field, was developed when the work on the concrete problem of dispersion, or the light-matter interaction in more general terms, was running high in Copenhagen. And the BKS theory should nonetheless be conceived of, in accordance to Bohr's personal conceptual convictions and his

[12] Peter Debye previously was assistant to Sommerfeld in Munich, successor of Albert Einstein at the University of Zürich and spent two years as Professor in Utrecht. In 1920, he left for a professorship at the ETH Zürich.

[13] This is not a statement about the acceptance of the light-quantum hypothesis in a larger part of the community. As Brush (2007, Especially p. 232) has shown, there were nearly as many supporters as opponents of the light-quantum hypothesis around 1923, with a number of physicists simply not taking sides. But both attempts at explaining the light-matter interaction and specifically the scattering of light from atoms discussed in Sects. 4.1 and 4.3 were explicitly framed either within a wave-theoretical conception or a light-quantum conception of the radiation field.

general approach to theoretical research, as a harmonious amalgamation of different concepts aiming to explain a vast amount of empirical material.

While in Copenhagen the BKS theory and a wave-theoretical conception of light was still en vogue, Arnold Sommerfeld, unsympathetic to the conceptual blending between classical and quantum-theoretical approaches as it was practised in Copenhagen, fully accepted the light-quantum conception of light.[14] From 1923 to 1925, at Sommerfeld's institute in Munich but also at the former focal points of quantum statistical research, physicists put forward proposals for an explanation of scattering and dispersion mostly within a light-quantum framework: Albert Einstein (Berlin) and Paul Ehrenfest (Leiden), Karl Ferdinand Herzfeld and Gregor Wentzel (both in Munich), and Adolf Smekal (Vienna resident, but also trained in Berlin).[15] Although these contributions constituted a side road to the overall development of quantum theory and were taken up only rarely (and were eventually refuted), they enlighten the close connection between the light-quantum concept and the conception of scattering as taking place through absorption and emission subprocesses (see Sect. 4.3).

While the light-quantum-theoretical approaches in terms of real transitions in the old quantum theory were not followed, the development of the correspondence principle approach in Copenhagen eventually led to Heisenberg's breakthrough of 1925.[16] Yet, the ascriptions of the chronologically first explanation of the light-matter interaction in terms of successive virtual emission and absorption subprocesses stretches from the dispersion theory of Kramers and Heisenberg (1925) to the one of Dirac (1927b) (see further below).[17] But it is unquestionably the case that the mathematical structures which Heisenberg and Kramers called "characteristic amplitudes" became the matrix elements of the newly established quantum mechanics. And even though the route to matrix mechanics went through the calculations of intensities of multiplet spectral lines (Blum et al. 2017), the dispersion-theoretical attempts certainly put the mathematical structures we refer to as virtual or intermediate states, and their combinatorics, into sharper relief.[18]

---

[14] On his 1923 journey to the United States, the most decisive experience Sommerfeld remembered was the presentation of the discovery of the Compton effect: "The most interesting scientific experience that I learned in America is a study by Arthur Compton in St. Louis. From now on, the wave theory for X-rays would have to be dropped once and for all." Sommerfeld to Bohr, January 21. 1923; as cited in Kojevnikov (2020, 54).

[15] Also contributions by Wolfgang Pauli (1923), Pascual Jordan (1924) and Paul Dirac (1924) fall into the category of such statistical and equilibrium research. But Jordan, as Born in Göttingen, followed the BKS theory and tried to proof that a strict conservation of momentum is not necessary in such accounts. Pauli finished his paper while spending a short stay in Hamburg, away from Bohr. Furthermore, although these papers fall into the line of statistical research, the decomposition into emission and absorption subprocesses was not spelled out, neither mathematically nor verbally.

[16] See the short discussion in Sect. 4.2.3.

[17] See my discussion in Sect. 1.3.1.

[18] That Heisenberg's reasoning at this point was analogous to the solution of the dispersion problem was also stated by Blum et al. (2017, 12). Yet, in Heisenberg's route to *Umdeutung* "the interpretation of these amplitudes is unclear".

3.2 The Spread of Quantum (Field) Theoretical Research and the Development of a Verbal Model

The establishment of matrix mechanics by Heisenberg, Born and Jordan in Göttingen and the development of wave theory by Erwin Schrödinger in Zürich spurred a considerable increase in the number of publications on quantum-theoretical topics from 1925/26 onwards.[19] And even though Heisenberg's 1925 *Umdeutung* has been described as a revolutionary act within theory, the community of practitioners remained relatively stable.[20] The geographical centre at the time was Central Europe (Germany, Austria, Switzerland, Netherlands, and Denmark) and the preferred language German.[21] The most important journal for technical articles was the *Zeischrift für Physik*, while overview articles or papers of a rather philosophical nature were often published in Arnold Berliner's *Die Naturwissenschaften*.

Especially in the first years after the advent of quantum mechanics, physicists close to the centres of the old quantum theory had an advantage due to their faster access to publications and preprints as well as their close communication network. Due to an extended funding policy, especially through the Rockefeller Foundation's International Education Board (established in 1923), physics pot-docs gathered mobility. The majority of the generation only a few years younger than Heisenberg and Pauli would regularly spend post-doc research fellowships at different institutes and it would take them a considerably longer period of time to secure a permanent position.

The competition amongst quantum theorists ran high in the initial years after 1925, a development which is also reflected in the creation of QED (see further below). It was strengthened through the quick pace of the development and the short time between reception and publication. At times, the same theoretical findings and techniques were published by different authors independently. Time-dependent perturbation theory, which played an important role in the development of the virtual particle concept, was developed and published independently by Max Born (1927), John Clarke Slater (1927b), Erwin Schrödinger (1926b), and Paul Dirac (1926).

Although Munich and Göttingen (see further below) remained among the places to learn quantum theory, Copenhagen took on a special role in the further development of the community. Without exaggeration, the Copenhagen institute can be called the social nexus of the quantum-theoretical community in the interwar period. The informal atmosphere at the institute, the freedom to follow one's own line of research, and the rather broad approach to scientific problems, all fostered by Niels Bohr, became trademarks of the so-called "Copenhagen spirit". The casual and stimulating Copenhagen conferences, institutionalized in 1929 but still without a fixed

---

[19] See Kojevnikov and Novik (1989) for a quantitative study of the development of the new research field.

[20] See Kojevnikov and Novik (1989, 128).

[21] Nevertheless, the internationalisation of quantum theory had gathered speed in the early 1920s. One may consult the list of visits of European physicists to America collected by Katherine Sopka (1988, Appendix B).

programme beforehand, brought established experts and young theorists into contact. According to the recollections either of Bohr's assistants or of visiting scholars, the Copenhagen institute remained a place vibrating with scientific creativity. All of this was granted by a very successful funding policy.[22]

Concerning the development of the concept of the virtual particle, the role of the Copenhagen institute and particularly of Niels Bohr is nevertheless ambiguous. On the one hand, Bohr succeeded in creating a stimulating atmosphere for physical research and idea exchange, and thereby fostered the networks and careers of many practitioners of quantum field theory. The personal acquaintance between Paul Dirac and Ivar Waller (see Sect. 7.3), just to name one important example, was first established in Copenhagen. The observation that negative energy states had to be included in calculations as virtual (modern terminology) states was first presented at a small conference in Copenhagen by Werner Heisenberg (see Sect. 7.1). The investigation of the Dirac sea, or rather its second quantization and its interpretation, in Berkeley in 1933 was spurred by a prior visit of Bohr (Sect. 10.1.2).

On the other hand, Bohr's own focus remained on questions of principle and rather philosophical aspects of theory. The solution of concrete problems and the development of the necessary techniques, although such works were certainly performed at the institute, were not prominent in Bohr's own publications or the output of physicists closely working with him.[23] Higher order calculations in perturbation theory, in which non-mathematical representations of virtual particles and related concepts regularly showed up, were simply not high on the agenda. Furthermore, as I have already argued in publication (Ehberger 2022; see also Sect. 10.1.1), there is a striking tendency towards the non-usage of explicit non-mathematical representations by physicists closely associated with Niels Bohr and the Copenhagen institute (at least so in the years directly after the development of QED).

3.2.1 Two Cultures of Theory and the Usage of Non-Mathematical Representations

While Niels Bohr's concentration on fundamentals made his few contributions to QED during the 1930s uninteresting for my research, distinctions in the fundamental conception and formulation of the theory are important for developing an appropriate perspective on the relevant social and epistemic communities. From its first

[22] For the atmosphere at Bohr's institute and the funding from the late 1920s throughout the 1930s, see Aaserud (1990).

[23] The nearly sole published contribution of Niels Bohr to quantum electrodynamics during the 1930s is the well-known, well-respected, but little read Bohr-Rosenfeld paper on the measurability of fields (Bohr and Rosenfeld 1933). Although the concepts in the paper were operationally defined and the argument rested on the discussion of diverse experimental arrangements, the paper is rather detached from any practical application, both from a theoretical and experimental viewpoint (for a historical discussion, see (Darrigol, 1991)).

development between 1927 and 1929, QED was formulated in two distinct ways.[24] The electrodynamic field was quantized in both strands of theoretical research. But the conception of the material part of the systems differed: In one conceptualization the fields were conceived of as fundamental and particles arose as excitations of these fields (field viewpoint). The other conceptualization took particles as its primary entities (particle viewpoint). Both viewpoints were developed nearly simultaneously, either viewpoint could be practised by either community, and the approaches proofed to be essentially equivalent. Nevertheless, this distinction is important for the development of the concept of the virtual particle and its representations.

The particle viewpoint and, after its acceptance in 1933/34, hole-theory became the prime way of doing calculations during the 1930s and, thus, deserves special attention in a history that tries to develop an understanding of conceptual development through theoretical practice. Moreover, practitioners in this tradition had a stronger tendency towards visualization and the development as well as application of non-mathematical representations. If we follow the lines of Sam Schweber's overview in his introduction (Schweber 1994, xxi–xxviii), the most prominent example of a non-mathematical representation of virtual particles, Feynman diagrams, was based on research in this tradition.

Although the stronger tendency of applying non-mathematical formats of representation within the particle viewpoint is striking (at least so for the first years after the development of QED and around the time of the development of renormalization), this does not mean that practitioners of the field viewpoint did not essentially contribute to the development of the virtual particle concept. At around the mid 1930s, the language of the particle viewpoint was more strongly used in this tradition as well. Furthermore, the field viewpoint could more easily incorporate different kinds of fields and particles: As such, it was the ground for Enrico Fermi's theory of β-decay (1933/34), Hideki Yukawa's meson theory (1934), and the quantum field theories in their follow up, which in their own respect had important influences on the conception of the virtual particle (mostly discussed in Sect. 10.2).

Although it is chronologically second in establishing a workable theory, let me first turn to the development of the field viewpoint. It had its roots in the contributions of Pascual Jordan to the so-called *Dreimänner-Arbeit*, i.e. one of the funding papers of matrix mechanics by Max Born, Werner Heisenberg and Pascual Jordan, and gathered speed after Dirac had developed his radiation theory (see further below). Pascual Jordan, still in Göttingen, together with Oskar Klein (Jordan and Klein 1927), Eugene Wigner (Jordan and Wigner 1928), and finally Wolfgang Pauli (Jordan and Pauli 1928), worked out a formalism for non-interacting fields.[25] In 1929, Wolfgang Pauli and Werner Heisenberg took the final step in developing a completely field theoretical QED.

[24] See Schweber (1994, xxi–xxvii) for a very concise description of the different strands of conceptualizations and research.

[25] Pascual Jordan's contribution to the quantization of radiation and matter waves is, for example, discussed in Schweber (1994, Sect. 1.2).

Werner Heisenberg received his call to Leipzig in 1927 and Wolfgang Pauli became Professor at the ETH Zürich just the year after. At the University of Zürich another former student of Arnold Sommerfeld and aspiring quantum field theorist, Gregor Wentzel, was appointed professor in 1928. Both Leipzig and Zürich became centres for learning and practising QED and QFT during the 1930s. *Physikalische Vortragswochen*, both in Leipzig and Zürich, were initiated by Pauli and Heisenberg and helped to disseminate quantum-theoretical ideas. The close personal and professional relation between Pauli and Heisenberg is well-known. Just scanning through their correspondence, one easily gets the impression that Heisenberg often used Pauli, the "Theorienfresser", as his soundboard.

The close interrelations between the institutes was further strengthened by mutual visits of assistants and research students. Yet, and quite strikingly, in the initial period of the development of QED physicists working directly with Heisenberg and Pauli, such as Robert Oppenheimer or Léon Rosenfeld, did not use verbalizations or non-mathematical representations of concepts we can connect to the virtual particle concept (see Sect. 10.1.1).[26] Only with the discovery of the neutron, Werner Heisenberg's turn to imagery and visualizable models in 1932,[27] and finally Fermi's field theoretical description of β-decay the explicit use of non-mathematical representations among the practitioners of this branch of theory became more common at around 1933/34.

But in the initial development and application of representational formats other than mathematics as well as for the calculation of potentially observable effects, the contributions in the tradition of the particle viewpoint are far richer. Due to this fact, the main part of this book will be concerned with work in this conceptualization. Researchers who performed a large part of their research on QED from this viewpoint include Enrico Fermi, Walter Heitler, Igor Tamm, and (partially) Lev Landau, Hans Bethe, Hans Euler as well as Bernhard Kockel. Paul Dirac was its "founding father and guiding spirit" (Schweber 1994, xxii).

Dirac[28] was a Bristol trained engineer (1921) and applied mathematician (1923).[29] In 1923, he came to Cambridge as a research student. At the time, Cambridge was a prestiges University in physical research, but it had not yet made a huge impact on quantum-theoretical research.[30] All the same, Dirac's supervisor, Ralph Fowler, had close connections to the German quantum physicists and to Niels Bohr in

[26] Especially Rosenfeld was mostly occupied with fundamentals of the theory and its mathematical formulation. An aspect that is also apparent in his contributions to QED together with Niels Bohr. Jacobsen (2012, 29–30) gives a concise description of Rosenfeld's approach to theoretical physics in the late 1920s and early 1930s. Oppenheimer, on the other hand, only applied non-mathematical representations when he was working in terms of Dirac's radiation theory, while in his work on Pauli's and Heisenberg's programme such were missing.

[27] See, for example, Miller (1984, Chap. 4, especially pp. 155–165).

[28] For Dirac's biography, Kragh (1990) is still the standard reference. Farmelo (2009), although popular in nature, is also a valuable resource for biographic detail.

[29] Especially his first education is regularly invoked to explain Dirac's pragmatic and philosophically rather flexible to uninterested approach to quantum theory.

[30] See for example Navarro (2015).

Copenhagen. It was Fowler who brought Dirac into contact with quantum theory and provided him with the preprints of Werner Heisenberg's 1925 paper.

Dirac had already gained some experience in scientific publishing, studied quantum theory himself and joined the $\Delta^2 V$- and the famous Kapitza-Club in 1924. The former was mostly concerned with mathematical physics and the members presented their own work. The latter was rather informal and a forum to discuss recent publications in physics. Dirac had also already heard some lectures of prestigious quantum theorists.[31] When Fowler provided Dirac with the preprints of Heisenberg's paper and thereby initialized Dirac's work on quantum mechanics, Dirac was still in relative isolation from the continental research centres. But through his diverse and often major contributions, Dirac made his name in the quantum theorists' community. By September 1926, when he started a year abroad, he had turned from "an English man Dirac"[32] to one of the most promising and prominent quantum physicists of the younger generation.[33] First, and due to a suggestion by Fowler, Dirac went to Copenhagen until February 1927 and then to Göttingen from March to June of the same year. Although Dirac made closer connections to the quantum-theoretical community, he, a man of few words, kept his habit of working mostly on its own.

Most important for our story, in Copenhagen he came up with his radiation theory, sort of the prototype of quantum electrodynamics during the following years, at least as far as applications were concerned. Under the auspices of Max Born in Göttingen, Dirac applied his new theory to dispersion. And this is the point at which Dirac would introduce his verbal model of successive (modern terminology) virtual emission and absorption subprocesses, which I will discuss in Chap. 5. As I will argue, this was a major step in the development of the virtual particle concept as Dirac introduced and utilized a new representational format, which became part of most applications of QED during the 1930s and is (in a modified form) still underlying the practice of QED to this day.

Even though Dirac kept his habit of working on his own, his stay in Göttingen certainly helped to disseminate his methods and ideas.[34] Normally, Göttingen's occurrence in the history of QED is mostly limited to Pascual Jordan's contributions and Dirac's dispersion theory.[35] But when it comes to initial applications and

---

[31] Fowler was capable of bringing, e.g., James Franck (October 1924) and Niels Bohr (May 1925) to lecture in Cambridge. Also Werner Heisenberg spoke at the Kapitza-Club, while Dirac most probably missed the meeting.

[32] "ein Engländer Dirac," Heisenberg to Pauli, November 24. 1925; as cited in Pauli (1979, 266).

[33] Dirac also started corresponding with other quantum theorists at the time. Although I deem these early contacts rather unimportant for the development of his verbal model (Chap. 5), the correspondence with Heisenberg was of major significance for Dirac's further research development: In spring 1926, Dirac became aware of Schrödinger's wave mechanics and at first refused to apply it. Only after persuasion by Heisenberg and the informally shared equivalence proof by Wolfgang Pauli, Dirac started to work with Schrödinger's symbols.

[34] Of course, Dirac's thinking fell on prepared grounds at the mathematically and perturbation theoretically inclined Göttingen institute.

the development of ideas connected to the virtual particle concept, Göttingen trained physicists played an important role. I do not consider it a coincidence that one of the earliest examinations and applications of Dirac's methods in doctoral dissertations were achieved by Maria Göppert-Mayer and Victor Weisskopf, both doctoral students of Max Born and also informally supervised by James Franck (see Sect. 8.1).

Göppert-Mayer introduced the *Schüttelwirkung* to the perturbative evaluation of QED which would play an important role in Walter Heitler's theory of radiation damping (see further below). Weisskopf continued theoretical research on radiation theory, especially on line breadths and resonant scattering, during his post-doctoral period spent in Leipzig, Berlin, Kharkiv, Cambridge and Copenhagen. In 1933, he gave the first general description of perturbation theory to an arbitrary order.[36] When Weisskopf left Göttingen in 1931, Walter Heitler, since late 1927 research assistant to Max Born, started to focus his work on quantum electrodynamics. Yet, the rise to power of the NS-regime put an end to the quantum field theoretical research in Göttingen. Nearly the complete theoretical staff had to flee Germany and went into exile in Great Britain or the US (see further below).

Other than the theoretical institute in Göttingen, the physics institute in Rome under the supervision of Enrico Fermi did not suffer personal loss nearly until the outbreak of WWII. Enrico Fermi's pragmatic tendency, his inclination for *anschaulich* formulations and calculations, as well as his close contact to experiment, often self-performed, are well-known.[37] Fermi had already contributed to the old quantum theory and had spent extended research stays in Göttingen (1922–1923) and Leiden (1924). In 1926, he became professor for theoretical physics at the university in Rome. Under his patronage, the experimental and theoretical physics in Rome flourished. Experimental results obtained by one of his co-workers, Franco Rasetti at the time at Caltech, proofed incremental for the discussion around the Raman-effect,[38] where the connection between Dirac's verbal model and the terminology of the "virtual" was, to the best of my knowledge, first established. Edoardo Amaldi and Emilio Segrè turned towards the explication of Rasetti's findings in the language of Dirac's radiation theory (see Sect. 6.2.1). Furthermore, as Enrico Fermi felt not content with Dirac's formulation of the quantization of the radiation field, he reworked it on his own in 1928 to 1929. Fermi came up with a pragmatic and easily accessible presentation in the tradition of the particle viewpoint. Its publication in English (Fermi 1932), based on a lecture series at the University of Michigan,

[35] The perceived lack of further development of QED in Göttingen is of course also due to Hitler's seizure of power in 1933 and the emigration of nearly the complete theoretical staff of Göttingen.

[36] It is noteworthy that Weisskopf's approach to quantum electrodynamics considerably changed in 1933/34 when, after his stay in Copenhagen, he became Wolfgang Pauli's assistant in Zürich.

[37] On his pragmatic and intuitive reformulation of QED, see for example, Schweber (1994, Chap. 1.8). For a concise biographical overview and the building up of experimental research on nuclear physics in Rome, see Stuewer (2018, 284–297).

[38] On the one hand, his observations indicated that the nitrogen nuclei obeyed Bose statistics and thereby contested the proton-electron model of the nucleus. On the other hand, and more important for our story, it definitely showed that infra-red absorption lines did not necessarily coincide with Raman lines (see Sect. 6.2.1).

brought quantum electrodynamical calculations, if not to the masses, at least to the younger generation of quantum field theorists.[39]

In the beginning of the 1930s, Fermi's institute attracted a growing number of future quantum (field) theorists. Hans Bethe, for example, co-authored a paper with Enrico Fermi, which traditionally received credit for the development of the idea of forces transmitted through (virtual) particle exchange, during his stay in Rome in 1932.[40] Fermi's theory of β-decay (Fermi 1934) was, besides QED, the first workable QFT and immediately inspired physicists to perform perturbative calculations and explain the attraction of nucleons through particle exchange field theoretically. Further, his assistant at the time, Gian-Carlo Wick, was the first to propose the idea that the magnetic moment of nucleons may be explained through virtual dissociation processes in terms of the Fermi theory of β-decay. This explanatory model was also applied to meson theory later on and was, until the development of the quark model, the most popular account of the phenomenon (see Sect. 10.2.2). The turn to the experimental evaluation of nuclear theory in the early 1930s would make the institute's output fundamental for the discovery of nuclear resonances, the so-called Breit-Wigner formula which is used to model resonance scattering, the liquid drop model and finally nuclear fission, while these developments will not be discussed.

The development of QED by Dirac, Jordan, Pauli, Heisenberg, and Fermi was in its initial phase rather competitive[41] and different versions of the theory existed, initially side by side.[42] Although eventually the different formalisms were shown to be equivalent, the different perspectives, one mostly developed in Zürich, Leipzig, and Göttingen (Jordan, Heisenberg, Pauli), the other one practised in Göttingen (Weisskopf and Göppert-Mayer), Cambridge (Dirac), and Rome (Fermi and co-workers), were used for different purposes and also operated with different concepts, or at least representations of them, initially.[43] While the field viewpoint was more relevant for what one might call research on the fundamentals of the theory and the inclusion of new particles in the conceptual framework, the particle viewpoint was

---

[39] I will discuss the use of intermediate states in Fermi's didactic publications and papers related to it in Sect. 8.2.1.

[40] That such credit was given a bit hastily is well-known (Carson 1996b). I will discuss this paper in particular in Sects. 8.2.1–8.2.1 and the concept of particle exchange in its connection to the virtual particle in Sect. 10.2.1.

[41] One might take Léon Rosenfeld's published attacks on Dirac's radiation theory as an example (e.g. Rosenfeld 1931). Similarly, Wolfgang Pauli commented in a letter to Rosenfeld, that a "Dirac quantum electrodynamics does not exist" ("Eine Diracsche Quantenelektrodynamik existiert nicht."). According to Pauli, Dirac had "insufficient knowledge about the previous works about quantum electrodynamics (since 1926)" ("ungenügende Kenntnis der bisherigen Arbeit (seit 1926) über Quantenelektrodynamik"). Pauli to Rosenfeld, May 2. 1932; as cited in Pauli (1993, 748–749).

[42] Besides the ones already mentioned up until 1932 also a correspondence-like method by Heisenberg and a version including the many-time formalism had been proposed.

[43] Again, this is a simplification but holds approximately at least until the early to mid 1930s.

mostly applied in (perturbative) evaluations of the theory in order to compare it with (potential) experiments. As indicated, the second line of research was initially far richer in the inclusion of new representational formats of concepts connected to the virtual particle.

Still, the practitioners had no problems understanding each other's research and the non-mathematical representation became part of the practice of field theorists in the course of the 1930s. In general, physicists used whatever approach suited the problem or, quite often, their personal (philosophical and conceptual) preferences best. Still, physicists in the tradition of the wave viewpoint also tended towards the evaluation of the structure of the theory and not as much towards its higher order evaluation and its comparison to experiment. At times, they openly opposed and questioned non-mathematical descriptions of concepts in the line of the virtual particle development.

3.2.2 Bringing Quantum Electrodynamics (and Dirac's Verbal Model) to Russia, the US, and Japan

Both Soviet and US-American scientific cultures were particularly receptive for the new trends in physical research. In both countries, theoretical physics was a rather young and emergent branch of science. Furthermore, Soviet and US physicists had contributed to the old quantum theory and to the initial elaborations on quantum mechanics. Quantum theory was already taught in the Soviet Union and the US, when QED and QFT was constructed. Through fellowships spent at the European centres, physicists from these regions directly took part in the development.

The guiding figures of Soviet theoretical physics in the interwar period, Yakov Frenkel, Vladmimir Fock, Lev Landau, and Igor Tamm, all spent some time as fellows in Central Europe (Kojevnikov 2004, 80–85).[44] Fock (Leningrad), Tamm (Moscow), and Landau (Kharkiv) will eventually become the main protagonists of quantum electrodynamics and field theory in the Soviet Union. As Fock's contributions, valuable as they were, aimed mostly towards the principles of QED and to a lesser extend its evaluation in respect to concrete effects, I will focus on Igor Tamm and Lev Landau in the following.

Igor Tamm's initial work on and conception of QED is closely connected to Paul Dirac. In 1928, Dirac went on his second research trip abroad.[45] He went to Copenhagen, Leiden, Göttingen, Leipzig, and finally to the Soviet Union. Already during his stay in Leiden, he met Tamm, at the time visiting Europe on a Lorentz fellowship. Dirac and Tamm bonded quickly and, although Tamm was older than Dirac, he soon considered Dirac his scientific mentor. They joined for the following

[44] Similarly, Karl Hall (2008, 8) has observed that "a period of study in central Europe was crucial for a Russian physicist in training".

[45] At the time, Dirac was employed at Cambridge, were he received a professor position in 1932.

journey and a long-lasting friendship developed between them.[46] Their destination in the Soviet Union was the Sixth Congress of Russian Physicists.

This congress, attended by several foreign physicists, definitely put Soviet science on the map of quantum theory, the presentation of the simultaneous discovery of the Raman effect by Russian physicists being one of the highlights of the event. The congress started in Moscow, mostly took place on a steam-ship on the Wolga and ended at the Ukranian Physico-Technical Insitute (UFTI) in Kharkiv, which was to become an important centre of Soviet theoretical physics.[47] At the congress, Dirac also made the acquaintance of Yakov Frenkel. This contact contributed to Frenkel's further development of the (eventually failing) perspective on the light-matter interaction in analogy to the photonic concepts of the time prior to quantum mechanics (see Sect. 6.3).

The correspondence between Dirac and Tamm in the subsequent years is especially interesting for the development of the concepts connected to the virtual particle. Besides Ivar Waller, it was also Igor Tamm that made Dirac aware that negative energy states had to be included as virtual states in the theoretical description of scattering as long as the electromagnetic field is quantized and the electron described relativistically. A fact that eventually led to the construction of the Dirac sea (see Sects. 7.1–7.4).[48] Furthermore, through their correspondence we can follow some of the thinking possibilities that were opened up through Dirac's verbal model, how they were discussed at following conferences in Leningrad, and how their non-occurrence as physical effects eventually led to the elaboration of the characteristics of the intermediate states (see Sect. 7.7). In 1934 and together with Dimitri Ivanenko, Tamm, always showing a great tendency towards visualisation,[49] would also be among the first theoreticians to describe the interaction between neutron and proton perturbatively in terms of Fermi's theory of β-decay and in the language of Dirac's verbal model.

As Igor Tamm, Lev Landau obtained a fellowship (in his case by the Rockefeller foundation), lasting from fall 1929 to spring 1931 and taking him to all the major

---

[46] It should be noted that, starting with his acquaintance with Piotr Kapitza in Cambridge, Dirac got along with Soviet physicists quite well in general. Dirac was an atheist, had leftist tendencies and was interested in the "Soviet experiment". From 1928 to 1936 (with the exception of 1931) he visited the Soviet Union every year.

[47] The establishment of the UFTI in 1928 was politically motivated in a move to bring science to more rural areas. It was closely modelled after the already established physics institutes in Leningrad and Moscow. Through annual conferences and the professorship of Lev Landau (from 1932 to 1936) Kharkiv became a prospering centre for quantum-theoretical research in the Soviet Union (see, e.g., Josephson et al. 2010). As soon as 1929, the first Soviet conference on theoretical physics was held in Kharkiv which would again attract a number of foreign theoretical researchers. See Ivanenko and Fock (1929) for the reports including research on quantum electrodynamics.

[48] As will be discussed in Sect. 7.5, Tamm and Dirac shared a rather physical reading of the intermediate states.

[49] In reference to Tamm's thinking in visualizable models, his introduction of the phonon (named by Frenkel) is also a case in point (Kojevnikov 2004, Sect. 3.5).

scenes of quantum-electrodynamical research at the time.[50] Already during his studies in Leningrad, Landau showed a great gift and interest in quantum theory and was part of the famous *Jazz band*, the core of which was comprised of him, George Gamow, Dimitri Ivanenko, and Matvei Bronstein, all of them promising young theorists with a taste for non-uniform behaviour. Due to his outstanding capabilities and on recommendation by Yakov Frenkel, the Rockefeller fellowship was organized. Landau's longer stays included Zürich and Copenhagen.[51]

During his visit to Europe, Landau still had a taste for deep conceptual problems as epitomized by his collaboration with Rudolf Peierls (at the time Pauli's assistant in Zürich) on the measurability of electromagnetic field strengths in QED (Landau and Peierls 1931).[52] Nevertheless, Landau also showed signs of digression from the philosophically motivated thinking of Bohr.[53] When Landau returned to the Soviet Union, he received the call to become the head of the theoretical department at the UFTI in Kharkiv in 1932. There he formed an important school of theoretical physics,[54] which focused on the practice of theory, "in which 'practical' did not mean 'applied' but rather 'not overly concerned with the physics of principles,' with all its attendant epistemological problems" (Hall 2008, 5). The specific approach Landau's students, Aleksandr Akhiezer and Isaak Pomeranchuk, took on light-by-light scattering and the Delbrück effect (see Sect. 9.2.3), modelling the second closely after the first by introducing longitudinal "pseudo photons" and making use of Hans Euler's diagrams, should be understood as arising from this pragmatic, hands-on approach to theoretical problems.

Pragmatism was a common trait not only of Soviet theoreticians, but also of theoretical physics in the US. The most important research centre for QED and QFT during the 1930s was in Berkeley and under the leadership of Robert Oppenheimer. Oppenheimer, trained under Edwin C. Kemble and Percy Williams Bridgman at Harvard (1925), first went to Cambridge (1925–26) and then to Göttingen (1926–1927), mostly to study under Max Born. In Göttingen, Oppenheimer obtained his doctorate and made the acquaintance of Paul Dirac and his recently developed radiation

[50] See Hall (2008) and Kojevnikov (2004, Chap. 4).

[51] In chronological order, Landau visited Berlin, Leipzig, Zürich, Copenhagen, Leipzig, Cambridge, and Copenhagen again.

[52] I will come back to this paper in Sect. 10.2.3 but not in respect to the measurability of fields.

[53] As Karl Hall (2008, 26) has argued "the image of the barely comprehensible genius [the philosophically minded Niels Bohr] provided a useful foil for Soviet theorists bent on portraying themselves as tough-minded pramatists by comparison". As a sidenote, one can also consult Hall's paper for a discussion of the political dimension of the pragmatic outlook on theoretical physics in the Soviet Union.

[54] Landau was particularly famous for his pedagogical program, including several textbooks and the so-called "theoretical minimum": a test on theoretical physics and applied mathematics, which his research students had to pass in order to be part of the Landau school.

theory.[55] After a short interlude at Harvard and Caltech, Oppenheimer came back to Europe in 1928. First he went to Ehrenfest and Kramers in the Netherlands and then to Zürich, exactly at the time when Pauli and Heisenberg ran high on their efforts of creating quantum electrodynamics.

Thus, Robert Oppenheimer was in direct contact with Dirac, when he developed his dispersion theory, and directly helped in formulating (the problems) of QED in the hands of Heisenberg and Pauli. Back in the US, Oppenheimer became professor for theoretical physics at the University of California (Berkeley) and at CalTech. His school of young theoreticians contributed to the further development of QED, QFT and also to the virtual particle concept. Oppenheimer's own contributions include the initial analysis of the self energy divergency (1929, Sect. 10.1.1) and a critique on Dirac's proposal to conceive of a hole in the Dirac sea as a proton (1930, see Sect. 7.6.2). Especially the second paper shows Oppenheimer's inclination towards practical application: It operated with Dirac's radiation theory and was nearly entirely based on the theoretical description of light scattering in QED.[56]

Oppenheimer's students and assistants in Berkeley felt relatively isolated from the community. The developments of the theory were mostly learned directly from publications. Nevertheless, the connections to the European centres were not cut and showed up in diverse places in Oppenheimer's and his students' research. Oppenheimer's, Milton Plesset's and Wendell H. Furry's investigations on the effects and properties of the Dirac sea, starting in 1933, were spurred by a visit of Niels Bohr. The paper introducing the concept of "nascent pairs" (see Sect. 10.1.2) with Furry is unusually philosophic for Oppenheimer and one senses Bohr's prior presence in Berkeley throughout the paper. Wolfgang Pauli visited the Berkeley group when the investigations on the Delbrück effect, performed both in Zürich and Berkeley, were carried out. Furry's theorem (see Sect. 9.2.4), published in 1937, was partially motivated by the theoretical description of this higher order effect.

The isolation from Europe felt in Berkeley was even stronger for the Japanese community. Quantum theory was brought to Japan by Yoshio Nishina.[57] Nishina had spent an extended research stay in Europe (1921–1928) at Cambridge, Göttingen but mostly in Copenhagen. There, just to name one example, Nishina and Klein developed a theory of light scattering from a relativistically described electron, the

---

[55] Robert Oppenheimer remembered Dirac warmly and remarked that "perhaps the most exciting time of my life was when Dirac arrived [in Göttingen] and gave me the proofs of his paper on the quantum theory of radiation" (cited in Farmelo (2009, 123)). While Oppenheimer was impressed with radiation theory, he did not consider Dirac's work on dispersion as an important breakthrough. After all, he was disappointed with Dirac's work in Göttingen (Farmelo 2009, 132).

[56] Already Sam Schweber (1986, 87) noticed the difference of approaches to this problem by Wolfgang Pauli and Robert Oppenheimer. While Pauli highlighted symmetry reasons, Oppenheimer chose to discuss the problem along the lines of an observable effect.

[57] Also Yoshikato Siguira played a role in the dissemination of quantum ideas in Japan, but his work was mostly concerned with quantum chemistry (Nakane 2019). Also Seishi Kikuchi studied for a period of time in Leipzig under Werner Heisenberg and brought further import to Japan.

famous Klein-Nishina formula.[58] When Nishina returned to Japan in December 1928 he initiated modern quantum-theoretical research in Japan and became its leading figure. Hideki Yukawa and Sin-Itiro Tomonaga were deeply influenced by Nishina. During the early 1930s, a network of quantum theorists developed within Japan. But besides a few post-doctoral fellows and visits by foreign physicists, the influx of quantum field theoretical ideas was mostly provided through published papers and textbooks.

Yukawa, for example, who was struggling to understand nuclear interaction in a neutron-proton model throughout the years 1932 to 1934, introduced the meson only after he had received news of Fermi's publication on β-decay.[59] The relative isolation of Japanese physicists at the time is also remarkably underscored by the fact that Yukawa's hypothesis only made its way to the Western community in 1937, when its merits in explaining cosmic ray showers were realized.[60] Tomonaga's relativistically invariant formulation of quantum electrodynamics also suffered from a lack of reception in the US and Europe due to the isolation of the Japanese community and, eventually, the outbreak of WWII. Furthermore and although Tomonaga himself remembered the period he stayed in Leipzig with Heisenberg (1937–1939) to be formative, he knew the key contribution for his investigation, Dirac's paper on the many time formalism (Dirac 1932), only from publication.

The initial dissemination of QED, its practical applications and representations around the globe show two general themes. On the one hand, the initial implementation of the theory into a theoretical community in the respective countries often went through learning phases of physicists in Central Europe. The specifically pragmatist tendencies in the Soviet Union and the US as well as concrete higher order calculations contributed to the usage of (representations of) concepts connected to the virtual particle. Direct contact between researchers could strengthen inclinations for the usage of non-mathematical representations and even a physical reading of (modern terminology) virtual entities, as was for example the case for Dirac and Tamm. On the other hand, the application of such visualizable models, as Dirac's verbal model of successive virtual emission and absorption subprocesses, did not necessitate direct interaction between researchers when problems were worked out.[61] Its practice could be learned from publication. Its (ontological) reading depended on the concrete problem to be solved and, to a large extent, the personal stance of the respective physicist. The same holds for a fundamental obstacle for the further development of QED which could only partially be solved until the late 1940s, the divergences.

---

[58] It should be noted that the problem of negative energy intermediate states did not show up in Klein and Nishina's paper as they applied a "correspondence like" method which Klein had developed just two years priorly.

[59] Yukawa was even reluctant to perform a research stay abroad before he would have finished his work. See Brown (1981, Especially Sect. 6) for a discussion of Yukawa's route to his meson theory.

[60] Incidentally, Yukawa visited Europe and the US as late as 1939.

[61] Although Yukawa's initial proposal did not include the concept of virtual particle exchange explicitly, in his further work with Soichi Sakata andd Mitsuo Taketani he eventually used the picture of successive transitions for the evaluation of QFT. By the time, both physicists had not yet made a trip abroad.

3.3 Crisis in Theory and Totalitarian Regimes

Nearly from its on-set, research on QED had to face infinities in every higher order calculation. These divergences could not simply be subtracted from the equations.[62] The problems first showed in the calculations by Oppenheimer (1930a) and Waller (1930a) but manifested rather quickly and stayed until the late 1940s. For parts of the community, these infinities indicated that something with the conceptual basis of QED and QFT was fundamentally wrong.[63] But not only problems of internal consistency threatened QED. Also the fit with experimental data was questionable.[64] During the 1930s, a wide-spread believe held that QED would breakdown at lengths comparable to the electron radius.

The reactions to the felt crisis in theory were diverging and the infinities certainly restricted the further development of concepts connected to the virtual particle. The "correspondence-like approach",[65] which did not invoke intermediate states explicitly (and hence nothing connected to the virtual particle concept), gathered some sympathy in the early 1930s, as it constituted a workable version of the theory which circumvented the complicated formalism of QED and came with a set of ready-made rules of application.[66] But even when applying representations of intermediate states and working out concrete problems, higher order calculations were not only enormously cumbersome but also rather questionable.

Divergences occurred in every higher order calculation and cut-off procedures, although often applied by pragmatically inclined physicists, spoiled relativistic invariance and were therefore seen at best as a preliminary and unsatisfactory solution. Furthermore, guiding quantum field theorists, such as Max Born, Werner Heisenberg, Wolfgang Pauli, Paul Dirac, or Hans Kramers believed that the basis of the theory was flawed and proposed different reformulations which did not aim at the application of the theory but at its fundamental conceptions. Their aim was to eventually find the "future correct theory". In a mixture of disappointed high hopes for QED, the longing for another revolution of the conceptual framework, lack of

[62] I will discuss the divergences and their relevance for the virtual particle concept in Sect. 10.1.

[63] Even after renormalization and even amongst the ones who introduced renormalization, this view remained. See for example Tomonaga's view on the procedure as cited in Schweber (1994, 271–272). But also Julian Schwinger expressed doubts (Bacelar Valente 2011b, 124–125). Both concerns are essentially grounded on an argument presented by Bohr as early as 1932, namely that perturbative QED is based on the idea that the interaction is small, which is in conflict to the infinities arising due to the interaction.

[64] See Galison (2008, Especially Sects. IV and V) and Cassidy (1981) for cosmic ray research. See Roqué (1997, Sect. 3) and Brown and Moyer (1984) for a discussion of the so-called Meitner-Hupfeld effect which caused some controversy until the mid 1930s.

[65] Heisenberg (1931a) developed this approach in strong analogy to a contribution by Oskar Klein (Klein 1927).

[66] See Rueger (1992, 317–320). Such correspondence-like approaches are also partially responsible for the fact, that phenomena we connect to intermediate states and virtual particles, were initially not explained by their invocation, as for example in Klein's and Nishina's work.

empirical evidence of how to proceed further, and diverse failed attempts at wrestling with the infinities a rather pessimistic outlook on QED prevailed.[67]

But this is not to mean that the occurrence of infinities completely blocked the conceptual development of virtual entities. First of all, these infinities are caused, in modern terminology, by successive virtual emission and absorption processes (although this was initially not spelled out verbally, see Sect. 10.1.1). Secondly, the phenomena connected to vacuum polarization studied from 1933 onwards, whether described in the language of Dirac's sea or in a completely field theoretical language, suggested the picture of a cloud of virtual particles or, in the terminology of Furry and Oppenheimer, "nascent pairs".[68] Similarly, the application of cut-off procedures, although known to break with principles of theory, would allow for a qualitative description of otherwise unexplained phenomena as the magnetic moments of nucleons.[69] And finally, the theoretical problems of QED and QFT could actually attract physicists to engage in this field of research: Walter Heitler, for example, explicitly remembered that his turn to QED in 1932 was due to the unsolved riddles in the theoretical framework.[70]

The perceived crisis in theory thus led from within to distinct theoretical research programs. The alterations on fundamental concepts, as epitomized by Werner Heisenberg's fundamental length or Max Born's and Leopold Infeld's non-linear field theory, were complemented by (and often rival to) pragmatic attempts to extract sensible information from higher order calculations by specifically pragmatically inclined theoretical physicists. As Peter Galison (2008) has shown, throughout the 1930s these different outlooks on theory showed a rather strong tendency to bind specific physicists.

While the infinities caused a perceived crisis in theory, the political situation and the possibilities of doing quantum field theoretical research worsened considerably in Central Europe. The rise to power of the NSDAP in 1933, the 1935 *Nürnberger Rassengesetze* and finally the outbreak of WWII caused a serious brain-drain from Central Europe, especially from Germany. About 25 percent of the German physics community of 1932 would emigrate and theoretical physics was hit particularly hard.[71] From 1926 to 1933, German publications on quantum theory made up 31 to

---

[67] The pessimistic attitude was widespread although physicists found ways to circumvent some of the divergences by procedures we would today call charge renormalization. The pessimistic tone on which Euler's publication on the successful calculation on light-by-light scattering ends is just one amongst many examples (see Sect. 9.1.3).

[68] See Sect. 10.1.2.

[69] See Sect. 10.2.2.

[70] Furthermore, questions on the validity of QED for high energies (or, equivalently, small lengths) brought virtual entities, unconnected to the virtual particle as we today know it, into being. "Virtual photons" were the cornerstone concept of the Weizsäcker-Williams theory, which proved that a breakdown of QED at high energies was questionable, at least so from within the theory.

[71] As Helge Kragh observed, "the more theoretically oriented the institute was, the more dismissals: Of Germany's 60 university teachers with positions in theoretical physics, no fewer than 26 went into exile" (Kragh 1999, 232).

50 percent of all publications on these themes in the respective years, while approximately a quarter of these publications were written by later emigrants. Furthermore, the publications by emigrants showed an above average citation frequency. In 1933, the break was particularly radical in Göttingen: Max Born, James Franck, Lothar Nordheim, and Walter Heitler, just to name those who were actively pursuing research connected to quantum electrodynamics, were or felt forced to leave Germany.

Nevertheless, also high class quantum field theorists remained in Germany or German speaking countries. Werner Heisenberg famously stayed in Leipzig. Certainly, political interventions and isolation increased also at Heisenberg's institute. But, through his reputation in the community, well-known to and exploited by the political authorities, he managed to keep close contact to other physicists, was allowed to attend foreign conferences even during WWII, and could maintain a high level of physical research.[72] Besides his own contributions on the fundamentals of the theory, the work of his doctoral students Hans Euler and Bernhard Kockel on light-by-light scattering and higher order corrections was performed (see Sect. 9.1) from 1934 to 1937. As I will argue, Heisenberg's taking up of the terminology of the virtual and the prominence of his students' work in the field led to a wider spread of the terminology and allows to draw a direct connection to the "virtuality" in the BKS theory.[73] Kockel's dissertation and the work of Edwin Góra, Polish refugee from 1941 onwards under Heisenberg's patronage, partially set the conceptual grounds for Heitler's theory of radiation damping (see further below).

But most of the centres of quantum field theory in Central Europe began to dissolve. Wolfgang Pauli stayed in Zürich until 1940. But, as the threat of war and the political restrictions by the NS regime grew, the institute's (quantum field) theorists left the country one after another. Victor Weisskopf fled through Copenhagen and Kharkiv to the US and found employment at the Rochester University in 1937. Nicholas Kemmer went to London in 1936. Pauli himself left for the Institute for Advanced Studies in Princeton.[74] Gregor Wentzel remained in Zürich. In fascist Italy, physics could be pursued mostly undisturbed until 1938. Although Italian fascism was ideologically similar to National Socialism in Germany, Italian physicists turned more strongly towards the British and US community after 1933.[75] The eventual disintegration of the Italian researchers began when the government imposed racial laws

---

[72] For Heisenberg's foreign visits and their political exploitation, see Walker (1992).

[73] The turn to *anschauliche* models of quantum theory in Heisenberg's work at around 1932 is certainly important for the explicit application of non-mathematical representations by his students.

[74] Pauli was raised catholic but had Jewish origins. He was an Austrian citizen and with the annexation of Austria by the German Reich in 1938, Pauli was forced to take the German citizenship (the Swiss citizenship was refused to him twice; once after the Austrian *Anschluss* and once after the outbreak of the war). After Germany had also invaded France, Swiss was completely surrounded by the NS regime and the threat was increasing for Pauli, who would have been treated as a Jew by the German authorities. See https://library.ethz.ch/standorte-und-medien/plattformen/virtuelle-ausstellungen/wolfgang-pauli-und-die-moderne-physik/gescheiterte-einbuergerung-und-ausreise-in-die-usa.html (visited on August 23. 2023).

[75] Fermi, for example, would tend to publish papers in English rather than in German. Nevertheless, he and his students still corresponded with Heisenberg.

similar to the ones in Germany. Fermi's wife, Laura Fermi, was Jewish and Fermi was attacked himself. He left for the US directly after receiving the Nobel Prize in December 1938.

The Russian community suffered under the uprising of the NS regime and a connected militarization due to a growing fear of war as well as under the political purges under the Stalinist regime. Matvei Bronstein, aspiring young theoretician who had proved that light-quantum splitting in empty space was impossible, was arrested in August 1936 and eventually executed in 1937. The worst blow to a pragmatically oriented quantum electrodynamics and quantum field theory in the Soviet Union came with Lev Landau's escape from Kharkiv and the following short-term dissolution of the Landau school.[76] In 1937, only shortly after Akhiezer, Pomeranchuck and Landau had worked on light-by-light scattering and the Delbrück effect, Landau, seemingly fearing his own detention after a series of arrests at the Kharkiv university, disappeared for a few weeks and found refuge at the Moscow institute of Pyotr Kapitza.

The political situation in Europe led to a well-known shift of research centres, mainly to the USA. The enormous number of emigrants and the economic depression rendered the job situation dire. In Great Britain, only a few and the most established physicists could secure a permanent position.[77] Amongst the refugees who found exile in Great Britain, Walter Heitler is the most important one for our story. From 1933 to 1940, he worked on a research fellowship in Bristol and, after a six month detention as an "enemy alien", at the Dublin institute of Advanced Studies from 1941 onwards.

In exile, Heitler kept close contact with experimentalists and followed his pragmatic inclination as well as his tendency towards visualization and verbalization.[78] His contributions to meson theory, together with the refugees Herbert Fröhlich and Nicholas Kemmer, as well as the development of the theory of radiation damping with Huanwu Peng (keeping virtual particles closely on-shell, in modern terminology) heavily relied on (partially newly developed) non-mathematical representations and on concepts already practised and introduced in Göttingen (i.e. the *Schüttelwirkung* of Maria Göppert-Mayer and Wigner-Weisskopf theory; see Sects. 8.1 and 9.3). Heitler's 1936 textbook, the *Quantum Theory of Radiation*, became one of the most popular sources to learn quantum electrodynamics and its practice (see Sect. 8.2.2). It was based on the particle viewpoint.

---

[76] Not to be misunderstood: The Great Purge under Stalin and WWII not only left its mark on the thriving Landau school, but also on the physical institutes in Moscow and Leningrad. Focus on Landau is due to the contributions of his school to the representations and concepts connected to the virtual particle.

[77] Even Max Born worked for three years on a research fellowship, considered leaving for the Soviet Union, but was eventually called to Edinburgh. For structural factors increasing the tendency of physicists to spend only a limited amount of time in Great Britain and to more easily obtain a permanent position in the US, see Hoch (1983, Sects. 3 and 4).

[78] Walter Heitler always showed a great tendency towards visualisable concepts which is, for example, evidenced by the introduction of exchange forces in the London-Heitler theory of valence bonding.

Similar to Great Britain, the USA had already developed a firm stance in the theoretical community when the Central European refugees started to arrive.[79] The summer schools at the University of Michigan, at which, for example, Fermi presented his approach to QED, the conferences on theoretical physics at the George Washington University, modelled closely after the Copenhagen Conferences at Niels Bohr's institute,[80] as well as conferences held by the American Physical Society all contributed to the dissemination of theoretical knowledge, also amongst an ever growing number of doctoral and post-doctoral students of theoretical physics. The pragmatist attitude of the experimentally-driven theoretical physics in the US has already been indicated and is well-known. Concerning the intake of refugees, this also meant that physicists with pragmatist inclinations found positions more quickly than their rather philosophically motivated peers. But also assimilation to the pragmatic attitude and the close interrelation between experiment and theory occurred.[81]

By the late 1930s, when the theoretical output of the community was slowly coming to a low due to war related work,[82] the geographical centres of quantum field theoretical research had shifted. Werner Heisenberg remained in Leipzig and Gregor Wentzel in Zürich. Yet, most of the main practitioners had left Central Europe and were distributed over British and mainly US institutes. Quantum theory in Europe would not recover from the severe blow of WWII and the NS regime until the early 1950s and the next huge steps, both in quantum electrodynamics in general but also concerning the development of the virtual particle concept, were taken in the pragmatic surrounding of the post-WWII USA.

3.4 The War and Its Aftermath

Although work on the problems of quantum field theory was still followed in Europe and already partially solved in Japan, the final chapter of the historical part will close in on the work of US physicists and especially the work of Richard Feynman. For the development of the community and the practice of physics, the wartime work had

[79] I refrain from giving a complete outlook on the theoretical development of physics in the US. For a quick overview of the theoretical staff in the early 1930s, one may consult Schweber (1986, 77, Table 1) and Sopka (1988, 245, Table 4.3).

[80] The Washington Conferences were initiated in 1935 by George Gamow, Edward Teller and Merle Tuve. At the first conference, for example, Hans Bethe informed the participants about Wick's explanation of the nuclear momenta through virtual dissociation (wrongly ascribing it to Heisenberg and von Weizsäcker). A short overview as well as reports on the conference of the 1930s are available under: https://home.gwu.edu/~kargaltsev/HEA/washington-conferences.html (visited on July 10. 2022).

[81] See the case of Eugene Wigner as discussed by Schweber (1986).

[82] This does not mean that research on non-war related issues was given up completely. Max Born in Edinburgh and Wolfgang Pauli in Princeton continued their work. Werner Heisenberg developed greater parts of his S-matrix theory during the war. In neutral Ireland, Walter Heitler worked on his theory of radiation damping.

several effects. First of all, the experimental capabilities in micro-wave technologies increased considerably, which set the ground for reliable experimental data on the Lamb shift and the anomalous magnetic moment of the electron, effects which are interpreted as being caused by virtual particles and which finally spurred the development of renormalization techniques. Secondly, the interrelation between experiment and theory was strengthened and the pragmatist attitude of getting the right numbers fostered. Especially theoretical physicists enjoyed a higher esteem both in the physical community and society after the war. Finally, Los Alamos and the other sites of war related work created a (career) network and even for young theorists professor positions opened in the expanding environment of physical institutes directly after the war.

American physicists started to (re-)organize and created the opportunities for direct idea exchange on the new developments in QED and particle physics directly after the war. The Shelter Island (June 1947), Pocono (April 1948) and Oldstone (April 1949) conferences, modelled after the ones in Copenhagen and Washington, brought the (mainly US centred) elite of quantum field theorists and experimentalists together. The influence of these gatherings on the further development of renormalization, as for example through Bethe's "trainride calculation" of the Lamb shift, has been discussed several times. And even though the development of renormalization and the reformulation of QED was an effort with many contributors, the final chapter will close in on Richard Feynman.

Richard Feynman portrayed himself and was portrayed as a "practical man". From his early years at MIT and his assistantship to John Archibald Wheeler in Princeton, through his war time work at Los Alamos and finally to his space-time approach to quantum electrodynamics, getting the numbers right was important to him.[83] But as Adrian Wüthrich (2010) has argued, getting the right picture, a suitable image of the theory, was driving Feynman's research and eventually gave way to the most prominent representation of a virtual particle in terms of his diagrammatic method. The relevance of perturbation theory, the input of his doctoral supervisor John Wheeler, and the general new outlook on QED as proposed by Feynman for the conception and the properties of the virtual particle concept will be studied in Chap. 11.[84]

---

[83] For the localization of Feynman in the American physicists' community and the prevailing pragmatist attitude, see Forstner (2007, Chap. 4).

[84] The historical literature on Feynman's QED and his diagrams is already extensive and I will heavily draw from Sam Schweber's assessment of Feynman's path to his formulation of QED (Schweber 1994, Chap. 8), from Adrian Wüthrich's study of the genesis of Feynman diagrams (Wüthrich 2010), from Olivier Darrigol's very accessible reconstruction (Darrigol 2019), and from Alexander Blum's description of the paradigmatic change taking place in QED at the late 1940s (Blum 2017), always highlighting how Feynman's approach shaped the practice with and the conception of the virtual particle concept.

Part I
From Virtual Oscillators to Virtual Transitions (1923–1929)

Chapter 4
The BKS Theory and the Light-Quantum Hypothesis: Virtual Entities and Transitions to Intermediate States, But in Different Conceptual Frameworks (1923–1925)

This chapter will focus on the theoretical description of the light-matter interaction, specifically of dispersion and the scattering of light, in the old quantum theory,[1] roughly from 1923 to 1925. For a reconstruction of the historical development of the concepts of virtual transitions and virtual particles there are plenty of reasons to do so: The concept of virtuality was prominently introduced into quantum physics in the BKS theory (an acronym based on the last names of its inventors: Niels Bohr, Hendrik Anthony "Hans" Kramers and John Clarke Slater)[2]; characteristic, later to be called quantum, amplitudes clearly came to the fore in the description of quantum phenomena; according to some historians (see further below) the shift from virtual oscillators to virtual transitions has already occurred during this time period; and the decomposition of scattering as two successive acts of emission and absorption accompanied by a transition of the atom to an "intermediate state" had been spelled out explicitly, but in a methodological and conceptual framework decidedly differing from the BKS theory.

This chapter will thus introduce prime aspects of concepts connected to the virtual particle, as the quantum-amplitude level of the theory and the terminology of the virtual. The focus on scattering and dispersion is furthermore motivated by the observation that Dirac would introduce his verbal model of successive transitions through intermediate states exactly in this problem context (see Chap. 5). Thus, one

[1] The term "old quantum theory" will be used here and onwards as a chronological umbrella term, capturing modes of physical explanation which applied quantum conceptions before the advent of quantum mechanics. The historical end-point for this period is, as a purely chronological distinction, the publication of the so-called *Umdeutung paper* by Werner Heisenberg in September 1925 (Heisenberg 1925), generally interpreted as having opened the field of matrix mechanics. Importantly, the term should not imply that there was any kind of generally accepted and practised theoretical framework shared by all involved actors, a conception that has increasingly and rightfully been rejected in the last decades.

[2] I will use the acronym BKS to refer to the theory as such as well as to the respective authors, as long as there is no possibility of misidentification.

M. Ehberger, *Representing the Unobservable*, Science Networks. Historical Studies 68,
https://doi.org/10.1007/978-3-032-09188-8_4

of the prime connections investigated in the following is in how far, in what way and in which conceptual framework a connection between transition processes and the process of scattering and dispersion had already been established in the old quantum theory.

Traditionally, dispersion played a particularly important role in the historical narratives on the old quantum theory: Until recently,[3] the theoretical description of dispersion, most prominently Hans Kramers' dispersion theory (Kramers 1924a, b) as spelled out in a paper by Kramers and Heisenberg (1925), had been conceived of as the most important source for Heisenberg's *Umdeutung* (Heisenberg 1925),[4] i.e. the paper generally acknowledged to have opened the field of matrix mechanics. In consequence, quite an extensive amount of secondary literature on quantum-theoretical accounts on dispersion and their relation to Heisenberg's *Umdeutung* is already extent and provides an important backdrop for parts of this chapter.[5] Importantly, Marta Jordi Taltavull, who has written her dissertation on the development of dispersion from 1870 to 1925 (Jordi Taltavull 2017), has covered a lot of the developments I am about to discuss but naturally from a different perspective.[6]

The empirical discovery and theoretical explanation of the Compton effect, another phenomenon of light scattering, enjoys a particularly important stance in the historical development of quantum theory as well: It is generally assumed to have tipped the balance towards the photon in the conceptual dispute over a wave- and quantum-theoretical conception of the radiation field.[7] BKS, as a theory or rather conceptual framework which was intended to explain the light-matter interaction and which is well-known both for the breach with energy conservation and its main theoretical concepts, "virtual oscillators" and "virtual radiation field", was developed and published in early 1924, exactly at the time when dispersion and the Compton effect became increasingly important in the theoretical discussions. As such, and

---

[3] In a detailed study of the archival material in connection to Heisenberg's *Umdeutung* paper, Blum et al. (2017) have highlighted the role of the problem of calculating spectroscopic intensities for Heisenberg's breakthrough in 1925 and thereby challenged the traditional narrative which conceived of dispersion as the most important problem context.

[4] With Max Jammer's *The Conceptual Development of Quantum Mechanics*, his discussion of dispersion and the rise of matrix mechanics, this standard narrative found a firm place in the historic literature on quantum mechanics (Jammer 1966, especially Sects. 4.3 and 5). This view was upheld until recently: Duncan and Janssen (2007a, 571) called dispersion theory "the bridge between the old quantum theory and matrix mechanics".

[5] I mostly engaged with the discussions as given by Dresden (1987, Especially Chap. 13), Darrigol (1992, Chaps. IX and X), Konno (1993), Duncan and Janssen (2007a); as well as Duncan and Janssen (2007b).

[6] Besides her dissertation, I also consulted her papers Jordi Taltavull (2013), Jordi Taltavull (2016), Jordi Taltavull (2020).

[7] Roger Stuewer (1975) has called the Compton effect a "turning point in physics", as the title of his monograph reads. See also Brush (2007) or Hentschel (2017, Sect. 5.2) for the framing of the Compton effect as a turning point or watershed.

surely also due to its (at first sight) perplexing terminology and basic assumptions, the BKS theory attracted a lot of attention by historians as well.[8]

Since the BKS theory is generally accredited with the introduction of the concept of virtuality into quantum physics, a discussion of this theory and the concepts applied is certainly in place, especially to judge in how far it influenced the further development of concepts carrying the respective epithet. Some historians have already proposed, at least implicitly, a direct connection. Max Dresden (1987, Chap. 13, Sect. IV) has argued that the description of dispersion in the Kramers-Heisenberg paper, which was originally presented in the BKS framework, rested on the concept of virtual transitions. Edward MacKinnon (1977) has argued that in Heisenberg's 1925 breakthrough the conceptual step from virtual oscillators to virtual transitions was taken. While it is unquestionable that Kramers and Heisenberg introduced "characteristic amplitudes", which we can retrospectively identify as probability or quantum amplitudes, and that they provided mathematical and diagrammatic representations, which make Dresden's point of view suggestive, I do not fully share his argument. It is also undoubtedly true that Heisenberg has stripped his physical interpretation of the virtual oscillators in his 1925 paper, that its basic building blocks were in modern parlance matrix elements and that a corresponding physical interpretation might read virtual transition into them. Looking closely at the representations provided and the use made of them, I want to argue for a more nuanced picture of this development and will review the conceptual differences but also connections between virtual oscillators and virtual transitions.

While the historical lines which led to Heisenberg's *Umdeutung* have found a lot of attention, a dead-end of the theoretical discussions remains under-investigated.[9] It had its roots in the theoretical elaborations by Wolfgang Pauli as well as by Albert Einstein and Paul Ehrenfest on the Compton effect and was fully developed at the institute of Arnold Sommerfeld in Munich. Based on Gregor Wentzel's work, Karl Herzfeld introduced a conception of scattering in which the atom performed transitions to "intermediate states" for a very short amount of time while first absorbing and then re-emitting an impinging photon. Although these "intermediate states", also called "meta-stationary states" by Adolf Smekal, were surely distinct from the intermediate or virtual states (modern terminology) of Dirac and there seems to be no generative connection, a discussion of this work is still suggested because of two reasons: This conception will resurface in the late 1920s as a possible explanation of the light-matter interaction in quantum theory, and its development points to a

[8] The BKS theory has been discussed from different angles and rather extensively. See, for example, Hendry (1981), Stolzenburg (1984, 3–96), Klein (1970, Sect. 7), Konno (1993), Dresden (1987, Chap. 13, Sect. III), Darrigol (1992, Chap. IX), Duncan and Janssen (2007a, Sect. 4), Jordi Taltavull (2017, Sect. 6.5.2), Blum and Jähnert (2022, 137–139) as well as Blum and Jähnert (2024).

[9] The work of Pauli as well as Einstein and Ehrenfest have only been touched upon in the secondary literature and, to the best of my knowledge, Klein (1964) provided for the most detailed discussion. The work on dispersion in Munich has only been discussed by Jordi Taltavull (2017, Sect. 7.3). Wentzel's quantum phase has been identified as a conceptual forerunner of Feynman's integral over path method by Antoci and Liebscher (1996).

strong conceptual entanglement of the decomposition of the scattering process into emission and absorption subprocesses and the light-quantum hypothesis.

I will start out my discussion with a concise discussion of the BKS theory, its emergence, its basic concepts, and the connotations the term "virtual" acquired thereby. This will not only serve as the basis for the further discussion of concepts carrying this epithet, but also for my discussion of the Kramers-Heisenberg paper and Heisenberg's *Umdeutung* paper (Sect. 4.2). These two papers have, as mentioned, been suggested as the first occurrences of virtual transitions. In Sect. 4.3, I will turn to the description of dispersion and scattering in a light-quantum framework in the years 1923 to 1925. Although this constituted a historical side road, it is not only interesting due to the above mentioned points. A discussion is, as I will show, necessary to understand Wolfgang Pauli's criticism of the BKS theory, and it will constitute a further explanation of the lack of verbal explications of successive transitions in the work of Kramers and Heisenberg.

Summarizing my discussion, I will argue that although physicists had provided mathematical and diagrammatic representations which we can retrospectively identify with virtual transitions, the historical actors did not yet (necessarily) think in such terms about their formulas. For the sake of further discussion, I will highlight the main points of the conceptual development that will either be taken over or reoccur in the chapters to come. Thereby, I will touch on an important point for my discussion but also on a central issue for quantum theory, namely the question: What is to be perceived of as a transition?

4.1 The BKS Theory (1923/24): Virtual Oscillators, (Virtual) Radiation, and Real Energy Non-Conservation

> All progress in science emphasizes difficulties. All progress in physics [entails] no proofs whatever but only simple connections of different conceptions.
>
> [...] I look on most general reasoning in science as opportunistic (success- or unsuccessful) relations between conceptions more or less defined by other conception[s] and helping to overlook things [...].[10]

In 1923, quantum theory was in a state of crisis–perceived so by the actors themselves and by historians who have engaged with this period. The concepts and methods of celestial mechanics, which had been widely applied to and refined in the atomic domain since Bohr's famous 1913 trilogy on his atomic model, seemed to have reached an impasse, especially for describing the spectra of atoms more complex than hydrogen, as epitomized by the "Helium catastrophe". Whether it was the problem of accounting for static phenomena with the methods of Bohr's atomic model (Darrigol 1992, 213) or the communication between Copenhagen and the

[10] Draft of a letter from Bohr to Darwin written in July 1919, as cited in Stolzenburg (1984, 15–16).

Breslau physicists Rudolf Ladenburg and Fritz Reiche, the development of a dynamical description of the light-matter interaction gathered prominence at Bohr's institute in mid 1923, exactly at the time when dispersion and the Compton effect also gathered more attention in the theoretical community. The time seemed ripe for a new description of the light-matter interaction, particularly since, as Blum and Jähnert (2022, 129) have put it, Bohr's idea of the emission of radiation by transitions between non-radiating states had "blackboxed the dynamics of the radiation process entirely".

Especially dispersion had posed a serious epistemic obstacle both from the perspective of a quantized radiation field and for Bohr's atomic model, and diverse (unsuccessful) attempts had been made to give a quantum-theoretical explanation.[11] According to classical theory, an atom–modelled as a set of harmonic oscillators–struck by impinging radiation would send out coherent spherical waves, and the interference between primary and secondary radiation would cause a frequency dependent digression of the velocity of light in the medium; this is the basis of the classical explanation of dispersion, a phenomenon which occurred for any frequency of light, also the ones not coinciding with an absorption or emission frequency of the respective scattering material.

How conceptual problems for a quantum-theoretical explanation of dispersion emerged can be seen quite easily.[12] The light-quantum hypothesis was generally seen as incompatible with the phenomena of interference and therefore it could not explain the most basic principle of dispersion, the interference of primary and secondary radiation. But also with a classical radiation field dispersion held conceptual difficulties when combined with Bohr's atomic model. How would an atom in a stationary state react to an impinging radiation with arbitrary frequency? How could the coherence between primary and secondary radiation be explained, when the concept of a stationary state itself came with a finite lifetime and, thus, a finite amount of time between emission and absorption of radiation? Moreover, it became increasingly clear in the mid to late 1910s that an application of classical electrodynamics to the movement of electrons in stationary states was empirically and conceptually incoherent.[13]

As we know, the BKS paper was written mostly (if not exclusively) by Bohr and Kramers. And Bohr was certainly the driving force behind the development of the connected conceptual framework. The BKS theory went beyond the explanation of

---

[11] In her comprehensive reconstruction, Marta Jordi Taltavull (2017, 595) has called dispersion "a conceptual laboratory for quantum physics". One might certainly borrow the analytical terms "epistemic challenge" of Friedrich Steinle (2009) or "empirical challenge" of Olivier Darrigol (2008) to describe the catalytic function of the phenomenon of dispersion for the conceptual development.

[12] The following paragraph is certainly retrospectively informed but, generally speaking, we find diverse instances where historical actors accentuated the following points and the problems for a quantum-theoretical explanation of dispersion.

[13] The conceptual incoherences were strikingly put to the fore by Carl Oseen (1915)–and Bohr agreed–in a reaction to the Debye-Sommerfeld theory of dispersion. See for example Duncan and Janssen (2007a, 579).

dispersion, but important concepts and ideas of this theory had emerged in the discussion of dispersion.[14] All the same, Bohr was opting for a more general description of the light-matter interaction. And although the quote opening this section is taken from a never sent letter from Niels Bohr to Charles Darwin, drafted in July 1919, and Niels Bohr, Hans Kramers and John Slater wrote and published their joint work in early 1924, so four and a half years later, I still hold that this quote gives a hint of how to best understand the development and content of the BKS framework.

In most general terms, I conceive of the BKS theory as Bohr's attempt to harmonious amalgamate different, but interrelated conceptions and the corresponding techniques in order to provide for a description of the light-matter interaction which could explain the available empirical material. In essence, Bohr tried to give an internally consistent interpretation of the empirical phenomena and the light-matter interaction. As such, BKS rested on the ideas and methods Bohr and others had developed in the prior years. One of the most important conceptual alterations, which took full form in BKS, was the partial decoupling between transition processes and the radiative reaction of atoms, which, pushed to its final conclusion, meant that atoms could also radiate while still in stationary states. The most important technical tools were certainly the adiabatic and correspondence principle.

4.1.1 The Methodological and Conceptual Pillars of the BKS Framework

Before laying the grounds for the conceptual pillars of BKS, let me shortly introduced the basic ideas behind the two guiding methodological principles[15] and start with the adiabatic principle. In his "Quantum Theory of Line-Spectra" (Bohr 1918), Bohr firmly connected Ehrenfest's adiabatic theorem to his atomic model. In short, the principle stated that a system remained in one particular state as long as a perturbation to the system is turned on adiabatically slowly. For Bohr, this principle was conceptually important:[16] The comparison between the energy of two stationary states only made sense to Bohr, if there was a way of *continuously* transforming one state into the other. This is exactly what the adiabatic principle allowed and thus,

[14] The influence of ideas developed in a dispersion-theoretical context is most clearly described by Duncan and Janssen (2007a) and Stolzenburg (1984, 3–96), but it is essentially part of any reconstruction.

[15] In "Die Grundpostulate der Quantentheorie", Bohr stressed that the adiabatic and correspondence principle seemed best suited to unlock the riddles of quantum theory and to finally "give hope for the future of a consistent theory" (Bohr 1923, 165).

[16] See Darrigol (1992, 132–137). See also Darrigol (1992, 220–221) for the relevance of the adiabatic principle for the identification of the "real" entities in BKS.

as Bohr himself had expressed it in a never sent letter to Charles Galton Darwin, it ensured that "we can in the worlds of stationary states obtain a rational definition of the energy".[17]

The correspondence principle,[18] in most general terms, was a means of establishing contact between the motion of the atom in a stationary state, or, more to the point, its Fourier decomposition, and the radiation emitted or absorbed by the atom in quantum-theoretical terms. In Bohr's 1913 trilogy, the movement of the electron and the radiation process had been disconnected, and the correspondence principle could reinstate this connection.[19] Initially it was used by Bohr to establish selection rules: A Fourier component of the motion of the atom corresponded to a possible quantum transition, hence the missing of such a component meant that the transition was forbidden. To understand the implication of the correspondence principle more fully, we shortly have to turn to Albert Einstein's work on black-body radiation.

In 1916, Einstein (1916b) studied the conditions under which an ensemble of quantum systems, described in terms of a state-transition model, would remain in thermal equilibrium with radiation. In order to do so, Einstein introduced probability coefficients for absorption, stimulated and spontaneous emission (modern terminology) and, in an extension, showed that such processes were connected to a momentum transfer between atom and radiation (Einstein 1916a).[20] Einstein considered this proof of the directed character of emission/absorption and thus of the light-quantum hypothesis. Bohr did not accept the latter, but he imported Einstein's statistical description and the probability coefficients (for spontaneous emission) to his atomic model in his "Quantum Theory of Line-Spectra". Einstein's argument could not provide any clues of how to calculate these coefficients. And this is where the correspondence principle comes in.

According to the correspondence principle, the probability for a transition to occur was assumed to be proportional to the intensity of the corresponding harmonic of the movement of the atom. Through this assumption, grounded on the classical limit, the Fourier decomposition of the motion of the atom

$$x(t) = \sum_{\tau} C_{\tau} \cos\left(2\pi\left(\tau\omega t + \gamma_{\tau}\right)\right) \tag{4.1}$$

where C_τ are the amplitudes of the harmonics, ω are the frequencies of the harmonic components, and τ an integer number (the summation stretches over all integer numbers), became a central theoretical entity in Bohr's atomic model. As the classically calculated intensity was proportional to the absolute square of the C_τ, the same assumption was put forward for the quantum-theoretical probabilities:

[17] Draft of a letter from Bohr to Darwin written in July 1919, as cited in Stolzenburg (1984, 16).

[18] For a detailed study of the application of the correspondence principle, see Jähnert (2019).

[19] This fact is most coherently expressed and followed in its consequences by Blum and Jähnert (2022). My shortened description follows Blum and Jähnert (2022) and the reader is referred to the paper for more context and details.

[20] For a short discussion of Einstein's first paper, see e.g. Darrigol (1992, 118–121).

$|C_\tau(n)|^2 \propto A^n_{n-\tau}$. The correspondence principle thus established contact between the motion of a single quantum system and its transitions accompanied by emission and absorption processes.

The use and extension of the correspondence principle is mostly connected to the direct scientific environment of Bohr, most importantly to Hans Kramers,[21] and with its application different new questions emerged, as for example what Martin Jähnert has called the "initial-final-state problem": It was unclear whether the harmonics of the initial state, the final state or some combination of the two should be used. Whatever the concrete answer to this question was, starting with the dissertational research of Hans Kramers (1919), a new, mostly implicit concept emerged. Historical actors often referred to it as the *Zwischenbahn*, Blum and Jähnert captured it under the analytical term *Ersatzstrahler* (substituted radiator): "the general interpretation of this amplitude [derived from a suitable averaging procedure] with a hypothetical motion, which was thought of as producing classical radiation and delivered explicit expressions for spectroscopic quantities [...]." (Blum and Jähnert 2022, 136)

With the concept of the substitute radiator we already enter the conceptual terrain of the BKS theory. And most of the grounding pillars of BKS have actually been entertained by Niels Bohr for quite a while. He had vocalized a clear preference for the wave theory of light, at least in free space.[22] As early as 1917, he had questioned strict energy and momentum conservation in the interaction of light and matter[23]; and by 1923 (Darrigol 1992, 214) the denial of strict energy conservation became entangled with a refutation of the light-quantum hypothesis. Bohr rejected the latter both on methodological[24] and conceptual[25] grounds. As early as 1916 and in the context of dispersion, Bohr had (initially quite tentatively) put forward the idea that the mechanism regulating emission and absorption would act in a way analogous to a classical oscillator.[26] This classical analogy, if not substitute model, acquired more prominence in Bohr's thinking in the following years and was several times vocalized in the context of dispersion.[27]

[21] This is a rather traditional and–due to the spatial restrictions–under-complex framing. For a comprehensive discussion of the correspondence principle and its applications, also outside of Copenhagen, see Jähnert (2019).

[22] See e.g. a never sent letter to Charles Galton Darwin, dating to July 1919, as cited in Stolzenburg (1984, 15–16).

[23] See Stolzenburg (1984, 14–15).

[24] According to Bohr, the application of both the light-quantum hypothesis and strict energy conservation might shed light on the processes that can occur but not on the quantum-theoretical nature of these processes. Bohr (1923, 164–165).

[25] According to Bohr, the light-quantum hypothesis entailed "insuperable difficulties, when applied to the phenomena of interference, which constitute our chief means of investigating the nature of radiation" (Bohr 1923, 157).

[26] This manuscript was published as a supplement to a German translation of a collection of Bohr's early work on quantum theory in 1921 (Bohr 1921). See also Duncan and Janssen (2007a, 579–580).

[27] For example, shortly before the development of the BKS theory in Bohr (1923, 161–162).

Finally, in mid 1923, Bohr had come to the conclusion that there was a "fundamental difficulty to relate the observable absorption phenomena directly with the occurrence of quantum jumps".[28] As he put it in a letter to Rudolf Ladenburg:

> Quantum transitions are not the direct cause of the absorption of radiation but rather constitute an effect that accompanies the continuous dispersing (and absorbing) effect of the atoms, without us being able to account for the close quantitative connection in detail by means of the usual concepts of physics.[29]

A purely formal comparison between quantum (probability coefficients) and classical (oscillator strength) parameters of dispersion by the Breslau physicist Rudolf Ladenburg (1921) was enriched–in an exchange of ideas between Copenhagen and Breslau–with the oscillator model in a joined paper with Fritz Reiche (Ladenburg and Reiche 1923).[30] They introduced the notion of an *Ersatzoszillator* (substitute oscillator), the oscillator strength of which was proportional to the transition probabilities. These oscillators, which Ladenburg and Reiche explicitly referred to as "formal", were connected to all *possible* transitions of the quantum systems and they would send out the coherent spherical wavelets necessary for dispersion. They provided for a substitute model of the atom in the case of dispersion, which was formally based on transition probabilities, while the atom rested in a stationary state[31] or–as Ladenburgand Reiche have put it–while "we do not assume that the transition actually takes place [...]."[32]

---

[28] "grundsätzliche Schwierigkeit, die beobachtbaren Absorptionsphänomene direkt mit dem Auftreten von Quantensprüngen in Verbindung zu bringen." Bohr to Ladenburg, May 17. 1923; as cited in Stolzenburg (1984, 399).

[29] "[dass die] Quantensprünge nicht die direkte Ursache der Strahlungsabsorption sind, sondern eine Wirkung darstellen, die die kontinuierliche dispergierende (und absorbierende) Wirkung der Atome auf die Strahlung begleitet, ohne dass wir von der engen quantitativen Verbindung mittels der gewöhnlichen Begriffe der Physik im Einzelnen Rechenschaft geben können." Bohr to Ladenburg, May 17. 1923, as cited in Stolzenburg (1984, 400).

[30] See Duncan and Janssen (2007a, Sect. 3.3), or Jordi Taltavull (2017, Chap. 5 and Sect. 6.4) and Jordi Taltavull (2020) for detailed elaborations.

[31] For clarity's sake: Here I digress from the reading of Marta Jordi Taltavull (2017, Sect. 6.4.4.3). She has argued that Ladenburg and Reiche connected the reaction of the atom to radiation which was not equivalent to an emission or absorption frequency with so-called "incomplete transitions" ["unvollständige Quantensprünge"] (Ladenburg and Reiche 1923, 591, Footnote 29) and, thus, provided for an explanation in terms of the orbital model of the atom. Close reading of the paper and the correspondence between the Breslau physicists and Bohr reveals, that Ladenburg and Reiche only refer to incomplete transitions in the context of resonant scattering and not in the case of regular dispersion. Thus, an explanation of dispersion in terms of Bohr's atomic model which went beyond tentative remarks in the application of the correspondence principle were not put forward by Ladenburg and Reiche.

[32] "Dabei nehmen wir nicht an, daß die Übergänge [...] wirklich zustande kommen." Ladenburg and Reiche (1923, 591).

Kramers, Bohr's assistant and important advocate of the correspondence principle, noticed a problem with the classical limit in the equations of Ladenburg and Reiche.[33] Whether directly motivated by this problem or not, Kramers' own quantum-theoretical take on dispersion resolved it. Kramers' theory was presented in a set of two short papers (Kramers 1924a, b), which–simply going by publication date–takes us beyond the BKS theory. And although Kramers used the BKS "terminology", as he put it (Kramers 1924b, 311), he had already derived the formulas when Slater, who gave the last input to the BKS theory, arrived in Copenhagen in 1923.[34]

For the further discussion two points in reference to Kramers' dispersion theory are particularly important.[35] First of all, Kramers included a new term in the dispersion formula which he referred to as "negative dispersion" in analogy to Einstein's "negative absorption", i.e. induced emission. For the corresponding "emission oscillators" the oscillator strength was negative. And although counter-intuitive–and according to a critique by Gregory Breit (1924) without a classical analogue of an oscillator–this term ensured that the classical limit was kept. Secondly, Kramers derived his dispersion formula through a perturbative approach to the orbital model of the atom and a subsequent translation of the formulas into quantum theory by a new instantiation of the correspondence principle, which is often referred to as the replacement of differentials by differences and was simultaneously developed by Max Born and John Van Vleck.[36] As for Ladenburg and Reiche, Kramers' oscillators stood in for all "possible transitions" the atom could perform, while it firmly rested in one stationary state during the interaction process.[37]

The concept of the substitute oscillators, which had emerged in the context of dispersion, and the concept of the substitute radiator, which had emerged in the context of the correspondence principle, were merged in BKS.[38] This step, together with a relaxation of energy conservation to a statistical law, opened up a path to a new

---

[33] For further discussion, I refer the reader to Jähnert (2019, Sects. 6.3 and 6.4) and Jordi Taltavull (2017, Sect. 6.5.1.1) (both focussing on Reiche's reactions and application of the correspondence principle).

[34] See Duncan and Janssen (2007a, 597–598), and the literature cited therein, for an elaboration on the temporal relation between BKS and Kramers' dispersion theory.

[35] For a (technically) detailed discussion of Kramers' dispersion theory, see Darrigol (1992, 225–259), Dresden (1987, 150–159), Jordi Taltavull (2017, Sect. 7.2.1.1), Duncan and Janssen (2007a, Sect. 3.4) and Konno (1993).

[36] See, for example, Darrigol (1992, 225–229) or Duncan and Janssen (2007b, 635–637), for technical details. I will shortly discuss both points in relation to Kramers and Heisenberg's paper on dispersion.

[37] Gregory Breit put the contrast between the modelling of the radiative reaction of the atom on impinging light, on the one hand, in terms of a classical radiation field and, on the other hand, in terms of Einstein's light-quantum hypothesis rather distinctively in a letter to Kramers: "Einstein's problem is statistical. Ours is not. His atoms change from one quantum state to another. Ours stay in one state." Breit to Kramers, September 26. 1924, AHQP, MF 8a, Sect. 2. Permission by the Niels Bohr Archive, Copenhagen.

[38] For a concise discussion see Blum and Jähnert (2022, Sects. 3.2 and 3.3).

conception of the light-matter interaction on the grounds of the correspondence principle. Kramers' extension of Ladenburg and Reiche's dispersion formula could have only strengthened Bohr's hope to find further insights into the light-matter interaction by exploiting the correspondence principle. The discovery of the Compton effect seemed to have tipped the balance towards light quanta in the physicists community. At least, the idea was taken much more seriously afterwards.[39] This might have stimulated Bohr to work out a theory of the light-matter interaction that would not rely on the light-quantum hypothesis.[40]

The last steps in the emergence of the BKS theory were taken in late December 1923 when John Clarke Slater arrived in Copenhagen.[41] Slater brought a proposal for a quantum theory of interaction with him and thereby provided the last missing piece of the puzzle for the BKS theory, at least so in Bohr's recollections. Bohr considered it "far more harmonious from the point of view of the correspondence principle to connect the spontaneous radiation with the stationary states themselves and not with the transitions".[42] According to Slater's initial proposal, atoms established a classically behaving guiding field during the time they were in excited states.[43] The emitted energy, clumped into photons, would travel along the Poynting vectors of this field. During a short period of intense discussion from late December 1923 to January 1924, Bohr and Kramers persuaded[44] Slater to give up the light quantum part of his theory. The radiative activity of the stationary states was kept as it cohered with Kramers', Bohr's as well as Reiche and Ladenburg's concept of an *Ersatzstrahler/-oszillator*.

The BKS theory combined these different, but interrelated conceptions Bohr and others had developed and entertained with Slater's innovation; BKS put emphasis on the correspondence and adiabatic principle for the development of theory; and it was at the same time the peak of a conceptual development that was already expressed in Ladenburg and Reiche's paper for dispersion: *Possible* transitions were active and effective, while the transitions actually performed by the material part of the system were not causally connected (see further below) to the radiation process.

---

39 For this standard narrative of the Compton effect as a turning point see, for example Stuewer (1975) or Hentschel (2017, Sect. 5.2).

40 Although I share large parts of Alexander Blum's and Martin Jähnert's outlook on the BKS theory, the relevance of the wave-theoretical description for Bohr is, in my conception, somewhat downplayed in their contribution.

41 For the work by Slater, its connection to the BKS paper, and his role in the writing process, see e.g. Duncan and Janssen (2007a, 597–601), Jordi Taltavull (2017, Sects. 6.5.2.1 and 6.5.2.2), Blum and Jähnert (2022, 137–138), Stolzenburg (1984, 6–12), Darrigol (1992, 218–219), or Dresden (1987, 159–163).

42 Bohr to Slater, January 10. 1925. BSC, MF 16, Sect. 2. Permission by the Niels Bohr Archive, Copenhagen.

43 A similar idea had been proposed, but not published as a so-called "Gespensterfeld" by Albert Einstein.

44 See for example the two snippets of the letters Bohr to Slater, January 28. 1926, and Slater to Bohr, May 27. 1926; as cited in Stolzenburg (1984, 68–69).

In the following, I will first give a general outline of the BKS theory, which Kramers and Bohr clearly perceived of as a working programme.[45] Then I will turn to the concept of virtuality in BKS. I will review both the attributes of the "virtual oscillators" and the "virtual radiation field", construct a meaning of "virtuality" thereby and also engage with broader, philosophical identifications of the term "virtual". Finally, I will shortly review the demise of the BKS framework, which occurred only a year after its proposal.

4.1.2 *Outline of the BKS Framework*

In late January 1924 Niels Bohr, Hendrik Anthony Kramers, and John Slater handed in "The Quantum Theory of Radiation" (Bohr, Kramers, and Slater 1924a) for publication in the *Philosophical Magazine*.[46] According to their own assessment, they had found a way of reconciling the quantum discontinuities with the wave theory of light by the application of the correspondence principle (Bohr, Kramers, and Slater 1924a, 785–786). BKS expressed the crucial new feature of their theory in the following terms:

> We will assume that a given atom in a certain stationary state will communicate continually with other atoms through a time-spatial mechanism which is virtually equivalent with the field of radiation which on the classical theory would originate from the virtual harmonic oscillators corresponding with the various possible transitions to other stationary states. (Bohr, Kramers, and Slater 1924a, 790)

Atoms radiated in stationary states. The frequencies of the emitted virtual radiation field were determined by the frequencies of the possible transitions via $h\nu = E' - E''$. The intensities were to be determined by an application of the correspondence principle, although how exactly this would be done was not described. This was how distant atoms came into contact with each other, how they "communicated".

The process of emission was thus intimately connected to the movement of an electron in the orbital model of the atom. The possible transitions were given as components of the Fourier decomposition of this movement (of course, in some way or another, translated into quantum theory). As the authors stated themselves:

> On the view presented in this paper [...] the origin of radiation is not directly sought in the occurrence of transitions but in the motion of the electron. (Bohr, Kramers, and Slater 1924a, footnote on p. 792)

The transitions themselves were not the reason for the emission of radiation. Yet, they involved a change of energy and momentum of the atom. The transitions were

---

[45] Kramers described the BKS theory in a letter to Sommerfeld as "an attempt to present the different observations as simply and correctly as possible". Kramers to Sommerfeld, September 6. 1924, AHQP, MF 8b, Sect. 11. Permission by the Niels Bohr Archive, Copenhagen.

[46] A German translation (revised by Wolfgang Pauli) was communicated in late February 1924 to the *Zeitschrift für Physik* (Bohr, Kramers, and Slater 1924b).

the endpoint of the emission process and the start of the establishment of a new virtual field in the respective final state. The connection between the transitions performed and the virtual radiation inducing them was not a causal but a probabilistic one (Bohr, Kramers, and Slater 1924a, 790–791).

A transition in a given atom depended on its own state and the state of the atoms with which it communicated, but not on the other atoms actually performing a transition. A causal connection between transitions in different atoms was no longer assumed. Energy and momentum were only conserved statistically (Bohr, Kramers, and Slater 1924a, 792).[47] All the same, BKS did not assume "any departure from the electrodynamic theory of light as regards the laws of propagation of radiation in free space, but [seek the cause for statistical conservation of energy] in the peculiarities of the interaction between the virtual field of radiation and the illuminated atoms" (Bohr, Kramers, and Slater 1924a, 793).

When irradiated, the atoms did not only gather a probability to perform a transition but also acted as sources for secondary radiation which could interfere with the impinging one. In effect, BKS assumed "a far reaching analogy with the classical theory of electrodynamics as regards the interaction of the virtual radiation field and the virtual harmonic oscillators [...]" (Bohr, Kramers, and Slater 1924a, 793). But for the atoms, such an assumption was explicitly excluded (Bohr, Kramers, and Slater 1924a, 792) because of the postulated stability of stationary states.

In turning to dispersion, BKS pointed to the "essentially formal character" of the analogy between the classical theory and quantum theory established through the correspondence principle, by invoking the "fact that on the quantum theory the absorption and emission of radiation are coupled to different processes of transition, and thereby to different virtual oscillators" (Bohr, Kramers, and Slater 1924a, 797). Although they already indicated that a quantitative description of dispersion will be worked out and mentioned Kramers' work on dispersion, they only emphasized "once more the continuous character of the optical phenomena, which seemingly does not permit an interpretation based on a simple causal connexion with transition processes in the propagating medium" (Bohr, Kramers, and Slater 1924a, 797–798).

4.1.3 Virtualities in BKS

As we will need it later on, a discussion of the term "virtual" in the BKS theory is necessary, and the previous outline should serve as a basis. We know for a fact that

[47] BKS accepted Einstein's idea of an alteration of energy and momentum in a transition processes, but did not connect it to the emission of a light quantum. The direction of the momentum in the case of transitions induced by other atoms was determined by the wave propagation of the virtual radiation field. If the transition was induced by the atom's own radiation field, the direction of the change in momentum was statistically distributed. See Bohr et al. (1924a, 793).

the term "virtual" was not introduced thoughtlessly by BKS.[48] Consequently, the attribution of the term "virtual" being "at best ambiguous and at worst meaningless" (Hendry 1981, 198) seems spurious. Yet, also BKS' contemporaries were wondering what the term "virtual" should mean.[49]

To entangle the meaning(s), let's first discuss each concept on its own. For the virtual oscillators, it is first to note that there is disagreement whether or not they behave purely classically. For absorption oscillators this is taken for granted, but for emission oscillators Darrigol (1992, 222 and 228) and Jordi Taltavull (2017, Chap. 6.5, especially 6.5.2.3) argued that they did not behave classically. In contrast, Blum and Jähnert (2022, 138) argued that their behaviour could be described entirely classically by assuming a particular phase relation between the impinging radiation and the already vibrating oscillator.

The BKS paper is notoriously vague and it is actually regardless for my argument whether the oscillators behaved entirely classically or not.[50] What can safely be stated is: BKS used an entirely classical language to describe oscillators and waves. And regardless of their (non-)classical nature, by merging the *Ersatzstrahler* (Kramers) and the *Ersatzoszillator* (Ladenburg and Reiche), BKS came up with a picture of the light-matter interaction, including dispersion, absorption, stimulated and spontaneous emission which rested on (nearly) classically behaving and oscillating entities conjugated to the motion of the electron *in a stationary state*. All the same, this conjugation was a purely formal one, which Blum and Jähnert clearly showed through their placing the virtual oscillator in the tradition of Kramers' *Ersatzstrahler*. As they noted: "The virtual oscillators of BKS [...] provided a classical substitute model for the transition process [...]." (Blum and Jähnert 2024, 343) And this modelling function, the formal character of the virtual oscillators, came with a certain freedom to describe their behaviour: They could move at different velocities than the atom itself.[51]

Let's turn to the virtual radiation field. It was emitted during the time the atom rested in a stationary state. The radiative activity was transferred from the

[48] When Bohr asked Pauli for a language-review of the German BKS publication, he asked Pauli to look favourably on the words "communicate [kommunizieren]" and "virtual [virtuell]", "for, after lengthy considerations, we have agreed here on these basic pillars of the exposition" ["denn um diese Grundpfeiler der Datstellung haben wir uns hier nach längerer Überlegung geeinigt."]. Bohr to Pauli, February 16. 1924, as cited in Pauli (1979, 146).

[49] See, for example, the letter of Oskar Klein to Bohr, June 30. 1924, as cited in Stolzenburg (1984, 29); or Erwin Schrödinger–generally approving of BKS–to Bohr, May 24. 1924, as cited in von Meyenn (2011, 92).

[50] It should still be noted that the question whether the virtual oscillators were entirely classically behaving entities or whether they had non-classical features has an influence on whether historians of science considered them to be a model of the atom or not. While Blum and Jähnert conceived of the virtual oscillators as a classical but formal substitute model, Olivier Darrigol has noted that "the virtual oscillators should be considered not a new classical model but rather (as Bohr had already asserted in the BKS paper) another formal product of the correspondence principle applied to the ordinary orbital model" (Darrigol 1992, 228–229).

[51] BKS needed this modelling freedom to describe the Compton effect (which is the most famous case), but also for dispersion.

transitions to the states. This also entailed that the emitted radiation did not have to be monochromatic: From one excited state there can be several possible transitions to other states.[52] In vacuum the radiation behaved entirely classically.[53] When it comes to the energy of the virtual radiation field and its relation to the regular radiation field, there is considerable disagreement in the secondary literature.[54] This disagreement has its roots in the ambiguities in the paper itself.[55]

In my eyes the most coherent interpretation as well as the one covered best by the historical material was proposed by Duncan and Janssen (2007a, Especially p. 607). They suggested that BKS did not introduce a new radiation field but only a new version of its interaction with matter. Already Duncan and Janssen (2007a, 605–607) gathered some sources published after the BKS paper by Kramers or Slater to back up their reading. But also in the BKS paper we find hints.[56] Actually, the epithet "virtual" was regularly dropped for the radiation field in later publications using the BKS framework, while it was kept for the oscillators. And at diverse points, both in publication, drafts, and in correspondence, Kramers and Slater stated that energy is radiated by atoms.[57]

While it is thus likely that BKS assumed that the radiation field carried energy, there is also an explanation why they did not point this out explicitly. The energy of the radiation field was definitely not observable. According to BKS all observations were based on the quantum jumps and quantum states of the material system.[58] For the quantum matter a reasonable definition of energy could be given through

---

[52] This was most clearly spelled out in Kramers and Holst (1925, 135–136).

[53] As stated by BKS themeselves (Bohr, Kramers, and Slater 1924a, 793).

[54] Dresden (1987, 179–182) conceived of the virtual radiation field as a probability field, which carried no energy, while an additional regular radiation field existed. Darrigol (1992, 219–221) and Blum and Jähnert (2022, Sect. 3.3) argue that the virtual radiation field carries no energy, is a probability field but that there is no other radiation field.

[55] BKS dropped the term virtual for the radiation field, they put it in brackets, they use it when it seems counter-intuitive (Compton effect), and eventually state that in the classical limit "the emphasizing of the 'virtual' character of the radiation field [...] automatically looses its importance" (Bohr, Kramers, and Slater 1924a, 801–802).

[56] "The cause of the observed statistical conservation of energy and momentum we shall not seek in any departure from the electrodynamic theory of light as regards the laws of propagation of radiation in free space, but in the peculiarities of the interaction between the virtual field of radiation and the illuminated atoms." Bohr et al. (1924a, 793).

[57] Compare for example: Slater in Anonymous (1925, 242), Kramers and Holst (1925, 135); or H. A. Kramers: An incomplete manuscript draft entitled "Über die Eigenschaften von Atomen in einem Strahlungsfelde." [This paper was intended for publication in *Naturwissenschaften*, but was not published], AHQP, MF 27, Sect. 16. The correspondence between Kramers and Fritz Arndt, Professor for chemistry at the Universtiy of Breslau and translator of Kramers' and Helge Holst's textbook, reveals that in the discussion on Kramers' talk in June 1924 in Breslau, Kramers already made a comment about the energy content of the radiation field. See Kramers to Arndt, October 22. 1924, AHQP, MF 8, Sect. 1; and especially Arndt to Kramers, October 24. 1924, AHQP, MF 8, Sect. 1.

[58] "[...] the time-spatial occurrence of the various transition processes *on which the observations of the optical phenomena ultimately depend*," Bohr et al. (1924a, 791, emphasis added).

the adiabatic principle.[59] As Bohr had put it five years before in a never-sent letter to Darwin, there was a way to rationally define energy in the "world of quantum states". But the interaction between matter and radiation field was statistical. Thus, the energy of the radiation field could not be measured. The radiation field may or may not have carried energy, but the definition of the energy of the radiation field for quantum theory was only of limited interest since it could not be measured and only was upheld statistically.

Having discussed the entities alone, let me try to synthesize a meaning of the "virtual" as given in BKS. This will constitute a rational reconstruction because a historical one in terms of sources BKS might have drawn from is hardly possible. None of the authors left a written comment on the sources for the term. Bohr had used it priorly,[60] other physicists like Kossel (1916, 357) or even Ladenburg and Reiche (1923, 585) used the term "virtual" in reference to different theoretical entities. The term was around and used, not only for classical concepts like the "virtual displacement" or the "virtual image", but also directly in quantum theory. Philosophical connotations are of course also possible, but hardly traceable. Thus, the search for generative connections of the terminology without any clues were to start was not followed.

To start out this reconstruction, let me first exclude one possible connotation. A strong connection between the energy content of the virtual radiation field and the term "virtual" seems not tenable (see my argument above).[61] Energy non-conservation was real, so this is also no possible candidate, although it is rather closely connected to the term today. My own interpretation of the virtuality of the theoretical entities of BKS rests on two prior historical reconstructions: the ones by Blum and Jähnert (2024) and Duncan and Janssen (2007a). In most general terms, I believe that Blum and Jähnert (2022, Sect. 3.3), as further spelled out in Blum and Jähnert (2024), provided the (nearly) correct interpretation: BKS actually introduced two kinds of worlds, a virtual one in which the oscillators and the radiation interacted, and a real one of quantum transitions and energy changes.[62]

Let me elaborate: There is a duplication, inherent in BKS, in the description of the phenomena.[63] The statistically governed quantum transitions were connected to observation, while there was another world in which the oscillators, conjugated to

[59] This fact was emphasized by Olivier Darrigol (1992, 220).

[60] In his initially unpublished, English manuscript of 1916, he invoked the term "virtual", in German "virtuell", in connection with certain transformations of systems connected to the adiabatic principle (Bohr 1921, 128, 130, 134).

[61] Ko (1997, 31) suggested this interpretation. The idea that the radiation field did not carry any energy was also part of Olivier Darrigol's interpretation (Darrigol 1992, 220). Besides this point, I believe Darrigol was on the right track but did not push his argument far enough.

[62] The only further historical reconstruction were such a reading has at least been touched upon was in the edition of Erwin Schrödinger's correspondence by von Meyenn (2011, 60–61, 104).

[63] Also Duncan and Janssen (2007a) stressed the duplicate nature of BKS, while not connecting it to a world-view of two ontologically different planes of description.

the possible transitions, interacted with the radiation in a (nearly) classical manner. Importantly, this duplication is not a retrospective reconstruction. Slater (1925, 399) himself spoke of this duplication in the theory, and Richard Becker, theoretical physicist at the Technische Hochschule in Berlin, noted that the virtual character of oscillators and radiation field is "justified by [the fact] that in principle only the quantum transitions taking place inside the atoms are accessible to observation."[64]

Gregor Wentzel, at the time Privatdozent in Munich, provided the clearest exemplification of this line of thought. In an overview article about the different ways of explaining the Compton effect theoretically, Wentzel presented the BKS theory under the title "The dualistic theory" ("Die dualistische Theorie"):

> According to this theory [BKS] one has in a sense to differentiate two worlds: In one the natural events are determined wavelike, in the other quantumlike. The wave world is the primary and the preferred one insofar as its events entail the ones of the quantum world. However, it is in no way accessible to our observations; neither the wave nor the particles (oscillators) creating the same are in any way observable; they are "virtual". What we observe and measure rather are electron velocities in the quantum world, or respectively their photoelectric and photochemical effects, which are coupled through probability laws to the field strengths of the "virtual" ones.[65]

I hold that this interpretation of the (virtuality in the) BKS theory is the most coherent one not only in respect to the historical source material, but also in respect to the later usage of the term. The most important (although not the first) invocation of the term "virtual" in quantum mechanics and quantum electrodynamics, which eventually led to a widespread use of the term in connection to transition processes, was made by Werner Heisenberg in the mid 1930s. According to the specific terminology he used ("virtual possibilities") and the conception of quantum theory and the measurement process he entertained at the time, it is this connotation of an unobservable/unobserved virtual world that Heisenberg will invoke.[66] Although I do agree with Blum and Jähnert (2024) that the concept of the virtual in BKS differed

[64] "dadurch begründet, daß grundsätzlich nur die in den Atomen erfolgenden Quantensprünge der Beobachtung zugänglich sind." Becker (1924, 174). Becker's paper is also interesting as Kramers reacted on a preprint and did not question Becker's definition of the virtual. Kramers to Becker, August 12. 1924. AHQP, MF 8, Sect. 2.

[65] "Nach dieser Theorie [BKS] muß man gewissermaßen zwei Welten unterscheiden: in der einen ist das Naturgeschehen wellenmäßig, in der anderen quantenmäßig bestimmt. Die Wellenwelt ist insofern die primäre und bevorzugte, als ihre Ereignisse diejenigen der Quantenwelt nach sich ziehen. Allerdings ist sie unserer Beobachtung in keiner Weise zugänglich; weder die Wellen noch die dieselben erzeugenden geladenen Partikeln (Oszillatoren) sind irgendwie beobachtbar; sie sind 'virtuell'. Was wir beobachten und messen, sind vielmehr Elektronengeschwindigkeiten in der Quantenwelt, bzw. deren photoelektrische und photochemische Wirkungen, welche durch Wahrscheinlichkeitsgesetze mit den Feldstärken der 'virtuellen' gekoppelt sind." Wentzel (1925, 445–446).

[66] See Sect. 9.1.2 for my argument about Heisenberg's use of the terminology.

markedly from the later usage of the term in quantum field theory,[67] the general distinction between unobservable and observable world (or, for QFT, unobserved and observed world)[68] still resonated in Heisenberg's mid 1930 invocation.

But BKS actually allows for another interpretation of the term "virtual". In empty space, the virtual radiation behaved equivalently to the regular radiation field. The non-classical features came out most pronounced at the location of the atoms, or rather in the interaction of the waves with the atoms. Wolfgang Pauli, who spent Easter 1924 in Copenhagen discussing the BKS framework with Bohr and who showed his mastery of the conceptions by his particularly critical assessment of the theory later on,[69] put the phrase "(because acting non-classically)" right behind the word "virtual" in his 1926 article in the *Handbuch der Physik*.[70]

But these two views on the "virtual" in BKS, the introduction of a wavelike unobservable world and the non-classical (i.e. energy non-conserving) light-matter interaction, are actually strongly interrelated. In reference to the virtual oscillators, the radiation field was virtual because it occupied the same plane of reality, i.e. the virtual one. Oscillators and radiation interacted directly, but were unobservable, formal, nearly a classical supplementary model of sorts. In reference to the real atoms, the radiation field was virtual because it interacted non-classically. It did not exchange energy directly, but only induced the probability of transitions, and thus energy changes, to occur.

According to BKS, all our observations rest on the quantum jumps of the real atoms. But these were only probabilistically connected to the virtual world. Thus, the latter was in principle unobservable. The probabilistic (instead of causal) connection between virtual and actual world introduced the cut which made the former unobservable in principle.

The Philosophical Connotation

I consider these two interrelated definitions in terms of physical attributes, modelling function and effectiveness decisive for the further use of the terms and concepts in the physicists' community. But for constructing a more general outlook on the theme and also for the sake of completeness, I want to discuss rather philosophical, more general connotations of the term "virtual" as implied in BKS. Up until now, such a

---

[67] This holds not only for the characteristics of the theoretical entities that were connected to the terminology of the virtual, but also to the ontological connotations the term came with. For BKS, the virtual entities seem to be, as expressed by Blum and Jähnert (2024, 329), "categorically distinct and ontologically separated" while this separation was blurred, at least for some actors, in quantum field theory.

[68] This differentiation between unobservable and unobserved is the point where the concept of "virtuality" in BKS and QFT differs according to Blum and Jähnert. It is also the point which was bridged by Heisenberg when invoking the term in the context of QFT.

[69] See e.g. Pauli to Bohr, October 2. 1924, as cited in Pauli (1979, 163–166).

[70] "in die Art der Wirksamkeit dieses 'virtuellen' (weil eben nicht klassisch wirkenden) Feldes", Pauli (1926, 83).

reading has only been discussed by Blum and Jähnert (2024) and Martinez (2024). I will focus on Blum's and Jähnert's definition, since Martinez aimed at a more general definition, which should also hold for the 1930s. Therefore, Martinez slightly altered the definition which Blum and Jähnert referred to.

Blum and Jähnert have argued that BKS' kind of virtuality coheres with the concept of virtuality as presented by Charles Sanders Peirce, which reads as follows:

> (1) A virtual X (where X is a common noun) is something, not an X, which has the efficiency (virtus) of an X.
> This is the proper meaning of the word; but (2) it has been seriously confounded with "potential," which is almost its contrary. For a potential X is of the nature of X, but is without actual efficiency. (Peirce 1902, 763)[71]

Blum and Jähnert's argument is based on the fact that the virtual radiation field did not transmit energy to the atom while a classical radiation field should. Thus, the virtual radiation field was not a radiation field (as it did not exchange energy with the atoms) but acted in (relation to the virtual oscillators)[72] as such.

I think this identification is doubtful. In fact, the virtual radiation had, besides interaction properties and potentially energy content, all the attributes of a regular radiation field. In free space, virtual radiation behaved like classical radiation. But it differed from its classical counterpart through its interaction with the real atomic systems, i.e. through its efficiency. Thus (compare the quote above), the virtual X (radiation field) is (of the nature of) X (radiation field), but it does *not* have the efficiency of an X (radiation field). The radiation field was certainly effective, but not in the regular way. This identification would rather be close to Pierce's definition of the "potential", which he opposed to the concept of the virtual. Hence, I do not share Blum and Jähnert's assertion that there is a close connection between the concept of virtuality in BKS and Pierce's definition, at least so if we start from the radiation field.

Taking the virtual oscillators as the primary theoretical, "virtual" entities and focusing on their relation to the real quantum atom, an identification with Peirce's definition seems more coherent. In terms of the light-matter interaction, the atom behaved as if it consisted of a set of virtual oscillators which were conjugated to the possible transitions it might perform. But the virtual oscillators were not oscillators: They were theoretical entities which replaced the atom, or rather the transition mechanism, in the BKS framework. Yet, they should in some way describe the reaction of the real atom on the radiation field. Hence, a virtual oscillator was not, in an ontological sense, an oscillator but still had the efficiency of an oscillator on the radiation field. Thus, conceiving of the virtual oscillators as the primary entities and the virtual nature of the radiation as a consequence of the fact, that radiation and oscillators inhabited the same theoretical space in the BKS framework, one might still (tentatively) apply Pierce's concept of virtuality as defined above.

---

[71] See also Steinle (2024) for a discussion of Peirce's definition.

[72] This insertion is my reading of their short comment.

While there are other philosophical traditions which we might turn to reconstruct the concept of virtuality in BKS,[73] I want to provide for my own, more general definition of the "virtual". It is the best general description I could come up with in order to connect the role of the virtual entities in quantum physics throughout the development I am about to sketch.

According to the framework of BKS, the virtual world was one of possibilities but it was still effective. The atom rested in a certain state when it reacted to the radiation field, but not through an alteration of its movement but through the virtual oscillators. Already Ladenburg and Reiche but also BKS and later on Kramers and Heisenberg emphasized, that the *Ersatzoszillatoren* and virtual oscillators were conjugated to all *possible*, but not actualized, transitions the atom could perform (unbound by energy conservation).[74] In BKS, but also in more concrete dispersion-theoretical accounts, the transition only potentially performed by the atom became effective agents in the description of the phenomena. In short:

The virtual entities extended the set of possible processes and rendered this new set of possibilities, although not actualized, physically effective.

Hence, through the virtual oscillators also energy non-conserving, merely potential transitions were effective in the BKS framework, as was particularly important in the case of dispersion. Even though virtual oscillators are thus connected to energy non-conservation indirectly, this was not the type of energy non-conservation BKS emphasized in their paper. The latter kind of energy non-conservation was real, observable, and due to the statistical connection between the real transitions induced in the atoms and the radiation impinging on them. Even though there was a connection between virtual oscillators and energy non-conserving transitions, the violation of the energy law was not yet the defining feature of virtuality in quantum physics.

4.1.4 An Intermediate Conclusion

In my interpretation of BKS, I combined insights from Blum and Jähnert (2022), Blum and Jähnert (2024) and Duncan and Janssen (2007a), grounded them by recourse to assessments by contemporaries of BKS, and showed how these insights are interrelated. Accordingly, I conceive of the virtual in the BKS paper in a twofold

[73] Since Max Jammer's work (e.g. Jammer 1966) the influence of Williams James on Bohr's thought has been identified. The "virtual" was also an important concept for the philosophy of Henri Bergson, which might have figured as a source. The Danish philosopher Harald Høffding, whose teaching of and personal contact to Bohr is well documented, had, for example, (critically) engaged with Bergson's philosophy. I am well aware that, because of his debate with Einstein at around 1922 and some retrospective attributions of Bergson as a forerunner of thought in modern physics, mentioning Bergson at this point is certainly controversial. I also admit that Bergson is not the most likely candidate, but his influence is certainly not ruled out.

[74] As Max Jammer (1966, 191) has expressed it: "[H]ypothetical possibilities had to be regarded as physically active efficient causes.".

way. On the one hand, BKS described a virtual world in which virtual oscillators–conjugated to all *possible* transitions–and virtual radiation interacted (nearly) classically. This allowed for the incorporation of dispersion into the Bohr model through an application of the correspondence principle. On the other hand, the most striking departure from classical physics occurred in the interaction between the actual atoms and virtual radiation, the latter inducing transition probabilities in the former. Thus, the non-classical character of the interaction was also invoked as a reason for the terminology. It was not energy non-conservation as such, but the induction of transition probabilities instead of energy exchange due to the radiation which made the latter virtual.

Although virtual oscillators were also, but not exclusively, conjugated to potentially energy non-conserving transitions, these transitions did not occur, neither virtually, nor actually. The concept of a virtual transition, even as a purely theoretical description, was foreign to the BKS framework. Quantum transitions were real. Virtual oscillators replaced all *possible* transitions for the light-matter interaction. In the same vein, the idea of energy non-conservation in the BKS paper was no reason for the invocation of the term virtual. It was real and observable. Contemplated for a long time by Niels Bohr as a possibly necessary part in the description of the light-matter interaction, the violation of the energy (and momentum) law rested on the statistical connection between the radiation field and the quantum atoms. For BKS, energy non-conservation was as real as the quantum atoms: It was observable and eventually the cause for the refutation of the BKS framework (see the next section).

In reference to the further and overall development of the concepts I investigate, I want to highlight three points. First of all and in terms of functionality, the virtual oscillators share a rather general feature with virtual transitions and virtual particles: They were intermediaries in the description of the light-matter interaction. And they performed this function by extending the set of possible processes (virtual oscillators stood in for all transitions, also energy non-conserving ones) and rendered this set of possibilities physically effective. For BKS this is particularly striking: In the paper as such, but also in the follow-up contributions applying the BKS framework as an interpretative scheme, the reaction of the atom was always framed as the reaction in a specific state (even in cases where a modern reader tends to see virtual transitions, see my discussion of the Kramers-Heisenberg paper in Sect. 4.2).

Secondly and in reference to the usage of the term "virtual", three connections between the BKS framework and the terminology will become important. First of all, and as Duncan and Janssen[75] have most strikingly argued, the virtual oscillator (concept and term) will not vanish, also after the advent of matrix and wave mechanics. It will survive as a classical substitute model (expressed in terms of the matrix elements of the position operator) in semi-classical approaches to the light-matter interaction. Secondly, although there was no direct relation between the invocation of the term virtual and energy non-conservation internal to the BKS framework, both

[75] Duncan and Janssen (2007a) as well as Duncan and Janssen (2007b).

the terminology and energy non-conservation were prominent features of the presentation of the theory. Energy non-conservation and the term "virtual" were thus already indirectly connected through their prominence in this work by Bohr, Kramers, and Slater. As I will argue, this indirect relation might have led to a first invocation of "virtual transitions" as energy non-conserving transitions by Chandrasekhara Venkata Raman in 1929.[76] Thirdly, the actual connotation of the term "virtual" in BKS, i.e. as referring to an unobserved, virtual world, was not lost. It will resurface in the mid 1930s: According to my reconstruction based on the then current views of Werner Heisenberg, he used the term "virtual" in exactly this sense and, through his status in the community, initiated the wider spread of the term.[77]

4.1.5 *The Demise of the BKS Framework (1925)*

The BKS theory was controversially discussed immediately after its publication and, as is well-known, empirically refuted within little more than a year.[78] Initially, BKS found a number of supporters, among them Rudolf Ladenburg and Fritz Reiche (Breslau), Richard Becker (Berlin), Max Born and James Franck (Göttingen), and Erwin Schrödinger (Zürich). The most prominent public opponents giving detailed critique of the conceptual framework were surely Wolfgang Pauli (at the time in Hamburg) and Albert Einstein (Berlin).

In the course of the year 1924, the theoretical critique, specifically by Einstein and Pauli, led Bohr to question the validity of the BKS theory. All the same, Bohr admitted that he was "nonetheless convinced, that the swindle of mixing the classical theory with quantum theory will still prove fruitful in many ways for seeking out the secrets of nature".[79] At the time, Kramers and Heisenberg were working on a further evaluation of the correspondence-like dispersion-theoretical formalism which was cast into the framework of the BKS theory. Due to the relevance of the collaboration of Kramers and Heisenberg for the conception of virtual transitions, I will devote the following section to its discussion.

Yet, the empirical refutation of the statistical conservation of energy and momentum came only a few months after the Kramers-Heisenberg paper was finished.

[76] See Chap. 6.

[77] See Sect. 9.1.2.

[78] For an extensive discussion, also of the papers that used BKS as an interpretation scheme, see Stolzenburg (1984, 23–93). See also Darrigol (1992, 246–255), Duncan and Janssen (2007a, Sect. 4.2), or Jordi Taltavull (2017, Sect. 6.5.2.4) for further detailed elaborations and the relation between the BKS interpretation and the actual methods employed.

[79] "[daß ich] nichtsdestoweniger davon überzeugt bin, daß der Schwindel des Vermischens der klassischen Theorie und der Quantentheorie sich noch auf viele Weisen beim Aufspüren der Geheimnisse der Natur als fruchtbar erweisen wird." Bohr to Pauli, December 11. 1924; as cited in Pauli (1979, 185).

In early 1925, results by Compton and Simon (1925) as well as by Bothe and Geiger (1925) showed that the correlation between scattered radiation and recoiling electrons in the Compton effect was much too high to be reconcilable with the BKS framework. As is well-known, the dispersion-theoretical accounts, which partially motivated BKS but were derived by methods independent of it, as well as the Kramers-Heisenberg formula remained untouched by the demise of BKS. In this respect, I certainly agree with Duncan and Janssen (2007a, 559) that "the rise and fall of BKS was largely a sideshow distracting from the main plot line".

All the same, the conceptual framework of BKS, in which the atom firmly rested in a stationary state during its interaction with radiation, remained an important ingredient in the interpretation of dispersion theory until its empirical refutation. It had considerable influence on the framing of formulas, which might in retrospect be conceived of as including virtual emission and absorption processes, as we will see in the following section. In a broader sense, and as Blum and Jähnert (2022) have argued, the failure of the BKS theory eventually led to different, not necessarily dynamical approaches to quantum-theoretical systems by Born, Heisenberg, and Schrödinger which would develop into quantum mechanics. But this is a different story.

4.2 The Kramers-Heisenberg Paper (1925), the BKS Theory and "Characteristic Amplitudes"

While BKS served as an interpretation scheme for the more popular accounts of the light-matter interaction (or perturbatively describable interactions) during 1924, the principle methodological path taken in such approaches by Kramers, Born, van Vleck, and later on also by Kramers and Heisenberg was independent of BKS and essentially the same for all papers: They all used classical perturbation theory on the movement of the atom, a periodic system expressed in action-angle variables. The classical expression was then translated into quantum theory by applying a three step procedure, which is known as the replacement of differentials by differences (and which I will shortly outline below). For sure, Born's, van Velck's, as well as Kramers and Heisenberg's calculations in classical mechanics differed,[80] yet the general structure was the same.

[80] I will not go into any details of the mathematical accounts on the classical level of theory, as my focus is on the concept of quantum transition processes. For technical details on van Vleck's method see Duncan and Janssen (2007b). A technically detailed description of Kramers and Heisenberg's paper is given in Dresden (1987, Part II, Chap. 13, Section A–C). Darrigol (1992, 231–234) provided a concise description of Born's methods, based on Kramers dispersion-theoretical account. Jordi Taltavull (2017, Sect. 7.2.1) also provided technical details to all of these approaches.

All the same, the paper by Kramers and Heisenberg (1925) traditionally holds a particularly important stance in the path towards quantum mechanics. The paper provided for a first detailed published elaboration of the methods applied by Kramers. Kramers and Heisenberg extended the procedure to include incoherently scattered radiation, proposed by Smekal (1923b) and later to be called Raman effect.[81] Yet, the most important theoretical, if not conceptual, advancement was the identification of the physical significance of new quantum objects. Kramers and Heisenberg called them "characteristic amplitudes".[82] This identification is, in turn, intimately related to the more general treatment of the theoretical description of dispersion-theoretical effects.

According to arguments brought forward in the secondary literature, the ascription of the amplitudes and their phases with quantum-theoretical significance led to a far more, if not the first explicit representation of what we today call an "intermediate" or "virtual state". I partially share this assessment. For the mathematical representation and even for a diagrammatic version, the prominence of such states is undeniable and their occurrence constituted an important input for Heisenberg's route to *Umdeutung*. Yet, according to an argument by Max Dresden (1987, 237–239), reproduced by Marta Jordi Taltavull (2017, 522–523), the framework of BKS and the nature of energy non-conservation of it allowed Kramers to conceive of these states as more than a symbolic representation. Furthermore, as Dresden argued, a physical interpretation of successive transitions guided the reasoning in the publication.[83]

In the following, I want to question these two conjectures by reconstructing the arguments of Kramers and Heisenberg. In order to do so, I will first discuss Heisenberg's background before he came to Copenhagen in autumn 1924. Then I will reconstruct the arguments of the Kramers-Heisenberg paper, closely following both the reasoning processes and nomenclature of the publication. Finally, I will locate my argument about the role of the intermediate states and characteristic amplitudes in reference to Heisenberg's *Umdeutung*, mostly by a backdrop to the already existing secondary literature.

---

[81] Smekal's prediction is well-known. For a detailed and context-sensitive discussion see Jordi Taltavull (2017, Sect. 6.3). I will turn to the interpretation of the effect directly after its experimental establishment in Chap. 6.

[82] "charakteristische Amplituden," see, for example, Kramers and Heisenberg (1925, 690).

[83] For the full discussion, see Dresden (1987, Part II, Chap. 13, Section A–C). Olivier Darrigol (1992, 229–230) was more careful in his judgement and did not imply any guiding function of a physical picture of successive transitions. He also warned his reader of a retrospective identification of matrix products and commutators.

4.2.1 Heisenberg's Route to the Kramers-Heisenberg Paper: Celestial Mechanics and Born's Discretization Programme

When Werner Heisenberg and Hans Kramers set to work on their co-authored paper in late 1924, two incredibly strong formalists, but also adherents of the BKS theory[84] joined forces. Bohr and Kramers had pioneered the then current application of perturbation theory to atomic systems, the further development of which eventually led to Kramers' theory of dispersion. Werner Heisenberg had studied under Arnold Sommerfeld in Munich and spent the Winter semester 1922/23 at Max Born's Göttingen institute. After his dissertation in Munich in September 1923, Heisenberg returned to Göttingen as Max Born's assistant. Heisenberg's time in Göttingen was, for the development of his analytical techniques, certainly formative.

Max Born, since 1921 Professor for theoretical physics in Göttingen, had launched a research programme in order to evaluate how far the mechanical model of the atom, especially for many-electron systems, could be pushed.[85] This task necessitated a mastery of the analytical took kit of celestial mechanics, the corresponding perturbation theory, and, eventually, its extension. The focus on and the further development of these methods for the application in the atomic domain is mirrored in a substantial output of papers dealing with this theme by Born and his co-workers.[86] Already in the Winter semester of 1922, Born established a "Poincaré-Lesezirkel" (Cassidy 1995, 186) in which Heisenberg learned the sophisticated methods of celestial mechanics and classical perturbation theory. The strong emphasis put on these perturbative approaches and their extensions is also reflected in the official curriculum in Göttingen: During the winter semester 1923/24 Born held a seminar "Höhere Mechanik (Störungstheorie mit Anwendung auf Atomphysik)" [Advanced Mechanics (perturbation theory with applications to atomic physics].[87]

[84] As is well-known, Werner Heisenberg was initially rather critical about the BKS theory and did not consider it to be a big step forward (see for example the letters Heisenberg to Pauli, March 4. 1924, and June 8. 1924; as cited in Pauli (1979, 149–150, 154–156)). But during his stay in Copenhagen and under the influence of Bohr, Heisenberg's attitude towards BKS changed. Although he never became as convinced as Kramers, his attitude towards BKS was certainly not unfavourable.

[85] For a short outline of this line of work and the location of its beginnings in crystal lattice dynamics, see Schirrmacher (2019, 86–89). Greenspan (2005, 107–121) provides an easily accessible description of the period from 1921 to mid 1924 in Göttingen. Cassidy (1995, 174–226) is still a standard reference for Werner Heisenberg's (scientific) biography from his arrival in Göttingen to his first longer research stay in Copenhagen. James and Joas (2015, Especially Sect. 4) discuss both the development of perturbation theory and its application to molecular theory in Göttingen in the years shortly before the advent of quantum mechanics.

[86] Most important were certainly Born and Pauli (1922) and Born and Heisenberg (1923). A small priority dispute emerged between Born/Pauli and Paul Epstein, who had independently developed similar methods. See Duncan and Janssen (2007a, 580).

[87] Assisted by Friedrich Hund, Lothar Nordheim, and Werner Heisenberg, Born wrote his textbook *Vorlesungen über Atommechanik* (Born 1925) partly based on these lecture notes. See Schirrmacher (2019, 81). On Max Born's *Vorlesungen über Atommechanik*, see also Giulini (2013).

This systematic evaluation eventually led to the so-called "Helium catastrophe", which caused Max Born to radically question the suitability of classical mechanics for calculating stationary states.[88] The following strive of Born to discretize mechanics itself, carried out with the help of Heisenberg, first bore fruits in Heisenberg's initially quite critically received paper on the Zeeman effect.[89] After communicating his results to Bohr, Heisenberg was invited to Copenhagen in March 1924. Bohr himself came for a one-day visit to Göttingen in early June 1924. Through this exchange, both the BKS framework as an interpretational scheme and the translation of classical theoretical components by the replacement of differentials by differences was followed and fostered in Göttingen and Copenhagen.

In the course of his discretization programme, Born, again with the help of Heisenberg, expanded the methods of Kramers' dispersion theory to a general theory of coupling, not only for the interaction of light and matter but for any atomic system that could be described perturbatively (Born 1924).[90] Yet, there is an important difference between Born's paper and the Kramers-Heisenberg paper in connection to the concepts we are interested in.

As in Kramers' communications on dispersion theory (Kramers 1924a, b), Born's quantum theory of coupling only represented the absolute squares of "characteristic amplitudes" as relevant quantum-theoretical entities: Although the classical expression was decomposed in a Fourier series with the respective amplitudes C_τ, in reference to the translation into a quantum-theoretical expression Max Born explicitly stated:

> In this process [of substituting the classical Fourier coefficients with their quantum counterparts] it does not seem reasonable to look for corresponding quantities for the Fourier coefficients C_τ themselves [...]; apparently only the quadratic forms $|C_\tau|^2 = C_\tau C_{-\tau}$ have quantum-theoretical meaning.[91]

Born called the corresponding theoretical quantity $\Gamma(n, n')$ but could not provide for means of calculating it. For dispersion, these quantities had of course close relations with Einstein coefficients but, for the time being, Born's programme remained of a purely formal nature.

Being a strong supporter of the BKS theory, Born framed his approach in terms of the virtual oscillators, which he referred to as "the real, primary".[92] But he did not have to buy into "the still disputed conceptualizations of this theory, such as the statistical

[88] For the "Helium catastrophe" and Born's radical reaction, see Darrigol (1992, Chap. VIII, Section "The Catastrophe of Helium", and "A New Quantum Principle in Göttingen"). See also Born (1923) for his own published opinion on the failure of his programme.

[89] For a summary of Heisenberg's paper and the contextualization in Born's urge for a new mechanics, see for example Darrigol (1992, 197–199). The development of the discretization programme by Heisenberg and Born was also discussed by Hendry (1981, 203–209).

[90] See Darrigol (1992, 230–234) for a technical reconstruction of Born's approach.

[91] "Dabei [im Prozess der Ersetzung der klassischen Fourierkoeffizienten mit den korrespondierenden quantentheoretischen Größen] scheint es nicht sinnvoll, für die Fourierkoeffizienten C_τ selbst korrespondierende Größen zu suchen [...]; quantentheoretische Bedeutung haben offenbar nur die quadratischen Bildungen $|C_\tau|^2 = C_\tau C_{-\tau}$." Born (1924, 388).

[92] "das reale, primäre," Born (1924, 387).

conception of the energy- and momentum transmission [...]."[93] Even though the $\Gamma(n, n')$ combined states which could not be reached according to the frequency of the impinging radiation, Born's approach was independent of the energy non-conservation as implied in BKS. Why? Because in Born's approach, as in BKS, the transition mechanism was, for the purpose of modelling the light-matter interaction, replaced by a set of oscillators which resonated at the right frequencies. But this model did not imply that anything like a transition to the higher state actually, or virtually, occurred.

When Heisenberg came to Copenhagen in September 1924, he was fully prepared to carry out the involved calculations that were to be performed in the derivation of the Kramers-Heisenberg paper and he had acquired first-hand knowledge of the BKS framework. Sadly, there is little historical source material on the actual writing process.[94] We know that Kramers announced an extension of his dispersion theory in a talk before the German Physical Society in Hamburg in June 1924. Bohr referred to a forthcoming publication of Kramers and Heisenberg in a paper handed in on November 1. 1924.[95] According to a letter to Arnold Sommerfeld, during the first months in Copenhagen Heisenberg was mostly concerned with interpreting the polarization of resonance fluorescence as observed in experiments. Only in the last, short paragraph of the letter, Heisenberg mentioned that "I want to write a note on the 'Smekal Compton jumps' with Kramers [....]."[96] It is thus very likely that Kramers did not only develop the general idea behind the paper, but also performed most of the calculations and the writing process. The insight to extend the wave-theoretical considerations to the incoherently scattered radiation is, for example, explicitly accredited to Kramers in the publication (Kramers and Heisenberg 1925, 687). And on January 2. 1925, Kramers sent a copy of the paper, at the same time, to Werner Heisenberg and the editor of the *Zeitschrift für Physik*.[97]

Taking into account that Kramers had already obtained results on the incoherently scattered radiation in June 1924, the delay in publication might be explained by the focus on the polarization of resonance fluorescence in Copenhagen and the dispute between Heisenberg and Kramers about the "false resonances" (I will turn to these below). These differences between Kramers and Heisenberg were emphasized in Max Dresden's reconstruction of the paper in order to develop an argument about the further trajectories of the two physicists. According to Dresden, Kramers took the general physical picture of BKS rather seriously and its empirical refutation hit

93 "noch umstrittenen Begriffsbildungen dieser Theorie, wie der statistischen Auffassung der Energie- und Impulsübertragung [...]." Born (1924, 387).

94 The reconstructions of Mehra and Rechenberg (1982, 170–189) and Dresden (1987, Part II, Chap. 13, Section A–C) rest to a larger extent on the AIP interview with Werner Heisenberg.

95 See the timeline in Dresden (1987, 221) and the comments on p. 223.

96 "Mit Kramers will ich eine Note über die 'Smekalschen Comptonsprünge' schreiben," Heisenberg to Sommerfeld, November 18. 1924; as cited in Eckert and Märker (2004, 174).

97 Bohr to Heisenberg, January 2. 1925; BSC, MF 11, Sect. 2. Heisenberg to Kramers, January 8. 1925; AHQP, MF 8, Sect. 6.

him hard. On the contrary, Heisenberg was far more concerned with the theoretical structures, not their interpretation in terms of the BKS framework, and could thus abstract features from the formulas that gathered prominence in his *Umdeutung* paper.[98] For the following evaluation, I will not differentiate between Heisenberg and Kramers' perspectives on the contents of the paper. Rather, I will focus on the representations given in the publication and the reasoning process as it was presented. The premise of this perspective is that the authors, regardless of favouring a more formal or a more physical approach, provided the readers with every aspect that was needed to reconstruct the derivation by themselves.

4.2.2 Smekal Jumps and Non-Verbal Representations of Intermediate States

In the following, I will provide for a detailed, at times technical reconstruction of the Kramers-Heisenberg paper on dispersion. I will closely follow the nomenclature and the reasoning as it was presented in the paper. I will put you through this detailed evaluation as I consider it necessary to coherently interpret the role of what we call "virtual transitions" and "virtual states" for Kramers and Heisenberg. Furthermore, without this detailed evaluation a judgement of the role of the BKS theory and the energy non-conservation assumed in it for the physical interpretation given by Kramers and Heisenberg would not be possible.

Introducing the Physical Interpretation: BKS as Interpretational Scheme

Even though Kramers and Heisenberg did not use the terms "virtual oscillators" or "virtual radiation field", the interpretation given and some of the reasoning provided are strongly related to the BKS framework. Already in the abstract, the results were framed as "an interesting support for this conception".[99] The decisive features of the BKS theory were, according to the authors, the wave-theoretical description of the radiation field, the guiding function of the correspondence principle, and the new interpretation of the reaction of the atom on the radiation field. The atom emitted spherical wavelets while it rested in one particular state. As Kramers and Heisenberg stated:

> What is above all typical for this theory [BKS], is the assumption that the reaction of the atom to the radiation field should primarily be understood as a reaction of an atom existing in a given stationary state; it is also assumed that the transitions between two stationary states

[98] See Dresden (1987, Part 2, Chap. IV, Sections D.–G.).

[99] "eine interessante Stütze für diese Auffassung," Kramers and Heisenberg (1925, 681).

> are of very short duration, and that the detailed nature of these transitions will not play any role in the description of the optical phenomena. (Kramers and Heisenberg 1925, 682)[100]

The essentially new effect Kramers and Heisenberg discussed was that the dipole moment of the atom $\mathfrak{P}$ had to be expressed as ($\Re$ indicates that the real part of the mathematical terms had to be taken):

$$\mathfrak{P}(t) = \Re\left(\mathfrak{P}e^{2\pi i\nu t} + \sum_k \mathfrak{P}_k e^{2\pi i(\nu+\nu_k)t} + \sum_l \mathfrak{P}_l e^{2\pi i(\nu-\nu_l)t}\right) \tag{4.2}$$

In words, Kramers and Heisenberg described this phenomenon as follows:

> Under the influence of irradiation with monochromatic light, an atom not only emits coherent spherical waves of the same frequency as that of the incident light [first term in brackets]: it also emits systems of incoherent spherical waves, whose frequencies can be represented as combinations of the incident frequency with other frequencies that correspond to possible[101] transitions to other stationary states [the two second terms in brackets]. (Kramers and Heisenberg 1925, 655)[102]

Kramers and Heisenberg framed Smekal's original proposal of this type of secondary radiation (Smekal 1923b) in terms of the light-quantum hypothesis, which was a misidentification on their part: Smekal had actually remained agnostic about the nature of the radiation field in his initial contribution[103]; all he assumed was strict conservation of energy. Kramers and Heisenberg followed the BKS framework and stated that "light of frequency $\nu + \nu_k$ or $\nu - \nu_l$ is emitted by the atom, while at the same time the atom acquires a probability for a decrease in its energy content by $h\nu_k$, or an increase by $h\nu_l$" (Kramers and Heisenberg 1925, 686).[104] But Kramers and Heisenberg even extended Smekal's conjecture. They predicted what today is called induced double emission: The simultaneous emission of light of frequency ν (of the impinging radiation) and of frequency ν' such that $h(\nu + \nu')$ corresponded to an energy difference between two states of the atom.

In total, Kramers and Heisenberg's introduction constituted a straightforward application of the BKS framework. They conceived of the reaction of the atoms, also for the case of incoherent secondary radiation and spontaneous emission, as the reaction of atoms in some specific state gathering a probability to perform a quantum jump. Although this might seem artificial, especially when we look at the

---

[100] Here and in the following, most translations are taken from van der Waerden (1967, 223–252). The location of the quotes in the original German version will be presented in parentheses in the text while the location of the translation will be presented in footnotes. Here: van der Waerden (1967, 224–225). If the translation is my own, I will provide the German original in the footnote.

[101] *Comment by the author*: The German original reads "denkbar", which is closer to "conceivable" than to "possible". Since the terminology of "möglich" ("possible") was common at the time and also used by Kramers and Heisenberg later on, I did not differentiate at this point.

[102] van der Waerden (1967, 227).

[103] See Smekal's own comments as discussed in Sect. 4.3.3, or the discussion by Jordi Taltavull (2017, Sect. 6.3).

[104] van der Waerden (1967, 228).

formal derivation of the results and its diagrammatical representation (see below), it was a straightforward consequence of the conceptions of BKS.

The Classical Perturbative Calculation

In the second part of their publication, Heisenberg and Kramers provided the classical perturbative calcualtion of the influence of an external radiation field on the atom, modelled as a mutli-periodic system. The Fourier series of the electric moment of the unperturbed system, expressed in action-angle variables, took the following form:

$$\mathfrak{M}(t) = \sum_{\tau_1,\ldots,\tau_s} \frac{1}{2} C_{\tau_1,\ldots,\tau_s} e^{2\pi i(\tau_1 w_1 + \ldots + \tau_s w_s)}$$

where the τ_i had integer values, the ω_i were the harmonic frequency of the movement, the amplitudes C depended on the action variables and fulfilled the relation $C_{\tau_1,\ldots,\tau_s} = \bar{C}_{-\tau_1,\ldots,-\tau_s}$ (the bar indicates complex conjugation).

This system was perturbed by an incoming radiation of frequency ν and with the electrical vector $\mathfrak{E}(t) = \mathfrak{R}(\mathfrak{E}e^{2\pi i\nu t})$. The entirely classical perturbative calculation led Kramers and Heisenberg to the following, still classical expression of the perturbed electric moment of the atom[105]:

$$\mathfrak{M}_1(t) = \mathfrak{R}\left\{ \sum_{\tau_1^0,\ldots,\tau_s^0} \sum_{\tau_1,\ldots,\tau_s} \frac{1}{4} \left[\tau' \frac{\partial C}{\partial J'} \frac{(\mathfrak{E}C')}{\omega' + \nu} - \right. \right.$$
$$\left. \left. -C\tau \frac{\partial}{\partial J} \left(\frac{(\mathfrak{E}C')}{\omega' + \nu} \right) \right] e^{2\pi i(\omega^0 + \nu)t)} \right\} \tag{4.3}$$

where the τ_s^0 were defined by the condition that the sums $\tau_1 + \tau_1' := \tau_1^0;\ \ldots\ ; \tau_s + \tau_s' := \tau_s^0$ had the same value. ω^0 was defined by $\omega^0 := \tau_1^0 \omega_0 + \ldots + \tau_s^0 \omega_s$.

From these equations, Heisenberg and Kramers already inferred some of the attributes of the scattered radiation: It will be proportional to the intensity of the incident radiation (Eq. (4.3) is proportional to $\mathfrak{E}$). The spectrum will contain the incident frequency ν as well as frequencies, which are either sums or differences between ν and ω^0. Importantly, the frequency ω^0 was not necessarily a frequency of the unperturbed motion of the atom, but was, according to the definition of ω^0 and τ_s^0 above, the sum or difference between two harmonic frequencies of the unperturbed motion.

---

105 For details of the classical calculation I refer the reader to the Kramers-Heisenberg paper itself (Kramers and Heisenberg 1925, 687–689) and the detailed reconstructions of Kramers' dispersion theory as well as the Kramers-Heisenberg paper by Max Dresden (1987, 150–154, 230–231) and Olivier Darrigol (1992, 225–230).

Translation Into Quantum Theory and Coherent Scattering

So far, this was a purely classical endeavour. For the transition to a quantum-theoretical expression, Heisenberg and Kramers considered the two cases of coherently and incoherently scattered radiation separately. In the case of coherent scattering ($\omega^0 = 0$ or $\tau_1^0 = ... = \tau_s^0 = 0$), the classical formula was simplified by Kramers and Heisenberg. After some formal alterations they arrived at:

$$\mathfrak{M}_{kl} = \mathfrak{R} \sum_{\tau}{}' \frac{1}{4} \tau \frac{\partial}{\partial J} \left\{ \frac{C(\mathfrak{E}\bar{C})}{\omega - \nu} + \frac{\bar{C}(\mathfrak{E}C)}{\omega + \nu} \right\} e^{2\pi i \nu t}$$

where the prime indicated that the summation only runs over positive values of ω and τ.

Kramers and Heisenberg then translated the classical expression into quantum theory by a three step procedure. (1) They translated the classical (ω) into the quantum-theoretical transition frequencies (ν). (2) They translated the classical amplitudes C into the "characteristic amplitudes" $\mathfrak{A}$ as "a symbolic average [...] extended over the region between two stationary states".[106] The relation between the characteristic amplitudes and the Einstein coefficients a_q was already known from the work on dispersion: $a_q h\nu_q = \frac{(2\pi\nu_q)^4}{3c^3}(\mathfrak{A}_q \bar{\mathfrak{A}}_q)$. (3) Finally, they applied the method of substituting differentials by differences. In most general terms, it can be represented by $\tau \frac{\partial F}{\partial J} \rightarrow \frac{1}{h}[F(n+\tau) - F(n)]$, where F is an arbitrary function.[107]

Kramers and Heisenberg explained the application of the last step with the diagram shown in Fig. 4.1. The atom under consideration was in the state P. The figure described a system with two degrees of freedom and, consequently, the diagram depicted the J_1, J_2-plane. As the stationary states were defined through their discretized J-values, they were depicted as dots in the diagram. Heisenberg and Kramers considered two states differing from P, namely Q and R. Their J values were $h\tau_1$ and $h\tau_2$ higher (Q) or lower (R) then the J values of P. The connected transitions were called a (for absorption) and e (for emission). The differential was then interpreted as the difference between a magnitude corresponding to the transition a and the same magnitude corresponding to the transition e. In effect, the classical expression $\tau \frac{\partial}{\partial J} F$ was substituted by the expression $\frac{1}{h}(F_a - F_e)$.[108]

As all possible transitions a and e had to be taken into account, Kramers and Heisenberg provided the following "quantum-theoretical expression for the scattering moment with frequency ν which acts in a given state"[109] (Kramers and Heisenberg 1925, 692):

[106] "ein symbolischer Mittelwert [...], erstreckt über den Bereich zwischen zwei sationären Zuständen," Kramers and Heisenberg (1925, 692).

[107] For the discussion of this procedures, see for example Darrigol (1992, 225–229) or Duncan and Janssen (2007b, 635–637).

[108] See also Jordi Taltavull (2017, 519–520) for a more detailed, but essentially equivalent discussion of the procedure by Kramers and Heisenberg.

[109] van der Waerden (1967, 234).

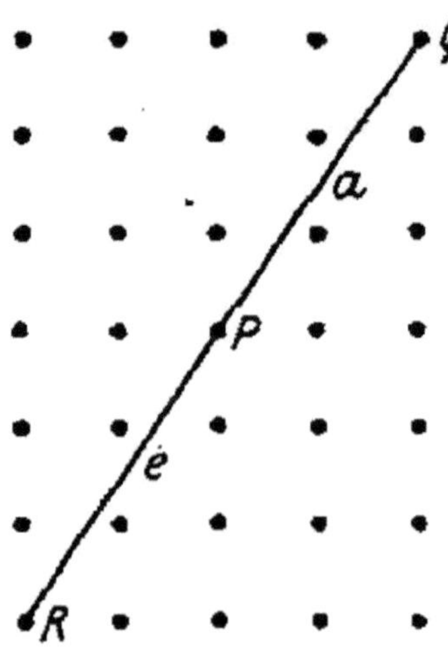

Fig. 4.1 Graphical display for the explanation of the method of substituting differentials by differences in the case of coherent scattering. Figure in Kramers, Hendrik, and Werner Heisenberg. 1925. "Über die Streuung von Strahlung durch Atome." *Zeitschrift für Physik* 31, 681–708, here 693, https://doi.org/10.1007/BF02980624 (cited as Kramers and Heisenberg (1925) in the bibliography). Reproduced with permission from Springer Nature

$$\mathfrak{M}_{qu} = \Re \left[\sum_a \frac{1}{4h} \left(\frac{\mathfrak{A}_a(\mathfrak{E}\bar{\mathfrak{A}}_a)}{\nu_a - \nu} + \frac{\bar{\mathfrak{A}}_a(\mathfrak{E}\mathfrak{A}_a)}{\nu_a + \nu} \right) - \right. \\ \left. - \sum_e \frac{1}{4h} \left(\frac{\mathfrak{A}_e(\mathfrak{E}\bar{\mathfrak{A}}_e)}{\nu_e - \nu} + \frac{\bar{\mathfrak{A}}_e(\mathfrak{E}\mathfrak{A}_e)}{\nu_e + \nu} \right) \right] e^{2\pi i \nu t} \tag{4.4}$$

For the case that $\mathfrak{A}_a$, $\mathfrak{A}_e$, and $\mathfrak{E}$ were real and parallel, Kramers and Heisenberg could easily reconstruct Kramers original dispersion formula. Importantly, upto this point, the sums given by Kramers and Heisenberg, although the relation to transition possibilities to states Q and R was obvious, were not expressed in terms of the upper or lower states, but, according to the labelling of the characteristic amplitudes, in terms of the transition oscillators a and e.

Translating the Incoherent Part

The last, most extensive, and essentially new part of Kramers and Heisenberg's derivation dealt with the incoherently scattered radiation. To start out their discussion, Kramers and Heisenberg singled out one specific $\omega^0 \neq 0$, i.e. (and in quantum-theoretical terms) they picked out a specific frequency of the atom (equal to the frequency shift of the scattered radiation), and, thus, two states P and Q with $h\nu^0 = H(Q) - H(P)$. Essentially, they singled out one specific term through its time-dependence $e^{2\pi i(\omega^0+\nu)t}$ which obviously translated to $e^{2\pi i(\nu^0+\nu)t}$ in quantum theory.

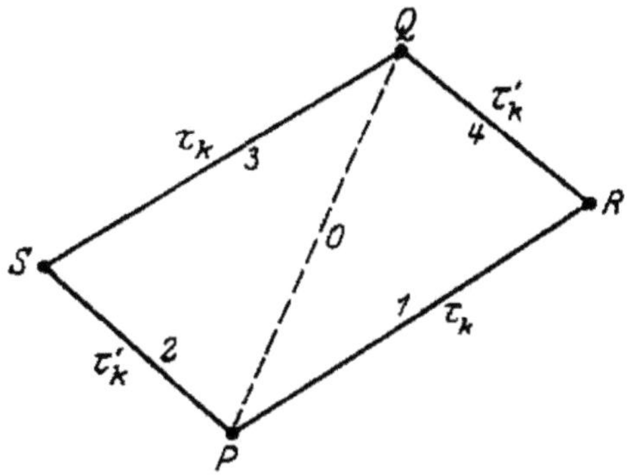

Fig. 4.2 Graphical display for explaining the method of substituting differentials by differences in the case of incoherent scattering. Figure in Kramers, Hendrik, and Werner Heisenberg. 1925. "Über die Streuung von Strahlung durch Atome." *Zeitschrift für Physik* 31, 681–708, here 694, https://doi.org/10.1007/BF02980624 (cited as Kramers and Heisenberg (1925) in the bibliography). Reproduced with permission from Springer Nature

For the translation of the differential expressions and the classical terms occurring in (4.3), Kramers and Heisenberg used the diagram shown in Fig. 4.2. As before, the picture shows the J_1, J_2-plane. Besides P and Q, Kramers and Heisenberg had to consider two further states, called R and S, which corresponded to the states Q in the case of the coherently scattered radiation. The quantum number of P was defined as $n_k = (n_k)_P$. The corresponding quantum numbers of the states Q, R, and S can be read off the diagram:

$$(n_k)_Q = (n_k)_P + \tau_k + \tau_k' = (n_k)_P + \tau_k^0; \ (n_k)_R = (n_k)_P + \tau_k; \ (n_k)_S = (n_k)_P + \tau_k'$$

The transitions between the states were given numbers zero to four.

In hindsight, I want to emphasize two points here. First of all, Kramers and Heisenberg noted that the transition PQ or 0 does not have to correspond to a possible transition of the unperturbed atom. Only the transition 1, 2, 3, and 4 have to correspond to possible transitions, i.e. their characteristic amplitudes were non-zero. This conclusion corresponded to the above mentioned insight that in the classical description ω^0 was not necessarily a harmonic frequency of the movement of the atom but a difference or a sum of such frequencies. It was a straightforward consequence of the classical description.[110]

[110] Such arguments were actually long known in relation to the correspondence principle. See for example Bohr (1918, 35): "From the above considerations [about the action of a small perturbing force on the classical behaviour on a periodic system] we shall therefore expect that, in addition to the main probabilities of such transitions between stationary states which are possible for the original system, there will for the perturbed system exist small probabilities of new transitions corresponding to the above mentioned combinations of the τ's." See also Darrigol (1992, 123–132) for a detailed reconstruction of this conclusion in the context of Bohr's correspondence principle and Bohr and Kramers' perturbation theory.

Secondly, I want to point the reader to the order of the numbering of the possible transitions in the diagram. The transitions were not numbered in terms of successive transitions from state P over R (or S) to Q (or the other way around). Rather, Kramers and Heisenberg attached the lower numbers (1 and 2) to state P and the higher numbers to state Q (3 and 4). Whether even or odd numbers were chosen, depended on the value of τ_k (1 and 3) and τ'_k (2 and 4).

The translation of Eq. (4.3) into quantum theory could be performed, according to Kramers and Heisenberg, in a "seemingly unambiguous manner".[111] It led them to the following identification:

$$\tau' \frac{\partial C}{\partial J'} \frac{(\mathfrak{E}C')}{\omega' + \nu} \rightarrow \frac{\mathfrak{A}_3 - \mathfrak{A}_1}{h} \frac{1}{2} \left(\frac{(\mathfrak{E}\mathfrak{A}_4)}{\nu_4 + \nu} + \frac{(\mathfrak{E}\mathfrak{A}_2)}{\nu_2 + \nu} \right) \tag{4.5}$$

and

$$C\tau \frac{\partial}{\partial J} \left(\frac{(\mathfrak{E}C')}{\omega' + \nu} \right) \rightarrow \frac{\mathfrak{A}_3 + \mathfrak{A}_1}{2} \frac{1}{h} \left(\frac{(\mathfrak{E}\mathfrak{A}_4)}{\nu_4 + \nu} - \frac{(\mathfrak{E}\mathfrak{A}_2)}{\nu_2 + \nu} \right) \tag{4.6}$$

Kramers and Heisenberg did not explicitly explain the translation procedure. But corresponding to the substitution of differentials by differences, an arbitrary classical function F had to be replaced according to $\tau^{(')} \frac{\partial F}{\partial J^{(')}} \rightarrow \frac{1}{h}[F(n + \tau^{(')}) - F(n)]$. This is exactly what Kramers and Heisenberg did for the differentials in the above two equations[112]: Expressed in terms of quantized J-values, all indexes of $\mathfrak{A}_3$ (or $\mathfrak{A}_4$) differed from the ones of $\mathfrak{A}_1$ (or $\mathfrak{A}_2$) by τ'_k (or τ_k).[113] The undifferentiated classical Fourier coefficients were replaced by the arithmetic mean of the characteristic amplitudes of the corresponding transitions (and, respectively, with the quantum-theoretical frequencies in the denominators).[114] Corresponding to their first translation, the characteristic amplitude was replaced by its complex conjugate whenever a quantum frequency turned out negative.

This translational procedure, and at this point my reconstruction differs from the one by Dresden and Jordi Taltavull, was not motivated by an interpretation in terms of successive transitions from state P over S (or R) to state Q. This is not only

111 "scheinbar eindeutigen Weise," Kramers and Heisenberg (1925, 696).

112 Of course, they also substituted the classical ω's with their quantum counterparts ν_i, and the classical Fourier coefficients with the "characteristic amplitudes".

113 To make this explicit, let's take $\mathfrak{A}_3$ and $\mathfrak{A}_1$ and change the nomenclature a little. The replacement of $C_{\tau_k} \rightarrow \mathfrak{A}_{n+\tau_k,n}$ was known. In the given example, $\mathfrak{A}_{n+\tau_k,n}$ corresponded to $\mathfrak{A}_1$ (see Fig. 4.2). Shifting all indexes by τ'_k, which you had to do for the first term in the course of replacing the differentials $\tau' \frac{\partial}{\partial J'}$ by differences, you get $\mathfrak{A}_{n+\tau_k+\tau'_k,n+\tau'_k}$ which is, in Kramers and Heisenberg's nomenclature $\mathfrak{A}_3$.

114 To be honest, it exceeded my capabilities to exactly reconstruct why Heisenberg and Kramers chose the arithmetic mean of these two amplitudes. It is probable that it was suggested from the translation of the priorly established case of coherent scattering, i.e. by comparing the prior quantum-theoretical formula with the full classical Eq. (4.3). As Kramers and Heisenberg (1925, 701) proved, the formulas they established merged with the coherent part if the states P and Q coincided.

evidenced by the numbering in the diagram and the replacement of differentials by differences as described above, but also by the combinations of amplitudes Kramers and Heisenberg arrived at. Both in Eqs. (4.5) and (4.6) terms such as $\mathfrak{A}_1(\mathfrak{E}\mathfrak{A}_2)$ or $\mathfrak{A}_3(\mathfrak{E}\mathfrak{A}_4)$ occurred. Each pair of amplitudes in the respective term was connected to either state P or to state Q, but it did not connect these two states. These terms only cancelled when the two Eqs. (4.5) and (4.6) were subtracted, which of course had to be done according to Eq. (4.3).

This subtraction and the observation, that switching τ_k with τ_k' referred to the same set of states,[115] eventually led Heisenberg and Kramers to the following quantum equivalent of the terms of Eq. (4.3) which corresponded to the set of states P, Q, R, and S:

$$\mathfrak{M}_{qu} \propto \frac{1}{h}\left\{-\frac{\mathfrak{A}_1(\mathfrak{E}\mathfrak{A}_4)}{\nu_4+\nu}+\frac{\mathfrak{A}_4(\mathfrak{E}\mathfrak{A}_1)}{\nu_1+\nu}-\frac{\mathfrak{A}_2(\mathfrak{E}\mathfrak{A}_3)}{\nu_3+\nu}+\frac{\mathfrak{A}_3(\mathfrak{E}\mathfrak{A}_2)}{\nu_2+\nu}\right\} \tag{4.7}$$

where the nomenclature can be inferred from Fig. 4.2.

Determining the Current State of the Atom

For the general case, Kramers and Heisenberg defined a complex vector for a triplet of states P, Q, R:

$$\mathfrak{M}(P, Q : R) = \frac{1}{4h}\left\{\frac{\mathfrak{A}_q(\mathfrak{E}\mathfrak{A}_p)}{\nu_p+\nu}-\frac{\mathfrak{A}_p(\mathfrak{E}\mathfrak{A}_q)}{\nu_q+\nu}\right\} e^{2\pi i(\nu_0+\nu)t} \tag{4.8}$$

where the indexes p, q, 0 refer to the transitions RP, QR, QP respectively (for the nomenclature, see Fig. 4.3). The frequencies are defined by the usual quantum conditions. With this definition, the right side of Formula (4.7) could be written as $\mathfrak{M}(P, Q : R) + \mathfrak{M}(P, Q : S)$. Through a comparison with the classical formula Kramers and Heisenberg inferred that the electrical moment of the atoms could be represented as the real part of a sum of such terms.

Yet, at this point, Kramers and Heisenberg hit upon a problem. Up until now, they had left the current state of the atom completely undefined. And, by correspondence considerations alone, they could not determine which state, P or Q, the atom was in when the electric moment was described by Eq. (4.8). The state R was not considered without giving any further reason. Max Dresden (1987, Specifically pp. 234–239) reconstructed Kramers and Heisenberg's derivation in total but also this specification process by invoking a set of successive transitions and modern energy level diagrams,

[115] Switching τ_k and τ_k' corresponded to switching the numbers 1 and 2 as well as 3 and 4 in Eqs. (4.5) and (4.6).

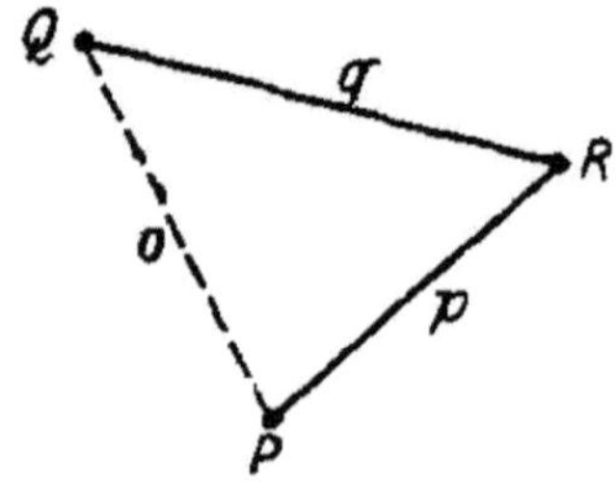

Fig. 4.3 Graphical display for the nomenclature of a electric moment corresponding to a triplet of states. Figure in Kramers, Hendrik, and Werner Heisenberg. 1925. "Über die Streuung von Strahlung durch Atome." *Zeitschrift für Physik* 31, 681–708, here 697, https://doi.org/10.1007/BF02980624 (cited as Kramers and Heisenberg (1925) in the bibliography). Reproduced with permission from Springer Nature

involving all states P, Q and R. Dresden also implied that specifically Kramers had a mental imagery of successive transitions in mind.[116] Following the reasoning process of the paper more closely, I want to call Dresden's emphasis on the mental imagery of successive transitions into doubt. The respective passage in the original publication is Kramers and Heisenberg (1925, 697–699).

Since Heisenberg and Kramers could not determine which state of the atom Eq. (4.8) referred to by correspondence considerations alone, they had to turn to the energy principle.[117] Their conclusion was that Formula (4.8) referred to Q, if the frequency $\nu_0 + \nu$ was positive, and that it referred to P, if $\nu_0 + \nu$ was negative. The reasoning went as follows.

If there was scattered radiation of frequency $\nu_0 + \nu$ then there was a certain probability that the atom would loose energy $h|\nu_0 + \nu|$ in a transition. On the other hand, the impinging radiation of frequency ν also resulted in a scattered radiation of frequency ν and, therefore, the atom had a probability of gaining or loosing energy $h\nu$ in a transition.[118] But the atom could only perform a transition to a different stationary state. All the same, the frequency of the impinging radiation ν was arbitrary.[119] Thus, $h|\nu_0 + \nu|$ did not (necessarily) correspond to a possible energy difference of the atom under consideration. This meant that the contributions of the scattered and impinging

[116] According to Dresden, also in Kramers' notebooks such a reading in terms of successive transitions is suggested. But Dresden did not give any indication which passages in the notebooks strengthen his conjecture. Having consulted the microfiches of Kramers' notebooks in the AHQP, I did not find any hard evidence for Dresden's conjecture.

[117] It is to note that this was not contrary to the general BKS framework. An example for the heuristic utility of the light-quantum hypothesis and energy considerations, which Kramers and Heisenberg (1925, 697–698) name themselves and which was proposed by Bohr in 1923, was the inference that spontaneous transitions always occur from higher to lower quantum states.

[118] It gathered a probability to loose the energy through the action of the scattered radiation and gathered a probability to gain the energy by the action of the impinging radiation.

[119] Actually, Kramers and Heisenberg's approach was only valid in the case that this frequency did not coincide with a possible emission or absorption frequency.

radiation had to be summed or subtracted in such a way that in total the energy amount gathered or lost by the atom was $h\nu_0$.

Kramers and Heisenberg considered two cases:

1. $(\nu + \nu_0) > 0$ (the frequency of the transition ν_0 might be positive or negative $(H(Q) > H(P)$ or $H(P) > H(Q))$, depending on the incident frequency)
2. $(\nu + \nu_0) < 0$ (for this case ν_0 is always negative, because ν is always positive $(H(Q) < H(P))$

In the first case, the atom would definitely loose the energy $h(\nu + \nu_0)$. This had to be compensated by the gain of the energy $h\nu$: In total the atom would loose the energy $h(\nu + \nu_0) - h\nu = h\nu_0$. Consequently, the atom had to be first in the state Q and then in the state P.[120] In the second case, the atom would loose the (positive) energy $h(-(\nu + \nu_0))$. To obtain a value which was proportional to ν_0 it had to loose at the same time $h\nu$. So, in total the energy $-h\nu_0$ was lost by the atom. Consequently, the atom was first in the state P (the energetically higher state for the second case) and then in the state Q.

I laid out this argument in some detail as I wanted to show that it did not depend on what we call the intermediate state, i.e. the state R. Heisenberg and Kramers nowhere mentioned this state in their explanation. Moreover, and in harmony with the BKS framework, the authors clearly noted that the atom emitted/absorbed multiple frequencies while it was in one stationary state. The whole consideration did not revolve around the idea that the incident radiation ν might induce some probability for a transition of the atom to a higher state for which the energy difference did not coincide with $h\nu$ (as Dresden suggested in his description). Rather on the contrary, it rested on the idea that the energy contributions of the different frequencies had to be combined in a way so that a transition to another stationary state could actually occur, i.e. that the entire result was $\pm h\nu_0$. According to the BKS framework, the atom did not have to perform these transitions but only gained a probability to do so.

From the above considerations, it followed that the dipole moment of an atom "in a stationary state P"[121] had to consist of two terms:

$$\mathfrak{M}(t) = \Re \left\{ \sum_Q \sum_R \mathfrak{M}(Q, P : R) + \sum_Q \sum_R \mathfrak{M}(P, Q : R) \right\} \tag{4.9}$$

where $\mathfrak{M}(P, Q : R)$ is given by Eq. (4.8). In both terms, the summation had to be extended over all states R, which were different from P and Q. The restrictions on the summation over the states Q can be inferred from the above line of thought: In the first term the sum ran over all states Q with $H(Q) < H(P) + h\nu$ and in the second term over all states Q with $H(Q) < H(P) - h\nu$.

---

[120] Since P might be energetically higher than Q, the loss of energy $h\nu_0$, which is then negative, corresponds to the gain of an energy $h\nu_0$.

[121] "in einem stationären Zustand P," Kramers and Heisenberg (1925, 699).

The above given formula, with the specific restriction regarding the summation over Q, was the most general formula of the electric moment that Kramers and Heisenberg provided in their paper. For my argument, it is important to note that only at this point, after most questions regarding the general nature of the electric moment had been settled, they changed the index of the summation. It did no longer run over the transitions themselves (as for the coherent case) but over the states R which connected the states Q and P. It is also quite obvious why Kramers and Heisenberg changed the summation index at this point: For the coherently scattered radiation, both characteristic amplitudes referred to the same transition (oscillator). For the incoherently scattered radiation, this was not the case. For given Q and P, the transition q was all the same defined by p (see Fig. 4.3 for the nomenclature), i.e. by the state R. Thus, choosing the summation over all possible states R was far more natural and unambiguous. For the characteristic amplitudes, Kramers and Heisenberg still kept the nomenclature referring to a transition (oscillator), such as q (in Fig. 4.3), and did not alter it to PR, i.e. they did not express it in terms of the states.

Specifying the Contributions to the Scattered Radiation

Having thus established a general formula for the electric moment of the atom in a given state P, Kramers and Heisenberg specified their formula in order to make the different contributions to the scattered radiation more accessible. Therefore, they introduced the frequency ν^*, which was considered to be positive and which contributed to the electric moment of the atom $\mathfrak{M}(|\nu \pm \nu^*|)$ (see Fig. 4.4 for the definition of ν^*). For the discussion of all possible energetic configurations between the states P, Q, and R, Kramers and Heisenberg referred their readers to diagrams as given in Fig. 4.4. It is to note, that only at this point Kramers and Heisenberg changed the numbering of the transitions in comparison to Fig. 4.2. In Fig. 4.4, each pair 1 and 2, 3 and 4, as well as 5 and 6 connected the states P and Q via the state R_a, R_b, or R_c. The lower numbers were always connected to the state P, i.e. the numerically first characteristic amplitude was connected to the state the atom was in.

With the rules thus far established, Kramers and Heisenberg could simply refer their readers to the diagrams and directly constructed the mathematical representation from Eq. (4.9). For example, the left of the two diagrams in Fig. 4.4 referred to the case where $H(Q) > H(P)$, so that $H(Q) - H(P) = h\nu^*$. The states R were divided in three groups according to the value of the energy with respect to the energy of P and Q. For R_a, $H(R_a)$ was bigger than $H(Q)$; for R_c, $H(R_c)$ was smaller than $H(P)$. The states R_b lied energetically in between Q and P: $H(P) < H(R_b) < H(Q)$.

One could straightforwardly infer that the moment had to take the following form (the attribution of numbers to the amplitudes and frequencies can be inferred from Fig. 4.4):

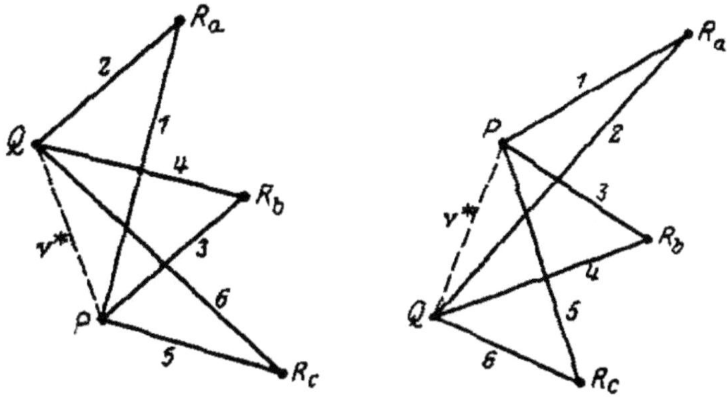

Fig. 4.4 Graphical representation for specifying the possible combinations for the induced moment for incoherent scattering. Figure in Kramers, Hendrik, and Werner Heisenberg. 1925. "Über die Streuung von Strahlung durch Atome." *Zeitschrift für Physik* 31, 681–708, here 699, https://doi.org/10.1007/BF02980624 (cited as Kramers and Heisenberg (1925) in the bibliography). Reproduced with permission from Springer Nature

$$
\begin{aligned}
\mathfrak{M}(\nu-\nu^*) = \Re\frac{1}{4h}\Bigg\{ & \sum_{R_a}\left(\frac{\mathfrak{A}_2(\mathfrak{E}\bar{\mathfrak{A}}_1)}{\nu_1-\nu}+\frac{\bar{\mathfrak{A}}_1(\mathfrak{E}\mathfrak{A}_2)}{\nu_2+\nu}\right)+ \\
& +\sum_{R_b}\left(\frac{\bar{\mathfrak{A}}_4(\mathfrak{E}\bar{\mathfrak{A}}_3)}{\nu_3-\nu}-\frac{\bar{\mathfrak{A}}_3(\mathfrak{E}\bar{\mathfrak{A}}_4)}{\nu_4-\nu}\right)+ \\
& +\sum_{R_c}\left(-\frac{\bar{\mathfrak{A}}_6(\mathfrak{E}\mathfrak{A}_5)}{\nu_5+\nu}-\frac{\mathfrak{A}_5(\mathfrak{E}\bar{\mathfrak{A}}_6)}{\nu_6-\nu}\right)\Bigg\} e^{2\pi i(\nu-\nu^*)t}
\end{aligned} \tag{4.10}
$$

Similarly, Kramers and Heisenberg constructed the moments of the atom in the cases of $H(Q) < H(P)$ and $H(Q) = H(P)$, where the last case coincided with coherent scattering if P and Q were equivalent.

I have copied Kramers and Heisenberg's nomenclature very closely in the discussion so far. For a modern version of the nomenclature, that will actually be applied in Heisenberg's 1925 breakthrough paper as well, see for example the discussion of the paper by Max Dresden. In this case, the simple indexes of the $\mathfrak{A}_i$ and the ν_i are substituted by a double index which refers to the two states of the transitions corresponding to the characteristic amplitudes, e.g. PR_a. Such a nomenclature makes the role of the states R more transparent for the mathematical representation. All the same, the formal role of the states R in the above given expressions, although not mirrored in the indexes of the characteristic amplitudes, was clearly perceived of by Kramers and Heisenberg: These states somehow connected the state P and Q.

Excluding False Resonances: BKS at Work

For the physical interpretation of the electric moment, I want to discuss one last part of the paper, namely Kramers and Heisenberg's description of resonances (Kramers and Heisenberg 1925, 701–707). As can easily be seen, in the case that the impinging frequency ν was close to ν_1, the first term in brackets of Eq. (4.10) got extremely large; when the frequencies were equal the expression actually diverged. Since the classical formula broke down in such cases, the quantum translation was also not to be trusted. Nevertheless, in analogy to the coherently scattered radiation, Kramers and Heisenberg assumed that in this case the equation was insofar correct as it would give rise to a strong increase in the intensity of secondary radiation of frequency ν_2.[122]

The occurrence of the secondary radiation of frequency ν_2 was, according to Kramers and Heisenberg, of a twofold nature. On the one hand, "the scattered radiation is emitted by all atoms in state P"[123] (Kramers and Heisenberg 1925, 702). On the other hand, because of the irradiation with frequency ν_1 the atom acquired a probability to perform a transition to state R_a. As soon as an atom arrived in this state, it spontaneously emitted radiation of frequency ν_2 because the atom had the possibility of spontaneously dropping to state Q.

To a modern reader this might seem as a rather cumbersome and counter-intuitive feature of the physical framework that Kramers and Heisenberg applied. Yet, it is the basis of a crucial argument in the paper: the exclusion of what was called "false resonances".[124] As one can easily see, Eq. (4.10) also has a sharp maximum if $\nu \approx \nu_6$ (compare the denominator of the last term on the right of Eq. (4.10)). The emitted radiation would have the frequency ν_5. This ran counter to the observation that only frequencies corresponding to an absorption frequency of the atom in the given state P would act as critical frequencies in the dispersion formula: The frequency ν_6 was connected to the transition from R_c to Q and not to P. A sharp maximum at frequency ν_6 was not expected.

Without going into any details, the basic argument of Kramers and Heisenberg to mitigate this problem was the following: Since P was energetically higher than R_c, the atom already had a probability of spontaneously dropping into the state R_c. Corresponding to the BKS framework, this meant that the atom already spontaneously emitted radiation of frequency ν_5 while resting in the state P. This radiation could interfere with the radiation emitted due to the false resonances. By establishing a

[122] The radiation emitted by the atom is of frequency $\nu - \nu^*$. Hence, from $\nu \approx \nu_1$ it follows that $\nu - \nu^* \approx \nu_1 - \nu^* = \nu_2$ (see also Fig. 4.4).

[123] van der Waerden (1967, 245).

[124] Max Dresden discussed this argument at length with a specific focus on the interpretations given by Kramers and Heisenberg. As Dresden's division between Kramers and Heisenberg's approach to the problem is exclusively based on the oral history interview with Heisenberg, I will not reproduce this argument here.

suitable phase relation between the two kinds of characteristic amplitudes, Kramers and Heisenberg could show that, under seemingly natural restrictions, the false resonances did not cause any further maxima in the scattered radiation.

Kramers and Heisenberg framed this solution to the riddle posed by the false resonances as proof for the conception of the BKS paper:

> It can be deduced from this [the solution of the false resonances by assuming that the atom spontaneously emits radiation in an excited state], that a conception [Auffassung] according to which the scattering effect of an atom under external irradiation is to be conceived of as the effect of the atom in a certain [bestimmten] stationary state, necessarily leads to the conception that also the spontaneous radiation of an atom has to be described as a radiation in a specific stationary state [of the atom], and not as an effectiveness that the atom only unfolds [entfaltet] in a transition between stationary states. But this is just the hypothesis introduced by Slater, which gave rise to the views further developed by Bohr, Kramers, and Slater.[125]

4.2.3 *Summary and Outlook: Mathematical/Diagrammatic Representations and Heisenberg's* Umdeutung *Paper*

At this point, you might rightfully ask: Why did I put you through all these details? My answer is: Because there are some misconceptions in the secondary literature about the Kramers-Heisenberg paper which need such a detailed account. Maybe more importantly, I also consider such a detailed elaboration necessary in order not to be misunderstood in the points I want to make in reference to these misconceptions.

The Role of Intermediate States in the Kramers-Heisenberg Paper

In line with prior historical arguments and in reference to the concepts I am investigating most importantly, Kramers and Heisenberg provided a mathematical and diagrammatic representation of what we today call "virtual transitions" and "virtual

[125] This is my own translation as the translation in van der Waerden (1967, 251) is misleading at different points. On the one hand, the passage which I translated as "necessarily leads to" was translated by van der Waerden as "need not necessarily lead to", which is simply wrong. Furthermore, in what I have translated as "is to be considered as the effect of the atom in a certain stationary state" is translated by van der Waerden as "as if it were in a stationary state". As Heisenberg and Kramers obviously assumed that the atom actually is in a stationary state, the clause "as if" is misleading in respect to the physical assumptions Kramers and Heisenberg made. The German original reads: "Man kann daraus schließen, daß eine Auffassung, nach der die streuende Wirkung eines Atoms bei äußerer Bestrahlung als die Wirkung des Atoms in einem bestimmten stationären Zustande anzusehen ist, notwendig die Auffassung mit sich führt, daß auch die spontane Strahlung eines Atoms als eine Strahlung in einem bestimmten stationären Zustande beschrieben werden muß, und nicht etwa als eine Wirksamkeit, die das Atom nur bei einem Übergang zwischen stationären Zuständen entfaltet. Das ist aber gerade die von Slater eingeführte Hypothese, die den Anlaß zu den von Bohr, Kramers und Slater näher entwickelten Ansichten gab." Kramers and Heisenberg (1925, 708).

states". The states R_a, R_b, and R_c, as well as the characteristic amplitudes A_i ($i = 1, ..., 6$), both in the summation of Eq. (4.10) and in the diagrams shown in Fig. 4.4, are examples of such. These states connected the initial and final state of the corresponding dipole moment, namely P and Q. Kramers and Heisenberg also clearly perceived of the characteristic amplitudes as being connected to transition possibilities. Maybe, but I have not seen any historical evidence for this conjecture, they thought of successive transitions through these states when figuring out the steps in their derivation.

Yet, it is important to note at which place these entities entered the theoretical discussion and specifically how they were referred to and treated by the authors themselves. Let's first turn to the derivation process. According to the above reconstruction and contrary to Max Dresden's argument, a physical picture of successive transitions was not guiding the derivation of the general electric moment. The two most crucial steps in the paper, the translation of the classical equation into quantum theory and the identification of the stationary state for a specific form of the electric moment, did not depend on such a mental picture. As laid out above, terms that contradicted such a conception occurred during the translation procedure (of course they would eventually cancel). But also for the identification of the momentary state of the atom, a sequence of energy non-conserving transition processes was not invoked. Rather on the contrary, Kramers and Heisenberg combined the energy amounts corresponding to the emitted and impinging radiation in order to arrive at processes which occurred only between stationary states. This argument was entirely independent of the intermediate state.

Only after having solved these essential and important steps, thus after having arrived at a theoretical representation of the electric moment, the intermediate states and an ordering of the characteristic amplitudes according to a conception that might be guided by a sequence of transitions was applied. Only in this last step, the summation ran over the intermediate states. Also the diagrammatic representations in the paper share this feature. While in Fig. 4.2, the one used for the translation into quantum theory, the numbers where not assigned according to a conception of successive transitions from initial over intermediate to final state, the numbers in Fig. 4.4, the one used to order the terms after the general formula had been established, can in principle be interpreted according to such a mental imagery.[126] Still it has to be noted, that the specific diagrammatic representation Kramers and Heisenberg provided, the representation in the J_1, J_2-plane, was of a rather abstract nature and a temporal dimension was, if at all, exclusively incorporated through the numbering of the respective lines connecting the states. At best, the interpretation of the formulas in terms of successive transitions was a product of the derivation process and not a tool used therein. But even then, I highly doubt that such a physical interpretation was underlying the discussion. Therefore, let me turn to the second point mentioned

[126] But this does not mean that such transition subprocesses were necessarily the reason for this nomenclature. The nomenclature might as well have been guided a purely mathematical insight: In formulas such as Eq. (4.9) only the characteristic amplitudes of transitions 1 and 2, 3 and 4, as well as 5 and 6 combined in products.

above: The way the authors referred to the theoretical entities of interest and their relation to the BKS framework.

Although Kramers and Heisenberg were clearly aware that there was a connection between the characteristic amplitudes and transition probabilities, a connection they could actually quantify, they are also very clear about the fact that these theoretical entities were not transitions. According to the conceptual framework the paper was presented in, the atom rested in one particular state when it reacted on the impinging radiation. This fact is repeatedly stressed by the authors. At no point an explicit verbal statement that the atom performed transitions through the intermediate states can be found. The combinations of the characteristic amplitudes were not conceived of as a sequence of transition processes, but rather as a suitable formal representation of the dipole moment when the classical Fourier coefficients were replaced by characteristic amplitudes. No more, no less.

As the BKS framework served as a blueprint for the interpretation of the formulas, this is not surprising after all. The BKS theory included the concept of a virtual oscillator, but not the concept of a virtual transition. Transitions were real processes on which our observations were based. The fact that the atom rested in a stationary state during its interaction with radiation was actually used by Kramers and Heisenberg to exclude the "false resonances" that came with their formal procedure. The scattered radiation and the radiation which was emitted due to the possibility of dropping into a state of lower energy interfered. Kramers and Heisenberg used this interference to mitigate the occurring problem. Hence, the conception that the atom rested in what we call the initial state was of prime importance for the given argument.

Furthermore, I doubt that energy non-conservation as it was implied by the BKS framework could have explained the occurrence of what we might conceive of "virtual transitions" in the Kramers-Heisenberg paper. This is one of the pillars of Max Dresden's conviction that Kramers conceived of the successive transitions as more than a formal representation. But, on the contrary, as laid out in the section on the BKS paper, energy non-conservation was real for BKS. And it rested on the idea that radiation of a frequency coinciding with a transition was emitted or could be absorbed by an atom in a stationary state, without actually performing the respective transition. This conception is evidently part of the paper by Kramers and Heisenberg and it was used in the solution of the "false resonances". But nowhere, neither in the BKS nor in the Kramers-Heisenberg publication, it was implied that a transition could be performed even though the impinging radiation had a frequency which did not coincide with the energy difference between two states.

But even if energy non-conservation as implied by BKS could have explained what we conceive of as "virtual transitions", I think Dresden's argument is spurious. Combining the fact, that the paper by Kramers and Heisenberg is thoroughly based on the BKS outlook, with Dresden's assumptions that Kramers conceived of the connection between the amplitudes as more than formal and that BKS could explain the implied energy non-conservation, one big question remains: Why shouldn't Kramers have spelled this out? Why did Kramers not refer to successive transitions when one could easily read such into the formulas, especially if this interpretation was warranted by the general outlook of the paper and would have made the discussion far more accessible? Given the material at hand, I would argue that Kramers did not do so because he did not conceive of the combinations as more than formal representations and because energy non-conservation as implied by BKS could not explain what we describe as "virtual transitions".

Hence, my argument goes: Kramers and Heisenberg had surely established a mathematical and a diagrammatic representation of what in modern terms would be referred to as "virtual transitions" through intermediate states. But a mental imagery of transition subprocesses was neither necessary for Kramers and Heisenberg's derivation, at best it was a product of the same, nor was it spelled out explicitly in verbal terms by the authors. And I think these two points are correlated: If the conception of successive transition processes was necessary for their evaluation, Kramers and Heisenberg would have spelled it out (as Dirac would do in 1927, see Chap. 5). In the BKS framework, energy non-conservation served a different purpose than to allow for what we conceive of "virtual transitions", and virtual transitions were not part of the conceptions of BKS. For the time being, the intermediate states in the Kramers-Heisenberg formula had a mostly (if not purely) formal character.

The Role of Intermediate States for Heisenberg's *Umdeutung*

As is well-known, the widely shared belief in the validity of the dispersion-theoretical accounts was not shaken by the empirical refutation of the BKS theory. This fact has in increasingly detailed elaborations been explained by the independence of the methods applied in dispersion theory and the BKS framework.[127] For Kramers' dispersion theory as well as for the Kramers-Heisenberg paper, the BKS theory remained the blueprint on which the interpretation was based but not a theoretical model guiding the general approach. As mentioned above, Born explicitly stated that his work was independent of the still disputed conceptions of the BKS paper. Also Wolfgang Pauli, besides Einstein the strongest critic of the BKS framework, applied Kramers and Heisenberg's dispersion formula. And he emphasized, both in a letter to Kramers and in publication, that the results were independent of the specific conceptions of the radiation processes, i.e. the BKS framework.[128]

With the demise of BKS the reference to virtual oscillators and virtual radiation field vanished from the publications (at least so for a while). At the same time, as Blum and Jähnert (2022, especially Sect. 4.2) have argued, it led to a shift in focus: away from the dynamical relations to a search for representations of the kinematics that could be connected to the theoretical structures representing transition probabilities. In these attempts, both the virtual oscillators and the classical orbits of the Bohr atom faded into the background. This vanishing of the oscillators and Heisenberg's

[127] The strongest argument in this direction was certainly given by Duncan and Janssen (2007a), who framed the BKS theory as a side road to the development of quantum theory.

[128] See Pauli to Kramers, July 27. 1925; as cited in Pauli (1979, 232–234); and Pauli (1925b, 5).

explicit refutation of the concept[129] led Edward MacKinnon (1977, especially Sect. 4) to argue that in Heisenberg's 1925 paper the transition from virtual oscillators to virtual transitions was performed.

I will not go into any detail about Heisenberg's path towards *Umdeutung* as this would require a considerable departure from the main theme of this chapter. As was clearly shown by Jähnert (2019, Chap. 7), Blum et al. (2017), and put into the context of radiation theory by Blum and Jähnert (2022), the development of a general framework for representing and calculating spectroscopic intensities in terms of purely quantum-theoretical quantities and the toy model of the anharmonic oscillator became the bedrock of Heisenberg's approach. Yet and as has long been argued,[130] the combination principles governing the characteristic amplitudes and the frequencies, as expressed in the Kramers-Heisenberg paper, were central ingredients in this problem context.

Regardless of the exact historical origins of the multiplication and summation rules for quantum amplitudes in Heisenberg's 1925 paper and in the development of the corresponding ideas the relations between quantum-theoretical amplitudes and frequencies took centre stage. Heisenberg had acquired deep insights in the combination principles by mid June 1925. In their detailed elaboration of a letter sent to Ralph Kronig, Blum et al. (2017, 17) noted that Heisenberg's translation procedure "relied on the physical insight that the transition from n to $n-3$ can be split up in two distinct ways", i.e. once over different states $n-1$ and once over $n-2$. This meant that Heisenberg had certainly developed an intuition about how quantum amplitudes would combine through intermediate states.

In the presentation of his publication, this fact boiled down to the translations displayed in Table 4.1. The first translation described how frequencies in quantum theory were to be combined. It was obviously based on Walter Ritz' long known combination principle. The second translation stated how the coefficients and the time dependence of the Fourier expansion of $x^2(t)$ should be translated into quantum theory. The $B_\beta(n)$ were defined through the equation $x^2(t) = \sum_{\beta=-\infty}^{\infty} B_\beta(n)e^{i\omega(n)\beta t}$. The above two identifications (as well as their generalizations) were translations of purely kinematic relations, which Heisenberg subsequently and with the necessary care plugged into the classical equations of motion.

In the context of *Umdeutung*, it has often been noted that Heisenberg's emphasis on observable quantities was, on the one hand, not necessarily guiding his evaluation (even though it was contemplated for a time priorly) and, on the other hand, to

[129] See the letter from Heisenberg to Ralph Kronig on May 21. 1925, as, for example, cited in Blum and Jähnert (2022, 141).

[130] The central role of dispersion for Heisenberg's *Umdeutung* is presented, for example, in MacKinnon (1977), Darrigol (1992, Chap. X), Duncan and Janssen (2007a, Especially Sect. 3.5), and Jordi Taltavull (2017, Chap. 7). For a short outline of this historical narrative and the questions it leaves unanswered, see Blum et al. (2017, Sect. 2).

Table 4.1 Translation of classical into quantum-theoretical quantities as proposed by Werner Heisenberg in 1925

Classical	Quantum
$\nu(n,\alpha)+\nu(n,\beta)=\nu(n,\alpha+\beta)$	$\nu(n,n-\alpha)+\nu(n-\alpha,n-\alpha-\beta)=$ $=\nu(n,n-\alpha-\beta)$ Or: $\nu(n-\beta,n-\alpha-\beta)+\nu(n,n-\beta)$ $=\nu(n,n-\alpha-\beta)$
$B_\beta(n)e^{i\omega(n)\beta t}=$ $=\sum_{\alpha=-\infty}^{\infty}A_\alpha A_{\beta-\alpha}e^{i\omega(n)(\alpha+\beta-\alpha)t}$	$B(n,n-\beta)e^{i\omega(n,n-\beta)t}=$ $=\sum_{\alpha=-\infty}^{\infty}A(n,n-\alpha)A(n-\alpha,n-\beta)\times$ $\times e^{i\omega(n,n-\beta)t}$

some extend an exaggeration.[131] Heisenberg's translation relied on the quantum amplitudes and their phases which are not directly, at best indirectly, observable. And, indeed, the summation rule for quantum-theoretical frequencies and the replacement of $B(n,n-\beta)=\sum_{\alpha=-\infty}^{\infty}A(n,n-\alpha)A(n-\alpha,n-\beta)$ retrospectively only make sense when we conceive of them, as MacKinnon (1977, Sect. 4) has most explicitly described, as virtual transitions (modern terminology). In contrast to the Kramers-Heisenberg paper, the intermediate states (e.g. $n-\alpha$) are also explicitly written out. In this sense, virtual transitions and intermediate states (again, modern terminology) formed an important ingredient of Heisenberg's 1925 publication, at least so in a formal, although well understood way.

Yet, it must have been obvious that the theoretical entities, the quantum amplitudes, Heisenberg operated with were related to transitions, but were not transitions in any manner of speaking (given the language the actors had at their disposal in 1925). As Blum et al. (2017, 13) noted, the evaluation of the anharmonic oscillator, Heisenberg's most important toy model, still relied on the conception that the system is in one particular state and that this conception is also mirrored in the published contribution. In a letter to Wolfgang Pauli in late June 1925, Heisenberg also wondered "what the equations of motion actually mean [bedeuten], when one conceives of them as a relation between the transition probabilities".[132] When sending his manuscript to Pauli on July 9. 1925, Heisenberg explicitly stated:

131 See for example MacKinnon (1977) and Darrigol (1992, 273–276). Blum et al. (2017, 19) framed this more positively as "the result of a faithful reflection on the heuristics that actually guided his research".

132 "[...] was eigentlich die Bewegungsgleichungen bedeuten, wenn man sie als Relation zwischen [den] Übergangswahrscheinlichkeiten auffaßt." Heisenberg to Pauli, June 24. 1925; as cited in Pauli (1979, 228).

> My own opinion about the scribbling [Geschreibsel; i.e. his *Umdeutung* paper], which I am not very happy about, is the [following]: That I am strongly convinced of the negative critical part [i.e. the introduction and the refutation of unobservable quantities], but that I consider the positive [part] rather formal and tenuous.[133]

As a matter of fact, there was neither a temporal dimension in Heisenberg's conception of the combinations of the quantum amplitudes, nor was there any explicit verbal reference to successive transition processes. And even though Born and Jordan (1925a) identified matrix products and commutators in Heisenberg's multiplication rules and grounded his approach on a theoretically more fundamental level (on the level of the Hamiltonian instead of on the equations of motion, which are derived from it),[134] there was no verbal explication of transitions through intermediate states. The same holds for the so-called Dreimänner-Arbeit (Born et al. 1926), one of the funding publications of matrix mechanics, and for most papers dealing with dispersion and matrix multiplication[135] up until Dirac's 1927 paper on a quantum-electrodynamical theory of dispersion.[136] For the time being, the concept of a virtual transition was, if it was developed at all, only represented mathematically.[137]

An explanation of the lack of verbal references to such a model of successive transitions, although formally underlying the theory, is in my eyes best explained by the fact that the physicists were clearly aware that they did not work with transitions. They worked with theoretical entities, quantum amplitudes, the squares of which referred to transition probabilities. For an emerging theory, the conceptual framework of which was still in flux, caution in accrediting certain theoretical structures with real world counterparts, when it is known that these were not the actual counterparts, is certainly advisable. And as a matter of fact, even after Dirac had introduced a verbal model of successive energy non-conserving transitions through intermediate states, especially physicists close to Copenhagen did not follow Dirac in making things verbally explicit.[138]

This observation makes a more detailed elaboration of Dirac's verbally explicit reference to energy non-conserving transitions through intermediate states necessary. I will provide it in the following chapter. It is also clear that this general line of the development of quantum mechanics should be followed in the further development. Yet, there is a historical side-road that deserves further elaboration. My attention was

---

[133] "Meine eigene Meinung über das Geschreibsel, über das ich gar nicht sehr glücklich bin, ist die: daß ich von dem negativen kritischen Teil fest überzeugt bin, daß ich aber den positiven für reichlich formal und dürftig halte." Heisenberg to Pauli, July 9. 1925; as cited in Pauli (1979, 231).

[134] See Blum et al. (2017, Sect. 7).

[135] Even the term "intermediate state", let alone "virtual state" is absent in the greater part of the theoretical literature on quantum mechanics from 1925 to 1927.

[136] That I did not find such attributions does not rule out that they existed. But it shows that they were not widespread and not part of the conceptual mainstream.

[137] I thus come to a similar conclusion as Marta Jordi Taltavull (2017, 546), who highlighted the formal part of Heisenberg's usage of the quantum amplitudes: "Yet, at the end, the reconciliation between the physical pictures of resonance and quantum transitions was only achieved at formal level, this giving rise to a highly abstract matrix mechanics."

[138] For further reflection on the reception of Dirac's verbal model, see Ehberger (2022, Sect. 2).

drawn to it by the most explicit verbal description of the virtual transition processes (modern terminology) of the Kramers-Heisenberg formula. Wolfgang Pauli, even before he turned to the work on matrix and wave mechanics, stated:

> The characteristic of these [the Kramers-Heisenberg] formulas is, that the occurrence of a transitions PQ in the electric field is bound to the possibility to get from level P to level Q by the detour through a third level R, such that PR [...] and QR [...] are allowed transitions in the absence of the field. Each path PRQ contributes a certain amount to the line PQ in the field and because of the denominators ν_{RP} and ν_{RQ} the paths PRQ provide the largest contributions, for which the term R is particularly close to P or Q.[139]

Although Pauli did not talk about transitions and would refrain from the above description in his 1926 article in the *Handbuch der Physik*, we must turn to a little-known side road of the development to understand why it was Wolfgang Pauli who provided for this most explicit comment on the Kramers-Heisenberg formula.

In the following, I will revisit quantum-theoretical accounts of dispersion that were based on the light-quantum hypothesis, or more generally, did not question energy conservation for the light-matter interaction. Within this conceptual framework, a conception of scattering was proposed which rested on real transitions through what was called "intermediate states", although this concept should not be confused with the modern concept of an intermediate state. I will discuss these conceptions not only to put Pauli's remarks into context. It will also serve to exemplify the strong interrelation between a light-quantum conception of the radiation field and the conception of successive transitions as a model for dispersion. Eventually, this conception of real transitions will shortly resurface but be refuted in the discussions around the Raman effect (see Chap. 6).

4.3 Scattering, the Light-Quantum Hypothesis, and Actual Intermediate States

Part of Bohr's motivation to work out the BKS theory, which exclusively rested on the wave-theoretical description of light, was the discovery of the Compton effect. The directed momentum transfer between electron and impinging radiation was seen as proof for Einstein's 1916 theoretical elaboration in which he introduced the probability coefficients, also by Bohr.[140] But while Bohr only accepted the conception that a directed momentum transfer was connected to a transition process, he refuted

---

[139] "Das charakteristische an diesen Formeln ist, daß das Auftreten eines Übergangs PQ im elektrischen Feld gebunden ist an die Möglichkeit vom Niveau P nach dem Niveau Q auf dem Umweg über ein drittes Niveau R zu gelangen, derart daß PR [...] und QR [...] bereits bei Abwesenheit des Feldes erlaubte Übergänge sind. Jeder Weg PRQ gibt einen bestimmten Beitrag zur Linie PQ im Feld und zwar geben wegen der Nenner ν_{RP} und ν_{RQ} diejenigen Wege PRQ die größten Beiträge, bei denen der Term R besonders nahe an P oder Q liegt." Pauli (1925b, 14).

[140] This does not mean that the Compton effect can be conceived of as an aim to proof the light-qunatum hypothesis. Such a linear connection was certainly not the case, as Stuewer (1975) proved.

Einstein's own conclusion, i.e. that a directed momentum transfer came with a refutation of the wave theory of light and was further evidence for a quantized description of the radiation field itself.

Up until Compton's discovery, the numbers of public supporters and opponents of the light-quantum hypothesis were nearly equal.[141] But the Compton effect and the ease of a light-quantum explanation of the corresponding frequency shift seemed to have tipped the balance towards the light quanta.[142] In the following, I will discuss theoretical attempts of explaining the Compton effect and dispersion which either explicitly rested on the light-quantum concept or applied methods which are related to it, i.e. equilibrium considerations between matter and radiation. I will do so with a specific focus on the interpretation in terms of successive transitions or emission and absorption processes.

First, I will shortly discuss the initial derivations by Compton and Debye to open the discussion (Sect. 4.3.1). Then I will turn to a theoretical description of the Compton effect by Pauli as well as by Einstein and Ehrenfest in which a first explicit decomposition of the emission and absorption processes was proposed, also in mathematical terms. But contrary to the standard narrative, I will argue that this decomposition might have solved a riddle initially posed by Pauli, but at the same time caused new conceptual problems (Sect. 4.3.2). Then I will turn to the work of a group of Munich physicists around Arnold Sommerfeld as well as its elaboration by Adolf Smekal in which a temporally ordered sequence of emission and absorption processes through intermediate states of the atom as a model for dispersion and scattering was proposed (Sect. 4.3.3). Finally, I will turn to Wolfgang Pauli's synthesis in his 1926 article in the *Handbuch der Physik*, in which such a conception was still defended (Sect. 4.3.4).[143]

4.3.1 *Compton and Debye (1923): The Billiard Ball Model and the Impossibility of Absorption*

Compton's struggle in accepting the light-quantum hypothesis as a possible explanation of the wavelength shift he himself discovered as well as the independence of

---

[141] See Brush (2007, 232).

[142] This statement is a simplification both in conceptual and historical terms. On the one hand, further theoretical and empirical work carried out during the 1910s considerably contributed to a renewal of the interest in the light-quantum hypothesis. For a concise summary, see Blum and Jähnert (2022, 133–134). Furthermore, Compton's experiments and his explanation have not been accepted as smoothly as the above phrase suggests. See, for example, the discussion of the "Duane-Compton Controversy" in Stuewer (1975, Chap. 6, Section F). On the other hand, the conceptualization of a light quantum was still in flux and found a coherent quantum-theoretical embedding as late as 1927 with the work of Paul Dirac. Yet, for the purpose of this chapter and especially for one of the main actors, Arnold Sommerfeld, the above statement certainly holds. For the identification of the Compton effect as a turning point for the conception of the radiation field, see for example Stuewer (1975, Chap. 7) or Hentschel (2017, Sect. 5.2).

[143] I will discuss the respective, rather sparse secondary literature in the respective sections.

Compton's empirical research from the light-quantum hypothesis are well known.[144] But in early December 1922, Compton fully embraced the light-quantum hypothesis, at least so for the purely kinematic explanation of the frequency shift of the secondary radiation.[145] In Compton's own words, the theoretical assumptions underlying his explanation were the following:

> The present theory depends essentially upon the assumption that each electron which is effective in the scattering scatters a complete quantum. It involves also the hypothesis that the quanta of radiation are received from definite directions and are scattered in definite directions. The experimental support of the theory indicates very convincingly that a radiation quantum carries with it directed momentum as well as energy. (Compton 1923, 501)

Compton's description of the change in wavelength of the secondary radiation exclusively rested on energy and momentum conservation. He modelled the phenomenon as an elastic collision between a quantum of radiation and a freely moving electron. This was clearly expressed in his mathematical elaborations but also in the diagrammatic display he provided. Compton represented the kinematic relations between the quantum and electron in terms of Fig. 4.5.[146] What Compton applied here can be considered a billiard ball model of the interaction process between quantum and electron.

The ascription of a billiard ball model is in fact not a retrospective identification. As Compton has expressed it after first empirical observations of the recoiling electrons had been published:

> We find that the wavelength and the intensity of the scattered rays are what they should be if a quantum of radiation bounced from an electron, just as one billiard ball bounces from another. Not only this, but we actually observe the recoiling billiard ball, or electron, from which the quantum has bounced, and we find that it moves with just the speed it should if a quantum had bumped into it. (Compton 1924, 70)

Simultaneously to and independent of Compton, Peter Debye (1923) developed a theory of the scattering of light from a free electron with the same basic conceptual components.[147] But Debye was already a little more explicit in reference to a verbal description of the decomposition of the process:

[144] See, for example, Stuewer (1975, Chaps. 5 and 6).

[145] Compton was well aware of the problems of reconciling such a model of the process with interference phenomena. See, for example, Stuewer (1975, 232–234).

[146] See also Wüthrich (2010, 30–32) for a discussion of Compton's diagrams in relation to diagrams used in QED during the 1930s and later on.

[147] For a discussion of Debye's paper, its relation to Compton's previous work and the slightly different conclusions they came to, see Stuewer (1975, Chap. 6, Section D). See also Brown (2002, 219–223) on both Compton's and Debye's calculations.

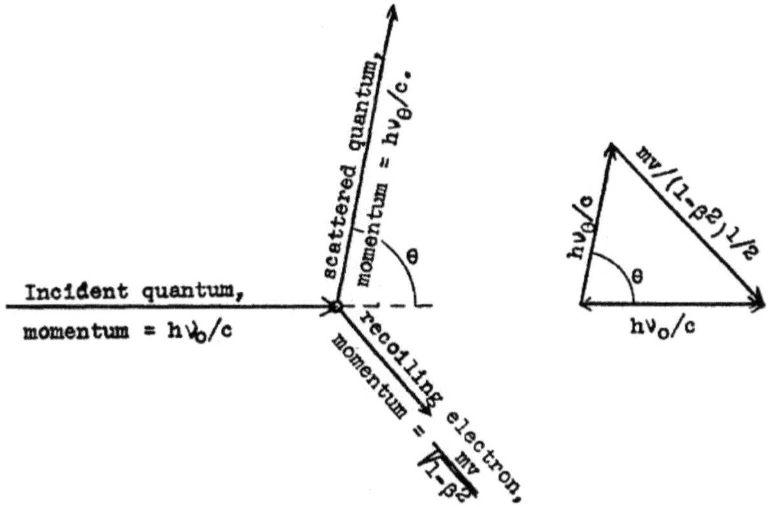

Fig. 4.5 Compton's diagram for the momentum conservation in the collision between a quantum of radiation and an electron. Reprinted figure with permission from Compton, Arthur, "A quantum theory of the scattering of X-rays by light elements," *Physical Review* 21, 483–502, here 486, https://doi.org/10.1103/PhysRev.21.483 (cited as Compton (1923) in the bibliography). Copyright 2025 by American Physical Society

> The primary X-ray of frequency ν_0 transfers the energy $h\nu_0$ to a free electron in a single process. This energy is converted quantitatively as it serves, on the one hand, to generate a secondary radiation of frequency ν and the energy $h\nu$ and, on the other hand, to give a velocity to the electron.[148]

Although this might seem as if Debye already decomposed the process of scattering into two distinct subprocesses of absorption of energy and its conversion into secondary radiation and kinetic energy of the electron, this was actually not the case. Debye even gave an argument, if only indirectly, why such a decomposition was not possible in the case of free electrons: He showed that a free electron cannot absorb a light quantum and simultaneously fulfil energy and momentum conservation. Debye did so in a comparison of the photoelectric effect, modelled as the absorption of a light quantum by a bound electron, and the Compton effect.[149]

For simplicity's sake, Debye assumed that the electron gathered momentum in the forward direction by absorbing a quantum of energy $h\nu_0$. If there was no further effect,[150] i.e. if the electron was completely free and did not radiate, then the energy of the absorbed photon would completely be transformed into the kinetic energy of

[148] "Die primäre Röntgenstrahlung der Frequenz ν_0 überträgt auf ein freies Elektron in einem Einzelprozeß die Energe $h\nu_0$. Die Energie setzt sich quantitativ um, indem sie dazu dient, erstens einen Sekundärstrahl von der Frequenz ν und dem Energiebetrag $h\nu$ zu erzeugen und zweitens dem Elektron eine Geschwindigkeit zu erteilen." Debye (1923, 162).

[149] The short argument can be found in Debye (1923, 164–165).

[150] The mass was assumed to be constant.

the electron, i.e. $\frac{mv^2}{2} = h\nu_0$. This would imply that a momentum of $mv = \sqrt{2mh\nu_0}$ was created, while the photon's momentum $\frac{h\nu_0}{c}$ vanished. In general, these two contributions do not cancel.[151] According to Debye, this led to a contradiction with momentum conservation that needed to be compensated. In the Compton effect, the compensation was due to a secondary radiation and in the photoelectric effect due to a momentum transfer to the atom from which the electron was expelled. Debye argued that there is still a connection between the two processes and that the photoelectric effect should slowly go over into the Compton effect, if either the energy of the impinging quantum got higher or the binding energy of the electron lower.

Both Compton and Debye did not analyse the process in terms of emission and absorption subprocesses: They treated the scattering process as an elementary act for which, in total, energy and momentum was conserved. Designations such as "incident quantum" and "scattered quantum" (Compton 1923, 485) by Compton or the differentiation between the energy of impinging and scattered radiation in the above cited quote by Debye rather refer to the overall process and, in modern terminology, the initial and final states. But it did not take long until the question of the composite character of the scattering process was discussed. These discussions had their origin in the realm of quantum equilibrium considerations which I will turn to in the following.

4.3.2 *The Compton Effect in the Hands of Pauli, Ehrenfest and Einstein (1923/24): A Composite Elementary Process?*

The discussions presented in the following,[152] if they were taken up in the historical literature at all, normally take little space and end with Einstein and Ehrenfest solving a riddle that Pauli had posed[153]: The mathematical representation of the equilibrium conditions for free electrons and radiation seemed to suggest that radiation of different frequencies interfered. By focussing on the nature of the process, especially on the question if it should be considered as a two-step phenomenon, I will argue that Einstein and Ehrenfest's paper rather created new conceptual issues that could not be directly resolved.[154]

---

[151] Simply equating the two momenta, one finds that this is only the case if $\nu_0 = \frac{2mc^2}{h}$.

[152] I will focus on the problems that arose in describing scattering from free electrons as a two-step process and the discussions between Pauli and Einstein/Ehrenfest. Other historical lines and possible explanations, as the propositions of a finite dimension of light quanta by Ornstein and Burger (see, e.g., Stuewer 1975, 306–307) or of "light molecules" as put forward by Mieczyslaw Wolfke and further elaborated by Walther Bothe (Fick and Kant 2013) will not be discussed.

[153] See, for example, Klein (1964, Section III, B), Stuewer (1975, 288–289), or Enz (2002, 108).

[154] As late as 1929, Placzek (1929) showed that Einstein and Ehrenfest's conception of the scattering process could not be upheld in general. See also Sect. 6.3.2.

Wolfgang Pauli and the Interference of Light of Different Frequencies

At the time Wolfgang Pauli engaged with the Compton effect and its relation to thermodynamic equilibrium between electrons and black body radiation, he was perfectly prepared for the job.[155] Shortly after graduating from Gymnasium in Vienna in 1918, he already published his first work on general relativity and, in 1922, gave a thorough review of this theory for the *Enzyklopädie der mathematischen Wissenschaften* which became a standard reference, praised by Einstein himself. Pauli soon focussed on quantum theory and received his PhD from the LMU in Munich, supervised by Arnold Sommerfeld, for a thesis on the quantum theory of the ionized diatomic hydrogen. From 1921 to 1922 Pauli was assistant to Max Born in Göttingen, where he contributed to the development of perturbation theory for atomic problems. From Göttingen, Pauli moved for one year to Copenhagen and worked at the institute of Niels Bohr, where he mainly focused his work on the Zeeman effect. This line of work would eventually lead to his postulation of the exclusion principle.[156] Consequently, Pauli was well-versed in relativity calculations and the Bohr-Sommerfeld quantum theory as it stood in 1923. Moreover, Pauli also had an interest in questions concerning the constitution of the radiation field,[157] which had gathered more attention in the community through the years 1917–1922.

Although Pauli's work in Göttingen and Copenhagen had focussed on the Helium problem and the Zeeman effect, on atomic stability and the theory of optical spectra, he found time to engage with the question of the equilibrium between the radiation field and free electrons. Pauli's elaboration was motivated by the recent discovery of the Compton effect as well as Debye's explanation of the same. Pauli wrote the paper on a short vacation in Hamburg.[158] By end of July he had finished his calculations, but still wanted to discuss some of the "principle questions" with Bohr before the publication.[159]

Pauli's paper "Über das thermische Gleichgewicht zwischen Stahlung und freien Elektronen", received in early August 1923, was explicitly independent of the constitution of the radiation field. All the same, Pauli himself denied attempts of harmonizing quantum theory with a classical conception of the radiation field in private communication but also publicly at the time.[160] The self-chosen topic of Pauli's Habilitationsvortrag, proposed in February 1924, read: "On the limits of the rep-

[155] For a biography of Wolfgang Pauli see e.g. Enz (2002) or Fierz (1974).

[156] See, for example, Serwer (1977) or Darrigol (1992, Chap. VIII).

[157] Pauli seemingly had criticized Schrödinger's derivation of the Doppler shift for photons, but not his results. Schrödinger to Pauli, November 8. 1922, cited in Pauli (1979, 69–71).

[158] Kojevnikov (2020, 54–55). As the Compton effect was conceived of as proof for the light-quantum hypothesis and Bohr defended the wave-theoretical description of light, Kojevnikov spoke of an act of "insubordination or hooliganism" by Pauli. At the time, academic practice still required the approval of the respective professor before publication.

[159] "prinzipielle Fragen," Pauli to Bohr, July 16. 1923, cited in Pauli (1979, 102–103). The main results as published in the *Zeitschrift für Physik* were communicated to Sommerfeld by Pauli in July 1923. Pauli to Sommerfeld, July 19. 1923, cited in Pauli (1979, 105–107).

[160] Pauli to Arthur Eddington, September 20. 1923, as cited in Pauli (1979, Footnote on p. 117).

resentability of the physical phenomena by continuous fields".[161] Pauli would also become one of the fiercest critics of the BKS theory and, even before its empirical refutation, he would write to Sommerfeld that "I am convinced that (independent of the outcome of Geiger's experiment) the future will rather belong to the light quanta".[162] Yet, the professor of the institute where Pauli was working in mid 1923, Niels Bohr, did not consider Einstein's thought or the Compton effect as proof of the light-quantum hypothesis; Bohr favoured the wave-theoretical description of light. In this vein, we see Bohr's influence when Pauli stated that no "final conclusions about the nature of the radiation field"[163] can be drawn from Einstein's 1916 considerations.[164]

Pauli's publication rested on the description that Einstein had given in 1916, especially on the concept of a directed momentum transfer, as well as the Compton effect. Thus far, the description of the thermal equilibrium between radiation and free electrons had resisted a consistent description. Both the calculations by Hendrik Lorentz and his student Adriaan Fokker, which rested on classical electrodynamics, had either led to a breakdown of Planck's law or, if this law was artificially upheld, implied that the mean value of the kinetic energy of the electrons would not be equal to $\frac{3}{2}kT$, in contrast to the kinetic theory of gases. Pauli concluded that also the description of the interaction of free electrons and radiation could not be described by classical laws. The problem now was "to find a quantum-theoretical mechanism for the interaction between radiation and free electrons, which satisfies the thermodynamic demand, that electrons with Maxwellian velocity distribution and a Planckian radiation field have to be in equilibrium with each other".[165]

After reviewing the general results of Einstein's 1916 publication, Pauli highlighted the differences between his and Einstein's approach. Einstein's evaluation rested on the assumption that the forces between the material part and the radiation field were small in comparison to the forces between the material parts themselves. This assumption could not be upheld because for free electrons only forces due to radiation were present. Consequently, no spontaneous processes occurred and only the translatory energy of the electrons had to be taken into account.

Pauli took the principle assumptions about the "elementary process"[166] of scattering from Compton and Debye, namely that in the interaction of light with a free electron, energy as well as momentum is transferred to the electron, while the quan-

[161] "Über die Grenzen der Darstellbarkeit der physikalischen Erscheinungen durch kontinuierliche Felder." Pauli's Habilitation is shortly discussed in Reich (2011, 113–114).

[162] "[...] ich bin überzeugt, daß (unabhängig vom Ausfall des Geigerschen Versuches) die Zukunft mehr den Lichtquanten gehört." Pauli to Sommerfeld, November 1924; as cited in Pauli (1979, 174).

[163] "endgültige Schlüsse über die Natur der Strahlung," Pauli (1923, 274).

[164] At this point, Pauli cited Bohr's critique of the light-quantum concept in Bohr (1923).

[165] "[...], einen quantentheoretischen Mechanismus der Wechselwirkung der Strahlung und freien Elektronen zu finden, welcher der thermodynamischen Forderung genügt, daß Elektronen mit Maxwellscher Geschwindigkeitsverteilung und ein Plancksches Strahlungsfeld miteinander im Gleichgewicht sein müssen." Pauli (1923, 272).

[166] "Elementarprozess," Pauli (1923, 284).

tum of energy $h\nu$ is transformed into a secondary one with energy $h\nu_1$. The total energy and momentum had to be conserved. The task was then to find the form of the probability function F, especially its dependence on the radiation density, for the respective scattering process. The Maxwell-Boltzmann distribution, named after James Clerk Maxwell and Ludwig Boltzmann, of the electrons and the Planckian form of the radiation density had to be upheld.

The main part of the paper[167] was concerned with finding a proper reference frame in which the processes could most easily be handled, the independent variables that should be considered, and the general conclusions that could be drawn from relativistic invariance about the form and the behaviour of the probability function F. As statistical equilibrium required that the inverse process would occur as often as the regular one, Pauli substantiated how this requirement could be treated relativistically.

After these considerations, Pauli studied different assumptions about the general dependence of the probability function F on the radiation density. First Pauli set F proportional to the radiation density before the process, so that $F = A\rho_\nu$. Pauli considered this the most natural assumption as the frequency of occurrence of the scattering process would then be proportional to the radiation which caused it. Yet, this assumption led him to (Max) Wien's instead of Planck's radiation law and therefore could not be valid in general.

Motivated by the classical analogue of interference fluctuations, which were proportional to ρ_ν^2, Pauli assumed that a second term had to be added into the ansatz for F. He showed that if one set

$$F = A\rho_\nu + B\rho_\nu\rho_{\nu_1} \tag{4.11}$$

where ρ_{ν_1} is the spectral density of the frequency after the scattering, Planck's law could be derived.

Pauli considered this to be an instance were radiation densities of different frequency could interact with each other and which implied "that the elementary process $h\nu \rightarrow h\nu_1$ happens more often, when radiation of the frequency ν as well as radiation of the frequency ν_1 is present, than if only the radiation of the primary frequency ν is present".[168] He mitigated this counter-intuitive finding, by the remark that already the assumption of a wavelength shift during scattering is unclassical and, therefore, a deviation from classical physics actually had to be expected.

Pauli then went one step further and proposed that a similar effect should be found for dispersion phenomena, namely a wavelength shift of the scattered radiation due to some recoil momentum of the atom. As the mass M of the atom is larger than the electron's, the wavelength shift would accordingly be much smaller. Pauli expected it to be so small as to be unobservable. To be sure, Pauli did not introduce Raman

---

[167] Pauli (1923, 277–281). Here Pauli could fully play out his mastery of relativity theory.

[168] "[Dieser Ansatz bedeutet], daß der Elementarprozeß $h\nu \rightarrow h\nu_1$ öfter vor sich geht, wenn sowohl Strahlung der Frequenz ν als auch Strahlung der Frequenz ν_1 vorhanden ist, als wenn nur die Strahlung der ersteren Frequenz ν allein vorhanden ist." Pauli (1923, 284).

scattering at this point. It was introduced by Adolf Smekal around the same time and even with very similar reasoning, but definitely independent of Pauli's remarks.

Einstein and Ehrenfest Respond: The Decomposition of the Compton Effect in Emission and Absorption Subprocesses

The counter-intuitive feature in Pauli's paper, the mutual interaction of waves that did not even nearly coincide in their frequencies and, moreover, the dependence on the frequency after the scattering process, was mitigated by Pauli but not resolved. But Pauli obviously did not only communicate his findings to Bohr and Sommerfeld, but also to Einstein or Ehrenfest. They handed in their reaction on Pauli's consideration even before Pauli's paper was published. Ehrenfest had been interested in the topic and discussed the issue with Einstein in September and October 1923.[169] They reacted on Pauli's contribution and gave a different conceptualization of the scattering process which was even more closely modelled after Einstein's initial account of 1916.

Einstein and Ehrenfest did not only generalize Pauli's consideration to molecules and atoms of quantized states but also "to the case that several radiation quanta are involved in the elementary process".[170] Einstein and Ehrenfest closely followed Einstein's original publication of 1916 and took emission and absorption processes as the basic building blocks of the analysis. Thereby, they wanted to arrive at a "certain deepening of our conception of the interaction between radiation and material particles".[171] The basic assumption and, according to Klein (1964, 24), the "key point" in their evaluation, was that in each elementary process radiation quanta $h\nu_1, h\nu_2, \ldots$ will be absorbed and $h\nu_1', h\nu_2', \ldots$ quanta will be emitted. These "partial processes of emission and absorption"[172] each had their probability coefficients $a_1b_1, a_2b_2, \ldots$ and $a_1'b_1', a_2'b_2', \ldots$ respectively (the a_i referred to spontaneous, the b_i to induced transitions). Einstein and Ehrenfest "naturally generalized"[173] Einstein's original probabilistic considerations by

$$dW = \prod_i b_i\rho_i \prod_j (a_j' + b_j'\rho_j')dt \tag{4.12}$$

The probability for the inverse process was given by

$$dW = \prod_i (a_i + b_i\rho_i) \prod_j b_j'\rho_j' dt \tag{4.13}$$

---

[169] See Kromos-Buchwald et al. (2015, xliii).

[170] "auf den Fall, daß bei dem Elementarprozeß mehrere Strahlungsquanten beteiligt sind," Einstein and Ehrenfest (1923, 304).

[171] "eine gewisse Vertiefung unserer Auffassung über die Wechselwirkung zwischen Strahlung und materiellen Teilchen," Einstein and Ehrenfest (1923, 302).

[172] "Absorptions- bzw. Emissions-Teilprozesse," Einstein and Ehrenfest (1923, 305).

[173] "naturgemäß verallgemeinernd," Einstein and Ehrenfest (1923, 305).

In Eq. (4.12) the index i refers to the absorbed quanta and j to the emitted ones. The primed radiation density refers to the radiation density at the frequency of the secondary quantum. Equation (4.13) has to be read in a similar way.

Einstein and Ehrenfest showed that for those products the Planck law could be upheld if the a_i and b_i were treated as regular Einstein coefficients. Pauli's result was a straightforward consequence of the above formulas, if only one quantum was absorbed and one was emitted. In Einstein's and Ehrenfest's hands, the corresponding formula took the form

$$dW = b\rho(a + b\rho') \tag{4.14}$$

For the first time, Einstein and Ehrenfest provided a formal decomposition of scattering in emission and absorption subprocesses in the quantum theory of radiation, symbolized by the products in Eqs. (4.12) and (4.13). Moreover, they explicated it verbally. To be sure, Einstein and Ehrenfest decomposed the processes in emission and absorption subprocesses, but did not imply any temporal order of the events. Yet, this "deepening of our conception of the interaction between radiation and material particles" was achieved by a direct incorporation of the priorly established coefficients of emission and absorption and a modelling of the processes resting directly on Einstein (1916a). Eintein's initial conception, already expressed in his 1905 article introducing the light-quantum hypothesis, that such quanta "can only be absorbed and created as a whole"[174] was obviously underlying this approach.

Pauli's Reaction: From Refutation to Argument Against BKS

Pauli's reaction[175] to a preprint of Einstein and Ehrenfest's paper[176] was in general rather sympathetic. Still, Pauli considered Einstein and Ehrenfest's introduction of probability coefficients for emission and absorption subprocesses untenable:

> Of course, these partial processes do not however have to be possible in isolation, in order that the composite process can occur. (For example, a free electron can never absorb a quantum $h\nu$ and transform it into translatory energy without a quantum $h\nu'$ being emitted at the same time, because in this case energy and momentum law could not be fulfilled at the same time.)[177]

[174] "nur als Ganze absorbiert und erzeugt werden können," Einstein (1905, 133).

[175] Pauli to Einstein, November 10. 1923; as cited in Pauli (1979, 128–130); and Kromos-Buchwald et al. (2015, 228–230).

[176] Einstein handled the publication process and sent a preprint to Pauli at around mid October 1923. See Einstein to Ehrenfest, November 5. 1923, as cited in Kromos-Buchwald et al. (2015, 224).

[177] "Natürlich brauchen jedoch diese Teilprozesse isoliert gar nicht möglich zu sein, damit der zusammengesetzte Prozeß vorkommen kann. (So kann ja z.B. ein freies Elektron niemals ein Quant $h\nu$ absorbieren und in Translationsenergie verwandeln, ohne daß gleichzeitig ein Quant $h\nu'$ emittiert wird, weil dabei Energie- und Impulssatz nicht zugleich befriedigt werden könnten.)" Pauli to Einstein, November 10. 1923; as cited in Pauli (1979, 129).

Therefore, Pauli concluded that the coefficients for the partial processes are "by themselves not physically defined".[178] One should rather write the probability for a "composite elementary process"[179] in the following way:

$$W = C^{\nu_1 \ldots \nu_p}_{\nu_{p+1} \ldots \nu_{p+q}} \frac{\rho_1}{\alpha \nu_1^3} \cdots \frac{\rho_p}{\alpha \nu_p^3} \left(1 + \frac{\rho_{p+1}}{\alpha \nu_{p+1}^3}\right) \cdots \left(1 + \frac{\rho_{p+q}}{\alpha \nu_{p+q}^3}\right)$$

where $C^{\nu_1 \ldots \nu_p}_{\nu_{p+1} \ldots \nu_{p+q}}$ is the probability coefficient for the whole elementary process in which the quanta $h\nu_1, \ldots, h\nu_p$ are absorbed and the quanta $h\nu_{p+1}, \ldots, h\nu_{p+q}$ are emitted.

For Pauli, the decomposition of the whole process into several steps was not physically justifiable. The whole process was, in his words, "physical". The definition of coefficients for the subprocesses (although they referred to the emission and absorption coefficients) was simply not reasonable because these processes did not exist for the primary case of interest, the free electron. What was generally framed as a solution to the riddle Pauli had posed, created new conceptual tensions: The decomposition of the scattering process, as it was formally proposed by Einstein and Ehrenfest, could not be established when energy and momentum conservation had to be fulfilled.

An answering letter of Einstein to Pauli's objections does not exist. But shortly afterwards Pauli further elaborated on the conception of scattering in a letter to Alfred Landé dating to November 28. 1923.[180] Pauli's reasoning is only indirectly related to the composite nature of the process, as he mainly discussed the relation of scattering processes in different energy regimes. But interestingly, Pauli did not relate the probability coefficients for emission and absorption, on the one hand, and for scattering, on the other, in any way. They were distinct entities which he referred to with distinct mathematical symbols: He used unprimed variables for the emission and absorption coefficients and primed ones for the scattering coefficients. And he did not construct scattering coefficients from emission and absorption coefficients as Einstein and Ehrenfest had.

For further discussion, Pauli's reasoning is of interest as it led him to a duplication in the theoretical description of resonance scattering and emission and absorption, thus (in a modified form) to the bedrock of his most important critique of the BKS theory (see further below). Pauli distinguished three different types of processes:

---

[178] "für sich nicht physikalisch definiert," Pauli to Einstein, November 10. 1923; as cited in Pauli (1979, 129).

[179] "zusammengesetzter Elemtentarprozess," Pauli to Einstein, November 10. 1923; as cited in Pauli (1979, 129).

[180] Pauli to Landé, November 28. 1923, as cited in Pauli (1979, 130–131). Seemingly Landé questioned some of Pauli's conclusions in his 1923 paper on the Compton effect in a postcard sent on November 15. 1923.

1. the photoelectric effect (or its inverse, electron capture) modelled as an absorption (or emission) ("Einstein-Prozesse", one quantum involved)
2. scattering with ionisation ("Compton-Debye Prozesse", two quanta involved)
3. scattering without ionisation (also "Compton-Debye Prozesse", two quanta involved)

Pauli explained that for loosely bound electrons, processes of type (1) and (2) could occur, but they had different outcomes. In processes of type (1) the atom would take up most of the recoil. For free electrons these processes cannot occur because of energy and momentum conservation. The processes (2) are different in so far as they involved two quanta and the momentum was, for loosely bound electrons, mainly taken up by the electron. These processes should slowly and, as Pauli suspected, continuously go over into the third kind of process. When the binding energy was higher than the energy of the impinging photon, processes of type (2) would no longer occur. The recoil had to be completely taken up by the atom.

If the energy of the impinging quantum was lower than the ionisation energy, Pauli, following his initial differentiation, distinguished the "Einstein-Prozesse" without ionisation, so simply emission or absorption of light, and the scattering without ionisation. The emission and absorption processes were, of course, the ones giving rise to the line spectrum. The third kind of processes must be possible for any frequency as otherwise, according to Pauli, the phenomenon of dispersion would be left unexplained. And at this point, Pauli noted a problem:

> According to this conceptualization, it might seem as if two processes were possible inside one [spectral] line (a) Einstein processes, for which the re-emission only occurs after a finite dwell time of the atom in the excited state. (b) Compton-Debye processes of the third kind, for which instantaneous scattering of the radiation occurs and the excited quantum state is not established at all. – I am however convinced that this is only a shortcoming of the current quantum theory and in reality the two types of processes will (in the region of one [spectral] line) turn out to be only two possible conceptions of one and the same process.[181]

Since the following will take us nearly a year into the future let me shortly recap where Pauli stood in late 1923. He had accepted the idea that two quanta were involved in the scattering process. Yet, he still denied the possibility of decomposing the scattering in one emission and one absorption subprocess. If he would have accepted this picture, the two processes he described in the last quote would have been the same. Therefore, the "key point" of Einstein and Ehrenfest, that the scattering was to be conceived of as a combination of an emission and an absorption subprocess, was not necessarily taken up without controversy. Rather, it caused conceptual problems that for the moment could not be resolved, at least not by Pauli.

---

[181] "Nach dieser Auffassung könnte es scheinen, als wären im Inneren einer Linie zweierlei Prozesse möglich (a) Einstein-Prozesse, bei denen die Rückstrahlung erst nach der endlichen Verweilzeit des Atoms im angeregten Zustand erfolgt. (b) Compton-Debye-Prozesse 3. Art, bei denen momentane Streuung der Strahlung stattfindet und der angeregte Quantenzustand gar nicht hergestellt wird. – Ich bin jedoch überzeugt, daß dies nur ein Mangel der jetzigen Fassung der Quantentheorie ist und in Wirklichkeit sich (im Gebiet einer Linie selber) beide Arten von Prozessen nur als zwei mögliche Auffassungen desselben Prozesses herausstellen werden." Pauli to Landé, November 28. 1923, as cited in Pauli (1979, 131).

The problem Pauli hit upon in the above letter, namely that apparently one and the same process was theoretically described by two different mechanisms, was not uncommon at the time.[182] And Pauli's main criticism of the BKS theory was directed in a similar direction. While in the above description there were two overlapping processes (Einstein processes and Compton-Debye processes) referring to one and the same observable phenomenon (the weakening of the intensity of the radiation inside a spectral line), the BKS theory also included two kinds of mechanism in the case of impinging resonant radiation: spontaneous emission by atoms in an excited states and the reaction of the oscillators conjugated to the initial quantum state.

In October 1924, when Pauli communicated both Einstein's and his criticism of the BKS theory to Bohr, he had found a solution to the above problem and could therefore criticize the BKS theory due to the duplication in the theoretical description.[183] Neither Einstein's critique nor the energy and momentum non-conservation were particular strong arguments against the BKS theory, at least so for Pauli. But he could not accept the duplication in the description of the resonance radiation. And in this connection the relation between the Compton processes, now called Compton jumps, and regular absorption, resurfaced. As Pauli stated:

> [...] I do not consider it possible to meaningfully differentiate Compton jumps and actual quantum transitions within one [spectral] line. The scattering processes outside and inside of a spectral line only differ in the *mean* dwell time of the atom in the excited state. If it [the excited state] is notably removed from the quantised state, it [the dwell time] is of the order of the period of oscillation (Compton process),[184] within the [spectral] line it [the dwell time] is of the order of $10^{-8} sec$ (actual quantum jump). Because these are only *mean* dwell times, there is no possibility to differentiate the two kinds of processes within one spectral line. The total frequency of occurrence of the process must be given by $B\rho_{\nu}$.[185]

Thus Pauli had accepted a decomposition of the scattering process in terms of successive emission and absorption processes. Even more, he connected these processes with transitions of the atom and introduced a temporal dimension. He even included a differentiation of this temporal dimension in respect to the energy relation between the possible quantised states and the frequency of the impinging radiation.

---

[182] Ladenburg and Reiche, for example, struggled with this duplication of scattering and absorption a bit, and Bohr put it into sharp relief in his "Grundpostulate der Quantentheorie" (Bohr 1923, 162–163).

[183] Pauli to Bohr, October 2. 1924; as cited in Pauli (1979, 163–167).

[184] *Footnote in the original*: at the same time the order of magnitude of the duration of the transitions process.

[185] "[...] halte ich es auch nicht für sinnvoll möglich, innerhalb einer Linie Compton-Sprünge und eigentliche Quantensprünge zu unterscheiden. Die Streuprozesse außerhalb und innerhalb der Spektrallinie unterscheiden sich nur durch die *mittlere* Verweilzeit des Atoms im angeregten Zustand. Wenn dieser merklich vom gequantelten Zustand entfernt ist, ist diese von der Ordnung der Schwingungsperiode (Compton-Prozess) [*Footnote in the original:* Zugleich Größenordnung der Dauer eines Übergangsprozesses überhaupt.], in der Linie ist sie jedoch von der Ordnung $10^{-8} sec$ (eigentlicher Quantensprung). Da dies *mittlere* Verweilzeiten sind, besteht keine Möglichkeit in der Spektrallinie zweierlei Arten von Prozessen zu unterscheiden. Die gesamte Häufigkeit der Prozesse muß durch das $B\rho_{\nu}$ gegeben sein." Pauli to Bohr, October 2. 1924; as cited in Pauli (1979, 166); emphasis in the original.

By late 1924, Pauli conceived of scattering as a sequence of transition processes connected to emission and absorption of radiation for which the mean duration of the occupation of the excited state depended on the frequency of the impinging radiation: It took the empirically established value of approximately 10^{-8} seconds for resonant radiation and a much shorter time if the excited state did not coincide with a quantised state of the atom, i.e. if the frequency of the impinging radiation did not coincide with a possible absorption frequency of the scatterer.[186]

But although Pauli had thought about the scattering mechanism in terms of a light-quantum picture in the meantime,[187] the conception of the scattering process he discussed in the letter to Bohr was developed in Munich, also shortly after the discovery of the Compton effect.[188] Thus, to follow the development of the conception of scattering as a sequence of two temporally separated emission and absorption processes, which first arose in the light-quantum framework, we have to turn our attention to the theoretical research performed at Arnold Sommerfeld's institute in Munich and its further development by Adolf Smekal, at the time in Vienna.

4.3.3 Dispersion and the Light-Quantum Hypothesis: Transitions to Short-Lived Intermediate States (1924–1925)

As discussed above, a first decomposition of the scattering process in terms of emission and absorption subprocesses was introduced into quantum theory by Einstein and Ehrenfest. But Einstein and Ehrenfest did not imply any temporal order of the subprocesses in their paper and Pauli had serious difficulties, at least so at first, to accept their reasoning. While Wolfgang Pauli was pondering about Einstein and Ehrenfest's conception, physicists in Munich, again motivated by the discovery of the Compton effect and independent of the above described theoretical endeavours, turned to the

---

186 For the optical spectrum ($\nu \approx 400-800$ THz) and according to Pauli's above mentioned connection between frequency and dwell time, the latter would be of the order $T \approx 10^{-15}$ seconds.

187 In a letter from November 1924 to Sommerfeld, Pauli also mentioned that he had concerned himself with the explanation of quantum-theoretical radiation theory in terms of light quanta, but that his ideas were not yet in a publishable form. "(Ich habe mir in dieser Richtung einiges überlegt, aber das ist noch nicht in mitteilba[r]er Form.)" Pauli to Sommerfeld, November 1924; as cited in Pauli (1979, 174). Also Smekal (1925a) indicated that Pauli and Wilhelm Lenz in Hamburg had thought about scattering in terms of the light-quantum hypothesis. See Sect. 4.3.3.

188 Pauli did not provide any sources in the letter but cited Karl Herzfeld's work in the same problem context in his article in the *Handbuch der Physik* of 1926 (Pauli 1926, 26). Due to the contact between Pauli and Sommerfeld, and the fact that they both just had been to the same conference in Innsbruck in late September 1924 (88. Versammlung deutscher Naturforscher und Ärzte, held from 21. to 27. September 1924), Pauli must have known about the research in Munich by October 1924.

evaluation of dispersion and, more generally, a theory of the light-matter interaction within a light-quantum framework. These attempts were subsequently further developed by Adolf Smekal, at the time in Vienna.[189]

The Light-Quantum Hypothesis in Munich: Quantum Phases and Dispersion as Successive Transitions

Karl Ferdinand Herzfeld and Gregor Wentzel were both Privatdozenten at the Munich institute. Prior to his arrival in Munich, Herzfeld had obtained his PhD on a theme of statistical physics in 1913 in Vienna and served in the army during WWI. In 1919, he was offered a double position in Munich: He was Privatdozent at Sommerfeld's institute and assistant to Kasimir Fajans, a physical chemist.[190] Wentzel had earned his PhD under the supervision of Sommerfeld for a study on X-ray phenomena in 1921 and took his Habilitation only a year later. Both Wentzel and Herzfeld stayed in Munich until 1926, when they could secure professor positions elsewhere.

The correspondence principle was also practised at Sommerfeld's institute in the years prior to the advent of quantum mechanics.[191] Nevertheless, the status of the principle cannot be compared to the role it played in Copenhagen. As Bohr did with light quanta, Sommerfeld would acknowledge the usefulness of the correspondence principle but he was reluctant to count it as a cornerstone of quantum theory. According to Suman Seth, Sommerfeld's well-known critique of the correspondence principle (Sommerfeld 1924, 1048) did not as much rest on practical uselessness, but on the "'problematic mixing of incommensurable worlds' [the quantum and the classical one]. [...] Distinct worlds were, for him [Sommerfeld], to be treated in distinctly different fashions." (Seth 2010, 347) And, indeed, Sommerfeld had connected the correspondence principle closely to the wave theory of light. When he came more and more to accept the failure of the latter, he also thought that the former became under increasing pressure.[192]

---

[189] The following line of work has found next to no attention in the secondary literature. Antoci and Liebscher (1996) identified Wentzel's approach, introducing a quantum phase for light quanta, as a precursor of Feynman's path integral approach. Marta Jordi Taltavull has studied Herzfeld's, Wentzel's and Smekal's work in some more detail than it will be presented in the following. For further information, especially on Herzfeld's and Wentzel's approach, I refer the reader to Sect. 7.3 of her dissertation (Jordi Taltavull 2017). I will focus on different aspects of the development and thus the following will not constitute a repetition of Jordi Taltavull's work.

[190] For a biography of Herzfeld, see Mulligan (2001).

[191] See for example Seth (2010, Especially pp. 232–241) and, for concrete applications, Jähnert (2019, Chap. 4). Even in Wentzel's paper, introducing a phase for light quanta, the correspondence principle was applied. See Jordi Taltavull (2017, 556–558).

[192] See, for example, the letter Sommerfeld to Max von Laue, November 27. 1923; as cited in Eckert and Märker (2004, 155–156); and the corresponding snippet as cited in Jordi Taltavull (2017, 551).

The light-quantum hypothesis, on the other hand, was taken rather seriously in Munich and the discovery of the Compton effect strengthened this trend.[193] Sommerfeld seriously questioned the wave theory of light even before the discovery of the Compton effect. As he stated in a letter to Einstein in early 1922:

> I am happy, if it [the ondulatory theory of light] gets a visible hole at any tip [Zipfel]. Things cannot go on as they are with the dualism of standpoints. If you [Einstein] say, that your experiment is decisive, I want to believe you, although I cannot comprehend it yet, in spite of the explanations by Geiger in Jena. [...] I can only promote the technique of the quanta, you have to do your philosophy. In private [innerlich], I do no longer believe in the spherical wave. [...] Do severely afflict it [Setzen Sie ihr nur ordentlich zu].[194]

Sommerfeld was in the US when Compton took the step to interpreting his empirical results according to the light-quantum hypothesis and Sommerfeld and Compton discussed the theoretical interpretation.[195] Although Sommerfeld felt not completely convinced initially, he wrote a letter to Niels Bohr on January 21. 1923, that if Compton's results proved correct "the wave theory of Röntgen-rays will become invalid."[196] Yet, the conceptual difficulties of the extreme light-quantum viewpoint were certainly not lost on Sommerfeld and, thus, the question of integrating interference and dispersion into a light-quantum framework became points on the agenda at the Munich institute. Wentzel and Herzfeld approached this problem together, based on ideas William Duane had developed.[197]

The Basis: Wentzel's Quantum Phase

In November 1923, Wentzel held a talk about the "Light-quantum hypothesis and lattice interference"[198] at the institute's colloquium and communicated his work

---

[193] When asked about the influence of Albert Einstein in Munich, Wentzel responded in his AIP oral history interview that Einstein's arguments and the light-quantum hypothesis were taken rather seriously in Munich, possibly more seriously than somewhere else. Interview of Gregor Wentzel by Thomas S. Kuhn, 1964 February 3 to 5, Niels Bohr Library & Archives, American Institute of Physics, College Park, MD USA. https://doi.org/10.1063/nbla.akbm.qngm. Pages 12–13 of the PDF.

[194] "Ich bin froh, wenn sie [die Ondulationstheorie] an irgend einem Zipfel ein sichtbares Loch bekommt. So wie bisher geht es mit dem Dualismus der Standpunkte nicht weiter. Wenn Sie [Einstein] sagen, dass Ihr Versuch entscheidend ist, will ich es glauben, obwohl ich es noch nicht begreife, trotz der Erläuterungen Geigers in Jena. [...] Ich kann nur die Technik der Quanten fördern, Sie müssen ihre Philosophie machen. Innerlich glaube ich übrigens auch nicht mehr an die Kugelwelle. [...] Setzen Sie ihr nur ordentlich zu." Sommerfeld to Einstein, January 11. 1922; as cited in Kromos-Buchwald et al. (2012, 76–77). The experiment Sommerfeld referred to should be an *experimentum crucis* for the decision between light-quantum hypothesis and the wave theory of light. It turned out to be non-decisive. See e.g. Klein (1970, Sect. 3).

[195] See Eckert (2013, 270).

[196] Sommerfeld to Bohr, January 21. 1923; as cited in Stuewer (1975, 241).

[197] See also Jordi Taltavull (2017, 551–553) for further details.

[198] See Jordi Taltavull (2017, 552).

on "Quantenoptik" in early February 1924 to the *Zeitschrift für Physik* (Wentzel 1924).[199] Wentzel showed a way of defining a quantum-theoretical phase using the light-quantum hypothesis. Wentzel's definition rested on the idea that, along its path, light caused unmechanical perturbations in the medium. His definition of a quantum phase was thus based on "the deviations of the inner atomic movement from Hamiltonian mechanics".[200]

Wentzel proposed to describe the movement of all atomic systems along the light-path by canonical momentum (α_k) and position (β_k) variables, where k is the number of degrees of freedom for the whole system.[201] With these definitions, a measure for the non-mechanical deviations could be written as $\int \sum_k \beta_k d\alpha_k$. The quantum phase was accordingly defined as this deviation divided by Planck's constant:

$$\varphi = \frac{1}{h} \int \sum_k \beta_k d\alpha_k \tag{4.15}$$

where the integration had to be performed over all "unmechanical processes"[202] along the path from emitting to absorbing system.

Singling out the energy W of all systems, which were considered to be "coupled to each other in principle",[203] and the canonically conjugated time t, Wentzel wrote the above expression as:

$$\varphi = \frac{1}{h} \left[\int t dW + \int \sum_{k=2} \beta_k d\alpha_k \right] \tag{4.16}$$

He could show that this definition was suitable to describe the propagation of light in the vacuum. By replacing the classical by his quantum definition of the phase in the classical definition coupling phase and refractive index, Wentzel arrived at the following quantum definition of the latter:

$$n = \frac{c}{\Delta W} \frac{d}{ds} \sum \int \beta_k d\alpha_k \tag{4.17}$$

[199] Wentzel published three notes on a light-quantum theory of radiative phenomena. I will only shortly discuss the relevant points of his first note. For a full discussion of Wentzel's work, see Jordi Taltavull (2017, Sects. 7.3.2 and 7.3.4).

[200] "die Abweichungen der inneratomaren Bewegungen von der Hamiltonschen Mechanik," Wentzel (1924, 193).

[201] The momentum variables were, for simplicity's sake and without loss of generality, considered to be constants of the motion: $\dot{\alpha}_k = -\frac{\partial H}{\partial \beta_k} = 0$.

[202] "unmechanischen Vorgänge," Wentzel (1924, 194).

[203] "prinzipiell miteinander gekoppelt," Wentzel (1924, 194).

As he noted: "The refractive index thus measures quantum-like the deviations from mechanics per unit path and unit energy."[204]

In his discussion of interference, Wentzel realized that a light quantum was in general not restricted to travel from the emitting (E) to the absorbing system (A) along one particular path. To each possible path s, Wentzel associated a wave $f_s e^{2\pi i \varphi_s}$, where f_s was the amplitude of the classical wave and φ_s the quantum phase. The probability for the light quantum to travel from A to E along an arbitrary path was not simply defined by the sum over the a-priori-probabilities over the single paths s but it was J-times as big, where J was defined as

$$J = \frac{F\tilde{F}}{|F_0|^2} = \frac{(\sum_s f_s e^{2\pi i \varphi_s})(\sum_s \tilde{f}_s e^{-2\pi i \varphi_s})}{(\sum_s f_s)(\sum_s \tilde{f}_s)} \tag{4.18}$$

The phase was connected to the probability of the light quantum to travel along each path, unrestricted by classical laws. The different probabilities interfered and thereby provided, in Wentzel's description, the total probability for the light quantum to travel from A to E.

Wentzel emphasized that this conception of interference and the corresponding quantum phase assumed a coupling between the emitting and absorbing system (and for that matter all systems in between). It implied a sense of non-locality.[205] The presence of the absorbing system, through which the paths and the quantum phases were defined, was essential in the construction of the interference formula. Interference in the vacuum, was not only not detectable but "in principle not extent".[206]

Through his definition of a quantum phase (4.15) and the probabilistic interpretation of interference as given in Eq. (4.18), Wentzel could rederive Bohr's frequency condition as well as the correspondence principle for selection rules and the polarization of emitted radiation.[207] Importantly, for the purpose of this section, Wentzel's introduction of the quantum phase allowed Karl Herzfeld to discuss dispersion in terms of light quanta.

Herzfeld's Elaboration: Dispersion as Transitions Through Short-Lived Intermediate States

In an "attempt of a quantum-like interpretation of dispersion",[208] submitted in the same month as Wentzel's quantum optics, Herzfeld tried to give a detailed account of dispersion in terms of the light quanta and Wentzel's quantum phase. Herzfeld split

---

[204] "Der Brechungsindex mißt also quantenmäßig die Abweichungen von der Mechanik pro Weg- und Energieeinheit." Wentzel (1924, 195).

[205] Such a conception was first introduced by Adolf Smekal (see for example Smekal 1923a), who will further elaborate on Wentzel's and Herzfeld's contributions.

[206] "prinzipiell nicht vorhanden," Wentzel (1924, 196).

[207] For details see, Jordi Taltavull (2017, 556–558).

[208] "Versuch einer quantenhaften Deutung der Dispersion," as the title of Herzfeld (1924) read.

the description of dispersion into two parts: On the one hand, one needed to explain interference; on the other hand, one needed to explain the different velocity of light in different materials. Wentzel had taken care of the first point. Hence, Herzfeld concentrated on the second one: How can one explain that the velocity of light was different in different media when working with the light-quantum hypothesis?

To start out his discussion and to make dispersion describable via Wentzel's quantum phase, Herzfeld reevaluated the classical theory of dispersion, both in terms of the phase and group velocity. While the details of Herzfeld's evaluation are not important for us, his interpretation of the same certainly is.[209] For the phase, Herzfeld came to the following conclusion:

> Consequently it looks [...] as if the primary wave performed a backward phase shift $2\pi\,\Delta\varphi$, or in other words, as if the phase was delayed in the layer for a time $2\pi\,\frac{\Delta\varphi}{\nu}$.[210]

For energy propagation within the medium, Herfeld arrived at a similar interpretation. The result could again be understood as if the energy was on average stopped for a certain amount of time within the medium before propagating further.

Herzfeld then reinterpreted the behaviour of the classical phase in terms of Wentzel's quantum-theoretical one. Consequently, the phase of a light quantum with a frequency which did not coincide with an absorption frequency would experience a phase shift when travelling through a molecular layer. In contrast to classical theory, this process should not occur with certainty but would be governed by probabilistic laws. Herzfeld reinterpreted the classical delay in energy propagation as a light quantum being stopped by an atom for a short amount of time before freely propagating through the vacuum again.[211]

To arrive at a possible interpretation of these features in terms of a mechanical atomic model, Herzfeld had to relax Bohr's postulates:

> The orbits calculated according to the usual quantum rules are the only stationary ones in the sense that they have a lifetime of the order of 10^{-8} sec. Besides these, also intermediate states [Zwischenzustände], which the atom can acquire in the case of the impact of a non-fitting [nichtpassenden] light quantum, are possible; these states only have a lifetime of a comparatively vanishing order, in the case of visible light $\propto 10^{-15}$ sec.[212]

---

[209] For details on Herzfeld's evaluation on the classical level, see either Herzfeld's own contribution (Herzfeld 1924, 342–348) or Marta Jordi Taltavull's discussion (Jordi Taltavull 2017, 559–561).

[210] "Es sieht also so aus, als ob die Primärwelle [...] in der Schicht einen Phasensprung $2\pi\,\Delta\varphi$ nach rückwärts gemacht hätte, oder mit anderen Worten, als ob die Phase in der Schicht die Zeit $2\pi\,\frac{\Delta\varphi}{\nu}$ aufgehalten worden wäre." Herzfeld (1924, 344).

[211] "Entsprechend muss die langsamere Energieforpflanzung bedeuten, daß die Quanten in den Atomen eine Zeitlang aufgehalten werden und so im Mittel langsamer fliegen als im Vakuum." Herzfeld (1924, 348).

[212] "Die nach der üblichen Quantenvorschrift berechneten Bahnen sind die einzigen stationären in dem Sinne, daß sie Lebensdauern von der Größenordnung 10^{-8} sec haben. Neben diesen sind auch noch Zwischenzustände möglich, die das Atom beim Auftreffen eines nichtpassenden Lichtquants annehmen kann; diese haben nur Lebensdauern von daneben verschwindender Größenordnung, bei sichtbarem Licht $\propto 10^{-15}$ sec." Herzfeld (1924, 349).

For a reinterpretation of the classical parameters in terms of this quantum-theoretical interpretation Herzfeld proposed the following. Imagine planes with N' atoms per unit area, which are ordered spatially so that N atoms per unit volume can be found. J light quanta, each of energy $\frac{h\nu}{2\pi}$, should impinge on a unit area per second, so that the intensity of light is $J\frac{h\nu}{2\pi}$. Per plane, the probability of a photon being absorbed was assumed to be qN'. The total amount of photons being absorbed per unit volume was thus NqJ. Herzfeld referred to q "as a kind of 'absorption cross-section'"[213] and to $\frac{1}{Nq}$ as "a kind of mean free path of the light quanta".[214] Of the absorbed quanta, the fraction w_E were assumed to be emitted in the direction of propagation after being stopped for the time τ_E on average. $1 - w_E$ was accordingly the fraction of light quanta scattered in different directions with the corresponding dwell time τ_s. $\Delta\varphi$ was the delay of Wentzel's quantum phase between emission and absorption for one light quantum.

Through this modelling, Herzfeld could formally connect the quantum-theoretical mean dwell times, phases and the fractions of light quanta being scattered either in the forward or any other direction to the classical parameters and formulas describing dispersion and scattering.[215] At this point, Herzfeld turned to a more detailed qualitative characterisation of the quantum-theoretical parameters, most importantly the time the quanta were stopped inside the atoms. According to Herzefeld, one had to assume that for quanta scattered in different directions, τ_s must be very small, in order that the states can be considered non-stationary. For the case of resonance, it was known from experiment that the dwell time was of the order 10^{-8} seconds. Herzfeld explicitly stated that, contrary to prior dispersion-theoretical accounts,[216] the proposed interpretation treated processes inside and outside a spectral line in essentially the same way: i.e. as two subprocesses of emission and absorption, while the mean dwell time of the photons differed in the two cases.[217] This was exactly the point which Pauli took up in the letter to Bohr (see above) and which he will discuss more thoroughly in his article in the *Handbuch der Physik* (see below).

For the case of quanta emitted in the forward direction, the ones responsible for dispersion, Herzfeld (1924, 350–351) could establish an upper boundary for the mean dwell time when the frequency of the light quantum did not coincide with an absorption frequency of the atom. Since the phase shift could not be greater than one,

[213] "eine Art 'absorbierender Querschnitt'," Herzfeld (1924, 349).

[214] "eine Art freie Weglänge der Lichtquanten," Herzfeld (1924, 349).

[215] For a discussion of the formulas, see either Herfeld's own paper (Herzfeld 1924, 349–350) or Jordi Taltavull (2017, 561–562).

[216] Herzfeld explicitly mentioned Ladenburg and Reiche (1923).

[217] At this point, I digress from Jordi Taltavull (2017, 563). Jordi Taltavull suggested that Herzfeld referred to a differentiation between complete and incomplete quantum transitions in Ladenburg and Reiche's paper. As argued above, Ladenburg and Reiche differentiated, on the quantum side of their explanation, between an atom resting in a stationary state for non-fitting quanta and incomplete quantum transitions as an explanation for the scattered radiation at an absorption frequency. I believe that Herzeld pointed towards the duplication in the description in terms of dispersing oscillator and quantum transitions in Ladenburg and Reiche's approach, which were responsible for the scattering for non-fitting frequencies.

he knew from the established connection between quantum-theoretical and classical variables that $\frac{\nu}{2\pi}\tau_E < 1$ or:

$$\tau_E < \frac{2\pi}{\nu} \tag{4.19}$$

Again, the mean life-time was vanishingly short in comparison to the stationary states. For the region of visible light, it was in the range of 10^{-15} seconds.

But for the case of quanta with a frequency equivalent to an absorption frequency and emitted in the direction of the impinging radiation, Herzfeld faced a difficulty. In the case of quanta scattered in different directions, the mean dwell time was finite, for fitting frequencies of the order of 10^{-8} seconds. But the equations for dispersion suggested that in the case of forward scattered radiation the dwell time was, again for fitting frequency and at least to first order, equal to zero. This could not be reconciled with a probabilistic modelling of the emission process,[218] at least as long as one did not want to provide for completely different probabilistic modellings of the emission processes for photons scattered in the forward direction and for the ones scattered in any other direction.

Herzfeld (1924, 352) proposed the following solution: The probability for emission should be governed by the law of radioactive decay, i.e. the probability of re-emission by the atom during the time dt was given by βdt, regardless of the direction of propagation after the emission process. But on top of this probabilistic modelling of the emission process, Herzfeld assumed another probabilistic connection: The lower the dwell time of the light quantum in the atom, the more likely the quantum was emitted in the forward direction. The fraction of atoms emitting their quantum in the forward direction was given by $e^{-\gamma t}$ and the fraction of atoms radiating in any other direction by $1 - e^{-\gamma t}$. As Herzfeld stated, "t corresponds to the time, which passes between the absorption [Aufnahme] and the re-emission [Wiederabgabe] of the quantum by the atom, $1/\gamma$ is, so to say, a measure for the memory of the atom or the quantum for the impinging direction."[219] The detailed elaborations on such a probabilistic model are irrelevant for the following. But it is important to note, that Herzfeld could show, although with some ambiguity in the interpretation, that the statistical energy balance was upheld within his new conceptualization.

In total, Herzfeld established a completely new model of how dispersion and scattering should be conceived. In wave-theoretical approaches to dispersion of the time, there was some kind of entity supposed to resonate at the right frequency, "substitute" or "virtual" oscillators, while the atom stayed in a stationary. Herzfeld's analysis, on the contrary, depended on a re-evaluation of the classical dispersion theory concentrating on the phase relation between impinging and resulting radiation. Herzfeld explained the stopping power of the atoms for light quanta with non-fitting energy in terms of transitions to "intermediate states" [Zwischenzustände], in this

---

218 Herzeld assumed, as was common, a modelling equivalent to the radioactive decay.

219 "[hier] bedeutet t die Zeit, die zwischen der Aufnahme und Wiederabgabe des Quants durch das Atom verfließt, $1/\gamma$ mißt sozusagen das Gedächtnis des Atoms oder Quants für die Auftreffrichtung." Herzfeld (1924, 352).

case lying both energetically and temporally between the emission and absorption processes. The time the quanta would be stopped by the atoms was proportional to $\frac{1}{\nu}$ for non-fitting light quanta and the empirically established value 10^{-8} seconds for fitting ones.

Herzfeld saw a remarkable advantage over prior contributions to dispersion, such as Ladenburg and Reiche's. Instead of assuming two different reaction mechanisms in- and outside of the region of spectral lines, Herzfeld could conceive of both types of processes as qualitatively equal.[220] He assumed both to be composed of emission and absorption processes. The prize he had to pay was nonetheless huge. Bohr's first postulate had to be relaxed (and with it necessarily the second). Non-fitting light quanta could be absorbed by atoms because the atom was capable of acquiring intermediate states for a very short amount of time. All the same, while the behaviour of the light quanta, being absorbed and reemitted, or "stopped" in Herzfeld's terminology, were concretely expressed in mathematical formulas, the behaviour of the atoms and the occupation of the intermediate states mostly remained of a verbal nature. Herzfeld did not base his reasoning in any kind of mechanistic model of the movement of the electrons.

Adolf Smekal: Intermediate States Without Light Quanta

Unsurprisingly, this new conception of scattering and dispersion was taken up favourably by two physicists: Adolf Smekal and Wolfgang Pauli.[221] I already discussed Pauli's reference to Herzfeld's modelling in private communication to Bohr and I will turn to his elaborations in his article for the *Handbuch der Physik* to conclude this chapter. But in the mean-time also Adolf Smekal had turned to Wentzel's and Herzfeld's approach.

Adolf Smekal, at the time in Vienna, had received his PhD in Graz (1917) and had spent two years in Berlin, where Max Planck and Albert Einstein worked at the time. Here, he came in contact with Bohr's atomic model and the light-quantum hypothesis, and the different methods of working with them. Although Smekal would gather some reputation through his work on X-ray spectra, the field of research he

[220] As Herzfeld's paper was communicated on February 21. 1924, he was certainly unaware of the details of the BKS theory. BKS essentially achieved the same goal but by a completely different assumption: They decoupled emission and absorption processes from the transition processes of the atoms, although in a rather sophisticated manner. While the BKS atoms remained in one stationary state during radiative processes, Herzfeld's atoms always jumped, even when the frequency of the impinging radiation did not fit the frequency of an absorption line.

[221] Also Erwin Schrödinger liked Wentzel's ideas, as he wrote in a letter to Sommerfeld as late as March 7. 1925: "By the way, Wentzel's attempts to arrive at a quantum theory of dispersion seem the most promising to me. But I reinterpret them in terms of Bohr-Kramers-Slater." [*Original*: "Übrigens scheinen mir die Wentzelschen Versuche zu einer Quantentheorie der Dispersion zu kommen, die meisversprechenden. Nur deute ich sie mir um nach Bohr-Kramers-Slater."] Schrödinger to Sommerfeld, March 7. 1925; as cited in von Meyenn (2011, 108).

was and is best-known for is classical and quantum statistics. And within this line of research, or rather with the respective theoretical tool kit, he had worked on dispersion problems and predicted (Smekal 1923b) the Raman effect, i.e. a discrete shift in the frequency of emitted radiation corresponding to an energy difference of the scattering material.

Obviously, Wentzel's and Herzfeld's contributions must have appealed to Smekal. Both Wentzel and Herzfeld had cited Smekal's contributions. Wentzel's concept of a quantum phase further incorporated two of Smekal's previous ideas. First of all, emitting and absorbing system (and all the systems in between) were necessarily connected, which had been spelled out by Smekal earlier on (Smekal 1923a). Secondly, in Wentzel's theory, as in Smekal's earlier evaluation, questions about the behaviour of light in vacuum were excluded. Both for Smekal and Wentzel interference was a concept that only made sense in reference to a material system which made it observable.

At the time Smekal came to discuss Wentzel's and Herzfeld's contribution in some detail in early 1925,[222] the investigations had already taken on speed, especially due to Kramers and Heisenberg's description of incoherent scattering in terms of Bohr's atomic model. The first time Smekal discussed Herzfeld's ideas was in a direct reaction to Kramers and Heisenberg's treatment of the incoherently scattered radiation. Smekal definitely felt bitter at the time: He stressed that "due to external circumstances the [his own] [...] more detailed treatment of the subject had to be postponed several times"[223] and that he had been aware of a classical correspondence analogue of the incoherently scattered radiation when he first proposed it in 1923.[224] In his second announcement Smekal even stated that Kramers' dispersion theory was essentially based on his ideas.[225]

In his reaction to the Kramers-Heisenberg paper, which was at the same time an outline of ideas he would more thoroughly discuss in a follow-up paper, Smekal (1925b) first defended himself against the attribution of applying the light-quantum hypothesis. On the contrary, his prediction of the incoherent scattering only relied on "the two so far *single*, truly *general* results of quantum theory, that in each (non-virtual) *interaction between matter and radiation* an energy exchange of $h\nu$ and a directed momentum exchange of $h\nu/c$ occurs, as Einstein has proven".[226] The assumption that the energy of light was transported in small packages was, according to Smekal, as ungrounded as the assumption of the wave theory of light in vacuum.

---

[222] Smekal had taken note of Wentzel's contribution by 1924 but, at that point, only mentioned it in a discussion of a paper by Epstein and Ehrenfest. See Jordi Taltavull (2017, 568).

[223] "Die [...] eingehendere Behandlung des Gegenstandes hat aus äußeren Gründen mehrfach hinausgeschoben werden müssen." Smekal (1925b, 241).

[224] See Smekal (1925b, 241, Footnote 2).

[225] Smekal (1925a, 84, Footnote 1). To be honest, I cannot see how this might be the case.

[226] "die beiden bisher *einzigen*, wirklich *allgemeinen* Ergebnisse der Quantentheorie, daß bei jeder (nicht-virtuellen) *Wechselwirkung zwischhen Strahlung und Materie* ein Energieumsatz von der Größe $h\nu$ und ein gerichteter Impulsumsatz vom Betrage $h\nu/c$ auftritt, wie Einstein nachgewiesen hat." Smekal (1925b, 242, emphasis in the original).

Smekal further argued that the new phenomenon of double emission, which Heisenberg and Kramers had introduced, could easily be established according to his initial derivation, simply by inverting a sign in his original equations of 1923.[227] But, as he went on and therefore borrowed and cited Herzfeld's conception of dispersion, such processes can easily be imagined "when one consistently conceives of the scattering process on a singular atom as decomposed into two temporally mutually *independently* occurring interactions of the atom with the radiation field, for which the regular Einstein probability approaches are valid".[228] According to Smekal, the assumption of a "*finite* time difference"[229] was the only way of interpreting dispersion without introducing wave-theoretical assumptions. In this context, Smekal took up Herzfeld's "intermediate states", renamed them to "meta-stationary quantum states"[230] and introduced, for the time being only verbally, Einstein coefficients for transitions between meta-stationary and stationary states. From this point of view, as Smekal argued, the processes he proposed[231] would correspond to absorption processes, the ones proposed by Kramers and Heisenberg[232] would correspond to spontaneous and induced emission processes.[233]

Only in a footonote, Smekal introduced a further conceptual piece to the puzzle. He argued that the concept of such short-lived states could be understood within the framework of Bohr's atomic model. The lifetime of the meta-stationary states was to be explained by what came to be known as weak quantization.[234] Bohr (1923, Chap. 2, Sects. 3 through 5) had already introduced the idea that his quantum postulates became invalid when the inverse of the frequency $1/\nu$ of a harmonic of the orbit is of the same order as the lifetime of the states. Due to the short lifetime of the meta-stationary states, Smekal's argument went, they could be integrated into Bohr's orbital model. The prime advantage of his interpretation, was according to Smekal, that it completely rested on the interaction between matter and monochromatic light.

Smekal further elaborated on these thoughts in a second communication (Smekal 1925a). His basic assumption was that

227 Smekal (1925b, 242, Footnote 2).

228 "wenn man sich den Streuvorgang am Einzelatom konsequent in zwei zeitlich voneinander *unabhängig* erfolgende Wechselwirkungen des Atoms mit dem Strahlungsfelde zerlegt denkt, für welche die gewöhnlichen Einsteinschen Wahrscheinlichkeitsgesetze gültig sind." Smekal (1925b, 242–243; emphasis in the original).

229 "*endlichen* Zeitdifferenz," Smekal (1925b, 243, emphasis in the original).

230 "metastationäre Quantenustände," Smekal (1925b, 243).

231 In modern terminology: the Raman effect.

232 In modern terminology: double emission processes.

233 For a more concrete description, both writing out the sign switch as well as the expression in terms of Einstein coefficients, see Smekal (1926a, 981–982).

234 Paul Ehrenfest and Richard Tolman (Ehrenfest and Tolman 1924) discussed such effects, but not in relation to dispersion, and also based them on Bohr's thoughts as expressed in Bohr (1923).

> a consistent decomposition of the unfamiliar Compton-Debye elementary process[235] in an absorption-like and an emission-like elementary process is possible by intersecting a finite time period of the order of a reciprocal radiation frequency, which corresponds to an average value between primary and scattered frequency.[236]

In this scheme Smekal saw a far more consistent connection of quantum theory and the classical theory of radiation. In both theories, the case of resonance was the exception to the more general case, i.e. interaction with non-fitting frequency. Scattering and dispersion faded harmoniously into the absorption and emission in the case of fitting frequencies. In a footnote, Smekal noted that Wilhelm Lenz and Wolfgang Pauli in Hamburg had been thinking about the same general description, but had not published their results because of the problem of fulfilling the energy and momentum law simultaneously for free electrons (Smekal, 1925a, 85, Footnote 1).

Again, Smekal argued for the existence of meta-stationary states on the grounds of weak quantization. Since the quantization rules might break down when the mean lifetime of a state is of the same order as the reciprocal of the frequencies of the harmonics, or, often equivalently, the frequency of the emitted radiation, there was conceptually enough room for the assumption that the meta-stationary states are actually established. Accordingly, Smekal introduced the "postulate" that

> the "meta-stationary" states should only differ from the "stationary" states in reference to their partial or complete independence of special quantum conditions and therefore also in reference to the order of their mean lifetime; for the newly introduced states, this postulate will therefore have to take the place of the first postulate of the Bohr theory, which concerns the existence of "stationary" states.[237]

Smekal extended the second postulate as well as Einstein's equilibrium and probability considerations to the transitions between stationary states and meta-stationary ones as well as for transitions only amongst meta-stationary states. The application of the adiabatic principle to meta-stationary states, on the other hand, was out of the question because of the contrast of slow parameter variation and the short lifetime of the states.

Smekal applied equilibrium considerations to show that his scheme was consistent. On the one hand, the statistical equilibrium between matter and radiation was not perturbed by the inclusion of such states. On the other hand, the probability of a transition from a stationary to a meta-stationary state was of the order of a

---

[235] *Comment by the author*: Smekal referred to the Compton effect as well as Raman scattering by the term "Compton-Debye process".

[236] "eine konsequente Zerlegung des fremdartigen Compton-Debyeschen Elementarvorganges in einen absorptionsähnlichen und einen emissionsähnlichen Elementarvorgang möglich ist unter Zwischenschaltung einer endlich Zeitdauer von der Größenordnung einer reziproken Strahlungsfrequenz, die einem Mittelwert von Primär- und Streufrequenz entspricht." Smekal (1925a, 84).

[237] "[daß] sich die 'metastationären' Zustände von den 'stationären' Quantenzuständen überhaupt nur hinsichtlich ihrer teilweisen oder vollständigen Unabhängigkeit von besonderen Quantenbedingungen und damit auch bezüglich ihrer mittleren Lebensdauergrößenordnung voneinander unterscheiden sollen; dieses Postulat wird für die neueingeführten Zustände demnach an Stelle des ersten, die Existenz der 'stationären' Zustände betreffenden Postulates der Bohrschen Theorie zu treten haben." Smekal (1925a, 86).

transition from a stationary to another stationary state, while the transition from a meta-stationary state to a stationary state was far more probable, i.e. occurred a lot more frequently than the transition from a stationary state. Although therefore the probability of finding an atom or molecule in a meta-stationary state was negligibly small, possible observable effects might occur: For a large amount of atoms, such as in a gas, a considerable amount of atoms would be in meta-stationary states.

Smekal had taken up Herzfeld's ideas and the connection between mean lifetime of meta-stationary or intermediate states as derived from dispersion theory. But Smekal went beyond Herzfeld's contribution in two important points. First of all, he avoided comments about the nature of radiation in the vacuum. Smekal neither based his approach on the wave theory of light nor on the light-quantum hypothesis. Secondly, he argued that the introduction of meta-stationary states, lying both energetically and temporally in between stationary states, could be argued for within the framework of Bohr's atomic model. Yet, Smekal did not concretize these ideas, i.e. he did not perform any calculations on a mechanical model of the atom. All the same, he could show that their introduction did not disturb thermal equilibrium. As is clear both from the introduction of Einstein coefficients as well as from Smekal's verbal statements, he conceived of such transitions as actually occurring and having a temporal nature. By bending (or supplementing) Bohr's first postulate, energy and momentum conservation could be upheld for each partial process.

One thing that is striking about Smekal's papers is the lack of any reference to Einstein and Ehrenfest's paper of 1923. Smekal was certainly aware of this publication, he cited it regularly in an overview article (Smekal 1926a) which he already referred to in the above described contributions. Turning to this overview article, we also find a reason why Smekal did not refer to Einstein and Ehrenfest's publication. He argued that Einstein and Ehrenfest's conception rested on the idea that there is no finite time interval between emission and absorption and hence a direct mutual influence of radiation of different frequencies. Only in the case of the completely free electron, which Smekal considered an idealization, Ehrenfest's and Einstein's modelling might, according to Smekal, be of use.[238]

Wentzel's, Herzfeld's, and Smekal's work is normally not considered in the historical literature on quantum theory. There is one obvious reason: Their elaborations merely constituted a side road which was (nearly) completely superseded by matrix and wave mechanics. Although this is essentially true, especially when it comes to the general theoretical development of quantum theory, the proposed conception of scattering did not fade entirely. It resurfaced in the discussions around the Raman effect.[239] Secondly, it was taken up in an important, but at the time of its publication already outdated overview of the old quantum theory by Wolfgang Pauli.[240]

---

[238] Compare Smekal (1926a, 1108–1109, Footnote 577).

[239] See Sect. 6.3 for my reconstruction.

[240] Jordi Taltavull (2017, 569) observed that Pauli only discussed Herzeld's conception of dispersion in a footnote. As we will see, this is not correct but the oversight is understandable: Pauli only shortly mentioned Herzfeld's contribution in the context of dispersion but actually defended a conception

4.3.4 *Pauli's Article in the* **Handbuch der Physik** *(1926): A Defence of Decomposition*

To conclude the historical discussion of this section, I want to turn attention to Wolfgang Pauli again. As already noted, Pauli had taken up Herzfeld's ideas to substantiated his criticism on the BKS theory and had obviously thought about the connection between absorption and scattering in terms of successive and temporally ordered emission and absorption subprocesses. And, as a matter of fact, Pauli would still defend this point of view in his article in the *Handbuch der Physik* published in 1926.

At the latest, Pauli started his work on the article in the *Handbuch der Physik* in February 1925[241] and seemed to have finished his contribution in mid to late October of the same year.[242] Although in the meantime quantum mechanics had been developed and Pauli was well aware of it, he did not discuss any of the corresponding papers.[243] All the same, he reviewed the description of dispersion in terms of the Kramers-Heisenberg paper (Pauli 1926, Sect. 16). And quite interestingly, he retracted his former "detour through a third state"-description for the non-resonant case. Only for resonance scattering, the third state should be occupied.[244] Yet, a decomposition into emission- and absorption-like processes was proposed by Pauli when he discussed a different approach: the one by Einstein, Ehrenfest, Herzfeld and Smekal. It is hard to judge when exactly Pauli finished the corresponding passage (Pauli 1926, Sect. 5). As Pauli cited Smekal's first but not his second contribution, it must have been before September 1925, but definitely at a time, when Pauli was already aware of Heisenberg's attempts at a new formulation of quantum theory.

In contrast to Smekal (see above), Pauli conceived of the formal decomposition of the scattering processes into emission and absorption subprocesses by Einstein and Ehrenfest as providing first signs for a deeper understanding, i.e. suggesting a temporal order and a finite intermediate period of time. As he commented:

---

of independently occurring emission and absorption subprocesses in the context of the thermal equilibrium of radiation and matter.

[241] "Ich muß überdies im Sommersemester einen zusammenfassenden Quantenartikel für das Springersche Handbuch schreiben [...]." Pauli to Heisenberg, February 28. 1925, as cited in Pauli (1979, 211).

[242] "Der erwähnte Artikel [für das Springersche Handbuch] konvergiert Gott sei Dank seinem Ende zu und ich hoffe in 2 bis 3 Wochen von dieser Last endgültig befreit zu sein!" Pauli to Kronig, October 2. 1925, as cited in Pauli (1979, 247).

[243] See also Pauli (1926, 1, Footnote 1).

[244] "Während bei den ins Auge gefassten Streuprozessen zweiter Art [Raman scattering and double emission] die Atome sonst *direkt vom Zustand P [initial state] nach dem Zustand Q [final state] übergeführt werden*, geschieht dies bei naher Übereinstimmung der Frequenz ν der auffallenden Strahlung mit einer Absorptionsfrequenz ν_{RP} auf dem Umweg über den höher angeregten Zustand *R*." Pauli (1926, 94–95); emphasis added.

> The product representation of the frequency of scattering processes by Einstein and Ehrenfest suggests the assumption, *that the scattering process actually consists of two partial processes with the frequency* $b\rho_\nu$ *and* $(a_1 + b_1\rho_{\nu_1})$, *the second of which succeeds the first by a time of the order of the period of light* $1/\nu$. For such a conception, which was proposed by Herzfeld (1924) in connection with the change of the propagation velocity of light for dispersion in transparent media and which was discussed in more detail by [Smekal (1925b)], a few reasons should be put forward.[245]

Pauli essentially provided two arguments and an exemplary modelling which should back up this conception (Pauli 1926, 27–29).

The first argument aimed at the relation between absorption and scattering, the problem Pauli emphasized in his critique of the BKS theory. In classical theory, Pauli argued, emission and absorption are limiting cases of scattering and this relation should also be taken over into quantum theory. To exemplify his point of view, Pauli constructed an example of N atoms, capable of two quantum states 1 and 2 with the respective number of atoms in ground (N_1) and excited (N_2) state. This system should interact with radiation with the frequency of emission/absorption, called ν. Working exclusively with the Einstein coefficients for emission and absorption processes, Pauli constructed the equilibrium conditions in which, since the number of atoms in the excited state and the ground state remained constant in the equilibrium, the atoms only performed a scattering function:

> But in this case we are certain, that this [scattering] consist of two subprocesses; the absorption of the quanta by the atoms in the ground state and the re-emission of the quanta by the atoms in the excited state. The time between absorption and re-emission process is here on average given by $\frac{1}{A_1^2}$, the mean lifespan of the excited state.[246]

Now, as Pauli's argument went, this was only a mean dwell time and, in some cases, the re-emission will occur immediately after the absorption. Hence and

[245] "Die Einstein-Ehrenfestsche Produktdarstellung der Häufigkeit der Streuprozesse legt die Vermutung nahe, *daß der Streuprozeß tatsächlich aus zwei Teilprozessen mit den Häufigkeiten* $b\rho_\nu$ *und* $(a_1 + b_1\rho_{\nu_1})$ *besteht, von denen der zweite dem ersten innerhalb einer Zeit von der Größenordnung der Lichtperiode* $1/\nu$ *nachfolgt*. Für eine solche Auffassung, die von Herzfeld [Herzfeld (1924)] in Verbindung mit der Änderung der Fortpflanzungsgeschwindigkeit des Lichtes bei der Dispersion in durchsichtigen Körpern vorgeschlagen und die von Smekal [Smekal (1925b)] näher ausgeführt wurde, sollen hier einige Gründe vorgebracht werden." Pauli (1926, 26); emphasis in the original. Pauli included the references to Herzfeld's and to Smekal's article in footnotes.

[246] "Hier sind wir aber sicher, daß diese [Zerstreuung] aus zwei Teilprozessen besteht; der Absorption der Quanten durch die Atome im Grundzustand und der Reemission der Quanten durch die Atome im angeregten Zustand. Die Zeit zwischen Absorption- und Reemissionsprozeß ist dabei im Mittel durch $\frac{1}{A_1^2}$, die mittlere Lebensdauer des angeregten Zustandes, gegeben." Pauli (1926, 28).

> in principle, there is no way to distinguish between scattering processes and 'true/actual' ['wahren'] absorption and emission processes within a spectral line. In order to retain the connecction between emission and absorption on the one hand and scattering on the other in quantum theory, it seems natural to assume that the scattering processes *always* consists of two suprocesses.[247]

The mean dwell time outside of the spectral line should, following Herzfeld's argument, be of the order of $1/\nu$. The reasoning was, as Pauli pointed out, still provisional but according to the given conceptualization the atom should remain in an "unstable intermediate region"[248] between two stationary states for this short amount of time. For the time being and without further arguing for it, Pauli simply stated that also in the case of free electrons, where emission and absorption cannot properly be defined, we should uphold this decomposition.

The second argument Pauli provided was based on the conception of a static field as the limiting case of a time-varying field with very small frequency $\nu \rightarrow 0$ and the reaction of free electrons on this field. A classical electron in a static field experienced an ongoing force, was thus continuously accelerated. For a slowly varying field, this behaviour went over into a slow undulatory motion. Thus, Pauli argued, the partial emission and absorption subprocesses in quantum theory must behave in exactly such a way as to slowly go over into this harmonic oscillatory motion of classical theory in the case of slowly varying fields and into the continuously accelerated motion of the static case. Thus, the connection to the classical limiting case seemed to suggest a decomposition into emission and absorption subprocesses.

Pauli then made an interesting compromise to account for emission and absorption subprocesses in the case of free electrons. He explicitly stated that one cannot yet oversee which characteristics such partial processes should have. Energy and momentum exchange, such as was proposed initially by Einstein for absorption and emission, cannot in general be upheld. For a free electron, as was well-known, energy and momentum conservation cannot be fulfilled for emission and absorption. Pauli extended this idea: For any process in which the time between emission and absorption was of the order of the period of the emitted light, energy and momentum transmission as initially proposed could not be upheld.

Even though Pauli thus discussed both the Kramers-Heisenberg formula and its derivation as well as the approaches by Smekal and Herzfeld, the references amongst the two sections are rare and vague. If at all, they are mostly concerned with the resonant case, the one the least interesting for our purposes. The discussion of the Kramers-Heisenberg formula was of course much more prominent and took a lot more space. Yet, also the conceptions of Smekal and Herzfeld were argued for by Pauli.

---

[247] "[es] besteht prinzipiell keine Möglichkeit, zwischen Streuprozessen und 'wahren' Absorptions- und Emissionsprozessen innerhalb einer Spektrallinie zu unterscheiden. Um den Zusammenhang zwischen Emission and Absorption auf der einen und Streuung auf der anderen Seite auch in der Quantentheorie beizubehalten, scheint es daher naturgemäß, die Streuprozesse *stets* als aus zwei Teilprozessen bestehend anzunehmen." Pauli (1926, 28); emphasis in the original.

[248] "instabile Zwischengebiet," Pauli (1926, 28).

4.3.5 An Intermediate Summary

The discovery of the Compton effect led to a conceptual development that has so far found little attention in the historical literature on quantum theory: Physicists proposed to decompose the theoretical description of scattering and dispersion into emission- and absorption-like processes. Pauli's first treatment on the equilibrium conditions between free electrons and radiation, in which Compton scattering served as the basic elementary process, led to a conceptual riddle: According to Pauli's interpretation it seemed as if radiation components of different frequencies interacted.

This astonishing result was at first approached by Einstein and Ehrenfest: They formally decomposed the scattering mechanism into subprocesses of emission and absorption, each with its corresponding probability coefficients. Thereby, they could reconstruct Pauli's result and mitigate the initial conceptual riddle. But thereby they also introduced a new one, at least so in Pauli's eyes: Since energy and momentum conservation could not be upheld simultaneously when a free electron absorbed or emitted a light quantum, the respective probability coefficients were, according to Pauli, not physically defined.

While Pauli pondered about this problem, the integration of interference and dispersion into a light-quantum framework was approached in Munich by Wentzel and Herzfeld. Wentzel could establish a quantum-theoretical description of the phase for light quanta, thus found a way of describing interference. Herzfeld, in his attempt of treating dispersion with this new concept, consequently turned to the propagation of light quanta in transparent media. By reconfiguring the classical theory of dispersion and establishing a formalistic comparison with a quantum-theoretical description, Herzfeld arrived at a completely new conceptualization of the light-matter interaction: Non-fitting light quanta, i.e. quanta, the frequency of which did not correspond to a possible emission or absorption frequency of the medium, should be stopped in their passage for a short amount of time. The atom occupied an "intermediate state", lying energetically in between two stationary states and temporally between emission and absorption process. Through his re-interpretation of the classical formulas, Herzfeld could even provide an upper boundary for the temporal dimension of the occupation of these "intermediate states": It was proportional to $1/\nu$, where ν is the frequency of the impinging radiation, and thus of the order of 10^{-15} seconds for the visible range, much shorter than the mean dwell time for actual emission and absorption processes.

This new conceptualization was taken up favourably by two physicists: Adolf Smekal and Wolfgang Pauli. Smekal provided two conceptual alterations. On the one hand and in line with his conviction that no judgement can be made about the nature of radiation in the vacuum, he removed the light quanta from the corresponding model of dispersion and scattering. On the other hand, he connected Herzfeld's argument with the concept of weak quantization: The mean dwell time, as estimated by Herzfeld, was short enough to be reconcilable with the general description of Bohr's atomic model and the respective rules for quantization. Yet, neither Herzfeld nor Smekal substantiated the description of intermediate states by actually applying a mathematical representation of the mechanical model of the atom.

Pauli, who had favoured the light-quantum hypothesis over a wave-theoretical description of light, saw an important advantage of Herzfeld's description of dispersion and scattering over the BKS theory: It did not introduce any duplication in the description of the phenomena. Thus, Pauli used it to back up his criticism on the BKS theory. And even though he was well aware that the description of dispersion and scattering in the light-quantum framework was still provisional, he defended this model in his monumental article in the *Handbuch der Physik* of 1926.

Of course, these considerations did not make it into the mainstream of theoretical physics and were superseded by the development of matrix and wave mechanics. Yet, I included them in my discussion for two reasons: First of all, the idea that scattering processes might be dissected into actual emission and absorption processes, i.e. connected to corresponding Einstein probability coefficients, will resurface in the discussions around the Raman effect (see Sect. 6.3). Yet, the nature of the intermediate states in these discussions would considerably have changed, both in reference to the conceptual framework and their characteristics: Herzfeld's and Smekal's intermediate states were conceived of in the framework of a mechanical model (although not spelled out mathematically) of the atom and they laid both temporally and energetically in between initial and final state. Energy was conserved in the transition processes. All of these aspects, were altered by Yakov Frenkel who reintroduced the idea that scattering could be conceived of as temporally separated and independently occurring emission and absorption subprocesses in mid 1929.

Secondly, the discussed theoretical attempts to describe scattering and dispersion working with the light-quantum hypothesis provide for an informative contrast to the dispersion-theoretical accounts followed in Breslau, Göttingen, and especially Copenhagen. In the following conclusion, I will first summarize the development of the respective strands of research and then turn to a discussion of their relevance for the historical development of the virtual particle concept.

4.4 Summary and Analysis: A Virtual World, Real Transitions, and Their Formats of Representation

The problem of the light-matter interaction as well as dispersion and scattering as their empirical focal points proved to be important obstacles for the conceptualization and methods of the old quantum theory. Although the route to Heisenberg's 1925 breakthrough would eventually lead through the intensity calculations of the Zeeman effect, important (and less important) conceptual innovations took place in attempts at explaining dispersion and scattering. I separated two lines of attempts at explaining these phenomena: in the first one, virtual classical entities were included; in the second one, scattering was conceived of as composed of actual and energy conserving emission, absorption and transition subprocesses. The watershed for both approaches was the year 1923.

4.4.1 The BKS Theory, Its Virtual Entities, and Their Relevance for Theoretical Progress

The BKS theory (Sect. 4.1) had its roots in thoughts that Bohr had entertained for quite a while and culminated, with the import of some recent developments on dispersion theory, in an harmonious amalgamation in order to provide for a theoretical explanation of the light-matter interaction. It rested on the conception that the atom rested in one stationary state during the interaction with light. That this description emerged in a theoretical tradition that rested on the correspondence principle is rather unsurprising: In its core, the correspondence principle provided a connection between the movement of the electrons in the atom and the radiation emitted. The harmonics of the classical movement or more to the point, the square of their quantum-theoretical counterparts, were identified with the Einstein coefficients which stood in for the transition probabilities.

The concept of virtuality in BKS had, according to my reconstruction (Sect. 4.1.3), two interrelated connotations. First of all, it referred to an *unobservable*, nearly classically governed and mainly model-like world, inhabited by virtual oscillators and virtual radiation. This world interacted with the actual and observable atoms in quantized states only statistically, i.e. by inducing transition probabilities. This non-classical interaction was the second reason for the terminology, which not only historians of science, but also contemporaries of Bohr, Kramers and Slater have already pointed to. Energy non-conservation was actual for BKS and the concept of virtual transitions was not included in this conceptual framework. Quantum entities were real; virtual ones were wave-theoretical.

In hindsight, I already identified four aspects of the BKS framework and the corresponding virtual entities that will become important in the further development:

1. The semi-classical concept of a virtual oscillator will survive the demise of the BKS theory as the matrix elements of the position operator or as a replacement of the atom by a classical dipole. But I do not believe, as seems to be suggested in some of the secondary literature, that there is any generative connection between virtual oscillators and virtual transitions (see my discussion in Sect. 5.2).
2. The concept of an unobservable, or at least unobserved world will be taken up in the mid 1930s by Werner Heisenberg when he referred to "virtual possibilities" and thereby led to a further spreading of the terminology in QED.
3. The prominence of energy non-conservation as well as the terminology of the virtual in BKS, although at best indirectly connected, could have led to a first invocation of the term "virtual transitions" by Chandrasekhara Venkata Raman as late as 1929.

4. From a more general point of view, the virtual oscillators had an important characteristic, which allowed for the description of dispersion within BKS and was regularly stressed by the historical actors: They stood in for all *possible* transitions of the atomic system, unbound by energy conservation. Thus, and this will become a common characteristic of all virtual entities we will encounter, they extended the set of possible processes and made this new set of possibilities, although not actualized, effective.

Although the meaning of the term virtual in BKS was in many respects different from the one we mostly have in mind today, and which was referred to by the historical actors after the terminology had been accepted in QED and QFT, there is a historical connection.

In the further development (Sect. 4.2), there were two prior suggestions by historians of science where the concept of a virtual transitions was applied: the Kramers-Heisenberg paper and Heisenberg's *Umdeutung* paper. I think these judgements are partially correct. Kramers and Heisenberg provided diagrammtic and mathematical representations which can easily be read as describing virtual transitions. And, indeed, the derivation of the Kramers-Heisenberg formula will be the point at which Paul Dirac will introduce a verbal representation of successive transitions through intermediate states. But neither Kramers and Heisenberg (1925) nor Heisenberg (1925) would do so. This is not only due to the conceptual framework of the BKS theory, or the idea that the atom rested in a stationary state during the interaction with light, which can still be found in Heisenberg's published contribution. Looking closely at the methods and representations Kramers and Heisenberg provided and used, I argued that the idea of successive transitions did not guide their evaluation. Certainly, it was not necessary. Otherwise, they would have reproduced a verbal statement somewhere in the paper. On the contrary, the idea that the atom rested in a stationary state became part of the argumentative structure. A conception of successive energy non-conserving transitions through intermediate states had not yet acquired a tool character. The concept of a "virtual transition", i.e. a virtual quantum entity, had not yet been fully developed.

And even though Heisenberg had definitely gathered some intuition about how to handle the combinations of what Kramers and Heisenberg had called characteristic amplitudes and the mathematical representation of such formed the basis for his *Umdeutung*, he did not explicate them verbally as transition subprocesses. Why? Because, my argument went, the concept of a virtual transition was simply not yet developed and the historical actors were certainly aware that the amplitudes were not equivalent to transition probabilities. But there is yet another reason: The explicit invocation of actual transition processes to short-lived intermediate states had been developed in a different approach.

4.4.2 Dispersion, Scattering, and the Light-Quantum Hypothesis

After the discovery of the Compton effect, the question of how to explain dispersion and scattering within a light-quantum framework and the corresponding methodology of equilibrium considerations was first discussed by Wolfgang Pauli. But Pauli came to a seemingly perplexing conclusion: Light of different frequencies interacted. This seemed counter-intuitive and the problem was taken up by Einstein and Ehrenfest (Sect. 4.3.2). They modelled the Compton effect even more closely after Einstein's initial 1916 paper in which he had introduced the probability coefficients for spontaneous and induced emission as well as absorption. In their paper, Einstein and Ehrenfest showed that the process could formally be decomposed into emission and absorption subprocesses. All the same, a temporal dimension was not implied by Einstein and Ehrenfest. At first, Pauli did not consider this formal decomposition to be meaningful, as the partial emission and absorption subprocesses were not physically defined in the prime case of interest, the free electron. By late 1924, Pauli seemed to have changed his mind due to approaches followed in Munich.

Building upon Wentzel's idea of integrating a phase into a light-quantum framework, Herzfeld elaborated on dispersion and scattering in transparent media (Sect. 4.3.3). According to Herzfeld's attempts, it seemed reasonable to assume that "non-fitting" light quanta, i.e. with a frequency not equivalent to an absorption frequency, would be stopped by the atoms on their path through matter. The atom would be in an "intermediate state", lying both energetically and temporally in between the initial and final state of the atom. Energy conservation was fulfilled in such a process and Einstein coefficients introduced for the subprocesses. Adolf Smekal took up this idea and developed it a little further. He argued that the short lifetime of the intermediate state, which Herzfeld had obtained from a formal comparison with and re-interpretation of the classical formula, was a striking example of what came to be known as weak quantization: When the dwell time of a state is short, it was known that the regular quantizaton rules could not be applied. Thus, the intermediate states could be explained within the mechanical model of the Bohr atom. But neither Herzfeld nor Smekal concretized such ideas by explicit calculations of atomic orbits.

Pauli, who had meanwhile already taken up Herzfeld's ideas in his critique of the BKS theory, had recognized some benefits of this approach. In his famous article in the *Handbuch der Physik* of 1926 (Sect. 4.3.4), he actually defended the idea that scattering should be conceived of as composed of independent subprocesses of emission and absorption. Still he acknowledged that one would probably have to assume that this modelling might come with a redefinition of the characteristics of such emission and absorption processes, since energy and momentum could not simultaneously be conserved in the emission or absorption of a quantum by a free electron. All the same, he also took up the Kramers-Heisenberg formula, gave it considerably more room, and finally retracted his earlier statements that dispersion occurred, in this mode of representation, due to "a detour" through a third state.

4.4.3 Virtual Entities and Actual Transitions Through Intermediate States, But in Different Conceptual Frameworks

Where does this leave us and the state of the concepts we are interested in at mid 1925? In the BKS conception (but stretching to Heisenberg's 1925 *Umdeutung* paper) the atom was conceived of as resting in one stationary state during the interaction with radiation. A set of theoretical entities, the virtual oscillators, were introduced to mediate the interaction with the classically conceived of radiation field. Even in cases for which we retrospectively but still nearly unavoidably read virtual transitions and intermediate states into the formulas and diagrams, i.e. the Kramers-Heisenberg paper, the historical actors did not provide verbal explication in terms of transition subprocesses. The concept of a virtual transition was expressed, at best, mathematically and diagrammatically. Within the correspondence principle tradition, from which quantum mechanics will eventually emerge, the concept of a virtual transition was at best implicitly developed.

On the other hand, suggestions were put forward to conceive of the scattering processes as two independent, actual, and energy conserving transitions. Within a light-quantum framework, such a conception seemed to have occurred more natural than within the correspondence tradition. All the same, this development did not have an important influence on the further development of quantum theory and thus is also very unlikely to have influenced Paul Dirac, who gave the first verbally explicit representation of transitions through energy non-conserving intermediate states.

To understand these diverging approaches and conceptualizations, and to make understandable the fact that a modified form of the verbal description of the light-quantum approach, *although conceptually far off from Semkal's and Herzfeld's attempts and without any generative connection*,[249] will be introduced into the mainstream of quantum theory by Paul Dirac in 1927 (see Chap. 5), we have to take into account both the conceptual framework but also the corresponding methods the historical actors applied.

From a conceptual point of view, the light-quantum hypothesis seemed to have lent itself far more easily to a conception of successive emission and absorption subprocesses and transitions. Starting with their introduction into quantum theory, the light quanta were entities "which move without dividing and can only be absorbed

[249] I really want to emphasize this point here. There is not only no generative relation between what I call Dirac's verbal model and the attempts of Smekal and Herzfeld. But the transitions to intermediate states of Dirac and the transitions to intermediate states of Smekal and Herzfeld are conceptually distinct entities which have next to nothing in common, contrary to what the name suggests.

and created as a whole".[250] The basic conceptual building blocks were emission and absorption processes. Viewed from the mathematical representational system, this is mirrored by the Einstein coefficients being first of all defined for (modern terminology) absorption, induced and spontaneous emission. For the equilibrium considerations the actors in this line nearly exclusively engaged with (starting with Einstein and Ehrenfest but stretching to Smekal's 1925 papers), the decomposition into emission and absorption subprocesses was not only more natural in terms of the conceptual building blocks they worked with; also in reference to the mathematical modelling, i.e. the Einstein coefficients on which such considerations were built, such a decomposition was rather suggestive.[251]

The Breslau/Göttingen/Copenhagen approaches rested primarily on the wave theory of light and a classical, mechanical model of the atom, which was then translated into quantum theory. According to the conceptual framework, the atom rested in a stationary state during dispersion and scattering, while this conception was even extended to absorption and emission within the BKS framework. Transition processes and the light-matter interaction were causally separated, they were only probabilistically connected. A wave-theoretical, unobservable world of vibrating oscillators and radiation was the cornerstone of these descriptions. The entities we can conceive of as virtual transitions were conceived of as a double sums over virtual oscillators replacing the reaction of the actual atom for the purpose of the theoretical description. Importantly, the amplitudes of these entities were not equivalent to Einstein coefficients, but only their squares.

Thus, in contrasting these approaches we can understand the differences in the respective verbal explication of transition processes for scattering and dispersion both emerging from their respective theoretical and conceptual traditions, but also from their concrete mathematical instantiations: In the light-quantum framework actual Einstein coefficients could consistently be introduced for the subprocesses, while the wave-theoretical descriptions introduced theoretical and unobservable entities, the amplitudes of which were not equivalent to Einstein coefficients. Since we know, that Kramers and Heisenberg were aware of Wentzel's approach, and thus most probably also informed about Herzfeld's explanation of dispersion,[252] we can identify

[250] "welche sich bewegen, ohne sich zu teilen und nur also Ganze absorbiert und erzeugt werden können." Einstein (1905, 133).

[251] As was proved by Smekal and Herzfeld, their conception did not disturb the thermal equilibrium between matter and radiation.

[252] This is even more likely as Pauli used Herzfeld's conceptualization in a letter to Bohr to argue against the BKS theory.

yet another reason why Kramers and Heisenberg might not have spoken about transitions: They were not only aware that the theoretical entities they handled were not transitions, there was another rival theory which worked with actual transitions explicitly.

The concept of virtual transitions as a description of scattering processes was not yet completely formed and the one theory which explicitly referred to scattering as a two-stage process of successive transitions would eventually prove inadequate. What physicists considered to be transitions in quantum theory in mid 1925 were in modern terminology actual transitions. But even if the actors referred to transitions, they mostly referred to their probabilities. Virtuality was only connected to wave-theoretic and nearly classical concepts. Quantized states and transitions between them were real, not virtual. It needed quite a far reaching re-definition of what a transitions is, it needed to be conceived of on a quantum-amplitude level, and in 1925 no one did do so, at least explicitly.[253]

As I will argue in the following chapter, it took the boldness of the young Paul Dirac, the development of time-dependent perturbation theory, the definition of the photon concept on a quantum-amplitude level, and finally the necessity of introducing a verbal representation of successive transitions to come into contact with the Kramers-Heisenberg formula to make transitions a concept of the quantum-amplitude level: Only when the verbal explication of successive energy non-conserving transitions acquired a tool character, it was explicitly spelled out and made transitions a quantum concept.

[253] Again, this does not mean that the formal connections between what we conceive of as quantum amplitudes were not clearly perceived. As argued, Heisenberg had acquired quite an intuitive handling of these entities. But, according to my reconstruction, accrediting Heisenberg with the development of the concept of virtual transitions is as false and as true as accrediting him with the concept of a matrix element.

Chapter 5
Dirac's Verbal Model: Making Transitions a Quantum Concept (1927)

In the previous chapter, I investigated two different lines of theoretical research of the old quantum theory and how the historical actors respectively described the light-matter interaction, or more to the point, scattering and dispersion. As argued (Sect. 4.1), we might retrospectively identify the diagrammatic and mathematical representations Kramers and Heisenberg (1925) or Heisenberg (1925) worked with as virtual transitions (modern terminology). Yet, they did not give any verbal description of these structures, at least not in terms of successive transition processes. This chapter will describe the origins of a verbal model for such entities.[1]

In the old quantum theory, BKS (Sect. 4.1) had introduced a virtual, unobservable, but purely wave-theoretical world. Yet, they did not introduce something like "virtual transitions": Only classical concepts, oscillators and radiation field, were connect to the term "virtual". Transitions, quantized states, and energy non-conservation were real. On the contrary, physicists working within a light-quantum framework (Sect. 4.3) in the old quantum theory explicitly described the scattering process as being composed of successive emission and absorption processes through intermediate states. But these "intermediate" or "meta-stationary" states lay in between the regular energy levels. The transition, emission and absorption processes were real and energy conserving ones, i.e. directly connected to Einstein coefficients. All the same, in the main line of research comprising matrix and wave mechanics as well as other semi-classical approaches, a verbally explicit description in terms of successive acts of emission and absorption processes connected to energy non-conserving transitions was neither given in the published papers nor in the correspondence I know of until 1927.

Corresponding to the analytical framework I introduced in Sect. 2.3, physicists had a mathematical and a corresponding diagrammatic representation of virtual transitions, at least for the description of the material part of the system. But a verbal

[1] The main part of this chapter (mainly its introduction (slightly modified) and Sect. 5.1) has already been published as Ehberger (2022, Sect. 1, pp. 536–554).

M. Ehberger, *Representing the Unobservable*, Science Networks. Historical Studies 68,
https://doi.org/10.1007/978-3-032-09188-8_5

representation was nowhere to be found in the research strand leading up to matrix and wave mechanics.[2] Surely, the connection between the characteristic amplitudes and transitions was well-known and clearly perceived, but this did not entail that these concepts were equivalent, i.e. that such equations as the Kramers-Heisenberg formula were cast in verbal expressions describing successive transition processes. Specifically, I argued that the mental imagery of successive, energy non-conserving transitions seemed not to have acquired a tool character in the theoretical evaluations.[3]

After matrix and wave mechanics as well as their equivalence had been established in the years 1925–1926, the interpretational debate around the quantum-theoretical formalism picked up pace. In this debate, physicists did not speak about scattering or dispersion in general, and the Kramers-Heisenberg formula in particular, in terms of successive (virtual) transitions.[4] This might at first sight seem all the more astonishing, as this formula had become, as Lacki et al. (1999, 462) phrased it, "un passage obligé" for all following versions of quantum mechanics: Born et al. (1926) (matrix mechanics), Schrödinger (1926b) (wave mechanics), and Klein (1927) all rederived this formula in the respective theoretical and conceptual framework without explicitly framing the formula in terms of successive transitions.

Schrödinger's non-invocation of a verbal description in terms of transitions, for example, is easily understandable: Schrödinger held that the wave function described an actual and vibrating charge configuration. For the incoherent scattering, Schrödinger thought that both initial and final state had to be occupied at the same time. In this view, neither the transition to the intermediate state is performed, nor the one to the final state.[5]

But also for matrix mechanics and the so-called "correspondence-like" approach by Klein (1927), it was always clearly stated that matrix elements do not directly refer to transitions or even their probabilities. Only the squares of the elements of the position operator referred to Einstein coefficients. Hence, there was no direct identification between the mathematical terms describing what we today interpret as virtual transitions and the transitions the historical actors referred to.

As Konno (1993, 161) has observed, "the real solution to these problems was given by P.A.M. Dirac. In his quantum theory of radiation [...], dispersion is regarded as one successive process; an atom absorbs radiation and simultaneously emits it via an

[2] The one exception, Wolfgang Pauli's reference to the "detour" of the atom through a third state, was, as discussed, retracted in his article for the *Handbuch der Physik*.

[3] At least the explanations given in the Kramers-Heisenberg paper (see Sect. 4.2) did not rely on such a picture.

[4] At least I did not find any explicit verbal statements in the most important papers of the time and the correspondence I consulted. If someone had already entertained this description, it was certainly not in the mainstream of the conceptualization of the scattering process.

[5] For Schrödinger's own description of his conception of the Kramers-Heisenberg formula, see Schrödinger (1926b, especially pp. 121–123). In retrospect, Heisenberg (1946) described the Raman effect as an experimentum crucis for the different conceptions of the wave functions at the time. But Schrödinger accepted a probabilistic interpretation of the latter already before the empirical observation of the Raman effect.

intermediate state." kragh (1990, 126) has similarly observed that "Dirac introduced a concept of great importance, later known as the concept of 'virtual states'." And indeed, in his second paper on his radiation theory (Dirac 1927b, communicated in April 1927), sort of the prototype of QED during the 1930s, Dirac introduced his reader to the scattering of light from an atom in the following way:

> [...] radiation that has apparently been scattered can appear by a double process in which a third state, n say, with different proper energy from m [final state] and k [initial state] plays a part. If initially all the b's [the quantum amplitudes of the states] vanish except b_k, b_n [the amplitude of the intermediate state] gets excited on account of transitions from state k by an amount proportional to v_{nk} [the matrix element of the interaction between radiation field and matter], and although it must itself always remain small, a calculation shows that it will cause b_m to grow continually with the time at a rate proportional to $v_{mn}v_{nk}$. The scattered radiation thus appears as the result of two processes $k \to n$ and $n \to m$, one of which must be an absorption and the other an emission, in neither of which is the total proper energy even approximately conserved.[6]

As discussed, such a verbally explicit and temporally ordered account of the scattering of light in terms of successive energy non-conserving or "virtual" (modern terminology) transitions to and from intermediate or "virtual" states (again, modern terminology) was (nearly) absent from the papers dealing with this effect.

But if quantum physicists did not describe their mathematical formulas of dispersion in terms of successive transition processes, why did Dirac eventually, and contrary to prior (and some later) descriptions, start to talk in a specific way about the structures he encountered? More to the point: Why did Dirac choose not only to name the structures, but to describe them in terms of temporally ordered subprocesses at least suggesting some kind of causal connection between them and thereby setting "the basic language and concepts characteristic of the modern conception [...]" (Lacki, Ruegg, and Telegdi 1999, 484)? In order to engage with this question, I will follow Dirac's path towards his account of dispersion within a second quantized radiation field.

The main technical and conceptual ingredients we need to understand Dirac's invocation of an explicit verbal representation of the model of the scattering processes (or in short the verbal model) are his time-dependent perturbation theory and the respective conception of the development of a system (Sect. 5.1.1), his radiation theory which worked with a quantized radiation field, i.e. the transversal part was described by photons (Sect. 5.1.2), and a way to arrive at sensible results in quantum electrodynamics which is today known as Fermi's Golden Rule No. II (Sect. 5.1.3). Having discussed the relevant aspects of these methods and conceptions leading up to Dirac's theory of dispersion, I will show how Dirac used the verbal model in his radiation theory to reconstruct the Kramers-Heisenberg formula, i.e. he applied the verbal model and the virtual transitions (modern terminology) as a tool (Sect. 5.1.4). Then I will turn to the question why it was specifically Dirac, and not another physicist, who introduced this kind of verbal description (Sect. 5.1.5). And, finally (Sect. 5.2), I will revisit the (dis-)connections between Dirac's verbal model, including what later on

[6] Dirac (1927b, 712). The notation has been changed slightly in comparison to the original.

will be called virtual transitions, and the research strands of the old quantum theory in general and the BKS theory as well as the virtual oscillators in particular.

5.1 Dirac's Technical and Conceptual Background and His Verbal Model

The output of Paul Dirac in the years 1925 through 1928 was, to say the least, outstanding. Dirac made a long lasting contribution to quantum theory in these years: from an alternative formulation of matrix mechanics (his q-number algebra); through his technical foundation of quantum mechanics (his transformation theory) and the first applications of quantum electrodynamics; to his relativistic description of the electron, just to name the most influential of his achievements. By early 1926, he had made a name for himself in the community of quantum physicists and his work had gathered wider attention.[7]

The following contextualization will focus on the aspects of Dirac's work which I deem important to understand his invocation of the verbal model. One of the most important aspects of his technical framework is his time-dependent perturbation theory. But, as I will argue, also the conceptual framework of his radiation theory and the specific way of its perturbative evaluation played an important role for his invocation.

5.1.1 Time-Dependent Perturbation Theory (Dirac 1926)

In mid 1926 and after some persuasion by Werner Heisenberg,[8] Dirac took up Schrödinger's wave function in his *On the Theory of Quantum Mechanics* and developed a version of time-dependent perturbation theory. In this mode of evaluating the formulas, the question was not how the energy levels of the system were altered due to the perturbation but how the occupation of unperturbed states changed over time.[9]

As was common for Dirac, he treated the problem first in most general terms.[10] Dirac evaluated an arbitrary system, described by the Hamiltonian H_0, and considered the respective problem to be solvable. Then he introduced a time-dependent external perturbation V starting to act at some moment $t = 0$. Using the most general solution,

[7] For Dirac's scientific biography, Kragh (1990) is still the standard reference. Although popular in nature, Farmelo (2009) is also a well-researched and valuable source for biographical details.

[8] See for example Hragh (1990, 32).

[9] The same procedure was independently and further developed by Born (1927). Slater (1927b) came up with this technique as well.

[10] The relevant passages can be found in Dirac (1926, 673–675).

a superposition of eigenstates $\psi = \sum_n c_n \psi_n$, Dirac showed that the wave function of the perturbed system could be expressed in the following way:

$$\psi(t) = \sum_n b_n(t)\psi_n \tag{5.1}$$

where the ψ_n are the unperturbed wave functions and the b_n, the new expansion coefficients, are time-dependent. Their initial values $b_n(0)$ are given by the coefficients of the unperturbed problem and their temporal development is governed by the following equation:

$$i\hbar\dot{b}_m = \sum_n V_{mn} b_n \tag{5.2}$$

where the V_{mn} are the matrix elements of the perturbation between the two unperturbed states m and n.

The unperturbed states of the system referred to energy eigenstates of the atoms under study. Yet, as Dirac was not interested in the behaviour of a single atom but an assembly of similar ones, he did not normalize the sum of the squares of the coefficients c_n and $b_n(t)$ to one but rather to the number of atoms in the respective state. Hence, in Dirac's initial presentation $\|b_n(t)\|^2$ corresponded to the number of atoms in the state n after the perturbation had been acting on this assembly for a time t.

Whether one follows Dirac's conception of an assembly of atoms or interprets the square of the coefficients as the probability of finding one atom or system in a given state,[11] this perturbative scheme came with a particular understanding: In contrast to time-independent perturbation theory, Dirac did not ask about the influence of the perturbation on the energy levels of the system. Rather, according to the mathematical modelling, the focus was on the temporal development of the coefficients of the unperturbed states. Which states coupled depended on the structure of the perturbation, i.e. which matrix elements of the perturbation V_{mn} were non-zero. The perturbation energy was further not directly part of the system's energy, but caused alterations in its behaviour. In his papers on radiation theory, Dirac would therefore introduce the term "proper energy" to refer to the energy of the unperturbed part of the system (or, equivalently, to the energy of the total system minus the interaction energy).

Dirac applied his new way of performing perturbative calculations directly in the same paper for a derivation of the Einstein coefficients for induced emission and absorption (Dirac 1926, 675–677). The specifics of this semi-classical evaluation are not important for our purpose. But there is a methodological step which Dirac incorporated here that I will refer to later on. He already indicated how a second order perturbative calculation must be carried out: A first approximation is derived by plugging in the initial values c_n into Eq. (5.2) and then integrating it with respect

[11] See for example Born (1927) or Dirac (1927a) for such a conceptualization.

to t. For a second approximation, these time-dependent values are re-introduced into Eq. (5.2). Hence, Dirac essentially described an iterative procedure for the construction of higher order terms.

5.1.2 Radiation Theory (Dirac 1927b, c)

Shortly after finishing the paper introducing his version of time-dependent perturbation theory, Dirac went to Copenhagen (September 1926 until early February 1927) and subsequently to Göttingen (until June 1927). Although Dirac was ever more strongly integrated into the circles of quantum physicists, he kept his habit of mostly working on his own. In Copenhagen, Dirac developed the so-called transformation theory,[12] which provided rules for quantizing any dynamical system and provided the formal basis for Heisenberg's uncertainty paper. The transformation theory was also one of the pillars of Dirac's radiation theory, which he developed in Copenhagen and which we shall address now.[13]

Essentially, when constructing quantum electrodynamics, Dirac applied quantization to the wave functions, or rather to the coefficients of his perturbation theory, again. Dirac started out with an assembly of bosons and showed that the temporal development of the coefficients could be cast in an Hamiltonian form with the canonical variables b_r and $b_r^\dagger$. As before, N_r, the number of the systems, was given by $N_r = \|b_r\|^2 = b_r b_r^*$.[14]

Then, to use Dirac's language, he did not treat these coefficients as c-numbers but as non-commuting q-numbers and imposed commutation relations on them, so that

$$[b_r, i\hbar b_s^\dagger] = i\hbar\delta_{rs} \tag{5.3}$$

[12] From a different perspective, the same theory was independently developed by Pascual Jordan at approximately the same time. For a historical reconstruction of Jordan's work, see Duncan and Janssen (2009).

[13] In the following, I will not go into too much detail about the derivation of what came to be called second quantization but rather pick out the aspects that are, in my opinion, necessary for the comprehension of Dirac's conceptualization of dispersion in this framework. For detailed studies of Dirac's development of radiation theory, see Jost (1972, Sects. 1–4), Darrigol (1986, 225–229), Lackis et al. (1999, Sects. 4.1–4.2), Bromberg (1977), Kragh (1990, Chap. 6), Schweber (1994, Sect. 1.3), or Pais (1986, Sect. 15c). For a historically sensitive modern reconstruction, see Duncan (2012, Sect. 2.1). My own reflection will most closely follow the paper by Lackis et al. (1999) due to the shared focus on perturbation theory.

[14] Dirac worked with the transformed variables $b_r = a_r e^{-iW_r t/\hbar}$ and $b_r^\dagger = a_r^\dagger e^{iW_r t/\hbar}$. But this is of no importance for the conceptualization I am concerned with.

These operators b_r and $b_r^\dagger$ are today known, and were interpreted by Dirac, as creation and annihilation operators. Their action was expressed in the following terms (Dirac 1927c, 252)[15]:

$$b_r^\dagger \psi(N_1, ..., N_r, ...) = \sqrt{N_r + 1}\psi(N_1, ..., N_r + 1, ...) \tag{5.4}$$

$$b_r \psi(N_1, ..., N_r, ...) = \sqrt{N_r}\psi(N_1, ..., N_r - 1, ...) \tag{5.5}$$

Hence, they raised or lowered the occupation number in the respective state r by 1.

Dirac could generalize this procedure: The second quantized bosonic assembly was coupled to an external perturbation, an atom. The Hamiltonian thereby constructed consisted, besides the proper energies, only of one additional term proportional to the product of each one annihilation and one creation operator: $H \propto H_0 + \sum_{r,s} v_{r,s} b_r^\dagger b_s$. Hence, it "will contribute only to those matrix elements that refer to transitions in which N_r decreases by unity and N_s increases by unity" (Dirac 1927c, 252). Without any detour through intermediate states, an initial photon was absorbed and the final photon created. In the same paper, Dirac already referred to such processes as "direct scattering processes" (Dirac 1927c, 263).[16]

In the last section, Dirac left his initial conceptualization of a perturbed assembly of bosons and turned towards "the wave point of view" (Dirac 1927c, 262). By resolving the radiation into its Fourier components, the interaction between radiation and matter was, in classical theory, proportional to $\sum_r A_r \dot{x}$, where A_r is the rth Fourier component of the vector potential and x the position variable at the location of the atom times the electric charge. After some manipulation, Dirac could describe the vector potential in terms of the number of light quanta N_r and the conjugate phases θ_r, which allowed him to impose the previously developed quantization rules.

The Hamiltonian Dirac derived in this "wave-theoretic" way only contained matrix elements which changed the occupation number by ± 1. As he himself noted, "it would seem that there are no direct scattering processes [from this wave-theoretic point of view], but this may be due to an incompleteness in the present wave theory" (Dirac 1927c, 263). Since the emission and absorption coefficients could now be treated by this Hamiltonian and the two points of view led to the same Hamiltonians, the missing direct scattering term in the wave point of view set aside for the moment, Dirac still concluded that "there is thus a complete harmony between the wave and light-quantum descriptions of the interaction" (Dirac 1927c, 245).

Hence, in his first paper on quantum electrodynamics, Dirac had already established the concepts of creation and annihilation operators acting on the occupation number of the wave functions. Furthermore, he had constructed an interaction Hamiltonian that consisted of a product of such operators. It was responsible for the "direct

[15] Dirac actually worked with number and the conjugated phase operators which have the same action on the wave functions. They are connected to the creation and annihilation operators by $b_r = \sqrt{N_r + 1}e^{-i\theta_r/\hbar} = e^{-i\theta_r/\hbar}\sqrt{N_r}$ and $b_r^\dagger = \sqrt{N_r}e^{i\theta_r/\hbar} = e^{i\theta_r/\hbar}\sqrt{N_r + 1}$.

[16] To treat emission and absorption processes with the above Hamiltonian, Dirac envisioned an infinity of photons in an unobservable zero state. For a discussion of this procedure, the interested reader is referred to the papers cited in Footnote 13.

scattering of light". Before turning to its counterpart, the scattering through intermediate states, I will revisit a technical aspect of the perturbative evaluation of Dirac's radiation theory as it provides important insights into Dirac's reasoning process.

5.1.3 *"Dirac's Mogelei" Or: How to Arrive at Sensible Results in Radiation Theory?*

Today, Dirac's method[17] is known as Fermi's Golden Rule No. 2.[18] During the late 1920s and 1930s, it did not necessarily have a name. Once, it was discussed in private communications between Werner Heisenberg and Wolfgang Pauli as "Dirac's Mogelei".[19] To find some balance between the "Golden Rule" and the "Mogelei", I will call it Dirac's trick. This procedure allowed Dirac and the physicists in his follow-up to construct probabilities in radiation theory that were proportional to the square of matrix elements and linear in time, as was expected by the actors.

Dirac's trick was actually one of many intricacies of the perturbative evaluation of radiation theory,[20] but studying it and its derivation will provide important insight into Dirac's way of reasoning in connection with dispersion theory.[21] Let's follow Dirac and discuss Eq. (5.2) in most general terms. We can simply integrate the equation with respect to t and assume that we know that the system was initially in a specific state $b_n(0)$. This results in

$$b_m(t) = v_{mn} b_n(0) \frac{1 - e^{i(W_m - W_n)t/\hbar}}{W_m - W_n} + \delta_{mn} \tag{5.6}$$

[17] It was independently discussed by Wentzel (1927).

[18] This is most probably due to the prominence of this rule in Fermi's lecture on nuclear physics from 1949/50 (Fermi 1950). Golden rule No. 1 is the application of second order perturbation theory when the first order matrix elements vanish.

[19] This may roughly be translated as "Dirac's cheat," while "mogeln" has in German a childish touch. Judging from Heisenberg's response to a lost letter from Pauli, it seems as if Pauli suspected some hidden averaging over the phases of the wave functions in Dirac's methods. Heisenberg came to a different conclusion: "In the course of this letter, in contrast to what I expected, I came to the result, that there is no cheat in Dirac's paper." (*Original*: "Ich [bin] im Laufe dieses Briefes, im Gegensatz zu dem, was ich im Beginn glaubte, zu dem Resultat gekommen, daß bei Dirac gar keine Mogelei steht.") Heisenberg to Pauli, June 13. 1928; as cited in Pauli (1979, 460–462).

[20] For example, Dirac applied time-dependent perturbation theory to an Hamiltonian actually independent of the time; or: Dirac conceptualized the interaction between radiation and matter as a perturbation switched on at some point $t = 0$.

[21] I will follow Dirac's description of the trick as he applied it in the later part of his paper on emission and absorption and his paper on dispersion. For the general derivation see Dirac (1927c, 257–259) from which some of the quotes are taken.

As Dirac noted, as long as the proper energies of the states m and n "differ appreciably", the amplitude of the state m varies periodically with time and is small, i.e. "these stationary states are not excited to any appreciable extent" (Dirac 1927c, 258).

While it might seem strange at first that states with energies different from the initial state could be excited, the real problem occurred when Dirac tried to simply look at the transition amplitudes to a state whose energy was exactly equal to the energy of the initial state. Then the coefficient b_m became proportional to t.[22] But since only the absolute square of this coefficient referred to the probability of finding the system in the respective state, this would render a quadratic time dependence. And this ran counter to any expectation. Dirac's trick was a solution to exactly this problem.

As Dirac noted, the probability of finding the system in a state with exactly the same energy "is of no importance, being infinitesimal" (Dirac 1927c, 258). Rather, he proposed to multiply the absolute square of this coefficient with the density of states $(\Delta W_m)^{-1}$ around the energy value of the final state and to integrate over this energy range. Expressed in formulas this meant that

$$P_m(t) = \int \frac{dW_m}{\Delta W_m} \|b_m(t)\|^2 = 2 \int \frac{dW_m}{\Delta W_m} \|v_{mn}\|^2 \frac{1 - cos[(W_m - W_n)t/\hbar]}{(W_m - W_n)^2} \tag{5.7}$$

Dirac now substituted $x = \frac{(W_m - W_n)t}{\hbar}$ and pushed t to infinity.[23] The point was, that by pushing t to infinity (which is the formalization of considering "large" t), the energy peaked around the value of the initial state (see Fig. 5.1 for Dirac's representations in his calculational notes)[24] and the integral could be evaluated to give π. This effectively rendered

$$\lim_{\text{large } t} P_m(t) \propto \frac{t\|v_{mn}\|^2}{\Delta W_m} \tag{5.8}$$

where $W_m = W_n$, the proper energy of the initial state. Hence, a transition probability between states of the same energy, proportional to t and the square of the respective matrix element was obtained.

In summary, there are essentially three things to note for Dirac's later conceptualization of the scattering of light when it comes to this procedure. First of all, Dirac

---

[22] One easily sees this by applying de l'Hôpital's rule for the value $W_m - W_n \approx 0$ twice on Equation (5.6).

[23] This might seem strange as Dirac's procedure was, from the start, only valid for short times. The restriction Dirac imposed for the method to give reliable results were: (a) the time had to be large in comparison to what Dirac called "the periods of the atom", i.e. the reciprocal absorption and emission frequencies; (b) the time had to be small in comparison to the life time of the states considered. According to Dirac "[t]here is no difficulty in T satisfying both these conditions simultaneously provided the perturbing energy is sufficiently small" (Dirac 1930c, 171).

[24] *General Calculations*. PAMDP, Box: 41, Folder 5. Digitized Version under: http://purl.flvc.org/fsu/fd/FSU_MSS_1989009_S02_B41_F05, p. 13 (visited on July 2. 2025), courtesy of the Florida State University Libraries, Special Collections and Archives. This material will be abbreviated by *Dirac. General Calculations* in the following. Page references refer to the pages in the PDF file..

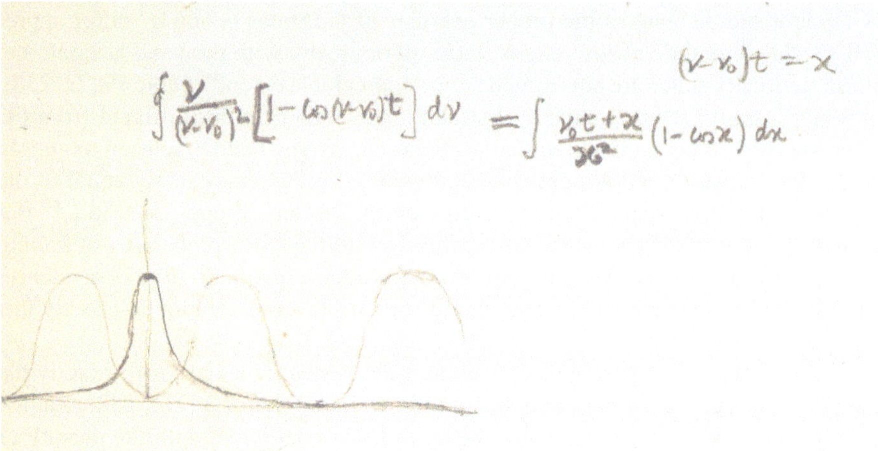

Fig. 5.1 Application and graphical representation of Dirac's trick in the Archival Material. Taken from *Dirac. General Calculations*, p. 13 (see Footnote 24), courtesy of the Florida State University Libraries, Special Collections and Archives

first imposed energy conservation directly and found that then the amplitude of the states would grow proportional to t. As we shall see in a moment, although Dirac correctly identified this as unexpected and reconfigured his calculation, he would use exactly this kind of reasoning in his dispersion paper to find the relevant matrix elements. Secondly, one can simply square Eq. (5.6), as Dirac did as well, to arrive at something which should be interpreted as a probability of finding the system in a state with different proper energy than the initial state. Hence, even the probability of finding the system in a state with different proper energy was, albeit fluctuating around a small value, non-zero. Only for longer times, this would reduce to energy conserving transitions.[25]

Thirdly, these states of different proper energy had to be incorporated into the procedure developed by Dirac to arrive at energy conservation in total and to arrive at a linear dependence on t. To some extent, Dirac had to use the energy non-conserving transitions already in first order approximations to arrive at physically sensible results.

[25] It should be noted that this was exactly how Dirac's procedure was at first connected to the energy-time uncertainty by others, as for example by Landau and Peierls (1931) (see also Sect. 10.2.3): namely, by referring to the probabilities of energy non-conserving transitions and not their amplitudes. Whether or not Dirac himself had something along the uncertainty relations of Heisenberg in mind at this point is hard to answer. He might have been aware of Heisenberg's uncertainty relations when finishing his dispersion paper, but Dirac did not directly refer to them.

5.1.4 Dirac's Quantum Theory of Dispersion *(Dirac 1927b)*

After these preliminaries, it is time to delve into Dirac's theory of dispersion. Dirac resolved some of the shortcomings of his prior wave-theoretic treatment by starting with the relativistic classical Hamiltonian for the interaction of a charged particle with the electromagnetic field.[26] Expanding it and applying his previous quantization procedure, i.e. treating the number operators and their conjugated phases as non-commuting q-numbers, he derived a three part Hamiltonian:

$$H = H_0 + V + D \tag{5.9}$$

H_0 described the proper energies of the material part of the system and the radiation field (in terms of $h\nu_r$). V and D were treated as perturbations. V was proportional to a sum over creation and annihilation operators which Dirac had already identified as connected, but not equivalent to emission and absorption processes. The last term, called D here, consisted of a product of such operators and, after some further reductions, boiled down to the direct or true scattering term Dirac had already partially discussed from his "bosonic point of view".

To derive a dispersion formula, Dirac first applied the iterative procedure he had already used when introducing his perturbation theory: The general solution of the first order equation was plugged back into Eq. (5.2). To second order, this yielded (Dirac 1927b, 721):

$$i\hbar\dot{b}_m = d_{mk}e^{i(W_m-W_k)t/\hbar} + \sum_n \frac{v_{mn}v_{nk}}{W_n - W_k}(1 - e^{i(W_n-W_k)t/\hbar})e^{i(W_m-W_n)t/\hbar} \tag{5.10}$$

Next, Dirac expanded the last term and reordered the whole equation with respect to the time-dependence.

After integration, the probability amplitude of the final state m was given by

$$\begin{aligned} b_m = \left(d_{mk} - \sum_n \frac{v_{mn}v_{nk}}{W_n - W_k}\right)\frac{1 - e^{i(W_m-W_k)t/\hbar}}{W_m - W_k} + \\ + \sum_n \frac{v_{mn}v_{nk}}{W_n - W_k}\frac{1 - e^{i(W_m-W_n)t/\hbar}}{W_m - W_n} \end{aligned} \tag{5.11}$$

Applying the exact same reasoning as he had done when developing and applying his trick in the previous paper, Dirac analysed the behaviour of this equation by simply imposing energy conservation for initial and final state at first. While the second term

[26] The Hamiltonian read: $H = c(mc^2 + (p + \frac{e}{c}A)^2)^{\frac{1}{2}} - e\Phi$. In the expansion of this Hamiltonian the vector potential occurred quadratically. The quadratic terms were responsible for what Dirac called "direct scattering," so exactly for those terms that had previously been missing in his wave-theoretic point of view.

remained periodic, the time dependence of the first term became linear. Dirac now interpreted:

> The rate of increase [of the first term] consists of a part, proportional to d_{mk}, that is due to direct transitions from state k, together with a sum of parts, each of which is proportional to $v_{mn}v_{nk}$, and is due to transitions first from k to n and then from n to m, although the amplitude b_n of the eigenfunction of the intermediate state always remains small. (Dirac 1927b, 721)

Having thus identified the relevant terms for the description of dispersion,[27] he could simply apply his trick and derive the probability for transitions to the set of final states. Expressed in formulas this led to[28]:

$$\sum \|b_m\|^2 = \frac{4\pi^2 t}{h\Delta W_m} \left\| d_{mk} - \sum_n \frac{v_{mn}v_{nk}}{W_n - W_k} \right\|^2 \tag{5.12}$$

So far, this was actually a rather general argument about perturbation theory: The only functional attribute of the perturbing terms was that d could coupled initial and final state directly, while v had to do so through the intermediate state. In order to connect this argument with physical processes, i.e. the scattering of light, and eventually to get into contact with the previously known and accepted Kramers-Heisenberg formula, Dirac had to consider two different sequences of events. I quote in full:

> We can now take the state n [intermediate state] to be either the state $J = J'',\ N_s = N_s' - 1,\ N_t = 0\ (t \neq s)$ for any J'',[29] which would make the process $k \to n$ an absorption of an s-quantum and $n \to m$ an emission of an r-quantum, or the state $J = J'',\ N_s = N_s',\ N_r = 1,\ N_t = 0\ (t \neq r, s)$, which would make $k \to n$ the emission and $n \to m$ the absorption. (Dirac 1927b, 722)

With this set of verbally expressed intermediate states and subprocesses Dirac was able to rederive the Kramers-Heisenberg formula in his new conceptual framework. By plugging in the specific form of the interaction matrix elements and after some smaller manipulations, the intensity of the scattered radiation was proportional to:

$$I_r \propto \left\| \sum_{J''} \left(\frac{x_r(J'''J'')x_s(J''J')}{\nu(J''J') - \nu_s} + \frac{x_s(J'''J'')x(J''J')}{\nu(J''J') + \nu_s} \right) \right\|^2 \tag{5.13}$$

The J'' referred to the variables of the atom in the intermediate state, $\nu(J''J')$ to the frequency corresponding to the energy difference between initial and intermediate state and ν_s to the frequency of the incident radiation. The two part structure, the

[27] The second term of Eq. (5.11) was only important in the case of resonant scattering which Dirac turned to separately in the last section of the paper.

[28] This equation is directly taken from Dirac (1927b, 722). The sum on the left hand side refers to the already performed summation/integration over the energy range of the final state.

[29] *Comment by the author*: The J are the variables describing the state of the atom.

addition of the two fractions in the brackets, was a direct consequence of Dirac's invocation of the verbal model as quoted above. The first fraction resulted from the imagined process of first an absorption and then an emission, the second fraction resulted from the inverse order of events.

Obviously, the quote of Dirac as cited in the beginning of this chapter, which Dirac provided in the introduction of his dispersion paper, was written after he had finished his calculations. This was what all of the reasoning he had applied boiled down to. As should have become clear, Dirac's description had shifted completely to the probability or quantum-amplitude level. He noted that the amplitudes would vary periodically with the time if the initial and final states did not have the same energy. The amplitudes were then "changed only by a small extent" (Dirac 1927b, 712). If initial and final state had the same energy, Dirac stated that the amplitude would increase linearly with the time and thus the transitions became "physically recognizable" (Dirac 1927b, 712). The amount of excitation was then, in Dirac's verbal description, proportional to the matrix element and not its square. While Dirac still applied his trick to impose energy conservation and to get reasonable results in agreement with priorly established formulas, his verbal description was no longer based on the level of probabilities, but on the level of probability amplitudes.

What Dirac did here was non-trivial and physicists clearly recognized that speaking about transitions when no actual transitions were concerned was not in harmony with the general interpretation of quantum theory at the time.[30] Dirac had made transitions, emission and absorption processes take place on the quantum-amplitude level. To some extent, he redefined what these terms actually referred to and thereby explicitly introduced what will later be called "virtual transitions" and "virtual states". And now it is time to provide an answer to the question guiding this first section: Why did Dirac invoke such a temporally ordered sequence of events when describing his theoretical procedure? And why was it specifically Dirac who did so?

5.1.5 So, Why Was It Dirac Who Introduced a Verbal Model of Successive Transitions?

A huge part of the material underlying my answer to the above question has already been presented. As should have become clear through the foregoing discussion, the technical and conceptual environment Dirac had developed was highly suggestive. But as will be exemplified in Sect. 10.1.1, Dirac's conception was not unanimously accepted and applied. And this certainly raises the question why it was specifically Dirac who introduced this specific kind of verbal model. In order to present an answer I will first look more closely at Dirac's introduction of the dispersion paper and the archival material connected to Dirac's verbal model. Secondly, I will contextualize Dirac's verbalization in respect to his philosophical stance towards quantum theory

[30] See for example the quote of Léon Rosenfeld as cited in Sect. 10.1.1, the respective Section or my discussion of the reception of the verbal model in Ehberger (2022, Sect. 2).

and the concepts involved in it. Concluding this subsection, I will draw together all of these aspects and provide a multifactorial answer to the question posed.

The Careful Introduction

At first we have to note that the direct identification of matrix elements with transitions only occurred in Dirac's paper on dispersion. In his first paper on radiation theory, mainly dealing with emission and absorption, Dirac spoke of "those matrix elements that *refer to* transitions" (Dirac 1927c, 252) or "the matrix elements *associated with* that transition" (Dirac 1927c, 259).[31] As he noted, "the probability of a transition [...] is proportional to the square of the modulus of that matrix element of the Hamiltonian which refers to this transition" (Dirac 1927c, 261). The connection between the matrix elements and transitions was a cornerstone of the interpretation of quantum theory from its beginning. But an identification suggesting a physical process standing behind these structures was not.

When introducing the verbal model in his paper on dispersion, Dirac was also rather careful. Shortly before engaging with the description of the processes on a probability-amplitude or quantum-amplitude level, he observed:

> If V_{mn} are the matrix elements of the perturbing energy V [...] then each V_{mn} gives rise to transitions from state n to state m; *more accurately*, it causes the eigenfunction representing state m to grow if that representing state n is already excited [...]. (Dirac 1927b, 711, emphasis added)

Dirac actually acknowledged himself that he was speaking somewhat loosely when invoking the transition terminology.

Yet, the verbal description and the differentiation between direct scattering and scattering through intermediate states were cornerstones of his exposition. In early notes on the dispersion paper, Dirac wrote (see Fig. 5.2): "state how coeft. [coefficient] v_{rs} in perturbation terms gives [emphasized and added] *direct* transitions $r \rightarrow s$ when there is no change of energy. Describe dispersion effect Scattering terms."[32] When drafting the relevant passage, Dirac would wrestle with the exact formulation for framing the "dispersion effect" (see Fig. 5.3). He noted that "a_n [the amplitude of the intermediate state] gets excited [*scribbled out*: at the expense of a_m; by transitions from m'] on account of the existence of $a_{m'}$ by an amount proportional to $v_{nm'}$ [...]."[33] Dirac certainly chose his words with care at this stage and the version he settled for in the draft ("on account of the existence") was rather uncontroversial. Since the amplitude of the initial state was set equal to one from the beginning, this is

[31] Emphasis added.

[32] *Early Work*. PAMDP, Box: 39, Folder 1. Digitized Version under: http://purl.flvc.org/fsu/fd/FSU_MSS_1989009_S02_B39_F01, p. 13 (visited on July 2. 2025), courtesy of the Florida State University Libraries, Special Collections and Archives. This material will be abbreviated by *Dirac. Early Work* in the following..

[33] *Dirac. Early Work*, p. 11 (see Footnote 32).

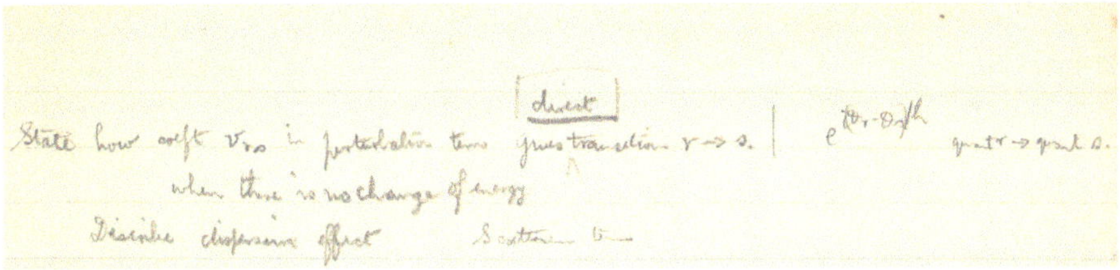

Fig. 5.2 Notes for a draft of Dirac (1927b), probably March 1927. Taken from *Dirac. Early Work*, p. 13 (see Footnote 32), courtesy of the Florida State University Libraries, Special Collections and Archives

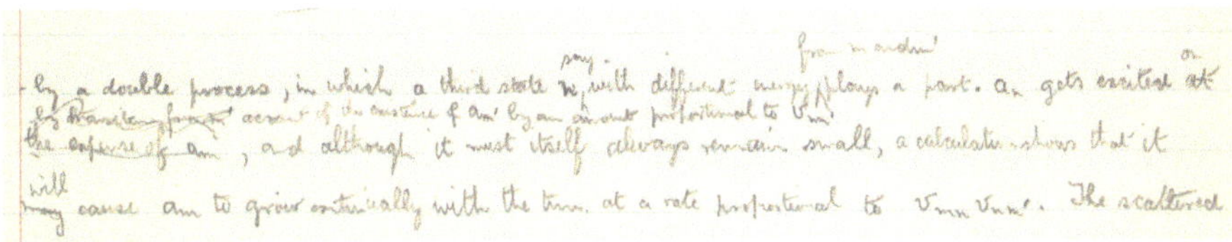

Fig. 5.3 Draft of Dirac (1927b), probably late March 1927. *Dirac, Early Work*, p. 11 (see Footnote 32), courtesy of the Florida State University Libraries, Special Collections and Archives

a straightforward extrapolation. In the published paper, contrary to this draft, Dirac would nevertheless choose the transition terminology.

When it comes to the specifics of his route towards the verbal model, the archival material only holds minor indications. Dirac's notes are scattered over different folders, temporally unordered while obviously stretching over several years. He only took sparse annotations to the calculations and some are on the back of prior drafts. Worst of all, there are bigger gaps in the material which make a coherent reconstruction next to impossible.

Some of the calculations and drafts are nearly equivalent to the published version.[34] Others simply back up what was clearly the case: For example, Dirac explicitly used the wave function in occupation number representation in his drafts while such a representation, although obviously guiding his work, is nearly completely lacking in the dispersion paper.[35] Yet, there are a few pages that actually point towards conceptual issues and exemplify the strong grasp the ideas Dirac had developed already had on his thinking. One of them is reproduced in Fig. 5.4.

Although the notation is a little bit different,[36] up until the third line of the sheet the calculation is equivalent to the presentation in the publication. The break with

[34] See for example the calculation of resonance scattering on p. 14 of *Dirac. Early Work* (Footnote 32).

[35] For example, p. 7 of *Dirac. General Calculations* (Footnote 24), but essentially to be found at diverse places.

[36] The matrix elements of the perturbation were marked by an α_{rs}. The initial state was called 0, the intermediate state s and the final state r. The energies of the state were denoted by E and, from the fifth line onwards, the α' denoted the dynamical variables of the system besides the energy.

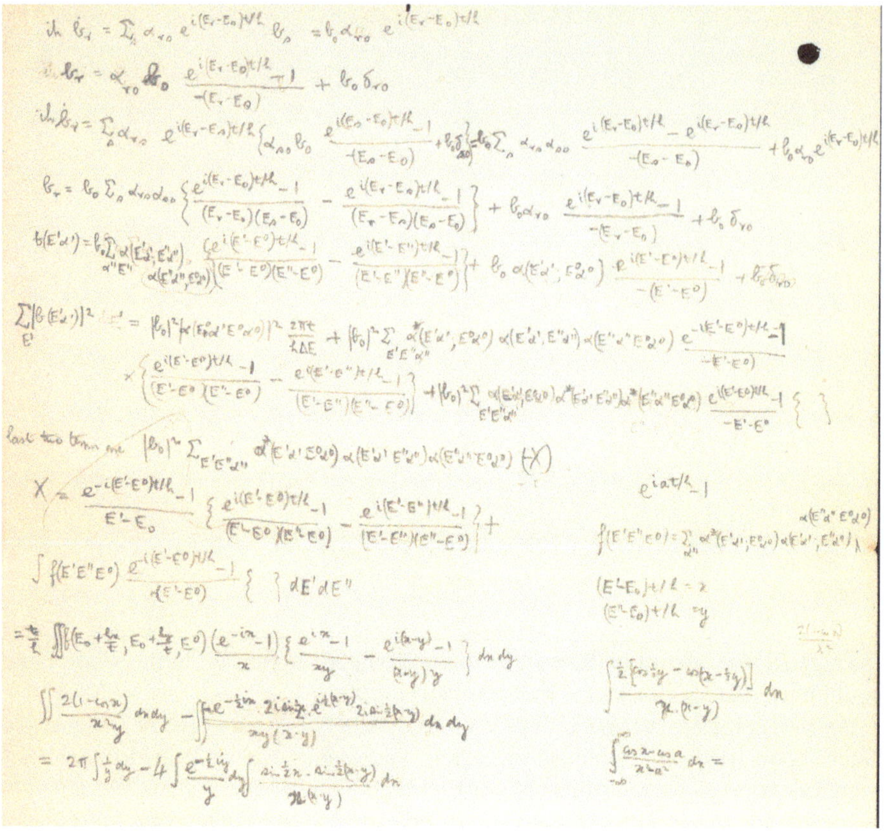

Fig. 5.4 Second order perturbative calculation. Taken from *Dirac, Early Work*, p. 92 (Footnote 32), courtesy of the Florida State University Libraries, Special Collections and Archives

his published paper occurs at the second equal sign in line three and is then carried through the whole calculation: Dirac did not split his formulas according to the time-dependence of the corresponding matrix elements, but rather in terms of direct scattering and processes through intermediate states.

There are two ways of temporally locating this calculation in Dirac's derivation process. On the one hand, we might interpret this calculation as an attempt at treating resonance scattering. Dirac kept all terms of Equation (5.11), including the ones he dropped in his calculation of dispersion. In the lower part of the draft, Dirac further applied his trick to the energy range of the intermediate states. Since this procedure was essentially a way of ensuring energy conservation, this points to an interpretation of the sheet as an attempt of treating resonant scattering.

On the other hand, had Dirac already derived the Kramers-Heisenberg formula, so had he finished his dispersion-theoretical treatment, which he nearly certainly did before developing an account of resonant scattering, he would have noted that he needed to group the terms differently, namely corresponding to their time-dependence. As already noted, this is how he would present his calculations in the published paper.

Although I am not able to temporally locate this calculation exactly, it nevertheless provides an important insight: The ordering of the terms according to the temporal development was not guiding Dirac's evaluation at this point, although he had already developed a firm grasp on the application of his trick. It was the distinction between direct scattering processes and scattering through intermediate states that led Dirac's derivation of the probabilities of observable results in his radiation theory. When Dirac performed the above calculation, the conception based on a quantum-amplitude level structured his evaluations.

The still existent notes by Dirac himself, taken in the period from early February to early April 1927, thus point towards the important role the verbal exposition of dispersion had for Dirac and to the influence of the conceptualization on early calculations of the effects. In particular, they highlight the role of the differentiation between direct scattering and scattering through intermediate states. Still, they do not provide any clues as to why Dirac chose to connect the calculations with his verbal model. For this reason, as well as to finally have all the arguments needed for a coherent interpretation in place, we need to engage with Dirac's philosophical stance towards theoretical physics and its concepts.

Dirac's Philosophical Stance Towards Quantum Theory

The best-known aspects of Dirac's approach to theoretical physics are certainly his striving for mathematical beauty[37] and the method of "playing with equations".[38] Both of these aspects were certainly part of Dirac's development of radiation theory. Dirac remembered that, in coming up with the quantization procedure, he was simply playing with equations. The archival material suggests that he often simply tried things out to see where they lead when evaluating his radiation theory as well. One aspect of mathematical beauty, the formulation of the problem in most general but equivalently simple terms and then shifting focus to the actual application, is also mirrored in Dirac's approach to perturbation theory. Yet, neither of these points help in answering the question guiding this section.

However, Dirac's methodology included a standpoint that is connected to the above two guidelines and which might actually be used for an explanation: Dirac himself has called it "Eddington's principle of identification".[39] In most general terms, it stated that the mathematics of the theory should be developed and evaluated before any physical interpretation and then, in a second step, the quantities referring to physical properties should be identified. Most explicitly, Dirac used this kind of

[37] See for example Kragh (1990, Chap. 14) or Wright (2016).

[38] See for example Pais (1987).

[39] Helge kragh (1990, 270–274) discussed it under the headlines of a "principle of plenitude"-argument. Olivier Darrigol (1992, 302–304) also identified the principle of identification as an important piece of Dirac's methodological framework.

reasoning in his argument for magnetic monopoles (Dirac 1931). But, as Olivier Darrigol (1992, 331–333) argued, it also guided Dirac's application and interpretation of Schrödinger's wave function in his 1926 paper. To understand how such a standpoint might be applied to explain Dirac's invocation of his verbal model, we need to briefly discuss the results of Dirac's evaluation of resonant scattering.[40]

In the last section of the paper on dispersion, Dirac demonstrated that in the case of resonance between the incoming frequency and the frequencies of the material system, the most relevant part of the scattered light was due to actual emission and absorption processes, connected to the squares of the respective probability amplitudes. Actual counterparts to Dirac's virtual transitions (modern terminology) existed that described the scattering of light when the energy of the impinging radiation was fitting. As is necessary for the principle of identification to apply, in the case of dispersion there was further no theoretical hindrance in conceiving of such counterintuitive events as energy non-conserving transitions. From this point of view, an "anything that can happen, will happen" attitude seems to fit Dirac's reasoning process.

And even though the identification with actual processes in the case of resonance might have enhanced the suggestiveness of Dirac's framework, there is one obstacle for invoking this kind of argument as an explanation of the emergence of the verbal model. Since the principle of identification aimed at associating mathematical structures with physical objects or processes, a close look at the structures Dirac identified with transitions in dispersion theory is important. And this is where a problem occurs: The mathematical elements describing dispersion on the one hand and the ones describing resonant scattering on the other were not equivalent. The starting point of Dirac's mathematical and verbal description of dispersion was the first part of Eq. (5.11) while resonant scattering was represented by its second part. Actually, the parts describing dispersion were also part of the final results on resonant scattering in Dirac's paper, but due to the dominance of the second part of Eq. (5.11) they could, according to Dirac's argument, be neglected. The two mathematical elements existed side by side. Although a direct identification of matrix elements with transitions might have become more suggestive, an implementation of the principle of identification seems (at least to me) on shaky grounds.

Rather, I would argue that Dirac's explicit verbalization was not due to a general philosophical stance but to Dirac's flexibility when it came to such. Dirac is often portrayed, and described himself, as a philosophically uninclined physicist. In the most prominent discussions about the interpretation of quantum mechanics, as for example the EPR-paradox, named after Albert Einstein, Boris Podolsky and Nathan Rosen, Dirac remained silent. The focus on philosophical issues, as it was most distinctly represented by Niels Bohr, and the obsessive care to keep the conceptual framework as clean as possible were to some extent foreign to Dirac's style of work.

Dirac's research practice shows a rather instrumentalistic outlook on quantum theory and the concepts involved in it. In most general terms, he did not care much about the ontological implications the handling of the theory had.[41] As long as the

---

[40] I would like to thank Alexander Blum, who pointed me towards the relevance of the resonant case for Dirac's interpretation in one of my presentations.

[41] This is certainly exaggerated but captures the general picture that is drawn of Dirac.

theory was internally consistent and its predictions cohered with the empirical results, everything was fine for Dirac. As Max Born, contemporary of Dirac and his host in Göttingen in 1927, once pointedly described Dirac's stance:

> They [physicists like Dirac] say: the existence of a mathematically consistent theory is all we want. It represents everything that can be said about the empirical world; we can predict with its help unobserved phenomena, and that is all we wish. What you mean by an objective world we don't know and don't care.[42]

Dirac's disinterest in issues he considered mostly philosophical came with a rather flexible attitude towards the principles guiding the interpretation of quantum mechanics from its start. This can best be exemplified by recourse to a principle that is directly connected to Dirac's introduction of the verbal model: the observability principle. It stated that only quantities in principle observable could be counted as genuine concepts of quantum theory. It was clearly expressed in Heisenberg's 1925, generally acknowledged to have opened the field of matrix mechanics, and was a basic pillar of the interpretation of quantum theory in the following years. Dirac himself referred to the principle all throughout his scientific career.[43] An analysis of his actual research practice nevertheless shows that this was not much more than lip service.

As Helge Kragh observed, "he [Dirac] did not hesitate to propose quantities that seemed to have only the slightest connection to observables" (Kragh 1990, 264). Kragh's prime example was the hypothetical negative energy world proposed by Dirac in 1942 to cure some of the problems QED was facing at the time.[44] Andrea Oldofredi and Michael Esfeld (Oldofredi and Esfeld 2019) also argued that Dirac's work was not guided by the observability doctrine. One of their examples is the sea of negative energy electrons introduced by Dirac in 1930. As their argument went, these entities were from a theoretical point of view unobservable by construction.

Some smaller annotations in Dirac's writings actually indicate that he considered a careful introduction of unobservable concepts tenable in theoretical physics. In his first lecture notes on quantum theory, we find a passage by Dirac which was an obvious attack on Schrödinger's conception of the wave function actually representing an electrical density. In this passage, Dirac explicitly stated that "one may introduce auxiliary quantities not directly observable for the purpose of calculation; but variables not observable should not be introduced merely because they are required for the description of the phenomena according to ordinary classical notions."[45] In a strange twist of reasoning, he made a similar statement in his paper on the many-time formalism (Dirac 1932). Therein, he upheld the observability doctrine as one of the basic pillars of the interpretation of quantum theory, yet, as he commented, "strictly speaking, it is not the observable quantities themselves (the Einstein A's and B's) that form the building stones of Heisenberg's algebraic scheme, but rather certain

[42] Born (1938, 12); as cited in Kragh (1990, 81).

[43] See Kragh (1990, 262–267) for a discussion of the observability doctrine and instrumentalism in Dirac's work.

[44] In the paper, Dirac would explicitly state that "negative energies and probabilities should not be considered as nonsense. They are well-defined concepts mathematically. [...] [they] should be considered simply as things which do not appear in experimental results." Dirac (1942, 8).

[45] Taken from lecture notes of Dirac from 1927, as cited in Kragh (1990, 80).

more elementary quantities, the matrix elements, having the observable quantities as the squares of their moduli" (Dirac 1932, 456, emphasis added).

All in all, Dirac was a physicist who was neither hindered in his methodology by some strict application of the observability principle, nor was he concerned with the ontological implications of his theoretical framework. The point is, that others were and I consider this one of the reasons why physicists would shy away from talking about transitions when no "actual transitions" occurred.

An Interim Conclusion

Now it is time to put everything together and present my whole argument as to why it was Dirac who introduced a verbal representation of the scattering of light in terms of temporally ordered and energy non-conserving subprocesses taking place on a quantum-amplitude level at least suggesting some kind of causal connection between these processes.

First of all, Dirac developed an extremely suggestive technical and conceptual framework. The amplitudes and probabilities of energy non-conserving transitions were, albeit small, non-zero and fluctuating. Energy was conserved for the whole process by application of Dirac's trick, but only by pushing time to infinity. The role of the interaction, serving to introduce alteration in the system's behaviour and not as altering the energy levels of the system, allowed for a possible explanation of energy non-conservation.[46] Shifting the focus on the quantum amplitudes was suggested by the creation and annihilation operators working on this level of theory and by the differentiation between direct scattering and scattering through intermediate states.

Secondly, Dirac did not only interpret the equations in terms of processes and leave it at that. He actually used the verbal model. To reproduce the structure of the Kramers-Heisenberg formula, Dirac had to invoke two different sequences of events: Either first an absorption occurred and then an emission or the other way around. Since energy conservation was not imposed on these subprocesses, there was no theoretical hindrance in conceiving of the second, quite counter-intuitive sequence of events. The storyline Dirac came up with was used, and through it the relevant matrix elements and their specific combinations could be constructed. From the first time it was explicitly expressed, Dirac's verbal model, emission and absorption subprocesses as well as the concepts figuring in it served as tools for the construction of mathematical representations, which were, in a second step, evaluated further. The verbalization was used to bridge the gap between an abstract mathematical representation of perturbation theory and the physical process to be calculated.

---

[46] This whole line of argumentation was inverted and connected to the energy-time uncertainty by others in the follow-up of Dirac: As long as short times were considered energy non-conserving transitions were not ruled out as such. Energy was then, strictly speaking, not not-conserved, but provided by the interaction energy. See, for example, Landau and Peierls (1931, 59) (as discussed in Sect. 10.2.3) or Yakov Frenkel's literal reading of these energy non-conserving transitions as discussed in Sect. 6.3.

Thirdly, this intricate interplay between mathematical techniques, previously established conceptualizations and, finally, the direct application from which, according to my argument, the verbal model emerged, would not have led any physicist to provide such an exposition. Even after the introduction by Dirac, some physicists would show reluctant in accepting this kind of framing. But Dirac was philosophically flexible enough (and I consider this, from a pragmatic point of view and in the given circumstances, a good thing) not to worry too much about strict philosophical guidelines of interpretation. After all, the reference to a sequential occurrence of transitions on a quantum-amplitude level, when no actual transitions occurred, was retrospectively a huge step, specifically in that it provided a language suitable for the theoretical treatment of quantum-electrodynmical processes and was consequently used in the majority of the evaluations of QED in the 1930s.

5.2 Analysis: Dirac's Double Transitions, the Light-Quantum Hypothesis, and the Virtual Oscillator, Or: Why Didn't Dirac Call Virtual Transitions "Virtual"?

Given that Dirac was the first to introduce the verbal model of QED and thus verbally explicitly used the concept of virtual transitions (modern terminology) in his dispersion paper, the question surely arises in how far he was influenced by the developments discussed in Chap. 4. As my own reconstruction in the prior sections went, I believe that we can understand Dirac's introduction of the verbal model through his research at the time. All the same, aspects of the former developments might have influenced Dirac. So, let's revisit both of the strands of research I discussed in the previous chapter separately and assess the (dis-)connections to Dirac's dispersion paper.

5.2.1 Dirac's Verbal Model and the Light-Quantum Hypothesis Before Quantum Mechanics

Let me first turn to the papers in the light-quantum conceptualization (Sect. 4.3). There is an obvious overlap as Einstein and Ehrenfest, Herzfeld, as well as Smekal argued that the scattering mechanism should be understood as a sequence of emission and absorption processes through intermediate states. The idea that photons can only be absorbed and emitted as a whole was still included in Dirac's formulation of the scattering mechanism. It was manifested mathematically in terms of the effect of the creation and annihilation operators on the wave functions of the system and in the verbal description in terms of the respective subprocesses contributing to the scattering amplitude. While I believe that a photonic conception of the radiation field was important for Dirac's description, I nevertheless want to argue that the import of the concepts of the old quantum theory into Dirac's radiation theory is questionable on different grounds.

In order to do so, we must note that physicists using the light-quantum approach of the old quantum theory and Dirac argued on different planes of theory and that the concepts they used decidedly differed. First of all, the authors of the old quantum theory worked on the level of what we today call actual transitions, while Dirac worked with quantum amplitudes. The latter concept was not developed in the light-quantum conception of the old quantum theory, but by the physicists decidedly digressing from a photonic conception of the radiation field. Dirac was clearly aware of the difference between quantum amplitudes and real processes and was rather careful in not talking about transitions when no actual transitions occurred initially (see Sect. 5.1.5). In a paper on the statistical equilibrium of radiative and non-radiative processes, written by Dirac in 1924, he actually spoke about emitted and absorbed quanta when discussing processes involving more than one quantum, as for example scattering (Dirac 1924, especially Sect. 5). Yet, and this is important for the question at hand, the analogy to the other papers in this strand of research is superficial: Dirac did not split the Einstein probability coefficients of scattering processes into products of coefficients referring to emission and absorption processes, even though he was aware of Einstein and Ehrenfest's paper.[47] In terms of probabilities and their mathematical representation, Dirac thought of scattering as one single process.

Secondly, the concept of an intermediate state as envisioned by Herzfeld or Smekal differed from the one Dirac proposed. As discussed above, the former authors worked with real processes and conceived of the intermediate state as actually occupied for a short amount of time, which is in general not warranted by Dirac's description.[48] In contrast to Dirac's conception, energy was conserved in each subprocess for Smekal and Herzefeld. Furthermore, the systems to which these states referred to differed: Herzfeld's and Smekal's intermediate states referred to the atom (in Bohr's orbital model) alone, while Dirac's intermediate states referred to the material part of the system and the radiation field together.

Finally, the intermediate character of the states does not cohere in the two cases. For Herzfeld and Smekal the states were intermediate in a twofold manner: The states laid temporally in between initial and final state and were placed in between the regularly allowed energy levels of the atom under consideration. Dirac's intermediate states provided for a connection of initial and final state, while their possible energy values were given by the unperturbed, thus normally allowed, energy values of the system. This conception was possible because of the digression from energy conservation for the subprocesses.[49]

---

[47] See, for example, Equation (4c) of Dirac (1924, 591).

[48] All the same, Dirac himself was vague in the published paper and Dirac's stance towards the temporal dimension of the intermediate state was not entirely clear (see Sect. 7.5).

[49] I will also revisit these differences in Sect. 6.3, when discussing the conception of intermediate states by Yakov Frenkel.

Even though I hold that the conception of the radiation field as composed of photons was important for a description in terms of successive emission and absorption subprocesses, I argue that there was no direct causal connection between the conceptions of the works applying light-quantum conceptions before the advent of quantum mechanics and Dirac's inception of the intermediate state. This is also clear from Dirac's own scientific trajectory. After he had received the preprint of Heisenberg's 1925 breakthrough paper, he worked on quantum and eventually wave mechanics. From this research strand Dirac's radiation theory and his theory of dispersion eventually emerged.

5.2.2 *Dirac's Verbal Model and the "Virtual" Entities of the Old Quantum Theory*

Let me now turn to the research strand which is obviously closer to Dirac's work and from which quantum mechanics, and eventually Dirac's radiation theory, emerged. Thereby, let me also engage with a question which might already have occurred to the reader: Why didn't Dirac call virtual transitions "virtual transitions"?

There are obvious structural and conceptual overlaps between concepts that lie in one line with the BKS theory, Heisenberg's *Umdeutung*, and the development of quantum mechanics, on the one hand, and the ones introduced in Dirac's radiation theory, on the other. First of all, virtual oscillators, matrix elements, and Dirac's description of the scattering processes played their part on the quantum-amplitude level of the theory. In this respect, they were what one might call genuine quantum objects. Secondly and connected to the first point, they were conceived of as unobservable. As argued in Sect. 4.1.3, the term "virtual" of the BKS theory can be understood as referring to an unobservable world in which the virtual oscillators and the virtual radiation field were placed. Also Dirac noted in his dispersion paper that only states of the same energy were connected by "physically recognisable transitions" (Dirac 1927b, 712). As such, Dirac's transitions to intermediate states (as well as the intermediate states themselves) and the virtual oscillators were located on the same plane of reality.

Thirdly, there were obvious overlaps in the mathematical representation. There was the sum over intermediate states (of the atom) at which one arrived when carrying out the calculation. In retrospect and as was most clearly argued by Duncan and Janssen (2007b), this sum over "virtual oscillators" can be identified with a sum over the matrix elements of the position operator, i.e. the elements figuring in the Kramers-Heisenberg formula. In this sense, the virtual oscillators were certainly taken over into quantum mechanics and even occurred in Dirac's paper on radiation theory (I will come back to this observation in a moment). Considering these two points together, the attributions of the inception of the concept of virtual transitions to Heisenberg and Kramers (Dresden 1987) or Heisenberg (MacKinnon 1977) is understandable.

All the same, these actors did not refer to virtual transitions explicitly in terms of a verbal representation of successive transition processes.

Finally, and this is an important point regarding the functional similarity of virtual oscillators, matrix elements, and Dirac's transitions to intermediate states (later to be called virtual), they all constituted a way of including *possible transitions* and the corresponding absorption frequencies in the formulas expressing dispersion. It was known that a representation of possible transitions, although forbidden by energy conservation and thus not actualized (or "physically recognisable"), had to play a role in the description of dispersion, at least since Reiche and Ladenburg's introduction of the *Ersatzoszillatoren* in 1923. The terminological variant of a "virtual possibility"[50] as introduced by Werner Heisenberg in the mid 1930s and used by his Leipzig students and assistants might be conceived of as exemplifying exactly this connection. Heisenberg, at least immediately after taking up the term "virtual", did not speak about "virtual processes" but mostly about the "virtual possibility of a process". When reading the term virtual as referring to an unobserved quantum world, then in this world the actually forbidden processes are possible, therefore their possibility is virtual, hence the "virtual possibility".[51]

Thus, there seems to be a lot of conceptual overlap between the concepts of virtual oscillators and virtual transitions (modern terminology) which makes the above mentioned question even more pressing: Why didn't Dirac call virtual transitions and virtual states "virtual transitions" and "virtual states"? I want to give a twofold answer: On the one hand, even if Dirac saw a conceptual connection to virtual oscillators, there are reasons why he would not have used the term "virtual" at the time. On the other hand, even though I consider all the above conceptual connections to be (at least retrospectively) reasonable, the concepts of virtual oscillators and the concept of virtual transitions are far from alike, and I believe that Dirac was aware of these differences.

First, let's assume that Dirac saw a connection between virtual oscillators and his energy non-conserving transitions to intermediate states. Then we can still note that Dirac simply seemed not to have liked the term "virtual". From the mid 1920s to the late 1940s, Dirac (to the best of my knowledge) did not use the word "virtual" in his papers, although the terminology seems to have been accepted widely in the mid to late 1930s. Even when Dirac referred to the BKS theory (Dirac 1936) to explain a seeming empirical proof of energy non-conservation, Dirac did not apply the term "virtual". In the third edition of his textbook *The Principles of Quantum Mechanics* published in 1947 (Dirac 1947) the term is nowhere to be found.

Furthermore, the word "virtual" was strongly connected to the BKS theory which so famously failed as an interpretation scheme. Some reluctance to use the word might thus be expected. But, it is still to note, that physicists did not shy away from using the term. They applied the term "virtual oscillator" after the emergence of quantum

[50] "virtuelle Möglichkeit," e.g. Heisenberg and Euler (1936, 714).

[51] For a more detailed discussion of the terminology of the "virtual possibilities" of Werner Heisenberg and the connection to BKS it introduced, see Sect. 9.1.2.

mechanics.[52] They even used the term "virtual" to refer to concepts unconnected to both virtual oscillators and virtual transitions, as for example Guido Beck did with the "virtual states" (Beck 1930).[53] It would not take long until the first physicist connected Dirac's concepts (although in my eyes wrongly, see Sect. 6.2.2) with the term "virtual". Even if we accept that Dirac was hesitant to connect his concepts with the word "virtual", given the usage of the terminology by other actors, it is hard to explain why a wide-spread reference to "virtual transitions" only occurred in the mid 1930s, so nearly eight years later.[54] Hence, let me turn to the dissimilarities of the two concepts.

First of all, a rather popular reason for connecting the term "virtual" with transitions will become the energy non-conservation of the latter. But as argued, energy non-conservation was not the reason for the invocation of the term in the BKS theory. Energy non-conservation was according to BKS observable, thus actual/real. Energy non-conservation in Dirac's treatment was not observable.

Secondly, the two concepts, virtual transitions and virtual oscillators, belonged to different conceptual frameworks. The virtual oscillators, both in the BKS theory and in their later invocation in semi-classical methods such as Klein's (Klein 1927), were connected to a classical description of the radiation field. Exactly for this reason the term oscillator (i.e. something vibrating in reaction to the wave field) was used. But the transitions to intermediate states of Dirac were embedded in a light-quantum framework, governed by successive emission and absorption processes.

Thirdly, both of these concepts performed the function of including the possibilities of transitions unbound by energy conservation in the calculations. But they did so in very different ways. In the BKS theory, as well as in Kramers and Heisenberg's paper, the atom stayed in one state during the dispersion process. The oscillators were explicitly connected to a stationary state and, if a transition would have occurred, another set of oscillators would have acted. The oscillators replaced the actual atom; they were wave-theoretical, "grafted on" entities, probably simply a non-dynamical analogy; they did not describe the behaviour of the actual atom. The transitions to intermediate states of Dirac, on the other hand, were a (possibly pseudo-physical) description of successive transition processes of the actual atom. The possible transitions could act as transitions to intermediate states since, according to Dirac's derivation, energy did not have to be conserved in the intermediate steps. Dirac's transitions did not replace the atom by another set of entities; they were a (possibly pseudo-physical) description of its behaviour.

This last point, although reconstructed retrospectively above, underlines the difference of how the virtual oscillators and the virtual transitions performed their task of including possible transitions into the respective conceptual frameworks. And to further underline this point of conceptual difference, I shortly need to anticipate Maria

[52] For example Slater (1927a, 110) and Klein (1927, 413).

[53] See also the discussion by Jean-Philippe Martinez (2024, Sect. 2).

[54] While Chandrasekhara Venkata Raman was the first to use the terminology of "virtual transitions", only with Werner Heisenberg's use of the term in the mid 1930s the terminology became accepted in wider circles.

Göppert-Mayer's contribution to Dirac's radiation theory (Göppert-Mayer 1931; see Sect. 8.1.1). As my argument goes, both Dirac and Göppert-Mayer actually used the concept of a virtual oscillator, while Göppert-Mayer explicitly named the respective theoretical entity.

Göppert-Mayer discussed, amongst other phenomena, double emission and absorption in terms of Dirac's radiation theory in her dissertational research. She applied Dirac's verbal model by noting that the discussed phenomena behaved "as if two processes, each of which does not obey the energy law, occur in one act".[55] But Göppert-Mayer did not use the term "virtual" to refer to these processes. Yet, she explicitly invoked the concept of a virtual oscillator. And she did so after applying the verbal model. She argued that the atom behaved under the influence of an external radiation "as if a virtual oscillator with moment $[P]$[56] [...] emits the frequency ν' spontaneously".[57] Although both of the concepts were used and actually connected to the term "as if", they differed in their function and in terms of the description: The virtual oscillator was a semi-classical replacement of the atom; the virtual transitions were a way of (pseudo-physically) describing the transition behaviour of the atom to arrive at the electric moment which Göppert-Mayer referred to as the virtual oscillator.

And in this sense also Dirac used the virtual oscillator concept in his dispersion paper (Dirac 1927b, 722–724). After having constructed the general expression of the perturbative evaluation for the non-resonant case, Dirac applied his verbal model and plugged in the respective matrix elements for the interaction between matter and radiation and restricted the emitted light quanta to have the same frequency as the absorbed ones, i.e. he discussed the case of dispersion (in contrast to the Raman effect). Then he stated: "The most convenient way of expressing this result is to find the amplitude P (a vector) of *the electric moment of that vibrating dipole of frequency* ν_s *that would, according to the classical theory*, emit the same distribution of radiation as that actually scattered by the atom." (Dirac 1927b, 723, emphasis added) This is the concept of a virtual oscillator as expressed by Göppert-Mayer. By some formal manipulations, Dirac could thereby show that his results were equivalent to the Kramers-Heisenberg formula.

[55] "[...] als ob zwei Vorgänge, von denen jeder nicht dem Energiesatz genügt, in einem Akt geschehen," Göppert-Mayer (1931, 284).

[56] *Note by the author*: This electric moment is accordingly given by a formula which resembles the Kramers-Heisenberg formula in structure.

[57] "als ob ein virtueller Oszillator mit dem Moment $[P]$ [...] die Frequenz ν' spontan emittiert." Göppert-Mayer (1931, 281–282).

5.2.3 Why Dirac Did Not Call Virtual Transitions "Virtual Transitions"

To recap, let me shortly rephrase the argument why Dirac didn't call virtual transitions "virtual transitions". Obviously, Dirac was not very fond of the term "virtual"; but there are further reasons. And to understand the fact, that at least in the mainstream of theoretical physics the term "virtual transitions" (or process, in more general terms) had to wait another eight years to become applied regularly, we should acknowledge them. Of course there were some important similarities of the concepts of virtual oscillators and virtual transitions and one might even understand the virtual oscillator (conceived of as matrix elements of the position operator) as arising from virtual transitions in terms of Dirac's theory.

Still, there was a considerable conceptual gap. A gap that, as discussed above, allowed for the invocation of Dirac's verbal model and the virtual oscillators in one and the same paper, but for different purposes. First of all, virtual oscillators and virtual transitions differed in their conceptual surroundings: The virtual oscillator were connected to a classical radiation field and the virtual transitions to a photonic conceptual space. The virtual oscillator was an after the fact replacement with the right eigen-frequency and it was grafted on the atom; it did not describe the behaviour of the atom. The virtual transition was a (pseudo-physical) quantum-theoretical description of the behaviour of the whole system, atom plus radiation field. Energy non-conservation was not the reason for the "virtual" terminology of BKS, while it will become the most prominent reason for the term "virtual" in connection to the virtual transition. And connected to the last point: Energy non-conservation was observable (thus not virtual) in the BKS framework, while Dirac only considered energy conserving transitions to be "physically recognisable".

Finally, BKS only called wave-theoretical, nearly classical concepts virtual (oscillators and radiation field), while a reference to a concept of quantum physics as virtual had not yet been established. This is again connected to the huge step which Dirac took in his quantum theory of dispersion: He spoke about transitions, which did not actually take place. He used them as a description of the behaviour of a system on the probability-amplitude or quantum-amplitude level: He made transitions a genuine quantum concept. And he did so because it served a purpose, because he could use this kind of language as a tool in his derivation process. We can, thus, understand the invocation of the verbal model, as argued in the prior sections, from within his research practice.

But if we acknowledge that there was a conceptual gap between virtual oscillators and the energy non-conserving transitions of Dirac, we certainly need to explain why physicists began to talk about virtual processes at all. As I will argue, there are two routes towards the connection between the term "virtual" and quantum-theoretical processes: I have already touched upon Heisenberg's conception above and I will turn to its historical contextualization in connection with the practice of time-dependent perturbation theory in Sect. 9.1.2. But there was another and earlier occurrence, which I want to historically investigate in the next chapter. In a speech in 1929, Chandrasekhara Venkata Raman was, to the best of my knowledge, the first to explicitly connect transition, emission and absorption subprocesses with the term "virtual". Thereby, he already proposed a reading of the term that will later on become popular: According to Raman, these subprocesses were virtual because they did not conserve energy and only arose in the mathematical treatment of the theory.

Chapter 6
The Raman Effect: How Virtual Transitions Became "Virtual" (for the First Time) and Real Transitions Were Excluded from the Conception of Scattering (1928–1929)

In the previous chapter I discussed how the verbal model of successive energy non-conserving emission and absorption processes each correlated with a transition of the material part of the system was introduced by Dirac as a description of the quantum-electrodynamical scattering process. As a tool in the calculation, Dirac used it to re-derive the Kramers-Heisenberg formula within the framework of his radiation theory. The following chapter will focus on the first empirical proof of this formula:[1] the discovery of the Raman effect and its aftermath.[2]

The discussions of the characteristics of the effect in the two years after its experimental realization has a twofold relevance for the history of the virtual particle concept. On the one hand, the term "virtual" was for the first time (as far as I know) directly connected to transitions (i.e. to a concept of quantum theory) in 1929. It was Chandrasekhara Venkata Raman himself who introduced this connection in direct response to the discussion around the interpretation of the effect named after him.[3] On the other hand, the explanation of scattering as real transitions through intermediate states (as discussed in Sect. 4.3) resurfaced in a modified form. Yakov Il'ich Frenkel connected and motivated this conception by directly referring to the mode of evaluation, i.e. Dirac's time-dependent perturbation theory.

The theoretical and empirical depth to which the said effect was investigated plays a crucial role in the following. Therefore, a short overview of the theoretical and

[1] The observation of the so-called "negative dispersion" predicted by Kramers in 1924 (Kramers 1924a) and observed by Hans Kopfermann and Rudolf Ladenburg in 1928 (Kopfermann and Ladenburg 1928) will not be discussed. This is due to the important role of the intermediate state for the Raman effect and its only implicit occurrence for the case of negative dispersion.

[2] Throughout this chapter, the terms "incoherent scattering", "Raman effect", or "Raman scattering" will be used synonymously to describe the scattering of radiation from molecules connected with a change of the internal energy of the molecule and the frequency of the emitted light. Peter Pringsheim (1928) coined the term "Raman effect" shortly after its experimental discovery.

[3] I have already shortly discussed Raman's speech as the first occurrence of the connection between the term "virtual" and the concepts of Dirac's radiation theory in Ehberger (2020, 267–269).

M. Ehberger, *Representing the Unobservable*, Science Networks. Historical Studies 68,
https://doi.org/10.1007/978-3-032-09188-8_6

conceptual basis of the Raman effect is certainly in place. This phenomenon was first proposed by Adolf Smekal in 1923[4] and put on firmer theoretical ground by Kramers and Heisenberg in their dispersion paper (see Sect. 4.2). In most general terms, the Raman effect corresponds to the scattering of light from an atom or molecule with an exchange of energy between radiation and matter. Both the frequency of the scattered radiation and the internal energy of the material part of the system is altered. Thus, the scattered radiation includes a frequency component which is discretely shifted by an amount $\Delta\nu$ determined by the difference of the energy of the atom in initial and final state:[5]

$$\Delta\nu = \frac{1}{h}(E_m - E_k) \tag{6.1}$$

This equation constitutes the essence of the Raman effect.[6] For the experimentalist the effect would consequently show itself in spectroscopic lines discretely shifted from the frequency of the primary radiation.[7]

This effect is included in the quantum-theoretical dispersion formula, i.e. the Kramers-Heisenberg formula. In Dirac's nomenclature (see Sect. 5.1.4) it reads:

$$P_r \propto \left| \sum_{J''} \left(\frac{x_r(J'''J'')x_s(J''J')}{\nu(J''J') - \nu_s} + \frac{x_s(J'''J'')x_r(J''J')}{\nu(J''J') + \nu_r} \right) \right| \tag{6.2}$$

where P_r is the electric moment of the atom or molecule. x refers to the unperturbed electric moment, the J's represent the quantum numbers of a given state of the material part of the system. The $\nu(J''J')$ represent the frequency corresponding to the transition from the state J' to J'', i.e. from initial to intermediate state. ν_s refers to the frequency of the impinging light and ν_r to the frequency of the secondary radiation. The sum has to be extended over all possible states J'' which combine with the initial and the final state (Dirac 1927b, 721–726).

It is important to note, that although the verbal model of Dirac was based on two acts of emission and absorption only the absolute square of P_r is directly connected to observations (i.e. the intensity of the scattered light). The decomposition of the scattering process was valid on the level of probability amplitudes and not on the level of probabilities. But through the intermediate state the initial and final state did not have to be connected directly. Also transitions actually forbidden for regular absorption played a role for the incoherent component of the scattered radiation,

[4] Smekal (1923b). For a context sensitive discussion, best see Jordi Taltavull (2017, Sect. 6.3).

[5] Either the initial or the final state might have higher energy. In the first case the frequency of the radiation increases (so-called Anti-Stokes case), while in the second case the frequency decreases (so-called Stokes case).

[6] Of course and importantly for the following discussion, this equation does not describe all of its aspects.

[7] That this part of the spectrum is not due to fluorescence, which would constitute two successive real (modern terminology) transitions and emission and absorption processes, was initially argued for by reference to the polarization of the secondary radiation.

as was already pointed out by Kramers and Heisenberg (1925, 695) as well as by Schrödinger (1926b, 122).[8] The selection rules, i.e. the rules describing which initial and final states couple in a respective situation, were not equivalent for absorption and the Raman effect. This last fact will play an important role in the discussion of the historical material.

With the experimental discovery of the Raman effect different possible conceptualizations of the scattering mechanism and their observable consequences were discussed among experimentalists and theoreticians. These discussions have, to the best of my knowledge, found next to no attention in the historical literature.[9] The secondary literature on the Raman effect is rather concerned with the preconditions of the empirical discovery, the priority dispute between the two experimental groups based in Moscow and Calcutta, the reception of the effect in Europe and the US, and its further empirical investigation.[10] The theoretical evaluation of the effect is often only touched upon and it is regularly simply framed as one of the experimental proofs of quantum-mechanical predictions. When it comes to theoretical considerations based on the empirical data, the work by Georg Plazcek on the selection rules in the early 1930s and the beginning doubts about the concept of intra-nuclear electrons are predominantly mentioned.[11]

The following chapter, will according to the nature of this investigation, focus on the role of the intermediate states and the transitions to and from them in the empirical and theoretical evaluations connected to the phenomenon of Raman scattering. To set the ground, the first section will deal with the discovery of the effect in Calcutta and the initial conceptualization given and defended by Raman himself (Sect. 6.1). In the second section, I will engage with the growing empirical evidence on the incoherently scattered radiation, which allowed more and more insights into the nature of the effect and the corresponding selection rules (Sect. 6.2.1). In what can only be called a defence of his initial "simple and easily understood picture of the process" (Raman 1929b, 789), Raman invoked the term "virtual" in respect to the involved transitions

---

[8] Of course, the respective conceptualization of Kramers and Heisenberg as well as by Schrödinger appreciably differed from Dirac's.

[9] One notable exception is Brand (1989).

[10] See for example Singh (2002) for a concise history of its discovery and the initial reception, Singh and Riess (2001) for the priority dispute and the award of the Nobel Prize for Raman, or Singh (2004) for a in depth study of Raman's biography and scientific work. Banerjee (2014) provided a socially sensitive study of Raman's discovery and its initial reception in Europe. Fabelinskii (1978) discussed the work of the physicists in Russia and also touched upon questions of priority. Naturally the effect was also discussed in the standard works on the history of quantum physics such as in Mehra and Rechenberg (2000, Sect. 3.1 (a)–(d)).

[11] Empirical data suggested that the nitrogen nucleus obeyed Bose statistics, i.e. it should be constituted of an even number of spin one-half particles. According to an electron-proton model of the nucleus, nitrogen consisted of 14 protons and 7 electrons (in total 21 spin one-half particles), thus Walter Heitler and Gerhard Herzberg stated that "it seems as if the electron looses with its spin also its co-determination right for the statistics of the nucleus" (Heitler and Herzberg 1929, 674). Of course, this only constituted one of the empirical and theoretical building blocks that would eventually lead to a refutation of the nuclear electron hypothesis. For overviews, see, e.g., Stuewer (2018, 116–119) or Mehra and Rechenberg (2000, Sect. 3.7 (c)).

to intermediate states and, as I will argue, (most probably unknowingly) altered the meaning of the term "virtual" thereby (Sect. 6.2.2). In the final section, I will discuss the re-occurrence of the proposal of scattering of light as composed of real transitions, as put forward by Adolf Smekal and Yakov Il'ich Frenkel, and its refutation by Georg Plazcek (Sect. 6.3).

6.1 The Discovery and Raman's Initial Explanation

The presence of a frequency component discretely shifted from the incident radiation was discovered independently and nearly simultaneously by the group around Chandrasekhara Venkata Raman in Calcutta and by Leonid Mandelstam and Grigory Landsberg in Moscow.[12] The line-shift was in principle already predicted by theory. Yet both of the experimental discoveries were not directly influenced by the prior quantum-theoretical work on dispersion. Landsberg and Mandelstam's research was motivated by Peter Debye's theory of the specific heat of solids which also predicted a frequency shift in the scattered radiation. The results turned out quite different from what they had expected. Consequently the authors interpreted the spectrum as a new kind of effect in the scattering of radiation.[13]

Also for the experimental work in Calcutta leading up to the discovery of the effect, quantum-theoretical accounts of dispersion initially played next to no role.[14] Raman's experimental interests were from the beginning of his career diverse, but acoustics and optics remained major themes of his investigations. His first paper on light diffraction was published in 1906, while still a physics student at the Presidency College of the University of Madras. Until his appointment as Palit Professor of Physics at the University of Calcutta in 1917, his research nevertheless had mainly focussed on acoustics and astronomy. Around 1921, Raman began concentrating more exclusively on the field of optical research.[15] In the period from 1921 to 1928, he worked intensively on the scattering of light from diverse materials.

Within this line of research, a lot of Raman's co-workers and employees established empirical findings which were retrospectively identified by Raman as

[12] The first announcement of the effect was delivered by Raman on March 16. 1928 in a lecture subsequently published on March 31. 1928 in the *Indian Journal of Physics* (Raman 1928). The first note to Nature was also published on March 31. 1928, while it was communicated on February 16. (Raman and Krishnan 1928b). Although the Russian's communication was only published in July 1928 in the *Naturwissenschaften* (Landsberg and Mandelstam 1928) and therein Raman's publication in *Nature* is already cited, the independence of the experimental work is generally unquestioned.

[13] See e.g. Landsberg and Mandelstam (1928); or Fabelinskii (1978).

[14] The following is mainly based on Singh (2002) and Singh (2004, Chap. 3).

[15] The motivation for Raman to go into this line of research was Lord Rayleigh's theory of the blue colour of the sea. According to Rayleigh, the colour of the sea was due to the reflection of the blue colour of the sky. Raman could show that this hypothesis was untenable and that the colour of the sea was actually caused by diffraction of light by water molecules.

decisive for the discovery of the effect. Kalpathi Ramakrishna Ramanathan observed in 1922 that the intensity of the light scattered from anisotropic molecules did not cohere with theoretical predictions. Ramanathan explained this by a "trace of fluorescence", the origin of which "has not yet been definitely ascertained".[16] Such a "feeble fluorescence" was again observed by another of Raman's employees, namely Kariamanikkam Srinivasa Krishnan in 1925, but was still left unexplained: At the time, it was not possible to record the spectrum of this radiation because of its weak intensity.

Raman and his co-worker Krishnan focused their experimental investigations on the "feeble fluorescence" in 1927, after Arthur Compton was awarded the Nobel Prize for the experimental discovery of the effect named after him.[17] The decisive steps towards the discovery of the new kind of secondary radiation were done in the first two months of 1928. Although the exposure times were incredibly long, Raman and Krishnan could establish more and more characteristics of the scattered radiation. They found that the "feeble fluorescence" was strongly polarized and could therefore not be actual fluorescence. The effect occurred in diverse materials, also in highly purified liquids. It was, thus, likely that the effect was universal and not due to impurities of the scattering material. Raman concluded that they actually discovered a new kind of secondary radiation.

Even though Raman's line of research was nearly independent of the quantum theory of dispersion up until early 1928,[18] he already pointed, although only vaguely, to the Kramers-Heisenberg formula in his initial publication:

> As a tentative explanation [of the scattered radiation], we may adopt the language of quantum theory, and say that the incident quantum of radiation is partially absorbed by the molecule, and that the unabsorbed part is scattered. The suggestion does not seem to be altogether absurd and indeed such a possibility is already contemplated in the Kramers-Heisenberg theory of dispersion. If we accept the idea indicated above, *then the difference between the incident and scattered quanta would correspond to a quantum of absorption of the molecule.* (Raman 1928, 373, emphasis added)

In a similar vein Raman discussed what is today called Anti-Stokes scattering, namely Raman scattering with an increase of frequency.

Raman's interpretation is quite noteworthy. First of all, he suggested that the photon, normally conceived of as emitted or absorbed as a whole, could supply

[16] Ramanathan (1923), as cited in Singh (2002, 408–409).

[17] For a discussion of Raman's comments on the Compton effect at the 1924 BAAS meeting and the dispute between Arthur Compton and William Duane, see Singh (2002, 409). Raman remembered that the Compton effect and Compton's 1927 Nobel Prize had important influence on his research programme.

[18] The first connection between the effect and the quantum theory of dispersion I know of is the diary entry of Krishnan on February 7. 1928: "After meals at night, Venkateswaran and myself were chatting together in our room when Prof. suddenly came to the house (about 9 p.m.) and called for me. When we went down, we found he was very excited and had come to tell me that what we had observed that morning must be the Kramers-Heisenberg effect we had been looking for all these days. We therefore agreed to call the effect *modified scattering.*" Excerpts from Krishnan's diary as cited in Venkataraman (1995, 45).

part of its energy to the molecule or atom from which it was scattered. Secondly, this certainly did not cohere with Kramers and Heisenberg's conception as they explicitly worked with a classical radiation field. Thirdly, already Kramers and Heisenberg had noted that for the effect Raman would discover the transition from initial and final state of the atom or molecule "does not have to represent a possible transition"[19]. While Raman identified the energy difference with "a quantum of absorption of the molecule", since the first detailed theoretical description of the effect it was known that such an identification was not universally valid. To some extent, Raman simply ignored the role of the intermediate state and the detailed theoretical description.

Raman's phenomenological and (to some extent pseudo-)photonic description seemingly did not have its roots in the quantum-theoretical evaluations of dispersion. It was, as Smekal's initial contribution (Smekal 1923b), only guided by the energy conservation for the whole process.[20] Even more curiously, it is generally argued that Raman favoured semi-classical approaches. Yet, the analogy between Raman's observation and the Compton effect might explain why Raman chose a simplified but still somewhat quantum-theoretical description of the phenomenon.[21] In his Nobel Lecture in December 1930, Raman even stated that "in interpreting the observed phenomena, the analogy with the Compton effect was adopted as the guiding principle" (Raman 1930, 27). He particularly emphasized the conservation of energy and momentum.[22]

The specific kind of interpretation and the identification of Raman lines with absorption lines can further be explained by backdrop to the empirical material at hand. Up until late 1928, most of the observations seemed to indicate that absorption lines and Raman lines coincided, although some anomalies were already known. In Raman's publications up until early 1929 we find the above given conceptualization of the scattering mechanism and the conjecture of the frequency shifts being proportional to absorption frequencies. In a communication to *Nature*, sent off on March 22. and published on May 5. 1928, Raman and Krishnan repeated the above given conceptualization and stated: "This explanation is supported by the fact that the diminution in frequency is of the same order of magnitude as the frequency of the molecular infra-red absorption line." (Raman and Krishnan 1928c, 711) More quantitative experiments on Benzene even strengthened Raman's conviction about the interpretation of the scattering mechanism. Raman stated, that "it is clear that the wavelength [shifts of the scattered radiation] correspond to certain fundamental frequencies of absorption of light by the benzene molecule" (Raman and Krishnan 1928a, 417).

---

[19] "braucht [...] keinen möglichen Übergang darzustellen," Kramers and Heisenberg (1925, 695).

[20] Quite curiously, there is no evidence that Raman was aware of Smekal's paper. See, e.g., Banerjee (2014, 163).

[21] In the above mentioned publication as well as in further communications Raman (see e.g. Raman and Krishnan 1928c) stressed the analogy between his discovery and the Compton effect.

[22] That the conservation principles alone were not sufficient to explain the phenomenon was known in 1930 and acknowledged by Raman.

By the end of 1928 Raman seemed to be very confident about his interpretation. In October of the same year, Charles Galton Darwin published a report about the Sixth Congress of Russian Physicists, held in early August 1928, in *Nature*. Darwin mentioned that the communications by "Profs. Mandelstam and Landsberger [sic!]" were especially interesting. "The latter described how they had independently discovered Raman's phenomenon, the scattering of light with changed frequency." (Darwin 1928, 630) In the race for priority and driven by the hope of being awarded the Nobel Prize,[23] Raman promptly responded. In a letter written on November 13. 1928 and published in January 1929, Raman stressed his (and his group's) priority for the discovery of the effect. One of his "observations [that] established and emphasized the fundamental character of the phenomenon" and therefore spoke, according to Raman, in his favour was "that the frequency differences between the incident and scattered radiations represent the absorption frequencies of the medium" (Raman 1929a, 50).

Also Raman's idea of scientific progress stressed the role of empirical evidence over theoretical considerations. The most explicit enunciation on this conviction can be found in a letter to Niels Bohr written on March 13. 1932. Therein, Raman stated that "theories must stand and fall according to as they agree with the facts, and not vice versa"[24]. At the time, empirical evidence pointed towards an identification of Raman lines with absorption lines, or at the least made such an interpretation possible.

Therefore, it can safely be assumed that at the beginning of 1929 Raman thought of the mechanism of scattering he had envisioned to universally account for the process under study. Guided by the empirical data, Raman upheld that the implication of his conception, namely that line-shifts are proportional to absorption lines of the medium, was one of the universal characteristics of his newly found effect. It must be noted that Raman was not the only one who conjectured that the line-shifts of the secondary radiation should be proportional to absorption lines. Up until 1929, this seemed to be the most widely upheld view on the observed spectrum, even though some digressions from this rule already occurred. All the same, in the publications of 1928 we mostly find Raman's identification of absorption lines with frequency shifts.

In most general terms, the reception of Raman's discovery was very favourable all over Europe and the US. The easy to replicate set-up, the novelty of the phenomenon, and the promise of a new means of studying molecular spectra led to an increasing number of experimentalists to engage with the study of the Raman effect in diverse materials.[25] But exactly this wide spread of empirical studies of the phenomenon more and more showed that Raman's initial identification could not be upheld.

---

[23] For Raman's strategies for ensuring his sole priority for the discovery of the effect, see e.g. Singh and Riess (2001), or Singh (2004, Chap. 4, especially Sect. 4.2).

[24] Raman to Bohr, March 13. 1932; as cited in Singh (2004); emphasis in the original.

[25] To get a detailed view of the empirical and theoretical work in the different countries see Long (1988). The reception of Raman's work is also shortly discussed in Singh (2002, 414). A bibliography of 160 papers published until August 1929 can be found in Ganesan (1929).

6.2 Selection Rules, Raman's Defence, and a New Meaning of Virtuality

6.2.1 Establishing Selection Rules and Explaining Them Theoretically

The spread of the news about the Raman effect eventually led to experimental studies in diverse countries and to the improvement of the empirical methods for studying the spectra. Experimentalists observed more and more deviations from the rule given by Raman, i.e. that the frequency shifts should correspond to absorption lines of the materials.[26] Observations by Robert Wood, professor at the John Hopkins university in Baltimore,[27] and by Franco Rasetti, based in Rome but at the time on a research stay at Caltech,[28] were particularly influential. Both Wood and Rasetti contributed immensely to the design and techniques involved in the experimental work on the Raman effect, which eventually led to the possibility of using gases constituted of simple molecules or atoms as the scattering medium. The most striking deviations from the simple model of Raman could be identified in these studies.[29]

In the first communication of his measurements published in February 1929 in *Nature*, Rasetti discussed his observations on carbon monoxide and carbon dioxide. For the latter molecules Rasetti noted, evidently with some surprise, that the absorption frequencies normally observed for the gases did not appear as line-shifts of the secondary radiation and, vice versa, that the observed line-shifts were as yet unknown from absorption measurements (Rasetti 1929c). Wood, on the other hand, observed transitions including the so-called "missing line" in experiments on HCl-gas. In his publication in *Nature* printed in February 1929, Wood communicated his observation of a line "almost exactly [at] the centre of the vibration rotation band. The line thus appears to be the so-called 'missing line', corresponding to a vibration transition unaccompanied by change of rotation, which does not appear in the absorption spectrum of the gas." (Wood 1929, 279).

---

[26] For an overview of the drive towards intensity measurements and selection rules, see Brand (1989, 14-18).

[27] Wood had made his name by experiments on fluorescent and resonance radiation in the 1910's and contributed immensely to the empirical knowledge on which quantum theory was built. The importance of his work and his high reputation is mirrored in a total amount of 39 nominations for the Nobel Prize (https://www.nobelprize.org/nomination/archive/show_people.php?id=10300; visited on April 21. 2023), one of which was made by Niels Bohr in 1929. Bohr proposed to award the Nobel Prize in physics for the year 1930 to both Raman and Wood (Singh 2004, 81).

[28] Rasetti eventually gathered international reputation by his work on the Raman effect. He continued this line of research when he came back to his home institution, the University of Rome.

[29] Further discrepancies between Raman's conjecture and empirical data were found by Carrelli, Pringsheim, and Rosen (1928), Czerny (1929), Ellis (1929), or McLennan and McLeod (1929). For the sake of brevity and without loss of significant information I will only discuss the work by Rasetti and Wood. All of the observations had in common that the observed line-shifts did not coincide with known absorption lines of the studied material.

The identification of absorption lines with frequency shifts in the Raman spectra was not tenable. The idea, as proposed by Raman, that a photon provided the molecule with part of its energy which was directly taken up did not coincide with the empirical results of the more refined experimental methods. As Langer[30] noted in a short communication in *Nature*:

> The idea suggested by Smekal[31] that a molecule may subtract or add to an incident quantum one of its characteristic energy quanta and scatter the resultant sum or difference in a single quantum is so neat and clear that it is accepted as the explanation of the scattering experiments of Raman and Krishnan and others. The most important characteristics of such a process are those just indicated [e.g. absorption lines were equivalent to Raman line-shifts] which are not realised in the experiments. (Langer 1929, 345)

In the following months the attempts were increased and from mid February to early May 1929 at least eight papers were communicate discussing the selection rules of the Raman effect in connection with the Kramers-Heisenberg formula.[32] In all of these publications, the observations made in simple, mostly diatomic gases were used as the starting point for the discussion. All of the authors came to the conclusion that neither the intensity nor the selections rules could be given by the simple picture of direct energy exchange. How firmly established this conception already was, can be seen from the introductions of some of these works.

Gerhard Dieke (1929, 564) observed that "several investigators have recently stated that in many cases Raman lines are found which do not correspond with infra-red absorption frequencies [...]. Several authors state that this is contrary to the theoretical expectations." In a similar vein, Charles Manneback opened his considerations on the intensities of the Raman lines with the following remark:

> Since the experimental discovery by C. V. Raman and K. S. Krishnan as well as by Gr. Landsberg and L. Mandelstamm, at least implicitly, the view has been upheld that a strong secondary scattered radiation has to correspond to a strong ultra red absorption line in the molecular spectra (and vice versa). That this is not the case, has become apparent by the examples of quartz and recently the gas CO_2.[33]

---

[30] Langer actually worked with transitions in vibrational and not electronic quantum numbers, which proofed untenable. He also applied Schrödigner's ideas about the quantum theory of dispersion, i.e. that both initial and final state had to be occupied at once, which was not uncommon at the time, see for example Frenkel's discussion in Sect. 6.3.1 or Pringsheim (1928) and the corresponding correction by Max Born (1928). Yet, for the point I am about to make these details are irrelevant. For followers of Langer's idea and its untenability, see Brand (1989, 16).

[31] *Note by the author*: In the following, we will see that Smekal already favoured a different interpretation of dispersion.

[32] Rasetti (1929b), Rasetti (1929a), Rasetti (1929d), Amaldi and Segrè (1929), Amaldi (1929), Dieke (1929), Manneback (1929), and Hill and Kemble (1929).

[33] "Seit der experimentellen Entdeckung von C. V. Raman und K. S. Krishnan und von Gr. Landsberg und L. Mandelstamm wurde, mindestens implizit, schon die Anschauung vertreten, eine starke sekundäre Streustrahlung müsse einer straken Ultrarotabsorptionsstelle im Molekülspektrum entsprechen (und umgekehrt). Daß dies nicht der Fall sein muß, ist am Beispiel des Quarzes und neuerdings des Gases CO_2 klar geworden." Manneback (1929, 364).

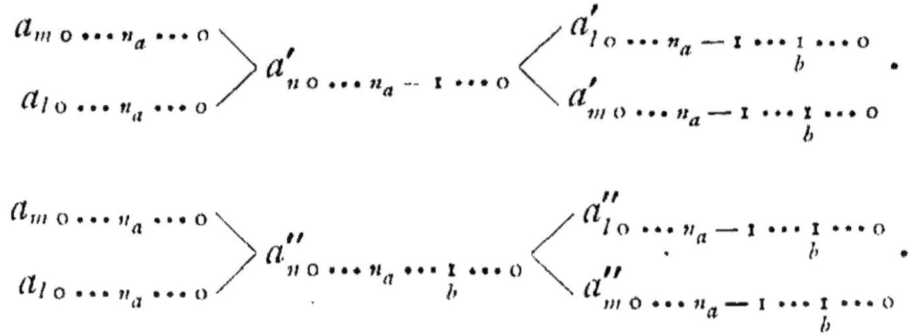

Fig. 6.1 Diagrams representing both the Stokes and Anti-Stokes case of the Raman effect. In the upper diagram, the absorption of the incident quantum occurs first. In the lower diagram, the emission of the final quantum occurs in the first step. Figures in Amaldi, Edoardo. 1929. "Sulla teoria quantistica dell'effetto Raman." *Rendiconti Accademia dei Lincei. Classe di Scienze Fisiche, Mathematiche e Naturali* 9, 876–881, here 878–879 (cited as Amaldi (1929) in the bibliography). Reprinted with kind permission by Accademia Nazionale dei Lincei

Also at Rassetti's home institute in Rome,[34] theorists noted and engaged with the seeming confusion amongst experimentalists. Edoardo Amaldi and Emilio Segrè (Amaldi and Segrè 1929, communicated on March 3. 1929) noted that the conception proposed by Raman, which they accredited to Smekal, as it was "currently accepted"[35], implied that the line-shifts should correspond (in selection rules and intensity) to infrared absorption lines. But as experiment had shown, this was not always the case. As the authors noted their motivation: "We seemed to notice that most experimenters wonder about this circumstance."[36] Following this introductory remarks, they provided a description in terms of the Kramers-Heisenberg formula and the concept of successive transitions.

In a follow-up paper, Amaldi (1929) even provided diagrams to make the conception of the processes according to Dirac's radiation theory and its verbal model accessible to his readers (see Fig. 6.1). The $a_{n0\ldots n_a\ldots 0}$ represented the probability amplitudes of the state of the total system. The state of the atom/molecule was represented by the first index (e.g. n) and the one of the radiation field by all following (e.g. $0\ldots n_a\ldots.0$). The diagrams were read from left to right. There were two initial and final states as Amaldi represented both the Stokes and Anti-Stokes case of the Raman effect in one diagram. Following the indexes, the upper diagram represented the sequence of first an absorption (see the index $n_a - 1$ of the intermediate state) and then an emission (1_b in the final state). The lower diagram represented first an emission (1_b in the intermediate state) and then an absorption. As should be noted, these diagrams strikingly differed from regular energy level diagrams: They repre-

[34] I shortly discussed the diagrams by Amaldi (see below) as an example of the influence of Dirac's verbal model on other representational formats in Ehberger (2022, 555–556).

[35] "correntemente accettato," Amaldi and Segrè (1929, 407).

[36] "Ci è sembrato di notare che la maggior parte degli sperimentatori si meravigli di questa circonstanza." Amaldi and Segrè (1929, 407).

sented the temporal conception on the quantum-amplitude level, as it was expressed in Dirac's verbal model, through the sequence from initial over intermediate to final state explicitly.

All of these publications, although most probably written independently, came to the overarching conclusion, that there was no direct connection between a possible absorption line and the line-shift of a Raman line, both in reference to the frequency shift and the intensity. The structure of the Kramers-Heisenberg formula in which a third state, Dirac's intermediate state, needed to be taken into account was guiding these considerations. According to the formula, the scattering was only allowed if initial and final state combined with at least one further state. They did not have to combine amongst themselves. As was noted, this third state normally corresponded to a transition in which the electronic configuration of the molecule changed.[37]

As the spectrum of these transitions regularly laid in the far ultraviolet, incident radiation in the visible or infrared spectrum could not lead to permanent transitions. In the framework of Dirac's radiation theory and the corresponding verbal model, energy non-conservation had to be taken into account for the intermediate step. Physicists not directly working with Dirac's conception were at times rather careful in their choice of words, but the range of comments was rather mixed. They often referred to the amplitudes corresponding to a transition but also directly to quantum jumps. Rasetti (1929b) used the term transition only for the complete process while he stated that initial and final state had to "combine" with at least one third state. Hill and Kemble (1929) also avoided talking about transitions to the intermediate states. John van Vleck (1929) used the term "combine" to speak about the connection between initial, final, and intermediate state. As he pointed out: "It is to be clearly understood that the term 'intermediate' relates merely to the position of the sate such as *b* in the products in Eq. (2) [the Kramers-Heisenberg formula], and does not at all imply that its energy is intermediate between those of the initial and final Raman states, as this is usually not the case." (van Vleck 1929, 755). A conceptualization in terms of temporal intermediate states was not even discussed by van Vleck.

6.2.2 *Raman's Defence: A New Meaning of "Virtual"*

Until mid 1929 Raman himself did not further publish on incoherent scattering and its conceptualization. In September 1929, he reacted to the new findings and the relevance of the quantum-mechanical description including the intermediate state. He did so at a symposium held by the Faraday Society in Bristol when summarizing the recent developments for the section concerning *The Raman Effect*. Raman was more careful in framing the frequency shift as equal "to a characteristic frequency of the molecule" (Raman 1929b, 783) and dated this insight back to his initial

[37] The matrix elements of transitions between purely vibrational states were very small in contrast to the matrix elements connecting different electronic configurations with each other. See e.g. van Vleck (1929, 755).

communication of the discovery in the *Indian Journal of Physics* in March 1928. He was evidently well informed about the experimental and theoretical work[38] on the newly established selection principles. In particular, he mentioned the work by McLennan, Wood, and Rasetti (Raman 1929b, 783), which were highly relevant for the establishment of the selection rules. Raman obviously accepted the empirical material gathered, the partial absence of infrared absorption spectra in the Raman effect, and the occurrence of line-shifts which were not proportional to infrared absorption frequencies.

Raman addressed the criticism of his conception, i.e. direct absorption of part of the energy of the impinging photon, and the relation between the incoherently scattered radiation, absorption spectra, and the theory of dispersion in more detail in the second part of his communication. When motivating the quantum-theoretical description by a short outline of the classical theory of dispersion, he emphasized that already therein "the refractivity of molecules for visible rays and for radiations in the near ultra-violet is determined by the existence of characteristic frequencies lying mainly in the far ultra-violet region of the spectrum" (Raman 1929b, 789). Nevertheless, Raman went on and defended his "simple and easily understood picture of the process involved in this new phenomenon":

> The process pictured [by Raman] is one which can actually occur, for the incident quantum of energy is greater than the quantum of energy absorbed by the molecule; that the interpretation is a valid one is shown by the fact that the difference between the frequencies of the incident and scattered radiations is characteristic for the molecule, and in not a few cases agrees with known infrared absorption frequencies. (Raman 1929b, 789)

According to Raman, the criticism of his initial conception was "based on a misapprehension". Raman acknowledged the role of the intermediate state in the quantum-mechanical scattering formula. Yet, he upheld that "the ultimate result is, of course, a transition of the system from A [initial state] to B [final state]". As Raman argued, the critics have "overlooked the fact that in the case with which we are actually concerned, the introduction of the third level *C* [intermediate state] is merely a *mathematical device*" as it corresponded to "a characteristic frequency of the molecule in the far ultra-violet". Since this transition could not be reached energetically when the incident radiation was in the visible spectral region "the transition of the molecule assumed *for the purpose of calculation* is a *purely virtual one which cannot actually occur*".[39]

Raman then asked how to further proceed in the evaluation of the effect and distinguished "two types of physical theory which experience proves to be useful in the exploration of a new field of research". He called the first one the "a priori or fundamental type of theory" which tried to reconstruct the empirical data and made new predictions by "a deep analysis of all the elements involved in the problem". The second type of theory was called "phenomenological theory" by Raman. It started from the observed phenomena and tried to establish from them a coherent picture of

---

[38] Raman referred to "Professor's Kemble's explanation of Wood's observation", which was most probably Hill and Kemble (1929), and to the paper by Amaldi (1929).

[39] All citations are taken from Raman (1929b, 789–790, emphasis added).

the effects under study. In line with his positivist attitude, Raman naturally favoured the second type of theory while "we may venture to hope that some day it will be possible to work it [the first type of theory] out in full detail, for our phenomenon".[40]

Before going into a deeper analysis of Raman's defence, I should point out that Raman's description of the phenomenon is an accepted way of understanding the quantum-theoretical dispersion formula. It is also true that the frequency shift depends only on the initial and final state and not on the energy of the intermediate state.[41] Furthermore, Raman actually has a valid point when noticing that some of his critics have overlooked the fact that energy is not conserved in the transitions to and from the intermediate states. Langer (1929) only considered combinations of infrared absorption frequencies in his initial contribution and actually attained an empirically untenable version of the processes. Also Amaldi and Segrè, who Raman explicitly mentioned, stated that one should "suppose that the molecule is in a state k [initial state] and consider a state j [intermediate state] so that the difference of energy $h\nu_{kj}$ is smaller than the energy of the incident quantum $h\nu$"[42]. Still, besides these two examples, the literature mentioned above did not restrict the intermediate state to be energetically in reach of the initial state and noticed that the transition was actually an electronic one.

The centre of Raman's argument was obviously the impossibility of reaching the intermediate state while upholding energy conservation. Yet, Raman did accept the selection rules that were suggested by the structure of the formula. The seeming contradiction was absorbed by Raman by a distinction drawn between a physical explanation of the phenomenon and its mathematical representation. The third level was only a "mathematical device"; the transition of the molecule to the third state C was only "assumed for the purpose of calculation". The line drawn by Raman ran between (observable and actual) physics and (unobservable and virtual) mathematics.

Raman was further very careful in not mentioning that his initial model of the process came with the conjecture that the frequency shifts would be proportional to absorption lines. In effect, Raman defended his initial model by shifting the third level C into the realm of mathematics. In harmony with his positivist outlook, Raman's initial interpretation only rested on the observable phenomenon of the frequency shift of the scattered radiation. Although Raman accepted the selection rules explainable by the quantum-theoretical dispersion formula, but more importantly established by experiment, he still argued that further theoretical research on the nature of the effect should progress along the lines of a "phenomenological" theory. The discrediting of the seemingly purely theoretical components of the scattering formula was in line with such an approach to physical theory.

---

40 All citations are taken from Raman (1929b, 790).

41 The difference of the energies involved are: $E_m - E_i$ and $E_m - E_f$, where i refers to the initial state, m to the intermediate state, and f to the final state (these are differences which occur if the initial and final state energetically both lie below the intermediate state). The difference between the energies give $(E_m - E_i) - (E_m - E_f) = E_f - E_i$.

42 "supponiamo che la molecola si trovi in uno stato k e consideriamo uno stato j tale che la differenza di energia $h\nu_{kj}$ sia minore dell'energia del quanto incidente $h\nu$." Amaldi and Segrè (1929, 408).

But we also have to take the historical situation Raman found himself in at the end of 1929 into account. When Raman gave the lecture before the Faraday Society, he was definitely hoping to be awarded the Nobel Prize in physics in the near future.[43] The admittance of an initial misinterpretation would definitely not fit Raman's personality[44] especially if Raman was able to identify theoretical lapses on the critics' side, while defending his initial interpretation within his positivist framework. The invocation of the term "virtual" can in this connection be understood as a rhetorical move.

There is, for the purpose of this investigation, still another aspect I want to emphasize in reference to the terminology. As outlined in Sect. 4.1.3, the term "virtual" as introduced in the BKS theory had two entangled meanings. On the one hand, it referred to an unobservable world of nearly classical entities. On the other hand, it was used to indicate the probabilistic connection between virtual and real world, on which the unobservability of the former was based. Energy non-conservation was real for BKS. While the reference to a "virtual oscillator" occurred in semi-classical approaches, the unobservable world seemed to have faded from sight as a possible connotation of the term "virtual" at around the time.

Raman's speech is the first instance I could locate in which the term "virtual" was connected to a concept of quantum theory, namely directly to transitions between states. Raman used the term "virtual" to point towards the energy non-conservation of the transitions and to locate them in the realm of mathematics. Quite in contrast, in the BKS theory energy non-conservation was an observable consequence of the theoretical framework and not the motivation for the introduction of the term. With Raman's connection of the term virtual to transitions an important shift in the meaning of the term "virtual" occurred. Energy non-conservation is to this day one prime aspect of the virtual particle concept, although of course in a modified form.

Raman, following his positivist and phenomenological attitude, identified the virtual transitions and states with "a mathematical device" which were "assumed for the purpose of the calculation" (see the quote above). An underlying kind of reality of the virtual entities was in Raman's invocation of the term excluded. Quite in contrast and as we will see in the Sect. 7.5, some physicists, as for example Dirac and Tamm, considered virtual transitions "physical", providing them with a more realistic reading.[45] In the mid 1930s, the group around Werner Heisenberg in Leipzig would as well use the term "virtual" to refer to such structures as Raman did. But, as I will argue in Sect. 9.1.2, their reading seems to cohere with the invocation of a virtual world rather closely, while still highlighting energy non-conservation.

Still, I believe, that Raman's resource for the term "virtual" was the BKS theory. While the usage of the term "virtual" is far more common in English than, for example, in German, there are indications. The distinguishing features of the BKS

---

[43] Only three months later Raman sent a letter to Niels Bohr asking him to propose him as a possible Nobel Laureate. Singh (2004, 79).

[44] One of his biographers, Singh (2002, 401), has noticed that "a mixture of self-confidence, pride, and arrogance were part of his [Raman's] personality".

[45] It is to note, that Dirac and Tamm did not use the term virtual at this point.

theory were the usage of the term "virtual" and energy non-conservation. Although initially distinct features, Raman blended these in the quote presented above. The structure of the Kramers-Heisenberg formula had not changed since its inception. In its original publication, the formula was still interpreted within the BKS framework. And Raman was evidently aware of this paper. Furthermore, even after quantum mechanics was established, the term "virtual oscillator" did not fade away completely but was, for example, used by Klein in his correspondence-like formulation, in which also the Kramers-Heisenberg formula was re-derived.

Hence in total my argument goes: BKS is the most probable source for Raman's usage of the term "virtual". Whether or not this conjecture is true, he definitely used the term as a rhetorical tool to mitigate the tension between his initial interpretation and the growing evidence that a more detailed picture of the process was necessary to explain the empirical data. Raman pushed the virtual transitions into the mathematical terrain. Given the historical situation, Raman's outlook on physics and his personality, the invocation of the term can be explained. All the same (and most probably unknowingly), Raman considerably altered the meaning of the term virtual (in comparison to BKS) as well as the theoretical entities it could be connected to: The term virtual was directly associated with energy non-conservation and, as far as I know for the first time, referred to a quantum concept, i.e. a transition. Thereby, not only the meaning but also the extension of the term "virtual" changed.

Raman's speech apparently led to the usage of the term "virtual" in connection with transitions or processes by some physicists. Robert Wood, who was present at the symposium, and Gerhard Dieke defended the picture implied by Dirac's radiation theory, but at the same time invoked "virtual transitions", "virtual absorption act" or "virtual reemission act" (Wood and Dieke 1930, 1357). In a context different from the Raman effect but still in connection with dispersion, it appeared that this terminology was first used by Henry Hulme (1932). He reminded the reader "that the transitions which occur in the dispersion formula are virtual ones only" (Hulme 1932, 238). In connection with problems of full quantum electrodynamics Heisenberg wrote to Pauli about the "infinitely many virtually possible transitions"[46] in reference to the self-energy of the electron. Rudolf Peierls used the term "virtual transitions" (Peierls 1934, 439) to describe the same phenomenon. As already indicated, these latter invocations, especially the ones by Werner Heisenberg and his student/assistant Hans Euler, will be further discussed in Sect. 9.1.2.

Returning to the development of Raman's thought, he admitted the relevance of the quantum-theoretical dispersion formula more fully in his Nobel Lecture. He emphasized "that though the conservation principle of Compton [which underlay his initial interpretation] is useful in interpreting the effect disclosed by experiment, it is by itself insufficient to explain the observed phenomena" (Raman 1930, 273). Nevertheless, Raman still upheld his distrust towards the quantum-theoretical explanation:

[46] Heisenberg to Pauli, February 5. 1934; as cited in Pauli (1985, 273).

> The work of Kramers and Heisenberg, and the newer developments in quantum mechanics [...] seem to offer a promising way of approach towards an understanding of the experimental results. But until we know much more than we do at present regarding the structure of molecules, and have sufficient quantitative experimental knowledge of the effect, it would be rash to suggest they afford a complete explanation of it. (Raman 1930, 274)

To conclude this subsection, I finally want to point the reader to an historical overview, published by one of Raman's associates in 1935, Suri Bhagavantam. It explicitly stated:

> The simplest picture that may be obtained of the production of a Raman line has been proposed by Raman himself and consists in regarding the collision between the molecule and a light quantum as satisfying the law of conservation of energy. The equation $h\nu +$ Molecule(normal) $\leftrightarrows h\nu' +$ Molecule(excited) represents the passage of a molecule from a normal state to an excited state the energy of the light quantum diminishing from $h\nu$ to $h\nu'$ if the transformation proceeds from left to right and *vice versa* if it proceeds in the opposite direction. [...]
> After a few preliminary investigations, it soon became clear that there is no exact one to one correspondence between Raman lines and the infrared absorption spectra and it was concluded that the simple equation, while indicating that an energy change similar to the one occurring in direct absorption is also responsible for the production of a Raman line, is not adequate enough to give a complete picture of the phenomenon. (Bhagavantam 1935, 527–528)

The first connection of the term "virtual" with transitions thus occurred in a historically entangled situation: Raman, hoping to be awarded the Nobel Prize in due time, used the term as a rhetorical tool in the defence of a conceptualization which was actually "not adequate" (using the words of his own co-worker). Most probably unknowingly, Raman changed the meaning and the extension of the term thereby.

6.3 A Former Conception Resurfaces: Actual Transitions to Intermediate States

Raman's initial interpretation coincided with Smekal's version of the effect as a direct energy exchange between the incoming radiation and the scattering molecule (Smekal 1923b).[47] Yet, there was a different conceptualization of the mechanism underlying dispersion, also followed by Adolf Semkal, still in play (but not in the mainstream of quantum theory) at the mid to late 1920s: the description of scattering as *independently* occurring acts of absorption and emission. This conception of scattering was first proposed by Einstein and Ehrenfest (1923). Adolf Smekal, building upon Karl Herzfeld's dispersion theory within a light-quantum framework, most explicitly formulated and defended such a conception.[48] In the following, I

[47] They did not coincide exactly: Smekal did not make any comments about the nature of the radiation field, while Raman used a photonic language. But this difference is unimportant for the further development.

[48] For more details, see my discussion in Sect. 4.3.

will show how this idea of dispersion was still upheld by Adolf Smekal until 1928, how it was adopted and reinterpreted by Yakov Il'ich Frenkel in 1929 and eventually refuted by Gregor Plazcek in the same year. I will further show how specific parts of the conceptualization given by Frenkel survived Placzek's disproof and were taken up in his textbook *Introduction to Wave Mechanics* published in 1934.

But let me first shortly recap this kind of interpretation: According to Herzfeld, dispersion and the necessary phase relation between the incoming and the scattered radiation should be explained by the concept of an "intermediate state" lying energetically in between the regular stationary states:[49] If the frequency of the impinging photon did not correspond to an energy difference between two stationary states of the scattering atom or molecule, the atom could still absorb the photon and reach such an intermediate state. After a short amount of time the atom would reemit the photon. This timespan should be much shorter than the lifetime of a regular stationary state.[50]

Adolf Smekal further worked with this idea in 1925 (Smekal 1925a, b). It should represent an alternative conceptualization to Kramers' and Heisenberg's derivation, which still had to assume a classical radiation field. According to Smekal, the scattering process should be conceived of as "two temporally *independently* occurring interactions of the atoms with the radiation field"[51]. Smekal defined absorption and emission coefficients for these "absorption- and emission-like" processes. The "meta-stationary state"[52] occupied by the atom after absorption should only differ from a regular stationary state by its shorter lifetime and a correspondingly lower probability to be occupied.[53]

Smekal's papers were based on Herzfeld's and Wentzel's work. Yet, Herzfeld or Wentzel did not further investigate the consequences of this conceptualization but rather turned to the work on quantum mechanics. Smekal himself elaborated on his view of independent acts of absorption- and emission-like processes and the concept of meta-stationary states in his compendium *Allgemeine Grundlagen der Quantenstatistik und Quantentheorie* (Smekal 1926a, Nr. 21; especially pp. 1108–1112). But also Wolfgang Pauli discussed the conception of scattering as a process of two independent acts of emission and absorption in his *Handbuch* article (Pauli 1926, 25–29). Yet, he did not invoke "meta-stationary states".

---

[49] Of course, this concept should not be confused with the intermediate states of Dirac. Dirac's intermediate states corresponded to stationary states and were only intermediate in reference to the processual modelling of the scattering phenomena.

[50] While the stationary states were supposed to have a lifetime of approximately 10^{-8} seconds, the intermediate state should only be occupied for 10^{-15} seconds in the case of visible light (Herzfeld 1924, 349).

[51] "[...] zwei zeitlich voneinander *unabhängig* efolgende Wechselwirkungen des Atoms mit dem Strahlungsfelde". Smekal (1925b, 242; emphasis in the original).

[52] This is what Smekal called the intermediate states of Herzfeld.

[53] Smekal noted that this constituted an effective breach with Bohr's first postulate (Smekal 1925a, 85).

Smekal used his conception even after the work on quantum mechanics started to gain speed.[54] After the discovery of the Raman effect, Smekal was quick in pointing to his initial theoretical prediction of the line-shift (Smekal 1928b), especially since Raman had not mentioned his name in the initial communication. But Smekal did not discuss how the scattering process should be conceived of in the short communication.[55] In a publication communicated one month later, mostly dealing with quantum processes without radiation, Smekal invoked the Raman effect as an analogy and noted:

> The Raman dispersion of light can be conceived of as decomposed into an absorption-like and an emission-like partial process, from which it follows, that real absorption and emission of radiation can be regarded as limiting cases of these elementary processes as in the classical theory of light.[56]

It should be noted that Smekal did not cite Dirac's quantum theory of dispersion at this point, but his set of 1925 articles on the "meta-stationary states" as well as Pauli's *Handbuch* article.

But Smekal was well aware of the new developments in quantum theory. He even cited Dirac's papers on radiation theory as a method for calculating emission and absorption coefficients (Smekal 1928a, 964–965). Seemingly, Smekal did not consider his conception to stand in contrast to the Kramers-Heisenberg formula or even Dirac's description of the interaction between radiation and matter.[57] And Smekal was not alone in conceiving of the scattering of radiation as an act of two independent emission and absorption processes: Yakov Il'ich Frenkel picked up Smekal's considerations and connected them to Dirac's time-dependent perturbation theory.

6.3.1 *Frenkel and a Third Concept of Intermediate States*

By 1929, Frenkel[58] was one of the most distinguished theoretical physicist of the Soviet Union and had already gathered a wide reputation in Europe and the US. Since 1921 Frenkel held the chair for theoretical physics at the Physico-Technical (Röntgen) Institute in Petrograd (from 1924 to1991: Leningrad; today: Saint Petersburg), which

---

[54] See for example, Smekal (1926b, especially p. 337).

[55] Nevertheless, he cited one of his 1925 papers.

[56] "[...] Raman-Zerstreuung des Lichts kann zerlegt gedacht werden in einen absorptionsähnlichen und einen emissionsähnlichen Teilprozeß, woraus sich ergibt, daß echte Strahlungsabsorption und -emission als Grenzfälle dieser Elementarprozesse anzusehen sind wie in der klassischen Lichttheorie." Smekal (1928a, 961–962).

[57] For the latter, this is not entirely clear. He cited the paper, but did not use it for the description of dispersion.

[58] If not otherwise indicated, the following biographical notes are taken from Frenkel (1974) and Frenkel (1996, Chaps. 2–4). For the (scientific) biography of Frenkel up until 1930 and his work on solid state physics, see also Kojevnikov (2004, Chap. 3, especially Sects. 2–4).

was to become one of the centres of physical research in the Soviet Union. During a research stay in Germany (late 1925 to late 1926) he could not only establish a wide network,[59] but also received first hand knowledge on the development of quantum mechanics.

Frenkel's research interests spread over many different areas. In the mid to late 1920s he mainly focused on the theory of electrons in metals and the application of quantum mechanics to the theory of the solid state.[60] At the time the Raman effect was discovered Frenkel was well aware of its quantum-theoretical explanation in terms of the Kramers-Heisenberg formula.[61]

Frenkel took up Smekal's concept of a "meta-stationary" state and the corresponding conception of the scattering process in a publication that dealt mainly with the absorption of light in gases and its subsequent transformation into heat (or kinetic energy) by collisions.[62] Frenkel used Smekal's interpretation either to motivate the specific theory he had in mind or to give an example of an extension of this mechanism. In fact, Frenkel's considerations about the Raman effect itself never made it into a full-length publication. Incidentally, this fits Frenkel's style of research in general. He used metaphors and analogies quite widely in order to develop new conceptions of processes and systems.[63] He freely developed ideas and put them forward without necessarily having developed them mathematically.[64] The rather casual introduction of further considerations in introductions or outlooks was also a trademark of Frenkel's papers: "Frequently, his ideas were expressed in an offhand manner, as certain variations on the main theme of the work in progress." (Frenkel 1974, 23).

The specific problem Frenkel had in mind, the transformation of light into heat in gases, was investigated by the application of Dirac's time-dependent perturbation theory. As Frenkel dealt with absorption only, he restricted himself to the first approximation for the interaction between light and matter. In his derivation, Frenkel particularly emphasized that "the frequency of the impinging light waves does *not* have to coincide with ν_{n0} [the frequency corresponding to the transition of the atom from the ground state 0 to an excited state n], so that this wave can induce the

---

[59] Among the physicists Frenkel had contacts with were Einstein, Pauli, Born and Jordan. Frenkel met Paul Dirac during the VI. International Congress of the Association of Russian Physicists in 1928.

[60] Until today, popular concepts of solid-state physics are still connected to Frenkel's name, as the Frenkel defect or the exciton. Frenkel's research focus was also due to the experimental work on these fields at the Physico-Technical Institute under the supervision of Abram Ioffé.

[61] Already in 1926, while staying in Göttingen, Frenkel published a paper on the Faraday effect in which he applied the Kramers-Heisenberg formula and shortly discussed the incoherent part of the secondary radiation (Frenkel 1926).

[62] I shortly mentioned Frenkel's conception discussed below as an example for an argument in favour of a realistic interpretation of virtual transitions (modern terminology) in Ehberger (2020, 265–266).

[63] See, for example, Kojevnikov's discussion of the development of the collectivist metaphor for electrons in metals, Kojevnikov (2004, Chap. 3, especially pp. 55–56).

[64] See the quotes by Abram Ioffé and Igor Tamm in Frenkel (1974, 2).

transition $0 \rightarrow n$"[65]. By invoking the well-known results of the perturbation theory, Frenkel found that for non-fitting radiation the coefficient for the occupation of the state n, $c_n(t)$, has a small and oscillating value.[66]

By averaging over the absolute square of the coefficient $c_n(t)$ Frenkel found a small, but finite value, which multiplied by the total number of atoms in a given volume was, according to Frenkel, equivalent to the number of atoms in the state n.[67] Frenkel used this value to calculate the change of the amount of heat which is created by collisions between excited and unexcited atoms. Although the state n could not have been reached energetically, it played, in Frenkel's theory, a decisive role in calculating the transformation of light into heat.

In his "closing remarks", which make up over a third of the paper, Frenkel invoked Smekal's conception of light scattering:[68]

> This process was conceived of by Smekal as a "double quantum transition" $0 \rightarrow n \rightarrow 0$ [this is for coherent scattering], in which the light quantum is first absorbed and then reemitted in a different direction without alteration of its energy. One conceives of the excited state n as an unstable "intermediate state" or "transition state", the lifetime of which can be disregarded. Therefore one does not have to worry about the breakdown of the energy law in the case $\nu \neq \nu_{n0}$. According to wave mechanics the lifetime of the transition state τ_{0n0} turns out to have a *finite* value, which is in general much smaller then the usual dwell time τ_{n0} of the excited state.[69]

Before further following Frenkel's reasoning, we must note that there are several important alterations to the original conception of Smekal in this short statement. Contrary to Frenkel's assessment, Smekal did not consider the time between absorption and emission negligible. The short, but finite lifetime of the meta-stationary states was their distinguishing feature. Frenkel further altered the concept of the intermediate state. For Smekal, the intermediate or meta-stationary state were energetically in between two stationary states. This was the reason, why Smekal did not have to worry about energy conservation. Energy was conserved in the transition to the intermediate state, while their existence was at odds with Bohr's first postulate. In

---

[65] "Die Frequenz ν der einfallenden Lichtwellen braucht *nicht* mit ν_{n0} zusammenfallen, damit diese Wellen Übergänge $0 \rightarrow n$ induzieren können." Frenkel (1929c, 799, emphasis in the original).

[66] This is exactly what Dirac and others found for the intermediate state. See for example Dirac's description of the intermediate state in his paper on dispersion theory (Dirac 1927b, 712), also discussed in Sect. 5.1.4.

[67] This can be seen analogous to Smekal's conjecture of 1925 that a small amount of molecules should occupy a meta-stationary state.

[68] Frenkel did not cite any of Smekal's paper but explicitly mentioned him in the text. Considering the conception Frenkel invoked, he most probably had read Smekal's 1925 papers.

[69] "Dieser Vorgang wurde nämlich von Smekal als ein 'doppelter Quantenübergang' $0 \rightarrow n \rightarrow 0$ aufgefaßt, bei welchem das Lichtquant zunächst absorbiert und dann in einer anderen Richtung ohne Änderung seiner Energie reemittiert wird. Den angeregten Zustand n stellte man sich dabei als einen unstabilen 'Zwischenzustand' oder 'Übergangszustand', dessen Lebensdauer vernachlässigt werden könnte, vor. Man brauchte sich deshalb um das Versagen des Energiesatzes im Falle $\nu \neq \nu_{n0}$ nicht zu kümmern. Nach der Wellenmechanik ergibt sich für die Lebensdauer des Übergangszustandes τ_{0n0} ein *endlicher* Wert, der im allgmeinen viel kleiner ist als die gewöhnliche Verweilzeit τ_{n0} des angeregten Zustandes n." Frenkel (1929c, 802; emphasis in the original).

Frenkel's interpretation of Smekal, the intermediate states corresponded to an excited state of the atom, which was intermediate in respect to the temporal order from intial over intermediate to final state. This redefinition of the intermediate state was most likely due to Frenkel's usage of Dirac's perturbation theory. Still, an important idea remained stable: The transitions were conceived of as independent.

In his further evaluation Frenkel provided two different ways of explaining or avoiding the problem of energy non-conservation for this, in his conception real processes. As a matter of fact, until today these two reasons are invoked as a motivation for the inclusion of the concept of virtual particles in the conceptual framework, i.e. as a narrative for the virtual particle as I defined it in Sect. 2.3.2.[70] At first, Frenkel (Frenkel 1929c, 802) pointed out that the problem of energy non-conservation can be mitigated by the short lifetime of the state in connection with the time-energy uncertainty. Since the state was assumed to have a rather short lifetime, the uncertainty in energy of this state was large and thus could explain the seeming breach with energy conservation.

But Frenkel further provided an explanation "in harmony with the energy law"[71]:

> If one conceives of the scattering of a light quantum as a collision process of this quantum and the respective atom – in the sense of Dirac's light-quantum theory [*footnote in the original*: I owe this remark to Mr. P. Dirac] –, then one further has to take their *mutual* [interaction] energy into account where the difference $h\nu_{n0} - h\nu$ [energy difference between impinging light quantum and state to be reached in the intermediate transition] has to be compensated at the expense of this mutual energy.[72]

As discussed in the previous chapter, in Dirac's time-dependent perturbation theory the interaction energy was not conceived of as part of the energy balance but as the cause for transitions. Frenkel used this fact to explain the seeming energy non-conservation. According to his own statement, Dirac made him aware of this idea. And as we will see in Sect. 10.2.3, Frenkel was not the last one to propose this reasoning.[73]

Frenkel (Frenkel 1929c, 802–803) even argued that this kind of explanation could easily take into account the observation that one and the same state could have two different lifetimes. If the state played the role of an intermediate state, the interaction with a light quantum actually present had to be taken into account. If the state was considered as a regular excited state, it only interacted with light quanta in the latent state, i.e. the photons introduced by Dirac to theoretically explain spontaneous emission in his first paper on radiation theory.

---

[70] Please note that Frenkel did not argue for the inclusion of virtual states in the conceptual framework, but for actual energy non-conserving transitions.

[71] "[...] im Einklang mit dem Energiesatz." Frenkel (1929c, 802).

[72] "Betrachtet man nämlich die Streuung eines Lichtquants als einen Stoßvorgang zwischen diesem Quant und dem betreffenden Atom – im Sinne der Diracschen Lichtquantentheorie [*Fußnote im Original*: Ich verdanke diese Bemerkung Herrn P. Dirac] –, so muß man noch ihre *gegenseitige* Energie in Betracht ziehen, wobei die Differenz $h\nu_{n0} - h\nu$ auf Kosten dieser gegenseitigen Energie zu kompensieren ist." Frenkel (1929c, 802).

[73] For a modern version of this argument, see the quote of Anthony Sudbery in Jaeger (2019, 13).

On a side note, but for clarity's sake, Frenkel actually stated that this description of the scattering process was unconnected with the Raman effect. But this was due to Frenkel's application of Schrödinger's conception of the dispersion formula, i.e. the idea that both initial and final state had to be occupied for a Raman line to occur.[74] This misconception was corrected in Frenkel's second communication published in *Nature* (Frenkel 1929b). Essentially, it constituted a summary of the paper in the *Zeitschrift für Physik* with some smaller alterations.

Frenkel now connected the incoherently scattered radiation with the concept of "double quantum switches" and opened his paper with his version of Smekal's ideas. According to Frenkel:

> The main point is that the intermediate states n of the double switch ($k \rightarrow n \rightarrow l$) have a definite *duration*, although a much shorter one than that of the two end states. This is contrary to the usual assumption, which does not regard the state n as "really" occuring, and it involves a temporary breach with the conservation of energy, though this is an objection that can be easily overcome formally. (Frenkel 1929b, 758)

Frenkel also mentioned, and thereby it is clear that he conceived of the transitions as actual ones, that one had to introduce two probability coefficients (one for each of the partial processes of the scattering) "which are a generalisation of Einstein's well-known B" (Frenkel 1929b, 758).

In Frenkel's work we therefore meet with a mixture of different conceptualizations that were often conceived of as mutually incompatible. On the one side, he gave a realistic interpretation of independent transition processes based quite freely on Smekal's work and formally connected it to the perturbative approach of Dirac; this resulted in the "meta-stationary" states of Smekal being identified with stationary states of the unperturbed system. In his first communication, Frenkel further discussed such a conception in connection with Schrödinger's interpretation of the wave function, which, according to Schrödinger, was initially presented as an alternative to quantum jumps. The gist of Frenkel's reasoning still was that the emission and absorption processes as well as the connected quantum jumps occurred independently. He even provided arguments why energy conservation might be expected. Although this might seem like a wild mixture of different conceptions, it fitted Frenkel's style of physical reasoning quite well.

As mentioned, Frenkel was known for a wide use of anthropomorphic language as well as analogical and metaphoric reasoning in his textbooks and his research publications.[75] Frenkel's reasoning was often led by model-like approaches. He showed a clear preference for intuitive ideas, that would guide his derivation of possible new

[74] Frenkel invoked Schrödinger's description of the dispersion formula in the discussed publication (Frenkel 1929c, 803). It was also part of the first edition of Frenkel's textbook on wave mechanics published shortly before (Frenkel 1929a, 182–183).

[75] The explanation of Pauli's exclusion principle by the picture of a hotel in which at most one married couple (spin-up and spin-down) was allowed to check into one room, or the invocation of the collectivist approach to the movement of electrons in metals, are only two out of many cases in point. See e.g. Frenkel (1996, 59, 112–113, 132) or Kojevnikov (2004, 52–56).

phenomena.[76] Frenkel's invocation of qualitative ideas rather than their exact formalization, was noted by his contemporaries and was at times seen rather critical.[77] Although this kind of imaginative and analogical thinking combined with Frenkel's intuition often bore fruit, Georg Placzek showed that, at this point, it had led Frenkel astray.

6.3.2 *Georg Placzek Shows that the Raman Effect Is Not Composed of Real Transitions*

Independently of and at the same time as Frenkel, Georg Placzek turned to the problem of scattering in the quantum theory and the (in)dependence of the involved transitions.[78] Placzek had studied physics in Prague and Vienna, where Smekal was Honorardozent for theoretical physics. After his dissertation in 1928, his first position as a post-doc was in Leiden with Hendrik Anthony Kramers. There, he started to engage with the Raman effect, a line of work which Placzek would follow up until the early to mid 1930s. Until today, the theoretical investigation of the Raman effect is closely connected with his name.[79]

In his first publication on the theory of the Raman effect, Placzek (1929) turned to the conception that was still entertained by Smekal and Frenkel: the modelling of the scattering process as two independent acts of emission and absorption. He rightly identified its roots in the works of Einstein and Ehrenfest, and observed that it was further elaborated by Smekal. Placzek even admitted that, in connection with the uncertainty relations, such a "decomposition does not seem to posses the character of a special hypothesis about the course of the scattering process"[80]. All the same, Placzek set out to prove that for incoherent scattering, the decomposition was not in accordance with the empirical material.

---

[76] Frenkel (1996, 132). Frenkel himself stated in a letter to Arnold Sommerfeld dating to March 8. 1928, that to him "it was always a pleasure to get simple results in the most simple and intuitive way". Frenkel to Sommerfeld, March 8. 1928, as cited in Frenkel (1996, 132).

[77] Arkady Migdal, one of Frenkel's post-doctoral students, remembered that theoretical physicist like Matvei Bronstein or Lev Landau rather "insisted on quantitative solutions and looked down on the qualitative ideas with which Frenkel was brimming". Cited in Frenkel and Gorelik (1994, 56–57).

[78] Both Frenkel and Placzek communicated their results in early October 1929.

[79] Placzek's article in the *Handbuch der Radiologie* of 1934 on Rayleigh and Raman scattering (English tranlsation: Placzek 1959) is still often cited and nearly every historical outline of the theoretical developments acknowledges his work with highest appraisal. Brand (1989, 22) even called Placzek "the dean of Raman theorists".

[80] "[will es jedoch zunächst scheinen, als ob der] Zerlegung hier nicht der Charakter einer besonderen Hypothese über den Verlauf des Streuprozesses zukäme." Placzek (1929, 585).

Placzek's main line of argument went as follows: The probability of scattering could, as was shown by Einstein and Ehrenfest (1923), at least formally be expressed in two different ways:

$$dW_{mn} = \rho(\nu)\Phi_{mn}(\nu, \nu')\left(1 + \frac{c^3}{8\pi\nu'^3}\rho'(\nu')\right)d\nu d\Omega dt \tag{6.3}$$

and

$$dW_{mn} = \rho(\nu)b(a_1 + b_1\rho'(\nu'))d\nu d\Omega dt \tag{6.4}$$

with $\frac{a}{b} = \frac{8\pi\nu^3}{c^3}$ and $\frac{a_1}{b_1} = \frac{8\pi\nu'^3}{c^3}$. $\rho(\nu)$ referred to the radiation density for the frequency ν, $\Phi_{mn}(\nu, \nu')$ was the probability coefficient for the atom/molecule in state m to scatter radiation of the frequency ν, while at the same to go over into the state n and sending out radiation of frequency ν'. Primed values, in general, refer to quantities after and unprimed values to quantities before the process took place. a and b refer to the probability coefficients for the subprocesses. Equations (6.3) and (6.4) are so far formally equivalent, but the first equation described one elementary process while the second represented a decomposition of the scattering process in terms of Einstein coefficients.

But, as Placzek showed, the formal equivalence could not be upheld when the coefficients a_1, b_1 and b were invested with the physical meaning that was defended by Smekal and Frenkel, namely as referring to actual emission and absorption probabilities. Then, as Placzek's argument went, it had to be assumed that b depended on the impinging frequency as well as the initial and an intermediate state, while a_1, b_1 depended on the frequency of the scattered radiation as well as the final and the intermediate state. As Placzek could show by invoking only very general arguments, the Formulas (6.3) and (6.4), invested with this physical interpretation, did not cohere with the predictions of the Kramers-Heisenberg formula. In particular, this conception would lead to different selection rules for the Raman effect.[81] Since the Kramers-Heisenberg formula had found empirical proof through the work of Wood and Rassetti, Placzek ruled out the idea that the two processes of emission and absorption occurred independently.

Thereby Placzek showed that while the process of scattering could be decomposed, according to Dirac's conceptualization, into an emission and absorption act on the quantum-amplitude level, the same decomposition had to break down if it was used for probabilities or probability coefficients. In consequence, such attempts as given by Smekal and Frenkel were deemed to fail when trying to reproduce the Kramers-Heisenberg formula and could not be upheld as viable candidates for further investigation.[82]

---

81 For scattering without change of wavelength the two conceptualizations coincided in their empirical testable results.

82 This was only the first part of Placzek's publication. In the second and third part he investigated the intensity relation between the Stokes and Anti-Stokes case of the Raman effect.

6.3.3 *Frenkel: Accepting the Consequences and Upholding the Narrative*

Seemingly Frenkel acknowledged these results.[83] At the same time, he still used the narrative initially provided for the interpretation of independently occurring acts of emission and absorption in his textbooks, now working on the probability-amplitude level. Take, for example, his 1934 textbook *Wave Mechanics. Advanced General Theory*. In a rather detailed discussion of the general time-dependent perturbation theory and its relation to Dirac's transformation theory (Frenkel 1934, Chap. 5, Sects. 21–24), Frenkel emphasized that the perturbation caused transitions between unperturbed states of the system.

Staying faithful to his preference for visualisable models, Frenkel even explicated the perturbative approach by reference to a set of infinitely many coupled pendulums: Each pendulum corresponded to one unperturbed quantum state (see Fig. 6.2). Frenkel suggested this analogy because the equations of the perturbative series described in classical theory a set of weakly coupled harmonic oscillators. According to Frenkel, "the 'pendelum model' [...] proves especially convenient" (Frenkel 1934, 206). He used it at diverse places throughout his discussion of the general theory of perturbations as well as in the concrete cases of the first and second order approximation.

In his explications, Frenkel again attributed the apparent violation of energy conservation to the role played by the part of the Hamiltonian used as a perturbation.[84] With this understanding Frenkel turned to the discussion of intermediate (or virtual) states in what he called "combined transitions":

> the fact, that in the combined transitions $H' \rightarrow H''' \rightarrow H''$ the particle has to pass through a state with an energy H''' different from the initial (and final) value does not prevent it from making such transitions. The apparent violation of the energy law for each of the two "legs" of the jump from H' to H'' can obviously be straightened out by taking into account the perturbation energy S not only as the cause of the transition but also as an invisible factor in the energy balance. If, for instance, $H''' > H'$, then we can imagine that the energy $H''' - H'$, which is required for the first step of the transition, is "borrowed" from the perturbation energy S and restored to it during the second step. The probability amplitude [...] of the state H''' will remain small, the corresponding probability [...] oscillating about [a] constant value [...], while the number $N_{H''}$, though initially much smaller, increases with the time, and may finally become very large. (Frenkel 1934, 226–227)

For the visualization of the process, Frenkel provided his readers with two possible models. On the one hand, one might imagine each state to be represented by a vessel.

---

[83] I did not find any reference by Frenkel to Placzek's paper. Yet, the empirical phenomenon discussed in 1929, the transformation of light into heat in gases, was treated by Frenkel in his textbook on wave mechanics on the probability-amplitude level and not on the level of actual probabilities (Frenkel 1934, 232).

[84] "It [the perturbation] can therefore [because it is not taken into account for the energy of the whole system] be regarded as the 'goat' responsible for the deviations from the conservation law [...] in the transitions for which this equation is not satisfied." Frenkel (1934, 226). See the surrounding paragraphs for the whole discussion of this fact by Frenkel (1934, 225–226).

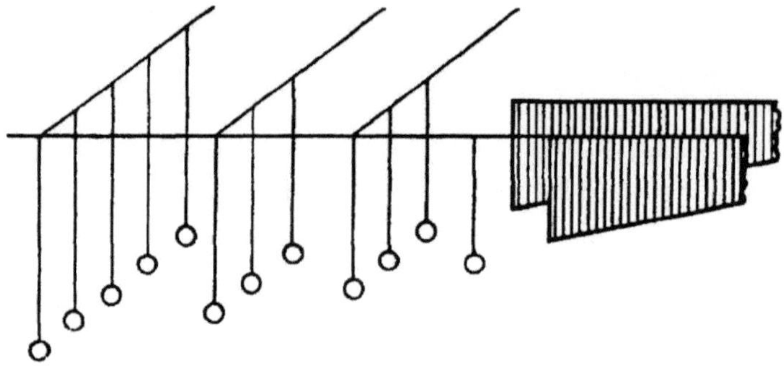

Fig. 6.2 Depiction of Frenkel's pendulum-model of the perturbation theory in quantum mechanics. Taken from Frenkel (1934, 207)

The liquid filled into a vessel should represent the probability of the number of copies in this particular state. According to this model, the perturbation would pump the liquid from the initial state to the final state, which are connected by the vessel representing the intermediate state. The liquid would not accumulate in the vessel corresponding to the intermediate state but only pass through it. Yet:

> A still better picture of this transition is provided by our pendulum model, the probability or number of copies being represented by the energy flowing from the pendulum H' to the pendulum H'' which is coupled with it through the pendulums H'''. The lack of resonance between the latter and H' results in these pendulums performing steady oscillations of small amplitude and functioning simply as carriers of energy from H' to H''. (Frenkel 1934, 227)

In this vein, Frenkel still argued for a quite literal reading of the specific subprocesses, backing it up with two different classical models. All the same, from a mathematical point of view, hc obviously acknowledged that a reading in terms of independent emission and absorption subprocesses was not tenable. The arguments he had previously provided for the conception of real and independent subprocesses were still upheld:[85] Now, they functioned as a narrative for the inclusion of energy non-conserving transitions into the conceptual framework of quantum theory, or more to the point, time-dependent perturbation theory, but on a quantum-amplitude level, i.e. for virtual transitions (modern terminology).

6.4 Virtual Transitions: Neither Actual, Nor Ineffective

In the prior sections, I discussed two different conceptualizations of the scattering mechanism in quantum theory which were provided or resurfaced shortly after the empirical discovery of the Raman effect: the direct energy exchange between photon and atom (Raman's initial conception) and a realistic conception of independent

[85] Quite noteworthy, Frenkel did no longer refer to the uncertainty relations.

acts of emission and absorption processes (Smekal and Frenkel). In reference to the intermediate (or virtual) states and transitions, both of these models can be considered as lying on two different sides of a spectrum: The first one denied the intermediate state to have any influence on the observation; Raman line-shifts were identified with absorption frequencies of the material part of the system. The second interpretation postulated specific Einstein coefficients for the two subprocesses, considering them consequently as real and independent. Both of these suggestions were found to be untenable in the face of the empirical material gathered in early to mid 1929. The respective empirical results showed that the mathematical structure of the Kramers-Heisenberg formula rightly reconstructed the observed selection rules, while the differing interpretations would lead to different observable consequences.

In my reconstruction of this episode, the empirical data on the Raman effect was important for the decision between the possible conceptualizations of the scattering mechanism. Yet, it was not the discovery of the line-shifts itself, but further empirical data that was gathered mostly in early 1929 by diverse physicists in Europe, Asia and the US. Only by reference to molecules simple enough and with known absorption spectra, the difference between the Raman line-shifts and the absorption lines could be identified. The establishment of the selection rules, i.e. the rules governing which two states could combine as initial and final state in the scattering process, became the focal point of my reconstruction. As of mid to late 1929, only the Kramers-Heisenberg formula, working with virtual and not-independent transitions through intermediate states could reproduce the selection rules that were empirically established.[86]

Still, both Raman and Frenkel did not have to give up all aspects of their models and arguments. Frenkel had to give up the independence of the different acts of emission and absorption. At the same time, he could take over his invocation of the perturbation Hamiltonian as an invisible factor in the energy balance and argue for a rather physical interpretation of the transitions on the quantum-amplitude level. For Frenkel, the narrative functioned both for real energy non-conserving as well as for virtual transitions. In his positivist attitude and contrary to Frenkel, Raman still denied the intermediate state any physical reality and pushed the description of the scattering mechanism completely into the realm of mathematics. Thereby, he defended his "simple and easily understood" picture of the phenomenon.

Concerning the different modes of representation of a concept, the above episode is rather insightful in respect to their mutual development: From a mathematical point of view, the structure of the Kramers-Heisenberg formula did not change. An explication of the different subprocesses, in terms of a verbal model, was only provided by Dirac two years later (see Chap. 5). Within his radiation and perturbation theory, such a reading was not only suggestive but also necessary for the derivation. The introduction of the verbal model directly influenced the diagrammatic representations

[86] Due to the complicated nature of the evaluations dealing with the most general formulation of dispersion in quantum theory, by mid to late 1929 a further development started in the theoretical development of the Raman effect: Physicists shifted focus on symmetry characteristics of molecules without a direct recourse to the Kramers-Heisenberg formula. But this is a different story and one in which the intermediate states lost their relevance.

of theoretical physicists: The diagrams of Amaldi (see Fig. 6.1) digressed from the more popular energy level diagrams and clearly indicated a temporal order of events. Now, again two years after the introduction of Dirac's verbalization and without change of either verbal model or mathematical structures, Raman introduced the term virtual to refer to the transition subprocesses.

Raman could have taken the term "virtual" from the BKS theory. At the same time, Raman altered both the extension and the intension of the term. For the first time, the term virtual was connected to a quantum concept, i.e. transitions. Within the BKS theory the term was exclusively connected to classical concepts, i.e. oscillators and radiation field. In Raman's usage, energy non-conservation and the realm of mathematics were clear connotations of the term. While the BKS theory was best known for only statistical energy conservation, the use of the term virtual therein was not a consequence of it or the distinguishing feature of the entities referred to as virtual. Thus an important shift in the meaning of the term occurred in Raman's lecture (but evidently also in later applications): Energy non-conservation became one of the most prominent differences between virtual and actual transitions. The connection of the term virtual to a virtual quantum world was missing in Raman's lecture as well (which is rather unsurprising given his phenomenological commitment); to Raman, virtual transitions were pure mathematics. This would change by a later invocation of the term in the mid 1930s, most importantly by Werner Heisenberg (see Sect. 9.1.2). But this would have to wait another five to six years.

Yet, with Raman's 1929 talk and the publication by Georg Placzek this first part on the history of the virtual particle concept comes to a conclusion. First of all, the connection between the terms "virtual" and "transitions" (or processes, in more general terms) was established. From roughly the mid 1930s to the late 1940s, this was one of the most common expressions to refer to entities connected to the historical trajectory of the virtual particle concept. Secondly, the relevance or, better, adequacy of these entities was supported by experimental results: With the discussion around the selection rules it was established that, on the one hand, a picture of direct energy exchange was not tenable. On the other hand, also a realistic reading of the intermediate states could not be upheld. Thirdly, two very important representations of the concepts I am investigating have been established and will become widely used in the course of the 1930s: the mathematical, i.e. perturbative, representation of Dirac's radiation theory, and the corresponding verbal model which allowed for its application in concrete situations. In the following three chapters, I will engage with themes that have also occurred in this section: The epistemic relevance of concepts connected to the virtual particle's historical trajectory and the role of the representational formats in the reasoning processes.

Part II
Theoretical Practice with Virtual Transitions (1928–1942)

Chapter 7
Scattering and the Sea: Antiparticles and Intermediate States (1928–1931)

In the following chapter I will revisit the development of one of the most prominent, but also most notorious ideas in the history of quantum electrodynamics: the Dirac sea and the corresponding concept of anti-matter as a hole therein. I will do so with the specific focus on the conception of scattering which was proposed by Paul Dirac, i.e. a sequence of successive, later to be called virtual, emission, absorption, and transition subprocesses. For the greater part of the following chapter, I will not discuss the introduction of new means of representation or, in a broader sense, the conceptual development of the theoretical entities I am investigating. Rather, the argument will focus on the role of virtual transitions (modern terminology) in the development of the idea of the Dirac sea and Dirac's identification of a hole in the sea with an anti-electron in 1931. I will investigate the epistemic relevance of virtual transitions for the development of the concept of anti-matter, the (re-)configuration of its conceptual surroundings, and the stances physicists took towards the physicality of virtual transitions[1] in this process.

Even though my historical discussion concentrates on the theoretical explanation of empirical phenomena, new experimental results played next to no role in the development I am about to discuss in this chapter.[2] But even when focusing on the theoretical modelling of the relativistic wave equation and its conceptual surroundings in the period from 1928 to 1931, the secondary literature has a considerable

[1] A short outline of the argument on the physicality of the intermediate states in the hands of Ivar Waller, Paul Dirac and Igor Tamm as presented in Sects. 7.3 and 7.5 can be found in Ehberger (2022, Sects. 2.1–2.2). Here, I will provide for a detailed evaluation and historical contextualization.

[2] The experimental discovery of the positron and its identification with a hole in the Dirac sea occurred in 1932–1934 and are, thus, outside the historical focus in this chapter. For a concise and thorough discussion, best see Roqué (1997).

M. Ehberger, *Representing the Unobservable*, Science Networks. Historical Studies 68,
https://doi.org/10.1007/978-3-032-09188-8_7

scope.[3] Historians of science, for example Pais (1986, 348–349), have already recognized that the scattering process and intermediate states played a crucial role in the inception of the Dirac sea.[4] On the basis of the correspondence between Paul Dirac and Ivar Waller, Karl Grandin (2008, 200–208) has most decidedly argued that the occurrence of negative energy intermediate states was the motivation for Dirac to turn to the general problem of negative energy states in late 1929.

I will focus on the role of the description of scattering, especially on the virtual states (modern terminology), throughout the whole historical episode, from the initial discussions on the negative energy solutions to the identification of a hole in the Dirac sea with an anti-electron. Thereby, I want to provide for a new perspective on the development of the most prominent, or at least most practised, conception of anti-matter until the late 1940s. I will study the impact of Dirac's mathematical and verbal description of the scattering process on its conceptual and theoretical surroundings, i.e. on the development and reconfiguration of theoretical entities and effects. At the same time, this historical episode allows for carving out the different stances which physicists took towards the physicality[5] of transitions to intermediate states for which "the usual assumption [...] does not regard the state *n* [intermediate state] as 'really' occurring" (Frenkel 1929b, 758). Hence, this chapter will be concerned with the epistemic consequences and the ontological connotations of virtual transitions (modern terminology).

I will first introduce what I will call the "general negative energy problem" (Sect. 7.1.1): Interpreted as a one-particle equation, Dirac's relativistic description of the electron necessarily included (real transitions to) negative energy states. But already in mid 1928 (Sect. 7.1.2), Werner Heisenberg recognized the necessity of

---

[3] Kragh (1981) is still the standard reference for the details of the development of the relativistic wave equation in the hands of Dirac. Moyer (1981b), Moyer (1981a), and Moyer (1981c) stretch over the whole period from the inception of Dirac's equation to the discovery of the positron, giving equal account of Dirac's contributions and those of others. Kragh (1990, Sect. 5) located the development of the idea of the Dirac sea and the inception of the anit-electron in Dirac's scientific biography and the criticism of his peers. Darrigol (1982, especially Sect. 2.2), Pais (1986, Chap. 13 (e) and 15 (f)), as well as Schweber (1994, Sect. 1.6) provided concise and technically informed reconstructions from the development of the Dirac equation to the discovery of the positron. Grandin (2008, 200–208) discussed Ivar Waller's contribution to the problem of negative energy intermediate states and his correspondence with Dirac shortly before the inception of the Dirac sea. This article provides the basis of Sect. 7.3. Wright (2016, especially Sect. 2) placed the evolution of Dirac's thinking about the negative energy states in his correspondence network, while discussion of the arguments connected to the scattering mechanism was not included. Wright (2024, Sects. 2.3–2.4) has reevaluated the corresponding period, particularly focussing on the impact of many-electron theory on Dirac's modelling of the vacuum. Monti (1995) focused on the theoretical reasons for a re-identification of a hole in the Dirac sea as a positron at least partly taking the conception of scattering into account.

[4] Also Wright (2024, Sect. 2.3.1) has noted the role of Waller's scattering calculations, while not being aware of both Karl Grandin's work (Grandin 2008) and the correspondence between Dirac and Waller.

[5] At times it is hard to determine whether or not the actors' term "physical", which especially Dirac and Tamm used, included an ontological connotation. I will comment on this concept in the respective sections.

including transitions to negative energy intermediate states when the scattering of light is described by a combination of Dirac's radiation theory and the Dirac equation. Since Heisenberg came to this conclusion nearly one and a half years prior to Dirac's proposition of the Dirac sea, its impact on the initial investigations of the scattering from relativistic electrons will be investigated in Sect. 7.2. Sects. 7.3 and 7.4 will focus on the correspondence between Ivar Waller and Paul Dirac in the route to the Dirac sea. First, I will follow Karl Grandin's argument, enriching it with a discussion of Ivar Waller's oscillations between a more or less physical reading of the intermediate states, both in semi-classical and quantum-electrodynamical descriptions of scattering. Secondly, I will thoroughly revisit the argument whether the occurrence of negative energy intermediate states really constituted the decisive motivation for Dirac's turning towards the general negative energy problem.

While Ivar Waller was oscillating between a more or less formal reading of the intermediate states, Paul Dirac and Igor Tamm, close associate of Dirac at the time, shared a "physical" reading of the intermediate states, to be discussed in Sect. 7.5. In Sect. 7.6, I will revisit the historical development from Dirac's inception of a filled vacuum to Dirac's identification of a hole therein with an anti-electron. My main aim is to show how Dirac's conception of the scattering mechanism, which had to be reconfigured in the Dirac sea framework, and the analogical theoretical structure of the annihilation process caused first major difficulties for Dirac's identification of hole with a proton. Through their structural analogy, the numerical results derived from the theoretical evaluations (once backed up by empirical results, once completely off-bounds) created a conceptual tension. Finally (Sect. 7.7), I will discuss the consequences of the new means of describing scattering in a filled vacuum for the inception of possible, as yet unforeseen processes through the example of a little known paper by Igor Tamm (1930a). In conclusion, I will reflect on the epistemological role of Dirac's description of the radiation field-matter interaction being composed of successive acts of emission and absorption processes and the connected concept of the intermediate state in the discussed episode.

7.1 The First Occurrence of Negative Energy Intermediate States

7.1.1 Background: The General Negative Energy Problem

In early 1928, Paul Dirac published his relativistic description of the electron (Dirac 1928a).[6] Throughout the year 1927, Dirac became more and more unsatisfied with the proposed relativistic generalizations of quantum theory, most importantly the Klein-Gordon equation, named after Oskar Klein and Walter Gordon. This equation

[6] For the historical genesis of the Dirac equation and the discussion of its precursors best see Kragh (1981), Kragh (1990, Chap. 3), or Moyer (1981b).

was quadratic in the energy operator or, equivalently, in the time derivative. But the linearity in these operators was the basis of Dirac's transformation theory and its physical interpretation (Dirac 1927a). Transformation theory had served Dirac well, especially in his derivation of radiation theory,[7] and had immediately become a major resource for the interpretation of quantum theory, as epitomized by its prominence in Heisenberg's uncertainty paper (Heisenberg 1927). Thus, when constructing a relativistic wave equation for the electron, Dirac started out with an Ansatz which was linear in the time-derivative and the Hamiltonian operator.

Dirac thus based his new relativistic formula on the following equation:

$$(p_0 + \alpha_1 p_1 + \alpha_2 p_2 + \alpha_3 p_3 + \beta)\psi = 0 \tag{7.1}$$

with $p_0 = \frac{E}{c} = i\hbar\frac{\partial}{c\partial t}$ and p_i; $i = 1, 2, 3$ referring to the momentum components, the αs being independent of t, x_1, x_2, x_3. Under the assumption, that this equation would reproduce the relativistic energy-momentum relation for high quantum numbers, Dirac showed that it could be written as:

$$[p_0 + \rho_1(\sigma, \mathbf{p}) + \rho_3 mc]\psi = 0 \tag{7.2}$$

where σ_i and ρ_i are four by four matrices built up out of the Pauli spin matrices. They obeyed the following relations:

$$\sigma_i^2 = \rho_i^2 = 1 \tag{7.3}$$

$$\rho_i\rho_j + \rho_j\rho_i = \sigma_i\sigma_j + \sigma_j\sigma_i = 0 \text{ for } i \neq j \tag{7.4}$$

$$\left[\sigma_i, \rho_j\right] = 0 \tag{7.5}$$

The usual procedure of minimal coupling allowed Dirac to introduce the radiation field into this equation ($p_0 \rightarrow p_0 + \frac{e}{c}A_0$ and $\mathbf{p} \rightarrow \mathbf{p} + \frac{e}{c}\mathbf{A}$). By construction, Eq. (7.2) could be interpreted according to the usual rules of quantum theory. Dirac further proved the invariance under Lorentz transformations and that spin was naturally incorporated in this scheme.

At this point I already want to mention an important alteration, which is a direct consequence of the formal structure of the Dirac equation. The vector potential enters linearly therein. As was observed diverse times and independently by different physicists,[8] this entailed that the "direct" or "true" scattering term of Dirac's 1927 papers on radiation theory vanished. Thus, all of the scattered radiation had to be somehow included in the term including intermediate states. As we will see, establishing the connection between Dirac's new description and the direct scattering term will be

[7] In the introduction to his first paper on radiation theory, Dirac explicitly mentioned that "the mathematical development of the theory has been made possible by the author's general transformation theory of quantum matrices" (Dirac 1927c, 245).

[8] See the discussions throughout this chapter. It was, at least, independently identified by Werner Heisenberg, Ivar Waller, and Igor Tamm.

the starting point for the arguments of the necessity of negative energy intermediate states. But how do negative energy states enter in Dirac's relativistic description in general?

The free solution of the Dirac equation,[9] is a four component wave function, which can be written in the form:

$$\psi(x) = u(p)e^{-\frac{i}{\hbar}(Et-px)} \tag{7.6}$$

The general negative energy problem now occurred, as two of these solutions referred to positive energies and two to negative energies. In classical theory negative energies could simply be ignored as no transitions between positive and negative energy values could occur.[10] In quantum theory, on the other hand, "a perturbation will cause transitions from states with W positive to states with W negative" (Dirac 1928a, 612). As the first set of solution corresponded to a particle with negative charge and the second one to a particle with positive charge, Dirac initially interpreted "such a transition [...] as the electron suddenly changing its charge from $-e$ to e, a phenomenon which has not been observed" (Dirac 1928a, 612).

As for Dirac's radiation theory, the novelty of the description of the electron was not connected to empirical predictions, but rather to the coherence and generality of the formalism. For example, spin, previously grafted on the equations, was naturally incorporated in the Dirac equation. News about Dirac's theory of the electron spread quickly amongst the community and it was generally taken up favourably.[11] Nevertheless, the problem of the negative energies remained and physicists conceived of it as a serious drawback. For example, Wolfgang Pauli immediately asked Dirac if he was sure about his claim that transitions between positive and negative energy states were included in his theory: "It could as well be possible, that the energy values *always* decompose into *non*-combining groups?"[12]

In April 1928, Dirac started a lecturing and research tour that went from Leiden, over Leipzig and Göttingen, to the Soviet Union. In Leiden, Dirac met Igor Tamm with whom he bonded quickly. Their friendship and the discussions on physics will become important in the second part of this chapter. Together with Tamm, Dirac first went to Leipzig, where Werner Heisenberg had become professor just a few months earlier. In his presentation at the "Leipziger Universitätswoche", Dirac definitely negated Pauli's question:

[9] The following presentation is somewhat anachronistic, as Dirac himself did not write down these equations. All the same, the problem of negative energy states was clearly recognized from the beginning.

[10] This attitude seemed to have prevailed priorly also for the Klein-Gordon equation, see Kragh (1981, 63–64).

[11] See e.g. Kragh (1990, 62–64). For an overview of the initial applications of the formula, see e.g. Moyer (1981a, 1055–1056).

[12] "Es könnte doch sein, daß die Energiewerte *stets* in *nicht* kombinierende Gruppen zerfallen?" Pauli to Dirac, February 17. 1928; as cited in Pauli (1979, 437–438); emphasis in the original.

> The theory allows for transitions for which $+e$ transforms into $-e$. Nevertheless, the probability of these transitions is extraordinarily small (of the order $(\frac{v}{c})^4$; c = velocity of light; v = velocity of the electron). Hence the current theory is only an approximation. It seems, that this difficulty can only be remedied by a fundamental change of our previous conceptions, and possibly is connected with the difference between past and future.[13]

In the absence of any solution, not only Dirac called for radical digression from the current conceptual framework. Also Heisenberg thought "that the problem will not be solved until one has a theory of the proton and electron together".[14] Although, or maybe because, physicists had high hopes, within half a year the general negative energy problem was conceived of as a huge problem for the interpretation of the quantum-mechanical description of a relativistic electron. Werner Heisenberg famously reported to Pauli that Pascual Jordan, who had also been working on a new relativistic formulation, had become "depressed [trübsinning]" over the problems with Dirac's equation. In Heisenberg's eyes the Dirac equation had become the "saddest chapter of modern physics".[15]

7.1.2 Heisenberg's Observation: Scattering Through Negative Energy States

Werner Heisenberg had actually engaged himself intensively with Dirac's equation and the general negative energy problem in the first half of 1928 (Rechenberg, 2010, 705-706). Together with Pauli, he had already launched their programme to establish a consistent quantum electrodynamics and was therefore also deeply involved with studying Dirac's radiation theory. Consequently, it is rather unsurprising that it was Heisenberg who first hit upon the problem that takes centre stage in this chapter: the negative energy intermediate states.

After Dirac had lectured in Leipzig, Heisenberg tried to find the roots of the problems of Dirac's theory of the electron by a deeper investigation of different physical processes. On July 23. 1928 he reported to Niels Bohr: "For example the spontaneous transitions from $+mc^2 \rightarrow -mc^2$ should probably occur much more

---

[13] "Die Theorie gestattet Übergänge, für welche $+e$ in $-e$ übergeht. Jedoch ist die Wahrscheinlichkeit für diese Übergänge außerordentlich klein (von der Größenordnung $(\frac{v}{c})^4$; c = Lichtgeschwindigkeit; v = Geschwindigkeit des Elektrons). Folglich ist die gegenwärtige Theorie eine Annäherung. Es scheint, daß diese Schwierigkeit nur durch eine fundamentale Änderung unserer bisherigen Vorstllungen behoben werden kann, und vielleicht im Zusammenhang steht mit dem Unterschied zwischen Vergangenheit und Zukunft." Dirac (1928b, 94). Such a radical stance might have been motivated by Niels Bohr's speculations about the time direction in quantum mechanics (Kragh 1990, 327, Note 60).

[14] Dirac to Klein, July 24. 1928; as cited in Kragh (1990, 88–89).

[15] "traurigste Kapitel der modernen Physik," Heisenberg to Pauli, July 31. 1928; as cited in Pauli (1979, 467).

frequently than any other spontaneous emissions."[16] But on the contrary, one also recognized, "in the application of the Dirac theory to other questions, e.g. scattering of light, especially for free electrons, that the ominous transitions are nevertheless an integrating part of the Dirac theory".[17] What Heisenberg meant by an "integrating part" becomes clear from a letter he wrote a week later to Wolfgang Pauli.[18]

Let me discuss Heisenberg's argument in a little bit of detail, as the same kind of argument will take prominence in Ivar Waller's correspondence with Dirac and the further development. Heisenberg reported to Pauli that, when combining Dirac's theory of radiation with Dirac's relativistic wave equation, negative energy intermediate states had to be taken into account in the description of coherent scattering. The first thing Heisenberg noted in his derivation was that the expression referring to direct scattering ["wahre Dispersion"] vanished in this combination. As mentioned, this was a straightforward consequence of the structure of the relativistic Hamiltonian because the vector potential entered linearly therein.

Heisenberg's argument was that the formula classically describing the coherently scattered radiation by a free electron, i.e. Thomson scattering (named after Joseph Thomson),[19] also had to be explained by the processes of "absorption and reemission",[20] i.e. according to the formula:

$$P_r \propto \left| \sum_{J''} \frac{\dot{x}_r(J'J'')\dot{x}_s(J''J')}{\nu(J''J') - \nu_s} + \frac{\dot{x}_s(J'J'')\dot{x}_r(J''J')}{\nu(J''J') + \nu_s} \right|^2 \tag{7.7}$$

where the $x(J'J'')$ referred to the matrix elements of the position operator between the states J' and J''. The J'', thus, referred to the configuration of the intermediate states of the free electron.

As Heisenberg laid out, for a non-relativistic free electron with no negative energy states, such a scattering mechanism would be impossible since the free electron had no absorption lines.[21] But according to the Dirac equation, the free electron has the absorption frequency $\frac{2mc^2}{h}$, corresponding to the "notorious transitions $mc^2 \rightarrow -mc^2$".[22]

---

[16] "Zum Beispiel sollten die spontanen Übergänge $+mc^2 \rightarrow -mc^2$ wohl viel häufiger vorkommen, als irgendwelche anderen spontanen Emissionen." Heisenberg to Bohr, July 23. 1928; as cited in Rechenberg (2010, 739).

[17] "[...] bei Anwendung der Diractheorie auf andere Fragen, z.B. die Streuung des Lichtes, besonders bei freien Elektronen, dass die ominösen Übergänge doch ein integrierender Bestandteil der Diractheorie sind." Heisenberg to Bohr, July 23. 1928; as cited in Rechenberg (2010, 739).

[18] Heisenberg to Pauli, July 31. 1928; as cited in Pauli (1979, 466–469).

[19] See Dirac (1927b, 719–720) for Dirac's derivation of the classical formula from his true scattering term.

[20] "Absorption und Reemission," Heisenberg to Pauli, July 31. 1928; as cited in Pauli (1979, 468).

[21] In the denominators of Eq. (7.7), the resonance frequencies $\nu(J''J')$ referred to possible absorption lines of the scattering material. Dirac (1927b, 719) had already noted in his paper on dispersion that "the true scattering is the only kind of scattering that can occur for a free electron".

[22] "berüchtigten Übergänge $mc^2 \rightarrow -mc^2$," Heisenberg to Pauli, July 31. 1928; as cited in Pauli (1979, 468).

Including these transitions in the dispersion formula, Heisenberg arrived at

$$P_r \propto \left| \sum_{J''} \frac{\dot{x}_r(J'J'')\dot{x}_s(J''J') + \dot{x}_s(J'J'')\dot{x}_r(J''J')}{2mc^2/h} \right|^2 \tag{7.8}$$

where the frequency of the incident radiation ν_S had been neglected compared to $2mc^2/h$. By some simple manipulations, Heisenberg could show that

$$P_r = E \frac{e^2 cos(\alpha_{rs})}{4\pi^2 m\nu_s^2} \tag{7.9}$$

which corresponded to the equation Dirac had himself derived in his paper on dispersion, i.e. to the well-known and in its limits experimentally confirmed Thomson scattering formula. Heisenberg's comment: "It is after all rather amusing, that this formula comes out only just because of the crazy transitions."[23]

In analysis, Heisenberg had observed that the part of the scattering which was previously described by the direct scattering term vanished and that this part also had to be described by processes including intermediate states. For a free electron, the only absorption lines, thus the only possible intermediate states, were the ones of negative energy. Following this reasoning, Heisenberg could show that the scattering was correctly described. Hence, negative energy states necessarily figured as virtual (modern terminology) states in the derivation of the Thomson formula when combining Dirac's theory of radiation and his relativistic description of the electron.

From a letter by Yoshio Nishina to Ivar Waller,[24] we know that Heisenberg presented this reasoning at a small, informal conference shortly afterwards in Copenhagen, the dating of which will become important to judge its relevance for the further development. It must have been given in early to mid September 1928. In mid July, Heisenberg established the above argument. Directly after communicating his result to Pauli he went to Munich and stayed there throughout the whole of August.[25] In early September 1928, Heisenberg and Pauli went to Copenhagen[26] and stayed roughly until September 16.[27] Since Nishina, who was present at Heisenberg's presentation, left Copenhagen in early October 1928 Heisenberg must have presented the argument of the necessity of including negative energy intermediate states somewhere between early and mid September 1928.

---

[23] "Es ist doch sehr amüsant, daß diese Formel gerade erst auf Grund der verrückten Übergänge herauskommt." Heisenberg to Pauli, July 31. 1928; as cited in Pauli (1979, 468).

[24] Nishina to Waller, February 22. 1930. IWA. See also below.

[25] Heisenberg to Pauli, July 31. 1928; as cited Pauli (1979, 469).

[26] See Rechenberg (2010, 626).

[27] Pauli to Ehrenfest, September 16. 1928; as cited in Pauli (1979, 468). Pauli sent this letter from Copenhagen but indicated that he will leave in the next days.

By mid September 1928 at least Heisenberg, Pauli, Bohr, Nishina, and Oskar Klein knew of the negative energy intermediate states. All the same, as I will argue in the following, Heisenberg's conclusion did not have any impact on the further development.

7.2 Initial Unimportance of Heisenberg's Observation

While Heisenberg was pondering about the combination of Dirac's radiation theory and its relation to the negative energy states, Yoshio Nishina and Oskar Klein were working out the scattering of light from free electrons in Copenhagen: Their result is today known as the Klein-Nishina formula. But Nishina and Klein did not apply Dirac's radiation theory. They followed a semi-classical, "correspondence-like" procedure which was most prominently used by Walter Gordon (1926) in a derivation of Compton scattering in terms of the Klein-Gordon equation and which had been elaborated on by Klein (1927) himself.[28] Such procedures did not rest on the idea of successive emission and absorption subprocesses as they were part of Dirac's radiation theory. Rather, the wave functions perturbed by the incoming radiation were used to construct charge and current densities which were then used as sources for the secondary radiation according to the classical Maxwell equations. This approach not only was at odds with an interpretation in terms of successive transition processes but even concepts as initial and final state were not as obvious as in Dirac's radiation theory.[29]

Due to the specific approach taken, Klein and Nishina did not have to take the negative energy states explicitly into account. In effect, they denied the negative energy solutions any physical interpretation: They called them "physically not sensible"[30]. This stance towards the negative energy solutions, namely that they were physically meaningless and should be somehow excluded from the considerations, was actually common until Dirac's inception of the negative energy electron sea. Moreover, the scepticism about negative energy solutions faded in a greater part of the community only with the empirical observation of the positron and the theoretical identification of a hole in the sea with the observed particle.[31]

---

[28] For a detailed description of the semi-classical methods as well as the historical framing of the same, see e.g. Lacki, Ruegg, and Telegdi (1999, Sect. 3), Brown (2002, Sects. 6 and 7). For a technically informed, historical reconstruction of the derivation by Klein and Nishina see Yazaki (2017).

[29] As for example Lacki, Ruegg, and Telegdi (1999, 472) described the state of affairs: "[...] a purely classical electromagnetic radiation affects the motion of the electron which radiates according to the classical laws, all this along a purely continuous course of events: by the nature of the approach, notions such as initial or final states do not appear naturally in this context."

[30] "physikalisch nicht sinnvoll", Klein and Nishina (1929, 856).

[31] See e.g. Roqué (1997).

Oskar Klein even worsened the problem of the negative energy solutions. In autumn 1928 he was not only working on the scattering of radiation from free electrons according to Dirac's theory of the electron but also on the effect of a potential barrier on an impinging relativistic electron (Klein 1929): Today, the result of this work is known as the Klein paradox. He found that, for values of the height of the potential barrier $V_0 > E + mc^2$, where E is the energy of the impinging electron and mc^2 its rest mass, the reflection coefficient was smaller than one; the wave was partially transmitted. The sign of the kinetic energy of the transmitted part changed, therefore this phenomenon showed that even when no interaction with the radiation field was taken into account the theory predicted transitions to negative energy states, "to which no physical meaning can be attributed".[32]

Historians of science have proposed that the problem of intermediate transitions to negative energies was one of the reasons why Klein and Nishina applied the correspondence-like method to the scattering problem and did not investigate it within the radiation theory of Dirac.[33] Besides the clear stance towards the unphysicality of the negative energy states taken by Nishina and especially Klein the argument rests on the following pillars: the letter from Heisenberg to Pauli in which he laid out the problem (see prior subsection) and a letter from Nishina to Ivar Waller in which Nishina mentioned that Heisenberg presented this conclusion in Copenhagen:

> From Heisenberg's lecture at Copenhagen, I thought that it would be necessary to take the state of negative energy into account in order to get the correct result [in Dirac's radiation theory], but I did not expect that you would obtain the same formula as ours.[34]

In order to judge the resilience of this argument, let's confront the chronology. Nishina came to Europe in 1921 and studied and worked in Cambridge (Ernest Rutherford), Göttingen (Max Born and James Franck), Hamburg (with Isidor Rabi and Wolfgang Pauli), but mainly in Copenhagen at Niels Bohr's institute where he first moved in 1923. We know that Nishina thought about calculating the scattering of radiation from electrons according to Dirac's theory of the electron as early as February 1928. After a short stay in Hamburg, he went again to Copenhagen in March of the same year where he started the work with Klein (Yazaki 2017, 405). Nishina and Klein communicated their paper on October 30. 1928, while the main part of the work was already finished when Nishina left Copenhagen in early October 1928.

As discussed, Heisenberg must have given his talk in early to mid September 1928. According to Nishina's letter, he was not aware of negative energy intermediate states prior to Heisenberg's talk, although Heisenberg also touched upon it when writing to Bohr in mid July 1928. Nishina was already thinking about calculating the scattering

---

[32] "denen ein physikalischer Sinn nicht zugesprochen werden kann." Klein (1929, 158).

[33] This argument was first made by Yazaki (2017) [originally published in 1991] and was taken up by Brown (2002). Klein and Nishina only stated that the same result might be expected from Dirac's radiation theory (Klein and Nishina 1929, 853). But no materials of Klein or Nishina applying Dirac's radiation theory have been found (Yazaki 2017, 414).

[34] Nishina to Waller, February 22. 1930. IWA.

of radiation from free electrons according to Dirac's equation in February 1928 and Klein also remembered that their joint work started in the spring of the same year.[35] Thus, it is highly unlikely that the two were aware of Heisenberg's conclusion when deciding on a particular treatment. When Klein reflected on the development of the calculations, he also stressed the problem of finding suitable boundary conditions. He explicitly stated that "we made a derivation, not by using Dirac radiation theory, which we could have done, but from the correspondence point of view, since I was more familiar with the correspondence point of view and I knew that in that order of approximation [it] would give the correct results".[36]

Thus, we can conclude with some certainty that Heisenberg's conclusion was not important for Klein or Nishina when deciding on a particular treatment. It simply does not fit the chronology of the events. Rather, their approach should be understood through the invocation of the two further arguments Yazaki (2017, 414–417) and Brown (2002, 236–237) provided: Klein's prior work on the semi-classical procedure and the problem of choosing a reasonable boundary condition for the electron wave function. Klein was not only one of the main proponents of the correspondence-like method. The work by Nishina and him was carried out at the Copenhagen institute of Niels Bohr at the time, the inventor of the correspondence principle. A similar turn to what people called correspondence-like approaches to quantum-electrodynamical effects can as well be observed in the case of Léon Rosenfeld and Victor Weisskopf when they were in closer contact with Bohr.[37]

While Heisenberg's finding was unimportant for Klein and Nishina's work, it could have entered the historical development through a different route. Both Dirac and Ivar Waller were regular guests in Copenhagen. After Heisenberg's talk in early to mid September 1928, physicists working at Bohr's institute were aware of the problem of the negative energy intermediate states. Nevertheless, as will become clear through their correspondence, both Waller and Dirac were unaware of the necessity of including negative energy intermediate states. Ivar Waller seems to have missed Heisenberg's presentation just by a few days. He was in Copenhagen shorty before in late August 1928.[38] Dirac was in the Soviet Union to attend the Sixth Congress of Russian physicists in September. He went straight back to Cambridge in October 1928. In March 1929, Dirac went to the USA for a two month visiting professorship at the University of Wisconsin, followed by a lecturing tour. And even

[35] Interview of Oskar Klein by J. L. Heilbron and L. Rosenfeld, 1962 September 25 to 16 July 1963. Session IV, Niels Bohr Library & Archives, American Institute of Physics, College Park, MD USA. https://doi.org/10.1063/nbla.vdax.mmhs. Page 25 of the PDF.

[36] Interview of Oskar Klein by T. S. Kuhn and J. L. Heilbron, 1962 September 25 to 16 July 1963. Session VI, Niels Bohr Library & Archives, American Institute of Physics, College Park, MD USA. https://doi.org/10.1063/nbla.dgih.rwum. Page 25 of the PDF.

[37] See also Sect. 10.1.1.

[38] Waller to Dirac, July 28. 1928. PAMDP, Box 23, Folder 1, courtesy of the Florida State University Libraries, Special Collections and Archives.

though he joined up with Heisenberg in August 1929 and went with him to Japan, Dirac made his way back to Cambridge in October 1929 seemingly without knowing about negative energy intermediate states.[39]

At first sight this seems astonishing. But Heisenberg's finding was obviously not taken very seriously at the time. At the Easter Conference in Copenhagen in early April 1929, problems connected to the Dirac equation were debated, especially the Klein paradox. But negative energy intermediate states were not part of the discussion.[40] Ivar Waller attended the conference[41] but had to establish the occurrence of these transitions for himself. And it was Waller who made Dirac aware of the negative energy intermediate states.

7.3 Ivar Waller, Intermediate States, and the Dirac Sea

While there is no evidence that Dirac was working on the general negative energy problem from late 1928 to November 1929, others definitely did. Hermann Weyl (1929) (communicated in May 1929) famously proposed that the four-component wave function should be split into two two-component wave functions where one set of solutions should refer to the electron and the other one to the proton.[42] Pauli had tried to solve the problem but failed.[43] Heisenberg mentioned negative energy states and a possible solution in a letter to Pauli in July 1929.[44] Still, the route to follow here is the incorporation of Dirac's relativistic equation into Dirac's radiation theory and the extension to short wavelengths of the same, which was most thoroughly pushed forward by Ivar Waller.[45]

Waller had studied physics in Uppsala where he received his PhD in 1925. At the time, Waller had already made the acquaintance of Max Born and had stayed for a short period in Göttingen. Afterwards, Waller spent a semester in Zürich and the academic year 1925/26 at Niels Bohr's institute in Copenhagen. Through his dissertation and his subsequent research, Waller became an expert in the quantum-theoretical description of X-ray scattering. Karl Grandin (2008) has characterized Ivar Waller as an "intermediate theoretical physicist", in the sense that he mediated between

[39] See, for example, Farmelo (2009, Chaps. 13–14) for the sojourns of Dirac in the respective time frame.

[40] For a partitioner's view, see Rosenfeld (1979).

[41] See Grandin (2008, 202).

[42] See e.g. Kragh (1990, 89–90) or Wright (2016, 230), who goes as far as attributing Weyl with the first physical interpretation of the negative energy solutions.

[43] "Ferner will ich [Pauli] dir [Klein] [...] einen mißglückten Versuch von mir vorführen, die $\pm m_0$-Schwierigkeit der Dirac-Theorie zu umgehen, in der Hoffnung, daß du ihn (wenigstens teilweise) doch wirst retten können." Pauli to Klein, March 16. 1929; as cited in Pauli (1979, 494).

[44] Heisenberg to Pauli, July 20. 1929; as cited in Pauli (1985, 514–516).

[45] For a condensed form of the historical episode discussed in this section see Ehberger (2022, 558–560).

high level theory and experiment (and the respective actors).[46] And in the concrete problem of deriving a cross section for the scattering of light from a relativistically described electron, Waller hit upon the negative energy intermediate states.

Waller and Dirac first met at Niels Bohr's institute in May 1928. At the time, Waller had already generalised the semi-classical method of Klein (1927) to high frequency radiation (Waller, 1927, 1928) and was therefore already well-informed about the usage of the correspondence-like methods. At the same time, Waller was already working on the treatment of the scattering of high frequency radiation according to Dirac's radiation theory. A footnote in Waller (1928, 214) already indicated the principle method of how to modify the interaction energy within this approach. On July 28. 1928, Waller took up the correspondence with Dirac to inform him about his work and eventually get Dirac's opinion on his paper.

7.3.1 Between Radiation Theory and the "Density Approach", Between Less and More Formal

The letters between Waller and Dirac have been discussed by Karl Grandin (2008, 202–208) and I refer the reader to this paper for a detailed evaluation. In the following, I will focus on the development of the discussion as far as it concerns the major theme of this chapter, the inclusion of negative energy states in the description of scattering and the stances the historical actors took towards the intermediate states. Waller worked on different topics at the time and it took him until early February 1929 to implement radiation theory. At that point he sent his manuscript to Dirac.[47] In the course of the calculation, Waller showed that what he called the "density method",[48] i.e. the correspondence-like calculation of Klein, would lead to the same results as a calculation utilizing radiation theory.

Since Waller was now in a position to operate with radiation theory on a non-relativistic system, he turned to its extension on the Dirac equation. Some time between February and July 1929, Waller must have come to the conclusion that the sum over intermediate states necessarily included negative energy states. In the last section of Waller (1929, 93–94), communicated on July 21. 1929, Waller already indicated that the terms normally associated with direct scattering processes did not occur in the scattering formula when the electron was treated relativistically. The effect of this term was taken up by transitions through intermediate states with

[46] The mediating function of Ivar Waller and his work has, according to Grandin, further dimensions but which are not of relevance for the problem to be discussed (Grandin 2008, 194).

[47] Waller to Dirac, July 28. 1928. PAMDP, Box 23, Folder 1; Waller to Dirac, January 19. 1929 and February 18. 1929. PAMDP, Box 23, Folder 2, courtesy of the Florida State University Libraries, Special Collections and Archives.

[48] The terminology reflects the method and conceptualization of this approach: The perturbed wave functions were used to construct charge and current densities which acted as sources for the secondary radiation according to Maxwell's equations.

"negative energy values", which Waller tentatively put into quotation marks.[49] This was the first published announcement of the necessity of negative energy intermediate states. While Heisenberg arrived at this conclusion for a free electron, Waller had found it in the case of bound electrons.

Waller took up the correspondence with Dirac on November 2. 1929,[50] now focusing on the calculations combining the Dirac equation and Dirac's radiation theory. Waller came again to the conclusion that the dipole moments of the atom which described the scattered radiation could both be derived from Dirac's theory of radiation and the "Schrödinger-Gordon-Klein density method", i.e. the semi-classical method he had worked with before turning to Dirac's radiation theory. Waller now provided a similar argument for a bound electron as Heisenberg had for the free one.[51] A calculation with the Schrödinger equation included the direct scattering term (referred to as (II)), while working with a relativistic wave equation one only obtained scattering through intermediate states (Waller referred to the corresponding equation as (I)):

> We must now be able to get (II) from (I) by neglecting relativity and spin corrections. It is then found that the first term in [brackets] of (II) which is the essential term for high frequencies ν, comes essentially from that part of $\sum_{J'''}$ [intermediate states] in (I) which refers to negative energies $E_{J'''}$, whereas the other part of $\sum_{J'''}$ in $[(I)]$ corresponds to (gives) the other terms in $[(II)]$. (I have made the calculation in detail but I shall not trouble you about it.)[52]

Immediately after the above argument, Waller questioned the justification of Dirac's radiation theory for the derivation of the result:

> Now I really find it difficult to see how far it is justified to use your radiation theory for deducing (I) because one has then to consider transitions $E_{J'} \rightarrow E_{J'''}$ to negative energy values $E_{J'''}$. By using the density method these states enter more formally into the calculation, so the difficulty is perhaps not so serious there. Another question is how far (I) can be valid, but I suppose it is not easy to say that.
> Have you perhaps found some way to avoid that difficulty about the negative energy?[53]

From the draft of the letter we can judge why Waller did not consider the problem as serious for the density method. The negative energy states "which do not have any real existence" entered "more formally" in the density method because "the corresponding eigenfunctions only play the mathematical role of certain terms in an expansion in eigenfunctions".[54] While the results of the two derivations were equivalent, for Waller, expert in both methods, the intermediate state in Dirac's radiation

[49] "negative Energiewerte," Waller (1929, 94).

[50] Waller to Dirac, November 2. 1929, PAMDP, Box 23, Folder 3, courtesy of the Florida State University Libraries, Special Collections and Archives.

[51] This argument is also sketched in the already mentioned public announcement in Waller (1929, 93–94).

[52] Waller to Dirac, November 2. 1929, PAMDP, Box 23, Folder 3, courtesy of the Florida State University Libraries, Special Collections and Archives.

[53] Waller to Dirac, November 2. 1929, PAMDP, Box 23, Folder 3, courtesy of the Florida State University Libraries, Special Collections and Archives.

[54] Draft of the letter Waller to Dirac, November 2. 1929, IWA.

theory played at least a less formal role than an expansion in eigenfunctions. Their function was, according to Waller, more than purely mathematical. The unphysicality of the negative energy states consequently posed a larger problem for Dirac's radiation theory than for the semi-classical procedure.

Waller's remark constitutes a rather interesting case of conceptual reasoning. He thought of the intermediate states in Dirac's radiation theory as less formal, i.e. more physical, than in the density methods. Yet, negative energy states were, according to Waller, unphysical. Hence, the concepts used in the derivation and the concepts needed to connect the result with the non-relativistic equation were at odds. Following Waller's line of reasoning, the occurrence of intermediate states as unphysical negative energy states created a conceptual tension which made the combination of Dirac's radiation theory and Dirac's theory of the electron questionable.

Waller's remark is further interesting as an expert in both approaches to scattering problems, semi-classical and Dirac's radiation theory, compared the respective role of the intermediate states in these theories. The conceptions and the procedures of application decidedly differed. Dirac's method was an iterative procedure in which one arrived straightforwardly at the probability for an event to occur by squaring the corresponding probability coefficient and applying Dirac's trick. The construction of the corresponding probability amplitude was guided by a picture of successive emission, absorption, and transition processes. Quite on the contrary, in the density method approach the first order solutions of the wave functions were used for the construction of charge and current densities that were then interpreted as sources for the secondary radiation according to Maxwell's equations. In this vein, the concept of successive acts of emission and absorption was never directly incorporated in the semi-classical approach, while it formed one of the conceptual pillars of Dirac's radiation theory. As we will see, there is yet another reason why Waller thought of Dirac's intermediate states to be less formal.

The answering letter from Dirac indicates that he was still unaware of the results derived by Heisenberg and Waller:

> Are you sure you are right about the important part of the sum [of (I)] for large ν coming from those terms J''' referring to a negative energy? If one worked out the corresponding expression for scattering with neglect of relativity and spin, I think one would find that it corresponds term by term, i.e. each term in the non-relativity expression would approximately equal one of those terms in the relativity expression [(I)] belonging to a positive energy. It therefore seems to me that [(I)] would be approximately correct if one neglected the negative energy terms. In any case I think the negative energy terms are very small when one takes the proper retardation effects into account in calculating the matrix elements.[55]

The next two letters by Waller and Dirac were written simultaneously and must have crossed each others way when sent from Cambridge to Uppsala and vice versa. Waller's letter is mostly concerned with the explication of the argument he gave in his previous letter. Yet, at the same time he reconsidered his stances towards the formal nature of the intermediate states in Dirac's theory of radiation:

[55] Dirac to Waller, November 18. 1929, IWA.

> The scattering formula can also be found from your spin theory and the Schrödinger-Gordon-Klein dispersion theory. Since in your dispersion theory the intermediate state J''' plays a rather formal role, as long as resonance effects do not occur, the presence of the $\sum_{J'''}^{E_{J'''}neg}$ may perhaps not be too serious.[56]

Waller, faced with Dirac's unwillingness to accept his result, had reconsidered his initial stance towards the intermediate states in Dirac's radiation theory. He had found a way to relax the above described tension. In the case of resonance, Dirac[57] had found that the two emission and absorption processes were governed by Einstein coefficients and therefore were independent of each other, i.e. the product of two squared matrix elements combined and the probability of scattering was proportional to t^2 instead of t.[58] They were real processes in contrast to virtual ones (modern terminology). Now, as resonances were not considered by Waller, he hoped to put the intermediate states of Dirac on the same, rather formal level as those figuring in the correspondence-like approach. Then, as Waller conjectured, the problem "may perhaps not be too serious".

In summary, the suggestiveness of Dirac's procedure and the results priorly arrived at first led Waller to consider the intermediate states as less formal than in another method which led to exactly the same result. Yet, the concrete problem at hand allowed Waller enough flexibility to refute the occurring intermediate state any physicality. By this conceptual, if not ontological move, Waller tried to safe the credibility of his results. Interestingly, Dirac took a different stance.

7.3.2 Dirac's Reconsideration and the Reinterpretation of the Scattering Process

Dirac's letter written on November 27. 1929 was for the most part a shortened version of the well-known and often discussed letter from Dirac to Bohr dating to November 26., in which Dirac outlined the basic physical picture of the Dirac sea and which we shall turn to in a minute.[59] But he also directly reacted to Waller's previous calculations and insights. Dirac first apologized for his doubts about Waller's results.

---

[56] Waller to Dirac, November 26. 1929, PAMDP, Box 23, Folder 3, courtesy of the Florida State University Libraries, Special Collections and Archives.

[57] Dirac (1927b, Sect. 6) and Dirac (1927d, Sect. 3).

[58] As discussed in Sect. 5.1.5, this statement has to be handled with care since not exactly the same mathematical structures were responsible for dispersion and resonance scattering and I do not believe that Dirac's invocation of the verbal model was to a greater part influenced by the resonance case. Nonetheless, as I also argued, it certainly made this identification more plausible.

[59] See, e.g., Kragh (1990, 90–93) for the discussion and a nearly complete reproduction of these letters.

He himself has looked at the relativistic scattering from free electrons[60] and found that "the greater part of the scattering comes from 'double transitions' in which the intermediate state is one of negative energy for the electron".[61]

Dirac explained the principle ideas for solving the general negative energy difficulty which he had "recently obtained", i.e. the filling of all the negative energy states with an unobservable, calm and infinite distribution of electrons. Due to the Pauli exclusion principle positive energy electrons could no longer drop into negative energy states. Initially, Dirac conceived of the hole in this sea as a proton. He made the proposition to account for the mass difference between the electron and the proton by "Austausch" or exchange interactions.[62]

In the letter to Waller, Dirac went on and discussed the scattering mechanism in the filled Dirac sea:

> If we apply these ideas to the problem of scattering, we see that double transitions in which for the intermediate state the electron has negative energy must be excluded by the Pauli principle. On the other hand we can have double transitions in which firstly one of the distribution of the $-ve$ electrons jumps to a state of $+ve$ energy and secondly the original $+ve$ electron jumps down and fills up the hole. These processes will just make up for those excluded above, (in which the electron jumps to a $-ve$ energy for the intermediate state) and justify our including the latter in our scattering formula.[63]

This is the first account of scattering described in the conceptual framework of the Dirac sea we know of. As we see, scattering had to be reinterpreted in Dirac's new framework.

I want to point at a peculiar feature of Dirac's reinterpretation: It involved two electrons. A negative energy electron which jumped to the final state in the first step and the initial electron which jumped to the hole in the second step. Furthermore, in the intermediate state, a proton-electron pair is present: Retrospectively one might say, that the first particle/anti-particle pair that occurred in QED was actually a virtual electron-proton pair.[64] But please note, that this statement has to be treated with caution since throughout the 1930s physicists did not necessarily refer to particles as virtual, but to the subprocesses as they did not conserve energy.

---

[60] Waller had so far only discussed scattering from bound electrons. In his publication, he also included scattering from free electrons on motivation by Pauli. Pauli to Waller, November 15. 1929, IWA. The calculations were communicated to Pauli on January 4. 1930 and he considered them to be "very satisfactory" ["Sehr befriedigend"]. Pauli to Waller, January 22. 1930, IWA.

[61] Dirac to Waller, November 27. 1929, IWA.

[62] The reference to the exchange interaction is both part of the letter to Ivar Waller and the one to Niels Bohr. Dirac to Waller, November 27. 1929, IWA; Dirac to Bohr, November 26. 1929, BSC, MF 9. It is to be noted that the "Austausch"-interaction Dirac mentioned is not the conception of "Austausch" or exchange in modern QED. Dirac referred here to the concept of an exchange that would switch the place of two electrons and not to the exchange of a photon between two electrons. For the history of the exchange forces before its incorporation into quantum field theory, best see Carson (1996a).

[63] Dirac to Waller, November 27. 1929, IWA.

[64] See e.g. Pais (1986, 350).

So, let me shortly recap: It was Ivar Waller who made Dirac aware of the necessity of taking negative energy intermediate states into account. For Waller, a (at least) more than mathematical reading of the intermediate states combined with the occurrence of the unphysical negative energy states led him to question the combination of Dirac's radiation theory and the Dirac electron. Confronted with Dirac's unwillingness to accept his results, Waller re-evaluated his reading of the intermediate states and considered them formal to safe his calculation. He pushed both intermediate and negative energy states into the mathematical terrain.

Dirac, quite on the contrary, initially questioned Waller's calculation. But after performing the calculation for free electrons, as Heisenberg had done nearly one and half years earlier, Dirac accepted Waller's argument. But his solution to the negative energy problem entailed a difficulty for the previously conceived of scattering mechanism: Transitions to negative energy states were, by construction, forbidden. Thus, Dirac reconfigured the process by describing the scattering through virtual pair creation (modern terminology). As I will argue more fully in Sect. 7.5, Dirac conceived of intermediate and negative energy states as physical.

7.4 Were Waller's Letters the Reason for Dirac's New Look on the Negative Energy Problem?

Before following the historical line any further, let me pose a question which I consider important in respect to the epistemic relevance of concepts connected to the virtual particle: Were Waller's letters really the reason for Dirac's new look on the negative energy problem? Karl Grandin (2008, 203–206) has argued that this is the case. Some historians of science have proposed that the scattering mechanism was an important ingredient for Dirac's reasoning without being aware of the correspondence between Waller and Dirac.[65] Although I share this assessment (see below), I also want to indicate that we can interpret this episode differently. But let me first revisit and discuss the argument.

At first we must note, that the above episode can only provide reasons *why* Dirac looked at the general negative energy problem again. It does not give any indication of *how* Dirac came up with the particular idea of filling up all the negative energy states with electrons and interpreting a hole therein as a proton. The "how" is normally answered by a combination of the following arguments[66]:

- Dirac had already used the idea of an infinity of unobservable zero state photons in his first paper on radiation theory (Dirac 1927c, 260–261).

[65] See e.g. Pais (1986, 348–349).

[66] See Kragh (1990, Chap. 5), where all of the following reasons are discussed.

- The theoretical identification of an empty state in an assembly of electrons as a positively charged particle was already established.[67]
- Hermann Weyl (1929, 332) had already proposed to think of the negative energy states as referring to protons.
- Dirac followed the general conception that electrons and protons are the only kinds of particles.
- The identification of a hole as a proton would provide for a reduction of all known particles to electrons alone and the interaction between the negative energy electrons could serve as an explanation of the mass difference.

Especially the second point figured prominently in Dirac's publications and letters. In his initial proposal of the idea in a letter to Bohr he mentioned the analogy to X-ray phenomena. Also in his first public presentations (e.g. Dirac 1930e), Dirac introduced his theory of the electron and proton with a discussion of X-ray levels in terms of Llewellyn Thomas and Enrico Fermi's electron model.

The interesting question for my investigation is not *how* but *why* Dirac took up the general negative energy problem again. To judge the resilience of Grandin's argument, let us look at the chronology of the events. Waller communicated his findings to Dirac on November 2. 1929. Dirac replied as late as November 18. 1929, still doubting the correctness of Waller's results. On November 24., so only one week later, Bohr wrote to Dirac, that he had heard from George Gamow (at the time in Cambridge) that Dirac had come up with a possible solution to the negative energy problem.[68] The letter from Gamow still exists and is part of the AHPQ. Sadly, it is obviously misdated by Gamow himself to November 25. 1929.[69] But also without an exact date on Gamow's letter, we can infer that time was short for Dirac. Even if we allow for a very fast postal connection between Copenhagen and Cambridge, Dirac had at most a few days to take a look at the scattering from free electrons, coming to the conclusion that the negative energy states had to be taken into account in this case as well (against his initial suspicion of November 18.). Then he had to come up with his idea of the Dirac sea, and finally find a way of how to describe the scattering through negative energy intermediate states in a filled vacuum. This would be incredibly fast, yet possible.

But there is a different reading that is not excluded by the above correspondence. One can argue, that Dirac was already working on, or at least thinking about, a solution similar or equivalent to the Dirac sea. In this scenario, Waller's results would have actually posed the first problem for Dirac's new conceptualization. Since in a filled Dirac vacuum the process of scattering through negative energy intermediate states

[67] See Wright (2024, Sects. 2.2–2.4) for a close discussion of the role of many-electron theory, where such a conception was entertained, for Dirac's invocation of the Dirac Sea and its further development.

[68] Bohr to Dirac, November 24. 1929, BSC, MF 9.

[69] Gamow to Bohr, November 25. 1929, BSC, MF 11, Sect. 1. There is also the possibility that Gamow telegraphed first informations about Dirac's breakthrough to Bohr and the letter only constitutes a fuller account of Dirac's proposal. In either case, we do not posses an exact date of Gamow's communication to Bohr.

was forbidden, it had to be replaced by a different set of transitions. This would as well explain Dirac's hesitation to accept Waller's results at first. It must further be noted, that Dirac did not exclude the negative energy states from playing a role in physics at all. In his reply to Waller, he only considered their contribution to be "very small" and that it "would be *approximately* correct if one neglected the negative energy terms".[70] Dirac neither answered Waller's question whether he had a solution to the general negative energy problem, nor did he exclude negative energy states completely.

Although this interpretation cannot conclusively be ruled out at the moment, I tend to follow Grandin's reasoning. We have no indication that Dirac was thinking about the negative energy states before Waller informed Dirac about his results. Waller's first mentioning of the negative energy intermediate states to Dirac was mailed on November 2. 1929, which would leave enough time if Dirac started thinking about the problem after receiving the letter. Furthermore, Waller asked Dirac directly about a solution to the negative energy state problem, which Dirac could not or was not willing to give. Although the timespan mentioned above would be short, it is possible. Dirac himself remembered that after he had arrived at the idea of filling up the sea and interpreting a hole as a proton, there would have been no reason to hold this inception back.[71]

Furthermore, as far as I know, and seemingly also Dirac knew (see below), the theoretical description of scattering was the only occurrence of negative energy states in a theoretical explanation of a process within the reach of experimental practices of the time. In addition, and as discussed in Sect. 7.3, without solving the problem of the negative energy states, both theories Dirac was celebrated for, the relativistic description of the electron and his radiation theory, would be under high pressure. This was certainly the case as long as one interpreted intermediate states physically, which Dirac seemingly did (see below). According to Waller's first assessment, Dirac's radiation theory would be less adequate than the density approach and, in essence, not combinable with the Dirac equation. Where the correspondence-like method succeeded, Dirac's radiation theory would have failed.[72] But as Dirac came up with a solution that allowed for an incorporation of negative energy states into physical reasoning, the conceptual tension Waller had pointed out dissolved.

This is hopefully not the last word on this question. But, I think, that at the moment the historical material at hand suggests that the problem of negative energy intermediate states has led Dirac to take up the general negative energy problem again.

---

[70] Dirac to Waller, November 18. 1929, IWA, emphasis added. At this point, I digress from Grandin's reading of Dirac's letter, regardless of whether Waller's letters were Dirac's motivation or not.

[71] Interview of P. A. M. Dirac by Thomas S. Kuhn, 1962 April 1, 1963 May 6, 7, 10 and 14. Session V, Niels Bohr Library & Archives, American Institute of Physics, College Park, MD USA. https://doi.org/10.1063/nbla.ykni.ovcj. Page 53 of the PDF.

[72] In this reconstruction we see, that it was not only the occurrence of negative energy states in scattering processes that drove Dirac to the recognition of the negative energy states as part of physical theory (see e.g. Wright 2024, 55). The problem was that a rival theory of the radiation field might have been better suited because it treated such states more formally.

7.5 Dirac, Igor Tamm, and the Physicality of Intermediate States

Waller's and Dirac's reasoning constitute two different strategies to handle the conceptual tension arising due to the occurrence of negative energy intermediate states: Against his initial conviction, Waller pushed the intermediate states in the formal, i.e. mathematical, terrain, where he had located the negative energy states. After Dirac had shown how to handle the general negative energy problem, he, as I will now argue, went a different route: Dirac conceived of both concepts physically.[73]

7.5.1 Dirac's Defence Against Niels Bohr's Doubts

The first one to question Dirac about his new conception was Niels Bohr.[74] Having heard from Gamow that Dirac came up with a possible solution, Bohr immediately wrote to Dirac. Bohr was convinced that "the difficulties in your [Dirac's] theory might be said to reveal a contrast between the claims of conservation of energy and momentum on one side and of the conservation of individual particles on the other side".[75] Bohr obviously expected a more drastic revision of the conceptual framework than Dirac would offer him.

In his answering letter, Dirac replied, on the contrary, that he would like to uphold energy conservation at all costs. He provided Bohr with "a simple way of avoiding the difficulty of electrons having negative kinetic energy".[76] Then he discussed his idea of the Dirac sea and described the filling of a hole in the sea as an annihilation of a proton and an electron. As in the letter to Waller, Dirac wanted to explain the mass difference between electron and proton by the exchange interaction between the negative energy electrons.

Bohr was rather unsatisfied with Dirac's explanation and opted for a more drastic revision: He still wanted to abandon energy conservation. Bohr was also convinced that the transitions from positive to negative energy states would not "present as serious a difficulty for [Dirac's] wave equation as it might appear":

> The question is, how much those features of the theory which claim the transitions in question are involved in the problems, where your theory has been found in so wonderful agreement with experiments. In this connection I must correct the statement in my former letter regarding the probability of these transitions which is not nearly so large as I believed. [...] My hope

[73] For a very condensed form of the following historical argument and episode see Ehberger (2022, 561–562).

[74] The correspondence between Dirac and Bohr is well-known. For a comprehensive discussion and a near to complete reproduction see Kragh (1990, 90–95).

[75] Bohr to Dirac, November 24. 1929, BSC, MF 9. Permission by the Niels Bohr Archive, Copenhagen.

[76] Dirac to Bohr, November 26. 1929, BSC, MF 9. Permission by the Niels Bohr Archive, Copenhagen.

> is that it should be possible to defend all the successful applications of your wave equation, but I suspect that the natural limitation of these applications prevents an extrapolation of the kind you describe in your letter.[77]

Why Bohr and Klein, with whom Bohr had discussed these matters, believed that transitions to negative energy states should play no role in the theory is, to be honest, not entirely clear to me. Bohr and Klein certainly knew about the relevance of negative energy intermediate states through Heisenberg. Yet, this problem was obviously not taken very seriously and occurred within the framework of Dirac's radiation theory and not within the semi-classical treatment, which was favoured in Copenhagen. Furthermore, Bohr and Klein might have shared Waller's later view that intermediate states were only formal parts of the treatment and should not be considered physically problematic.

Be that as it may, Dirac was not aware of Heisenberg's talk and the fact, that Bohr must have known about the description of scattering in terms of Dirac's radiation theory. Thus, he answered Bohr:

> There is one case where transitions of electrons from positive to negative levels does [sic!] give rise to serious practical difficulties, as has been pointed out to me by Waller. This is the case of the scattering of radiation by an electron, free or bound. *A scattering process is really a double transition*, consisting of first an absorption of a photon with the electron jumping to any state and then an emission with the electron jumping to its final state (as in the Raman effect) (or else of first the emission and then the absorption). The initial and final states of the whole system have the same energy, but not the intermediate state, *which lasts only a very short time*. One now finds, for radiation whose frequency is small compared with $\frac{mc^2}{\hbar}$, that practically the whole of the scattering comes from double transitions in which the intermediate state is one of negative energy for the electron. Detailed calculations of this have been made by Waller. *If one says the states of negative energy have no physical meaning, then one cannot see how the scattering can occur.* On my new theory the state of negative energy has a physical meaning, but the electron cannot jump down into it because it is already occupied.[78]

In this letter, Dirac not only expressed his conviction that the negative energy states should have a physical meaning. He connected this conclusion with the intermediate states. Dirac proposed a physical, if not realistic, reading of these two concepts. He stated that scattering "is really a double transition" and ascribed a temporal dimension to the intermediate state: It "lasts only a very short time". Dirac, in effect, did not follow Waller in locating both the intermediate and the negative energy states in the formal, mathematical realm. On the contrary, Dirac considered both of them to be physical.

In one of his first publications on the theme, received on December 6. 1929 (Dirac 1930f), Dirac essentially provided the same reasoning as laid out above. When discussing scattering, Dirac stated that the intermediate state "lasts only a very short time" (Dirac 1930f, 365) and went on:

---

[77] Bohr to Dirac, December 5. 1929, BSC, MF 9. Permission by the Niels Bohr Archive, Copenhagen.

[78] Dirac to Bohr, December 9. 1929, BSC, MF 9; emphasis added. Permission by the Niels Bohr Archive, Copenhagen.

> According to previous ideas these [negative energy] intermediate states had no real physical meaning, so it was doubtful whether scattering processes that arise through their agency should be included in the formula for the scattering coefficient. This gave rise to a serious difficulty, since in some important cases nearly all the scattering comes from intermediate states with negative energy for the electron. [*Footnote in the original*: I am indebted to I. Waller for calling my attention to this difficulty.] In fact for a free electron and radiation of low frequency, where the classical formula holds, the whole of the scattering comes from such intermediate states. (Dirac 1930f, 365)

While these statements, especially Dirac's reference to a temporal dimension of the intermediate state, suggest that Dirac conceived of this concept rather realistically, his stance is not entirely coherent.

Take, for example, Dirac's influential textbook *The Principles of Quantum Mechanics* (Dirac 1930c).[79] The textbook has a considerable overlap with Dirac's lectures from 1927 to 1928 and was written from 1928 to early 1930. Having derived the general scattering formula for a non-relativistic particle, including both the direct scattering term and scattering through intermediate states,[80] Dirac explicitly noted:

> The scattering may now be considered as composed of two parts [...]. The first part [...] may be called the true scattering. The second part may be considered as arising from an absorption of the incident particle into some state k, followed immediately by a re-emission in a different direction. The fact that we have to add the two terms before taking the square of the modulus denotes interference between the two kinds of scattering. There is no experimental way of separating the two kinds, the distinction between them being only mathematical. (Dirac 1930c, 197)[81]

Thus, whether or not Dirac conceived of the intermediate states as real physical entities is hard to answer. The explicit reference to the temporal dimension of the same suggests that a physical conception came, at least at this point, with a realistic one. The reference to the experimental situation and to the distinction between direct scattering and scattering through intermediate states in the above quote might further weaken the perception of a non-realistic reading. In general, Dirac did not share the positivist attitude of the Copenhagen circles. Thus, the fact that the processes could not be separated experimentally (in the given situation) did not necessarily involve the refutation of an ontological status of the intermediate state on Dirac's side. Furthermore, the terms referring to direct scattering disappeared in a relativistic treatment, which led to an emphasis on the scattering through intermediate states.

---

[79] For a discussion of the emergence, dissemination, and content of Dirac (1930c) embedded in Dirac's research and stances towards quantum mechanics, see Kragh (2013).

[80] The distinction between a non-relativistic particle and the treatment with the Dirac equation did not alter the way in which the intermediate states enter formally in the perturbative equations. Still, for the above quote this distinction is important. Only in the case of a non-relativistic particle, the direct scattering processes occurred.

[81] Dirac made rather similar comments on the mathematical nature of the distinction between direct scattering and scattering through intermediate states when discussing the case in a second-quantized framework (Dirac 1930c, 233).

Thus, I tend to ascribe him a realist conception, while according to Dirac's general stance towards concepts of (quantum) theory ontological questions were, at best, a side issue.[82]

7.5.2 Tamm: Transitions to Intermediate States as "Elementary Processes"

Also a close, if not the closest, associate of Dirac at the time expressed a rather physical conception of the intermediate states within the same problem situation. Igor Tamm and Dirac had met in Leiden in 1928 and were from the following trip from Leipzig over Göttingen to the Soviet Union connected through a close friendship. I will discuss parts of Dirac's and Tamm's correspondence in the later sections but want to, for the sake of coherence, already present indications for Tamm's physical interpretation of the intermediate states.

Tamm was very favourable of Dirac's solution of the general negative energy problem.[83] Independently of Heisenberg, Waller, and Dirac, Tamm had discovered that negative energy states needed to be included as intermediate states in the scattering from a relativistically described electron. He concluded that "this proves the physical meaning/importance [Bedeutung] of these states [of negative energy]"[84]. As for Dirac, the import of physical relevance was from the intermediate states to the negative energy states.[85]

In a publication, the emergence of which will be discussed later on, Tamm also made an interesting connection between the virtual subprocesses (modern terminology) and an expression which was priorly only used for actual processes. He opened his paper by noting:

---

[82] See also the discussion in Sect. 5.1.5.

[83] After having heard about it through a copy of the letter Dirac wrote to Bohr and through the published paper, Tamm commented: "The idea to turn the whole question of the negative energy states upside down and to create a unitary theory of electricity out of a presumed difficulty – once one hears about it [the theory], then it is immediately clear [here Tamm uses a rather idiosyncratic expression: "leuchtet sie wie ein Blitz ein!" which combines German terms of "einleuchten" which translates to "make sense" but also "enlighten" and the term "Blitz" which is "lightning."]! I sincerely hope that you will succeed to calculate the mass of the proton and thereby support your whole theory." [*Original*: "Die Idee, die ganze Frage nach der negativen Energie auf den Kopf zu stellen und aus der vermutlichen Schwierigkeit eine unitäre Theorie der Elektrizität zu schaffen—wenn man sie einmal erfährt, dann leuchtet sie wie ein Blitz ein! Ich hoffe innigst, dass es Ihnen gelingen wird die Masse des Protons zu berechnen und dadurch Ihre ganze Theorie zu stützen."] Tamm to Dirac, February 2., 1930; as cited in Kojevnikov (1993, 30).

[84] "was die physikalische Bedeutung dieser Zustände beweist," Tamm to Dirac February 5. 1930; as cited in Kojevnikov (1993, 30).

[85] Tamm's argument that the mass difference between proton and electron should somehow be taken into account also for scattering (see the discussion in the next section) also indicates a rather physical reading of the intermediate states.

> As is well known, the scattering of a light quantum by matter consists according to Dirac's theory of *a sequence of two elementary processes*, namely the absorption of the impinging and the emission of the scattered light quantum.[86]

The emphasized term "elementary process" was normally used for actual processes, i.e. such processes which could directly be connected with an Einstein coefficient. This terminological choice by Tamm is certainly no proof of a physical reading. But it is indicative of the status the intermediate state had acquired in his reasoning.

To conclude: There are clear indications that both Tamm and Dirac inferred that the negative energy states were physical because they occurred as intermediate states. It is still to note that this conclusion was not necessary. Ivar Waller's approach of considering both concepts as formal was afterwards also put forward by Léon Rosenfeld. Rosenfeld was at the time closely working with Niels Bohr. In a lecture held in February 1931 at L'Institute Henri Poincaré he noted that "negative energy states play an important, although purely formal role in the theory of dispersion". Dirac verbally described the formula, according to Rosenfeld, in "a pictorial fashion, [...] which is not justified by the general precepts of the physical interpretation of quantum mechanics".[87] The conception of both the intermediate and the negative energy states as theoretical fictions did not unavoidably lead to an exclusion of these concepts from theoretical considerations. According to Waller's and Rosenfeld's standpoint, both concepts were necessary, but purely formal.

7.6 From Protons to Positrons: Scattering Problematics Revisited

The discussion so far focussed on the role of the conception of scattering for the inception of the Dirac sea and the different strategies to avoid the conceptual tension due to the status of intermediate and negative energy states. But, as is well-known, with Dirac's proposition of filling all the negative energy states with electrons the problems were far from settled. In the first version of the Dirac sea a hole was interpreted as a proton. This interpretation, which I will call the proton-hole model, was retracted by Dirac one and half years later (Dirac 1931) and replaced by the conception of a hole as an anti-electron. In the following, I will discuss this episode with a focus on the role of the conception of scattering and its relation to the annihilation process of electron and proton. According to my argument, the similarity of the theoretical description of these two effects created a conceptual tension for the

---

[86] "Die Streuung eines Lichtquants durch die Materie besteht nach der Diracschen Theorie bekanntlich aus *einer Folge von Elementarprozessen*, nämlich aus der Absorption des einfallenden und der Emission des gestreuten Lichtquants." Tamm (1930a, 705, emphasis added).

[87] "états d'énergie négative jouent un rôle important, bien que purement formel, dans la théorie de la dispersion." And: "une manière imagée, mais qui n'est pas justifiée par les précepts de l'interprétation physicque de la méchanique quantique [...]." Rosenfeld (1932a, 56).

proton-hole model. But let me first revisit the reasons for Dirac's re-identification of a hole as discussed in the secondary literature.

The historical reconstructions have focused on two arguments by Dirac's contemporaries.[88] The first one is the much too high annihilation rate of electrons and protons (Dirac 1930b; Tamm 1930b; and Oppenheimer 1930b). The mean lifetime would, according to these estimations, lie between 10^{-10} and 10^{-3} seconds. Hermann Weyl (1931, 233–234) was most regularly accredited with the second argument based on a second-quantized matter field: Due to the symmetry of the theory, the masses of the holes and the electrons have to be equal also when interactions are incorporated in the theoretical description. This observation is most regularly referred to as the decisive reason for Dirac to abandon the proton-hole model. The focus on these two arguments is understandable: Dirac himself referred to them when he introduced the hypothesis of a hole not being a proton but an anti-electron, a particle "unknown to experimental physics" (Dirac 1931, 61).

Dalida Monti (1995, Sect. 6) has partially discussed arguments based on the scattering mechanism, focusing on Robert Oppenheimer's contribution. But she did not stress the conceptual tensions arising from these considerations. Thomas Pashby (2012, especially Sect. 3) has reinforced the importance of Hermann Weyl's argument due to its formal stringency. On the contrary, Wright (2016, especially Sect. 2.2) has shown through detailed analysis of the archival materials of the P.A.M. Dirac papers, that Dirac's correspondence network was well aware, and made Dirac aware, of the problems connected to the proton-hole model.[89] Dirac was even aware of the symmetry argument, which was also put forward by Wolfgang Pauli and Werner Heisenberg, before Herman Weyl published his results.[90] Wright therefore argued that Weyl's importance in the story of the positron is due to his initial identification of two components of the wave function with the proton in 1929, thereby providing the first physical interpretation of negative energy states. Wright (2024, Sects. 2.3.2–2.3.4) focused on the theory of many-electron systems as well as the concept of exchange force both for the inception of the Dirac Sea and Dirac's shift from proton to anti-electron, thereby providing for another perspective on this episode in terms of the practice of theory.

For the following, I want to take a middle position: Essentially I accept the prominence of the symmetry arguments. Yet, I will show how the conception of scattering actually played a crucial role in pointing towards conceptual, and not only numerical or empirical,[91] problems with Dirac's proton-hole model. The initial proposal of the

---

[88] See e.g. Kragh (1990, 98–103), Schweber (1994, 65-67), Pais (1986, 349–351), or Miller (1994b, 38–40).

[89] A first analysis of the reaction of Dirac's correspondence network, the spread of his idea as well as the criticism it faced can be found in Kragh (1990, 94–103).

[90] Tamm wrote to Dirac that Pauli made this argument at a physical congress in Odessa (held August 20.–24. 1930). Tamm to Dirac, September 13. 1930, as cited in Kojevnikov (1993, 43). Heisenberg communicated this insight, without any detailed proof, directly to Dirac: Heisenberg to Dirac, January 16. 1930, as cited in Kragh (1990, 94–95).

[91] As Helge Kragh (1990, 102) has stated, "Dirac was not much impressed by the empirically based objections raised by Oppenheimer and others".

Dirac sea allowed physicists to take negative energy states seriously as constituents of physical processes, but due to the way holes figured in the theoretical modelling a tension arose which could only be settled by the identification of a hole as an anti-electron.

7.6.1 Scattering and Annhihilation I: Tamm's Unresolved Problems

According to Dirac, the scattering processes through intermediate negative energy states had to be replaced by a different set of transitions including a virtual (modern terminology) proton-electron pair. As he noted in his publication[92]:

> these new processes [creation of an intermediate proton-electron pair] just make up for those [electron jumps] of the more direct processes that are excluded on account of the intermediate state having negative energy for the electron, since the matrix elements that determine the transition probabilities are just the same in the two cases, though they come into play in the reverse order. In this way the old scattering formula, in which no intermediate states are excluded, can be justified. (Dirac 1930f, 365)

Ivar Waller simply accepted this kind of argument when he noted that "in the problem considered here [scattering of light from an electron (free or bound)] this new conception only provides, as far as it is considered in this work, essentially formal alterations, and we can avail ourselves for brevity's sake of the old way of expression".[93]

Igor Tamm, although in general very favourable of Dirac's new conception, was on the contrary not as easily persuaded. He had severe problems in coherently interpreting the scattering mechanism in the Dirac sea picture. At first, Tamm was unable to arrive at the new conceptualization of scattering with filled negative energy solutions and he was very pleased to find Dirac's explanation including an intermediate electron-proton pair. But at the same time Tamm raised a concern in a letter to Dirac:

> Nevertheless, I cannot understand well how the in any case very significant interaction of the electrons of negative energy can be left unconsidered in the calculations of the transitions with an intermediate step of negative energy.[94]

Tamm had a copy of the letter Dirac had sent to Bohr. Therein, Dirac had argued that the main part of the interaction contributing to the mass difference between proton

---

[92] The same statement was made in his second letter to Bohr and his letter to Waller (see above).

[93] "Beim hier betrachteten Problem gibt diese neue Auffassung, insoweit, als sie in dieser Arbeit in Betracht kommt, wesentlich nur formale Abänderungen, und wir können uns deshalb der Kürze wegen der alten Ausdrucksweise bedienen." Waller (1930b, 837).

[94] "Ich kann nur nicht gut verstehen, wie die übrigens sehr wesentliche Wechselwirkung der Elektronen negativer Energie bei der Berechnung der Übergänge mit einer Zwischenstufe negativer Energie unberücksichtigt bleiben kann." Tamm to Dirac, February 2. 1930; as cited in Kojevnikov (1993, 30).

and electron was due to the exchange interaction. Tamm had extrapolated that these interactions must also play an important role in the scattering process when negative energy states were included because an additional electron-proton pair was present in the intermediate state.

In his calculations, Tamm initially derived a result which differed from the Klein-Nishina formula due to a calculational mistake. Yet, Dirac had worked out the problem and was convinced that the Klein-Nishina formula was correct. At first, Dirac did not comment on Tamm's above mentioned worry.[95] Rather, Dirac pointed out that he was also working on the annihilation of electron and proton. He noted that "[t]he mathematics is very much the same as in the scattering process – in one case there are two emissions and in the other one emission and one absorption".[96]

In his response, Tamm acknowledged his calculational mistake and noted that he was as well working on the annihilation process. As Dirac, Tamm found that the calculation was "closely analogous to that made for the scattering".[97] Both Tamm and Dirac thereby hit on the problem of the much too high annihilation rate of proton and electron. Yet, Tamm raised another worry that was connected to the analogy of the calculation of annihilation and scattering. I quote in full:

> But the main difficulties [with the formula describing the annihilation of electron and proton] are: 1) if one (tentatively & approximately) applies the formula (1) [describing the annihilation cross section] to the case of bound electrons, one gets a ridiculously short [life] value for the life-time of an atom. & 2) the frequency of the radiation emitted, when an electron drops in a hole is of the order of magnitude of mc^2/h, where m is the mass of the electrons & not of the proton, & that can't be reconciled with the existence of the cosmic radiation.
> Of course these difficulties are of the same kind as the difficulty with the mass of the proton and are connected with the necessity to take into account the interaction of the negative-energy electrons, the disappearance of the field of the electron & of the hole [proton] when the electron drops in the hole and so on. *But I can't understand why one gets the correct scattering formula although one neglects all the complicating circumstances of this kind, connected with transitions of the electron through the states of negative energy.*
> Well, my conscience tells me that it is a shame to trouble you with all kinds of boyish questions.[98]

The annihilation process was modelled by Tamm, Dirac, and also Oppenheimer (see below), as an electron going from a positive energy initial state through an intermediate state to a negative energy final state. It was effectively described as a one-particle process, because the interaction purportedly responsible for the mass difference could not be included. But exactly this lack opened up the possibility for shoving the problems connected with the much too annihilation rate on this neglect.

But this is exactly where in Tamm's reasoning the derivation of the Klein-Nishina formula entered. The two processes, scattering and annihilation, were described by analogical theoretical structures, i.e. being built up of transitions between positive and negative energy states coupled to emission and absorption subprocesses. One

---

[95] Dirac to Tamm, February 21. 1930; as cited in Kojevnikov (1993, 35).

[96] Dirac to Tamm, February 21. 1930; as cited in Kojevnikov (1993, 35).

[97] Tamm to Dirac, March 3. 1930; as cited in Kojevnikov (1993, 37).

[98] Tamm to Dirac, March 3. 1930; as cited in Kojevnikov (1993, 37, emphasis added).

arrived, on the one hand, at quite a reasonable result, i.e. the Klein-Nishina formula. On the other hand, the rate of the annihilation of electrons and protons was completely absurd. Since Tamm believed that Dirac was on the right track, he hoped to explain the mass difficulty, and thereby the problems with the annihilation process by taking into account further interactions. But exactly this might ruin, in Tamm's eyes, the correctness of the scattering calculation.

Dirac could only tentatively answer Tamm's worries. He conjectured that one should take some mean value of the proton and the electron for the annihilation process since the theory was actually symmetric between these two entities. For the connected scattering problem Dirac pointed out:

> The theory at present predicts that an electron and proton should have the same mass, and is therefore inaccurate and unreliable in all questions where one has to take the different masses into account. I suppose the reason why the Klein-Nishina formula is right is because the process concerns only one particle, and not two interacting particles.[99]

For Dirac, the neglect of the interaction between two particles present in the initial state, i.e. an electron and a proton, should explain the difference in the results. For scattering, there was no proton in the initial configuration, it "concerns only one particle", while for the annihilation process there definitely was: There were "two interacting" particles.

To take full account of Dirac's reasoning here, it is important to note, that under the term interaction actually three well-known but decidedly different concepts hide: (1) the interaction between radiation field and matter; (2) the Coulomb interaction between charged particles; (3) and the exchange interaction between indistinguishable particles. The scattering process was described by the invocation of the interaction between radiation field and matter alone and rendered reasonable results. The exchange interaction should have explained, according to Dirac's letters to Bohr and Waller, the greater part of the mass difference of electrons and protons.[100] The neglect of the Coulomb interaction and the exchange interaction was used by Dirac to explain the much too high annihilation rate.[101]

The difference between the two processes of scattering and annihilation was, according to Dirac, that in the former process one had to deal only with the interaction between radiation field and matter while in the latter one also had to, but at the current state of the theory could not, include the Coulomb and the exchange interaction. For the time being, Dirac still had the possibility to explain the contradiction between reasonable and completely bogus results derived from completely analog-

---

[99] Dirac to Tamm, March 20. 1930; as cited in Kojevnikov (1993, 39).

[100] This identification also explains Dirac's next research paper on "the exchange phenomena in the Thomas atom", communicated on May 19. 1930 (Dirac 1930a).

[101] In the publication of a presentation Dirac gave to the British Association of Science on September 8. 1930, Dirac noted that "the frequency of occurrence of these processes [annihilation of proton and electron] has been calculated independently by several investigators, with neglect of the interaction between the electron and proton (that is the Coulomb force between them). The calculation gives a result much too large to be true. [...] Possibly the neglect of the interaction is not justifiable, but it is difficult to see how it could cause such a very big error." (Dirac 1930d, 606).

ical mathematical structures by shoving it on the neglect of parts of the interaction that did not have to be considered in one case but were important in the other.

Dirac's explanation obviously did not convince Tamm. Tamm included the replacement of the mass of the electron by a mean value of the masses of electron and proton for the process of annihilation in his publication (Tamm 1930b, 563). Yet, he still commented that those processes can only be discussed coherently

> when we succeed in theoretically taking into account the interaction of the negative energy electrons, the disappearance of the field of the electron and of the "hole" in case of annihilation etc. [*Footnote in the original*: It is not clear to me, why the calculation [...] of the scattering of radiation performed under neglect of these circumstances still results in the correct scattering formula.][102]

Tamm was still convinced that curing the problem of the much too high annihilation rate might lead to deviations in the scattering formula, since the two processes were described by analogical theoretical and conceptual structures. In this way, at least for Tamm, the numerical problem of the much too high annihilation rate was more than merely numerical. It was a problem of internal coherence in the application of Dirac's proton-hole model.

7.6.2 *Scattering and Annihilation II: Oppenheimer's Criticism*

The contradiction between scattering and annihilation was not only noticed by Tamm. Also for the criticism of the proton-hole model by Robert Oppenheimer, who had worked with Pauli and Heisenberg on the development of their version of QED, the theoretical description of the scattering mechanism was central. In his first note *On the theory of Electrons and Protons*, communicated in February and eventually published in May 1930, Robert Oppenheimer introduced the much too high annihilation rate as a "nummerical discrepancy". Yet, he further mentioned exactly the same point which was bothering Tamm:

> Now again it is difficult to see what large errors could be involved in the computation, since the matrix components which give [the annihilation rate between electrons and protons] are of precisely the same type as those which give correctly the Thomson formula and the optical transition probability of the electron. (Oppenheimer 1930b, 563)

The much too high annihilation rate was at odds with reasonable results derived from the same theoretical structures. But Oppenheimer provided two further arguments that directly rested on the conception of scattering in the Dirac sea.

---

[102] "wenn es gelingt, die Wechselwirkung der Elektronen negativer Energie, das Verschwinden des Feldes des Elektrons und des "Loches" bei der Zerstrahlung usw. theoretisch zu erfassen. [*Fußnote im Original*: Es ist mir nicht klar, warum die unter Vernachlässigung dieser Umstände [...] ausgeführte Berechnung der Streuung der Strahlung dennoch die richtige Streuformel ergibt.]" Tamm (1930b, 547).

Oppenheimer inferred that Dirac's theory actually required an infinite positive charge density as well. For the explanation, I quote Oppenheimer in full:

> If the explanation of the scattering of an electron is to be tenable, a negative [energy] electron must interact with the electromagnetic field in the way predicted by Dirac's theory of the electron; for otherwise the scheme proposed would not give the Thomson formula. But this means that there must be a term involving the current and charge vector of the negative [energy] electrons in the total energy momentum tensor for matter and radiation. Thus by [the Maxwell equation for the electric field], the divergence of the electric field will be everywhere infinite unless there is an infinite density of positive electricity to compensate the negative [energy] electrons. (Oppenheimer 1930b, 562)

To paraphrase this argument: The conceptualization of scattering implied that a negative energy electron emitted or absorbed a photon when jumping to the final positive energy state. Hence, there must be some interaction between the negative energy electrons and the electromagnetic field. Thereby, the restriction of one of the equations describing this interaction, $\nabla E = -4\pi\rho$, to positive energy particles seemed spurious, at least so to Oppenheimer.

For his third argument, Oppenheimer (1930b, 562–563) discussed the scattering of light by a proton or a hole in the Dirac sea. In Dirac's conception the process would be described as follows: There was an empty state in the vacuum, i.e. the initial proton, and an impinging light quantum. The transition to the intermediate state would, according to Oppenheimer, correspond to any negative energy electron jumping to some positive energy state. The transition to the final state would then be achieved by the electron filling the hole which corresponded to the initial proton.

Oppenheimer pointed out that the matrix elements involved in the mathematical representation of the above process are "precisely the same matrix components as those which give the electron scattering, and that the present theory gives equal scattering coefficients for electrons and protons" (Oppenheimer 1930b, 563). Although the interaction of the negative energy electrons have been omitted in this calculation,

> the difficulty is this, that such interaction would affect electron scattering and proton scattering in precisely the same way; whereas the Thomson formula requires the latter to be smaller by a factor proportional to the square of the ratio of the masses. (Oppenheimer 1930b, 563)

Essentially, all of Oppenheimer's arguments, as well as Tamm's, rested on the description of scattering in Dirac's sea. As the description of scattering was the one point where negative energies entered the calculations of a phenomena that had already been observed, this is rather unsurprising. Following the above arguments, Oppenheimer concluded that one should hardly find any hole in the distribution of negative energy electrons and that one should introduce a similar distribution for protons to cancel the fields produced by each. "In this way, we may accept Dirac's reconciliation of the absence of these transitions [to negative energies] with the validity of the scattering formulae." (Oppenheimer 1930b, 563).

7.6.3 Dirac Retracts the Proton-Hole Model

As I tried to show through the above discussion of the contributions of Tamm and Oppenheimer, the scattering mechanism as constituted of two acts of emission and absorption connected to an intermediate electron-proton pair production already pointed to several conceptual inconsistencies of the proton-hole model. First of all, the negative energy electrons had to interact with the radiation field and therefore the truncation of the Maxwell equations seemed spurious. Secondly, the scattering coefficients of electrons and protons came out equal in Dirac's description and, importantly, the inclusion of interactions, which might have explained the mass differences, would alter both scattering coefficients in the same way. Thirdly, scattering and annihilation were described by analogical mathematical structures which effectively meant that both should give, to some extent, reasonable results, which was certainly not the case: The credibility of the scattering formula directly reflected on the credibility of the annihilation formula due to its structural and conceptual analogy. In this vein, the scattering mechanism in terms of successive transitions to intermediate states resulted in the identification of conceptual tensions within the proton-hole model.

By the end of 1930, Dirac's faith in his proton-hole model was shaken, but not lost. In a talk given before the British Association of Science in Bristol, Dirac cited Oppenheimer's results as well as his proposal of filling up all the negative energy states and introducing another Dirac sea for the protons. At that time, Dirac did not reject this idea in total but stated that

> [a]t present it is too early to decide what the ultimate theory of the proton will be. One would like, if possible, to preserve the connexion between the proton and electron, in spite of the difficulties it leads to, as it accounts in a very satisfactory way for the fact that the electron and proton havc chargcs cqual in magnitude and opposite in sign. Further advances in the theory of quantum electrodynamics will have to be made before one can deal accurately with the interaction and see whether it will settle the difficulties, or whether, perhaps, a new idea can be introduced which will answer this purpose. (Dirac 1930d, 606)

Even though the concerns raised by Oppenheimer and Tamm restricted the space of possible solutions to the problems of the proton-hole model, Dirac still held on to "the dream of philosophers to have all matter built up from one fundamental kind of particle" (Dirac 1930d, 605). He still hoped that advances in quantum electrodynamics would result in a form of the interaction that could incorporate all the differences. Hermann Weyl's symmetry argument conclusively negated these hopes. Weyl proved that an additional interaction energy could not explain the mass difference between electrons and holes. In mid 1931, Dirac (1931, 61) eventually gave up his electron-proton model and introduced the idea of an anti-electron. The holes and the electrons now had the same mass.

The above line of argument is not meant to replace the usual narrative told around Dirac's identification of the hole with an anti-electron, but to enrich it. Following this line of argument, we can understand Dirac's giving up the "dream of philosophers" as a story of lost alternatives.[103] The scattering mechanism entailed strong doubts about the proton-hole model even before Weyl's argument arrived on the scene. As a matter of fact, it led Robert Oppenheimer to propose that protons and electrons are unconnected. The arguments by Oppenheimer and Tamm were partially based on the scattering mechanism itself, but further resulted from the analogical description of the annihilation of protons and electrons. Tamm and Oppenheimer pointed towards problems of internal consistency based on the idea that scattering as well as pair annihilation consisted of two successive emission, absorption and transition processes. In the years 1929 to 1930, Dirac first proposed to shove the mass difference on the neglect of exchange interactions, then on the neglect of the Coulomb interaction, and finally on the vague idea of "further advances in the theory of quantum electrodynamics". That this hope was in vein, was then fully established by Hermann Weyl.

7.7 Tamm (1930): Light Scattering from a Pair of Electrons and the Pauli Exclusion Principle

In the above sections I have argued that the scattering mechanism as proposed by Dirac had epistemic relevance for the inception of the Dirac sea and the evaluation of the proton-hole model. But Dirac's reconceptualization of the scattering process had a further, as far unnoticed, consequence. It led Igor Tamm not only to question the identification of a hole as a proton, but also to propose a new scattering process and to evaluate the role of the Pauli exclusion principle for intermediate states in more general terms.

I will discuss Tamm's little known contribution not because of its relevance for the further development of theory. Seemingly it went mostly unnoticed by Tamm's contemporaries; at least, I did not find any citation in the contemporary literature, even when the same topics were discussed.[104] But for my general approach, Tamm's publication has a twofold relevance: (1) Tamm's reasoning led him to an instance of what I defined as a "false prediction";[105] (2) it constituted the first reflection on the relevance of the Pauli exclusion principle for intermediate states, a theme that will reoccur in Bernhard Kockel's dissertation (Sect. 9.1.4) and constitutes one of the

[103] And in this vein, this reconstruction nicely coheres with Aaron Wright's perspective (Wright 2024, especially Sect. 2.3.4) on this episode. Wright discussed the inability of Dirac to reconstruct a mass difference between electron and proton through the exchange and Coulomb interaction by the investigation of archival material of the time, finally contributing to Dirac giving up his proton-hole model.

[104] This is for example the case for Kockel (1937) who also discussed the relevance of the Pauli exclusion principle for intermediate states.

[105] See Sect. 2.3.3 for the introduction of this analytical term and Sect. 12.2 for the analysis of the instances of false predictions I came across in my research on the historical development of the concept of virtual particles.

basic differences between the concepts of intermediate states during the 1930s and the methods introduced by Richard Feynman (see especially Sect. 11.2.3).

7.7.1 *The False Prediction, Its Background, and Its Refutation*

After Tamm communicated his difficulties in connection with annihilation and scattering processes in the proton-hole model to Dirac, he seemed to have further pondered about the description of scattering in Dirac's new conception of the vacuum. He had two realizations: Firstly, Dirac had found a workaround to the Pauli exclusion principle for the scattering process. This was obviously by design, but led Tamm to further investigate this problem. Secondly, as also Robert Oppenheimer noticed in his critique of the proton-hole model, the scattering process in the Dirac sea involved two different electrons: one which absorbed and one which emitted a photon. In June 1930, at a small conference in Kharkiv, Tamm discussed the generalization of these two observations with Dirac, seemingly without any results.[106] The corresponding questions were: Under which circumstances can a pair of electrons scatter one photon (to second order) and can the Pauli exclusion principle generally be ignored for intermediate states?

In September 1930, Tamm informed Dirac about his results:

> In Kharkov I discussed with you the possibility of the scattering of radiation by a pair of electrons, one electron absorbing the incident photon & the other emitting the scattered one. I see now that such processes are impossible, since the terms of the sum[107]
>
> $$\left|\sum_{\alpha''} \frac{(\alpha'|v|\alpha'')(\alpha''|v|\alpha')}{E_{\alpha''} - E_{\alpha'}}\right|^2$$
>
> cancel out pair by pair if the final states of the electrons are different from their initial states.[108] Only in the case, considered by you in the "Theory of protons", i.e. in the case, when the final state of an electron coincides with the initial state of the other, can such processes [light scattering from a pair of electrons] take place. I was very interested to find, that your statement: "the fact of an intermediate state of a double transition being occupied

[106] Tamm to Dirac, July 6. 1930; and Tamm to Dirac, September 13. 1930; as cited in Kojevnikov (1993, 40, 43–44).

[107] *Comment by the author*: I only had the reproduction of the letter in Kojevnikov (1993, 42–44) at hand. It is possible that the nomenclature of the given formula is retrospectively informed. Yet, for my argument, the exact nomenclature is irrelevant.

[108] *Comment by the author*: v refers to the interaction of the radiation field and matter and the α's to the respective states. α'', in particular, is the configuration of the intermediate state. $(\alpha'|v|\alpha'')$ refers to the matrix element between the states α' and α''. E_i refers to the energy of the state.

by a second electron does not alter the transition probability" can be proved only if one takes into account, that the wave functions of the electron contain "den Jordan-Wignerschen Vorzeichenfaktor".
Why did you not mention that in your paper?[109]

The sentence in quotation marks was actually not contained in this way in Dirac's paper introducing the Dirac sea. It constituted an extrapolation from Dirac's initial communication. While Dirac had outlined how occupied negative energy intermediate states could be taken into account, Tamm generalised this explanations to the case of any scattering process.

In his answering letter, Dirac stated: "What you say about the impossibility of scattering by a pair of electrons is quite correct. I had not previously noticed it."[110] Nevertheless, Dirac was reluctant to accept Tamm's way of reasoning in terms of the Jordan-Wigner sign-function. He noted that one could simply use anti-symmetric wave functions for the same derivation. Dirac's response to Tamm's usage of the Jordan-Wigner sign-function can rather simply be explained: This function occurred in matter-wave quantization as proposed by Pascual Jordan, which Dirac rejected and did not apply until the 1950s.[111] But in general, Dirac agreed with Tamm's results.

Shortly after having received Dirac's approval, Tamm sent in his little-known article to the *Zeitschrift für Physik* (Tamm 1930a). He opened with the following statement:

> The scattering of radiation is usually considered by one single electron. But one can pose the question, if in the presence of a number of electrons, let's say a pair of electrons, also other processes might not occur, in which the impinging light quantum is absorbed by one electron, but the scattered light quantum is emitted by the *other* electron. [...]
> In fact, Dirac has already considered a special case of such scattering by a pair of electrons, namely the case that the initial state of one of the scattering electrons is identical with the final state of the other.[112]

Tamm showed that the more general processes in which all initial states (m and n) and final states (m' and n') of the two electrons differed could not occur. As Tamm noted, there were eight possible ways in which such a process might come about but he did not discuss all of them. Rather, he paired the possibilities for which the order of the two subprocesses was reversed and showed that their contributions cancelled. In the discussion, I will follow Tamm's notation in which an electron was represented by the respective state it was in (e.g. m) and a photon by its frequency (e.g. ν).

---

[109] Tamm to Dirac, September 13. 1930; as cited in Kojevnikov (1993, 43–44).

[110] Dirac to Tamm, October 4. 1930; as cited in Kojevnikov (1993, 46).

[111] See e.g. Darrigol (1986, 239–241).

[112] "Gewöhnlich wird die Streuung der Strahlung durch ein einzelnes Elektron betrachtet. Es kann aber auch die Frage aufgeworfen werden, ob beim Vorhandensein von mehreren, sagen wir einem Paar Elektronen, nicht auch weitere Streuprozesse vorkommen können, bei denen das einfallende Lichtquant durch das eine Elektron absorbiert, das gestreute Quant aber durch das *andere* Elektron emittiert wird. [...] Tatsächlich hat Dirac bereits einen Spezialfall solcher Streuung an einem Paar von Elektronen betrachtet, nämlich den Fall, daß der Anfangszustand des einen der streuenden Elektronen mit dem Endzustand des anderen identisch ist." Tamm (1930a, 705, emphasis in the original).

Tamm considered the following process as an example. First the impinging photon ν was absorbed by one electron, initially in state m, and then the scattered photon ν' was emitted by the other electron, initially in state n. The two processes might be interchanged and both result in an overall transition from the initial state $a = (n, m, \nu)$ to the final state $c = (n', m', \nu')$ of the whole system. This was represented by Tamm in the following way:

$$\begin{aligned} &(m, \nu) \rightarrow m'; \quad n \rightarrow (n', \nu') \\ &n \rightarrow (n', \nu'); \quad (m, \nu) \rightarrow m' \end{aligned} \tag{7.10}$$

In the first case, the intermediate state b of the system is (m', n) and in the second case (m, n', ν, ν').

The total probability of the scattering process was proportional to

$$\left| \sum_b \frac{V_{cb} V_{ba}}{W_b - W_a} \right|^2 \tag{7.11}$$

where the V_{xy} referred to the matrix element of the transition between the states x and y and W_x to the energy of the state x. One knew, Tamm argued, that the energy of initial and final state had to be equal (conservation of energy for the whole process). Thus, the alteration of the energy of the whole system in the transitions from initial to intermediate state and from intermediate to final state had to compensate each other. In consequence, if one turned the order of the two subprocesses around, the denominators of Eq. (7.11) were equal but had opposite signs. Hence, the contributions of the two opposite arrangements of the two subprocesses (as for example in Eq. (7.10)) would cancel in Eq. (7.11) if the magnitude and the sign of the numerators stayed the same. Tamm proved that this was actually the case by the invocation of the Jordan-Wigner sign-function.

Although the matrix elements were non-zero, the process composed of the different mechanisms and orderings of the subprocesses turned out to be non-existent as it had a vanishing probability.

7.7.2 Tamm's Reflections on the Characteristics of the Process and a Short Analysis

Tamm's second aim was to prove that the process which Dirac had discussed, namely the one in which the initial state of one electron was equal to the final state of the other, had a non-zero probability and that the occupation of the intermediate state was irrelevant in this case. Therefore, he envisioned the following scenario. The initial state consisted of an electron in state m, while the state m' was unoccupied. Further there were a number of intermediate states $n_1, n_2, \ldots$ which combined with m and m'. Dirac's claim in his theory of electrons and protons, or at least Tamm's reading of it, was that the probability of the scattering was independent of the occupation of the intermediate state $n_1, n_2, \ldots$. The reasoning of Dirac, represented in Tamm's scheme, went as follows: If e.g. n_1 was unoccupied the double process $(m, \nu) \rightarrow n_1;\ \ n_1 \rightarrow (m', \nu')$ was possible. If this state was occupied, the following process, resulting in the same final state, was possible: $n_1 \rightarrow (m', \nu');\ \ (m, \nu) \rightarrow n_1$. Tamm now had to proof that these two processes did not cancel each other, but actually had the same probability. Yet, the actual proof of this statement is not of particular relevance for us as it was completely based on the behaviour of the Jordan-Wigner sign-function, which compensated the switch of the sign of the energy difference for the two processes.

In analysis, Tamm's reasoning process, although next to not recognized at the time, shows some interesting features. Tamm's proposal was based on Dirac's envisioning of scattering in the Dirac sea. Following Dirac's reconceptualization, Tamm imagined more general processes in which a pair of electrons scattered one photon. To show that such processes were not possible (as long as all the initial and final states were different), Tamm had to translate these processes into a mathematical format. Within this representation, Tamm could use the general behaviour of the system (energy non-conservation in intermediate steps but energy conservation for the whole process) and the fact, that the order of events was, within a theory of quantized matter fields, only reflected in the Jordan-Wigner sign-function, to exclude the possibility of such processes. On the same level of the mathematical formalism, he could further prove Dirac's conjecture, that for possible scattering processes the occupation of the intermediate states was irrelevant.

One should note that the overall conceptualization of these processes was very different from the modern conception. Tamm was well aware of the fact that energy and momentum conservation could not be upheld for free electrons in a single emission or absorption event (at least so if all particles were on shell, while this concept was not yet developed). But as the theoretical description had it, energy did not have to be conserved in the transitions to intermediate states. Therefore, the energy for the overall process of the whole system could only be conserved if the energy difference between first and second process cancelled. This effectively allowed for conceiving of processes as discussed by Tamm in principle, while their occurrence could be excluded without having to go into a detailed calculation. All the same it should be noted, that these processes of zero probability could be conceived of and could not directly be excluded both by Dirac and Tamm, who we can certainly attest a deep knowledge of the theoretical description, before going into a mathematical elaboration.

Tamm's description of the scattering of light from a pair of electrons is an instance of what I called a "false prediction". Dirac's verbal model, together with the idea that momentum was and energy was not conserved in interactions, made the conception of such *real*[113] processes possible. And even though Tamm did not have to perform a

[113] Real is used here as an antonym to virtual. It should not imply that the process existed, which, as discussed, Tamm proved is not the case.

detailed calculation, he had to transform the verbally described process into another format, in this case a mathematical one, to show that these processes did not occur.

From a modern vantage point, we do not need any mathematics to come to this conclusion. Representing such processes in terms of Feynman diagrams, we immediately see that the processes are not possible. They would be represented by two separate basic building blocks, each of which is unphysical by itself, standing side by side. Even without Feynman diagrams but only by invoking energy and momentum conservation at the interaction points, the processes are impossible to imagine. Since energy and momentum is conserved, at least one of the particles in each processes must be off shell. Since there is no connection between the emitting and absorbing electrons, we would be left with two off shell, i.e. not real, particles. Tamm, on the contrary, only invoking energy non-conservation and to some extent considering the pair of electrons as one quantum-mechanical system, could at least conceive of the processes.

Hence, the scattering of light from a pair of electrons (to second order) is a false prediction of Dirac's verbal model which, retrospectively, can be excluded by the application of Feynman diagrams or the conception that energy and momentum is conserved at the interaction points.[114]

7.8 Summary and Conclusion: The Epistemic Relevance of Intermediate States in the Emergence and Initial Evaluation of Dirac's Sea

In the foregoing discussion I revisited the development of the inception of the Dirac sea by focusing on the concepts involved in the description of scattering in Dirac's radiation theory. Thereby, I did not aim at substituting the standard historiography on this subject but rather at enriching it by a shift in focus. In this conclusion, I particularly want to stress the epistemic relevance of the theoretical description of scattering for the inception and evaluation of the most prominent description of anti-matter up until the late 1940s, i.e. the idea that an anti-particle is a hole in the Dirac sea.

At first I argued that Heisenberg's initial observation, that negative energy intermediate states had to be taken into account in the description of scattering, played, contrary to arguments presented in the secondary literature, no role in the evaluation of scattering by Klein and Nishina. Importantly, also Dirac was seemingly unaware of Heisenberg's conclusion until Ivar Waller pointed him towards the fact that negative energy intermediate states had to be included in the theoretical description of

[114] I want to stress here that the historiographical concept of a false prediction, as I introduced it in Sect. 2.3.3, does not necessarily rely on retrospective reasoning. I defined it as a prediction of a phenomenon which was found to be non-existent *by the historical actors themselves* (after the transformation into another representational format).

scattering. This communication by Waller in turn, and here I follow Karl Grandin, played a crucial role for Dirac taking up the general negative energy problem again.

As I argued, Waller's calculation (both in terms of the density method and in terms of Dirac's radiation theory), put the combination of radiation theory and the Dirac equation under pressure. As was clearly stated by Waller, the intermediate states in Dirac's radiation theory played a "less formal" role than in semi-classical methods, thereby showing that the picture of successive emission and absorption processes was most strongly connected to Dirac's time-dependent perturbation theory in his quantized conception of the radiation field. While in a second letter, Waller, in order to harmonize the above mentioned combination, pushed the intermediate states into the mathematical terrain, Dirac (and also Igor Tamm) took the other route and found a way of interpreting (the intermediate and) the negative energy states physically. Their occurrence as intermediate states led Dirac to reconsider the problem of negative energy states. In this vein, the virtual states (modern terminology), although Waller and Dirac fluctuated between a rather physical and more formal interpretation of the same, played a crucial epistemic role in a first step towards the creation of the idea of anti-matter in QED.

Underlying this whole developments was a shift in the conception of scattering through the alteration of the interaction Hamiltonian. While Dirac's interaction Hamiltonian of 1927 included a term proportional to the square of the vector potential responsible for direct scattering processes, this term was no longer included in the relativistic Hamiltonian. The actors recognized the consequences of the vanishing of this quadratic term: The interaction between matter and radiation was brought about exclusively through emission and absorption subprocesses in which the momentum was conserved also for the transitions to intermediate states, while energy was conserved only in total. The processes that were priorly described through direct scattering, consequently, had to be described through double processes. This was the place where negative energy intermediate states entered, both in Heisenberg's and Waller's derivation

The theoretical description of scattering further served to point towards conceptual inconsistencies of the proton-hole model. According to the arguments of Tamm and Oppenheimer, the analogy of the scattering mechanism to the calculation of the much too high annihilation rate did not only pose a numerical problem: As was clearly stated, the similarity in the theoretical description resulted in a conceptual tension which was only resolved by Dirac's introduction of the anti-electron in 1931.

Dirac's new description of scattering in a Dirac sea included two further characteristics, which were most clearly studied by Igor Tamm: Dirac found a workaround to the exclusion principle and he proposed a corresponding scattering process involving two electrons. Following Dirac's description, Tamm (and Dirac) could envision the process of light scattering from a pair (to second order). But Tamm proved that the more general processes involving two electrons with different initial and final states were actually forbidden and that the occupation of the intermediate state was unimportant; at least for scattering the exclusion principle could be ignored in the intermediate states. Although he had to transform the verbal representation into a mathematical one, Tamm did not have to go into detailed calculations. He could group the respective subprocesses in order to show that the involved matrix elements cancelled pair by pair.

I already indicated that Tamm's conclusion, although facilitated by the conception of successive acts of emission and absorption, is from a modern vantage point straightforward. Using Feynman diagrams, one immediately and literally sees that the process is impossible. It is further interesting to compare the reasoning of Dirac and Tamm to modern arguments involving the virtual particle concept[115]: Dirac's way of reasoning in the inception of the Dirac sea took an interesting further development. Heisenberg, Waller, Tamm, and Dirac were to some extent forced by their theory to take virtual negative energy states (modern terminology) into account in the explanation of scattering. This, in turn, led Dirac to take negative energy states seriously as constituents of physical reasoning.

In the course of the 1930s, physicists started to introduce new, as yet unobserved particles or processes as virtual ones into the theory to explain discrepancies between theory and experiment.[116] In these cases, the occurrence of the virtual entities was not forced on the historical actors by the theory itself, but through its comparison with experiment. Today, physicists introduce new fields into QFT which, in the form of virtual particles, allow for as yet unobserved processes and thereby probe into the domain of Beyond Standard Model Physics.[117] While today we do not know of any empirical deviations from the Standard Model and the search for "new physics" is still ongoing, the negative energy intermediate states needed to be taken into account in Dirac's case for the description of an empirically well established formula. As argued, only when the negative energy states showed up in theoretical description of the empirically accessible case of high frequency scattering their need to be taken seriously arose for Dirac. In this sense, the combination of experimental results and theoretical structures often not considered to be "real" fuelled the development of the description of anti-matter that was most common and a general building block of the conception of quantum electrodynamics and quantum field theory throughout the 1930s and 1940s. And the structure of reasoning applied by Dirac still resonates in Beyond Standard Model physics today.

---

[115] I already pointed at this argumentative structure of the historical actors and its further development in Ehberger (2022, 561–562).

[116] Gregor Wentzel's inclusion of hypothetical spin zero and one states of the proton and neutron (see Sect. 10.2.4) or the introduction of higher-spin states by Homi J. Bhabha and Walter Heitler (they will shortly be discussed in Sect. 9.3.2) are cases in point.

[117] See for example https://www.physik.uzh.ch/groups/serra/RareDecays.html (visited on March 31. 2020).

Chapter 8
The Practice of Time-Dependent Perturbation Theory (Part I): Formal and Conceptual Extensions (1929–1936)

In the previous chapter I focused on the relevance of the concept of an intermediate state and the verbal model of successive emission and absorption processes for the development of the concept of anti-matter in the framework of quantum electrodynamics. My argument was based on a detailed study of a small set of publications and of the correspondence of Dirac with Igor Tamm and Ivar Waller. In the following two chapters, I want to switch gears and take a broader look on the application of time-dependent perturbation theory during the late 1920s and all through the 1930s.

As I argued in the introductory chapters, the virtual particle in particular, but also the concepts connected to its historical development, are concepts very closely related to the practice of quantum electrodynamics and especially to the method of perturbation theory. In order to get a grasp on the development of these concepts a focus on the actual practice of the working theorist is necessary. In the following two chapters, I will concentrate on time-dependent perturbation and scattering phenomena because Dirac's verbal model of successive emission and absorption processes as well as the concept of an intermediate state was at the beginning most firmly connected to this method and these phenomena. The incorporation of verbally explicit references to intermediate states, the transitions to and from them, and virtual entities more generally into the theoretical evaluation of static effects will be discussed in Chap. 10.

Through the focus on the practice of theorists, I do not only aim at a deeper understanding of the usage of the concepts but also on the general practice of QED during the 1930s. As I will show in the following discussion, physical and mathematical reasoning as well as their respective representations had to be merged harmoniously in the actual application of QED in general and perturbation theory in particular. The exact nature of this interaction and the use of different modes of representation is of prime importance to gather an understanding of the practices of the historical actors, but also of the relevance they ascribed to the respective concepts.

This focus on practice is connected to an important conceptual point. One regularly encounters the concept of virtual particles in historical discussions of quantum

M. Ehberger, *Representing the Unobservable*, Science Networks. Historical Studies 68,
https://doi.org/10.1007/978-3-032-09188-8_8

electrodynamics of the 1930s.[1] By a focus on the application of the theory and the respective forms of representations, I will argue that it was not so much the concept of a virtual particle that was of prime importance. Rather the idea of energy non-conserving transitions to intermediate states coupled to emission or absorption processes was guiding the practice of theoretical physicists. As I will argue, the intermediate state as the interlink of the processes was in the centre of attention and, contrary to some historical judgements, something like a general algorithm for calculations in QED emerged during the 1930s.

I will investigate the applications of time-dependent perturbation theory throughout the 1930s up until Walter Heitler's and Huanwu Peng's theory of radiation damping (1942) in the following two chapters. I explicitly do not aim at comprehensiveness. Rather, the discussion will focus on representative applications of QED during the 1930s in order to probe the conceptual framework connected to time-dependent perturbation theory. Consequently, the following will be a snapshot story of the working physicists' application of time-dependent perturbation theory.

There is a considerable amount of secondary literature on QED during the 1930s and the general development of the theory is well-known.[2] Time-dependent perturbation theory has been discussed by Lacki et al. (1999) who focused, after its application by Dirac, on its development by Ernst Stueckelberg which had next to no impact during the 1930s. The theoretical practice of the 1930s has most extensively been discussed in the works of Alexander Blum and Adrian Wüthrich.[3] Through these works we know that QED during the 1930s was based on a conception of successive state transitions and the focal point of the evaluation were the characteristics of these states. But the actual theoretical practice of time-dependent perturbation theory, its modifications, the representations used by the historical actors as well as their place in the reasoning process, and their development throughout the 1930s have never been the focal point of the respective discussions. Thus, instead of reviewing the arguments so far presented, I will engage with (specific arguments of) the secondary literature in the respective sections.

In Sect. 8.1, I will engage with the further development of the usage of Dirac's radiation theory in Göttingen, where Dirac wrote his paper on dispersion and introduced his verbal model. I will focus on the work by two doctoral students of Max Born: Maria Göppert-Mayer (Sect. 8.1.1) and Victor Weisskopf (Sects. 8.1.2–8.1.4).

Göppert-Mayer introduced a new terminology, the "as if", and included a temporal dimension into energy level diagrams. She also introduced a new concept to the evaluation of time-dependent perturbation theory in QED, the *Schüttelwirkung*. The general mode of reasoning behind this concept, the adiabatic switching on of the perturbation, and the effect it had on the perturbative series will allow Victor Weisskopf to provide for a first general perturbative expansion of radiation theory to an arbitrary order in 1933. Both, Göppert-Mayer and Victor Weisskopf, will present

[1] See my discussion in Sect. 1.3.

[2] See the literature cited in my general overview of the historical development in Sect. 1.3 as well as in Chap. 3.

[3] Blum (2015, 2017), and Wüthrich (2010, Chap. 2).

arguments why processes one might consider to be independent due to the nature of the mathematical representation of the involved matrix elements were dependent: They further blurred the lines between actual and virtual (modern terminology) processes. In his joint work with Eugene Wigner, Weisskopf will extend the time-scope for which the perturbative expansion was to be trusted, thus providing for an important elaboration on the perturbative evaluation. Taken together, Weisskopf's work and the introduction of the *Schüttelwirkung* by Maria Göppert-Mayer will form the basis of Heitler's theory of radiation damping, which I will turn to in the final Section of the next chapter.

Section 8.2 will be devoted to the two most important didactic works on quantum electrodynamics of the 1930s: Enrico Fermi's 1932 "Quantum theory of radiation" and Walter Heitler's 1936 textbook of the same title. Didactic literature, as a particular important means of distributing theoretical knowledge and practice, deserves special attention when we want to judge in how far specific concepts were applied and accepted by the theoretical community. But I will also discuss papers closely connected to these works as they show how physicists had already become used to the handling of the concepts connected to perturbation theory and allow for a more detailed evaluation of the question in how far the concept of a "virtual particle" (as a concept of theoretical practice) had already been developed. I will conclude this chapter (Sect. 8.3) with the reconstruction of something like a general and iterative algorithm that had developed in the first half of the 1930s. This general scheme, although never spelled out explicitly as such, was underlying a great part of the theoretical calculations of scattering phenomena up until the advent of Feynman's methods in the late 1940s.

8.1 *Schüttelwirkung* and "Weisskopf's Formula": The Extension and Modification of Dirac's Perturbative Method in Göttingen

Dirac had finished his paper on dispersion and introduced the explicit temporal description of transitions through intermediate states as well as the differentiation between direct scattering and the processes involving intermediate states during a research stay in Göttingen in early 1927. One of the centres during the development of quantum theory from 1920 onwards, Göttingen stayed a vibrant place for experimental as well as theoretical physics until Hitler's rise to power in 1933. Under the auspices of Max Born and James Franck the institute attracted a great number of young researches. Besides Paul Dirac, Enrico Fermi, Walter Heitler, and Eugene Wigner had longer research stays in Göttingen throughout the 1920s and early 1930s.[4]

[4] For the scientists who spent research stays at the Göttingen institute (but also for the corresponding listing for Copenhagen and Munich) see the table in Eckert (2001, 156–157).

Among the doctoral students in the late 1920s to the early 1930s were Max Delbrück, Robert Oppenheimer, Maria Göppert-Mayer,[5] and Victor Weisskopf. The breadth of topics studied in Göttingen was as rich as the scientific personnel: Solid state physics (experimental work mainly by Robert Pohl) and quantum chemistry (most notably by Walter Heitler) were investigated as well as rather mathematical approaches to quantum physics, such as group theory.

The following section is devoted to the work of Maria Göppert-Mayer and Victor Weisskopf mostly carried out at the Göttingen institute. Their work is strongly connected: Both published their dissertation in 1931, thanked each other and cited their work in the publications. We can certainly assume a close exchange of ideas between the two physicists during their PhD-phase.

There are several reasons why the study of the contributions of these two physicists is suggested. Göppert-Mayer's work can be considered exemplary for the application of Dirac's radiation theory. But it certainly also exceeded the prior application through formal but also conceptual alterations. Weisskopf, together with Wigner, extended the time for which the perturbative account could be trusted to investigate the line-breadth and applied this method to the case of resonant scattering. Thereby, he gathered further insight into this mechanism, the transitions involved and its connection to regular scattering. Furthermore, Weisskopf gave a first derivation of Dirac's perturbation theory to all orders, a derivation that rested crucially on Göppert-Mayer's contributions.

Therefore, the work studied in the following section can be seen as an important extension of Dirac's methods and decidedly contributed to the understanding of the same. I will start out with the dissertational research of Maria Göppert-Mayer as it is more firmly rooted in the perturbative approach to radiation theory that was initially proposed by Dirac. Afterwards, I will turn to Victor Weisskopf's dissertation. In order to understand his work, the extension of Dirac's perturbative approach to longer times as proposed by him and Eugene Wigner has to be evaluated shortly. The extension to an arbitrary order as given by Weisskopf in 1933 will conclude this section.

8.1.1 Göppert-Mayer's Dissertation (1931): Exclusion of Indirect Processes and Inclusion of the Schüttelwirkung

Maria Göppert-Mayer,[6] born in 1906 in Kattowitz, was raised in Göttingen, where her family moved in 1910. The family soon became well-connected to the academic circles of the town: David Hilbert, Max Born, and James Franck were among the family's friends. The successful completion of the admission exams to the University,

[5] I will exclusively use her married name Göppert-Mayer. She altered her maiden name Göppert to Göppert-Mayer after marrying Joseph E. Mayer, a chemist, in early 1930. Sachs (1979, 311).

[6] This short biographic sketch is based on Sachs (1979, 311–314).

which Göppert-Mayer had prepared for by visiting the so-called "Frauenstudium", allowed her to enrol as a student of mathematics in 1924. Göppert-Mayer joined the University when the development of quantum theory had just entered a new stage. She was also able to gain first-hand knowledge about the theoretical development, as Max Born invited her to his physics seminar. Her interest shifted in the following years from mathematics to physics.

In 1929, Max Born asked Göppert-Mayer to write a chapter on radiation and dispersion theory for his and Pascual Jordan's textbook *Elementare Quantenmechanik* (Born and Jordan 1930, Chap. 6). In her first scientific publication (Göppert 1929), Göppert-Mayer attributed the development of the ideas of her dissertational research to this assignment. Furthermore, as will be laid out below, the experimental research previously carried out in Göttingen, and so far only qualitatively and semi-classically explained, formed an important motivation for the topic of her dissertation. In the following, I will mostly focus on her published dissertation "On elementary acts with two quantum jumps".[7] Yet, it should be noted that Göppert-Mayer was already in possession of most of the results of the first part of her dissertation in 1929. The terminology she used and the graphical representations she provided are already part of her above mentioned publication.

Göppert-Mayer's dissertation nicely falls into two parts. They are connected through the general theoretical approach and the mathematical as well as conceptual structures used for describing the different phenomena.[8] In the first part, Göppert-Mayer discussed processes which involve two external light quanta under the headline of "the synergy/combined action [Zusammenwirken] of two light quanta in one elementary act".[9] In the second part, she focussed on two processes which involve the simultaneous action of the impact of a material particle on an atom and the emission of light, i.e. "the synergy of light and collision in one elementary act".[10] Before going into a deeper analysis of Göppert-Mayer's work, her motivation for taking the specific theoretical approach and the processes studied with it will be outlined.

Theoretical and Empirical Roots of Göppert-Mayer's Dissertation

The application of Dirac's radiation theory was to some extent quite natural in Göppert-Mayer's dissertation. Dirac had finished his work on radiation theory at the Göttingen institute in early 1927. The perturbative approach used therein had

---

[7] "Über Elementarakte mit zwei Quantensprüngen," Göppert-Mayer (1931).

[8] The only analysis of Göppert-Mayer's dissertation I know of is Masters (2013). Part of the analysis of the theoretical framework and the origins of Göppert-Mayer's dissertation is based on this contribution. Yet, I differ decidedly from Masters analysis of the dissertation itself. While Masters seems mostly to question the originality of Göppert-Mayer's theoretical work, I focus exactly on those points where she went beyond Dirac's original communication. Secondly, other than Masters, I will also include a discussion of the second part of her dissertation.

[9] "Das Zusammenwirken zweier Lichtquanten in einem Elementarakt," Göppert-Mayer (1931, 275).

[10] "Das Zusammenwirken von Licht und Stoß in einem Elementarakt," Göppert-Mayer (1931, 284).

been developed by Dirac and by Max Born independently earlier on. Up until early 1929, Dirac's radiation theory was the only approach to the quantization of the electromagnetic field and its interaction with matter ready for application. Furthermore, the theory was simultaneously studied by Victor Weisskopf (and Eugene Wigner) at the Göttingen instintute (see below). As already mentioned above, the assignment for writing a chapter on this theory for the textbook of Born and Jordan eventually led Göppert-Mayer to first insights.

But Göppert-Mayer's dissertation also had its roots in the empirical research carried out in Göttingen. The simultaneous action of radiation and impinging particles on molecules and atoms had theoretically been discussed by Born and Franck (1925). Later on, Oldenberg (1928) experimentally studied such phenomena.[11] Oldenberg observed that the breadth of the resonance line of mercury atoms was increased, i.e. he found a continuous low frequency band directly joining the resonance line, when a different gas, in his case Helium, was infused. He interpreted this effect by invoking the simultaneous action of the emission of radiation and an in- or decrease of the kinetic energy of the atoms due to the impinging particles. The de-excitation energy of the mercury atoms was divided into kinetic and radiation energy, a process of the "synergy of collision and radiation".[12] Oldenberg then tentatively generalized his findings by considering elementary acts in which three energy sources (radiation quanta or material particles) were involved (Oldenberg 1928, 609–612). Within this line of thought he already discussed the "synergy/combined action of two light quanta in one elementary process".[13] Oldenberg qualitatively discussed the idea of the simultaneous absorption and emission of two light quanta. In Oldenberg's publication we thus encounter three of the four effects studied by Göppert-Mayer in her dissertation: double emission, double absorption, and the broadening of the emission line due to the simultaneous action of the radiation field and impacting particles.

The fourth phenomenon investigated by Göppert-Mayer was studied and semi-classically explained by Franck (1928). The intensity of spectral lines stimulated by electronic collisions not only showed a sharp maximum at the resonance energy, as was to be expected, but also a large flat maximum at about 100 Volts above the excitation energy. Since this energy already corresponded to an ionization of the atom, the phenomenon had been regularly interpreted as a recombination of the electron and the molecule or atom under study. Franck (1928), in contrast, gave reasons why such an interpretation was untenable.[14] Franck interpreted the phenomenon,

[11] See also Oldenberg (1930) for a discussion of the simultaneous action of different kinds of energies (kinetic, excitation and radiation).

[12] "Zusammenwirken von Stoß und Strahlung," Oldenberg (1928, 605).

[13] "Zusammenwirken zweier Lichtquanten in einem Elementarvorgang," Oldenberg (1928, 611).

[14] The main reason was, that if an actual ionization took place the fields present during the experiments would cause the secondary electrons to have a much too high velocity to make a recombination plausible.

in the language of the old Bohr theory, as a recombination that takes place under the emission of light before the electron "has left a certain zone of influence".[15] Translating this conjecture into the language of Dirac's radiation theory, Göppert-Mayer considered the possibility that the atom took up energy from the impinging electron while at the same time it emitted a radiation quantum so that it would eventually be found in a stationary state.

The main result of Göppert-Mayer's dissertation was the translation of the above mentioned qualitative reasoning by Oldenberg and Franck into the language and mathematics of Dirac's radiation theory and the proof, as far as empirical data was extent, of the harmony of the theoretical predictions and the empirical findings. The calculation in Göppert-Mayer's dissertation followed the procedures given by Dirac very closely. Many of the interpretations and applications are directly taken from Dirac (1927b). But we find also differences in interpretation, partly due to mathematical reformulations provided by Göppert-Mayer, partly due to the specific processes under study.

The Exclusion of Direct and "Abnorm" Transitions

The first observation I want to mention is concerned with the term in the interaction energy which was responsible for "direct scattering". As was already pointed out in the prior chapter, this term vanished automatically when the electron was described by the Dirac equation. Göppert-Mayer (1931, 277–278) showed that this was also the case in the non-relativistic theory if the so-called dipole approximation was applied, i.e. when the wavelength of the radiation was large in comparison to the atomic dimensions.

Göppert-Mayer worked directly with the Hamiltonian of Dirac (1927b). The perturbation H' consisted of two terms:

$$H' = \frac{e}{mc}(pA) + \frac{e^2}{2mc^2}A^2. \tag{8.1}$$

where p referred to the momentum of the electron and A to the vector potential. The first term of H' (linear in A) was responsible for emission and absorption and, to second order, for scattering via intermediate states. The second term (proportional to A^2) combined initial and final state of the system directly, i.e. was responsible for "true" scattering.

Göppert-Mayer transformed the Lagrangian

$$L = \dot{q}p - H_{el} \tag{8.2}$$

[15] "[...] eine bestimmte Einflußzone verlassen hat," Göppert-Mayer (1931, 275).

corresponding to the Hamiltonian of the electron by the addition of the total time derivative $\frac{d}{dt}(qA)$. Thereby, she could derive a new, but equivalent Hamiltonian $\bar{H}_{el}$[16]:

$$\bar{H}_{el} = \frac{\bar{p}^2}{2m} + V(q) - \frac{e}{c}\left(q\frac{dA}{dt}\right) \tag{8.3}$$

With the definition of the electric moment $P = \sum_r eq_r$ and the approximation of wavelength of light long compared to the radius of the atom[17] the interaction energy read:

$$H' = -\frac{1}{c}(P\dot{A}) \tag{8.4}$$

This interaction energy was independent of the terms proportional to the square of the vector potential. Since $-\frac{1}{c}\dot{A}$ was equal to the electric field strength, the interaction energy boiled down to the potential energy of the electric moment of the atom in the radiation field.

Although the procedure described above constituted a formal alteration of the interaction Hamiltonian,[18] only indirect scattering processes had to be considered. Processes referring to "direct" or "true" scattering did no longer occur, also in the non-relativistic theory (at least so under the given approximations). As will be seen below, Victor Weisskopf would apply this transformation in his dissertation and his subsequent work. It forms one of the pillars of the first extension of the perturbative formalism to higher orders and ensured that the general perturbative expansion could be taken over to a relativistic description of the material part as well.

For the second point I want to discuss we have to go into some detail of the perturbative expansion. To second order, the equation for the time-development of the coefficients was given by Göppert-Mayer as:

$$a_{ns} = \sum_{n's'} \frac{H'_{nn'ss'} H'_{n'n_0s's_0}}{h(\nu_{n'n_0} + \nu_{s's_0})} \left[\frac{1 - e^{2\pi i(\nu_{nn'} + \nu_{ss'})t}}{h(\nu_{nn'} + \nu_{ss'})} - \frac{1 - e^{2\pi i(\nu_{nn_0} + \nu_{ss_0})t}}{h(\nu_{nn_0} + \nu_{ss_0})} \right] \tag{8.5}$$

n referred to the state of the material system, s to the state of the radiation field. Primed, unprimed and indexes carrying a 0 corresponded to the intermediate, final and initial state respectively. Equation (8.5) formed the basic equation from which all of the physical considerations of Göppert-Mayer's dissertation started from.

For the following discussion the important part of Eq. (8.5) is the time-development of the coefficient a_{ns}, i.e. the terms in brackets. Dirac had shown that through an

---

[16] The addition of a total time derivative to a Lagrangian does not alter the equations of motion so that the transformed and the initial Lagrangian/Hamiltonian describe the behaviour of an equivalent physical system.

[17] With this approximation the variation of vector potential over the area of the atom can be neglected so that: $\frac{dA}{dt} \approx \frac{\partial A}{\partial t} =: \dot{A}$.

[18] See also Milonni (1994, Sect. 4.3) for a modern derivation of the vanishing of the A^2-term in the dipole approximation.

integration of the square of the second term of Eq. (8.5) one arrived at energy conservation for the whole process: This is what I have previously called Dirac's trick.[19] Further, Dirac had argued that the contribution of the first term in brackets "may be neglected since it always remains small" (Dirac 1927b, 722) for the non-resonant case. In her discussion of double emission and double absorption Göppert-Mayer explicitly focussed on non-resonant cases and therefore could have taken up Dirac's reasoning directly.[20] And even though Göppert-Mayer's result is essentially the same as Dirac's, namely the exclusion of the contribution of the first term in brackets, she provided an argument based on the specific way in which the perturbation was treated in Dirac's radiation theory.

To see how she developed her argument, I will outline her treatment of induced double emission.[21] Göppert-Mayer assumed a non-resonant impinging radiation of frequency ν_α ($\nu_\alpha \neq \nu_{nn'}$). The frequency of the secondary photon was ν_β. In the initial state of the system, the atom was assumed to be in an excited state n_0 and one of the oscillators of the radiation field was excited (ν_α).

By plugging in the explicit form of the perturbation and the intermediate states while keeping in mind that either light quantum may be involved in either subprocess, Göppert-Mayer found that the time-development of the final state was constituted of four factors:

$$\begin{aligned} &\frac{1-e^{2\pi i(\nu_{nn'}\mp\nu_\alpha)t}}{h(\nu_{nn'}\mp\nu_\alpha)}; \quad \frac{1-e^{2\pi i(\nu_{nn_0}+\nu_\beta\mp\nu_\alpha)t}}{h(\nu_{nn_0}+\nu_\beta\mp\nu_\alpha)}; \\ &\frac{1-e^{2\pi i(\nu_{nn'}+\nu_\beta)t}}{h(\nu_{nn'}+\nu_\beta)}; \quad \frac{1-e^{2\pi i(\nu_{nn_0}+\nu_\beta\mp\nu_\alpha)t}}{h(\nu_{nn_0}+\nu_\beta\mp\nu_\alpha)} \end{aligned} \tag{8.6}$$

where the upper sign of $\pm$ referred to the Raman effect and the lower sign to double emission. $h\nu_{nn'} = E_n - E_{n'}$ referred to the energy difference between two states of the scattering atom.

Göppert-Mayer then argued that the excitation of a_{ns} is only appreciable if one of the denominators of the terms in Formula (8.6) vanished; this is essentially Dirac's argument. Since $\nu_{nn'} \neq \nu_\alpha$ was assumed from the start, only $\nu_{n'n_0} + \nu_\beta$,[22] $\nu_{nn'} + \nu_\beta$, and $\nu_{nn_0} + \nu_\beta \mp \nu_\alpha$ could be considered. Yet, as Göppert-Mayer argued:

[19] See Sect. 5.1.3.

[20] She actually did this in the chapter of the textbook by Born and Jordan. See Born and Jordan (1930, 405).

[21] The case of double absorption was not discussed by Göppert-Mayer at length as she found that "the calculations proceed in this case essentially analogously." (*Original*: "Die Rechnungen verlaufen in diesem Fall im wesentlichen analog.") Göppert-Mayer (1931, 282).

[22] This denominator does not occur in Eq. (8.6) but, as one can easily check, figures as one of the resonance denominators of the matrix elements in Eq. (8.5).

> The neighbourhood of the zeros of the first two denominators render probabilities for processes in which the energy law is not obeyed, namely for transitions of the atom from state n_0 to n under absorption or emission of a quantum of the impinging frequency ν_α and emission of one of the eigen-frequencies of the atom.[23]

So far, this was not particularly new but rather constituted an interpretation of the terms normally dropped. But Göppert-Mayer also pointed out why these transition possibilities occurred and thereby gave a better grounded argument of why they could be neglected:

> These transitions [for which energy is not conserved in the overall process] do not correspond to real/actual ["wirkliche"] processes, they are based on the peculiarity of the method of the variation of constants applied here.[24] In the application of this method it is assumed, that the perturbation energy H' starts to act at time zero, while in reality it acts continuously. This process of switching on [the perturbation] is the cause of the occurrence of the abnorm transitions.[25]

Göppert-Mayer did not provide a derivation or references for the conclusion that the energy non-conserving terms, the "abnorm transitions", are due to the sudden switching on of the perturbation. But Göppert-Mayer's above conclusion most likely had its roots in the research of her doctoral advisor Max Born.[26] The invoked action of the sudden switching on of the perturbation has been called *Schüttelwirkung* (a direct translation would read "shaking action") by Born and Jordan (1925b, 490) in a different theoretical framework. This terminology was taken up afterwards in reference to Dirac's radiation theory as well. Furthermore, one of the best-known contributions of Max Born to early quantum mechanics was his derivation of the adiabatic principle in the new conceptual framework (e.g. Born 1927).[27] In this work, the concept of the *Schüttelwirkung* is used as well.[28] Since it was the sudden switching on of the perturbation that was made responsible for the energy non-conserving terms,

---

[23] "Die Umgebung der Nullstellen der ersten beiden Nenner liefern Wahrscheinlichkeiten für Prozesse, bei denen nicht der Energiesatz gewahrt ist, nämlich für Übergänge des Atoms von n_0 nach n unter Absorption bzw. Emission eines Quantums der eingesandten Frequenz ν_α und Emission einer Eigenfrequenz des Atoms." Göppert-Mayer (1931, 279).

[24] *Note by the author*: The term "variation of constants" was soon used for the procedure of time-dependent perturbation theory as envisioned by Dirac as it resembled a perturbative treatment for celestial problems. It was introduced by Leonhard Euler and further developed by Joseph-Louis Lagrange to discuss the movement of celestial bodies.

[25] "Diesen Übergängen entsprechen aber nicht wirkliche Prozesse, sie beruhen auf der Eigenart der hier angewendeten Methode der Variation der Konstanten. Hierbei wird nämlich angenommen, daß die Störungsenergie H' zur Zeit Null zu wirken beginnt, während sie in Wirklichkeit ununterbrochen wirkt. Dieser Einschaltvorgang ist die Ursache des Auftretens der abnormen Übergänge." Göppert-Mayer (1931, 280).

[26] See also Im (1996) for Max Born's usage of the *Schüttelwirkung* or shaking action, specifically in his work with Pascual Jordan.

[27] The adiabatic principle states that, if a perturbation is switched on very slowly, i.e. adiabatically, a quantum-mechanical system does not undergo state transitions.

[28] Naturally, there is a strong interrelation between time-dependent perturbation theory and the adiabatic principle. Although this connection still might lead to interesting insights into the applications of quantum theory, a complete evaluation would go beyond the scope of this book.

they should vanish when the perturbation is introduced adiabatically. This conclusion would only be spelled out explicitly in publication later on.

Although the argument of Dirac, namely the smallness of the contribution of such terms, and of Göppert-Mayer, by referring to the conceptualization of the perturbation, led to the same result and both would be used subsequently without specific references, I laid out Göppert-Mayer's reasoning in some detail for three reasons. First of all, it provides an example of physical reasoning in the evaluation of perturbation theory that seems to be generally accepted. The modelling of the perturbation as switched on at some particular instance led to physically unreasonable transitions. These transitions were excluded by Göppert-Mayer's reasoning. Secondly, Weisskopf would use the above conclusions about the structure of the interaction energy and the *Schüttelwirkung* in a first derivation of the perturbative expansion to an arbitrary order. Thirdly, the concept of the *Schüttelwirkung* will be discussed more thoroughly by the historical actors in the late 1930s and became, through some detours, one of the theoretical motivations of Heitler and Peng's theory of radiation damping, a development I will turn to in Sect. 9.3.

Incorporating Time Into Energy Level Diagrams: "As If"

The actual evaluation of the probability of the processes of double emission and double absorption is not of particular relevance for us but there are two further points that deserve mentioning. In her dissertation as well as in the prior communication Göppert-Mayer provided graphical representations of the processes she discussed.[29] The diagrams of her first communication are displayed in Fig. 8.1, while (nearly) the same diagrams can be found in her dissertation (Göppert-Mayer 1931, 284). Göppert-Mayer's representation was certainly rooted in the usual format of energy level diagrams, yet a clear temporal order was indicated in the behaviour of the material part of the system.

From left to right, the Stokes and Anti-Stokes case of Raman scattering as well as the processes of double absorption and double emission were represented. n and m referred to initial or final state of the atom. k was chosen as a representative of an intermediate state (in the above formulas it was referenced as n'). ν and ν', represented by solid lines, referred to the photons. The arrows on the solid lines indicated whether a photon was absorbed (pointing upwards) or emitted (pointing downwards), while no succession of emission and absorption processes was specified.[30] This was actually

[29] I shortly discussed the diagrams by Göppert-Mayer as an alteration of the diagrammatic representations in response to Dirac's verbal model and the terminology of the "as if" in Ehberger (2022, 556–557).

[30] The apparent order in the two diagrams on the left side is most probably due to the two-dimensional format. If you look at the third and fourth diagram, you see that the arrows are actually aligned. For clarity's sake: The diagrams in the dissertation only differ from the represented ones by a small displacement of the photon lines in the fourth diagram. I consider this a error in printing.

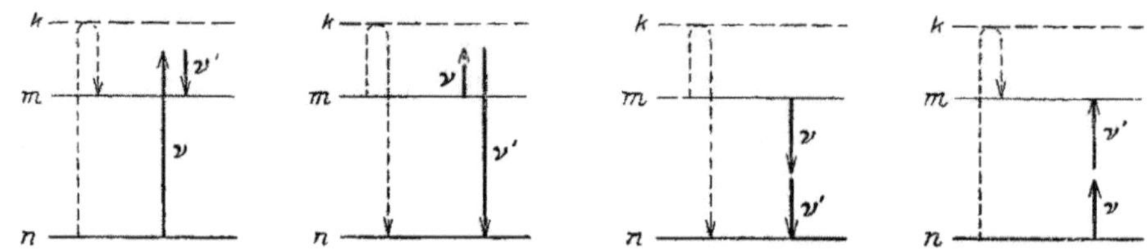

Fig. 8.1 Energy level diagram describing, from left to right, Raman scattering (Stokes and Anti-Stokes case), double absorption, and double emission. Diagrams in Göppert, Maria. 1929. "Über die Wahrscheinlichkeit des Zusammenwirkens zweier Lichtquanten in einem Elementarakt." *Naturwissenschaften* 17, 932, here 932. https://doi.org/10.1007/BF01506585 (cited as Göppert (1929) in the bibliography). Reprinted with permission from Springer Nature

in harmony with the conceptualization of Dirac's radiation theory: The sequence of emission and absorption processes was not fixed and both orders had to be taken into account.

The dashed lines referred to the behaviour of the material part of the system. They clearly indicated a temporal order of events. Take, for example, the third diagram: The atom was initially in the state m. It performed a transition to the intermediate state k and eventually, note the arrow, arrived in the final state n. Certainly, these diagrams were retrospective visualizations of the mathematical structures and their interpretation, but they served a specific, in this case a didactic purpose: They visualized the analogical conceptualization of the different effects (Stokes and Anti-Stokes case of the Raman effect, double emission and double absorption) through a representational format (energy level diagrams) which was familiar to the great majority of the community. Göppert-Mayer emphasized the analogical structure of the processes throughout her two communications. In her first publication she made this use of the diagrams particularly clear by stating: "For comparison, the two processes are schematically put side by side with the Raman effect."[31] The same reason for choosing this representational format can be inferred from the introduction of the diagrams in her dissertation:

> It is further to note, that the two discussed processes [double emission and double absorption] as well as the Raman effect behave *as if* two processes, each of which does not obey the energy law, occur in one act: The atom goes under emission or absorption of a frequency ν from the state n_0 into an intermediate state n' and from there under the emission or absorption of a frequency ν' into the end state n.[32]

Göppert-Mayer introduced a specific kind of terminology in the above quote. She noted that the processes behave "as if" they are constituted of two successive

[31] "Es seien hier zum Vergleich die beiden Prozesse schematisch dem Raman-Effekt gegenübergestellt." Göppert (1929, 932).

[32] "Zu bemerken ist noch, daß sich beide besprochenen Prozesse ebenso wie der Ramaneffekt so verhalten, *als ob* zwei Vorgänge, von denen jeder nicht dem Energiesatz genügt, in einem Akt geschehen: Das Atom geht unter Emission oder Absorption einer Frequenz ν vom Zustand n_0 in einen Zwischenzustand n' und von da unter Emission oder Absorption einer Frequenz ν' in den Endzustand n über." Göppert-Mayer (1931, 284, emphasis added).

subprocesses. This might be a reference to Hans Vaihinger's philosophy of the "As if".[33] In most general terms and concerning the natural science in particular, Vaihinger argued that scientists used fictional entities in their arguments. Thereby they might come to pragmatically justifiable conclusions, but not to genuine explanations. Since the early 2010's, the use of fictions in science and also Hans Vaihinger's work has gathered some attention, especially through Alisa Bokulich's work on explanatory fictions.[34] While it is likely that Hans Vaihinger's work was known in Göttingen, Göppert-Mayer did not provide for any indication that she took this concept from Vaihinger.

The "Virtual Oscillator" as a Classical Substitute

Whether or not Göppert-Mayer referred to Vaihinger at this point, the terminological twist of the "as if" is still insightful for our historical object: the intermediate states and the transitions to and from them. First of all, Göppert-Mayer obviously did not conceive of these transitions as something corresponding to actual subprocesses. All the same, she considered the description certainly suitable and insightful enough to explain the analogical structure and provide for a diagrammatical representation of the same. Secondly, she created a connection, albeit indirect, between the transitions to intermediate states and the virtual oscillators. As she noted in connection to induced double emission:

> If light of frequency ν is irradiated, then the portion of the induced emission of frequency ν' behaves, in reference to its intensity, as if a virtual oscillator [...] emits the frequency ν' spontaneously.[35]

The electric moment of this virtual oscillator was, as always to second order, composed of the matrix elements corresponding to transitions to intermediate states (Göppert-Mayer 1931, 282, Eq. (8)). Transitions to intermediate states and virtual oscillators were seemingly on a par when it came to their physical reality, both being referred to by an "as if". But Göppert-Mayer obviously did not conceive of them as equivalent.

The virtual oscillator was a classical object which stood in for the quantum atom as a whole. Göppert-Mayer did not use the term to refer to the individual components of the formula, but only to its complete result. The transitions to intermediate states were a theoretical description of the behaviour of a quantum atom, although

[33] Vaihinger (1911). For a discussion of the (mostly negative) reception of the Vaihinger's fictionalism in the physics community by way of a few examples from the early 1920s to the early 1930s, see Hentschel (2014).

[34] See, for example, Bokulich (2012). For the reader interested in philosophical reflections on the use of fictions in science, the edited volume "Fictions in science. Philosophical essays on modelling and idealization" (Suárez 2009) can form a good starting point.

[35] "Wird Licht der Frequenz ν eingesandt, so verhält sich der Anteil der erzwungenen Emission der Frequenz ν' hinsichtlich seiner Intensität so, als ob ein virtueller Oszillator [...] die Frequenz ν' spontan emittiert." Göppert-Mayer (1931, 281).

not corresponding to actual subprocesses. They were a means of describing the components, or better, to discover their structure through physical reasoning. What Göppert-Mayer called a "virtual oscillator" referred to a real transition. More to the point, it conceptually replaced the atom by a classically behaving oscillator sending out the radiation with the electric moment, which was derived by the application of the verbal model.

In this connection it must be noted, that Göppert-Mayer used the term "virtual" exclusively to refer to the oscillator, i.e. the replacement of the atom by a semi-classical entity. For the transitions to the intermediate states, Göppert-Mayer did not invoke the term "virtual" or another term. Although both "virtual oscillator" and the transitions to intermediate states were, for Göppert-Mayer, theoretical fictions, they were not equivalent and performed different functions. Even though Göppert-Mayer did not shy away from using the term "virtual" or the admittance of the fictional character of such entities, she did not connect the term "virtual" to intermediate states. As Göppert-Mayer wrote her dissertation at one of the centres of quantum mechanics (Göttingen) and under the supervision of Max Born (who had made important contributions to the old quantum theory and was certainly aware of the current developments of radiation theory), we can infer that at least in these (rather informed) circles, the difference between virtual oscillators and what was later to be called virtual transitions was still clearly perceived.[36] All the same, this indirect connection can be seen as an early indication of the terminological coherence that would manifest in the work of the group around Werner Heisenberg in Leipzig, i.e. as a reference to a virtual world.[37]

(In-)Depdence of Processes and Their Time Proportionality

There is one last aspect of Göppert-Mayer's dissertation I want to discuss and which is directly connected with the interpretation and application of Dirac's perturbation theory. She provided an argument for not conceiving of transitions through intermediate states as independent even when energy was conserved in the intermediate steps. This argument was part of the second part of her dissertation. Therein she treated "the synergy of light and collision in one elementary act".[38] The Hamiltonian of the system had to be altered in two respects. On the one hand, the kinetic energy of the impinging particle had to be included in the unperturbed part of the Hamiltonian. On the other hand, the interaction now included two individual perturbations: the interaction between the atom and the radiation field V and the interaction between atom and the impinging electron U.

For simplicity's sake, Göppert-Mayer assumed that the impinging electron did not interact with the radiation field. Accordingly, V was independent of the coordinates

[36] See also my discussion of the relation between virtual oscillators and virtual transitions in connection with Dirac's first invocation of the verbal model in Sect. 5.2.

[37] See my discussion in Sect. 9.1.2.

[38] "Das Zusammenwirken von Licht und Stoß in einem Elementarakt", Göppert-Mayer (1931, 284).

of the electron and U was independent of the coordinates of the radiation field. Thus, processes in which both the state of the electron and the radiation field changed had to be built up of a combination of the two perturbations. As Göppert-Mayer found, such processes first occurred in the second order of the perturbative expansion, where the matrix elements of U and V shared a common intermediate state.[39]

Göppert-Mayer could show that the treatment of these processes was in rough agreement with the empirical observations at hand. Of particular interest here is the process proposed by Franck and translated by Göppert-Mayer into the language of Dirac's radiation theory. Since this process included a state in the continuous spectrum of the atom, there were further resonances to be considered. Therefore, as in Dirac's initial discussion of the resonance case, also terms that were previously excluded had to be taken into account. This essentially led to further restrictions on the possible energy values of the states, i.e. created dependencies between the values of the energy of the photon and the energies of the atom as well as the electron, also for the intermediate state. In short: Since resonance occurred between the energy of the impinging electron and a transition to the continuous spectrum of the atom, the probability for this process was considerably higher than for any other processes which were therefore neglected. Consequently, energy conservation also held for the transition to the most important intermediate states.

Göppert-Mayer found that the probability for the respective process was:

$$\sum_{\kappa} |a_{ns\kappa}|^2 \propto |P_{nE} A^{\alpha}|^2 \int |U_{En_0\kappa\kappa_0}|^2 d\omega_{p_\kappa}\, t \tag{8.7}$$

The integral expressed the interaction between the impinging electron and the atom. It was taken over the possible directions of movement of the electrons and involved the absolute square of the interaction term between atom and electron. The first factor on the right side of the equation referred to the interaction of the radiation field and the atom. Here also the absolute square of the matrix element was taken. t referred to the time.

In reference to these equations Göppert-Mayer stated "that this process behaves as if two separated events, both of which obey the energy law, occur in one act".[40] Both partial processes obeyed energy conservation and they entered the final equation as products of squares, so as actual probabilities, rather than squares of sums of probability amplitudes. They were "getrennte Vorgänge". Nevertheless, Göppert-Mayer stated, that the two processes occurred in one elementary act. Her explanation went as follows:

[39] The most general expression for the time development of the final state was given in Göppert-Mayer (1931, 286, Eq. (11)).

[40] "daß dieser Prozeß sich so verhält, als ob zwei getrennte Vorgänge, deren jeder dem Energiesatz genügt, in einem Akt geschehen," Göppert-Mayer (1931, 291).

> That it is really one uniform act and not two mutually independent processes, comes from the proportionality of the frequency of occurrence to t. The frequency of occurrence of two independent processes—for example the excitation by collision and subsequent light emission after a dwell time of 10^{-8} s—has always to be proportional to t^2 for times which are small compared to 10^{-8}, and for those the calculation is valid.[41]

Although, when considering the composition of the matrix elements and energy conservation alone, the processes seemed to occur independently, Göppert-Mayer argued for their dependence on the ground of their time proportionality. Thereby she gave a further argument against the conception of independently occurring processes through intermediate states.

In connection to the broader picture, there are two things to mention here: Neither the interference of probability amplitudes nor the violation of the energy principle can be seen as a necessary condition for, in modern terminology, a virtual transition. They are clear indicators and possible instantiations of virtual processes,[42] but they do not cover the whole scope of such processes. Further arguments for the conception of one unitary process were based on the physical consequences and results: the frequency of occurrence, selection rules, or the width of spectral lines (see below). Mathematical such conclusions had their roots in the procedure outlined by Dirac, which operated from the beginning on the quantum-amplitude level.

Intermediate Conclusion

As already mentioned, Göppert-Mayer's dissertation can be seen as an exemplary instance of the application of Dirac's quantum theory of radiation to scattering processes. Although she closely followed Dirac's reasoning, his derivations and his application of the verbal model, she not only extended the scope of the processes discussed by his method but also put forward insights that will gain relevance for the future development. According to her transformation of the Hamiltonian (and the approximations applied), scattering was exclusively based on transitions through intermediate states, also for a non-relativistic description. The use of the concept of the *Schüttelwirkung*, although Göppert-Mayer did not use the term explicitly, not only constituted physical reasoning in the application of perturbation theory, but will

[41] "Daß es sich wirklich um einen einheitlichen Akt und nicht um zwei voneinander unabhängige Prozesse handelt, geht aus der Proportionalität der Häufigkeit mit t hervor. Denn die Häufigkeit der Aufeinanderfolge zweier unabhängiger Prozesse—etwa der Anregung durch Stoß und späterer Lichtemission nach einer Verweilzeit von 10^{-8} s—muß für Zeiten, die klein sind gegen 10^{-8}, und für diese gilt die Rechnung, immer proportional t^2 sein." Göppert-Mayer (1931, 292).

[42] Interference of probability amplitudes needs phase relations between the different processes and, as e.g. Werner Heisenberg stated in his uncertainty paper, observation destroyed the phase relations between different processes. Hence, at least it was a necessary condition that no observation took place for interference of probability amplitudes to occur. The violation of the energy law was also a clear indicator, as the formalism of quantum theory ensured energy conservation for the whole process. Hence, energy non-conserving processes could not be actual.

be prominently taken up in the second half of the 1930s when physicists tried to find ways of including radiation damping into the description of scattering of otherwise freely moving particles (see Sect. 9.3).

Göppert-Mayer particularly stressed the analogy between different effects throughout her dissertation and exemplified it through a diagrammatical means of representation. Although still rooted in the traditional format of energy level diagrams, a clear order of events was indicated in the behaviour of the material part of the system. The choice of this format may be explained by the familiarity of a great number of physicists with term schemes, but it also reflected the conceptual framework of Dirac's radiation theory. The important concepts for theoretical practice were the intermediate state and the transitions to and from them, i.e. a conception based on energy levels. Further, Göppert-Mayer's representations exemplify in which way a procedural description was incorporated in Dirac's radiation theory. The sequence of events was indicated in the representation of the material part of the system. The involvement of the photons in any of the subprocesses was not fixed and all different possibilities had to be taken into account. Consequently, no order of events was indicated in the lines describing the photons.

From a terminological point of view, Göppert-Mayer clearly perceived of virtual oscillators and the transitions to intermediate states as fictional, albeit helpful entities: She referred to both by the term "as if". Yet, the term "virtual" was exclusively used for the oscillator, i.e. the replacement of the atom by a (pseudo-)classical entity. It was not used for the transitions to and from intermediate states, which constituted a (pseudo-)quantum-theoretical description of the behaviour of the radiation field and atom.

Lastly, Göppert-Mayer gave another reason why processes through intermediate states were not independent of each other. Neither the structure of matrix elements nor the concept of energy non-conservation was important at this point. It was the temporal development, the proportionality of the probability to t, that led her to this conclusion. A very similar argument will be used by another doctoral student at Göttingen, Victor Weisskopf. Weisskopf's work was partially based on insights Göppert-Mayer had gained and partially on a different approach to the perturbative treatment which he had developed with Eugene Wigner and which we shall turn to now.

8.1.2 A Short Introduction to Weisskopf-Wigner Theory (1929)

Victor Weisskopf[43] was born into a Viennese upper middle-class family and enjoyed a classical education. Already in his younger years at the Gymnasium Weisskopf developed a taste for science, undertook first astronomical studies with a friend,

[43] This short biographical sketch is based on Jackson and Gottfried (2004).

and published a first report in the *Astronomische Nachrichten* in 1924 even before enrolling as student of physics at the University of Vienna. After two years of study in Vienna (1926–1928), Weisskopf followed the recommendation of Hans Thirring to go to Göttingen. At the time Weisskopf arrived in Göttingen, Born was in pure health. Weisskopf learned quantum physics first from Gerhard Herzberg and then immersed himself with the help of Eugene Wigner (based in Berlin but regular visitor in Göttingen) into the study of the width of spectral lines according to Dirac's radiation theory. According to Weisskopf's later recollections, discussions with the experimentalist James Franck were highly influential both for Weisskopf's work on the width of spectral lines together with Wigner as well as for his dissertational research on resonance scattering.[44]

Weisskopf and Wigner (1930a) were interested in the breadth and shape of emission lines, phenomena for which the lifetime of the excited state needed to be taken into account explicitly.[45] In essence, Weisskopf and Wigner's approach was based on Dirac's perturbative treatment but it constituted an alteration of the approximation taken. Dirac's method rested on the assumption that the temporal dimensions involved are short compared to the natural lifetime of an atom in an excited state (approximately 10^{-8} seconds).[46] But due to the question Weisskopf and Wigner posed they needed to take longer times into account. To still being able to solve the equations, they had to introduce the assumption that all the processes (approximately) conserved energy.

Consequently, Weisskopf and Wigner dropped terms not corresponding to energy conservation from the start. In essence, Weisskopf and Wigner truncated their Hilbert space to states that had (nearly) the same energy as the initial state. By construction, energy conservation also held for intermediate states.

The states therefore considered by Weisskopf and Wigner in the case of emission boiled down to (1) the atom being an excited state A and no radiation present (initial state a) and (2) the atom being in state B and one photon with frequency ν_ρ present, $N_\rho = 1$, (final state b_ρ). For treating the breadth of the emitted line, the energy of the states was assumed to lie in a small range $h\epsilon$: $E_A - E_B - h\epsilon < h\nu_\rho < E_A - E_B + h\epsilon$. The differential equations governing the temporal development of the initial and final state of the system were then given by:

---

[44] Just to give an example, I quote Weisskopf's memories from his AIP interview: "I can tell you what kind of problems excited me at this time, and they were more or less Franck's problems. I remember endless discussions, but deep discussions, about the difference between absorption and dispersion. [...] These questions excited me enormously, and in fact I think most of my deep thinking after I absorbed quantum mechanics was related to problems like this under the direct influence of James Franck." Interview of Victor Weisskopf by Thomas S. Kuhn and John L. Heilbron, 1963 July 10, Niels Bohr Library & Archives, American Institute of Physics, College Park, MD USA. https://doi.org/10.1063/nbla.ajsc.jkth. Page 12 of the PDF.

[45] I will not discuss prior approaches to the breadth of spectral lines as the alterations of the perturbative approach shall take centre stage here. A discussion of different approaches to the linewidth can nicely be extrapolated from the introduction of Weisskopf and Wigner (1930a).

[46] See, for example, Wentzel (1933c, 752), or Heitler (1936, Sect. 12) for explicit, contemporary acknowledgements of this fact.

$$\frac{h}{2\pi i}\frac{\partial a}{\partial t} = \sum_{\rho} w_{AB}^{(\rho)} b_\rho \tag{8.8}$$

$$\frac{h}{2\pi i}\frac{\partial b_\rho}{\partial t} = h(\nu_\rho - \nu_B^A) b_\rho + w_{BA}^{(\rho)} a \tag{8.9}$$

where $w_{XY}^{(\sigma)}$ referred to the matrix element of the interaction between radiation field and atom, i.e. it described the transition of the atom from state X to Y under the emission or absorption of a light quantum of frequency ν_σ. $h\nu_B^A$ referred to the energy difference between the energy levels A and B. The sum in the first equation resulted from the small energy range of the emitted radiation which had to be taken into account to make the natural breadth of the line accessible. Weisskopf and Wigner then showed that these equations could be solved by the Ansatz of an exponential decay of the initial state:

$$a(t) = e^{-2\pi\Gamma t} \tag{8.10}$$

where $4\pi\Gamma = \gamma$ is the decay or damping constant. As expected, it was found to be proportional to the transition probability between the two states and the corresponding frequency.

By neglecting all states that did not conserve energy, an explicit function for the temporal development of the probability coefficient of the states, modelled after the nuclear decay, could be implemented and it was shown that it constituted a solution to the set of differential equations studied. Weisskopf and Wigner could further establish equations for γ which only depended, besides constants, on the transition matrix element between the two states of the atom and the corresponding energy difference, and a formula for the intensity distribution of the emitted radiation, which was their initial aim (Weisskopf and Wigner 1930a, 66, Eqs. (21) and (22)).

But Weisskopf and Wigner also discussed cascade processes (see Fig. 8.2).[47] They considered the case of a double emission in which first the atom made a transition from state A to state B emitting a photon of frequency ν_ρ. Then a transition occurred in which the atom went from state B to state C while a photon ν_σ was emitted. Weisskopf and Wigner also allowed for a transition from state A to a metastable state B' under the emission of a photon of frequency $\nu_{\rho'}$.

The equations corresponding to Eq. (8.9) had to be extended for this process to include the temporal development of the third state of the whole system signified by $c_{\rho\sigma}$. But also the possible transitions to state B' (second line of the equations below) had to be taken into account. The equations then read:

[47] In a follow up paper, Weisskopf and Wigner (1930b) also discussed two cascade processes which were excluded in their first paper: the case in which the energy differences $E_A - E_B$ and $E_B - E_C$ were approximately equal and the case in which the state B' (see Fig. 8.2) could also decay into the state C so that different possibilities of successive transitions had to be included.

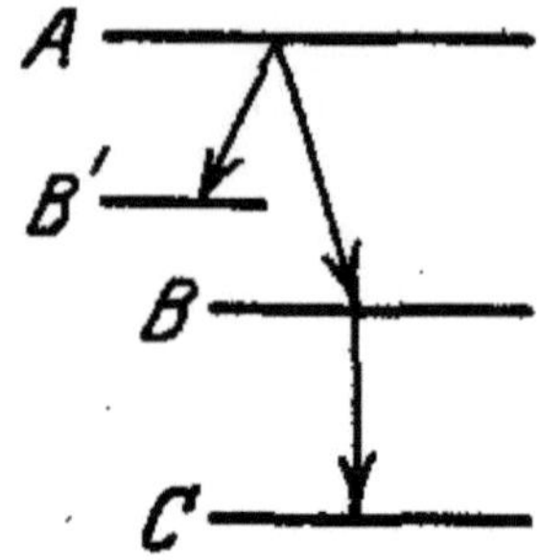

Fig. 8.2 The situation corresponding to Eq. (8.11–8.14). The atom was allowed to jump to a (meta)stable level B' or to make two successive transitions from state A over state B to state C. Figure in Weisskopf, Victor, and Eugene Wigner. 1930a. "Berechnung der natürlichen Linienbreite auf Grund der Diracschen Lichttheorie." *Zeitschrift für Physik* 63, 54–73, here 67. https://doi.org/10.1007/BF01336768 (cited as Weisskopf and Wigner (1930a) in the bibliography). Reproduced with permission from Springer Nature

$$\frac{h}{2\pi i}\frac{\partial a}{\partial t} = \sum_{\rho} w_{AB}^{(\rho)} b_\rho \tag{8.11}$$

$$\frac{h}{2\pi i}\frac{\partial b'_{\rho'}}{\partial t} = (h\nu_{\rho'} + E'_B) b'_{\rho'} + w_{B'A}^{(\rho')} a \tag{8.12}$$

$$\frac{h}{2\pi i}\frac{\partial b_\rho}{\partial t} = (h\nu_\rho + E_B) b_\rho + w_{BA}^{(\rho)} a + \sum_{\sigma} w_{BC}^{(\sigma)} c_{\sigma\rho} \tag{8.13}$$

$$\frac{h}{2\pi i}\frac{\partial c_{\rho\sigma}}{\partial t} = (E_C + h(\nu_\rho + \nu_\sigma)) c_{\rho\sigma} + w_{CB}^{(\sigma)} b_\rho \tag{8.14}$$

This system of equations was again solved by assuming an exponential decay rate for the states. The solution of these equations was a straightforward extrapolation from the prior case and mostly constituted a proof of the already expected results.

But as we see, the differential equation for the temporal development of the intermediate state (Eq. (8.13)) was written out explicitly. Dirac's treatment was iterative in its nature and the intermediate state showed up in the second step of the iteration. The time-development of this state was essentially written out but was not explicitly indicated as such, in contrast to Weisskopf and Wigner's approach. This definitely was due to inclusion of the radiation reaction in the equations and it might seem as a small alteration in notation. Nevertheless, I would argue that this also has influence of the perception of the perturbative equations. Especially physicists who studied Weisskopf's and Wigner's treatment of radiation phenomena, such as Fermi and Heitler (see below), used the same kind of notation, even when applying Dirac's approximation rather than Weisskopf's and Wigner's. The way of writing the equations did not alter the arrived at results, but it put more emphasis on the successive nature of events: Dirac's prescription was guided by an iterative procedure that was subsequently interpreted. In Weisskopf's and Wigner's equations the possible transitions to some other state were constructed first, which then turned out to be the intermediate state in a bigger process that came about when the transitions from the intermediate state to the final state were studied.

8.1.3 Weisskopf's Dissertation (1931): Resonance Scattering as a Uniform Process

While still in Göttingen, Weisskopf started to work on his dissertation. The publications with Wigner, as they were collaborative, could not be used. The choice of the topic for Weisskopf's dissertation was again heavily influenced by the experimentalists in Göttingen, foremost James Franck. Weisskopf remembered that it "[...] was simple work. This was written mainly as a sort of 'house theory' for Franck's experiments."[48] In the same interview, Weisskopf mentioned discussions between him and Franck about the relation between resonance scattering and emission and absorption processes, a theme that occurred several times in Weisskopf's dissertation about resonance scattering (Weisskopf 1931).[49]

In fact, Weisskopf's dissertation was a nearly straightforward application of Weisskopf's and Wigner's theory to the case of resonant radiation. Weisskopf took over the structure of the interaction energy from Göppert-Mayer's dissertation. The choice of the initial, intermediate and final state were naturally suggested to Weisskopf by the phenomenon studied. In the initial state the atom was supposed to be in the ground state and a photon should be present (a), in the intermediate state the atom was raised to an excited state and the photon was absorbed (b_ρ) and in the final state a photon had been emitted and the atom was in the ground state again ($a_{\rho\sigma}$). Accordingly, the differential equations read:

$$\frac{i}{2\pi}\dot{a} = \frac{1}{h}\sum_\rho H_{AB}^{(-\rho)} b_\rho \tag{8.15}$$

$$\frac{i}{2\pi}\dot{b}_\rho = \frac{1}{h} H_{AB}^{*(-\rho)} a + \frac{1}{h}\sum_\sigma H_{BA}^{(\sigma)} a_{\rho\sigma} + (\nu_A^B - \nu_\rho) b_\rho \tag{8.16}$$

$$\frac{i}{h}\dot{a}_{\rho\sigma} = \frac{1}{h} H_{BA}^{*(\sigma)} b_\rho + (\nu_\sigma - \nu_\rho) a_{\rho\sigma} \tag{8.17}$$

where $H_{XY}^{(\pm\tau)}$ referred to the matrix element of the perturbation connecting the states X and Y of the atom through the emission (+) or absorption (-) of a photon of frequency ν_τ. As Weisskopf noticed himself, the equations for the temporal evolution of the states "are formally completely identical to Eqs. [(8.11–8.14) of Weisskopf and

[48] Interview of Victor Weisskopf by Thomas S. Kuhn and John L. Heilbron, 1963 July 10, Niels Bohr Library and Archives, American Institute of Physics, College Park, MD USA. https://doi.org/10.1063/nbla.ajsc.jkth. Page 16 of the PDF.

[49] In the following, I will only discuss small parts of Weisskopf's dissertation that directly connect with the theme of this chapter.

Wigner (1930a)], if $w_{AB'}^{(\rho)} = 0$. The absorption and reemission corresponds formally to a sequence of two emissions with opposite and equal frequency."[50] This allowed Weisskopf to choose the same Ansatz as for the double emission discussed by him and Wigner and the solution of the equation was analogous.

Connecting Dispersion and Resonance Scattering

The specific calculations are not important for our purpose but the derived results are interesting. In the case of a broad incident frequency spectrum the probability of absorption, and therefore also of the scattering process, was equivalent to the one of previous studies, such as Dirac's. This made the conception of resonance scattering as a two step process rather suggestive.[51] Weisskopf further showed that for the case of nearly monochromatic incident radiation (ν_0), the result could be interpreted as a generalization of the Kramers-Heisenberg equation. The probability was in this case proportional to:

$$(\nu_A^B)^3 \frac{|(X_{AB}, e_j)|^2 |X_{BA}|^2}{(\gamma^B)^2 + (\nu_A^B - \nu_0)} \tag{8.18}$$

while the Kramers-Heisenberg equation, in this nomenclature was proportional to:

$$(\nu_0)^3 \left| \sum_N \frac{(X_{AN}, e_j) X_{NA}}{\nu_A^N - \nu_0} + \frac{(X_{NA}, e_j) X_{AN}}{\nu_A^N + \nu_0} \right|^2 \tag{8.19}$$

As Weisskopf argued, if ν_0 gets closer to ν_A^B, only the term proportional to $\frac{1}{\nu_A^B - \nu_0}$ in Eq. (8.19) is relevant but the small damping constant γ^B can no longer be neglected. Weisskopf consequently showed that his evaluation of resonance scattering nicely merged with the Kramers-Heisenberg formula. This was certainly expected and would have also resulted from semi-classical consideration. Yet, thereby he did not only connect resonance scattering, for which the general conceptualization involved two successive and independent acts of emission and absorption, and dispersion, which was known not to involve independent processes, but also the methods of him and Wigner and the more usual ones of Dirac.

---

[50] "Sie sind formal vollkommen identisch mit Gleichungen [(8.11–8.14) Weisskopf and Wigner (1930a)], wenn dort $w_{AB'}^{(\rho)} = 0$ gesetzt wird. Die Absorption und Reemission entspricht formal einer Aufeinanderfolge zweier Emissionen mi entgegengesetzt gleicher Frequenz." Weisskopf (1931, 37).

[51] Weisskopf did not explicitly spell out the above point. A further evaluation and the argument for conceiving of resonance scattering as two independent acts of absorption and emission in the case of a broad impinging frequency range can be found, e.g. in Heitler (1944, 141–143).

Why Resonance Scattering Is Not a Sequence of Independently Occurring Transitions

As pointed out above, resonance scattering was generally conceived of as two independent acts of emission and absorption. Weisskopf questioned this assumption. The total probability for the absorption of a light quantum in the frequency range $d\nu$ and the emission of a light quantum in the frequency range $d\nu'$ was according to Weisskopf given by:

$$W(\nu, \nu')d\nu d\nu' = \frac{1}{h^4} \frac{F_{AB}(-\nu)F_{BA}(\nu')d\nu d\nu'}{[(\gamma^A)^2 + (\nu - \nu')^2][(\gamma^B)^2 + (\nu_A^B - \nu')^2]} \tag{8.20}$$

Here, $F_{AB}(-\nu)$ symbolises the sums over the small frequency ranges $d\nu$ of the squares of the interaction matrix elements.

For Weisskopf's argument, the important part of Eq. (8.20) is the first bracket in the denominator. It showed that the probability of the process had a sharp maximum for $\nu = \nu'$ with a half width of γ^A which is $\frac{1}{4\pi}$ times the probability of absorption. As Weisskopf argued, for a process of independent absorption and emission one would expect that the emitted line would have the natural line shape for spontaneous emission with half width γ^B. Since, for regular radiation densities the probability of (spontaneous) emission $4\pi\gamma^B$ was a lot bigger than $4\pi\gamma^A$, the spectral distribution of the emitted radiation was a lot sharper than one would expect from a conception that rested on two independent processes. The calculated breadth of the emitted line, as Weisskopf noticed, was due to the finite lifetime of the initial state (not of the excited state) and could be explained by the breadth of the initial state due to the corresponding uncertainty in its energy. Consequently: "The resonance radiation coincides as sharply as possible with the frequency of the incident light."[52]

Weisskopf commented on this result as follows:

> The result is only insofar of theoretical interest, as it discards the idea that the resonance fluorescence is an independent sequence of absorption and emission. In this case the emitted radiation would have to be equal to the natural line breadth, regardless of the kind of the impinging radiation.[53]

Thereby, Weisskopf effectively gave another reason why one cannot conceive of scattering events as independently occurring acts of emission and absorption. Although, by construction, energy was conserved in the transitions and the structure of the matrix elements, i.e. the combination of squares of them (no interference

[52] "Die Resonanzstrahlung stimmt also so scharf wie irgend möglich mit der Frequenz des eingestrahlten Lichtes überein." Weisskopf (1931, 43).

[53] "Das Resultat hat nur insofern theoretisches Interesse, als es die Vorstellung verwirft, daß die Resonanzfluoreszenz eine unabhängige Aufeinanderfolge von Absorption und Emission sei. In diesem Falle müßte ja, unabhängig von der Art der Einstrahlung, die Ausstrahlung natürliche Linienbreit haben." Weisskopf (1931, 24).

terms), suggested that in terms of probabilities the events could be considered as independent, the line shape of the reemitted light depended crucially on the impinging one.[54]

In analysis,we can abstract two insights from Weisskopf's dissertation which are directly connected to the theoretical structures we refer to as virtual transitions. On the one hand, Weisskopf provided arguments that disagreed with the conception of resonance scattering being composed of two independent acts of emission and absorption. This is particularly striking as the phenomenon studied was priorly exactly conceived of in such a way and in the theoretical description of the process only the squares of matrix elements and not the matrix elements themselves entered. Also energy conservation was imposed from the beginning. Nevertheless, the shape of the emitted radiation depended crucially on the impinging one.

On the other hand, the way he argued for the second conclusion is a reversal of reasoning we have met in the previous chapter. For Ivar Waller (Sect. 7.3) the case of resonance was important for considering the intermediate state in Dirac's theory rather realistically, at least so at first. Weisskopf now turned things around. He had found that the mechanism of the excitation had influence on the emitted radiation, therefore the conception of independently occurring emission and absorption processes could not be upheld. The line between actual and virtual (modern terminology) processes was further blurred.

8.1.4 A First General Extension to Higher Orders: Weisskopf's Formula

After his dissertational research Weisskopf left Göttingen[55] for a sequence of post-doctoral stays in Leipzig with Heisenberg, in Kharkiv with Landau, in Berlin with Schrödinger, in Copenhagen with Bohr, and in Cambridge with Dirac.[56] Up until late 1933, when Weisskopf could secure the position as assistant to Pauli at the ETH Zürich, his research focussed on the scattering of light, especially on dispersion, resonance scattering, and the line-width. During his stay in Copenhagen and Zürich, Weisskopf's style of doing quantum electrodynamics, in particular which formulation of it he applied, changed considerably. His work on the self-energy of the electron, which was already strongly influenced by Bohr and Pauli, will be discussed

---

[54] Weisskopf provided another reason why resonance scattering should not be conceived of as a sequence of two independently occurring acts of emission and absorption based on the so-called Hanle effect, named after Wilhelm Hanle. The interested reader is referred to Weisskopf's discussion as given in Weisskopf (1931, 44–55).

[55] Starting in late 1932, Walter Heitler continued the work on radiation theory in Göttingen. I will turn to his work in Sect. 8.2.2.

[56] For the personal recollections of Victor Weisskopf on his post-doctoral years before coming to the ETH Zürich, see Weisskopf (1991, Chaters 3 and 4).

in Sect. 10.1.1. For the moment, I want to focus on one of the last papers in which Victor Weisskopf's explicitly used Dirac's radiation theory and discussed scattering phenomena. He finished it in Copenhagen in mid 1933 and developed a first formula for the perturbative expansion in terms of Dirac's theory to an arbitrary order of approximation.

To be sure, the extension of the perturbative approximation to all orders was not a particularly pressing problem from the point of view of experimental verifiability. Even second order effects, such as those studied by Göppert-Mayer, could not be studied with the empirical methods at hand. Maybe worse, the ultraviolet divergences cast serious doubts on the application of quantum electrodynamics and radiation theory. Efforts were often directed towards finding new means of establishing a quantum treatment of the electromagnetic field rather than to the extension of the known methods to higher orders. Yet, if theorists turned to the study of higher order effects, they often engaged with the evaluation of the perturbative expansion.[57] Weisskopf was, to the best of my knowledge, the first one to provide an expansion to all orders. He eventually found, although only rarely, credit for this endeavour: The expansion was also referred to as "Weisskopf's formula".[58]

Weisskopf's treatment was part of a little known paper on "the scattering of light from excited atoms".[59] It was a direct continuation of his work on resonance scattering and the Kramers-Heisenberg formula. As he had already established in his dissertation, important aspects of the emitted radiation, as the shape of the emitted spectral line, depended on the nature of the excitation process in the case of resonant scattering. Hence, for the discussion of the scattering from excited states the discussion of the excitation and the de-excitation process was important. Thus (loosely speaking) at least four different subprocesses were involved (excitation and de-excitation as well as the scattering mechanism constituted of (virtual) absorption and emission): Weisskopf needed to discuss fourth order processes in the interaction of light and matter or with other perturbations.

But instead of simply deriving the fourth order term, Weisskopf chose to develop a generalized scattering formula to an arbitrary order.[60] In order to do so, Weisskopf first of all defined a multiple scattering process:

> Under a multiple scattering process we understand a process, in which an atom transforms a number of light quanta into different ones with the same or different frequencies and directions, while the total energy of the atom and the light quanta is conserved. This process can also be connected with an alteration of the atomic state.[61]

---

[57] As for example Hans Euler and Bernhard Kockel did in 1935 to 1936, see Sects. 9.1.3 and 9.1.4.

[58] "Weisskopfsche Formel," Akhiezer and Pomeranchuck (1937, 480).

[59] "Die Streuung von Licht an angeregten Atomen," Weisskopf (1933).

[60] At the time, processes of third order had already been discussed by Paul Güttinger (1932) in Zürich with Pauli.

[61] "Wir verstehen unter einem mehrfachen Streuprozeß einen Vorgang, bei dem ein Atom eine Anzahl von Lichtquanten in andere mit gleicher oder verschiedener Frequenz und Richtung verwandelt, wobei die Gesamtenergie von Atom und Lichtquanten erhalten bleibt. Dieser Prozeß kann auch mit einer Änderung des Atomzustandes verbunden sein." Weisskopf (1933, 460).

Weisskopf discussed the general formula for such a "multi-stage scattering process"[62] through an intuitive generalization and a concrete mathematical derivation according to Dirac's radiation theory. I will only discuss the latter.

For simplicity's sake, Weisskopf considered the non-resonant cases, for which the damping constants Γ_{R_k} could be neglected. Weisskopf started out from the well-known formula for the second approximation:

$$a_n^{(2)} = \sum_{n'} \frac{V_{nn'}V_{n'0}}{h\nu_{n'0}} \frac{e^{2\pi i \nu_{n0}t} - 1}{h\nu_{n0}} - \sum_{n'} \frac{V_{nn'}V_{n'0}}{h\nu_{n'0}} \frac{e^{2\pi i \nu_{nn'}t} - 1}{h\nu_{nn'}} \tag{8.21}$$

where the indexes n, n', 0 referred to the state of the whole system, atom plus radiation field, and V_{XY} to the interaction.

Following Dirac and Göppert-Mayer, Weisskopf identified the denominator $h\nu_{n0}$ as referring to energy conservation for the process leading from the state 0 to state n and neglected the second term. Echoing Göppert-Mayer, he argued that "it [the second term] essentially rests on the sudden switching on of the interaction at the time $t = 0$ inherent in the method of the variation of constant".[63] Only taking the first term into account, Weisskopf pushed the procedure to the third order and arrived at:

$$\begin{aligned} a_n^{(3)} = &\sum_{n'n''} \frac{V_{nn'}V_{n'n''}V_{n''0}}{h^2\nu_{n''0}\nu_{n'0}} \frac{e^{2\pi i \nu_{n0}t} - 1}{h\nu_{n0}} - \sum_{n'n''} \frac{V_{nn'}V_{n'n''}V_{n''0}}{h^2\nu_{n''0}\nu_{n'0}} \frac{e^{2\pi i \nu_{nn'}t} - 1}{h\nu_{nn'}} \\ &- \sum_{n'n''} \frac{V_{nn'}V_{n'n''}V_{n''0}}{h^2\nu_{n''0}\nu_{n'n''}} \frac{e^{2\pi i \nu_{nn''}t} - 1}{h\nu_{nn''}} + \sum_{n'n''} \frac{V_{nn'}V_{n'n''}V_{n''0}}{h^2\nu_{n''0}\nu_{n'n'}} \frac{e^{2\pi i \nu_{nn'}t} - 1}{h\nu_{nn'}} \end{aligned} \tag{8.22}$$

Again, the terms which did not correspond to energy conservation, namely all terms but the first one, were neglected. Inductively, Weisskopf reasoned that for the non-resonant case the equation for the zth order had to read:

$$a_n^{(z)} = \sum_{n_1} \dots \sum_{n_{z-1}} \frac{V_{nn_1}V_{n_1n_2}\dots V_{n_{z-1}0}}{h^{z-1}\nu_{n_10}\nu_{n_20}\dots\nu_{n_{z-1}0}} \frac{e^{2\pi i \nu_{n0}t} - 1}{h\nu_{n0}} \tag{8.23}$$

This Eq. (8.23) constituted the general formula for the perturbative series in terms of Dirac's procedure for non-resonant cases.

For the time being, Weisskopf had made no assumptions about the nature of the interaction (besides it being non-resonant with the energy levels of the material system). To obtain a formula which corresponded to the interaction of light and matter, Weisskopf needed to take into account the explicit form of the perturbation in

---

[62] "mehrstufiger Streuprozess," Weisskopf (1933, 460).

[63] "Er [der zweite Term] beruht im wesentlichen nur auf dem in der Methode der Variation der Konstanten liegenden plötzlichen Einschalten der Wechselwirkung zur Zeit $t = 0$." Weisskopf (1933, 464).

Dirac's radiation theory. At this point, Maria Göppert-Mayer's simplification of the interaction Hamiltonian came in. Since V was (also in the non-relativistic case) linear in A, the matrix element $V_{nn'}$ always coupled states which differed in the occupation of the radiation field by one quantum. In consequence, the zth order term included z emission and absorption processes.

Now Weisskopf brought in the fact that the interaction could be written as a product of terms referring to the state of the radiation field and to states of the atom:

$$\frac{1}{h}V_{nn'} := \frac{1}{h}V_{RR'}^{(\pm\rho)} = (M_{RR'}e_\rho)f(\pm\nu_\rho) \tag{8.24}$$

where R and R' referred to stationary states of the atom, e_ρ to the polarization vector of the radiation and $f(\pm\nu_\rho)$ depended only on the properties of the ρth component of the radiation field and not on the states of the atom. The $+$ $(-)$ sign had to be used when the state n was reached from state n' by the emission (absorption) of the photon of frequency ν_ρ.

With this identification Weisskopf could write the second order term in the following way:

$$\begin{aligned} a_n^{(2)} &= \frac{e^{2\pi(\pm\nu_\rho\pm\nu_\sigma-\nu_{AM})t}-1}{\pm\nu_\rho\pm\nu_\sigma-\nu_{AM}}f(\pm\nu_\rho)f(\pm\nu_\sigma)\times \\ &\times\sum_R\left[\frac{(M_{AR}e_\rho)(M_{RM}e_\sigma)}{\nu_{RA}\pm\nu_\rho}+\frac{(M_{AR}e_\sigma)(M_{RM}e_\rho)}{\nu_{RA}\pm\nu_\sigma}\right] \end{aligned} \tag{8.25}$$

Again, the plus and minus signs had to be taken corresponding to whether the photon was absorbed or emitted. The two fractions in brackets reflected that either of the light quanta could be involved in either process. As Weisskopf stated in general terms: "$|a_n^{(2)}|^2$ represents the probability that after the time t two light quanta of type ρ and σ have been converted."[64]

The exact nature of the process (double emission, double absorption, or scattering) depended on which quanta were emitted or absorbed. Through the summation over a small frequency range around $d\nu_1$ and $d\nu_2$, including ν_ρ and ν_σ respectively, as well as the corresponding solid angles $d\Omega_1$ and $d\Omega_2$ while pushing towards large values of t Weisskopf arrived at the Kramers-Heisenberg formula. The factors $f(\pm\nu)$ thereby merged into the radiation densities. The square of the fraction at the beginning of Eq. (8.25) ensured energy conservation through Dirac's trick.

Now Weisskopf showed how this procedure could be generalized. Since each intermediate state n_1 to n_{z-1} could arise by the emission or absorption of 1 to $z-1$ photons, for an arbitrary intermediate state s the following equation held: $\nu_{n_s 0} = \nu_{R_s A} + \Theta_s^{(p)}$, where $\Theta_s^{(p)}$ referred to the partial sums of the first k-frequencies for a possible sequence p of emission and absorption subprocesses:

64 "$|a_n^{(2)}|^2$ stellt die Wahrscheinlichkeit dar, daß nach der Zeit t zwei Lichtquanten der Art ρ und σ umgesetzt werden." Weisskopf (1933, 465).

$$\Theta_s^{(p)} = \sum_{s=1}^{k} \pm\nu_{s(p)} \tag{8.26}$$

The partial sum over the frequencies corresponded to the part of the proper energy that was due to the occupation of the radiation field for the intermediate state s in the particular order of events p. Its structure was due to the fact that in each subprocess only one photon was involved and all permutations of emission and absorption subprocesses had to be included. For z photons, there were $z!$ possibilities of ordering them. Weisskopf consequently arrived at the following probability coefficient for a transition of the atom from state A to state M:

$$\begin{aligned} a_n^{(z)} &= \frac{e^{2\pi i(\pm\nu_1 \pm \ldots \pm\nu_z - \nu_{AM})}}{\pm\nu_1 \pm \ldots \pm \nu_z - \nu_{AM}} f(\pm\nu_1)\ldots f(\pm\nu_z)\times \\ &\times \sum_{R_1}\cdots\sum_{R_{z-1}}\sum_{p=1}^{z!} \frac{(M_{AR_1}e_{1(p)})\ldots(M_{R_{z-1}M}e_{z(p)})}{(\nu_{R_1A}+\Theta_1^{(p)})\ldots(\nu_{R_{z-1}A}+\Theta_{z-1}^{(p)})} \end{aligned} \tag{8.27}$$

The damping constants Γ_i were included by Weisskopf inductively, simply mentioning that this inclusion could be proved in Dirac's radiation theory. Similarly, the generalization to different causes of excitations, such as electron impacts, was discussed by Weisskopf only shortly.

To conclude, this first extension of the scattering formula to an arbitrary order still rested on a non-relativistic treatment of the electron. But through the insights gathered by Göppert-Mayer and Weisskopf previously, namely that the interaction could be written in a form that only involved emission and absorption processes and the neglect of the energy non-conserving terms for the total process, as they rested on the peculiarities of the methods chosen, it could later on also be applied in a fully relativistic treatment.

It is important to note that this formula was designed for the non-resonant case, forming the prerequisite for neglecting the energy non-conserving terms, and the particular form of the interaction between light and mater. Only through these assumptions the perturbative expansion could be cast in such a concise form as it is given in Eqs. (8.23) or (8.27). The interested reader is referred to the work of Paul Güttinger (1932, especially p. 243), where the perturbative expansion is given in most general terms for the third order approximation and spreads over approximately half of a printed page.

I discussed Weisskopf's derivation in some detail not only because it was the first generalization of the perturbative expansion to an arbitrary order. It further rested on, and made explicit, a lot of its most important traits. First of all, it clearly was based on the conception of successive transition processes. Secondly, in each step one light quantum was emitted or absorbed. Thirdly, having a general mathematical structure as Eq. (8.27), all one had to do was to fix the initial and final state or, equivalently, the numbers of photons emitted or absorbed in total to discuss a specific scattering process. Fourthly, the behaviour of the material system and the radiation field was, in this description and for the transitions to intermediate states, to some extent independent (see Eq. (8.24)).

Of course, each emission or absorption was connected to a transition of the material part of the system. But the order of the photons was completely arbitrary and the sum over its permutation had to be taken. As will be seen in this and the following chapter, physicists regularly only represented the material part of the system as the electromagnetic field was, to quote Adrian Wüthrich (2010, 20), "conceived of as a reservoir of photons". As Wüthrich has argued (Wüthrich 2010, especially Chap. 2.3), comparing the conceptions of processes during the 1930s with Feynman diagrams, not only the idea of what constituted a physical process was altered (energy level transitions vs. free movement of particles with interruptions through interaction). But Wüthrich further argued that the inability of one of the most prominent diagrammatic representations before Feynman diagrams, Euler's term schemes (see also Sect. 9.1.3) and its furhter developments, to explicitly depict photons was one of its particular shortcomings.

Having discussed Weisskopf's derivation, we see why such a representation must still have seemed suitable for a physicist working during the 1930s. The mathematical structures suggested that the material system was *the* component to focus on. As long as one knew how many photons were involved, their involvement could simply be permuted over all possibilities. Besides their number, which was equivalent to the order of the effect, emission and absorption processes did not interfere with the modelling of the process in total.

Before going over to the didactic literature, let me shortly recap, as Weisskopf's little known paper can be conceived of as a first conclusion to the application of perturbation theory in the study of the light-matter interaction. Through the prior work of Göppert-Mayer and Weisskopf himself not only a first general extension to all orders but also an extension of the perturbative approach to include non-radiative perturbations and longer times was established. Weisskopf could show how the case of non-resonant and resonant scattering merged in a quantum-theoretical description. Both Göppert-Mayer and Weisskopf provided new means of arguing for the conception of scattering being one elementary process, although the intermediate states fulfilled the energy law and only squares of matrix elements were involved. They blurred the lines between virtual (modern terminology) and actual processes. Although the subprocesses could therefore not be conceived of as independent, the stepwise theoretical description was represented verbally as well as diagrammatically and was heuristically applied by the two physicists. Specifically in the work of Weisskopf and Wigner we saw how also the mathematical representation changed although only slightly: Instead of closely following Dirac's iterative procedure, the time-development of the states, including the intermediate state, was written out explicitly. This strongly pronounced the relevance of the intermediate state and the stepwise conception of the processes.

8.2 Energy Non-Conserving Transitions and Intermediate States in the Most Influential Instructional Literature of the 1930s

With the discussion of Göppert-Mayer's and Weisskopf's work also the focus on non-relativistic quantum mechanics will end. From now on, in the theoretical work I am about to discuss the material part of the system will be described relativistically as well. In Chap. 7 we have already encountered three important alterations to the conception of the scattering mechanism when the relativistic theory of the electron was taken into account. First of all, the direct scattering term (proportional to A^2) vanished and only processes through intermediate states were included in the theoretical description. Secondly, for free particles momentum was conserved in every step of the process. Thirdly, negative energy states played a crucial role for the general description of the scattering mechanism.

The study of scattering and related phenomena (such as Bremsstrahlung or pair creation and annihilation) in relativistic terms became one of the focal points of theoretical research, especially for cosmic ray research but also for effects created with laboratory equipment. Rather than going through the corresponding publications of the 1930s[65] and the usage as well as the conception of perturbation theory applied therein, I will look more closely on the role of time-dependent perturbation theory in prominent examples of the didactic literature of the time. The following section will be concerned with Enrico Fermi's "Quantum Theory of Radiation" (Fermi 1932), Walter Heitler's textbook with nearly the same title (Heitler 1936) and publications closely related to these works. It is safe to say that these constituted the two most widely used instructional texts to QED during the 1930s.

There is a twofold motivation for the focus on this kind of literature. First of all, when investigating the practice and the relevant conceptions of a theory, didactic works have a particular important role: They teach the following generation(s) how to understand and use the theory. But they also aim at a representation that is most likely shared by a great number of fellow researchers. They are a means of communicating generally accepted practices, concepts, and their representations to a younger generation of scientists.

Secondly, the reception of these instructional texts and papers closely related to them in the secondary literature motivates a closer investigation in reference to the concept of the virtual particle. On the one hand, David Kaiser (2005, 30) observed that Enrico Fermi and Hans Bethe's study of electron-electron scattering (Bethe and Fermi 1932), which was methodologically based on Fermi's version of QED, has

[65] For an outlook on the general discussions surrounding the description of scattering and the involved phenomena, see e.g. Cassidy (1981), Galison (2008) (especially on cosmic ray research during the 1930s); Roqué (1997, Sect. 3) (detailed account of the so-called Meitner-Hupfeld effect); or Pais (1986, 374–376) (very concise overview of different scattering effects studied until 1935). For a contemporary assessment, which also includes detailed bibliographic information about the relevant studies, the 1936 textbook by Walter Heitler is an excellent source (Heitler 1936, especially Part III).

"put virtual particles to work". Also Sam Schweber (1994, 78) accredited Bethe and Fermi with "the description of the interaction between charged particles as mediated by the exchange of photons", according to Schweber one of the "landmarks in the development of quantum field theory during the 1930s". By a closer look at the actual practice of the authors I will argue that Kaiser's identification might be sensible in retrospect but is besides the point when we want to understand the conceptual framework the physicists were working in and with.[66]

On the other hand, seemingly motivated by Heitler's presentation of Dirac's perturbation theory, Jan Lacki (2009) conjectured that in the application of time-dependent perturbation theory during the 1930s "there is a lack of an iterative algorithm and, as a result, the intermediate states must be 'put by hand' to obtain [the second order equation]" (Lacki 2009, 30). While I share the conclusion that the physical instantiation of the intermediate states had to be introduced through a non-mathematical procedure, i.e. through the verbal model, I will argue that something like a general algorithm emerged in the first half of the 1930s. Certainly, this algorithm was based on non-covariant techniques but it was iterative (or at least backed up by an iterative mathematical procedure), while the physical concretization of the intermediate state was governed by the theoretical structures, especially the interaction term, with which the physicists worked.

8.2.1 Fermi, Bethe, and the Intermediate State

Starting in late 1928, Fermi worked out, independently from Heisenberg and Pauli, an approach to the quantization of the whole electromagnetic field.[67] It was first published in 1929 and 1930 in the *Rendiconti Lincei* and presented in lectures at the Michigan University in the summer of 1930. Through the publication of an overview article in the *Physical Review* in 1932 based on these lectures (Fermi 1932), Fermi's approach gathered prominence and was widely used, especially by pragmatically inclined physicists.

Fermi's approach to the quantization of the electromagnetic field became highly popular throughout the 1930s: "It taught an entire generation of physicists how to think about quantum electrodynamic effects in atomic phenomena." (Schweber 2002, 31). It stood out for its "intuitive, transparent formulation" and its "matter of fact, pragmatic approach" (Schweber 1994, 73, 74). Most of the examples discussed by

---

[66] Already Cathryn Carson (1996b) has argued that the concept of exchange forces was not fully formed in the work of Fermi and Bethe. See also my discussion in the following Sections and in Sect. 10.2.1.

[67] Enrico Fermi's early work on quantum electrodynamics has already found some attention in the historical literature. I will mostly follow Schweber (1994, Sect. 1.8). But see also: Darrigol (1982, Sect. 2.3.b), Cini (2001) or Schweber (2002). For a biography of Fermi by one of his students and co-workers, see Segrè (1970).

Fermi for illustration and justification were based on a straightforward application of time-dependent perturbation theory, which was re-derived by Fermi as well (Fermi 1932, 93–94).

The Intermediate State in Fermi's "Quantum Theory of Radiation" (1932)

Fermi's QED rested on an Hamiltonian approach. The material part of the system was not second-quantized. Fermi's approach is mostly credited for the conclusion that the Lorentz condition should not be read as an operator identity but as defining a supplementary condition for the physically realizable states of the system. This in turn allowed the implementation of coherent commutation relations for the field excitations.[68] Furthermore, Fermi showed how the scalar and longitudinal parts of the electromagnetic field could be subsumed to give the Coulomb interaction between the particles (Fermi 1932, 130–131).[69] In essence, this fact justified Dirac's initial approach in which only the transversal part of the electromagnetic field was quantized and the Coulomb interaction simply added to the Hamiltonian. As Fermi himself stated:

> The problem which we meet now in quantum electrodynamics is identical with that of radiation theory since our new Hamilton function [...] is the Hamiltonian function of radiation theory plus the electrostatic energy. (Fermi 1932, 131)

Fermi discussed the concept of interest for us, the intermediate state and the transitions to and from it, in connection to the propagation of light in the vacuum (I will reserve the discussion of this effect for the following subsection), the scattering of light from free electrons and pair annihilation/creation in terms of Dirac's relativistic description of the electron (Fermi 1932, Paragraphs 15 and 16).[70] As already discussed in Chap. 7, for the latter two processes intermediate states of negative energy had to be taken into account. Fermi's approach went as follows: Knowing the initial and final state, he first envisioned the possible intermediate states of the system (Fermi 1932, 120, Eq. (128)). Then he constructed the matrix elements corresponding to the transitions from initial to intermediate state (Fermi 1932, 121, Eq. (130)) and from intermediate to final state (Fermi 1932, 121, Eq. (131)). In the next step, Fermi explicitly derived the time-dependence of the intermediate state (Fermi 1932, 122, Eq. (132)) and, finally, plugged this result into the general formula for the time-development of the final state (Fermi 1932, 122, Eq. (134)).

---

[68] See e.g. Schweber (1994, 73–75). For the connected struggles of Heisenberg and Pauli in implementing the canonical quantization conditions for the electromagnetic field see Schweber (1994, 39–41).

[69] In Zürich, this approach was developed in the works of Léon Rosenfeld (1929) and Robert Oppenheimer (1930a), together with Wolfgang Pauli. They arrived at the same conclusion in Heisenberg's and Pauli's version of QED. See also Carson (1996b, Sect. 3.1).

[70] Fermi still used the model of either an empty Dirac sea or of a hole as a proton, which he both found to be unacceptable.

In his derivation, Fermi emphasized the role of negative energy intermediate states and the general properties of these states. As he noted in connection to the annihilation process:

> This intermediate state [of energy $\pm\sqrt{2}mc^2$] does not satisfy also conservation of (unperturbed) energy; as we have often seen in the preceding instances, the amplitude of probability for this state can not continuously increase with time, but it can nevertheless be different from zero though having very small oscillating values. (Fermi 1932, 124)

Also for the scattering processes through negative energy states he remarked:

> (notice the very big change in energy due to the transition of the electron from state 1 [initial state] with positive energy $+mc^2$, to state 3 [intermediate state] with negative energy $-mc^2$). (Fermi 1932, 122)

The reference to the "(unperturbed) energy" might point to a justification of these states in terms of the argument provided by Yakov Il'ich Frenkel (originally due to Dirac), i.e. that the energy non-conservation was only apparent as one neglected the interaction energy in the energy balance.[71] All the same, Fermi did not explicitly explain the energy non-conservation involved in the transitions. He did not provide a narrative for the inclusion of the intermediate states. They were used as a tool, but their irregular characteristics were not justified.

Quite on the contrary, Fermi questioned the applicability of time-independent perturbation theory for QED. At the time, the divergence of the self-energy was already well-known and Fermi directly referred to it.[72] And in this connection he commented:

> It could be noticed however that the application of the perturbation method is for this problem extremely uncertain, since the differences between the quantum states of the unperturbed problem are very small in comparison with the perturbation. (Fermi 1932, 132)

Fermi provided an explanation of the divergences physicists encountered in the perturbative evaluation of QED by directly blaming the application of perturbation theory.[73]

But all of this did not interfere with Fermi's application of the intermediate states and energy non-conserving transitions in the calculation of observable effects. In essence, Fermi's derivation was as well iterative (it entailed the repeated application of the fundamental equation of time-dependent perturbation theory). Yet, there is a subtle but I think important difference in the usage of the concept of an intermediate

---

[71] See Sect. 6.3 for the discussion of this argument in the hands of Frenkel.

[72] See also Sect. 10.1.1 and the literature cited therein.

[73] As a side-note: Fermi was not alone in the conjecture that the method of perturbation theory was to blame for the divergences: Ivar Waller (1930a, 675), Rosenfeld (1931, 461–462) and also Paul Dirac (in his third edition of his textbook on the *Principles of Quantum Theory* and his comments at the Solvay Conference in 1948 (Institut International de Physique Solvay 1950, 282)) noted that the method of calculation might be responsible for the divergence (see also Darrigol (1982, 143)). Also Hans Euler (1936, 447–448) (most probably due to Werner Heisenberg, see Sect. 9.1.3) and André Mercier (1939, 75) (see Sect. 10.1.1) explicitly made the perturbative expansion responsible for the divergence.

state. Fermi put the identification of the possible intermediate states of the system at the beginning of his calculation and explicitly derived and integrated the time-development of the intermediate state. As a matter of fact, Fermi did not explicitly write down the general second order equation of time-dependent perturbation theory anywhere in the paper. In Fermi's approach, the theoretical description of scattering as a stepwise process of emission and absorption was even more strongly reflected in the calculational procedure and the intermediate states were spelled out more distinctly.[74] As an application of this procedure, let me turn to the already invoked paper by Bethe and Fermi on the interaction of free electrons.

Fermi, Bethe, and the Relativistic Corrections to Electron-Electron Scattering (1932)

Hans Bethe received his doctorate degree in physics in 1928 under the supervision of Arnold Sommerfeld in Munich, where Bethe was employed for the larger part of the time until 1933. He first came to Rome and the institute of Fermi in early 1931 during his post-doctoral phase.[75] Reporting his first impressions of Fermi to Sommerfeld, Bethe stated: "The best thing in Rome is unquestionably Fermi. It is absolutely fabulous how he immediately sees the solution to every problem that is put to him, and his ability to present such complicated things as quantum electrodynamics simply."[76] Bethe learned quantum electrodynamics from Fermi's articles and their joint paper was Bethe's first practical application of the theory.[77]

In spring 1932, when Bethe came to Rome for the second time, Fermi had already launched a programme for experimentally and theoretically studying nuclear interactions. Bethe himself was mainly involved with writing an article for the *Handbuch der Physik*. Yet, they still found time to write a paper about the relativistic interaction of two electrons comparing the previously arrived at results of Gregory Breit and Christian Møller.[78] Bethe had already worked with Møller's equation to calculate the stopping power of relativistic particles. Fermi had found some interest in this work and proposed to Bethe to write an article about the relativistic collision of two electrons using his own formulation of QED. Judging from Bethe's recollections,[79] the work progressed quickly. The paper could be handed in June 1932.

As noted in the introduction to this section, this paper is often framed as the first instance of the "virtual particle" as a force-carrier and I will discuss this conception

[74] As a matter of fact, Fermi explicitly mentioned that the integration of the temporal development allowed for choosing the integration variables in a specific way so as to allow for a direct calculation (Fermi 1932, 122).

[75] For an excellent historical discussion of Bethe's post-doctoral years, see Schweber (2012, Chap. 4); especially pp. 193–223 about Bethe's work and life in Rome.

[76] Bethe to Sommerfeld, April 9. 1931; as cited in Schweber (2012, 193).

[77] See Schweber (2012, 211) and Dyson (2007, 49).

[78] For further information, especially on Møller's derivation, see Roqué (1992) and Kragh (1992).

[79] Schweber (2012, 221).

with a backdrop to the actual practices in a moment. Following Cathryn Carson (1996b), I do not fully agree with this judgement, but for slightly different reasons. I want to further point out that there was a process investigated by Fermi in which the concept of the interaction between two particles carried by an excitation of the electromagnetic field showed up priorly.[80] It was part of his "Quantum Theory of Radiation" article which was finished before Bethe and Fermi set to work. Actually, Fermi had already published a similar description in Italian (Fermi 1929).

The first concrete problem which Fermi discussed, which went beyond emission, was the "Propagation of light in vacuum" (Fermi 1932, Paragraph 9). To prove that the theory gave the correct speed of propagation of the interaction, the velocity of light, and that the intensity of the emitted light decreased with $\frac{1}{r^2}$, Fermi considered a system of two atoms, A and B each with two states, 1 and 2. The general question to answer was: If atom A was initially in state 2 and atom B in the lower state 1, what was the probability to find atom B in the excited state 2 and atom A in the state 1? In short: What was the probability that the two atoms exchanged some energy?

Fermi modelled the process as a real emission of atom A, the probability of which he had already calculated in the prior paragraph. The corresponding amplitude was the amplitude of the intermediate state, in which both atoms were in the ground state and one oscillation of the radiation field was excited. This amplitude was then connected to the final state of the process, i.e. the one in which the atom B was in state 2 and no component of the radiation field was excited. The square of this amplitude was the probability Fermi was looking for and behaved as it should, i.e. it was proportional to $\frac{1}{r^2}$ and equal to zero for $t < \frac{r}{c}$.

Of course, this modelling was not relativistic in terms of the material part of the system, it rested on real emission and absorption processes, it only held for distances r between the atoms A and B which were large compared to the frequency of the emitted radiation, and it rested on the specific lifetime Fermi chose for the respective excited states of the two atoms. But, and this is important, the general gist of the argument and even the mathematical modelling of the process was very similar to (parts of) Bethe and Fermi's article. It was clear that initial and final state could only be connected through an intermediate state in which an oscillator of the radiation field was excited.[81] Thus, something like a force carrier, although for non-relativistic atoms with only two states, at considerable distance, and as a real photon already existed in Fermi's earlier papers.

With this background, let's turn to Fermi and Bethe's work on the scattering of two electrons. At first, Fermi and Bethe verbally introduced the reader to the concepts involved in their calculation. The "deviation from the Coulomb law" (Bethe and Fermi 1932, 296) due to the interaction between radiation field and matter was expressed as follows:

---

[80] See also Schweber (2002) for a short discussion of this effect.

[81] Kikuchi (1930, 570), at the time in Leipzig with Werner Heisenberg, also turned to the propagation of light in the vacuum in Heisenberg and Pauli's QED. Seishi Kikuchi directly calculated the intensity of the emitted light. Only as a side note, he pointed towards the modelling Fermi studied.

> The coupling between matter and radiation field causes in *first* approximation only such transitions, in which the quantum state of the electrons is changed and a light quantum is absorbed or emitted. But we are interested in the matrix elements of the interaction energy which correspond to an alteration of the quantum state of the electrons alone without an alteration of the state of the radiation field, where we can specifically assume that both in the initial and final state no radiation is present. One gets such transitions of the electrons alone apparently only through a double process, in which first a quantum is emitted and then the same quantum is absorbed again.[82]

In the quantitative part of the paper, Fermi and Bethe first showed how Breit's formula can be constructed from Møller's scattering formula. Of greater interest for us is the derivation of both Breit's and Møller's formula from Fermi's QED. For the derivation of Breit's formula, the Coulomb interaction was conceived of as part of the unperturbed system, i.e. the perturbation only consisted of the interaction between matter and radiation field. Bethe and Fermi, did not restrict themselves to the case of two electrons but designated one quantum state of an arbitrary number of electron pairs by n. The state of the system was written as $nn_1n_2...n_i...$, where first the state of the material system and then the diverse occupation numbers of the radiation field are indicated.

At the beginning of their derivation Fermi and Bethe stated:

> Now the perturbation matrix has no element which couples the two states $n000...$ and $n'000...$ directly. Such a transition can only occur indirectly through a detour through an interlink ["Zwischenglied"]; as such, according to [the form of the perturbation matrix] only the states $m00...1_s...$ which combine with initial and final state can be considered.[83]

Fermi and Bethe then composed the scattering process, as Fermi did priorly, by explicitly considering the two successive processes. If the system was initially in state $n000...$, the time-development of the probability amplitude $m00...1_s...$ was:

$$\dot{a}_{m...1_s...} = \frac{-2\pi i}{h} H^{n00...0...}_{m00...1_s...} a_{n00...0...} e^{2\pi i(\nu_{nm}+\nu_s)t} \tag{8.28}$$

where $H^{n00...0...}_{m00...1_s...}$ was the matrix element of the perturbation and $h\nu_{nm} = E_m - E_n$ corresponded to the energy difference between the two states of the material part of the system.

---

[82] "Die Kopplung zwischen Materie und Strahlungsfeld verursacht in *erster* Näherung nur solche Übergänge, bei denen sich der Quantenzustand der Elektronen ändert und ein Lichtquant absorbiert oder emittiert wird. Es interessieren uns aber die Matrixelemente der Wechselwirkungsenergie der Elektronen, welche eine Änderung des Quantenzustandes der Elektronen allein, ohne Änderung des Zustandes des Strahlungsfeldes, entsprechen, wobei wir insbesondere annehmen können, daß sowohl im Anfangs- wie im Endzustand keine Strahlung vorhanden sein soll. Solche Übergänge der Elektronen allein bekommt man offenbar erst durch einen Doppelprozeß, wobei zuerst ein Quant emittiert und dann dasselbe wieder absorbiert wird." Bethe and Fermi (1932, 496–497); emphasis in the original.

[83] "Nun hat die Störungsmatrix kein Element, welches die beiden Zustände $n000....$ und $n'000...$ direkt koppelt. Ein solcher Übergang kann nur indirekt auf dem Umweg über ein Zwischenglied vorkommen; als solche kommen nach [der Form der Störungsmatrix] die Zustände $m00...1_s...$ in Frage, welche mit Anfangs- und Endzustand kombinieren." Bethe and Fermi (1932, 301).

Applying the approximation that the $a_{n000...}$ are practically constant, Fermi and Bethe integrated with respect to t and arrived at:

$$a_{m...1_s...} = -\frac{H^{n...0...}_{m...1_s...}}{h(\nu_{mn} - \nu_s)} a_{n00...} e^{2\pi i(\nu_{nm}+\nu_s)t} \tag{8.29}$$

Plugging this into the basic equation for the time development of the probability amplitude again, now considering transitions to the state $a_{n'...0...}$ they found:

$$\begin{aligned} a_{n'...0...} &= -\frac{2\pi i}{h} \sum_{ms} H^{m...1_s...}_{n'...0...} a_{m...1_s...} e^{2\pi i(\nu_{n'm}-\nu_s)t} \\ &= \frac{-2\pi i}{h} \sum_{ms} \Big(-\frac{1}{h} \frac{H^{n...0...}_{m...1_s...} H^{m...1_s...}_{n'...0...}}{\nu_{mn} + \nu_s} \Big) a_{n...0...} e^{2\pi i \nu_{n'n} t} := \\ &:= \frac{-2\pi i}{h} K_{n'n} a_{n...0...} e^{2\pi i \nu_{n'n} t} \end{aligned} \tag{8.30}$$

Fermi and Bethe then had to calculate $K_{n'n}$, "which we can conceive of as a matrix element that mediates the transitions between the states $n0...0... \rightarrow n'0...0...$ directly".[84] This matrix element described the correction to the Coulomb interaction that they were after. Fermi and Bethe could show that it merged, under certain approximations, with Breit's formula.

For the rederivation of Møller's formula the procedure was a little bit different. Fermi and Bethe restricted the process to involve only two electrons. They treated the Coulomb interaction as a perturbation. The unperturbed states, thus, contained two individual electrons and the excitations of the radiation field. The total transition under consideration was consequently $n_1 n_2 0...0... \rightarrow n'_1 n'_2 0...0...$, where the index 1 or 2 indicated which of the two electrons was considered.

The Coulomb interaction coupled initial and final states directly. But also the interaction with the radiation field allowed for transitions "through the detour through an intermediate state".[85] The total matrix element was, in analogy to Eq. (8.30):

$$K^{n'_1 n'_2}_{n_1 n_2} = \frac{-1}{h} \sum_{m_1 m_2 s} \frac{H^{n'_1 n'_2...0...}_{m_1 m_2...1_s...} H^{m_1 m_2...1_s...}_{n_1 n_2...0...}}{\nu_{m_1 n_1} + \nu_{m_2 n_2} + \nu_s} \tag{8.31}$$

Since the unperturbed electrons were independent of each other, each matrix element only depended on the coordinates of one electron. Thus, each matrix element corresponded to a transition of a single electron. Consequently, only the intermediate states $m_1 m_2 = n_1 n'_2$ and $m_1 m_2 = n'_1 n_2$ needed consideration. On the one hand, this

---

[84] "[...] als ein Matrixelement auffassen, das die Übergänge zwischen den Zuständen $n0...0... \rightarrow n'0...0...$ direkt vermittelt." Bethe and Fermi (1932, 302).

[85] "auf dem Umweg über einen Zwischenzustand," Bethe and Fermi (1932, 304).

reduced the matrix elements to be calculated decidedly. On the other hand and as a result, it showed that one electron emitted and the other electron absorbed the photon with frequency ν_s.

Bethe and Fermi then demonstrated how to derive Møller's scattering formula, while taking momentum conservation for the intermediate states and energy conservation for the whole process into account. In conclusion they noted

> It is satisfying that in our approximation no radiation is emitted. One could imagine that processes are possible in which both electrons change their states and two quanta are emitted. For such processes the momentum law would have to be satisfied twice [...] and the energy law once. But this is impossible, as long as none of the electrons has negative energy in the final state.[86]

Did Fermi and Bethe Put Virtual Particles to Work?

In their recalculation of Møller's formula in terms of Fermi's version of QED, Bethe and Fermi could clearly establish that in one transition a photon was emitted by one electron and the same photon was subsequently absorbed by the other. This insight essentially rested on treating both electrons as independently moving constituents of the system while their mutual influence, both the Coulomb force and the one due to the interaction with the radiation field, was treated as a perturbation. This fact effectively led to an important historical framing: Some historians consider Bethe's and Fermi's paper the first instance of the application of the idea that electromagnetic forces are due to photon exchange. As David Kaiser Kaiser (2005, 30) expressed it, Bethe and Fermi "put virtual particles to work".

Cathryn Carson (1996b) has already argued against this conjecture. The concept of exchange forces was at the time still connected to the exchange of one or two electrons to explain molecular forces or terms in electronic collisions but not to electromagnetic forces and photons as force carriers.[87] The authors also only considered the alteration to the Coulomb interaction as being due to the emission and reabsorption of a photon and not the entire electromagnetic interaction. Further, Bethe and Fermi did not speak about the exchange of photons but rather, and this is most common in the 1930s, about successive acts of emission, absorption, and transitions.

Looking at Bethe's and Fermi's procedure, I further argue that they did not apply the concept of a virtual particle. First of all, they lack the terminology. As has already been indicated priorly, the terminology of virtuality in connection with time-dependent perturbation theory will be mostly used for processes (transitions, emissions, absorptions, and so on) and not in connection with particles. But this

---

[86] "Es ist befriedigend, daß in unserer Näherung keine Strahlung emittiert wird. Man könnte an sich denken, daß Prozesse möglich sind, bei denen die beiden Elektronen ihre Zustände ändern und zwei Quanten emittiert werden. Bei solchen Prozessen müßte man aber zweimal den Impulssatz befriedigen [...] und einmal den Energiesatz. Das ist aber unmöglich, sofern im Endzustand keines der Elektronen eine negative Energie hat." Bethe and Fermi (1932, 306).

[87] The shift to force carriers taking place in nuclear theory in the mid 1930s as discussed by Carson will also be discussed in Sect. 10.2.1.

terminology only spread widely in the second half of the 1930s. Fermi and Bethe actually never used the term themselves. But this is, of course, only a terminological, not a clear-cut conceptual point.

Secondly, rather than back projecting modern concepts on this paper, I think it is far more adequate to conceive of it as an exemplary case of how scattering processes were treated during the 1930s, i.e. as a sequence of transitions connected to emission and absorption processes. Fermi and Bethe applied time-dependent perturbation theory à la Dirac, which entailed that energy conservation was violated in the transition, while conservation of momentum held in every step for the interactions of free particles. The structure of the interaction energy indicated that the process could not take place in one step but one needed to take into account intermediate states "which combine with initial and final state" as an "interlink" (Bethe and Fermi 1932, 301). Bethe and Fermi did not use the concept of a virtual particle to connect the two electrons and the respective final and initial state of the system. It was the concept of the intermediate state which performed this function. The idea of a (transversal) photon being emitted by one electron and being absorbed by the other was a consequence of a particular way of theoretical treatment, i.e. the conception of the Coulomb interaction as a perturbation, within such an approach rather than a tool in the derivation.

But even if we were to accept that the emergence of the intermediary particle in the calculational procedure allows for accrediting Bethe and Fermi the concept of a virtual particle, there is the problem what "virtual" means in this circumstance. Is it the mediating function of the photon? Then we might go further back in time and eventually state that Fermi applied this concept in 1929 in his discussion of the propagation of light in the vacuum although the processes were real. Or is it the properties of the processes, i.e. that each of the subprocesses cannot occur due to accepted physical principles, that makes the photon "virtual"? But then one is faced with the fact that the transitions to and from the intermediate state did not conserve energy, thus digressed from known physical principles and were thus "virtual".

Not to be misunderstood, this is not an argument against the relevance of Bethe and Fermi's finding that the interaction between freely moving electrons was (partially) mediated by photons for the further development of QED, QFT and the virtual particle concept as such. I believe and will argue in Chap. 10, that such effects, in which the particles were only present in intermediate or "virtual" states, and specific forms of mathematical representations will lead to the terminological variants of "virtual particles" to be applied during the late 1930s. And with this terminological variant certainly a slow shift in perception is diagnosable. While Carson decidedly argued against an important influence of QED on the conception of exchange forces, I will argue for an intricate interplay between suggestive mathematical structures, their results and heuristic imagery from other fields of research.

All the same, I want to stress that projecting the modern concept of a virtual particle on Bethe and Fermi's paper obscures the intricate differences between the practice of QED during the 1930s and the late 1940s. In the concluding chapter and with the help of the historiographical concept of a false prediction, I will argue that this difference was not only one of calculational ease but actually structured, through the modes of representations, what real processes could be conceived.

The difference to the modern techniques is also strikingly exemplified by the conjecture with which Fermi and Bethe closed their considerations. The possibility of a process in which two electrons change their states by emitting two light quanta was ruled out by Fermi and Bethe due to the impossibility of fulfilling the energy law once (for the whole process) and momentum conservation twice (for each single process). The intermediate states for such a process would essentially be the same as the ones in Bethe and Fermi's derivation of Møller's formula (one quantum being emitted while one electron changed its state to the corresponding final state). With the above conjecture, Bethe and Fermi (most probably unknowingly) provided for a false prediction which is very similar to Tamm's scattering of light from a pair of electrons (see Sect. 7.7). From a modern vantage point, using virtual particles and Feynman diagrams or energy and momentum conservation at the interaction points, we directly see that such a process is impossible since one would simply put side by side two (unphysical) vertices with no connection.[88]

8.2.2 Walter Heitler and a "General Scheme to Deal with All Kinds of Radiative Processes"

As laid out above, I think we should not conceive of Bethe and Fermi as applying the concept of a virtual particle; at best, such a concept might have emerged in their theoretical practice. According to the above argument, the idea of a photon only present in an intermediate state, which we might retrospectively identify as a virtual particle, was a result of a particular treatment and not a tool in the derivation. It was rather the concept of an intermediate state and the successive transitions to and from it that led the way in their calculation and conceptualization. This was in harmony with the perturbative methods they applied and the focus on transitions between stationary states in QED up until the late 1940s. Bethe and Fermi modelled their mathematical description closely after Fermi's review article: Instead of giving a mathematical derivation of higher orders in perturbation theory, they constructed the process of successive emission and absorption subprocesses in a stepwise manner.

That Fermi showed a rather pragmatic approach to QED and that especially pragmatically inclined theorists, like Bethe, took up Fermi's approach for the explanation of empirical data is well-known. In the following, I want to focus on another theorist who was more interested in the application of the theory than in investigating its basis. I will shift attention to Walter Heitler's early work on QED, as he would publish the second major didactic work on QED during the 1930s.

[88] The only possibility Fermi and Bethe leave open, namely that one electron is in a negative energy state after the scattering process, is, to the best of my knowledge, not expressible in a modern conception. The closest conception is the coupled process of pair annihilation with electron scattering, but since Fermi and Bethe work with an empty Dirac sea, i.e. they did not have the concept of pair creation, their process was essentially different.

Heitler's Route to QED (1932–1936)

Heitler had studied chemistry, mathematics and physics in Karlsruhe, Berlin and Munich.[89] In Munich, supervised by Karl Herzfeld, he wrote his PhD thesis on a subject of statistical mechanics in 1926. His career in physics began just when the new quantum mechanics was developed. Heitler spent the years 1926 and 1927 partly in Copenhagen with Bohr and partly in Zürich with Schrödinger, where he made his first major contribution to physics: the so-called Heitler-London theory of the covalent bond. In the same year, Max Born offered him an assistant position at the University in Göttingen which Heitler held until 1933. At first, Heitler's work focused on quantum chemistry, but took a major turn to the quantum theory of radiation in 1932.

Heitler himself remembered a twofold reason for changing his research subject. On the one hand, he felt that his job in quantum chemistry was done and he searched for other problems to engage with. On the other hand, quantum electrodynamics was "the fundamental unsolved problem" and he felt "that high energy phenomena would give some key to the further development of quantum electrodynamics".[90] Although this is certainly retrospectively informed, it fits the historical development quite well. At the time physicists already speculated that QED might break down for high energy phenomena or, equivalently, small distances. Simultaneously, empirical research showed inconsistencies with theory in terrestrial phenomena, most importantly the Meitner-Hupfeld effect, and in the realms of cosmic ray research,[91] which had also gathered more attention through the first hints at the positron in 1932. Furthermore, Weisskopf, who had priorly worked on radiation theory and QED in Göttingen, had just left for his post-doctoral fellowships.

Before actually turning to the evaluation of high energy phenomena, such as the inspection of Bremsstrahlung and pair creation processes for which Heitler is still best known, he calculated the line shape of the emitted radiation when the atoms were excited by the impact of electrons. This small piece of work was motivated through discussions with James Franck (Heitler 1933a, 150) and can be seen as a warm-up for the following contributions. Heitler followed Weisskopf's prescription and interpretation as given in his dissertation very closely. The main difference to Weisskopf's work was the fact that Heitler had to consider two different kinds of perturbations: the one between atom and impinging electron V, and the one between

[89] The short biographical sketch in this first paragraph is based on Mott (1982). The important role of Heitler's work on high energy phenomena is recognisable from its prominence in Cassidy's discussion of cosmic ray physics during the 1930s (Cassidy 1981). Heitler's work on the scattering phenomena and Bremsstrahlung as well as his social location in a group of young, striving theorists (partially German emigrants in 1933) was also discussed by Roqué (1997, Sect. 4).

[90] Interview of Walter Heitler by John L. Heilbron, 1963 March 18 and 19. Session II, Niels Bohr Library & Archives, American Institute of Physics, College Park, MD USA. https://doi.org/10.1063/nbla.fxxq.txtw. Quotes on page 8 of the PDF.

[91] For a contemporary assessment, see for example the conclusion of Heisenberg (1932a). The incoherences of QED during the 1930s and the connected, often rather pessimistic outlook on QED are well-known and I refer to the short introduction I gave in Sect. 3.3 and the literature cited therein.

atom and radiation field H. The interaction between electron and radiation field was assumed to vanish. This simplification was dropped later on, as it was exactly this kind of interaction that led to the phenomenon of Bremsstrahlung.[92] Knowing the structure of the interaction and the initial and final states of the system, Heitler argued that only through an "intermediärer Zustand" ["intermediate/intermediary state"] (Heitler 1933a, 147) the process could take place.

Heitler followed closely the prescription of Weisskopf (and Wigner). He wrote down the time development of the initial, intermediate, and final state, evaluated them for the given initial conditions, and introduced the Ansatz of an exponentially decaying probability amplitude. The result was also essentially the same as Weisskopf's for resonance radiation. Since "the relative phases of these three states were always taken into account"[93] the line shape of the emitted radiation depended crucially on the energy of the impinging electrons: If the energy was within the order of the line breadth, the line shape of the emitted radiation differed from the one of spontaneously emitted radiation.

While still in Göttingen, Heitler also worked out a first estimation of Bremsstrahlung in Dirac's radiation theory. For this process the interaction of the electron and the radiation field had to be included in the description. Consequently, the two perturbations consisted of the Coulomb interaction between atom and the electron and the interaction between radiation field and electron. The transitions from initial to final state were brought about by an "*intermediary intermediate state* [intermediärer Zwischenzustand] [...], which is not directly observed, but has *strict* phase relations to initial and final state, and which does not have to fulfil the energy law".[94] Two such intermediate states were possible, corresponding to whether the Coulomb interaction or the interaction between radiation field and electron acted first. In the paper, the respective processes were verbally constructed by Heitler at first and then translated into the mathematical format. As Fermi, Heitler first evaluated the corresponding matrix elements and then studied the temporal behaviour of all states. In the last step, Heitler put all things together and kept only terms corresponding to energy conservation.

Due to the rise to power of the NS-regime, Heitler left Göttingen and went to Bristol in 1933. There he further followed the discussion of Bremsstrahlungs phenomena. The papers with Sauter (Heitler and Sauter 1933) and Bethe (Bethe and Heitler 1934), in which the famous Bethe-Heitler formula was derived, also took processes of pair creation and annihilation, that were implicit in the prior calculation, as well as the effect of screening into account. The perturbative formula as derived in Heitler's prior papers were directly taken over.

But Heitler also engaged in the evaluation of higher order processes together with Lothar Nordheim. Nordheim, as Heitler, was a Göttingen theoretical physicists.

---

92 Heitler indicated this in a footnote in Heitler (1933a, 146).

93 "die relativen Phasen dieser drei Zustände stets berücksichtigen," Heitler (1933a, 147).

94 "[ein] *intermediäre[r] Zwischenzustand* [...], der nicht direkt beobachtet wird, sondern *strenge* Phasenbeziehungen zum Anfangs- und Endzustand hat, und der nicht dem Energiesatz genügen muß." Heitler (1933b, 152, emphasis in the original).

With the rise to power of the NS-regime, he first fled to Paris and then over Holland to the United States. In a rather little known paper, Heitler and Nordheim (1934b) investigated the process of double Compton scattering and, through the usage of the Weizsäcker-Williams method,[95] extrapolated the calculation to the case of double Bremsstrahlung, a process possibly important for the explanation of cosmic ray showers: Heitler and Nordheim explained cosmic ray showers as "multiple processes, in which a whole amount of electron pairs is created at once".[96] Following this reasoning they argued that also "Photonshower" (Heitler and Nordheim 1934b, 1060) should occur. From Heitler's prior calculation it was suspected that the process of Bremsstrahlung ($\propto e^6$) became more important than processes as the Compton effect ($\propto e^4$) in the high energy regime. Consequently, higher order processes might become increasingly important for higher energies, "[which] would have clearly brought the failure of exactly this expansion in e^2 and therefore of the usual radiation theory to light".[97]

To check the above conjecture, Heitler and Nordheim discussed the process of double Compton scattering and double Bremsstrahlung and compared them to Compton scattering and Bremsstrahlung. They found that the higher order processes always remained smaller by a factor $1/137$ in comparison to the simpler processes, which effectively, for the time being and in Heitler and Nordheim's eyes, saved the perturbative expansion.

For the processes discussed, "*two successive* intermediate states"[98] had to be considered. Rather then deriving the result mathematically, Heitler and Nordheim provided the reader directly with the possible intermediate states of the electron and then provided "the matrix element in analogy to [the second order equation]" as:

$$H = \sum \frac{H_{AI} H_{I\,II} H_{EII}}{(E_A - E_I)(E_A - E_{II}} + \dots \tag{8.32}$$

where also the permutation of the light quanta needed to be taken into account, which is indicated by the three dots in Eq. (8.32). This rather straightforward manner of

---

[95] The Weizsäcker-Williams method is a procedure for high velocity electrons. It rested on a calculation in the rest frame of the electron for which the Coulomb field of the scattering nucleus (in the case of high velocities) could be decomposed into plane waves and thus the regular Klein-Nishina formula could be applied for the calculation of Bremsstrahlung as well (von Weizsäcker 1934; Williams 1933, 1935). It showed that there were no theoretical reasons why the formulas for Bremsstrahlung should break down at high velocities, even though such a breakdown was expected from deviations between experiment and theory at the time (Cassidy 1981, 14–17). As this method treated the Coulomb field of the nucleus as a radiation field, thus as photons, the terminology of equivalent, but also of "virtual quanta" (especially in Williams 1935) was used. I will comment on this terminology in the Conclusion.

[96] "Mehrfachprozesse, bei denen eine ganze Anzahl von Elektronenpaaren auf einmal erzeugt werden," Heitler and Nordheim (1934b, 1059).

[97] "Dies hätte das Versagen eben dieser Entwicklung nach e^2, und damit der gewöhnlichen Strahlungstheorie klar zu Tage treten lassen." Heitler and Nordheim (1934b, 1060).

[98] "*zwei aufeinanderfolgende* Zwischenzustände," Heitler and Nordheim (1934b, 1066, emphasis in the original).

$$(\mathbf{p}_1^0\,\mathbf{p}_2^0\,\mathbf{p}_3^0) \rightarrow \left\{\begin{array}{ll} \mathrm{I} & \mathbf{p}_1'\,\mathbf{p}_2\,\mathbf{p}_3^0 \\ \mathrm{II} & \mathbf{p}_1''\,\mathbf{p}_2^0\,\mathbf{p}_3 \\ \mathrm{III} & \mathbf{p}_1\,\mathbf{p}_2'\,\mathbf{p}_3^0 \\ \mathrm{IV} & \mathbf{p}_1^0\,\mathbf{p}_2''\,\mathbf{p}_3 \\ \mathrm{V} & \mathbf{p}_1\,\mathbf{p}_2^0\,\mathbf{p}_3' \\ \mathrm{IV} & \mathbf{p}_1^0\,\mathbf{p}_2\,\mathbf{p}_3'' \end{array}\right\} \rightarrow (\mathbf{p}_1\,\mathbf{p}_2\,\mathbf{p}_3) \quad (5)$$

Fig. 8.3 Semi-diagrammatic or tabularic display of the different possible processes through intermediate states for pair production in collisions. Taken from: "Sur la production des paires par des choces de particules lourdes"—J. Phys. Rdaium—Volume 5, Numéro 9, septembre 1934—https://doi.org/10.1051/jphysrad:0193400509044900—W. Heitler et L. Nordheim, p. 449–454, here p. 450 (cited as Heitler and Nordheim (1934a) in the bibliography). Reproduced with kind permission from EDP Sciences

introducing higher orders and intermediate states showed that Heitler and Nordheim could assume their readers to be familiar with the concepts.

In a second paper written jointly by Heitler and Nordheim on the production of pairs by high energy collisions, the perturbative series was introduced in a similarly offhand manner. It was simply written down without any prior discussion or derivation.[99] To establish the possible intermediate states for the concrete case, according to Heitler and Nordheim "the first task"[100] in the derivation, they focused on the momenta of the three involved particles and first verbally explicated how such transitions can be developed. But they also put their reasoning in a semi-diagrammatic or tabularic form (see Fig. 8.3).

The table or diagram is read from left to right in the sequence initial → intermediate → final state. They assumed an interaction (relativistically corrected version of the Coulomb interaction as derived by Fermi and Bethe) for which the momenta of two electrons changed in one interaction. To exemplify the reading of the diagrams, let's take the third intermediate state. From the initial state $(p_1^0 p_2^0 p_3^0)$ the system passed into a state where the first particle was already in its final state p_1 and the second electron was in an intermediate state p_2'. The third electron was not involved in this subprocess. In the transition to the final state $(p_1 p_2 p_3)$ the second electron performed a transition from the intermediate to its final state p_2 and the third electron from its initial p_3^0 to its final state p_3. In this subprocess, the first electron was not involved.

As the diagrams by Amaldi (see Fig. 6.1) and Göppert-Mayer (see Fig. 8.1), this display clearly indicated the ordered fashion in which the processes were to come about according to the modelling in time-dependent perturbation theory. But it should be noted that this was more or less simply a shorthand notation to include all the transition possibilities into one and the same format. The reasoning process through

99 See Heitler and Nordheim (1934a, 450, Eq. (3)).

100 "La première tâche," Heitler and Nordheim (1934a, 450).

which Heitler and Nordheim (1934a, 450) arrived at these configuration was a physical one, which they exemplified in a combination of verbal terms and the short hand notation for one process.

And this was exactly the method Heitler will apply in the derivation of the intermediate states in his famous work with Bethe[101] and, importantly, in his influential textbook of 1936. Therein, the physical instantiations of the intermediate states are always constructed by the application of what I called Dirac's verbal model.

Perturbation Theory in Heitler's Textbook (1936)

The structure of Heitler's textbook (Heitler 1936) was heavily influenced by the perturbative approach. The introduction of the third and central chapter "Interaction of Radiation with Matter" included a discussion of perturbation theory and the general structure of the interaction of light and matter. Following this outline, Heitler arranged his presentation of physical processes according to the order in which they occur: "A. Processes of the First Order", "B. Processes of Second Order", "C. Processes of Third and Higher Order".[102]

Heitler actually argued for this structure in a short paragraph called "§10.4 Classification of radiation processes". Therein, he explained that radiation theory as presented in his textbook "is based on an expansion in powers of H' [the interaction energy]". He emphasized that all radiation processes with more than one light quantum involved have to go through an intermediate state which, for second order processes, "has to be chosen so that, in the transition from the initial state to the intermediate state, only one quantum is absorbed (or emitted), whereas the second quantum is emitted (or absorbed) during the second step, intermediate state → final state". Since $H' \propto e$ and the transition probability is proportional to H'^2 "we can [...] classify all radiation processes according to the power of e^2 or to the number of light quanta involved. The transition probability for a process in which n quanta are emitted or absorbed will be proportional to e^{2n}."[103] Naturally, Heitler reminded the reader that the Coulomb interaction, if treated as a perturbation, was already of second order.

The introduction of the perturbative expansion itself consisted of a rather standard derivation of time-dependent perturbation theory, closely following Dirac's initial presentation. Heitler simply dropped terms not corresponding to energy conservation for the whole process and applied Dirac's trick to ensure energy conservation for the whole process. Without further argumentation, Heitler simply stated that the "generalization of [the second order equation] when two successive intermediate state are necessary obviously gives" (Heitler 1936, 90):

---

[101] See Bethe and Heitler (1934, 84).

[102] See the table of contents of Heitler's textbook: Heitler (1936, vii).

[103] All quotes taken from Heitler (1936, 97).

$$H'(E_0) = \sum_{n'n''} \frac{H'_{n_0 n'} H'_{n'n''} H_{n''n}}{(E_{n_0} - E_{n'})(E_{n_0} - E_{n''})} \tag{8.33}$$

As Heitler noted in an offhand manner, "for the transitions from or to the intermediate states n' the energy is, of course, in general not conserved" (Heitler 1936, 90).

In the evaluation of concrete effects, Heitler always used Dirac's verbal model (or a corresponding short-hand notation)[104] to construct the different possible ways of combining initial and final state "through the agency of *intermediate states*" (Heitler 1936, 131, emphasis in the original). Since each interaction term corresponded to the emission or absorption of one light quantum, it was clear how many transitions of the material part of the system had to be taken into account at each order. The important point to figure out for each process was the respective intermediate state(s) of the material part of the system. The involved photons had to be permuted.

As in Fermi's and Bethe's case, there is a particular historiographic judgement in respect to the application of the theory that I want to argue against in connection to Heitler's 1936 textbook. Jan Lacki (2009, 30) has stated that for time-dependent perturbation theory as practised in the 1930s "there is a lack of an iterative algorithm and, as a result, the intermediate states must be 'put by hand' to obtain [the second order equation]". I hold that such a judgement is understandable in connection to the handling of perturbation theory in the hands of Heitler and Fermi, but does not hold against historical scrutiny.

The extension to higher orders was, from a mathematical point of view, an iterative procedure as exemplified in Weisskopf's derivation (see Sect. 8.1.4). Also in the case of the research publications of Fermi and Heitler, as laid out above, the procedure was certainly iterative as it consisted of a repeated application of the fundamental equation of time-dependent perturbation theory. As we will see in the following chapter, mathematical proofs, as envisioned by Hans Euler and Bernhard Kockel, in the derivation of the general structure of the matrix elements in time-dependent perturbation theory made use of the iterative nature of the procedure to construct a proof by induction.

But Lacki might have the way how a specific physical problem was treated in mind when referring to the intermediate state being "put by hand". For example, in the case of scattering the two orders of emission and absorption processes corresponding two different intermediate states (either no photon present or two photons present) had to be taken into account. It is true that here the iterative procedure of the general perturbative expansion was enriched by physical considerations. As argued in Chap. 5, this was one of the main reasons why Dirac introduced the verbal model, namely as a tool in the derivation process. But this reasoning, in turn, was guided by the nature of the perturbations involved, the transitions it might cause, and the idea that energy was not conserved in the transitions. While the process studied defined

[104] Once, Heitler used a diagram to represent the two different intermediate states in the theoretical description of the Compton effect (Heitler 1936, 190). This energy level diagram was, corresponding to its place in the textbook and the general perspective provided, based on the verbal model and was exclusively used for illustration.

the initial and final state of the system, the nature of the perturbation allowed for the construction of the route that would connect the two. The process of finding the different physical manifestations of intermediate states may not have been iterative, but it was certainly rule governed through the structure of the perturbations involved.

Consequently, I argue that the practice of QED of the time and of perturbation theory was iterative in its mathematical manifestation[105] and rule governed in its physical implementation. The general structure of the perturbative expansion and the general structure of the matrix element of the interaction between radiation and matter did render a generalized scheme that was filled with physical content in a given evaluation by Dirac's verbal model. To show that this is not a retrospective interpretation of the methods of QED of the 1930s but that the generality of the methods applied were actually grounds on which to argue for the application of QED, I want to draw the attention of the reader to a review of Heitler's textbook published in 1936.

Judging from the initials given, R. P., it was most likely written by Rudolf Peierls, also one of the experts in quantum electrodynamics of the time (R. P. 1936).[106] As already indicated a few times, semi-classical methods were equivalent in their predictive power and therefore often used when calculating observable phenomena. In discussing which method to use, semi-classical, "simple" ones or quantum electrodynamic ones, the reviewer stated:

> Which of these methods one prefers is a matter of taste, but quantum electrodynamics has the advantage that, instead of requiring a new 'recipe' for each new type of problem, one can develop a general scheme to deal with all kinds of radiative processes and, from this, solve special problems by merely inserting appropriate values into the general formulae. The importance of such a general scheme is emphasized by Dr. Heitler, and this will make his book very valuable in dealing with practical problems. (R. P. 1936, 483)

In contrast to Lacki's judgement there was something like a "recipe", a "general scheme" in the practice of QED during the 1930s which was clearly perceived of by the historical actors.

As a matter of fact, both the perturbative expansion and the matrix elements of the light-matter interaction could be derived before treating a specific problem. The task was to fill this framework with physical content for a specific process. Also this last step was rule governed, backed up by the mathematical structure of the interaction terms, and it applied verbal modelling or short-hand notations as tools. The whole process of finding the transition possibilities in this way, provided the "recipe" or "general scheme", which I want to outline in the following.

---

[105] This iteration process could even be enriched by physical interpretations as the *Schüttelwirkung*.

[106] The disagreement between the author and Heitler about the connection between field fluctuations and zero-point energy (R. P. 1936, 484) is a further hint at Rudolf Peierls. Compare, for example, the statements about the zero-point energy made by Heitler (1936, 67) and Landau and Peierls (1931, 56).

8.3 Conclusion and Outlook: A General Algorithm for the Evaluation of Scattering Processes

Although I did not find the following spelled out explicitly,[107] I want to provide for something like a general algorithm for the application of QED in the 1930s. To be sure, I do not claim complete generality of the following procedure, as there are surely deviations from this form of reasoning. Still, I hold that it exemplifies the general trend of doing perturbative calculations in QED and later on in QFT during the 1930s and early 1940s. Importantly, for my approach to conceptual development and for the discussion to follow in the next chapter, spelling out this procedure allows to locate the different representational formats in the reasoning process.

Step one: Split the Hamiltonian
First, one had to establish which part of the Hamiltonian described the behaviour of the unperturbed system and which part was conceived of as the perturbation. For example, Fermi and Bethe considered the Coulomb interaction as part of the unperturbed Hamiltonian in one case, while they treated it as a perturbation in the other.

Step two: Initial and final state
The structure of the Hamiltonian decidedly influenced the structure of the states of the unperturbed system. This choice influenced the representation of the material part of the system while the occupation number representation for the radiation field was common. For the process under study, one had to establish which of these states figured as initial and final states and thus should be combined.

Step three: Coupling two different states
The perturbation governed the way in which states of the unperturbed system could couple. The interaction terms, through their dependence on variables describing various components of the system, clearly showed which part of the system would alter its state in a transition of the system. The interaction Hamiltonian between radiation field and matter, for example, coupled states of the system in which (a component of) the material part performed a state transition and the radiation field changed its occupation number by one, i.e. they referred to emission and absorption processes. The Coulomb interaction, on the other hand, coupled states in which the occupation number of the radiation field did not change. But it combined states of the material part of the system that could not directly be connected by the perturbation due to the radiation field.

[107] Heitler comes close to it, especially when he argued for the ordering of the radiative processes.

Step four: Coupling initial and final state: the sequence of intermediate states
In general it was clear by the early 1930s that higher order calculations diverged, so that often only the first order in which a process occurred was studied. As such, this essentially constituted an evaluation of the general perturbation formula as given by Weisskopf (see Eq. 8.23). As soon as one knew which states would be coupled through the action of the perturbation, one could envision the set of possible intermediate states which would lead from initial to final state. Due to the structure of perturbation theory and the interaction energy, the transitions to intermediate states did not have to fulfil energy conservation while for the interaction between freely moving relativistic electrons and the radiation field momentum was conserved in each instance. At this point, the nearly purely mathematical structure of the perturbative series was filled with physical content in order to construct all the possible paths through virtual transitions (modern terminology) which coupled initial and final state. This is exactly the point where Dirac's verbal model, short-hand notations, or diagrammatic representations of the processes were applied by the historical actors.[108]

Step five: Dirac's trick
In general, the application of perturbation theory rendered probability coefficients the time evolution of which depended on the energy difference between unperturbed states of the whole system. Through the application of Dirac's trick one eventually found that only transitions corresponding to energy conservation for the whole process would lead to a probability proportional to t.

Step six: Evaluate the process by plugging in the explicit structure of the matrix elements
This last step normally took most of the room in the papers and was certainly the most time consuming one. The calculations were lengthy and further approximations had to be applied.

This way of approaching quantum-electrodynamical effects essentially rested on a model of successive state transitions. Virtual particles were not among the core concepts of this procedure. They might show up as particles only present in an intermediate state. But the most important concepts were transitions to intermediate states in which the energy did not have to be conserved. The guiding question, when it comes to the application of physical reasoning in this perturbative treatment, was: Through which paths of intermediate states can initial and final state be coupled?

---

[108] On a side note, this is also exactly the place where Feynman diagrams enter the calculation.

This function of the intermediate states was also reflected in the terminology of the historical actors. Fermi and Bethe, for example, called the intermediate state an "interlink [Zwischenglied]". Walter Heitler used a specific terminological variant, which strikingly exemplifies the above point. He once called the intermediate state "intermediärer Zwischenzustand" which can be translated as "intermediary intermediate state". At this point, we should be careful in not going over this potential doubling to quickly: "Intermediate" refers to something in between but does not necessarily give any further indication of a relation. But "intermediary" does: It means something like a connecting piece, something that is mediating and which is not simply in between. It was exactly this function of the intermediate states which secured its central position, which provided for its essential role in the conceptual framework of the practice of QED and QFT of the 1930s. It had, to use Heitler's words, "agency".

Comparing the procedure to the modern approaches to QED, there are a few important points for the history of the concepts I am studying. First of all, the verbal model of Dirac, diagrammatic, or tabularic representations describing virtual transitions enter exactly at the point where Feynman diagrams enter the procedure today. As soon as the mathematical basis of perturbation theory and the interaction was established in general terms, (pseudo-)physical reasoning was applied to construct the way of how the process was theoretically described. These representations served and serve to bridge the gap between abstract perturbative expansion and mathematical representation of the specific effect to be evaluated.[109]

Secondly, and in contrast to the modern approaches, it was the intermediate states and the transitions to and from them which broke with accepted physical principles, i.e. energy conservation. This was also reflected in the terminology used in the application of time-dependent perturbation theory: As soon as the term "virtual" became widespread in the mid 1930s, it was mostly connected to the individual subprocesses or the intermediate states.

Finally, there is an important conceptual difference between the modern practice and the practice of the 1930s. The connecting piece between initial and final configuration, the mediating entity in the theoretical description is in modern terms always a set of virtual particles. In the 1930s, particles in intermediate states might have arisen in the description, but the interlink or intermediary was the intermediate or virtual state.

---

[109] I already indicated this bridging function of the verbal model of Dirac and its representations in the conclusion of Ehberger (2022).

Chapter 9
The Practice of Time-Dependent Perturbation Theory (Part II): Virtual Possibilities, Modes of Representation, and the Reprise of the *Schüttelwirkung* (1934–1942)

In the prior chapter I set the stage for the discussion of the practice of time-dependent perturbation theory to follow in this chapter. In Sect. 8.2, I argued that intermediate states as well as the transition to and from them (with a specific focus on the configuration of the material part of the system) were the guiding concepts for the evaluation of QED during the 1930s, at least so in the most important didactic literature and for scattering processes. The corresponding algorithm of how to apply QED on specific effects within this conceptual framework was reconstructed in Sect. 8.3. As pointed out, the verbal model of Dirac served as the bridging element between the rather abstract perturbative series and the concrete instantiation of the light-matter interaction to be studied. In Sects. 9.1 and 9.2, I will further engage with this kind of reasoning and the representations the physicists came up with to express it.

In Sect. 8.1, I discussed important alterations to the concepts of perturbation theory and its practice. The inclusion of the *Schüttelwirkung* by Maria Göppert-Mayer and the extension of the perturbative formalism to longer time periods by Weisskopf and Wigner. These two contributions will, through some historical detours, serve as the methodological basis of Heitler's theory of radiation damping, which I will turn to in Sect. 9.3.

The in-depth study of the theoretical elaborations in this chapter will allow for a deeper understanding of the concepts of virtual transitions and virtual states (modern terminology) when conceiving of them as genuine concepts of theoretical practice, i.e. the reasoning which they were used for and the thinking possibilities they opened. The theoretical discussions in this chapter will also embed these conceptual structures in the developing conceptual framework of QED and QFT itself: At around 1934 the Dirac sea, the concept of anti-matter and processes such as pair creation became widely accepted,[1] which formed the basis for the elaborations of light-by-light scattering and the Delbrück effect. The acceptance of Yukawa's meson theory and the

[1] For a very well informed study of the acceptance of the Dirac sea as a suitable description of anti-matter, see Roqué (1997).

M. Ehberger, *Representing the Unobservable*, Science Networks. Historical Studies 68,
https://doi.org/10.1007/978-3-032-09188-8_9

identification of cosmic ray components with this particle provided the empirical context of the development of Heitler's theory of radiation damping.[2]

Light-by-light scattering (Sects. 9.1.1–9.1.3), the Delbrück effect (Sect. 9.2), and the theory of radiation damping (Sect. 9.3) have found some attention in the secondary literature. The context of cosmic ray research for the emergence of the theory of radiation damping has been identified. This theory is still best known as a possible candidate for filling the frame of Heisenberg's S-matrix and its eventual failure due its reintroduction of the infrared divergence.[3] Light-by-light scattering, as one of the great (still purely theoretical) successes of QED during the 1930s, is part of nearly any review of QED during this time period. It has shortly been discussed by David Kaiser (2005, 34–36), mainly to point out the extreme cumbersomeness of the calculational methods of the 1930s. The most detailed elaboration on the calculational methods has been given by Wüthrich (2010, Section 2.1), focusing on the diagrammatic representation in Hans Euler's publication and their comparison to Feynman diagrams, thereby highlighting the shortcomings of energy level diagrams. Wright (2024, Sect. 3.4) focussed on the terminological idiosyncrasy of "virtual possibilities" of the Leipzig physicists at the time, tentatively connecting it to neo-Kantian philosophy. I will give my own take on this terminological twist in Sect. 9.1.2.

The roots of the Delbrück effect as an explanation of the so-called Meitner-Hupfeld effect, an apparent anomaly in the scattering of hard γ-rays from heavy elements, is well-known.[4] Yet, a detailed elaboration of the methods applied by the different groups involved is still not available. Finally, the dissertation of Bernhard Kockel (Sect. 9.1.4) has found next to no attention in the secondary literature. It is well-known that he contributed to the calculations of light-by-light scattering,[5] but his dissertational research has only found some attention in connection to the infra-red divergence: Kockel stumbled over this theoretical problem, but did not discuss it in detail.[6]

In the following, I will order these contributions mostly in terms of the respective institutes where they were investigated. The first section will be concerned with the evaluation of light-by-light scattering and higher order processes in Leipzig. The qualitative introduction of this effect will show the merits of the Dirac's verbal model but also its short-comings: Otto Halpern's (at the time at the New York University) first communication included what I called a false prediction (Sect. 9.1.1). The investigation of light-by-light scattering is also of terminological interest: In 1934/35

[2] For this empirical background of radiation damping theories (both classical and quantum-theoretical ones), see, for example, Brown and Rechenberg (1996, Sects. 10.3 and 11.5) or Darrigol (1982, 301).

[3] See the discussions in Darrigol (1982, Sect. 7.4), Blum (2015, 86–87), and Blum (2017, 56–58). I will shortly review the infra-red divergence in Sect. 10.1.3.

[4] See e.g. Brown and Moyer (1984, especially p. 133) or Roqué (1997, especially pp. 100–101).

[5] The first communication of the results were given by Hans Euler and Bernhard Kockel together (Euler and Kockel, 1935) and Euler explicitly acknowledged Kockel's help, especially for the perturbative evaluation, in his dissertation (Euler 1936, 398, Footnote 1).

[6] See Blum (2015, 74).

Werner Heisenberg (and Hans Euler following him) (re)introduced the term "virtual" in connection to the concepts we call "virtual transitions" (Sect. 9.1.2). Euler's publication is, as mentioned, best known for its energy-level diagrams, but also constitutes a prime example of the application of the algorithm of QED I discussed in the prior chapter and allows for a comparison of the invoked concepts and the respective levels of theory he worked with (Sect. 9.1.3). Finally, Kockel's dissertation (Sect. 9.1.4) did not only introduce another false prediction and highlighted the role of the "picture of 'successive' transitions", as he called Dirac's verbal model, through the usage of tables instead of diagrams. His work is also one of the instances in which physicists engaged with the physical characteristics of the intermediate states (in this case the Pauli exclusion principle) and it included the first in-depth discussion of the so-called *Schüttelwirkung*.

The Delbrück effect (Sect. 9.2), introduced in Berlin and theoretically studied in Berkeley, Zürich, and Kharkiv, exemplifies the strong interrelation with established forms of evaluation at the respective institutes which all share a common feature: They all introduced the false prediction of the Delbrück effect as a third order process. This in turn will (partially) motivate Wendell H. Furry to prove the theorem named after him (Sect. 9.2.4). False predictions could thus directly lead to the elaboration of the perturbative techniques and the establishment of new rules for the application of perturbation theory in QED.

In my discussion of the emergence of the theory of radiation damping, I will focus on the remodelling of the perturbative series through the inclusion of the *Schüttelwirkung*. Edwin Góra was motivated by Bernhard Kockel's discussion of the same (Sect. 9.3.1). Building upon this conception, Walter Heitler (Sect. 9.3.2) developed a theory which excluded the ultraviolet divergences of QED and QFT but could still take into account the radiation damping necessary to understand the perceived of scattering anomalies in cosmic ray phenomena. Yet, as is well-known, it reintroduced the infra-red divergence.

Finally, I will draw the insights gathered in this and the previous chapter together and provide for an analysis of the different modes of representation, terminological variants, the characteristics of and rules for the practice with virtual transitions and states, the concepts connected to them, as well as the role of false predictions during the 1930s. Thus, I will analyse the practice of time-dependent perturbation theory with specific focus on the representations and applications of the concepts under study.

9.1 Light-by-Light Scattering and Radiation Theory in Leipzig (1934–1937)

So far we did not have to take into account a lot of the major changes in the conceptual framework of relativistic quantum theory of the 1930s. I already discussed the role of negative energy intermediate states for the emergence of the idea of the

Dirac sea (Chap. 7). Still, this conception was at first mostly seen as a workaround to the problem of negative energy transitions and was not taken seriously by the majority of physicists.[7] This changed radically in the years 1932–1934. Through the experimental observation of the positron and the subsequent theoretical identification of the same with a hole in the Dirac sea, the infinity of negative energy electrons gathered more and more physical meaning and the study of their behaviour came to the forefront for the physicists community.[8]

It was soon realized that the distribution of negative energy electrons was not as calm as initially proposed by Dirac. Vacuum polarization, a new divergence in QED, was first explicitly spelled out by himself in the contribution to the Solvay Conference in 1933. Therein he investigated the action of an external potential on the distribution of electrons in negative energy states. In modern parlance this effect is due to the creation and polarization of virtual pairs formed around a static charge.[9] Although this seemed at first to make the theory unhandable, Dirac and in his follow-up Werner Heisenberg devised ways of subtracting terms due to this distribution. These methods, although heavily involved,[10] made it possible to consistently treat problems in hole theory that were, without the application of the subtraction terms, formally divergent.

In the following sections I will focus on the probably most advanced studies on scattering phenomena in QED of the 1930s which rendered finite results in their evaluation: light-by-light scattering and the Delbrück effect. Although these two processes are closely related, I decided to make a separation according to the places and institutes where these effects were studied. At first I will take a look at the handling of radiation theory in Leipzig in the hands of two doctoral students of Werner Heisenberg, namely Hans Euler and Bernhard Kockel, and then turn to the evaluation of Delbrück scattering in Zürich, Kharkiv, and Berkeley.

The separation is, on the one hand, due to the close relation between the two dissertations of Euler and Kockel. Although mostly Hans Euler's name is connected to the phenomenon of light-by-light scattering, Bernhard Kockel was heavily involved in the perturbative evaluation. The two dissertations also show a rather similar structure that corresponds to the general algorithm I outlined in the last chapter. First a general and mostly mathematical evaluation of Dirac's perturbation theory was developed and then, in a subsequent step, this scheme was filled with physical content in the evaluation of the specific process under study. Although the authors did not make

---

[7] Heitler's first paper on Bremsstrahlung (Heitler, 1933b) is a very nice example.

[8] See Roqué (1997) for an in-depth study of the acceptance of the positron through its controlled production and its theoretical identification as a hole in the Dirac sea.

[9] See also my discussion in Sect. 10.1.2 and the literature cited therein.

[10] Wolfgang Pauli mockingly referred to these procedures as "Subtraktionsphysik" and "Limes-Akkrobatik". Pauli to Heisenberg, June 14. 1934; as cited in Pauli (1985, 327). Wolfgang Pauli had a rather fierce stance towards hole theory as well as the complications and non-rigourosities of the applied methods.

this explicit, a new mathematical evaluation of perturbation theory was in place since the subtraction terms of Heisenberg (which were of higher order by construction) also needed to be taken into account.[11]

On the other hand, I also decided to discuss Euler's and Kockel's work first since the next two sections will depend on these two dissertations. The study of light-by-light scattering had important influence on the work on the Delbrück effect in Zürich and Kharkiv; Bernhard Kockel's study of the *Schüttelwirkung* will lead the way to one of the most influential alterations of perturbation theory and the concept of the intermediate state up until the late 1940s: the theory of radiation damping by Heitler (and Peng), Góra, Wilson, and Sokolov.

I will at first revisit Otto Halpern's introduction of light-by-light scattering published in 1933, because it constitutes a very instructive example of the application of the physical reasoning encoded in Dirac's verbal model. Furthermore, Halpern proposed a different scattering possibility which has found very little attention in the secondary literature so far[12]: light-quantum splitting. After following the fate of this proposal, I will turn to Hans Euler's (and Bernhard Kockel's) work on light-by-light scattering and study its evaluation in Euler's published dissertation. Euler's dissertation is very rich when it comes to the concepts and procedures studied in this section: It included a specific terminology of the virtual ("virtual possibilities") and a general mathematical evaluation of perturbation theory; it introduced an influential diagrammatic representation; and, although positive in the result, involved serious doubts about the applicability of perturbation theory. In the next to final subsection, I turn my attention to Kockel's dissertation. To the best of my knowledge, Kockel developed the most advanced mathematical study of Dirac's perturbation theory during the 1930s. But Kockel did not only study the mathematical side of the perturbative approach but also elaborated on the concept of the *Schüttelwirkung* and on the physical rules that governed the intermediate state: He investigated under which circumstances the Pauli principle had to be taken into account for intermediate states.

9.1.1 The Qualitative Introduction of Light-by-Light Scattering and Light-Quantum Splitting

It is well-known that Peter Debye[13] (Leipzig) and Otto Halpern (1933) (New York) predicted the phenomenon of light-by-light scattering independently in the second half of 1933. Yet, it is less often noted that Halpern actually first introduced light-

[11] I could not reconstruct whether or not the Leipzig physicists were aware of Weisskopf's work (as discussed in Sect. 8.1.4).

[12] To the best of my knowledge, it is only deeply studied in the biography of Matvei Bronstein by Frenkel and Gorelik (1994), while naturally, the focus therein lies on Bronstein's contribution. It is also shortly mentioned as an explanation of the astronomical observation of a red-shift in Kragh (2017, 8).

[13] In private communication with Heisenberg. See, for example, Euler (1936, 398, Footnote 3).

quantum splitting as an explanation of the Hubble constant. Halpern's reasoning is a prime example of the application of Dirac's verbal model, which Halpern himself called "the language of Dirac's theory of radiation" (Halpern 1933, 856).

Halpern clearly stated that at the time, late 1933, "the insurmountable difficulties which the infinite charge-density without field offers to our physical understanding", i.e. the Dirac sea, needed further testing to be acceptable. In particular, "purely radiation phenomena", i.e. effects for which the energy of the photons was not high enough to cause actual pair production, might offer grounds for such an elaboration: "We are seeking, then, scattering properties of the 'vacuum'."[14] To qualitatively study the behaviour of light in the Dirac sea, Halpern distinguished two different initial states: (1) all incident quanta had the same direction of propagation and (2) they had different directions of propagation.

The second kind of initial conditions corresponded to light-by-light scattering. This process might change the frequency and the direction of the photons according to Halpern. It has gathered most attention in the physics community of the time and the secondary literature. I will turn my attention to this effect more closely in the following subsections. For now, I want to follow Halpern's reasoning in reference to the first kind of initial conditions more closely and show how this effect eventually was found to be non-existent.

As Halpern noticed, in the case that all quanta have the same direction of propagation the interaction between light quanta and negative energy electrons can only lead to a change in the frequency of the photons but not in their direction of propagation. Halpern introduced the particular sequence of events that would lead to this frequency reduction in the following way:

> In the language of Dirac's theory of radiation such splitting of the incident quantum occurs in processes of the following type: An electron in a negative energy state passes by absorption of the incident quantum into a state of positive energy; the electron then returns in several steps under emission of $h\nu$ in toto to its original state. At each step total momentum should be conserved. A scattering process of this type can only *reduce* the frequency; the reduction, if small, would on average be proportional to the distance travelled by the quantum through "vacuum." In this connection the hypothesis may be mentioned that Hubble's constant might be reducible to atomic constants without utilizing gravitational theories. (Halpern 1933, 856, emphasis in the original)

This is Dirac's verbal model, "the language of Dirac's theory of radiation" in application. Halpern could not provide any quantitative evaluation of the process and actually ended his paper with the outlook that "[t]he writer will supply shortly the necessary quantitative considerations lacking at present" (Halpern 1933, 856). Yet, through the usage of the verbal representations Halpern conjectured that light-quantum splitting should be possible and proposed it as an explanation of the Hubble constant and the corresponding red-shift.[15]

---

[14] Quotes taken from Halpern (1933, 855–856).

[15] For the contextualization of Halpern's proposal in the historical discussion around the explanation of the Hubble constant, see Kragh (2017, 8).

Although Halpern's proposal did not attract a lot of attention in research publications at first, it made it into Heitler's popular textbook of 1936:

> The *creation of pairs in intermediate states* gives rise to some processes which are impossible in principle according to ordinary electrodynamics. For instance, it can easily be seen that energy and momentum conservation allow two light quanta k_1 and k_2 to be scattered by each other, giving rise to two new quanta k_1' and k_2'. Or a single light quantum k_0 may break up into three quanta k_1, k_2, and k_3 with the same direction as k_0 (because of conservation laws) and an energy $k_1 + k_2 + k_3 = k_0$. [Footnote in the original: The probability for k_0 to break up in two light quanta vanishes.][16] (Heitler 1936, 193, emphasis in the original)

It was apparently this inclusion that triggered the interest of Mattvei Petrovich Bronstein and led to a first closer study of the effect.

Bronstein[17] had studied in Leningrad. He entered research in theoretical physics in 1930 at the Leningrad Physicotechnical Institute, where among others Yakov Il'ich Frenkel, Lev Landau, and Vladimir Fock were working. Bronstein specifically focused on the combination of quantum theory and (general) relativity. He was therefore well equipped to approach the question if the cosmological red-shift could be explained through the gradual loss of energy by photons. This hypothesis was favoured amongst Soviet officials at the time (Frenkel and Gorelik 1994, 114).

In the last two articles before his death[18] (Bronstein 1936; Bronstein 1937) Bronstein discussed the matter in some detail.[19] Bronstein differentiated two parts of Halpern's argument: (1) the process of light-quantum splitting is possible, and (2) the redshift can be explained thereby. To show that the second conjecture was wrong, Bronstein did not have to investigate the actual mechanism of light-quantum splitting. He only had to invoke principles of special relativity and quantum mechanics to show that it was untenable.

The rejection of the first hypothesis involved the calculation on quantum-electrodynamical grounds. Both in his German outline and his Russian publication, Bronstein did not give the calculations that led him to the refusal of the possibility of light-quantum splitting. Bronstein only mentioned that without considering the subtraction terms of Heisenberg, Heitler's conjecture, as taken up from Halpen, that a photon can split in three photons was correct. But, as Bronstein only pointed out but did not prove in the publication, through the subtraction terms of Heisenberg

[16] *Comment by the author*: This last conclusion is today most easily explained by reference to Furry's theorem (Furry, 1937). Heitler does not give any indication how he arrived at this result.

[17] For a rich biography of Matvei Bronstein, see Frenkel and Gorelik (1994).

[18] Bronstein was arrested and sentenced to death by the Military Tribunal of the USSR Supreme Court in February 1938. The reasons for his persecution are still not entirely known; the indictment accused him of "practical anti-Soviet activity", "preparing terrorists acts against the CPSU(B) leadership and the Soviet Government" and of being a "foreign spy". According to his biographers, these statements only correspond to the "flights of fantasy" of his interrogators. The only document found about his interrogations, stretching from August 1937 to February 1938, that carried Bronstein's signature is a sheet, where he denied all of the accusations. Frenkel and Gorelik (1994, 144–146).

[19] For a full discussion of the more elaborate Russian article of 1937, see Frenkel and Gorelik (1994, 117–121).

> the probability of spontaneous decomposition/decay [Zerfall] is strictly equal to zero. From a physical point of view this is natural and satisfying because otherwise this probability would increase with decreasing frequency, which would be in contradiction to the correspondence principle and experience.[20]

Since the calculations were not published, I have no means of judging the resilience of the result or the details of Bronstein's derivation.[21] But if we accept Bronstein's conclusion without knowing its specific derivation, the proposed process of light-quantum splitting gets cancelled by Heisenberg's subtraction terms. In essence, this would mean that Halpern's conjecture was correct within a non-charge-renormalized QED and that at the time he had published his article the proposition of such an effect would have been correct. It would only be cancelled by a procedure that was proposed by Heisenberg half a year later.[22]

9.1.2 The Terminology: Heisenberg, "Virtual Possibilities", and a Recourse to BKS

Bronstein's investigation on light-quantum splitting seemed not to have made a huge impact. On the contrary, the quantitative evaluation of Halpern's and Debye's proposal of light-by-light scattering performed in Leipzig by two doctoral students of Werner Heisenberg, Hans Euler and Bernhard Kockel, was widely received.[23] It resulted in the dissertation of Euler and, for a comprehensive evaluation of the corresponding calculations, I will turn to Euler's publication (Euler, 1936). But before turning to his methods, I want to discuss the terminology Euler used. He invoked the term "virtual" nearly exclusively in connection with processes (Euler 1936, 400), possibilities (Euler 1936, 403) of such, or intermediate states (Euler 1936, 415).

The reference to intermediate states and processes as "virtual" was consistent with the method of perturbation theory and also with a lot of the invocations of the word "virtual" up until the late 1940s: It reflected the fact that energy was not conserved in the transitions. Hence, the processes broke with known physical laws and were

---

[20] "[...] die Wahrscheinlichkeit des spontanen Zerfalls streng gleich Null wird. Vom physikalischen Standpunkte aus ist es natürlich und befriedigend, denn andernfalls würde diese Wahrscheinlichkeit mit abnehmender Frequenz zunehmen, was mit dem Korrespondenzprinzip und der Erfahrung im Widerspruch wäre." Bronstein (1936, 687–688).

[21] There might exist archival material but I had no means of accessing it. Considering the little archival material that is cited in the biography of Mattvei Bronstein, I consider it unlikely that there are any manuscript notes in archives. But even if there were any, my inability of reading Russian would certainly have made the search a near to impossible endeavour.

[22] Bronstein's argument seemed not to be of high relevance for the community. In the second edition of Heitler's textbook (Heitler 1944, 193), for example, the possibility of spontaneous light-quantum splitting is still included.

[23] Also in Kharkiv and under the supervision of Lev Landau physicists turned to the calculation of light-by-light scattering. See Sect. 9.2.3.

thus virtual. The terminology of "virtual possibilities" (of processes), on the other hand, is rather idiosyncratic. But it can be traced back to Euler's supervisor, Werner Heisenberg.

Heisenberg first used the term "virtual" in connection with QED in reference to the self-energy problem.[24] In a letter to Pauli dating to February 5. 1934, Heisenberg described the self-energy as being due to the "infinitely many virtually possible transitions to different states"[25]. Similarly and in publication, he connected van-der-Waals forces to "the possibility of a virtual exchange".[26] Heisenberg and Euler would again frame vacuum polarization as being due to the "virtual possibility"[27] of the creation of matter.

As Hans Euler in his dissertation, Heisenberg did not exclusively invoke the combination "virtual possibility". A rather nice example of the mixture of terminological variants is Heisenberg's contribution to the *Zeeman Festschrift* (Heisenberg, 1935a). According to Heisenberg, the "possibility of β-decay"[28] (note that the term "virtual" is missing) led to the exchange forces between nucleons. The "virtual creation [Entstehen] and disappearance [Wiederverschwinden] of a pair"[29] was responsible for the anomalous magnetic moments. But Heisenberg also referred to the energies "of the virtual electron-neutrino pair"[30] in the same context. We can find such an overlap of terminological combinations, but also different representational formats, in the publications of different physicists starting in the mid to late 1930s and I will turn to an explanation in Sect. 10.3. For the time being, I still want to engage with the "virtual possibility".

Let me first state the straightforward interpretation of the term "virtual possibility" for the discussed phenomenon of light-by-light scattering. In a first step, it is sensible to separate the two terms "virtual" and "possible". One could anticipate that it was in essence possible that two photons would create an electron positron pair, but as the energy of the photons was actually not high enough this possibility was only virtual.[31] The general idea behind this terminology would accordingly be that every in principle possible process, which is not forbidden by symmetries or conservation laws other than energy conservation (as e.g. charge conservation), had to be taken into account in the calculation as a virtual process. Hence, the "virtualness" of the possibility of the processes is due to the breach with energy conservation. This is certainly one connotation of the terminology and fits the usage of Raman (left aside

---

[24] For the role of virtual entities in the description of the self-energy, see Sect. 10.1.1 and the literature cited therein.

[25] "unendlich vielen virtuell möglichen Übergänge zu anderen Zuständen," Heisenberg to Pauli, February 5. 1934; as cited in Pauli (1985, 273).

[26] "Möglichkeit eines virtuellen Austausches," Heisenberg (1935b, 482).

[27] "virtuelle Möglichkeit," Heisenberg and Euler (1936, 714).

[28] "Möglichkeit des β-Zerfalls," Heisenberg (1935a, 112).

[29] "virtuellen Entstehen und Wiederverschwinden eines Paares," Heisenberg (1935a, 113).

[30] "des virtuellen Elektron-Neutrino-paares [sic!]," Heisenberg (1935a, 115).

[31] For clarity's sake, Euler and Kockel only discussed processes in which the energy of the photons was too low to produce actual pairs.

the non-realistic connotations the latter implied, see Sect. 6.2.2). And as such, energy non-conservation will become the most important connotation of the term "virtual" for concepts directly connected with the virtual particle. But there is another layer to the term "virtual possibility" which connects Heisenberg's invocation directly to one of the core meanings of the term "virtual" in the BKS theory: the conception of a "virtual world".

The "Possible" and the "Virtual" in Heisenberg's Writings Before 1934

In order to argue for this point, let me first and shortly contextualize Heisenberg's usage of the terms "possible" and "virtual".[32] Heisenberg rarely used the word "possible", in German *möglich* or *Möglichkeit*,[33] but for our purposes at important points of his writings.[34] For example, when interpreting the role of the measuring process in his paper on the uncertainty relations he noted that "all perception [Wahrnehmen] is a selection out of a wealth of possibilities and a restriction on the prospectively possible".[35] He also invoked the possible in his evaluation of fluctuation phenomena in 1931, where he concluded "that the *possibility* of the excitation of very short waves leads to infinitely large energy fluctuations".[36]

The term "virtual" played only a minor role in Heisenberg's reasoning on quantum mechanics up until 1934. And if Heisenberg used the term, he exclusively referred to virtual oscillators or, similarly, to semi-classical concepts which replaced the quantum systems by classically behaving entities. For example, in the English version of his 1930 textbook, the semi-classical description of emission, absorption, and dispersion was discussed under the headlines "Correspondence principle and the method of virtual charges" (Heisenberg 1930a, 82–84). In the German edition (Heisenberg, 1930b) the corresponding section is missing. In the English version, a "virtual source" is invoked in the discussion of interference, again as a semi-classical replacement of the radiating atom (Heisenberg 1930a, 90). In the German version a "virtual" entity

[32] Especially the contextualization of the concept of the "possible" in Heisenberg's interpretation of quantum mechanics, but also in the development of the theory as such, deserves further attention. I am personally not aware of any genuine historical works on the concept of the possible in Heisenberg's thinking.

[33] I will always translate the term "möglich" as "possible" since it seems to me the most natural connotation of the term in the historical situations. I still want to note that also "potential" is a possible translation.

[34] It is also possible that the terminology of the "possible" did not occur frequently in Heisenberg's writing since Pascual Jordan introduced the term "interference of probabilities" [*Original*: "Interferenz der Wahrscheinlichkeiten"] (Jordan 1927, 812), which was also used by Heisenberg, for what was actually an interference of probability amplitudes.

[35] "[...] alles Wahrnehmen ist eine Auswahl aus einer Fülle an Möglichkeiten und eine Beschränkung des zukünfitg Möglichen." Heisenberg (1927, 197).

[36] "daß die *Möglichkeit* der Anregung sehr kurzer Wellen zu unendlich großen Energieschwankungen führt." Heisenberg (1931b, 4, emphasis in the original).

is not mentioned (Heisenberg 1930b, 66). Whether or not the English or the German version better reflects the lectures held in Chicago on which the book was based is regardless for the point I want to make: Heisenberg was very cautious in the use of the term "virtual", either in German or in English, and if he used it, he did not connect it to transition processes.

The *Heisenberg Cut* and the Quantum World

There is an important aspect of Heisenberg's understanding of quantum theory which must shortly be discussed to understand the connection to BKS I am about to make: the so-called Heisenberg cut.[37] As described in his uncertainty paper (Heisenberg, 1927), Heisenberg was convinced that the perturbation introduced by the measuring process and the observer played an important role in the transition from a quantum-theoretical description of possibilities to the description of measured quantities. The latter, in turn, could only be expressed in classical terms. In 1927, Heisenberg already explicitly spoke about the reduction of the wave packet and the selection of one out of many overlapping possibilities due to the measuring process (Heisenberg 1927, 197).[38] In this conceptualization, as was for example observed by Gregor Schiemann, a separation of the physical world into two spheres is created through the measuring process: "Thereby, the physical world is split in the realms of the very small and the medium to large objects."[39]

This split and the measuring process were more explicitly discussed starting at around 1932. John von Neumann (1932, Chap. IV) famously gave a detailed discussion of the measuring process in quantum mechanics, wherein he also discussed the concept today known as the "reduction [or collapse] of the wave function". The important point for the following argument is yet another one. Von Neumann spelled out a sharp separation, a cut, between two parts of the world, the observed one and the quantum-theoretically described part. And von Neumann came to the conclusion that it was regardless where the cut between these two parts was drawn.

Heisenberg seems to have immediately noticed the importance of von Neumann's discussion. After having read the draft of Pauli's second article for the *Handbuch der Physik* (Pauli, 1933), Heisenberg recommended to Pauli to take up von Neumann's conclusion.[40] Pauli did so and described it in the following terms:

---

[37] This conception is directly connected to an interpretation and a formalisation of the measurement process in quantum theory which has gathered a lot of philosophical but also historical attention right from the start of such investigations, see for example Jammer (1966, Chap. 9.2). Quite recently, a short overview has been published by Pessoa JR (2022). For the following, only general aspects of these discussions are important and therefore I did not go into an in-depth study of the literature.

[38] For Heisenberg's views on possibility waves and the collapse of the wave function, see, for example, Camilleri (2006, especially Section 3).

[39] "Die physikalische Welt ist damit zweigeteilt in die Bereiche der ganz kleinen und der mittleren bis großen Gegenstände." Schiemann (2008, 52).

[40] See Pauli to Heisenberg, January 18. 1933; as cited in Pauli (1985, 148–151; especially p. 149).

> Because it [a previous derivation modelled after von Neumann's ideas] shows, that one arrives at consistent results, regardless of the point at which one makes the cut between the system to be observed and described by a wave function, and the measuring apparatus.[41]

Heisenberg himself further reflected on the cut between observer and quantum-mechanical system, and the arbitrariness of where the cut is exactly made, in publication (Heisenberg 1934d, 670–671) but also invoked the concept in his unpublished response to the EPR paradox in mid 1935.[42] As Heisenberg himself framed it, he engaged with the question: "At which point should the cut between the description in terms of wave functions and the classical-intuitive [klassisch-anschaulichen] description be drawn?"[43] The exact reasoning of Heisenberg is not of importance for us but the conceptual differentiation the cut implies is. There was a clear-cut differentiation between the system described in terms of wave functions and the intuitive level of the description. It separated a "quantum world" (with all the possibilities still in play) and the classical or "intuitive" one (where the decision has been made). This separation, between intuitive (often also semi-classical) description and the fully quantum-theoretical one can also be observed in and actually structured the dissertation by Euler (1936) and the joint work of Heisenberg and Euler (1936) (I will come back to this differentiation in the following section).

The *Cut* and the "Virtual"

With this background, let me turn to the terminology of the "virtual possibilities" again. As discussed in Sect. 5.2.2, the "virtual oscillators" and the "virtual transitions" have different points of conceptual overlap. They were both located on the probability- or quantum-amplitude level of theory and, in their respective theoretical and conceptual framework, they both allowed for conceiving of the possibilities of energy non-conserving transitions as physically effective entities. The replacement of all *possible* transitions (unbound by energy conservation) through virtual oscillators or *Ersatzoszillatoren* was clearly noted by the historical actors at the time. Heisenberg was in Copenhagen from September 1924 to April 1925. He himself applied virtual oscillators implicitly and explicitly. He must have been well aware of this fact.

---

[41] "Denn es zeigt, daß man zu übereinstimmenden Resultaten gelangt, an welchen Stellen immer man den Schnitt zwischen dem zu beobachtenden, durch Wellenfunktionen bechriebenen System und dem Meßapparat zieht." Pauli (1933, 148).

[42] See the appendix to Heisenberg to Pauli, July 2. 1935; as cited in Pauli (1985, 409–418). For a short and concise description of Heisenberg's reasoning, see Cassidy (1995, 322–324). Heisenberg also discussed the cut between observed and quantum world in a lecture at the university of Vienna in 1935, as cited in Heisenberg (1949, 35–46).

[43] "An welcher Stelle soll der Schnitt zwischen der Beschreibung durch Wellenfunktionen und der klassisch-anschaulichen Beschreibung gezogen werden?" Appendix to Heisenberg to Pauli, July 2. 1935; as cited in Pauli (1985, 411).

As early as 1927, Heisenberg described the probability or quantum amplitudes and their interference as expressing the possible, while the measurement chose among these possibilities. Von Neumann's clear proposition of a cut sharply differentiated between the quantum world and the classical world, an idea which we can observe in Heisenberg's conceptions of the measuring process early on. I argue that Heisenberg, and Hans Euler following him, invoked the term "virtual" as a reference to the quantum world in which all possible processes and transitions had to be taken into account.[44] They referred to a virtual, unobserved world in which possibilities were effective, exactly as (according to my reconstruction and the one by Blum and Jähnert) BKS did.

Of course, the virtually possible transitions often did not conserve energy.[45] In the given case, light-by-light scattering was only studied by restricting the processes to energies as low as $\sum_{i=1}^{2} h\nu_i < 2\ \mathrm{mc}^2$, i.e. by only considering energy non-conserving transitions. But according to the above reconstruction, energy conservation was not the primary characteristic of the transitions discussed. Rather it was their description on a quantum-amplitude level, i.e. as part of the "virtual world". And thereby the concept of the effectiveness of only possible, but not actualized, transitions in the quantum world connects Heisenberg's invocation directly to the BKS theory.

Neo-Kantian Philosophy, Retrospective Interpretations and the Spread of the "Virtual"

In my above reconstruction, an important thread of philosophical analysis which was discussed amongst Leipzig physicists at the time is missing: the influence of Neo-Kantian philosophy.[46] Werner Heisenberg had been interested in this strand of philosophy since the 1920s but it rose to prominence in Leipzig when Grete Hermann arrived in early 1934.[47]

Whether or not this philosophical tradition had impact on the particular framing of "virtual possibilities" is hard to judge.[48] An answer would require an in-depth study of the correspondence especially of Grete Hermann, Werner Heisenberg, and Carl

[44] In how far this differentiation also coheres with or is connected to Heisenberg's concepts of "anschaulich" (i.e. semi-classical) and "unanschauliche" (fully quantum-theoretical) theoretical description at around the time (see also below) needs further investigation.

[45] It is to be noted that a breach with energy conservation is not a necessary prerequisite for description in terms of quantum amplitudes. See, for example, the points made in this connection in Sect. 8.1, as Weisskopf's consideration of resonance scattering.

[46] A connection between the "virtual possibilities" of the Leipzig physicists and Neo-Kantian philosophy has been (tentatively) suggested by Sidney Aaron Wright (2024, Sect. 3.4). In my opinion, Wright downplayed the significance of the terminology of the virtual in BKS for the further development of such concepts in quantum theory, and thus filled the explanatory gap left in the reconstruction by the rise of Neo-Kantian philosophy in Leipzig at around the same time.

[47] A historical discussion of Grete Hermann's work on the philosophy of physics is given by Crull and Baccuagaluppi (2016).

[48] Although, all things taken into account, I do not consider it very likely.

Friedrich von Weizsäcker at the turn from 1933 to 1934.[49] Yet, since the "possible" already showed up in Heisenberg's writing before and a very prominent instance of the "virtual" existed as a recourse which also fits Heisenberg's view on quantum theory of the time—at least according to the reconstruction above—I consider it unlikely.

All besides the terminology, there is historical evidence that the conception of virtual processes Heisenberg entertained at the time, i.e. that they were not real or at least not on the same plane of reality as classical objects, was not influenced by the rise of Neo-Kantian philosophy in Leipzig in 1934. From 1932, we can find visualisable/intuitive models in Heisenberg's reasoning and papers, prominently exemplified by Heisenberg's 1932/33 trilogy on the nucleus, but of course also important for his contribution to the *Zeemann-Festschrift*.[50] While Heisenberg (as well as Pauli and Bohr) had initially not described the theoretical objects of quantum theory in an intuitive fashion, this tendency was loosing its grip on Heisenberg in the early 1930s.

And even prior to Grete Hermann's arrival in Leipzig, she received a letter from Gustav Heckmann in which Heisenberg's stance towards the concepts of quantum physics and its relation to Hermann's prior philosophical work is described as follows:

> Hei[senberg] spoke also about a certain difficult-to-acquire resignedness in the face of physics: one just has to give up the wish to state something about *real processes* in the atom – then one is able to work more fruitfully. Also your philos[ophical] considerations would be more fruitful once in regard to atomc theory you had managed to fight your way to this st[an]dp[oin]t. "It requires a great deal of strength to get there in the first place. Grete H[ermann] seems not to have quite got there yet, also Schrödinger has not yet arrived there". I do not understand this st[an]dp[oin]t, perhaps you can see what he means.[51]

Before a closer contact between Hermann and Heisenberg was established, Heisenberg had rejected the idea that the talk about "real processes" in quantum theory was sensible. All the same, at the time, Heisenberg spoke and wrote about quantum-theoretical processes, which were separated from the classical world by the Heisenberg cut.

Yet, we do not know if Heisenberg had already developed the terminology of the "virtual" for these not-real processes. But, given the fact that he used the phrase "infinitely many virtually possible transitions"[52] to describe the self-energy of the electron in a letter to Pauli shortly afterwards (February 5. 1934), and that Pauli did

---

[49] Since this possibility came to my mind to late during my research, this piece of the puzzle is left for future investigation.

[50] For the development of the "intuitive" ("anschaulich") in Heisenberg's reasoning see especially Miller (1984). For Heisenberg's trilogy on the nucleus see Sect. 10.2.1 and the literature cited therein.

[51] Gustav Heckmann to Grete Hermann, December 17. 1933, as cited in Crull and Baccuagaluppi (2016, 222, emphasis in the original).

[52] "unendlich vielen virtuell möglichen Übergängen", Heisenberg to Pauli, February 5. 1934, as cited in Pauli (1985, 273).

not comment on the terminology at all, I believe that the "virtual possibilities" were already in place, when neo-Kantian philosophy was en vogue in the physics circles of Leipzig.

Concerning the retrospective identifications one might have in mind, let me first turn to Werner Heisenberg's *Der Teil und das Ganze*. Therein, a fictional conversation between Heisenberg and Euler was constructed (Heisenberg 1972, 220–225). In this "conversation", Heisenberg stated that "the word 'virtual' indicates, that a possibility is concerned"[53] and that the possible processes were restricted only by the symmetries of the system under study.[54] In Heisenberg's reconstruction a photon consisted virtually of an electron-positron pair which meant "that the light quantum can be decomposed into an electron and a positron in certain experiments" (Heisenberg 1972, 220). Gerald Feinberg (1972) expounded a very similar understanding of the "virtual", most probably influenced by Heisenberg's considerations,[55] under the headlines "Effects of the Possible on the Real".

With this connection in mind, a reader acquainted with Heisenberg's thinking and popular writings might have already thought of another connotation: the Aristotelian "potentia", which Heisenberg used in his *Physics and Philosophy* to describe the quantum world as "something standing in the middle between the idea of an event and the actual event, a strange kind of physical reality just in the middle between possibility and reality" (Heisenberg 1958, 41). Right before this passage, Heisenberg discussed the BKS theory. Although this seems to cohere with my reconstruction it should, as a retrospective identification, obviously be handled with care. And it must also be mentioned that Heisenberg most probably had not yet developed a connection between Aristotle's concept and the possibility as referring to quantum amplitudes at around the mid 1930s.

For example, there is no trace of the Aristotelian potentia in Heisenberg's *Gedanken der antiken Naturphilosophie in der modernen Physik* (Heisenberg, 1937). Also in other, rather popular or philosophical articles of the time, as edited in the *Wandlungen in den Grundlagen der Naturwissenschaften*, I did not find the Aristotelian concept of a potentia.[56] I want to emphasize that these statements should not cause any doubt about the importance of the concept of "possibilities" in Heisenberg's reasoning about quantum physics at the time. Rather, it should point towards the fact that Heisenberg's reasoning about possibilities in quantum theory was not yet connected with the ancient concept of potentia. If the above reconstruction is correct, or at least tends in the right direction, we might actually turn things around:

---

[53] "Das Wort 'virtuell' deutet an, dass es sich um eine Möglichkeit handelt." Heisenberg (1972, 220).

[54] Interestingly, Heisenberg referred directly to a philosophy of the as if ("diese Philosophie des 'als ob'," Heisenberg (1972, 222)) but without mentioning Hans Vaihinger.

[55] Feinberg used the same example as Heisenberg, a classical double star system, to exemplify some consequences and boundaries of these ideas. Interestingly, Feinberg connected this conception with Cheffrey Chew's bootstrap programme.

[56] I consulted the eighth edition of Heisenberg's *Wandlungen in den Grundlagen der Naturwissenschaften* (Heisenberg, 1949). Therein, some of the popular and more philosophical articles by Heisenberg are gathered.

The virtual world of the BKS theory which Heisenberg appealed to with his terminology might be one of the roots of the concept which Heisenberg later described with potentia. Thus, instead of back-projecting the potentia on Heisenberg's thought in the 1930s, further historio-philosophical work might look for the roots of this potentia in Heisenberg's writings about virtualities in quantum theory.

Whether or not the above reconstruction is correct, we can safely state that with the usage of the term "virtual" by Heisenberg, prominently in his contribution to the *Zeeman Festschrift*, but also by Hans Euler in his widely received dissertation the term became firmly established in the community. In the follow-up, more and more physicists used the term in connection with energy non-conserving transitions to intermediate states (and at times in connection to the particles in these states). Among the most prominent physicists who used the term "virtual" from 1934 onwards[57] and mostly directly in connection with states or processes (transitions, emission, absorption, ...)[58] are: Hans Bethe,[59] Hans Euler (see the following Sections), Wendell H. Furry (see Sect. 9.2.4), George Gamow,[60] Werner Heisenberg (see above), Walter Heitler,[61] Rudolf Peierls,[62] Hideki Yukawa,[63] Carl Friedrich von Weizsäcker,[64] Gregor Wentzel (see Sect. 10.2.4), and Gian-Carlo Wick.[65]

The status of Heisenberg in the community and the popularity of the work on light-by-light scattering, which I shall turn to now, led to a wider spread of the terminology of the virtual in QED and QFT.

9.1.3 The Quantitative Evaluation of Light-by-Light Scattering: "Virtual Possibilities" on the Quantum, "Virtual Matter" on the Intuitive Level

Although light-quantum splitting as suggested by Halpern turned out to be what I called a "false prediction", the phenomenon of light-by-light scattering was quantitatively evaluated and found to have a non-vanishing probability. Breit and Wheeler

[57] I only list occurrences of the term for which I can safely state that the concepts the historical actors referred to stand in one historical line with the virtual particle concept. Nonetheless, the following invocations of the term were made in different strands of research.

[58] There are also deviations from this terminology which will already show up in this chapter but will be discussed more closely in the following chapter.

[59] E.g. Bethe and Bacher (1936, 204): "Each of the two neutrons emits 'virtually' (intermediate state [...]) an electron and a neutrino [...]."

[60] E.g. Gamow (1937, 148): "virtual creation of electrons".

[61] E.g. Fröhlich and Heitler (1938, 37): "virtual state".

[62] E.g. Peierls (1934, 439): "virtual transitions".

[63] E.g. Yukawa and Sakata (1937, 1088): "The quantum of negative or positive charge thus emitted virtually [...]".

[64] E.g. von Weizsäcker (1936, 573): "virtuelle Emission und Reabsorption".

[65] E.g. Wick (1938): "virtual transitions".

(1934) discussed the process in the case that the pairs were actually created and then, in a second step, the electron and positron recombined. The quantitative evaluation of the case that the energy of the two light quanta was not high enough to produce actual pairs (Halpern's initial proposal) was carried out in Leipzig by Hans Euler and Bernhard Kockel. The successive development of the results and the general approach can be followed through the Heisenberg-Pauli correspondence and the first communication by Euler and Kockel (1935).

As early as June 1934, Heisenberg informed Pauli that Euler and Kockel were working on light-by-light scattering and that they found qualitative agreement with a proposal by Peter Debye: namely, that the corona of the sun might be explained thereby.[66] On July 28., Heisenberg told Pauli that the calculations of light-by-light scattering were in progress and the results were still subject to change.[67] In the same letter, Heisenberg proposed that the Lagrangian for the Maxwell field will be altered by an additional term of the form

$$\frac{\hbar c}{e^2}\frac{1}{E_0^2}[\alpha(E^2-H^2)^2+\beta(EH)(E^2-H^2)+\gamma(EH)^2] \tag{9.1}$$

where E and H referred to the electric and the magnetic field strengths and α, β, γ were numbers, the values of which still had to be calculated.

Finally, on November 4. 1934 Heisenberg informed Pauli that "the scattering of light by light is now quite completely calculated".[68] The scattering can be expressed, according to Heisenberg's letter, as an addition to the Lagrangian density of the Maxwell theory in terms of:

$$\begin{aligned} L = {} & \frac{1}{8\pi}(E^2-H^2)+\frac{e^4\hbar}{m^4c^7}\left[\alpha_1(E^2-H^2)^2+\alpha_2(EH)(E^2-H^2)+\right. \\ & \left.+\alpha_3(EH)^2\right]+\frac{e^4\hbar}{m^4c^7}\left(\frac{h}{mc}\right)^2\beta_1(E^2-H^2)\left(\frac{\partial E}{\partial x}\frac{\partial H}{\partial x}+\ldots\right)+\ldots \end{aligned} \tag{9.2}$$

So far only the term $\alpha_1=\frac{1}{360\pi^2}$ had been calculated, while the other constants should follow soon. According to Heisenberg, the last line of the equation could be neglected due to the long wavelength considered in the calculation.

Pauli replied that due to invariance considerations the coefficient α_2 had to be equal to zero, as otherwise the full Lagrangian would not be invariant under spatial reflections.[69] The initial communication of the results by Euler and Kockel, handed

[66] Heisenberg to Pauli, June 16. 1934; as cited in Pauli (1985, 331–333).

[67] "Ich verpreche mir einiges von den immer noch nicht fertigen und im Resultat sich ändernden Rechnungen über die Streuung von Licht an Licht." Heisenberg to Pauli, July 28. 1934; as cited in Pauli (1985, 339–340).

[68] "Die Streuung von Licht an Licht ist jetzt ziemlich fertig gerechnet." Heisenberg to Pauli, November 4. 1934; as cited in Pauli (1985, 358–360).

[69] Pauli to Heisenberg, November 7. 1934; as cited in Pauli (1985, 360–363).

in in Febraury 1935, confirmed Pauli's reasoning (Euler and Kockel 1935). The term proportional to α_2 did not appear.

As they stated in the publication, Euler and Kockel followed a perturbative approach to the scattering problem. For small frequencies ($\sum_{i=1}^{2} h\nu_i < 2\,\mathrm{mc}^2$) of the interacting light quanta, there was still the possibility that they "virtually create a pair (positron and electron), which then immediately annihilates".[70] To calculate the cross section, "at first the matrix element of 4. order for this process was calculated according to the usual perturbative calculation of the Dirac theory"[71] which was then developed in terms of the energy of the light quanta $\frac{h\nu}{\mathrm{mc}^2}$. Through the subtraction physics of Heisenberg the 0th order of this development cancelled with a subtraction term, while the first, second, and third order terms vanished. The resulting fourth order term could be interpreted, as already laid out above, as an additional non-linear term of the Hamiltonian or the Lagrangian.

Analogies on the "Intuitive" Level of Theory

As already indicated above, there is, besides the "virtual", another term which Euler borrowed from Heisenberg, namely "anschaulich" which can roughly be translated as "intuitive".[72] Heisenberg used this term to differentiate between two approaches in his paper on subtraction physics (Heisenberg, 1934a). For the "anschauliche Theorie", following Dirac, Heisenberg defined a density matrix $(x't'k'|R|x''t''k'') = \sum_n \psi_n^*(x't'k')\psi_n(x''t''k'')$ from which

> all physically important properties of the quantum-mechanical system like charge density, current density, energy density, etc. can be read off. This however only holds in the approximation, in which the interaction of the electrons can be ignored, i.e. in which the typical quantum-theoretical unintuitive ["unanschaulichen"] characteristics of the events do not occur; the density matrix therefore conveys an intuitive, correspondence-like picture of the actual processes.[73]

The counterpart to Heisenberg's "anschauliche" theory in his 1934 paper was the theory of quantized matter and radiation fields. The "anschauliche" part corresponded to a non-second quantized theory in which the behaviour of the system was studied

---

70 "virtuell ein Paar (Positron und Elektron) erzeugen, das dann sofort wieder zerstrahlt." Euler and Kockel (1935, 246).

71 "[Dazu] wurde zunächst nach der gewöhnlichen Störungsrechnung der Diracschen Theorie das Matrixelement 4. Ordnung für diesen Prozeß berechnet." Euler and Kockel (1935, 246).

72 For a discussion of the development of the concept of "anschaulich" in Heisenberg's thought see Miller (1984, Chap. 4). For a detailed discussion of the concept of "Anschaulichkeit" in Heisenberg's uncertainty paper, see Beller (1999, Chaps. 4 and 5).

73 "[Aus der Dichtematrix] können alle physikalisch wichtigen Eigenschaften des quantenmechanischen Systems wie Ladungsdichte, Stromdichte, Energiedichte usw. abgelesen werden. Allerdings gilt dies immer nur in der Näherung, in der von der Wechselwirkung der Elektronen abgesehen werden kann, d.h. in der die typischen quantentheoretischen unanschaulichen Züge des Geschehens nicht vorkommen; die Dichtematrix vermittelt also ein anschauliches, korrespondenzmäßiges Bild des wirklichen Vorgangs." Heisenberg (1934a, 210).

through the equations of motion for the electromagnetic field, the Maxwell equations, and the single particle Dirac equation. The physically important characteristics of the system were constructed from the density matrix.

Miller (1984) discussed "Anschaulichkeit" mostly in connection with visualizability in Heisenberg's reasoning, yet noted that "in this method [referring to Heisenberg's 1934 paper] no mental images were involved" (Miller 1984, 167). In my own framing of Heisenberg's understanding of this "intuitive approach", a rather strong connection to correspondence-like, semi-classical methods and concepts was pronounced. Quite a similar connotation was evidently underlying the term in Heisenberg (1931b). There also seems to be a close connection to the representation in ordinary space. In a letter dating to June 14. 1934 from Pauli to Heisenberg, Pauli referred to "'anschauliche' (??!) ψ-waves in ordinary space".[74] A closer study of these connections and the concept of "anschaulich" in this case is surely interesting, but for our purpose the distinction Heisenberg himself set up, namely between a correspondence-like, semi-classical method (intuitive) and a fully quantum-electrodynamical description (unintuitive) shall suffice.

Hans Euler used the term "intuitive" in this vein when constructing, in explicit analogy to the Coulomb interaction between two electrons, additional terms to the Maxwell equations due to the interaction between two light quanta. The reasoning and the analogy ran as follows: Since two electrons "can create light quanta",[75] they can interact, e.g. scatter from each other, and to a certain approximation, this interaction can be represented by an "intuitive expression",[76] namely the Coulomb interaction. In Dirac's hole theory, the analogy went, photons can "virtually create an amount of pairs"[77] and through them an interaction takes place which should be brought into "a simple, intuitive term analogous to the Coulomb law".[78]

Euler's first step was the derivation of the general structure of an "analogous interaction of the light quanta"[79] which he would add to the Hamiltonian or the Lagrangian equations of the Maxwell field. Euler referred to this term as a "correspondence-like expression for the scattering of light by light".[80] The details of the derivation, based on the approximations of long wavelengths and invariance arguments as the one by Pauli quoted above, are not of particular interest for us. Yet, the result is of importance as it constituted a Lagrangian (see Eq.(9.2)), or Hamiltonian function, with

---

[74] Pauli to Heisenberg, June 14. 1934; as cited in Pauli (1985, 330). In a footnote, Pauli added that he considered classical light waves as "anschaulich" but not "quantum-ψ-waves" even when the latter were defined in ordinary space. Interestingly, Heisenberg would set "anschaulich" also in quotation marks when answering Pauli. Heisenberg to Pauli, June 16. 1934; as cited in Pauli (1985, 332).

[75] "können Lichtquanten erzeugen," Euler (1936, 400).

[76] "anschaulichen Ausdruck," Euler (1936, 400).

[77] "[...] erzeugen zwei Lichtquanten virtuell eine Menge von Paaren," Euler (1936, 400).

[78] "einem dem Coulombschen Gesetz analogen, einfachen, anschaulichen Ausdruck," Euler (1936, 400).

[79] "analoge Wechselwirkung der Lichtquanten," Euler (1936, 400).

[80] "korrespondenzmäßigen Ausdruck für die Streuung von Licht and Licht," Euler (1936, 436).

only two unknown constants (α_1 and α_3 in Eq. (9.2)). This entailed that only two specific cases of the scattering of light by light had to be calculated; it reduced the subsequent calculational efforts appreciably.

Before delving into the evaluation of the fourth order matrix element in Dirac's radiation and perturbation theory, Euler included a further analogy. This analogy was, again, most probably motivated by a paper of his supervisor (Heisenberg, 1935b): Therein, Heisenberg framed van-der-Waals forces as a virtual exchange interaction. Euler used the analogy to explain a possibly puzzling feature of the arrived at equations. He had started out from a formulation in which Maxwell's equations held, i.e. in which the conjugate momenta to the vector potential were the electric field strength E. Yet, the actual conjugate momenta he constructed D differed from E, i.e. there were additional terms to the Maxwell equations.

To remedy this apparent contradiction, Euler discussed the analogy between the scattering of light by light in the case in which no real pairs can be created and the van-der-Waals forces between two atoms. Euler filled three pages with two columns in which he put side by side the explanation of the two effects in a Lagrangian and Hamiltonian form (Euler 1936, 408–410). The essential feature he wanted to explicate thereby was that the reduction of the possibility space to only virtual and not real pair creation brought about the alteration in the Maxwell equations. There was no alteration in the basic equations of the field theory. Euler invoked the similar case of van-der-Walls forces, in which the mutual energy between two atoms was actual not large enough to induce real transitions but nevertheless the possibility of such transitions created a force which could be described by an additional term in the Lagrangian or Hamiltonian of the system, as an example which he considered to be better accessible (and better known) to most physicists.

The Perturbative Treatment: Mathematical In(tro)duction

Euler carried out the fully quantum-theoretical discussion in terms of Dirac's radiation theory, the corresponding time-dependent perturbation theory, and the evaluation of the arrived at matrix element for light-by-light scattering in collaboration with Kockel. Euler's starting point was the basic equation of time-dependent perturbation theory:

$$i\hbar\frac{\partial}{\partial t}a_{\mu'}(t) = \sum_{\mu} e^{\frac{1}{\hbar}(E_{\mu'}-E_{\mu})t}V_{\mu\mu'}a_{\mu}(t) \tag{9.3}$$

where a_μ referred to the probability coefficient of the state μ, E_μ to the corresponding energy, and $V_{\mu\mu'}$ to the matrix element of the perturbation, i.e. the interaction between matter and radiation.

Euler developed the probability coefficients as well as the perturbation:

$$a_\mu = a^0_\mu + a^1_\mu + a^2_\mu + a^3_\mu + a^4_\mu + ... \tag{9.4}$$

$$V = V^1 + V^2 + V^3 + V^4 + ... \tag{9.5}$$

To start out the calculation Euler assumed that the system was in a specific state i at the beginning ($a_i = 1;\ a_{j(\neq i)} = 0;$ for $t = 0$). After integration, Euler provided the following, usual first order equation:

$$a_k^1(t) = \left(\frac{e^{\frac{1}{\hbar}(E_k - E_i)t} - 1}{E_i - E_k}\right) V_{ik} \tag{9.6}$$

Euler noticed that the first factor of this equation only has an appreciable amount

> within the region of indeterminacy/uncertainty [Ungenauigkeitsbereich] $|E_k - E_i| \leq \frac{\hbar}{t}$. I.e. for small t $a^1(t)$ is always of importance, but for large t only if the system can go from the state i to the state k under conservation of energy $E_i = E_k$.[81]

The explicit reference to an "Ungenauigkeit" and the temporal dimension, and thereby indirectly to Heisenberg's uncertainty relation, was not regularly made at the time (see also Sect. 10.2.3). Interestingly, the invocation had no direct influence on the further development of the derivation. As always, Euler arrived at a transition probability which was proportional to the square of the matrix element and linear in time through the application of Dirac's trick (see Sect. 5.1.3).

So far, this was basic first order time-dependent perturbation theory. What Euler did next differed from the standard accounts: He constructed an inductive proof for the general structure of the matrix elements and the corresponding time-dependence. Therefore, he explicitly assumed that the arrived at order is the first in which the process could occur. For the details of the mathematical proof, I refer the reader to the respective passages in Euler's published paper (Euler 1936, 412–415). I simply want to point out a few of its features which directly relate to the main theme of this chapter.

First of all, Euler made no assumptions about the matrix element of the perturbation energy. The proof was true for any kind of perturbation and it was not limited to radiation theory. Euler applied next to no physical reasoning within this purely mathematical proof. Secondly, Euler's proof explicitly focused on the time-dependence of the probability coefficients. This focus allowed Euler to apply Dirac's trick to impose energy conservation, or put differently: It regulated which terms had to be considered at all. Thirdly, Euler assumed that he investigated the first order in which a transition could actually take place, so that no intermediate state had the same energy as the final or the initial state: by construction virtual states could not be reached through actual transitions.

According to Euler's proof, the explicit form of the matrix element to fourth order could be represented as:

$$H_{in}^4 = \frac{V_{ik}^1 V_{kl}^1 V_{lm}^1 V_{mn}^1}{(E_i - E_k)(E_i - E_l)(E_i - E_m)} + \ldots + V_{in}^4 \tag{9.7}$$

---

[81] "innerhalb des Ungenauigkeitsbereichs $|E_k - E_i| \leq \frac{\hbar}{t}$. D.h. für kleine t ist $a^1(t)$ immer von Bedeutung, für große t aber nur, falls das System unter Energieerhaltung $E_i = E_k$ vom Zustand i in den Zustand k übergehen kann." Euler (1936, 412).

where i referred to the initial and n to the final state of the system. The indexes on the mathematical representation of the perturbation V_{mn} described the successive transitions ($i \rightarrow k,\ k \rightarrow l,\ l \rightarrow m,\ m \rightarrow n$) which Euler described as follows:

> These partial processes to the "*virtual*" intermediate states do not demand energy conservation like the whole process ($i \rightarrow n$) to the *actual/real* [wirklichen] final state. Although therefore the state $k,\ l,\ m$ cannot really/actually be occupied by the system due to the energy law [...], their virtual possibility nevertheless causes the considered transition $i \rightarrow n$. The scattering of light by light is based on this circumstance.[82]

Up to this point, the discussion on the quantum-amplitude level was actually a nearly purely mathematical exercise. Euler did not in any way restrict the perturbations or the kind of transitions they could induce. The only thing he assumed was that the process could not occur in a lower order. Physically, only the energies and the time periodicity of the states were important, not the structure of the interaction energy. The physical implementation of this scheme as subsequently given by Euler, i.e. the derivation of the structure of the matrix element H^4_{in} and the order to which it had to be calculated, is a prime example of the application of Dirac's verbal model whilst providing for a new means of representation.

Filling the Frame: Tables and Diagrams

Besides the interaction of matter and radiation, Euler had to include one of the subtraction terms of Heisenberg, designated by V^4, in his physical evaluation. Since the matrix elements of the light-matter interaction coupled two states of the same momentum by an alteration of the number of photons in a particular state by ± 1 and as in total four photon emission/absorption processes were involved, the entire process had to be of fourth order. Euler consequently provided the general structure of the matrix elements as:

$$H^4_{in} = \sum_{klm} \frac{V^1_{ik} V^1_{kl} V^1_{lm} V^1_{mn}}{(E_i - E_k)(E_i - E_l)(E_i - E_m)} + V^4_{in} \tag{9.8}$$

As always for free particles, momentum conservation held for each subprocess and energy conservation for the total process.

In the next step, the possible intermediate states coupling initial and final state had to be found. Euler provided them, or rather the transition leading to them, in two different representations (see Fig. 9.1). How exactly Euler arrived at these transition possibilities was not stated. But according to his presentation and the then current

[82] "Diese Teilprozesse zu den '*virtuellen*' Zwischenzuständen verlangen keine Energieerhaltung wie der Gesamtprozeß ($i \rightarrow n$) zum *wirklichen* Endzustand. Obwohl also die Zustände $k,\ l,\ m$ wegen des Energiesatzes gar nicht wirklich vom System angenommen werden können [...], bewirkt doch ihre virtuelle Möglichkeit den betrachteten Übergang $i \rightarrow n$. Auf diesem Umstand beruht die Streuung von Licht an Licht." Euler (1936, 415–416, emphasis in the original).

Lichtquanten / Elektronen	Absorption von g^1	Absorption von g^2	Emission von $-g^3$	Emission von $-g^4$
$\mu = 1$	Paarerzeugung	Elektronensprung	Elektronensprung	Paarvernichtung
$\mu = 2$	Paarerzeugung	Elektronensprung	Positronensprung	Paarvernichtung
$\mu = 3$	Paarerzeugung	Positronensprung	Elektronensprung	Paarvernichtung
$\mu = 4$	Paarerzeugung	Positronensprung	Positronensprung	Paarvernichtung
$\mu = 5$	Paarerzeugung	Paarerzeugung	Zerstrahlung des Elektrons vom 1. mit dem Positron vom 2. Paar	Zerstrahlung des Positrons vom 1. mit dem Elektron vom 2. Paar
$\mu = 6$	Paarerzeugung	Paarerzeugung	Zerstrahlung des Positrons vom 1. mit dem Elektron vom 2. Paar	Zerstrahlung des Elektrons vom 1. mit dem Positron vom 2. Paar

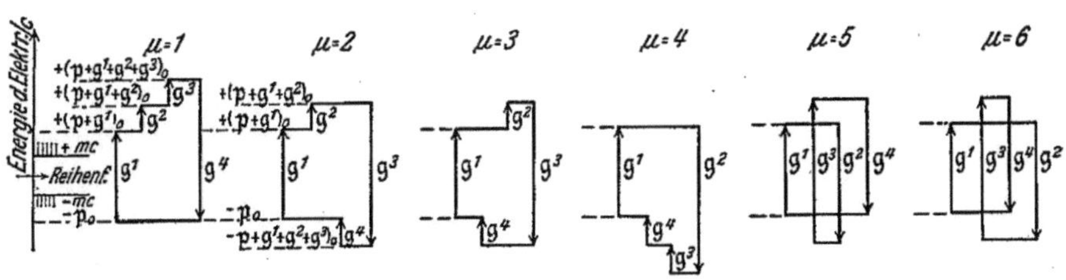

Fig. 9.1 Euler's graphical and tabularic description of the sequences of intermediate states. Figure and Table taken from Euler (1936, 419)

standards, physical reasoning in terms of successive transition processes by the material system, which were restricted by the Pauli principle, guided his construction.

Let's first take a look at the diagrams. On the outermost left side, there was a small energy scale indicating positive and negative energies. Also the way how to read the diagrams, namely from left to right, was indicated by a small arrow and the word "Reihenf[olge]" [order/sequence]. The arrows on the vertical lines did not indicate whether an absorption or an emission occurred but referred to the direction of the transitions of the electron. All of the diagrams had the structure of a closed loop because the energy of the *material part of the system* was conserved, i.e. there was in total no energy transfer from material system to the radiation field or vice versa.

To illustrate the reading of the diagrams, let's take the third one ($\mu = 3$) as an example (also compare the table).[83] First a pair was created (this was the case for every possible sequence) by lifting a negative energy electron into a positive energy state. Then a negative energy electron performed a transition to the unoccupied state in the sea created in the first process, which was conceived of as a positron jump.

[83] For the verbal description of all possible processes the reader is referred to Wüthrich (2010, 26).

In the third step, the electron of the initially created pair performed a state transition. Finally, it jumped to the vacant place in the Dirac sea: electron-positron annihilation occurred.

It should be noted that the assignment of photon momenta (g^i) in the diagrams did not coincide with the one in the table. In the diagrams, the photon momenta were ordered so that the structure of a closed circle is followed. g^{i+1} was always put where the electron subprocess involving g^i ended. Although the structure of the diagrams became somewhat more intuitive, it would involve processes that were forbidden due to the Pauli exclusion principle.[84] If we actually read the diagram $\mu = 3$ following the arrows and the enumeration of the photon momenta, the transition involving g^3 would be forbidden as the negative energy state would still be occupied by an electron.

Since this diagrammatical representation was not developed any further by Euler (or Kockel) it is hard to give reasons why the particular order was chosen.[85] On the one hand, it was almost definitely chosen on purpose, as it was consistently used throughout the diagrams. On the other hand, Euler was definitely aware of the problems due to the Pauli exclusion principle when following the processes in this way as the verbal statement in the table gave the correct order of events, i.e. read the diagrams from left to right. Furthermore, the horizontal arrow on the energy scale indicated the correct reading. Without further material I will refrain from trying to give an explanation but rather take it as an instance of the indeterminacy of the graphical representation and the methods applied, since the order of the photons, as was well-known, still had to be permuted for arriving at the total matrix element.

For the actual construction of all possible processes, Euler referred to the tabularic display. As any emission or absorption process of the photons might be involved in any of the state transitions of the electrons, the entries of the headline in the table still had to be permuted. Euler could then write the matrix element as:

$$H_{in}^4 = C \int dp \sum_{\text{Perm}} \sum_{\mu=1}^{6} \frac{Z_\mu}{N_\mu} + V_{in}^4 \tag{9.9}$$

where the integral was over the momentum involved in the first process; the first sum went over the permutations of light quanta; μ referred to the particular sequence of intermediate states; Z_μ to the corresponding product of first order matrix elements (which in itself also included a summation over the spin variables of the virtual electrons and positrons); N_μ referred to the product of the energy differences between initial and intermediate state; V_{in}^4 was the subtraction term of Heisenberg.

---

[84] As we will see in Sect. 9.1.4, Kockel argued that the exclusion principle had to be taken into account for the intermediate states in the case of light-by-light scattering.

[85] The diagrams were nevertheless taken up by researchers in Japan in the late 1940s (Wüthrich 2010, Sects. 2.2–2.4) and in another paper on light-by-light scattering by Akhiezer (1937) (see also Sect. 9.2.3).

After the construction of the matrix element, it came to real and painstaking calculational work for Euler (and Kockel). Having established the connection between the mathematical abstract formalism of time-dependent perturbation theory and the matrix elements actually to be taken into account, the evaluation of these terms, as usual, took up most of the space of the paper: In total, 144 terms had to be evaluated. Since my focus lies on the construction of the matrix elements, I will not discuss the involved calculation that will eventually render definite numerical values for the two unknown constants.

Representations of the "Virtual Matter" on the Intuitive Level and Doubts About the Perturbative Treatment

In the discussion of the result, Euler invoked the analogy between Coulomb interaction of two electrons and the interaction of two light quanta, as before, to motivate the incorporation of an additional term in the Lagrangian. This term "is an interaction between the light quanta, which indicates the virtual creation of matter and replaces the energy of the matter field, which surrounds the light quanta".[86] As Euler discussed, the final equations could be written in two forms. In the first formulation (Euler 1936, 444, Eq. (10,4)), there was no additional source or current density. The "virtual matter" (Euler 1936, 444) caused a difference between the electric field strength E and the electric displacement D. This way of writing the equations corresponded to what is today called macroscopic Maxwell equations in matter. But the additional source and current densities could also be written out explicitly in terms of their dependence on the electric and magnetic field (Euler 1936, 444, Eq. (10,5)).

Both of these mathematical representations, equivalent by construction, pointed to the fact that the vacuum was no longer empty: There was a "virtual matter coupled to the field".[87] In the first form, this was represented by the difference between E and D, so that the vacuum was actually considered as a material object. In the second form, this circumstance was explicitly written out as additional, only "apparent"[88] source and current densities. The second way of writing had the particular strength, that it clearly indicated that the Maxwell equations were actually not altered, but that there were additional sources and currents that depended on the field strengths. Yet, the first way of writing could be used to indicate the non-linearity of the equations, i.e. to show more clearly that there was a breach with the superposition principle.

Although the results of Euler's dissertation were in general rather positive (he could establish a way of calculating the cross section for light-by-light scattering and a way of incorporating the results in an intuitive form into the Lagrangian of the system), the closing paragraphs were rather pessimistic in tone. Euler discussed the reliance of the perturbative treatment. On the one hand, Euler argued that it might

[86] "[...] ist eine Wechselwirkung der Lichtquanten, welcher die virtuelle Materieerzeugung andeutet und die Energie des Materiefeldes ersetzt, welches die Lichtquanten umgibt." Euler (1936, 444).

[87] "die mit dem Feld gekoppelte virtuelle Materie," Euler (1936, 444).

[88] "scheinbare", Euler (1936, 444).

not be suitable to break off the expansion with the first non-vanishing term, as he did. If the field strengths were high enough, the terms calculated by Euler might as well give contributions of the same order of magnitude as the lower order terms, "i.e. the expansion in the coupling e of light and matter does not have to converge, if (for example for fast electrons or in the vicinity of an electron) that coupling becomes to strong by the creation of matter".[89]

On the other hand, higher order contributions to light-by-light scattering, and any other process, were known to diverge. Following a suggestion by Heisenberg, Euler framed this problem in terms of a mathematical expansion:

> One nearly gets the impression [*footnote in the original*: According to Prof. Heisenberg] that with the expansion in the electronic charge one carries out something similar to the calculation of a finite expression
>
> $$\int_0^1 \sqrt{1+\frac{e}{r}dr}$$
>
> in terms of an expansion
>
> $$\int_0^1 \sqrt{1+\frac{e}{r}dr} = \int_0^1 dr + e\int_0^1 \frac{dr}{2r} + e^2 \int_0^1 \frac{dr}{-8r^2} + \ldots$$
>
> the single terms of which diverge the more strongly the higher they are.
> The results of this work thus indicate, that one has first to try to replace the expansion in $e^2/\hbar c$ by something different to progress in the principle difficulties of the radiation theory.[90]

Even though the calculation eventually led to a reasonable and finite result, the problem of the infinities still laid heavy on an interpretation of the same. But it was not only the divergences that were used to question the appropriateness of the perturbative methods. Especially in the realm of high energies, it was uncertain if higher

[89] "d.h. die Entwicklung nach der Kopplung e von Licht und Materie braucht nicht zu konvergieren, wenn (etwa bei schnellen Elektronen oder in der Nähe eines Elektrons) diese Kopplung durch Materieerzeugung zu innig wird." Euler (1936, 447).

[90] "Man bekommt beinahe den Eindruck [*Fußnote im Original*: Nach Prof. Heisenberg], bei der Entwicklung nach der Elektronenladung etwas Ähnliches vorzunehmen, wie wenn man einen endlichen Ausdrcuk

$$\int_0^1 \sqrt{1+\frac{e}{r}dr}$$

durch eine Entwicklung berechnen würde,

$$\int_0^1 \sqrt{1+\frac{e}{r}dr} = \int_0^1 dr + e\int_0^1 \frac{dr}{2r} + e^2 \int_0^1 \frac{dr}{-8r^2} + \ldots$$

deren einzelne Glieder um so stärker divergieren, je höher sie sind.
Die Ergebnisse dieser Arbeit deuten also darauf hin, daß man, um in den prinzipiellen Schwierigkeiten der Strahlungstheorie weiter zu kommen, zunächst versuchen muß, die Entwicklung nach $e^2/\hbar c$ durch etwas anderes zu ersetzen." Euler (1936, 448).

approximations became as or even more influential than lower order approximations. This question would motivate Bernhard Kockel's dissertation, which we shall turn to now.

9.1.4 *Kockel's Dissertation:* **Schüttelwirkung,** *Pauli's Principle, and a False Prediction of the "Picture of 'Successive' Transitions"*

Bernhard Kockel had largely contributed to the evaluation of light-by-light scattering. Euler mentioned that the actual calculation in terms of Dirac's radiation theory had been carried out with the help of Bernhard Kockel. Paragraph five of Euler's publication, involving the construction of the particular matrix element to be calculated, was mostly due to Kockel (Euler 1936, 398, Footnote 1). In his own doctoral research Bernhard Kockel turned to the evaluation of third order processes in Dirac's radiation theory.

At the time Kockel was working on his dissertation, cosmic ray phenomena had gathered a lot of attention in the theoretical research. Although it was generally accepted that the formulas of quantum electrodynamics describing scattering effects should not break down in the high-energy regime,[91] the influence of higher order processes was uncertain in the corresponding energy realm.[92] As Kockel explicitly stated himself: "The investigation of third order processes is of interest for the question, if higher order processes can play a role in the cosmic radiation."[93]

The main result of Kockel's dissertation was the comparison between the cross-sections of two processes: the pair creation by two photons Φ^{II} and the same process with an additional emission of a photon Φ^{III}.[94] The ratio of the two cross sections increased with the square of the logarithm of the momentum of the involved light quanta and, according to Kockel, around the range of 10^8 eV the contribution of the third order process might be considerable. To arrive at this result, Kockel developed a general evaluation of Dirac's radiation theory.

[91] See Cassidy (1981) for a historical discussion of the validity of the theoretical predictions for cosmic rays.

[92] This is what Euler's comments in the conclusions of his paper (see above) gesture at.

[93] "Die Untersuchung von Prozessen dritter Ordnung ist von Interesse für die Frage, ob in der Höhenstrahlung Prozesse höherer Ordnung eine Rolle spielen können." Kockel (1937, 154).

[94] In the calculation of this process Kockel also hit upon the infrared divergence in early 1936. Heisenberg to Pauli, February 18. 1936; as cited in Pauli (1985, 439). Thus far this was the only reason why Kockel's dissertation was taken up in the secondary literature, as in Blum (2015, 74).

A General Mathematical Evaluation of Time-Dependent Perturbation Theory and the "Picture of 'Successive' Transitions"

Euler had inductively derived the general structure of matrix elements for the first order in which a process could occur. Kockel (1937, 157–164) also initiated his calculation by a general derivation of the structure of matrix elements in Dirac's perturbation theory. Kockel's approach was as general as possible: He discussed degenerate initial states, perturbations which could directly couple initial and final state to an arbitrary order, and it was not restricted to the first order in which the process was to occur.

Although Kockel's approach was thus certainly the most detailed evaluation of Dirac's time-dependent perturbation theory upto that moment, he would eventually apply different approximations and simplifying assumptions. Before taking the next to final step, Kockel had developed the probability for a transition from initial state i to final state s to νth order as:

$$W^{(\nu)} = \frac{2\pi t}{\hbar \Delta E_s} \left| \int \frac{dE_k}{\Delta E_k} \int \frac{dE_l}{\Delta E_l} \cdots \int \frac{dE_r}{\Delta E_r} \frac{V_{sr}^{(1)} V_{rq}^{(1)} \ldots V_{lk}^{(1)} V_{ki}^{(1)}}{(E_i - E_k) \ldots (E_i - E_r)} + \text{terms involving } V^{(2)} \text{to } V^{(\nu)} \right|^2 \quad (E_i = E_s) \tag{9.10}$$

Division by t (probability per unit time) and restriction to the third order, the one Kockel was investigating, led him to:

$$W^{(3)}_{\text{per sec}} = \frac{2\pi}{h \Delta E_s} \left| \sum_k \sum_l \frac{V_{ik}^{(1)} V_{kl}^{(1)} V_{ls}^{(1)}}{(E_i - E_k)(E_i - E_l)} + \sum_k \frac{V_{ik}^{(1)} V_{ks}^{(2)} + V_{ik}^{(2)} V_{ks}^{(1)}}{E_i - E_k} + V_{is}^{(3)} \right|^2 \tag{9.11}$$

As may have been noticed, Kockel altered the order of the matrix elements as well as the order of the indexes in the step from Eqs. (9.10) and (9.11). The initial state i was on the left, followed by two intermediate states, while the final state s was put to the right. Since, the matrix elements were assumed to be Hermitian this only constituted a formal change but Kockel gave a particular motivation for it:

> This way of writing has the advantage, that now one can assign to the order i, k, l, s of the indexes the picture of "successive" transition from the initial state i over the intermediate state k to a second intermediate state l and finally to the final state s.[95]

[95] "Diese Schreibweise hat den Vorteil, daß man jetzt der Reihenfolge i, k, l, s der Indizes das Bild 'aufeinander folgender' Übergänge vom Anfangszustand i über den Zwischenzustand k zu einem zweiten Zwischenzustand l und schließlich zum Endzustand s zuordnen kann." Kockel (1937, 162–163).

Although, obviously, Kockel did not consider this "picture of 'successive' transitions" to correspond to physical reality, he nevertheless chose to order the terms in the specific way, so that the picture can unfold its heuristic utility.

As in Euler's work, the derivation of the general structure of the matrix elements involved was nearly exclusively mathematical. There were next to no assumptions about the explicit nature of the states of the systems or the perturbation energy. But before applying this scheme to the calculation in radiation theory and thereby filling the matrix elements with physical content, Kockel provided a discussion of the *Schüttelwirkung* and Dirac's trick.

A First Study of the *Schüttelwirkung*

The application of Dirac's trick (Sect. 5.1.3) involved an integral over the energy range which had to be extended over all final state energies E_s. To give an explanation of this circumstance, Kockel studied the first order approximation and the influence of the different E_s-regions on the final result.[96] Kockel (1937, 163–164) identified two disturbances in the calculational scheme.

First of all, after the switching on of the perturbation, the u_i, the eigenfunctions of the unperturbed system, were no longer eigenfunctions of the whole system. The new solution was rather a superposition of different $a_i u_i$. Therefore, $|a_s(t)|^2$ included terms which did not correspond to transition probabilities but only occurred due to the expansion in eigensolutions of the unperturbed system. Secondly, it was assumed that the interaction was turned on "infinitely suddenly"[97] which constituted a heavy shock to the system: the "Schüttelwirkung des Einschaltens" ["Shaking action of the turning on"] (Kockel 1937, 163).

To take care of these two factors Kockel proposed to study the action of the perturbation a bit differently. On the one hand, the perturbation should be acting only for a limited amount of time $-t_0 < t < t_0$. Consequently, after the time t_0 the u_i were eigenfunctions of the system again. Thereby, the first disturbance in the calculation was remedied. On the other hand, following a suggestion by Heisenberg, Kockel introduced a function which turned the perturbation slowly on and off.[98] With the specific function chosen by Kockel he arrived at the probability for a transition to be given by:

[96] The restriction to a first order process is only applied for simplicity's sake. Since the question only concerns the relation between the energies of initial and final states Kockel's evaluation equally holds for any higher order process.

[97] "unendlich plötzlich," Kockel (1937, 163).

[98] The function chosen by Kockel was: $\Psi = \Phi(t + t_0) - \Phi(t - t_0)$ where $\Phi(t) = \frac{\alpha}{\sqrt{\pi}} \int_{-\infty}^{t} e^{-\alpha^2\tau^2} d\tau$.

$$W^{(1)} = \int \frac{dE_k}{\Delta E_k} |V_{ki}^{(1)}|^2 \frac{4e^{-\frac{1}{2\alpha^2\hbar^2}(E_i - E_k)^2} \sin^2 \frac{t_0}{\hbar}(E_i - E_k)}{(E_i - E_k)^2} \tag{9.12}$$

where $\frac{1}{\alpha}$ is the time the perturbation took to be switched on or off.

Due to the first exponential function in the integral, only for $|E_i - E_k| < \hbar\alpha$ considerable contributions arose from the integral. By decreasing α, i.e. turning on the perturbation more carefully, this energy region became smaller and smaller. In effect, when turning on the perturbation adiabatically the contribution of the energy non-conserving terms vanished. Kockel provided the following probability of transitions in relation to the energy difference between the two states and the time of the switching on process:

$$|a_k^{(1)}(t)| \propto 0 \text{ for } |E_i - E_k| > \hbar\alpha \tag{9.13}$$

$$\propto |V_{ki}^{(1)}|^2 \frac{2}{(E_i - E_k)^2} \text{ for } |E_i - E_k| < \hbar\alpha \tag{9.14}$$

$$\propto |V_{ki}^{(1)}|^2 \frac{4t_0^2}{\hbar^2} \text{ for } |E_i - E_k| < \frac{\hbar}{t_0} \tag{9.15}$$

The second equation described terms due to the *Schüttelwirkung*. The possible energy differences depended on the way how the perturbation was switched on through the dependence on α. The third line, in contrast, was independent of α. According to Kockel, these were "actual transitions, caused by the perturbation and obeying the energy law"[99].

Kockel did not further argue why these transitions actually are in harmony with energy conservation although the difference of the energies of initial and final state is finite. To make this better accessible to the reader I propose a heuristic argument based on the energy-time uncertainty.[100] If we connect to t_0, which is the time the perturbation is acting,[101] an energy indeterminacy of ΔE_0 we get that $t_0 \Delta E_0 \approx \hbar$. This directly renders $|E_i - E_k| < \frac{\hbar}{t_0} \approx \Delta E_0$. The energy difference between initial and final state is smaller than the uncertainty in the energy connected to the finite time the perturbation was actually acting.[102] Still, one should note that this derivation was a discussion of actual transitions and not of virtual ones, although the states might figure as virtual states in multi-step processes.

Kockel's discussion of the *Schüttelwirkung* is the first one I could find in which the problem was evaluated in quantitative terms. I discussed it here in some detail, because it does not only give credence to the arguments brought forward

[99] "wirklichen, durch die Störung hervorgerufenen, dem Energiesatz gehorchenden Übergängen." Kockel (1937, 164).

[100] The following argument can also be used to see why Eq. (9.13) has to be zero.

[101] Actually, it is $2t_0$, but this is unimportant for the qualitative argument given.

[102] See Landau and Peierls (1931, 59) for a similar argument, if not spelled out in this detail. See also my discussion of the narrative for virtual transitions in Sect. 10.2.3.

by Göppert-Mayer and Weisskopf (see Sect. 8.1), but Kockel's investigation of the *Schüttelwirkung* will motivate a different approach towards the study of quantum-electrodynamical processes. It motivated the work of the Polish physicists and later doctoral student of Heisenberg, Edwin Gorà (see Sect. 9.3.1), and thereby initialized the work on the theory of radiation damping.

Physical Manifestation: Intermediate States with and Without the Pauli Principle and Another False Prediction

Up until this point, Kockel's study did not involve the explicit nature of the physical system studied or the perturbation involved. For the calculation of radiation processes in hole theory Kockel introduced, in standard fashion, the interaction between matter and radiation field (each term referred to an emission or absorption; momentum is conserved in each transition) and noted that Eq. (9.11) can be further reduced. The terms $V^{(2)}$ or $V^{(3)}$ did not have to be considered. Kockel consequently arrived at:

$$W^{(3)}_{\text{per sec}} = \frac{2\pi}{\hbar \Delta E_m} \left| \sum_{k,l} \frac{V^{(1)}_{ik} V^{(1)}_{kl} V^{(1)}_{lm}}{(E_i - E_k)(E_i - E_l)} \right|^2 \tag{9.16}$$

To represent the possible combinations of subprocesses and intermediate states contributing to the total process, Kockel did not provide a diagrammatical representation, like Euler. He used a tabuaric one (see Figs. 9.2, 9.3, and 9.4). Because the order of the light quanta involved in each subprocess had to be permuted, only the material part of the system restricted the way in which the processes might be envisioned. In Fig. 9.2 you find the different sequences of events which lead to the creation of a pair under the absorption of two and the emission of one photon.

The table was read from left to right. The Roman numerals each referred to one possible sequence, the $p_i^{\pm}$ to the momenta of negative (−) or positive (+) energy

	1. Übergang	2. Übergang	3. Übergang
I	Paarerzeugung ($\mathfrak{p}_0^- \to \mathfrak{p}_1^+$)	Elektronensprung ($\mathfrak{p}_1^+ \to \mathfrak{p}_2^+$)	Elektronensprung ($\mathfrak{p}_2^+ \to \mathfrak{p}^+$)
II	„ ($\mathfrak{p}_1^- \to \mathfrak{p}_2^+$)	„ ($\mathfrak{p}_2^+ \to \mathfrak{p}^+$)	Positronensprung ($\mathfrak{p}_0^- \to \mathfrak{p}_1^-$)
III	„ ($\mathfrak{p}_1^- \to \mathfrak{p}_2^+$)	Positronensprung ($\mathfrak{p}_0^- \to \mathfrak{p}_1^-$)	Elektronensprung ($\mathfrak{p}_2^+ \to \mathfrak{p}^+$)
IV	„ ($\mathfrak{p}_2^- \to \mathfrak{p}^+$)	„ ($\mathfrak{p}_1^- \to \mathfrak{p}_2^-$)	Positronensprung ($\mathfrak{p}_0^- \to \mathfrak{p}_1^-$)
V	„ ($\mathfrak{p}_0^- \to \mathfrak{p}_1^+$)	Paarerzeugung ($\mathfrak{p}_2^- \to \mathfrak{p}^+$)	Paarvernichtung ($\mathfrak{p}_1^+ \to \mathfrak{p}_2^-$)
VI	„ ($\mathfrak{p}_2^- \to \mathfrak{p}^+$)	„ ($\mathfrak{p}_0^- \to \mathfrak{p}_1^+$)	„ ($\mathfrak{p}_1^+ \to \mathfrak{p}_2^-$)

Fig. 9.2 The successive transitions involved in the creation of an electron-positron pair and a photon from two photons. Table in Kockel, Bernhard. 1937. "Über einige Mehrfachprozesse zwischen Elektronen, Positronen und Lichtquanten." *Zeitschrift für Physik* 107, 153–176, here 167. https://doi.org/10.1007/BF01330357 (cited as Kockel (1937) in the bibliography). Reproduced with permission from Springer Nature

	1. Übergang	2. Übergang	3. Übergang
A	$\mathfrak{p}_0^- \to \mathfrak{p}_1^+$	$\mathfrak{p}_1^+ \to \mathfrak{p}_2^+$	$\mathfrak{p}_2^+ \to \mathfrak{p}^+$
B	$\mathfrak{p}_0^- \to \mathfrak{p}_1^+$	$\mathfrak{p}_1^+ \to \mathfrak{p}_2^-$	$\mathfrak{p}_2^- \to \mathfrak{p}^+$
C	$\mathfrak{p}_0^- \to \mathfrak{p}_1^-$	$\mathfrak{p}_1^- \to \mathfrak{p}_2^+$	$\mathfrak{p}_2^+ \to \mathfrak{p}^+$
D	$\mathfrak{p}_0^- \to \mathfrak{p}_1^-$	$\mathfrak{p}_1^- \to \mathfrak{p}_2^-$	$\mathfrak{p}_2^- \to \mathfrak{p}^+$

Fig. 9.3 The successive transitions involved in the creation of an electron-positron pair and a photon from two photons without taking account of the exclusion principle for intermediate states. Kockel, Bernhard. 1937. "Über einige Mehrfachprozesse zwischen Elektronen, Positronen und Lichtquanten." *Zeitschrift für Physik* 107, 153–176, here 168. https://doi.org/10.1007/BF01330357 (cited as Kockel (1937) in the bibliography). Reproduced with permission from Springer Nature

electrons. The momentum of the final state electron was indicated by p and the momentum of the final state positron by p_0. To exemplify the reading let's look at the third process (III): First, a pair was created ("Paarerzeugung") through an electronic transition from the state p_1^- to the state p_2^+: We have an electron in the state p_2^+ and a hole, i.e. a positron, with momentum p_1^-. In the second step, an electron of negative energy jumped from the state p_0^- to the state p_1^-, which was interpreted as a positron jump ("Positronensprung"): The hole in the sea or the positron now had the correct momentum p_0^-. Finally, the electron of the initial pair creation p_2^+ made a transition to its final state p^+ ("Elektronensprung"). In total, a pair was created and the positron (or the electron, depending on the process) performed an additional transition. By construction, two absorption and one emission processes were involved but not displayed in the table.

The transitions represented in Fig. 9.2 took the Pauli exclusion principle for the intermediate states explicitly into account. Yet, Kockel (1937, 167–169) also provided a scheme for the same process, shown in Fig. 9.3, in which he did not apply the exclusion principle to the intermediate states. Again, Kockel started out with an electron in a negative energy state. But disregarding the exclusion principle he could consider all other negative energy states to be empty (although, of course, they were not) or, equivalently, to be capable of being occupied by more than one electron. Kockel then simply followed the single initial electron around. The second transition in process *B* or the first ones of processes *C* and *D*, for example, would be forbidden in a filled Dirac sea if the exclusion principle held for intermediate states.

Kockel established that the processes represented in Fig. 9.2 and in Fig. 9.3 led to the same matrix elements, hence they were equivalent and both suitable for describing the process studied. By direct evaluation of processes upto the fifth order, Kockel could establish the following rule about the Pauli exclusion principle for intermediate states:

> If a process can be realized under the neglect of the Pauli exclusion principle for the intermediate states in such a way that a *single* electron arrives through successive jumps from

	1. Übergang	2. Übergang	3. Übergang	Π_χ
I.$_1$	Paarerzeugung ($\mathfrak{p}_1 \to \mathfrak{p}_1 + \mathfrak{g}_1$)	Paarerzeugung ($\mathfrak{p}_2 \to \mathfrak{p}_2 + \mathfrak{g}_2$)	Elektronensprung ($\mathfrak{p}_2 + \mathfrak{g}_2 \to \mathfrak{p}_2 + \mathfrak{g}_2 + \mathfrak{g}_3$)	1
I.$_2$	Paarerzeugung ($\mathfrak{p}_2 \to \mathfrak{p}_2 + \mathfrak{g}_2$)	Paarerzeugung ($\mathfrak{p}_1 \to \mathfrak{p}_1 + \mathfrak{g}_1$)	Elektronensprung ($\mathfrak{p}_2 + \mathfrak{g}_2 \to \mathfrak{p}_2 + \mathfrak{g}_2 + \mathfrak{g}_3$)	1
I.$_3$	Paarerzeugung ($\mathfrak{p}_2 \to \mathfrak{p}_2 + \mathfrak{g}_2$)	Elektronensprung ($\mathfrak{p}_2 + \mathfrak{g}_2 \to \mathfrak{p}_2 + \mathfrak{g}_2 + \mathfrak{g}_3$)	Paarerzeugung ($\mathfrak{p}_1 \to \mathfrak{p}_1 + \mathfrak{g}_1$)	1
II.$_1$	Paarerzeugung ($\mathfrak{p}_1 \to \mathfrak{p}_1 + \mathfrak{g}_1$)	Paarerzeugung ($\mathfrak{p}_2 + \mathfrak{g}_2 \to \mathfrak{p}_2 + \mathfrak{g}_2 + \mathfrak{g}_3$)	Positronensprung ($\mathfrak{p}_2 \to \mathfrak{p}_2 + \mathfrak{g}_2$)	-1
II.$_2$	Paarerzeugung ($\mathfrak{p}_2 + \mathfrak{g}_2 \to \mathfrak{p}_2 + \mathfrak{g}_3 + \mathfrak{g}_3$)	Paarerzeugung ($\mathfrak{p}_1 \to \mathfrak{p}_1 + \mathfrak{g}_1$)	Positronensprung ($\mathfrak{p}_2 \to \mathfrak{p}_2 + \mathfrak{g}_2$)	-1
II.$_3$	Paarerzeugung ($\mathfrak{p}_2 + \mathfrak{g}_2 \to \mathfrak{p}_2 + \mathfrak{g}_2 + \mathfrak{g}_3$)	Positronensprung ($\mathfrak{p}_2 \to \mathfrak{p}_2 + \mathfrak{g}_2$)	Paarerzeugung ($\mathfrak{p}_1 \to \mathfrak{p}_1 + \mathfrak{g}_1$)	-1

Fig. 9.4 The possible transitions leading to the creation of two pairs from three light quanta. Kockel, Bernhard. 1937. "Über einige Mehrfachprozesse zwischen Elektronen, Positronen und Lichtquanten." *Zeitschrift für Physik* 107, 153–176, here 176. https://doi.org/10.1007/BF01330357 (cited as Kockel (1937) in the bibliography). Reproduced with permission from Springer Nature

> an initial state at a *different* final state (like in table [9.3]), then both ways for finding the transition possibilities are equivalent.[103]

As Kockel remarked, this condition is fulfilled for Compton scattering or pair creation but not for light-by-light scattering (same initial and final state) or the process of the creation of two pairs from three light quanta (more than one electron is involved).

As was discussed in Chap. 7, the role of the exclusion principle for the intermediate states was crucial in the discussion of scattering in an empty and filled Dirac sea and eventually motivated the inception of the corresponding concept of anti-matter. Dirac could show that also in a filled Dirac sea one could still reconstruct the results with empty negative energy states. Igor Tamm consequently generalized this conclusion by showing that the probability of scattering from two electrons is independent of the occupation of an intermediate state as long as one of the initial states of an electron coincided with the final state of the other (see Sect. 7.7.2). Yet, at the time Kockel discussed the third order processes, Weisskopf had already shown that the occupation of the possible intermediate states could have decisive influence on the result: The self-energy in the Dirac sea was only logarithmic and not quadratic (see also Sect. 10.1.1). Kockel's conclusion was in harmony with both observations. Quite contrary, and as will be seen in Sect. 11.2.3, in Feynman's approach to QED the exclusion principle could be neglected in all kinds of processes for the intermediate states.

---

[103] "Wenn man einen Prozeß unter Verzicht auf die Paulische Besetzungsregel für die Zwischenzustände dadurch verwirklichen kann, daß ein *einziges* Elektron durch aufeinanderfolgende Sprünge von einem Anfangszustand zu einem von ihm *verschiedenen* Endzustand gelangt (wie in Tabelle [9.3]), so sind beide Wege zum Auffinden aller Übergangsmöglichkeiten gleichberechtigt." Kockel (1937, 169); emphasis in the original.

As for Euler's work on light-by-light scattering, the subprocesses established through the physical reasoning in terms of Dirac's verbal model, now represented in a tabularic display, formed the basis for the quantitative evaluation. And as in Euler's case, I will not discuss this evaluation in detail, as I believe the most important function of the concept of intermediate states and transitions to and from them had already been performed, i.e. the translation from the general perturbative expansion to the concrete physical process under study.

Yet, I want to point the reader also to the second process which Kockel investigated in his dissertation: the creation of two pairs from three light quanta. The transition possibilities are given in Fig. 9.4. Kockel's evaluation is rather short. Within one page Kockel finds that the two sets of transitions I_{1-3} and II_{1-3} (see outermost left of Fig. 9.4) each combine to give zero. Hence, although in principle allowed by the conservation laws applied in the theoretical treatment,[104] the process did not occur. It was a process which could be envisioned by the theoretical restrictions known for the application of Dirac's verbal model. But it turned out to have a vanishing cross section. It was conceivable but non-existent. It was a typical example of a false prediction. Retrospectively, we can rule such a process out by simply attempting to draw a corresponding connected Feynman diagram: Such a diagram does not exist.

9.1.5 Intermediate Conclusion

In these two dissertations a clean separation was made between a rather mathematical evaluation of perturbation theory and the filling of this frame with physical content according to the actual form of the perturbation and the effect studied. The first kind of elaborations, the ones on the mathematical structure of perturbation theory, were necessary since Heisenberg had derived higher order subtraction terms that could have been included in the evaluation. The second kind of elaboration, the concretization in terms of physical processes, was guided by Dirac's verbal model, i.e. by physical reasoning about the different virtual transition possibilities. The structure of the matrix elements (combining a transition of the material part of the system with an emission or absorption process) together with the lifting of energy conservation for the virtual subprocesses guided this implementation.

As a matter of fact, the first public announcement of light-by-light scattering rested on Dirac's verbal model or, as Halpern had framed it, "the language of Dirac's theory of radiation". But in the evaluation both Euler and Kockel considered it suitable to provide further representations of this "picture of 'successive' transitions": Kockel used tables and Euler a combination of a table and diagrams. But judging from the application of these representational formats in the papers, they seem to be constructed after the fact. The main principle guiding the reasoning processes was

[104] As Kockel stated: "The creation of two electron-positron pairs from three light quanta is not forbidden by the energy or the momentum law." [*Original*: "Durch Energie- und Impulssatz nicht verboten ist eine Entstehung von zwei Elektron-positronpaaren aus drei Lichtquanten."] Kockel (1937, 175).

the verbal model of Dirac, while the tables and diagrams allowed for a condensed representation of all possible processes.

It is noteworthy that a diagrammatic display of the processes was seemingly as appealing as a tabularic one at the time. To name another example, in the report of a seminar held at the Institute for Advanced Study (Princeton) by Pauli in the Winter semester 1935/36, Maurice Pryce discussed light-by-light scattering (Pauli 1936b, 41–47). Pryce also used tables, not diagrams, to present the different transition possibilities (Pauli 1936b, 44). Although Pryce only referenced the communication of the results by Euler and Kockel (1935) it is very likely that he had further information on the calculations through Pauli. Pryce already mentioned the analogy to van-der-Waals forces, which is not mentioned in Euler and Kockel's communication. Moreover, Pryce used exactly the same ordering of the subprocesses (in all details) as they were given in the table by Euler (Fig. 9.1). All the same, diagrams were not represented.

The fact that tables were used and seemingly seen to be as suitable as diagrams might be due to the conceptualization of QED of the 1930s, i.e. as a theory of transitions between unperturbed states, or the fact that the photons in the diagrams were only partially indicated: They only showed up as connected to a transition but did not have another form of representation. Adrian Wüthrich (2010, Chap. 2) identified this last point as a specific shortcoming of the representational technique initiated in Euler's paper. Through the study of its application in the late 1940s in Japan, Wüthrich noted that for some processes it is of importance to discern which light quantum is involved in which process; a task this mode of representation is not capable of.

Besides this shortcoming of the representational format, Wüthrich has already correctly pointed out that the diagrams and tables as given in Fig. 9.1 take the same place in the reasoning processes as Feynman diagrams today.[105] In response to Wüthrich's consideration and with backdrop to my own reconstruction, I have three points that will enrich the understanding of these diagrams and the concepts involved.

The first point concerns the place were the diagrams, tables or the verbal model enter the theory: Euler used diagrams (or rather the underlying physical model) to bridge the gap between mathematical evaluation and physical consideration. The diagrams were a format in which the different orders of events that led from initial to final state of the process could be represented in a condensed form. To some this might sound unsurprising but I think it is worth noting. It shows that diagrams (and other forms of representation of the verbal model), although still rooted in a conception of transitions between stationary states, were used in order to physically reason about the behaviour of the system.

Secondly, whether the virtual transitions were conceived of as corresponding to some kind of physical reality[106] or whether they simply were used as a suitable

---

[105] Wüthrich was motivated by David Kaiser's misidentification of tables displaying the evaluation of specific matrix elements as corresponding to Feynman diagrams today (Kaiser 2005, 34–36).

[106] At least, this is the gist of the terminology "virtual possibilities" in Euler's paper, if we accept them to have a historical connection to Heisenberg's later concept of potentia.

but unrealistic model,[107] they were used. The application of the physical reasoning was independent of any ontological connotation.[108] Bernhard Kockel, for whom the successive transitions seemed not to correspond to any kind of physical reality, even investigated the physical restrictions of the intermediate states, i.e. for which processes the Pauli exclusion principle had to be taken into account.

My last argument is concerned with the usage of the term virtual. In connection to Euler's description of light-by-light scattering, Wüthrich explained that "the *hypothetical possibility* of the creation of the pairs does affect the cross section of the scattering, even if the possibility does not actualize" (Wüthrich 2010, 25, emphasis added). This has overlaps with my reading of the term in the reconstruction given in Sect. 9.1.2, but is not equivalent to it. And even though Wüthrich rightly pointed to the difference in conception of processes in QED during the 1930s (state transitions) and the one implied by Feynman diagrams (interrupted movement of freely travelling particles), Wüthrich also referred to the particles as virtual in connection to Euler's description. This does not cohere with the publication of Euler itself. Although the particles were only present in the intermediate states, when we focus on the actual practices, the transitions were virtual in the sense that they broke with energy conservation. Euler referred to "virtual matter" only once but did so on the intuitive level of the theory, where the results of the previous discussion were represented in terms of additional source and current densities.

Euler mostly spoke of "virtual possibilities (of processes)". According to my reconstruction, the concept of "hypothetical" might be retrospectively adequate, but the reason for the inclusion was another. The terminology of "virtual possibilities" in Euler's dissertation points, through Heisenberg, to a key aspect of the BKS invocation of the term, namely as referring to an unobserved world which was separated from the actual world of observed entities, while the virtual entities were still effective. Whether or not it is possible to create a consistent historical line from this connotation of the term "virtual" to Heisenberg's invocation of the Aristotelian concept of "potentia" remains to be investigated. Regardless of this historical question, the possibility is in this reading not imagined, but has some degree of reality, although a different kind than the world we observe. The connotation of the "virtually possible" is not one of hypothetical possibilities but rather one of effective possibilities.

Before turning towards the investigation of the Delbrück effect, let me finally point out that the differentiation between the practice of the 1930s and late 1940s which both Wüthrich and I emphasized is not only one of complication of the respective calculation. According to the model of the light-matter interaction, whether framed verbally, diagrammatically, or tabularic, physicists could imagine processes which

---

[107] This corresponds to my reading of Kockel's invocation of the "picture of 'successive' transitions".

[108] We already encountered such a phenomenon in the discussions around the Dirac sea (Chap. 7). Ivar Waller's later opinion and Rosenfeld explicitly denied the intermediate states any kind of reality. Yet, they still used them and the corresponding verbal model. Dirac and Tamm, who seemed to have a more realistic or at least physical reading of the intermediate states, of course used them as well.

actually had a vanishing cross section. Halpern's idea of light-quantum splitting and Kockel's discussion of the creation of two pairs from three light quanta (to third order) are cases in point. They were found to be non-existent effects when transformed into a mathematical language. The relation to a modern formulation in terms of Feynman diagrams will be discussed in Sect. 12.2.

9.2 The "Delbrück-Nordsieck-Weisskopf-Kemmer-Ludwig-[Akhiezer-Pomeranchuck] Effect" and Furry's Theorem (1934–1937)

The work on light-by-light scattering had decisive influence on the evaluation of another scattering mechanism which was proposed during the initial reception of the Dirac sea: the Delbrück effect.[109] This effect, expressed in most general terms, refers to the scattering of light from an electrostatic potential. Therefore, from a modern point of view, the connection to light-by-light scattering is quite obvious. The Delbrück effect is described by the same Feynman diagram while one of the incoming and one of the outgoing real photons are replaced by virtual ones. But from a historical perspective this connection was not clear, at least at first.

I will discuss the different attempts at calculating the Delbrück effect in some detail for three reasons. First of all, to the best of my knowledge no detailed study of the historical development exists and therefore this study has a right of its own. Secondly, in the evaluation of Delbrück scattering we will encounter different conceptualizations which show that the concept of a virtual photon, as modern physics would use it to express the process, was not yet completely developed. Thirdly, a reoccurring false prediction in the treatments of the Delbrück effect (partially) motivated Furry's theorem: a theorem exclusively concerned with the perturbative evaluation of QED.

To show how the treatment of the effect developed and how the conceptualization and the corresponding evaluation could be cast in different forms, I will first follow Max Delbrück's initial proposal and his failed attempt at a quantitative evaluation (Sect. 9.2.1). Then I will go over to the calculations carried out by the Zürich physicists around Wolfgang Pauli. Their starting point was the initial conception of Delbrück. The work of the Zürich group still proved to be relevant for the trustworthiness of the calculation of light-by-light scattering (Sect. 9.2.2). In contrast, Kharkiv physicists investigated the problem in strong analogy to the scattering of light by light from the start. This calculation complemented the previous work carried out in Zürich (Sect. 9.2.3). A general theme we will meet throughout this discussion is the possibility of conceiving of Delbrück scattering as a third order process. This conjecture was rejected by the historical actors themselves on different grounds. Yet,

---

[109] Due to the large number of calculating physicists involved, Pauli called the Delbrück effect the "Delbrück-Nordsieck-Weisskopf-Kemmer-Ludwig-Effekt". Pauli to Weisskopf, April 8. 1936; as cited in Pauli (1993, 788). I added the names of Akhiezer and Pomeranhuck in the section heading.

only after these evaluations had been published Wendell H. Furry developed a general proof which excluded all odd order contributions for similar processes. A short evaluation of Furry's publication will conclude this section (Sect. 9.2.4).

9.2.1 *The Empirical Background and Delbrück's Qualitative Proposal*

The conceptual and empirical background of the proposal of Delbrück scattering has already been described in the historical literature:[110] Delbrück's conjecture was based on a long known problem, often referred to as the Meitner-Hupfeld effect. Beginning in 1929, different experimental groups set out to test the theoretical predictions of the Klein-Nishina formula and the corresponding absorption coefficient. Although the empirical findings diverged, there was one shared conclusion: The absorption coefficient varied with the atomic number Z of the scattering material. For low values of Z, the Klein-Nishina formula proved suitable. For large values, considerable deviations from the expected results occurred which could not be explained by the then current theoretical and conceptual framework. In the course of these empirical studies a number of investigators, amongst them Lise Meitner's group at the Kaiser Wilhelm Institut in Berlin-Dahlem, observed an unshifted line in the scattered radiation (i.e. having the same frequency as the incident γ-rays).

Although this phenomenon would subsequently be explained by real pair creation and annihilation in the vicinity of the nucleus, the Delbrück effect was originally intended as a theoretical interpretation of the unshifted line. At the time, Delbrück was the "Haustheoretiker" of the Berlin group around Lise Meitner. Delbrück proposed the mechanism in the appendix of an experimental paper of Meitner and Kösters (1933, 144). As he observed, according to hole theory pair production could be expressed as a photoelectric effect on a negative energy electron. He inferred that, therefore, these electrons should also be capable of coherently scattering light in the vicinity of a nucleus. Delbrück's conjectures in the short appendix were vague and qualitative. He made no indication how the process could be calculated and he did not make any attempt at an estimation of its intensity. He could only infer that the scattering cross section no longer depended on the constitution of the nucleus[111] but on its charge and the frequency of the incoming light.

---

[110] Brown and Moyer (1984) rightly pointed to the Meitner-Hupfeld effect. Roqué (1997) both discussed the empirical background of Delbrück scattering and frames its initial inception in terms of "hole heuristics", i.e. as an instance where the concept of the Dirac sea was applied because of its heuristic appeal. My own discussion of the initial proposal of Delbrück in Meitner and Kösters (1933) is based on these two papers, but goes beyond them in evaluating Delbrück's attempts as far as the historical material allows.

[111] The unshifted line was initially explained as scattering from the nucleus itself or from intranuclear electrons, see e.g. Heisenberg (1932c, 160–163).

Lise Meitner indicated progress in a short note communicated on February 23. 1934: "Mister Delbrück has meanwhile performed the calculation of these processes, although only qualitatively, which will appear shortly."[112] According to Meitner, Delbrück's calculations were in harmony with the angular dependence of the scattered radiation. But these calculations were never published. Meitner was most probably referring to Delbrück's Habilationsschrift "Beiträge zu Diracs Theorie des positiven Elektrons", which was first handed in in February 1934 and, due to the political circumstances, was only accepted in 1936. This work is sadly lost.[113] The best glimpse we can get on Delbrück's calculations is Max von Laue's report on the Habilitationsschrift (dated to April 29. 1934).

As von Laue reported, Delbrück could estimate the angular distribution of the scattered radiation and the total scattering intensity. Yet, these estimations were rather qualitative:

> Naturally, the exact discussion of the formulas also fails in this case due to mathematical difficulties. The calculation, sadly only represented broadly, shows at first far reaching similarity with Schrödinger's dispersion theory and like it includes certain matrix elements. But their evaluation seems to be impossible at the moment.[114]

These comments, the most concrete in the report, are so vague that thereby we can actually not infer how Delbrück had treated his own effect. The only further indication I know of is a letter from Pauli to Heisenberg. Pauli mentioned that Delbrück used the Born approximation.[115] Delbrück himself remembered that the above mentioned estimations could only be derived with "some advice by Hans Bethe" (Delbrück 1979, 46). Effectively, by mid 1934, Delbrück had given up on the quantitative evaluation. At that time, the effect was taken up in Zürich and with a little delay in Kharkiv.[116]

---

[112] "Herr Delbrück hat inzwischen eine allerdings nur qualitative Durchrechnung dieses Prozesses durchgeführt, die demnächst erscheinen wird." Meitner (1934, 174).

[113] The Habilitationsschrift is neither in the Archive of the Humboldt University nor in the Caltech Archives. In 1978, Delbrück himself requested a copy of his Habiltationsschrift from the Humboldt University. The University replied that they had already sent it in 1936. I have to thank Loma Karklins of the Caltech Archives for this information.

[114] "Freilich scheitert auch hier die genaue Diskussion der Formeln an mathematischen Schwierigkeiten. Die leider nur in großen Zügen wiedergegebene Rechnung zeigt zunächst weitgehende Ähnlichkeit mit der Schrödingerschen Dispersionstheorie und enthält wie diese gewisse Matrixelemente. Deren Auswertung scheint aber zurzeit unmöglich zu sein." *Gutachten über das Habilitationsgesuch Herrn Dr. Max Delbrück's* von Max von Laue, 29.4.1934. AHUB, HU UA, UK Personalakten bis 1945, Delbrück, Max, D 040.

[115] Pauli to Heisenberg, June 14. 1934; as cited in Pauli (1985, 327–331).

[116] Also in Berkeley the effect was studied by Arnold Nordsieck, but I came across this study to late in my research as to comprehensively discuss it in this section.

9.2.2 Zürich: Delbrück Scattering as Scattering from a Perturbed Negative Energy Electron

Wolfgang Pauli initiated the work on Delbrück scattering in Zürich. Pauli must have become aware of Delbrück's conjecture at the "Physikalische Vortragswoche" held in Zürich from June 28. until July 1. 1933. The newly discovered positron and its relation to high energy phenomena were part of the agenda.[117] Pauli was famously sceptical about Dirac's hole-theory and the connected "abuse of the infinity concept".[118] "But", as he wrote to Heisenberg two weeks after the conference, "of particular interest is the hole theory with regard to the anomalous nuclear scattering of γ-rays,"[119] i.e. the Meitner-Hupfeld effect. Pauli mentioned that Delbrück and Rudolf Peierls have independently arrived at a possible explanation of the unaltered frequency component:

> First expressed in the infinity terminology of the hole theory, in the presence of a nucleus the occupied infinite continuum of negative energy states corresponding to the hyperbolic orbits will always cause a scattered radiation of unaltered wavelength with appreciable intensity if the states deviate appreciably from the force-free ones.
> I believe, that besides the infinity terminology (which I do not take seriously), such an effect has to exist, because it so to say corresponds to the unobservable intermediate state of the birth of a $+e$, $-e$ pair and its following recombination, and in the presence of a nucleus (and only then) the momentum law can be fulfilled.[120]

Quite ironically, a conceptualization that is closer to the first of the two above mentioned will be followed at Pauli's institute. And Pauli himself (see below) might have initiated it nearly a year later.

Pauli, Weisskopf (Pauli's assistant at the time) and Heisenberg kept up close correspondence about the subtraction procedures of Heisenberg and connected processes, likc light-by-light scattering and the self-energy of an electron in the Dirac sea. Since they suspected that Delbrück would quantitatively evaluate the effect, they did not

[117] For a report on the presentations, see Bretscher (1933).

[118] "Mißbrauch des Unendlichkeitsbegriffes," Pauli to Heisenberg, July 14. 1933; as cited in Pauli (1985, 186).

[119] "Von besonderem Interesse ist aber die Löchertheorie im Hinblick auf die anomale Kernstreuung der γ-Strahlen." Pauli to Heisenberg, July 14. 1933; as cited in Pauli (1985, 186).

[120] "Nun wird, zunächst in der Unendlichkeitsterminologie der Löchertheorie ausgedrückt, bei Anwesenheit eines Kernes das besetzte unendliche Kontinuum der den hyperbolischen Bahnen entsprechenden Zustände negativer Energie eine Streustrahlung unveränderter Wellenlänge mit merklicher Intensität immer dann ergeben, wenn die Zustände merklich von den kräftefreien abweichen.
Ich glaube, daß auch abgesehen von der (von mir nicht ernst genommenen) Unendlichkeitsterminologie, ein solcher Effekt vorhanden sein muß, da er sozusagen dem unbeobachtbaren Zwischenzustand der Geburt eines $+e$, $-e$ Paares und seiner darauf folgenden Wiedervereinigung entspricht und da bei Vorhandensein eines Kernes (und nur dann) der Impulssatz dabei erfüllt werden kann." Pauli to Heisenberg, July 14. 1933; as cited in Pauli (1985, 186–187).

get deeply involved in the calculations at first. In April 1934, however, Delbrück met Pauli and Weisskopf in Copenhagen and informed them about his difficulties.[121] By the time, Heisenberg also knew of Delbrück's struggling and proposed that Weisskopf should assist Delbrück.[122]

Weisskopf had already arrived at rough estimations. Pauli suspected that the effect would not converge, even with Heisenberg's subtraction terms. Nevertheless, Pauli laid out how he thought that the problem should be handled. The "bad Ansatz" of Delbrück consisted, according to Pauli, in applying the Born approximation and therefore considering free particle solutions for initial, intermediate and final states. On the contrary, Pauli was convinced: "One should rather calculate with (unrelativistic) hydrogen-eigenfunctions in the ground state [i.e. the initial state] and with plane waves only for intermediate and end states [...]."[123] In early September 1934, Pauli finally proposed to Weisskopf that the effect should be investigated more closely in Zürich.[124]

Until June 1935 not much progress had been made. As Pauli wrote to Heisenberg, the problem was still investigated by Weisskopf and "the occurring questions of the subtraction physics become very unpleasant".[125] During the Winter semester 1935/36, while Pauli was in Princeton, the major part of the calculations on the Delbrück effect in Zürich was carried out. Weisskopf had gathered support from Nicholas Kemmer. Kemmer had studied in Göttingen under Max Born and finished his dissertation in 1935 under Gregor Wentzel at the University of Zürich. Kemmer's dissertation was concerned with the self-energy problem in relativistic quantum theory. He had established close contact to the ETH group while still working at the University of Zürich.[126]

It is hard to judge why the evaluation and a first publication took so long. Certainly, the initial Ansatz Pauli wanted to follow was not applied exactly: The problem was studied within Dirac's hole theory and the Born approximation was applied. Furthermore, only the case of low frequency incident radiation and slowly varying potential fields could be quantitatively evaluated. Since the effect was initially proposed to explain deviations from the Klein-Nishina formula in the case of high frequency radiation and near the nucleus, circumstances which were not taken into account by

---

[121] Pauli to Heisenberg, June 14. 1934; as cited in Pauli (1985, 328).

[122] "Delbrück war neulich hier und erzählte von seinen vergeblichen Versuchen, die kohärente Streuung zu rechnen. Er wäre sicher froh, wenn er die Arbeit gemeinsam mit einem anderen z.B. Weisskopf machen könnte [...]." Heisenberg to Pauli, June 16. 1934, as cited in Pauli (1985, 332).

[123] "Schlechter Ansatz"; "Man soll aber im Grundzustand mit (unrelativistischen) Wasserstoffeigenfunktionen und nur in Zwischen- und Endzuständen mit ebenen Wellen rechnen." Pauli to Heisenberg, June 14. 1934; as cited in Pauli (1985, 328).

[124] "Es wäre doch sehr wünschenswert, wenn im Wintersemester in unserem Institut darüber etwas geixt werden könnte." Pauli to Weisskopf, September 5. 1934; as cited in Pauli (1993, 763).

[125] "die dabei auftretenden Fragen der Subtraktionsphysik werden sehr unschön." Pauli to Heisenberg, June 15. 1935; as cited in Pauli (1985, 402).

[126] See Pauli (1985, XXVI) for this short outline of Kemmer's path to the ETH Zürich.

the above assumptions, the relaxation of the initial modelling might have taken some time.[127]

Although the calculation of Kemmer and Weisskopf was therefore empirically uninteresting it had some inner-theoretical value. The results as communicated by Kemmer and Weisskopf (1936) supported the usage of Heisenberg's subtraction terms. The coefficients of the Lagrangian of Euler could be rederived without invoking any kind of subtraction procedures. Consequently, "all the results following from the non-linearity of Maxwell equations, for example, the cross-section for scattering of light by light as found by Euler and Kockel, seem to be of the same trustworthiness as the cross-section for pair production derived by Bethe and Heitler" (Kemmer and Weisskopf 1936, 659). Hearing of the results through Weisskopf, Heisenberg himself acknowledged the "very nice support for the uniqueness of the subtracted terms".[128]

Kemmer (1937) published the derivation of the results separately. He did not conceptualize Delbrück scattering as analogous to the scattering of light by light but rather developed an approach that was close to the original proposal by Delbrück. At first, Kemmer worked with the wave functions of a relativistic electron in a potential field and plugged them into the scattering amplitude for the scattering of light by an electron. This effectively meant that the problem was treated, at least so when approaching the subject, as proposed by Pauli initially: It was described as the scattering of light from a negative energy electron in the field of the nucleus; no approximation was made for the second part of the description, at least so at first.

But this calculation could not be performed. Kemmer had to expand the wave function of the electrons in the electrostatic case, i.e. he applied the Born approximation, thereby deviating from Pauli's initial proposal. Consequently, only unperturbed relativistic eigenfunctions were involved in the calculation and the potential V was considered as a perturbation, similar to the light-matter interaction H. I reproduced the full equation for the matrix element $M^-_{\nu\nu'}$ Kemmer arrived at in Fig. 9.5.[129] In its theoretical structure, the effect was a combination of matrix elements of the different perturbations, the static potential and the radiation field. Still from a modern point of view the first two lines seem curious as they suggest that Delbrück scattering could occur through second or third order processes. But Kemmer excluded these possibilities.

---

[127] In this connection, it is noteworthy that, while in Princeton, Pauli together with Maurice Pryce tried to calculate light-by-light scattering without the low frequency assumption of Euler and Kockel, but eventually failed. See Pauli to Weisskopf, September 27. 1935; as cited in Pauli (1993, 769); and Pauli to Weisskopf, January 20. 1936; as cited in Pauli (1993, 782): "Das Problem der Streuung von Licht an Licht bei kurzen Wellen habe ich nun, nach Besprechung mit Oppenheimer, definitiv als zu kompliziert aufgegeben."

[128] "eine ganz schöne Stütze für die Eindeutigkeit der subtrahierten Glieder." Heisenberg to Pauli, February 18. 1936; as cited in Pauli (1985, 475).

[129] Actually, it was half of the perturbative expansion. A similar term $M^+_{\nu\nu'}$ also had to be calculated. $M^-_{\nu\nu'}$ and $M^+_{\nu\nu'}$ differed in the order of the emission and absorption processes of the light quanta.

$$\frac{1}{\hbar c} M_{\nu\nu} = \sum_{\substack{A \\ (E_A<0)}} \sum_{\substack{S \\ (E_S>0)}} \frac{1}{(E_A - E_S - \hbar\nu)} \Bigg[\tilde{H}_{AS} H_{SA}$$
$$+ e \sum_K \left(\frac{\tilde{H}_{AS} H_{SK} V_{KA}}{(E_A - E_K)} + \frac{V_{AK} \tilde{H}_{KS} H_{SA}}{(E_A - E_K)} + \frac{\tilde{H}_{AS} V_{SK} H_{KA}}{(E_S - E_K)} + \frac{\tilde{H}_{AK} V_{KS} H_{SA}}{(E_S - E_K)} \right)$$
$$+ e^2 \sum_K \sum_L \left(\frac{\tilde{H}_{AS} H_{SK} V_{KL} V_{LA}}{(E_A-E_K)(E_A-E_L)} + \frac{V_{AK} \tilde{H}_{KS} H_{SL} V_{LA}}{(E_A-E_K)(E_A-E_L)} + \frac{V_{AK} V_{KL} \tilde{H}_{LS} H_{SA}}{(E_A-E_K)(E_A-E_L)} \right.$$
$$+ \frac{\tilde{H}_{AS} V_{SK} V_{KL} H_{LA}}{(E_S-E_K)(E_S-E_L)} + \frac{\tilde{H}_{AK} V_{KS} V_{SL} H_{LA}}{(E_S-E_K)(E_S-E_L)} + \frac{\tilde{H}_{AK} V_{KL} V_{LS} H_{SA}}{(E_S-E_K)(E_S-E_L)}$$
$$+ \frac{\tilde{H}_{AS} V_{SK} H_{KL} V_{LA}}{(E_S-E_K)(E_A-E_L)} + \frac{\tilde{H}_{AK} V_{KS} H_{SL} V_{LA}}{(E_S-E_K)(E_A-E_L)}$$
$$\left. + \frac{V_{AK} \tilde{H}_{KS} V_{SL} H_{LA}}{(E_A-E_K)(E_S-E_L)} + \frac{V_{AK} \tilde{H}_{KL} V_{LS} H_{SA}}{(E_A-E_K)(E_S-E_L)} \right) \Bigg]. \qquad (10)$$

Fig. 9.5 The full equation for one of the matrix elements describing Delbrück scattering within the Born approximation according to Nicholas Kemmer. Equation in Kemmer, Nicholas. 1937. "Über die Lichtstreuung an elektrischen Feldern nach der Theorie des Positrons." *Helvetica Physica Acta* 10, 112–122, here 115 (cited as Kemmer (1937) in the bibliography). ETH-Bibliothek Zürich, E-periodica, https://www.e-periodica.ch. Reproduced with kind permission from the Swiss Physical Society

The first expression was non-zero only if the momentum of initial and final photon was equal in magnitude and direction. Therefore, it was not of interest in the case considered by Kemmer. The contribution of the second line also vanished: The terms exactly cancelled in the summation. Kemmer only mentioned that one easily sees this when using the explicit form of the matrix elements. This effectively left Kemmer with the calculation of the terms proportional to e^2. Interestingly, Kemmer mentioned that a symmetric treatment of the two perturbations constituted a simpler way of arriving at the conclusion that no subtraction procedures had to be applied. But he only did so on a side note (Kemmer 1937, 115–116).[130]

From this point onwards the calculation went along with the incorporation of the explicit form of the matrix elements and by the application of the two approximations mentioned earlier. By comparing his final result with the ones of Euler and Kockel, Kemmer found agreement in the coefficients of the Lagrangian, without having to apply subtraction physics at all.

In Zürich, Kemmer and Ludwig (1937) published a second paper on the effect. Yet, when Pauli came back from Princeton he was quite disappointed to see that the

[130] As we will see, this symmetric treatment was the basis of the approach followed in Kharkiv.

calculations went on quite slowly.[131] At that point it was already clear that the effect was out of reach of the then current empirical procedures. Thus, Pauli did not want to invest too much time in the calculations.[132] The publication only involved qualitative conclusions and shall not be discussed here in any more detail. Rather, I will turn my attention to Lev Landau's institute in Kharkiv.

9.2.3 *Kharkiv: Delbrück Scattering as Scattering of Photons from "Pseudo-Photons"*

In Kharkiv, the research on the Delbrück effect, as well as on light-by-light scattering, had been initiated by Lev Landau.[133] Landau had studied physics at the University of Baku (1922–1924) and at Leningrad (1924–1927), where he met and studied with George Gamow and Dimitri Ivanenko: Together they formed the core of the "physical jazz band" and learned the evolving quantum mechanics partly on their own from the research articles in the *Zeitschrift für Physik*. From 1927 onwards Landau held his first position as "aspirant" at the Leningrad institute and made first major impacts on theoretical physics through his publications on the density matrix and the relativistic theory of the electron.

In October 1929, Landau left for his first foreign trip including visits to the major centres of the development of quantum mechanics. In Zürich he met Rudolf Peierls and together they published two papers on quantum electrodynamics. One of them (Landau and Peierls, 1931) treated the uncertainty relations and the measurement in quantum electrodynamics. This piece of work would initiate, but was also overshadowed by the paper of Léon Rosenfeld and Niels Bohr on the same topic.[134] In 1932, Landau changed from Leningrad to Kharkiv. While he had been interested in foundational problems priorly, his research interests shifted towards applications at around that time. Landau had an excellent reputation and could attract a number of talented young physicists: Each had to pass a special set of exams, later called Landau's theoretical minimum. Thereby, Landau created the famous and specifically pragmatist Landau school of theoretical physics.

---

[131] "Ich bin sehr ärgerlich, daß ich die Arbeit über die kohärente Licht-Streuung hier nicht fertig vorfand. Im Gegenteil versichern mit die Herren [most probably Ludwig and Kemmer], daß sie noch viele Monate brauchen würden, um sie fertigzustellen. Ich bin dagegen, so lange Zeit mit dieser Rechnung zu verbringen, da der Effekt doch nicht beobachtbar sein wird (wovon die Herren sich duch Abschätzungen noch des näheren versichern sollen)." Pauli to Weisskopf, May 19. 1936; as cited in Pauli (1993, 788).

[132] In the published paper this reads: "insbesondere ist ausser Zweifel, dass der Wirkungsquerschnitt für alle nach (I,1) zulässigen Wellenlängen [langwelliges Licht] weit unter der Grenze des Beobachtbaren liegt. Die Durchführung wurde darum unterlassen." Kemmer and Ludwig (1937, 182).

[133] The following short recapitulation of Landau's early career is based on Kojevnikov (2004, Chapter 4). See also Hall (2008).

[134] For a detailed history of these developments see Darrigol (1991).

Light-by-Light Scattering in the High Frequency Domain

Among the young talents Landau could win were Alexandr Akhiezer, who had studied in Kiev, and Isaak Pomeranchuck from Leningrad.[135] As soon as Akhiezer entered the institute and passed Landau's theoretical minimum, Landau assigned to him the task of calculating the effect of light-by-light scattering. Landau had already set out to calculate the low-frequency range of the phenomenon but was outran by Euler, Kockel, and Heisenberg. Akhiezer and Landau, later to be joined by Isaak Pomeranchuck, consequently investigated the opposite approximation, the ultra-relativistic limit.

According to Akhiezer's memories on the calculations, Landau was particularly troubled by the non-covariance of the perturbative methods[136] and the possibly gauge dependent result. The last problem could only be solved by direct inspection, a tedious task of calculating and adding 144 terms, just to check that in sum they combined to zero. Still, the result satisfied Landau and they sent off a three author communication to *Nature* (Akhiezer, Landau, and Pomeranchuck, 1936). For the full evaluation I will turn my attention to Akhiezer's publication in the *Physikalische Zeitschrift der Sowjetunion* (Akhiezer, 1937).

Akhiezer took the perturbative expansion directly from Weisskopf's 1933 paper (see Sect. 8.1.4). He used the regular theoretical underpinning of Dirac's radiation and hole theory. Although it is likely that Akhiezer already arrived at the possible intermediate states which coupled initial and final states before becoming aware of Euler's dissertation, his presentation of the sequence of events was certainly motivated by Euler's diagrammatic representation. In Fig. 9.6 you find the diagrams Akhiezer provided.

Each diagram was accompanied by a verbal description of the process. In contrast to Euler's representation, the processes are not ordered from left to right, but through the number of the photons $\mathbf{K}_i$ involved. Take for example the diagram III: First a pair was created (K_1), then another pair was created (K_2, at the outermost right). Afterwards, the electron dropped into the hole formed by the second pair creation under the action of the third quantum (pair annihilation of the electron of the first pair with the electron of the second pair; K_3). Finally, the electron created in the second step dropped into the hole created in the first process (annihilation of the electron of the second pair with the positron of the first; K_4). The numbers 1, 2, and 3 on the sides of the horizontal lines were not explained by Akhiezer himself. Judging from

[135] For Akhiezer's memories on Landau and the calculations, see Akhiezer (1989).

[136] On a side, it is quite remarkable that even physicists who were particularly interested in a covariant perturbation theory and who even began to argue with each other about the trustworthiness of the results of regular perturbation theory, did not hit upon Stueckelberg's covariant form. In this respect it is also noteworthy that Wolfgang Pauli mentioned Stueckelberg's methods to Werner Heisenberg when the latter was trying to restructure the perturbative account (Pauli to Heisenberg, February 5. 1937; as cited in Pauli (1985, 513)). Nevertheless, Heisenberg never mentioned Stueckelberg's work in publication. For a discussion of Stueckelberg's covariant perturbation theory, see Lacki, Ruegg, and Telegdi (1999).

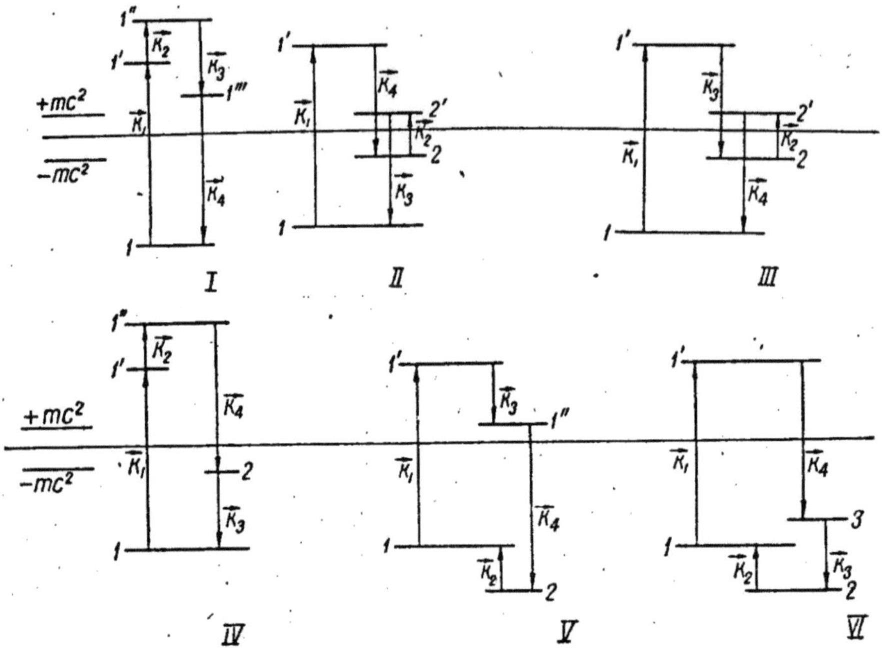

Fig. 9.6 The different possible processes that lead to the scattering of light by light according to Akhiezer. Taken from Akhiezer (1937, 268)

the way they were used, these numbers referred to the different electrons involved in the process, while primes on them referred to the second state, double primes to the third state they occupied, and so on.

A general quantitative evaluation was found to be impossible. The problem was only considered for small angles and the dependence of the total cross section on the frequency was derived. One of the prime results, derived through the comparison of Akhiezer's and Euler's results, was that the maximum of the cross section laid around $h\omega \propto \mathrm{mc}^2$, which was rather reasonable since in this case real pair production set in (Akhiezer 1937, 281).

The Delbrück Effect and "Pseudo-Photons"

As soon as the calculations on light-by-light scattering were finished, Akhiezer and Pomeranchuck again joined forces and turned towards the calculation of Delbrück scattering. As laid out above, Kemmer, rather then treating the two perturbations symmetrically from the start, used the scattering formula for photons from electrons to initialize his calculations. In contrast, Akhiezer and Pomeranchuck started their calculation with a symmetric treatment: They explicitly modelled the effect in analogy to light-by-light scattering.

Akhiezer and Pomeranchuck opened their paper with the following words:

> From Dirac's theory of the positron the possibility of coherent scattering of γ-rays by nuclei follows. This possibility comes about because the photon can create a virtual pair in the Coulomb field of the nucleus, which then annihilates again under the emission of a photon of the same frequency. The momentum difference of the photons is transferred to the nucleus.[137]

This introduction is not only interesting as the authors did not speak of virtual transitions but actually of virtual pairs, but also because the process of Delbrück scattering was conceived of, at first, as a third order effect. I quote in full:

> One could think, that the following scheme of intermediate states should be taken as a basis for the calculation of the probability amplitude of our process in the perturbation theory. Some electron in a negative energy state goes into a state of positive energy under the action of the photon, then it emits a photon, the energy of which is the same as the energy of the first photon, and in doing so passes over to another state with positive energy, and finally it goes over into its initial state under the action of the Coulomb potential of the nucleus.[138]

Kemmer also included such terms (second line of Fig. 9.5). By direct inspection Kemmer showed that this contribution should vanish. Akhiezer and Pomeranchuck invoked a symmetry argument for that purpose. The matrix elements in their treatment are all proportional to e so that the total probability amplitude would be proportional to e^3. Since it was well known, as Akhiezer and Pomeranchuck argued, that Dirac's theory was invariant under charge conjugation and the scattering amplitude of their calculation referred, due to the coherence of the effect, to an "observable quantity"[139] it could not have an odd value as then it would depend on the sign of the charge. Hence, the third order terms had to vanish. Akhiezer and Poermanchuck also proved this conjecture by direct calculation of the terms.

In the further development, Akhiezer and Pomeranchuck modelled Delbrük scattering closely after their prior calculation of light-by-light scattering. In order to apply "Weisskopf's formula" (Akhiezer and Pomeranchuck 1937, 480), i.e. the general perturbative expansion for a fourth order effect composed of first order matrix elements, they Fourier analysed the Coulomb potential of the nucleus and consequently came to the following conceptualization:

---

[137] "Aus der Diracschen Theorie des Positrons folgt die Möglichkeit der kohärenten Streuung von γ-Strahlen an Kernen. Diese Möglichkeit kommt dadurch zustande, dass das Photon im Coulombfeld des Kernes ein virtuelles Paar erzeugt, welches dann unter Ausstrahlung eines Photons derselben Frequenz wieder vernichtet wird. Die Impulsdifferenz der Photone [sic!] wird dabei dem Kerne übergeben." Akhiezer and Pomeranchuck (1937, 478).

[138] "Man könnte denken, dass bei der Ausrechnung der Wahrscheinlichkeitsamplitude unseres Prozesses bei der Störungsrechnung folgendes Schema von Zwischenzuständen zugrunde gelegt werden muss. Irgendein Elektron im Zustande negativer Energie geht unter dem Einfluss des Photons in einen Zustand positiver Energie über, strahlt dann ein Photon aus, dessen Energie gleich der Energie des ersten Photons ist und geht dabei in einen anderen Zustand mit positiver Energie über, und schliesslich geht es unter dem Einfluss des Coulombfeldes des Kerns in seinen Anfangszustand zurück." Akhiezer and Pomeranchuck (1937, 478–479).

[139] "beobachtbare Größe," Akhiezer and Pomeranchuck (1937, 479).

> Through the representation of the field of the nucleus in the form of plane waves, which we shall call for brevity's sake pseudo-photons, we can use the following scheme of intermediate states in the calculation of the probability amplitude of the coherent scattering of γ-rays on nuclei. The electron in the state of negative energy goes under the action of the quantum of momentum k_1 over to a state of positive energy, further [it passes over] into a different state of positive energy under the influence of a pseudo-photon with momentum g_2. Through the emission of a photon of momentum k_3 and a pseudo-photon of momentum g_4 the electron falls back into its initial state.
> Consequently, the coherent scattering [...] of γ-rays on nuclei can be considered as the collision of a photon k_1 with the pseudo-photon g_2 whereby a photon k_3 and a pseudo-photon g_4 are created.[140]

Although Akhiezer and Pomeranchuck used the term "pseudo-photon", they clearly point out that these entities are unconnected to the radiation field. The number of quanta of the radiation field is only changed by the interaction between transversally polarized photons; the transitions due to the interaction with the nucleus arose "without alteration of the number of quanta".[141] Nevertheless, their conceptualization allowed them to model their calculation closely after the case of light-by-light scattering: "In fact, we have here the same intermediate states as in the scattering of light by light with the difference, that now instead of the photons with momenta k_2 and k_4 and the frequencies ω_2 and ω_4 the pseudo-photons with the momenta g_2 and g_4 and frequencies [equal to] zero occur."[142]

9.2.4 Delbrück Scattering, the Reoccurrence of a False Prediction, and Furry's Theorem

Besides the particular method and conceptualization chosen by the authors, a general theme can be identified in the derivations laid out in the previous subsections: Delbrück scattering was conceptualized as a third order process but it was found that the third order contribution identically vanished. A similar conclusion was arrived at for the process of light-quantum splitting: Halpern had not specified the numbers

140 "Indem wir das Feld des Kerns in der Form von ebenen Wellen darstellen, die wir der Kürze halber Pseudophotonen nennen werden, können wir bei der Berechnung der Wahrscheinlichkeitsamplitude der kohärenten Streuung von γ-Strahlen an Kernen folgendes Schema der Zwischenzustände benutzen. Das Elektron im Zustande negativer Energie geht unter dem Einfluss des Quants von Impuls k_1 in einen Zustand positiver Energie, ferner unter dem Einfluss eines Pseudophotons mit dem Impuls g_2 in einen anderen Zustand positiver Energie über. Mit der Ausstrahlung eines Photons vom Impuls k_3 und eines Pseudophotons vom Impuls g_4 fällt das Elektron in seinen Anfangszustand zurück. Die kohärente Streuung [...] von γ-Strahlen an Kernen kann also aufgefasst werden, als der Stoss eines Photons k_1 mit dem Pseudophoton g_2, wobei ein Photon k_3 und ein Pseudophoton g_4 entstehen." Akhiezer and Pomeranchuck (1937, 483–484).

141 "ohne Änderung der Anzahl von Quanten," Akhiezer and Pomeranchuck (1937, 482).

142 "In der Tat haben wir hier dieselben Zwischenzustände, wie bei der Streuung von Licht and Licht mit dem Unterschied, dass statt der Photonen mit Impulsen k_2 und k_4 und den Frequenzen ω_2 und ω_4 jetzt die Pseudophotonen mit den Impulsen g_2 und g_4 und den Frequenzen Null auftreten." Akhiezer and Pomeranchuck (1937, 485).

of photons involved in his initial contribution, but Heitler (1936, 193) remarked in his textbook, that the cross-section vanished if only three photons were involved. In the case of light-by-light scattering such considerations were out of the question. By definition this process involved in its lowest order contribution four light quanta and therefore had to be at least of fourth order in the perturbation-theoretical account.[143]

Delbrück himself arrived at the conclusion that the third order terms should vanish and we can be sure that he had also informed Weisskopf and Pauli about this fact when he met them in Copenhagen in April 1934. In the letter exchange between Weisskopf and Pauli in late 1935, there are diverse references to the vanishing of the third order terms, when they discussed possible deviations from the Klein-Nishina formula, theoretically a similar phenomenon.[144] Nevertheless, Kemmer did not invoke any general principle in his publication but came to this conclusion by direct inspection. In contrast, Akhiezer and Pomeranchuck referred to symmetry principles when concluding that the third order term should be equal to zero.

Besides Zürich and Kharkiv, the Delbrück effect was also investigated in Berkeley. Robert Oppenheimer had put one of his students, Arnold Nordsieck, to calculate the Delbrück effect for his PhD-thesis: "The scattering of Radiation by an Electric Field" (defended May 1935; see Kelly (2006, 137)). In Berkeley, physicists were also well aware of the fact that the third order contribution should vanish: The first mentioning of this conclusion by Oppenheimer in print I could locate can be found in Lauritsen and Oppenheimer (1934, 80). They used it to point towards the unimportance of the Delbrück effect ("a coherent radiation, highly anisotropic, scattered by the electron positron cloud[145] about nuclei") for the explanation of deviations between the theoretical and experimental data at hand.

The work on Delbrück scattering in Berkeley, Zürich, and Kharkiv was mostly carried out independently. Between Kharkiv and Berkeley no connection had been established at all, while communication between Zürich and Berkeley started when Pauli was in Princeton and Nordsieck had already finished his dissertation. I also do not know of a direct connection between Zürich and Kharkiv when the calculations were

[143] One might argue that light-by-light scattering can lead to the fusion of two photons into one. But according to the differentiation already applied by Halpern, namely into light-by-light scattering and light-quantum splitting, this would refer to the reversal of light quantum splitting and not to light-by-light scattering.

[144] Just consider the following examples: "[...] das Verschwinden der Terme proportional zu Z^2 in der Intensität der Delbrückstreuung [...]," Pauli to Weisskopf, September 27. 1935; as cited in Pauli (1993, 769). Or: "Die Ordnung der Delbrück-Streuung ist *nicht* 'gerade die, deren Verschwinden wir zu beweisen glaubten', sondern eben die nächst-höhere, deren Nicht-Verschwinden ohnedies trivial vorasugesetzt wurde." Pauli to Weisskopf, December 5. 1935; as cited in Pauli (1993, 777).

[145] *Comment by the author*: I will further engage with the concept of an electron-positron cloud in Sect. 10.1.2.

carried out. Only later on, Pauli and Weisskopf approved of the results of Akhiezer and Pomeranchuck.[146] Without knowing that also Pomeranchuck and Akhiezer were working on the Delbrück effect, Pauli referred to this phenomenon, due to the great number of physicists involved, as the "Delbrück-Nordsieck-Weisskopf-Kemmer-Ludwig-Effekt".[147]

All of these groups came to the conclusion that the third order effect should vanish, but without generalizing this result. Wendell H. Furry was the first one to give such a general proof up to an arbitrary. Furry was a former National research fellow at Oppenheimer's institute in Berkeley[148] and was at the time employed at Harvard University. His result is still known as "Furry's Theorem" and was originally stated as:

> In calculations using plane wave functions as a basis ("Born approximation") for processes in which the appearance of electrons and positrons is transitory only, the odd order contributions vanish identically. (Furry 1937, 125)

Furry motivated his investigation by the effects of light-by-light scattering and Delbrück scattering "in which electrons and positrons, though they are not actually created so as to be experimentally observable, are able to play the part of catalytic agents producing effects which otherwise could not appear": I.e., "the transitory or virtual presence of charged particles provides a means of mediating indirect interactions". Delbrück scattering was Furry's prime example in the introduction and he mentioned: "Immediately after the Delbrück effect was suggested, it was remarked by Oppenheimer that, on account of such a cancellation [due to the symmetry between positively and negatively charged particles], the transition probability for such scattering is of order $(Z\alpha)^4$ instead of $(Z\alpha)^2$."[149] A further motivation to investigate the question and to provide a general proof of this conjecture mentioned by Furry was his recent interest in non-electromagnetic forces and the connected question if such forces would also be subject to the same restrictions (Furry 1937, 125).

Furry's proof rested on the general structure of the perturbative expansion as it was applied in the 1930s. Furry considered a general process A brought about through intermediate states in the regular way and the same process under reversal of the roles of electrons and positrons. This process was called A'. First Furry showed, through the introduction of creation and annihilation operators, that for even orders the numerators of the perturbative expansion have the same sign and for any odd order term it holds that: $A = -A'$. Secondly, he showed that no other changes of the sign occur if the interaction term is due to the radiation field or the Coulomb interaction. Since both kinds of processes had to be taken into account in the summation over

---

[146] Victor Weisskopf was in the Soviet Union, and also in Kharkiv, during a short visit in late 1936. But, as Akhiezer himself remembered, the calculations were already performed and Pauli and Weisskopf only approved of them. See "In memorium Isaak Yakovlevich Pomeranchuk (A tribute on the 80th anniversary of his birth)" under https://www.kipt.kharkov.ua/itp/akhiezer/en/recollections/pomeranchuk/index.html (visited September 1. 2023).

[147] Pauli to Weisskopf, April 8. 1936; as cited in Pauli (1993, 788).

[148] I will turn attention to the work of Oppnheimer and Furry in Sect. 10.1.2.

[149] All citations taken from Furry (1937, 125).

the possible intermediate states, the odd order terms must be identical to zero: Each contributing matrix element had a negative counterpart and cancelled.[150]

9.2.5 Intermediate Conclusion

As was emphasized, in Zürich and in Kharkiv two distinct ways of conceptualizing the problem, that still resulted in the same theoretical structures, were given. While the Zürich physicists closely stuck to the original proposal of Delbrück (scattering phenomena on a perturbed negative electron), in Kharkiv the process was modelled in close analogy to light-by-light scattering from the start. In Zürich, the effect was subsequently connected to light-by-light scattering through the Lagrangian that Euler and Kockel had derived. In fact, Kemmer's results were seen as proof for the uniqueness of the subtraction terms. In Kharkiv, the complete structure of the perturbative evaluation of light-by-light scattering was taken over directly through the incorporation of the concept of a "pseudo-photon". Both evaluations, in turn, show that the differentiation between static fields and photons was still upheld and formed an integral part of the conceptual framework: A further argument against the conjecture that physicists had yet completely developed the concept of a virtual photon.[151]

The study of the effects had a direct effect on the structure of the perturbative evaluation and on the particular sequence of intermediate states necessary in the calculation. Furry's theorem, the proof that only even orders contributed to processes in which pairs are only present in intermediate states, was (partially) motivated by the reoccuring false prediction of Delbrück scattering as a third order effect. As Kockel's study of the application of the Pauli principle for intermediate states, it provided insight into the theoretical structures of prime importance for the evaluation of QED processes in the 1930s: the number and sequence of the intermediate states that connected initial and final state of the system. Contrary to Kockel's conclusion,[152] Furry's theorem survived the transition to the covariant methods of Feynman.

From a terminological point of view, it is certainly interesting to note that Furry spoke of the "virtual presence" of pairs; he called them the "catalytic agents producing effects which otherwise could not appear" (Furry 1937, 1215). Akhiezer and Pomeranchuck spoke directly of "virtual pairs". As discussed in Sect. 9.1.3, Euler exclusively used the term "virtual matter" on the "intuitive" level of the theory where the respective contributions of the pairs were mathematically represented by additional charge and current densities. At the same time, Heisenberg already connected

[150] Although this seems rather straightforward, Furry's paper stretches over seven two columned pages and the proof still included some ambiguities (because of the divergences of the contributing integrals).

[151] See also my argument in reference to Fermi and Bethe's paper on electron-electron scattering (Sect. 8.2.1), as well as the arguments by Carson (1996b) mentioned there.

[152] Feynman explicitly stated that within his theory, "no account need to be taken of the exclusion principle in intermediate states." Feynman (1949b, 755).

the term "virtual" directly to particles. This step, from the terminological variants of "virtual (possibilities of) processes" to "virtual particles", will be discussed in more detail and also by drawing on different kinds of representational systems in Chap. 10.

For the time being, and with direct focus on the practices of (time-dependent) perturbation theory, I still want to emphasize that this terminology was an extrapolation from the particles only being present in an intermediate state which they occupied due to energy non-conserving transitions. If one looks at the concepts involved in the perturbative evaluation and, indeed, Furry's proof of his theorem, intermediate states and unobserved, energy non-conserving (i.e. virtual) transitions played the central role. That it was especially the concept of an intermediate state that, when altered, held the promise to lead to a consistent evaluation of QED will be argued for in the following section.

9.3 The Theory of Radiation Damping, or: Reprise of the *Schüttelwirkung* (1938–1942)

Up until now I have focused on the standard way of doing time-dependent perturbation theory during the 1930s as well as on the application of the concepts involved in it. In the following, I will revisit the development of the so-called Heitler-Peng theory of radiation damping, one of the candidates for "filling the frame" of Heisenberg's S-matrix during and right after the Second World War (Blum 2017, Sect. 2.5.1).[153] I will focus on the role of the *Schüttelwirkung*, a genuine concept of perturbation theory, and how the mathematical representation of the concept of the intermediate states was altered through its application.

The basic problem faced by the historical actors was the incorporation of radiation damping, i.e. the back reaction of the emitting particle on the emitted radiation, into the scattering of otherwise freely moving particles. For the non-relativistic case of quantized atoms, Weisskopf and Wigner had provided a solution (Sect. 8.1.2). For freely moving, relativistic particles, the *Schüttelwirkung* became central. The idea behind the inclusion of the *Schüttelwirkung* can be stated as follows: Through the sudden switching on of the perturbation at some time t_0 (mostly set equal to zero) transition probabilities occurred in the mathematical scheme which did not correspond to physical reality. Early on, physicists noticed that these terms vanished when the perturbation was switched on adiabatically.[154]

As discussed in Sect. 9.1.4, Bernhard Kockel published a first detailed study of the *Schüttelwirkung* in his dissertation. Although Kockel's evaluation was published as

---

[153] A historical contextualization of radiation damping has already been given by Olivier Darrigol (1982, Sect. 7.4).

[154] As for example Maria Göppert-Mayer, who introduced the respective concept into radiation theory. See Sect. 8.1.1.

a sort of appendix to his general assessment of perturbation theory[155] and Kockel's dissertation was rarely cited at the time (as well as in the historiographic literature on QED), it had an important influence on the further development of QED. As I will show, it stood at the beginning of what came to be known as the Heitler-Peng theory of radiation damping.

Although the incorporation of radiation damping into the scattering of free particles was first studied by Ivar Waller (Waller, 1934), I will start out with Edwin Góra's treatment (Sect. 9.3.1). Waller's calculations only came to be known to the actors in the early 1940's and, as Heitler had realized and in contrast to Góra'ss treatment, it could not be generalized. In Sect. 9.3.2, I will turn to Wilson's and Heitler's elaborations on the theme and finally provide for an outlook and reflection on their further development.

9.3.1 Edwin Góra Takes Up Kockel's Treatment

Edwin Góra,[156] a Polish physicist with German ancestors, graduated from the Jagiellonian University in Krakow with a masters degree in physics in 1934. He spent the years 1935 to 1937 in India, teaching at the St. Xavier's College. In 1937 he went back to Europe and held a post as an assistant at the Department for Theoretical Physics at the Johann-Casimir University (Lviv; today: Ivan-Franko University) and switched to Warsaw in 1938. Due to the occupation of Poland by Germany in September 1939, Góra could no longer work at the university. After consulting Heisenberg for a position at the Leipzig institute, Góra studied and worked there for a short time before he was banned from the classes by the Gestapo. His education went further through private lessons by Heisenberg. In 1942, in spite of the political repression, he successfully defended his dissertation on the quantum theory of radiation damping. After a further stop in Munich (1944–1946) he eventually went to the US after the war.

The publication I want to discuss was produced by Góra during his time in Lviv and formed the basis of his later dissertation under Heisenberg in 1942. In this paper, the basic ideas for the consistent treatment of radiation damping for freely moving particles were laid out.

[155] The short paragraph in which Kockel discussed the *Schüttelwirkung* was printed in smaller font size, indicating that it was not necessary for the reader to apprehend this part in order to understand the rest of the paper.

[156] This short biographic outline is based on Rovenchak (2013, 3002-3003) and an oral history interview with Góra to be found on the family webpage of his daughter: http://www.gora.us/ (visited on September 1. 2023).

Introducing the Problem

Góra's aim was to provide a theory of radiation damping for the Compton effect which was upto that point not obtained in general terms. But as Góra discussed in his introduction:

> On the contrary, if one assumes, following a suggestion of Heisenberg,[157] a gradual onset of the perturbation, the duration of which comprises many periods of the oscillation corresponding to the energy difference between initial and intermediate state, then a sensible more exact solution of the initial equations can be given. For the limiting case of an adiabatic switching on this constitutes an exact solution.[158]

To introduce the reader to the problem Góra laid out both Dirac's as well as Weisskopf and Wigner's approach to perturbation theory in terms of second order processes. The basic equations of time-dependent perturbation theory in the second approximation read:

$$i\hbar\dot{b}_0 = \sum_I H_{0I} b_I e^{\frac{i}{\hbar}(E_0-E_I)t} \tag{9.17}$$

$$i\hbar\dot{b}_I = H_{I0} b_0 e^{\frac{i}{\hbar}(E_I-E_0)t} + \sum_\lambda H_{I\lambda} b_\lambda e^{\frac{i}{\hbar}(E_I-E_\lambda)t} \tag{9.18}$$

$$i\hbar\dot{b}_\lambda = \sum_I H_{\lambda I} b_I e^{\frac{i}{\hbar}(E_\lambda-E_I)t} \tag{9.19}$$

where the indexes $0,\ I,\ \lambda$ referred to the initial, intermediate, and final state respectively, b_i to the probability amplitude, and E_i to the corresponding proper energy. H_{0I} was the matrix element of a transition from 0 to I with absorption/emission of a light quantum. As Góra laid out, Dirac's approximation consisted in taking $b_0(0) = 1,\ b_I(0) = 0,\ b_\lambda = 0$ and keeping $b_0(t) = 1$. In contrast, Weisskopf and Wigner set $b_0 = e^{-\gamma t}$ and showed that the first order equation could thereby be solved exactly.

For the indirect processes, such as Compton scattering, Góra observed that the Ansatz of Weisskopf and Wigner with the corresponding initial conditions could not be used. The problem had its roots in setting $b_I(0)$, the initial amplitude of the intermediate state, equal to zero. This rendered, as Góra explained, a non-compensable time dependence on the energy difference between initial and intermediate state in

[157] *Note by the author*: Góra will later on refer to Kockel's paper. Therein, Kockel mentioned that the study of the *Schüttelwirkung* was originally suggested by Heisenberg.

[158] "Nimmt man hingegen, einem Vorschlag Heisenbergs folgend, ein allmähliches Einsetzen der Störung an, dessen Dauer viele Perioden der den Energiedifferezen zwischen Anfangs- und Zwischenzustand entsprechenden Schwingungen umfasst, so lässt sich eine sinnvolle genauere Lösung der Ausgangsgleichungen des Problems angeben. Für den Grenzfall eines adiabatischen Einschaltvorgangs stellt dies eine exakte Lösung dar." Góra (1938, 159–160).

the solution of the initial state.[159] Consequently, in the way perturbation theory was regularly practised the problem could not be handled.

Proposal of a General Solution

As Góra noted, there was a simple formal solution to this problem:

> This difficulty can be circumvented, when one abandons the initial condition $b_I(0) = 0$ and sets b_I proportional to b_0, so that only forced vibrations are considered. [...] But with such an Ansatz one would presuppose, that the representatives of the intermediate states permanently coexist with b_0, which is, of course, meaningless. This assumption would be entirely arbitrary and would render no indications whatsoever about the possibly to be questioned approximate validity of results arrived at in this way.[160]

The question was: How to create such initial conditions, which allowed for setting the b_I proportional to b_0 for all times, in order to consistently follow the Ansatz of Weisskopf and Wigner? In a footnote, Góra indicated that such initial conditions were actually applied by Dirac in his first calculation of pair annihilation "but without discussing the associated conceptual difficulties".[161]

Góra found a possible solution in Kockel's dissertation: the adiabatic switching on of the perturbation, as Heisenberg had proposed to Kockel to study the action of the *Schüttelwirkung*. The prime aim of the following considerations of Góra was then to show that the above equations could exactly be solved if such a temporal setting in of the perturbation was taken into account. Since Góra's calculations for the case of scattering from a free particle neglected a specific intermediate state that was actually needed for a correct evaluation, I will not discuss his contribution any further.

Yet, I do want to point out that a shift in the usage of the *Schüttelwirkung* occurred. Normally, the argument went: There are terms in the expression of the probability coefficient that depend on t_0 and therefore they should be dropped. Certainly, it was also known that these terms would vanish if the perturbation was switched on adiabatically. But Góra, on the other hand, did not want to drop these terms: He looked for suitable physical conditions that would allow him to coherently exclude these dependencies from the start. Through this reasoning, Góra altered the mathematical representation of the intermediate state. Where regular perturbation theory (with $t_0 = 0$) would give

---

159 For the mathematical representation of this problem the reader is referred to Equation (8) of Wilson (1941, 305) and the connected elaborations.

160 "Diese Schwierigkeit lässt sich umgehen, wenn mann [sic] die Anfangsbedingung $b_I(0) = 0$ aufgibt und b_I proportional zu b_0 ansetzt, also nur die erzwungenen Schwingungen in Betracht zieht. [...] Mit einem derartigen Ansatz würde man aber voraussetzen, dass die Darsteller der Zwischenzustände dauernd mit b_0 koexistieren, was natürlich sinnlos ist. Dieses Vorgehen wäre also ganz willkürlich und gäbe auch keinerlei Aufschluss über eine etwa in Frage kommende näherungsweise Gültigkeit der so erhaltenen Resultate." Góra (1938, 163).

161 "ohne jedoch auf die damit verbundenen begrifflichen Schwierigkeiten einzugehen." Góra (1938, 163).

$$b_I \propto H_{0I} \frac{e^{i(E_i - E_0)t/\hbar} - 1}{E_i - E_0} b_0 \tag{9.20}$$

the exclusion of the *Schüttelwirkung* through the adiabatic switching on of the perturbation rendered[162]

$$b_I \propto H_{0I} \frac{e^{i(E_i - E_0)t/\hbar}}{E_i - E_0} b_0 \tag{9.21}$$

Góra argued that we can coherently envision physical circumstances in which the state b_I was coexisting with the state b_0. In effect, this constituted a physically motivated shift in the mathematical representation of the intermediate state.

9.3.2 Theory of Radiation Damping and Meson Physics

The proposal of the inclusion of radiation damping had its major impact on the theory of meson scattering. The empirical background and the theoretical discussion leading up to Heitler's theory of radiation damping have already been treated in the historical literature.[163] In the late 1930s, European physicists had already noticed Yukawa's meson theory and applied different modellings of it to explain cosmic ray phenomena.[164] There were diverse suggestions of how to explicitly model the meson field and its interactions with other particles. But the common problem, in retrospect, was that physicists still thought that the particles observed in cosmic rays were the mesons connected to the nuclear interaction, i.e. the differentiation between pion and muon was not yet available.

This misidentification effectively led to diverse troubles in explaining the empirical data. One of the most pressing ones around 1940 was that the theoretical cross section for scattering of mesons from nuclei turned out much too large and increased with increasing energy. In the possible explanations of this deviation, the intermediate or virtual states played a crucial role. For example, Bhabha (1939) proposed a classical theory with neutral mesons[165] that would eventually yield low enough cross sections. In a letter to Bhabha, Pauli pointed out that the decrease of the cross section

---

[162] See for example Góra (1938, 166, Eq. (17)).

[163] See Cassidy (1981) and Mukherji (1974, Section 16). Cassidy focused on the different stances towards the theoretical interpretation of the empirical material through a distinction between field theorists and more pragmatically inclined physicists as well as through the relevance of "explosions" in the theoretical description. Yet, the theory of radiation damping was referred to in the concluding sections of his paper in the respective empirical context. Blum (2017, Sect. 2.5.1) and Mukherji (1974, Section 16) have more thoroughly located Heitler's approach in the empirical background discussed below.

[164] For a discussion of the emergence of the theoretical work on meson physics in Britain shortly before the Second World War, see, for example, Brown and Rechenberg (1996, Chap. 7).

[165] For the introduction of the idea of neutral mesons, see e.g. Mukherji (1974, Sects. 6, 10, 11).

was effectively due to an increase of the different possible intermediate states: While for the scattering of a neutral meson two intermediate states were possible ((i) two mesons and the proton; (ii) only the proton) for charged mesons only one intermediate state was possible due to charge conservation (in the case of positive mesons $\rightarrow$ (i); for negative mesons $\rightarrow$ (ii)). The two possible intermediate states in the case of neutral mesons effectively cancelled their contribution and therefore the cross section was reduced.[166]

This insight, in turn, led Bhabha (1940a) (also Bhabha (1940b)) to consider the possibility of higher charge states and Heitler (1940) (also Heitler and Ma (1940)) to propose higher spin and charge states for the nucleons, which would effectively allow for different intermediate states in the calculation. The fact, that such particles were as yet not observed, and the lesser influence on structures inside nuclei was in turn explained by a considerably higher rest mass of these particles. With these additional degrees of freedom, the charged mesons could to some extent behave as neutral ones in the interaction with nucleons. Different intermediate states were allowed which in turn led to a smaller cross section (and only a logarithmic divergence of the magnetic momenta).[167]

Nevertheless, Bhabha himself pointed out, that the introduction of higher spin and charge states alone would not suffice.[168] Based on a classical theory of radiation reaction by Dirac, that was envisioned to mitigate the divergence difficulties in QED, Bhabha suggested that radiation reaction would have decisive influence on the scattering cross section of mesons with nucleons at high energies. Bhabha's as well as Dirac's theories were completely classical, nevertheless they showed a way of how to avoid infinities occurring in the classical theory while incorporating radiation reaction. Consequently, this allowed to arrive at a formulation of, in Bhabha's case, meson physics that would render finite and seemingly sensible results. These insights in turn triggered Alan Wilson and Heitler (again) to look for a possibility of incorporating radiation damping into the quantum-theoretical description.

Wilson's Approach to Radiation Damping

As I will follow Heitler's method afterwards, I will first shortly discuss Alan Wilson's contribution.[169] In Wilson's paper first mention was made of the discussion of radiation damping by Ivar Waller (1934). Unsurprisingly, Waller hit upon the

---

[166] Pauli to Bhabha, May 4. 1939; as cited in Pauli (1985, 643). See also Bhabha (1939, 400–401).

[167] It is noteworthy that here a similar, but even more speculative reasoning is applied as in Dirac's initial proposal of the Dirac sea. An as yet unobserved particle (or state of particles) was proposed because it could be coherently used in the intermediate states in the theoretical description. Eventually, Bhabha (1940b) will actually state that particles with higher charge states might be accessible to observation.

[168] See Bhabha (1940a) and Bhabha (1940b).

[169] For a contextualization of Wilson's work in his biography, see Sondheimer (1999, 554–555).

same problem as Góra but simply dropped the terms and therefore implicitly introduced different initial conditions for the intermediate states without arguing for them (Waller 1934, 444). Wilson, on the other hand, pointed out that Góra neglected some of the intermediate states that were necessary for a coherent treatment of the radiation reaction. Consequently, Wilson saw in his contribution rather a generalization of Waller's treatment, especially since the procedure to get rid of the troubling terms was different from Góra's.

Instead of introducing the perturbation adiabatically, Wilson chose to study the case of a bound electron first[170] and then to go over to the case of a freely moving particle. Although there was a different motivation for dropping the terms involving the time development proportional to the difference between initial/final and intermediate state the two procedures "really [amount] to the same thing," as Wilson explicated:[171]

> The difficulty [...] is due to there being in general uncompensated terms involving $e^{i(E_f - E_n)t/\hbar}$ [f (n) corresponded to the final (intermediate) state], which arise from the necessity of imposing the initial condition $a_n = 0$ at $t = 0$. Waller neglected these terms without giving any reason for doing so, but Góra pointed out that these terms do not occur if the perturbation is considered as being introduced adiabatically, that is, over a time which is large compared with $\hbar/(E_n - E_i)$ and $\hbar/(E_n - E_f)$. Góra's procedure and ours are therefore two different methods of introducing the averaging to eliminate the unwanted terms: in Góra's method the average is introduced via the perturbation energy, and in ours via the initial wave function. (Wilson 1941, 306)

Wilson could construct a theory which took damping effects into account and the most important case, namely the scattering cross section of mesons from nucleons, became appreciably suppressed. Still, Wilson pointed out that the self-energy terms have simply been neglected in his approach which rendered the theory "probably, but not certainly, correct" (Wilson 1941, 315).

Heitler's Reconsideration of Góra's Contribution

At the same time as Wilson, Heitler followed Bhabha's idea of incorporating radiation damping into the description of scattering phenomena. Heitler's first paper was developed independently of Wilson and was more firmly connected to Góra's ideas. As mentioned above, Heitler did not know of Waller's paper until Wilson made him aware of it. Even after hearing of Waller's contribution he did not put much emphasis on it as "Waller's method does not, however, lead to a general result which could be applied to other cases as well" (Heitler 1941a, 300). A defect that was in Heitler's eyes

[170] In this case, as Wilson argued, the troublesome terms in the equations could be neglected.

[171] These two different possibilities of getting to the same result, namely the neglect of the *Schüttelwirkung*-terms, have also been pointed out by Heisenberg in a letter to Pauli: "[...], dass all diese anderen Beiträge auf das Konto der 'Schüttelwirkung' gesetzt werden können (sie hängen von t_0 ab) und bei geeigneten Annahmen (z.B. adiabatisches Einschalten der Wechselwirkung oder: große Wellenpakete, die z.Z. [zur Zeit] t_0 noch nicht überlappen etc.) beliebig klein gemacht werden können." Heisenberg to Pauli, February 2. 1937; as cited in Pauli (1985, 511).

cured by Góra's approach. Although Heitler acknowledged the mistakes in Góra's paper, he stated that:

> In fact, the problem [of the incorporation of radiation damping] has already been attacked and solved in principle by Góra. [...] The following method of solution is largely based on Góra's ideas. (Heitler 1941a, 292)

Heitler initialized his discussion with the relativistic description of scattering of light as composed of two successive acts of emission and absorption.

> In order to calculate correctly the reaction of the radiation it is, however, necessary to take into account the fact that the once scattered quantum can be rescattered into another direction. Thus we have to introduce a third intermediate state (iii), corresponding to the emission of a second scattered quantum $\hbar\nu_\lambda$ from the state where one scattered light quantum $\hbar\nu$ is present.[172]

Denoting the initial, final, and intermediate states by $A,\ F_\nu,\ I,\ II_\nu,\ III_{\nu\lambda}$ and correspondingly their amplitudes and their energies, Heitler arrived at the following set of equations ($\hbar$ was set equal to 1 and a summation over indexes appearing twice was implied):

$$-i\dot{b}_A = H_{AI}b'e^{i(E'-E_A)t} + \sum_\nu H_{AII}b''_\nu e^{i(E''_\nu - E_A)t} \tag{9.22}$$

$$-i\dot{b}' = H_{IA}b_A e^{i(E_A-E')t} + \sum_\lambda H_{IF_\lambda}b_\lambda e^{i(E_\lambda - E')t} \tag{9.23}$$

$$-i\dot{b}''_\nu = H_{IIA}b_A e^{i(E_A-E''_\nu)t} + H_{IIF_\nu}b_\nu e^{i(E_\nu - E''_\nu)t} \tag{9.24}$$

$$\begin{aligned}-i\dot{b}_\nu = {} & H_{F_\nu I}b'e^{i(E'-E_\nu)t} + H_{F_\nu II}b''_\nu e^{i(E''_\nu - E_\nu)t} + \\ & + \sum_\lambda H_{F_\nu III}b'''_{\nu\lambda} e^{i(E'''_{\nu\lambda} - E_\nu)t}\end{aligned} \tag{9.25}$$

$$-i\dot{b}'''_{\nu\lambda} = H_{IIIF_\nu}b_\nu e^{i(E_\nu - E'''_{\nu\lambda})t} + H_{IIIF_\lambda}b_\lambda e^{i(E_\lambda - E'''_{\nu\lambda})t} \tag{9.26}$$

These equations were also given by Wilson and Waller and were not entirely without neglect. Equations (9.22), (9.23), and (9.25) were exact. In Eqs. (9.24) and (9.26) terms had been omitted which would correspond to an additional emission of a light quantum.

Noting that the solution of these equations in terms of Weisskopf's and Wigner's methods was impossible, Heitler mentioned that "substantial progress was made by Góra who showed that a much simpler solution of [(9.22)–(9.26)] can be obtained if we assume that *the interaction H is switched on adiabatically*" (Heitler 1941a, 294, emphasis in the original). The simplification Heitler referred to was essentially the

---

172 Heitler (1941a, 293). For the motivation of the introduction of the third intermediate state, see also Wilson (1941, 306–307).

disregarding of any further initial conditions of the intermediate states. As Heitler put it: “They may be assumed to be coexistent with b_A or b_ν.” (Heitler 1941a, 294)

With this assumption at hand, Heitler was able to solve the equation by standard procedures. He arrived at a general representation of a scattering formula which took into account the radiation damping effects. The final equations read:

$$\begin{aligned} \Gamma &= 2\pi \rho_{F_\nu} H_{AF_\nu} U_\nu \\ U_\nu &= H_{F_\nu A} + i\pi \rho_{F_\lambda} H_{F_\nu F_\lambda} U_\lambda \end{aligned} \tag{9.27}$$

where the matrix elements were constituted of sums over the intermediate states coupling the respective initial and final state indicated by the index. As Heitler pointed out, the second of the equations represented an integral equation from which one could calculate the physically interesting quantity U_ν.

Heitler (1941a, 296–299) then used his theory to compare the scattering cross section for different processes and models. For the case of non-relativistic scattering of a photon Heitler found that it merged with the classical formula. The most important motivation for taking up the theory of radiation damping, the scattering of mesons, was also discussed by Heitler. For low energies, his and Bhabha’s introduction of higher charge states still needed to be taken into account since the radiation damping only became important for high energies. But for this regime, Heitler showed that radiation damping effectively introduced a decrease in the cross section.

To provide for a short analysis and contextualization, it is to note that in Heitler’s, Góra’s, and Wilson’s discussion of radiation damping we encounter a mechanism which is complementary to Dirac’s introduction of the infinity of negative energy electrons (see Chap. 7). While the occurrence of negative energy intermediate states motivated Dirac to take a closer look at these theoretical entities again, the theoretical modelling he came up with, i.e. the Dirac sea, was informed by different concepts and areas of investigation. For Heitler and Wilson the motivation for taking up the question of radiation damping came from inconsistencies between theoretical predictions for the scattering of mesons and the empirical data. But the solution they proposed went through a reconfiguration of the temporal development of the intermediate, i.e. virtual, states. Certainly, there were different ways of getting rid of the specific terms that needed to be cancelled to apply Wigner’s and Weisskopf’s procedure. Yet, even after becoming aware of these different possibilities, Heitler used the adiabatic switching on of the perturbation, i.e. the exclusion of the *Schüttelwirkung*, to motivate the alteration of the time-dependence of the intermediate state.[173]

Corresponding to the general algorithm laid out in Sect. 8.3 we can also understand why this approach was general and in fact applicable to any interaction between different fields, as long as they were studied perturbatively. The method altered the behaviour of the intermediate state before any assumptions about the specifics of the interaction were introduced. It altered the perturbative series in a rather general

[173] See e.g. Heitler and Peng (1942, 299).

manner. Through the concept of the *Schüttelwirkung*, a general scheme was envisioned that could be implemented for the study of different phenomena and areas of theoretical physics, as meson physics or the light-matter interaction. This treatment rested on a physically motivated alteration of the concept of an intermediate state: They were assumed to coexist with the initial state.

9.3.3 Outlook and Intermediate Conclusion

Wilson's and Heitler's papers were the first steps of a more general scheme that was developed by Heitler and Peng Huanwu, a post-doctoral student at the Dublin Institute for Advanced Studies. They aimed at a "*general and unambiguous treatment of the radiation problem* [...], including the reaction of the field, without coming in conflict with the diverging effects" (Heitler and Peng 1942, 296, emphasis in the original). This scheme rested on the priorly developed incorporation of radiation damping and the observation that, so far, all the empirical predictions of quantum electrodynamics to first order "are found to be in excellent agreement with the experiments, without exception" (Heitler and Peng 1942, 296). Thus, higher order contributions, the ones causing the infinities, were simply disregarded.

The details of the calculations are not of particular relevance at this point. But, first of all, a differentiation Heitler and Peng, and also Heitler in his second edition of his textbook (Heitler, 1944) used, is interesting: The whole procedure was based on a differentiation between intermediate states and states with the same energy as the initial state (a bit confusingly all called "initial states" by Heitler and Peng). These two groups of states were set apart and structured the evaluation. Secondly, Heitler and Peng still referred to the adiabatic switching on of the perturbation as a reason for the alteration of the initial conditions. Heitler, in his 1944 textbook, rather tried to mitigate the conceptual problem of using non-zero initial amplitudes for intermediate states: He argued that, as these contributions were small and rapidly varying functions of the time, "there can [...] be no physical meaning at all in demanding any initial conditions for b_i, b_α, ..." (Heitler 1944, 242).

At around the same time, also Góra (1943) followed the topic and actually gave the most influential representation of the same.[174] Not all of the contributions by Waller, Góra, Wilson, Heitler and Peng made use of the adiabatic switching of the perturbation. Therefore, the above discussion was not an argument that radiation damping could only be incorporated through this idea.[175] But, as a matter of fact, the discussion of the *Schüttelwirkung* by Bernhard Kockel, the connected adiabatic onset of the perturbation, and the connected alteration of the concept of an intermediate state

[174] Also Sokolov (1941) introduced radiation damping into the scattering mechanism.

[175] It should still be noted, that the differentiation between final/initial state and intermediate state was important for all of these contributions and each altered the temporal evolution of the intermediate states.

as discussed by Góra stood at the beginning of Heitler's own theoretical evaluation in terms of time-dependent perturbation theory.

The above historical connection and the alteration of the concept of the intermediate state as proposed by Góra and Heitler have so far not been spelled out in the historical literature. Yet, we know very well how these elaborations on radiation theory ended.[176] Heitler's integral equation was for a time discussed as a possible candidate for filling the frame of Heisenberg's S-matrix theory, especially after static phenomena could be included in the discussion through the method of analytic continuation. Yet, there were two other problems which proved unsolvable.

On the one hand, as Heitler simply excluded all higher order contributions, the calculation of magnetic moments of nucleons faced serious difficulties. On the other hand and even more importantly, Heitler's theory reintroduced the infrared divergence (see also Sect. 10.1.3). First hints at this fact can be found in Pauli's letters. As early as November 27. 1944, Pauli wrote to Wentzel that Dirac's proposal of photons with negative energy but also Heitler's theory of radiation damping "results in something wrong for the scattering of many photons of low frequencies".[177] Effectively, shortly after the end of the war this defect of the theory led to its decline and Heitler's approach was eventually overshadowed by the new covariant theories of Schwinger, Feynman, and Dyson.

The above study of its development was, thus, not due to the relevance of the method for the further development of the theory. It should rather show that the concept of an intermediate state was one of the focal points of the discussions around scattering theory in terms of perturbation theory. In the original framing of Heitler and Peng, the theory rested on the dissection between initial and intermediate states, i.e. of states with nearly the same energy as the initial state and states that differed appreciably in energy. The temporal development of the states falling in the second category was manipulated to allow for a specific kind of modelling: The mathematical representation of the intermediate states was flexible, at least to a certain extent.

To use Theodore Arabatzis' terminology, "the recalcitrance of theoretical entities" (Arabatzis 2006, 37) can also be exemplified in the case of the intermediate state. Its mathematical representation could (through reasonable assumptions) be altered, but not at will: The theory allowed the incorporation of radiation damping through the manipulation of the mathematical representation of the intermediate or virtual states, but the exclusion of any higher order terms, i.e. processes through further intermediate or virtual states, proved to be untenable as it (re-)introduced the infrared divergence.

[176] The most detailed analyses, which I will follow here, were given by Alexander Blum (2017, Sect. 2.5.1) and Darrigol (1982, Sect. 7.4).

[177] "[...] gibt sie etwas Falsches für die Streuung von vielen Photonen kleiner Frequenzen." Pauli to Wentzel, November 27. 1944; as cited in Pauli (1993, 249). Similarly, Pauli to Bhabha, March 29. 1945; as cited in Pauli (1993, 260–264).

9.4 Analysis

9.4.1 *Terminological Variants, Their Spread and Their Meaning*

As should have become clear through the discussion of the practice of time-dependent perturbation theory throughout the last two chapters, the virtual or intermediate states were the connecting pieces, the mediating entities in the theoretical description of 1930's QED. Fermi and Bethe called the virtual states (and not the intermediary photon) the "interlink" in their theoretical description of the scattering of two electrons. Heitler referred to an "intermediary intermediate state". And this should not be conceived of as a doubling in terminology. "Intermediary" has a mediating connotation, while "intermediate" merely describes the position of an entity in relation to others. The above two terminological variants thus reflect the centrality of this concept. According to Heitler, the intermediate states had "agency".

As discussed in Sect. 6.2.2, C. V. Raman first explicitly established the connection between the intermediate states or the energy non-conserving transitions to and from them with the term virtual in 1929. Yet, initially theoretical physicists applied the terminology only rarely. This changed at around 1934/35. At this point, the usage of the term "virtual" in direct relation to the concepts involved in perturbation theory[178] became more regular. As the concepts themselves did not change, this must be explained otherwise. Certainly, at around this time these concepts gathered more attention through the acceptance and evaluation of the Dirac sea but also in the field theoretical description of nuclear physics (see also the following chapter). But probably even more importantly, Werner Heisenberg, the standing of whom in the community can hardly be overestimated, started to use the term at around 1934. And Hans Euler, his protégé, invoked the term in his widely received work on light-by-light scattering. The prominence of the concepts seemed to have necessitated a specific terminology and an influential actor provided it.

But, as discussed in Sect. 9.1.2, Heisenberg and Euler introduced the rather idiosyncratic terminological variant of the virtual possibility (of processes). The connotation of energy non-conservation for the processes was of course particularly prominent in Euler's evaluation: The restriction to low frequency photons was the starting point of their investigation. Yet, the contextualization of this terminology within Heisenberg's conception of quantum physics and the sharp cut between actual and quantum world through the measurement process suggests another layer of meaning: I argued that Heisenberg had one of the most important connotations of the term "virtual" of the BKS paper in mind, i.e. the reference to a virtual world which is by definition unobserved.

[178] There were also other invocations of the term "virtual" which I will shortly discuss in the concluding chapters. The "virtual states" as introduced terminologically in 1930 by Guido Beck have been studied by Martinez (2024, Sect. 2).

Heisenberg's and Euler's terminological variant did not prevail. They themselves dropped the possibilities regularly. Eventually, and in harmony with the methods of perturbation theory of the time, physicists started to refer to virtual processes (transitions, emission, absorption, creation, ...): According to the mathematical representation, the transitions to and from the intermediate states did not conserve energy and thereby broke with generally accepted laws of physics. But, as we have seen, physicists started to create the terminological combination of the "virtual" and the "particle". They did so when the particles were exclusively present in intermediate states, when the particles were the "catalytic agents", as Furry has expressed it.

It is certainly reasonable not to paint a black and white picture at this point. Rather one should regard the historical passage from the conception of a "virtual transition" to the "virtual particle" as a gradual process which manifested itself in the different representational formats the actors had at their disposal. I will engage with this process more fully in the next chapters. For the moment, I still want to emphasize that, from the point of view of the practice of perturbation theory, the "virtual particles" were not the primary entities but rather a consequence of its application which rested on the concepts of virtual transitions and states.

9.4.2 Physical Analogies and the Conception of Non-Relativistic Quantum Mechanics

Hans Euler, who was rather careful in not referring to "virtual particles" in his dissertation, used the term "virtual matter" once. He did so on the "intuitive" level of the theory, where the quantum-theoretical effects were represented as additional terms to the Maxwell equations. Euler actually used two different mathematical representations on the intuitive level: In one of these representations, additional source and current densities were included in the canonical equations. From this point of view, the terminology of the virtual matter can be better understood. It was represented by an only "apparent" additional charge density. But the intuitive level of the theory was also the place at which Euler most explicitly referred to analogical phenomena. And these analogies indicate how strongly the conception of non-relativistic quantum mechanics and QED still overlapped.[179]

Euler filled several pages putting side-by-side the description of light-by-light scattering and van der Waals forces, essentially a dispersive phenomenon, to make the additional contributions to the Maxwell equations accessible to his readers. Also Wolfang Pauli, when arguing for the existence of light-by-light scattering at a seminar held at the Institute for Advanced Study, invoked dispersion: "This process [light-by-light scattering] has the same relation to the generation of pairs as dispersion to absorption, so it is probable that such a process takes place." (Pauli 1936b, 1) Also

[179] As already mentioned, this general conceptual overlap of non-relativistic quantum mechanics and QED/QFT during the 1930s has most strikingly been identified and investigated by Wüthrich (2010, Chap. 2) and Blum (2017).

Maurice Pryce, reporting on light-by-light scattering at the seminar, invoked the van der Waals forces:

> A close analogy exists between the present work [light-by-light scattering] and the theory of Van der Waals forces between two atoms since these forces arise from virtual transitions of electrons in the atoms from low energy states to higher and then back. The ordinary Coulomb interaction is not sufficient to let the electrons stay in the excited states, just as the photons of visible light will not permit actual pair formation, but the virtual electron transitions caused by the Coulomb interaction lead to second order interaction terms which are just the Van der Waals forces. (Pauli 1936b, 46, emphasis in the original)

This invocation is exemplary for the strong conceptual overlap of non-relativistic quantum mechanics and QED: Both rested on the idea that processes took place through transitions between stationary states of the systems.

Such analogies had an obvious purpose: They made the processes studied understandable, maybe conceivable, through the invocation of already known phenomena. The reference to resonance fluorescence by Amaldi and Segrè (1929, 408) in order to explain the Raman effect is a case in which the corresponding real process was invoked to explain its virtual counterpart. The diagrams by Maria Göppert-Mayer (Fig. 8.1) were explicitly introduced to represent the analogy between Raman effect and double absorption/emission.

But physicists also used analogies in order to derive the mathematical structures describing the investigated effect. Tamm and Dirac noted the analogical structure of pair annihilation and scattering.[180] Victor Weisskopf (1931, 37) invoked the theoretical description of (real) double emission as a formal analogy for his derivation of resonance scattering. The invocation of the process of light-by-light scattering in the evaluation of the Delbrück effect by Akhiezer and Pomeranchuck (see Sect. 9.2.3) is another quite striking example. In order to draw this analogy and to invoke the theoretical structures involved in the evaluation of light-by-light scattering, Akhiezer and Pomeranchuck had to introduce a concept which was (at best) only implicit in prior calculations: the concept of a "pseudo-photon" represented by the Fourier components of the Coulomb potential. In this case, the invocation of the analogy certainly contributed to the concept formation process.

9.4.3 The Schüttelwirkung *and the Flexibility of the Intermediate States*

The close study of the practice of perturbation theory unravelled a little-known concept only applied within perturbation theory: the *Schüttelwirkung* and the connected proposal of switching on the perturbation adiabatically. When Maria Göppert-Mayer introduced the concept to Dirac's radiation theory (Sect. 8.1.1), it served to exclude transitions that did not conserve energy in the overall process. Weisskopf applied it

[180] As discussed in Sect. 7.6, this analogy would lead to conceptual incoherences as long as a hole in the Dirac sea was conceived of as a proton.

in each step of his first derivation of a general perturbative formula (see Sect. 8.1.4). Kockel engaged with the *Schüttelwirkung* to study which kind of processes corresponded to actual ones and which were artefacts of the perturbative method (see Sect. 9.1.4). This led Góra to apply the connected adiabatic switching on of the perturbation to arrive at a specific modelling of the intermediate state, namely as already initially excited. Why did he do so? Because only in this case the Ansatz of Weisskopf and Wigner (as shortly discussed in Sect. 8.1.2) could be applied to the problem he studied. Góra's paper as well as the alteration of the concept of the intermediate states and their boundary conditions was in the further development the entrance point for Walter Heitler in his description of radiation damping.

The concept of the *Schüttelwirkung* is not only of interest because it has a strong physical connotation yet was exclusively used in the evaluation of (the often unphysically framed) perturbation theory. It also shows that physicists could manipulate the central concepts of perturbation theory, or rather their behaviour. The concept of the virtual or intermediate state proved to be flexible. The idea that one might turn on the perturbation adiabatically led to an alteration of the mathematical representation of the concept[181] which could simply be interpreted: The boundary conditions of these states changed (or were simply disregarded); the states could be conceived of as excited initially. Thereby, the mathematical modelling of Weisskopf and Wigner and radiation damping could be incorporated.

9.4.4 The Role of Different Formats of Representation in the Quantitative Evaluation

In Sect. 8.3 of the previous Chapter, I already discussed how the implicit algorithm of the perturbative evaluation of QED during the 1930s can be conceived of. In this chapter, some further representational formats and methods were presented which allow for a deeper study of this scheme and the relevance of specific representations. Euler's and Kockel's dissertations most strikingly exemplify that the derivation of the general perturbative series was nearly purely mathematical: Next to no assumptions about the structures of the perturbation or the intermediate states were introduced.

When filling this mathematical framework with physical content in order to study a specific process, other, non-mathematical representations were applied. This is the place where Dirac's verbal model, as I called it, "the language of Dirac's theory of radiation", as Halpern called it, or "the picture of 'successive' transitions", as Bernhard Kockel called it, was applied. Language functioned as a tool for quantitative evaluation and took the same place as Feynman diagrams do today: It bridged the gap

[181] The causal relation implied above is only concerned with the usage of the adiabatic process. Considering the motivation, things were probably the other way around: The exponential decay could not be incorporated as long as one assumed the usual initial conditions and it was obvious which mathematical terms blocked the application. The adiabatic switching on of the perturbation provided a way of avoiding these terms.

between the abstract mathematical form of the perturbative series and the concrete combinations of matrix elements to be calculated in a quantitative evaluation of the physical process under investigation.[182]

Physicists did not only represent the underlying conception of successive transitions verbally but did so in terms of diagrams, tables and other schemes. As I discussed more fully in the respective sections, the diagrams by Maria Göppert-Mayer (Fig. 8.1), Eduardo Amaldi (Fig. 6.1), or Walter Heitler and Lothar Nordheim (Fig. 8.3)[183] clearly indicated a temporal order of the subprocesses. Also Hans Euler indicated the order in which his diagrams had to be read (Fig. 9.1). Akhiezer and Pomeranchuck used labels on the transitions to make the order of the processes visible (Fig. 9.6). Also in tables a clear sequential representation was implied: They were read from left to right.

Considering the functionality of the diagrams more specifically, they certainly were a concise representation of the transitions implied in the verbal model of Dirac. They might have been intuitive and could serve a didactic purpose, but I mostly regard them (at least so in the hands of the physicists of the 1930s) as a condensed version of the verbal statements. For the most part, I believe that they were after the fact representations. They did not provide for new inferential possibilities but allowed a concise and intuitive fashion of display.[184]

Let me argue for the above point by considering the usage of the diagrams by Euler and compare them to a different representational format of the same process, the tables. The one time Euler specifically referred to the diagrams, he did so for connecting (the indexes of) variables with the specific electrons involved in the process. The table could have performed this task equally well. But Euler also referred directly to the table: The necessary permutation of the photons could be represented by a permutation of the entries in the headline of the table. Also in this case, the permutation might simply be invoked verbally and mathematically represented by a sum.

At around the mid 1930s, diagrams were not seen as epistemologically superior to tables as such. Bernhard Kockel who had been deeply involved in the evaluation of light-by-light scattering did not use diagrams in his dissertation. He used tables. Also Maurice Pryce represented the processes leading to light-by-light scattering in a

[182] I already indicated this bridging function of the verbal model of Dirac and its representations in the conclusion of Ehberger (2022).

[183] See also my discussion in Ehberger (2022, Sect. 2.1) for the first two of these three examples.

[184] A possible exception to this rule, which interestingly has no connection to energy level diagrams and which comes closest to being a "paper tool", were Wentzels Schemata i, to be discussed in Sect. 10.2.4. For the sake of clarity, the above argument does not contradict Adrian Wüthrich's observation that Japanese physicists built upon Euler's diagrams (Wüthrich 2010, Chapter 2). While the Japanese researchers spelled out the inferential possibilities the diagrams allowed for, Euler did not.

table in his presentation at a seminar at the Institute of Advanced Study (Pauli 1936b, 44). Although Euler's dissertation was most probably not yet published when Pryce gave his presentation, its contents allow for the conclusion that Pryce had first-hand knowledge (most probably through Pauli) about the paper as it was written.

The fact that tables were as efficient as diagrams to represent the sequences of virtual subprocesses is connected to the focus of the physical modelling: Only the material part of the system restricted the way in which this sequence manifested. Each light quantum could be involved in each subprocesses. As Adrian Wüthrich (2010, 20) has expressed it, the electromagnetic field was "conceived of as a reservoir of photons". This partial dissection of transitions and the involved photons was already mirrored in Weisskopf's first derivation of a general perturbative expansion (see Sect. 8.1.4). In general, there was no necessity of representing the photons explicitly. Thus, the stepwise behaviour of the material part could as well be listed in the table. And in this way the tables served the same purpose as the diagrams: They were a concise representation of the underlying physical model of the perturbative evaluation of the 1930s. No more, no less.

9.4.5 Rules for Intermediate States and Perturbation Theory

While the development of the theory of radiation damping showed how specific representations of the intermediate state could be manipulated, physicists also tried to get a better grasp on the rules governing the application of perturbation theory. Igor Tamm (see Sect. 7.7.2) and, seemingly unaware of Tamm's contribution, Bernhard Kockel (see Sect. 9.1.4) evaluated to which extent the Pauli exclusion principle needed to be taken into account for the intermediate states. Kockel came up with a concrete rule which he proved by direct inspection to the fifth order. Kockel's investigation had a (potential) practical benefit:[185] Disregarding the exclusion principle (if possible) reduced the number of routes leading from initial to final state. The calculational efforts necessary for a quantitative evaluation were thereby reduced.

In 1937, Wendell H. Furry proved a theorem which still holds his name and is applied in perturbative evaluations: Furry's theorem. It states that the probability of odd order processes vanishes identically if the electrons and positrons are only present virtually, i.e. if the process exclusively includes photons as real particles. Furry's theorem, as Kockel's conclusion, reduced the calculational efforts involved:

[185] The benefit was only potential insofar as physicists seemed not to have used Kockel's conclusion in the further development.

Some in principle conceivable processes could be excluded before engaging with the concrete mathematical evaluation. But in contrast to Kockel's work, Furry's theorem still holds and is applied today. In a post-Feynman application of perturbation theory, the exclusion principle can be neglected for all possible configurations.[186]

9.4.6 *False Predictions and the Difference Between Virtual Transitions and Virtual Particles*

I will more fully engage with the false predictions I came across in my research in Sect. 12.2, yet I will provide for a rather shortened analysis at this point. According to my definition, a "false prediction" is a process which was conceivable according to the rules governing the usage of specific representations of the model of QED, but which turned out to have a vanishing probability when actually calculated.[187] And we have already encountered four instances of false predictions: Tamm's proposal of scattering of light from a pair of electrons (Sect. 7.7), Halpern's prediction of light-quantum splitting (Sect. 9.1.1), Kockel's proposal of the creation of two pairs from three photons (Sect. 9.1.4), and the (re-occurring) conception of the Delbrück effect as a third order process (Sect. 9.2).

The historiographic concept of a false prediction has a twofold purpose in my study. On the one hand, false predictions can figure as an analytical tool to frame and understand the further development of the practices, representations, and concepts. On the other hand, false predictions allow us to probe the practical, representational, and effectively conceptual differences between QED of the 1930s and the "same" theory in the late 1940s in some more detail.

Regarding the first point, although theoretical progress was no necessary consequence of a false prediction, a false prediction could motivate the evaluation of the conceptual and methodological framework. The derivation of Furry's theorem was obviously motivated by the fact that physicists stumbled over the false prediction of a third order Delbrück scattering diverse times. Tamm's false prediction, together with Dirac's reconceptualization of scattering in a filled Dirac sea, led him to study the possibility of light scattering from a pair of electrons in more general terms and to a first direct evaluation of the relevance of the Pauli exclusion principle for the intermediate states.

Regarding the second point, i.e. the differences between QED during the 1930s and the late 1940s, Tamm's and Kockel's false predictions are particularly enlightening because they cannot be represented in terms of Feynman diagrams. Tamm's case is pretty clear but also Kockel's conjecture can easily be seen not to be representable through Feynman diagrams by trying to draw a corresponding closed diagram: It

[186] For the discussion of these two aspects of Feynman's version of QED, see Sects. 11.2.3 and 11.3.2.

[187] This is already a specialised framing of the "false prediction". For my most general understanding of this analytical term, see Sect. 2.3.3.

doesn't work.[188] While Tamm and Kockel had to transform their verbal statements (or tables) into mathematics to find out that specific matrix elements exactly cancelled, physicists applying Feynman diagrams can rule out the corresponding processes by imagining the processes with pen and paper.[189]

The point why these processes turn out to be non-existent is also interesting for our purpose. In the representation in terms of Feynman diagrams, you get one (or two) unconnected single and unphysical building blocks. All the same, at least Kockel explicitly stated that his process should be allowed according to the general principles of energy and momentum conservation of the then current perturbative methods, i.e. momentum conservation at each interaction, energy conservation only for the whole process. Using virtual transitions, in the sense that energy was only conserved for the overall process, leads to different conclusions than using Feynman diagrams, i.e. virtual particles (which do not obey the relativistic energy-momentum relation) but where energy *and* momentum are conserved at each interaction point.

The examples of Tamm's and Kockel's false predictions show that the difference between the methods of the 1930s (with virtual transitions) and of the 1940s (with virtual particles and Feynman diagrams) is deeper than one of calculational effort.[190] In their conceptual framework, Tamm and Kockel could conceive of processes which are directly ruled out by the representational system of Feynman diagrams. Understanding concepts as manifested in their representations, this points to the fact that the differences of the (representation of) *virtual* entities in QED in the 1930s and the QED of the late 1940s reflected on the thinking possibilities of the historical actors, i.e. which *actual* processes were conceivable. The difference, often framed as one of calculational effort, is one of conception: Physicists in the 1930s and the late 1940s did not only conceive of the same processes differently, but they could conceive of different processes.

[188] We know how many external electron/positron lines (two each) and how many external photon lines (three) we have. Starting from there, one will pretty soon recognizes that no diagram corresponds to such a process (at least to third order).

[189] Turning things around, a physicist working with Feynman diagrams would most probably not think of the scattering processes Tamm and Kockel had envisioned.

[190] With the following point, I also want to contribute to the identified conceptual differences between QED of the 1930s and the late 1940s, as discussed by Wüthrich (2010), Blum (2015), or Blum (2017) (see also Sect. 1.3.2).

Part III
From Virtual Transitions to Virtual Particles (1930–1949)

Chapter 10
In Between: Traces of the Virtual Particle During the 1930s

In the prior chapters, I nearly exclusively engaged with the practice of time-dependent perturbation theory, its representations, and phenomena, such as scattering, which can be considered as temporal in nature. I thereby broke with chronology. But the discussion thus far will serve as an important backdrop for this final chapter on QFT during the 1930s. In the following, I will engage with the role and the representations of theoretical entities we can reconstruct as lying in one historical line with the virtual particle concept in the calculations of the divergences physicists encountered during the 1930s, on the one hand, and the interrelation between these concepts and their representations with quantum field theoretical approaches to nuclear physics, on the other.

Two observations motivate this arrangement. First of all, the phenomena to be discussed in this chapter were often described by time-independent perturbation theory. As I will argue, the difference to time-dependent perturbation theory initially made a notable impact on the representational practices and the conceptions of the physicists. Secondly, the theoretical description of the effects contained particles which were only present, according to the verbal model of radiation theory, in intermediate states, which made the attribution of the virtual character to the particles themselves more suggestive. Obviously, I discussed such phenomena before[1] but focused on the method of calculation. Now, by drawing such phenomena together, I want to discuss how their theoretical investigation altered the perception of physicists, while the general method of the perturbative evaluation remained stable during the considered period of time. According to the following reconstruction, there was a considerable overlap of the concepts of virtual *processes* and virtual *particles* which started to show in the mid 1930s and manifested itself in the different representational formats the actors used.

This trend from virtual transitions to virtual particles is, on the one hand, connected to the usage of a temporal language on phenomena theoretical physicists themselves

[1] Most importantly light-by-light scattering and the Delbrück effect in Sects. 9.1 and 9.2.

M. Ehberger, *Representing the Unobservable*, Science Networks. Historical Studies 68,
https://doi.org/10.1007/978-3-032-09188-8_10

considered to be static. On the other hand, it is related to the emergence of prominent instantiations of the virtual particle concept, namely as force carriers, as vacuum fluctuations, and as a particle cloud surrounding another charged particle. At the same time and, as I will argue, due to the uncertainties in the specific modelling of quantum field theoretical approaches to nuclear physics, the invocation of an argument for including such concepts as virtual transitions and virtual particles in the conceptual framework of quantum field theory became prominent. The argumentative structures I called narratives in my methodological introduction gathered force in this entangled epistemic constellation.

To approach these developments, I will, as in the previous chapters, discuss the evaluation and representations of different effects and ideas connected to the virtual particle concept in the respective subsections. The contents of these subsections will thus be interrelated in their connection to the development of the virtual particle concept.[2] I will present a chronologically ordered picture of the development and an analysis which identifies the conceptual overlap of *virtual transitions* and *virtual particles* at the end of the 1930s in the concluding section.

The (non-)invocation of verbal and diagrammatic representations in the historical investigations of the self-energy of the electron (Sect. 10.1.1) will serve as an entrance point and motivation. It exemplifies the strong interrelation of concepts connected to the virtual particle with cultures of theory, methodology as well as the tool character of the different representations. Vacuum polarization and the emergence of the connected concept of a cloud of (charged) virtual particles (modern terminology) will be discussed in Sect. 10.1.2. By a close look at the different conceptual frameworks in which this effect was proposed, the theoretical entities respectively made responsible for it, and the terminology used, its connection to both the anomalous magnetic momenta of the nucleons and the narrative of nuclear theory will be indicated. That the concept of a "dressed" particle, including the effects of virtual particles, and a "bare" particle (modern terminology) had not yet been fully developed, or at least followed to its consequences, is evidenced by the state of the infrared divergence in the later half of the 1930s (Sect. 10.1.3).

Turning to nuclear physics, the emergence of the idea that forces are carried by intermediary particles will be discussed in Sect. 10.2.1. Although I mostly share Cathryn Carson's assessment (Carson 1996b) that this conception was first coherently formulated in nuclear theory and nearly independently from quantum field theoretical considerations, I will examine it more closely according to its methodological preconditions and consequences. This discussion, as the qualitative explanations of nuclear magnetic momenta, described in Sect. 10.2.2, will provide for the necessary background to understand the prominence of the narrative for the inclusion of virtual particles and transitions in the conceptual framework of QFT in nuclear theory in the late 1930s (Sect. 10.2.3). In the last subsection and contrasting the mostly qualitative assessments of the historical actors priorly discussed, I will turn to a representational

[2] As those processes stretch over different fields of theoretical investigation, which have mostly been studied in distinct historical reconstructions, I will engage with the corresponding secondary literature in the respective subsections.

system that had already found some historical interest because of its striking similarity to Feynman diagrams: the "Schemata" of Gregor Wentzel. Evaluating their historical emergence, I will argue against a close historical relationship between Feynman diagrams and Wentzel's schemata. Yet, through their similarity to Feynman diagrams these schemata constitute a prime instance of the general theme of this chapter: the conceptual overlap of virtual particles and virtual processes through their representation in different formats.

10.1 The Divergences During the 1930s

The self-energy of the electron and, more generally, the divergences encountered in quantum field theory are probably the best studied phenomena in the history of QED during the 1930s.[3] This focus on the divergences is quite understandable as they constituted the main theoretical obstacle in the formulation of a consistent theory and this attribution was generally shared by the historical actors. During the 1930s, physicists devoted an appreciable amount of time on often too radical propositions for a solution. Only the renormalization procedures developed during the late 1940s allowed calculations beyond the first order, thus made QED a workable theory, and the main protagonists of these endeavours, Feynman, Schwinger, and Tomonaga, received a Noble Prize for their efforts in 1965.[4]

But comparing the representations of virtual entities (modern terminology) in the historiographic literature on 1930s QED with the historical papers themselves one encounters a striking dissimilarity. In most cases, if historians reconstructed the calculations, they applied verbally explicit forms of representations, including the concept of a virtual quantum,[5] and even diagrammatic representations[6] to make the calculational procedures accessible to their readers. Such representations are absent in the papers of the historical actors, at least so in the first contributions to the self-energy problem. This should not be misunderstood as a critique of the respective

[3] See, for example, Darrigol (1982, Chap. 3), Pais (1986, Chap. 16), Rueger (1992), Brown (1993), Schweber (1994, especially Sect. 2.5), Miller (1994b, Sects. 3 and 4), or Cao (1997, especially Sect. 7.6). Aramaki (1987, especially Sect. 2) and Blum (2015) provided the most detailed discussions of the infrared divergence, its connection to the ultraviolet divergences, and the later development of renormalization.

[4] The high esteem renormalization enjoys in connection with the development of QED can be judged from the title of the most popular and comprehensive historical work on QED: According to Sam Schweber (1994), Dyson, Feynman, Schwinger, and Tomonaga were the men who made QED.

[5] See for example, Pais (1986, 373) for explicitly spelling out such a model of the emission and reabsorption of a virtual photon in connection to Oppenheimer's work. But see also Darrigol (1982, 132) (in connection to Oppenheimer) or Blum (2015, 83) (in connection to Weisskopf). Miller (1994b, Sect. 3) is more careful in Oppenheimer's case. In total one must still note that Darrigol (1982, Chap. 3) is amongst the more careful reconstructions when it comes to the general development of the calculational procedures, especially Weisskopf's 1934 paper, and the involved concepts.

[6] Especially Sam Schweber (1994, Sect. 2.5.3) relied on a verbally and diagrammatically explicit reconstruction. See below.

historians: Such verbal and diagrammatic formats are often a lot more intuitive and transparent concerning the physical reasons for the occurrence of divergences or their mitigation.

All the same, the anachronistic use of representational formats can be problematic. Sam Schweber's application of Feynman diagrams to describe the reduction of the self-energy divergence to a logarithmic one in the Dirac sea constitutes such an example (Schweber 1994, 122–125). He included an unconnected bubble diagram to exemplify how the self-energy was reduced in a filled Dirac sea.[7] Yet, in terms of the most common usage of Feynman diagrams, unconnected diagrams do not alter the self-energy divergence, or the calculation of any other process for that matter. Instead of back-projecting representational formats on the practices of the historical actors, focusing on exactly these representations highlights conceptual differences in the practice of QED that have so far not been discussed.

10.1.1 The Self-Energy of the Electron and the Late Application of Non-Mathematical Representations in Its Evaluation

The problem of a divergent energy of a charged particle, a self-force, and even the idea of renormalization are older than QED.[8] But with the quantization of the electromagnetic field new and essentially quantum theoretical infinities occurred. The zero point energy of a quantized radiation field already showed up in the *three-man paper* (Born, Heisenberg, and Jordan 1926, Chaps. 4, Sect. 3), but as an additional constant it could largely be ignored. By 1931, Rosenfeld and Solomon (1931) figured out a way of formulating the theory so that this infinity would no longer occur.[9] Like the divergences connected to the mean square fluctuations of the radiation or the electron-positron field, it did not rely on the interaction between radiation and matter.[10]

In the following, I will exclusively discuss contributions which are due to the interaction between radiation field and matter and which were perturbatively investigated during the late 1920s and the 1930s. As the main line of my argument goes, the initial lack of non-mathematical representations of virtual (modern terminology) processes or particles in these publications (stretching to Weisskopf's calculation of the self-energy in a filled Dirac sea) needs a multi-factorial explanation, combining sociological and methodological reasons. But when physicists turned to the investigation of higher order contributions, the mathematical formalism had to or, at least,

---

[7] I will discuss this reduction more closely below.

[8] See e.g. Darrigol (1982, Sect. 3.1) and Dresden (1993).

[9] See, for example, Schweber (1994, 108–110) for an explanation of this divergence and a way of getting rid of it.

[10] See Heisenberg (1931b), Heisenberg (1934c), and Oppenheimer (1935) for contemporary discussions.

was enriched with verbal and diagrammatic representations. Such representations figured as tools in the theoretical investigations.

Initial Lack of Non-Mathematical Representations (1929/30)

For the following description of the self-energy calculations two things have to be kept in mind. First of all, Dirac's verbal model could have been used as a resource to describe the respective perturbative evaluations from the start. Dirac proposed it in 1927 (see Chap. 5), so two years prior to the first investigation of the self-energy and all involved actors were certainly aware of it. Secondly, when the transversal contribution to the self-energy was first calculated, the division of the quantized electromagnetic field into a part corresponding to the Coulomb interaction and a part corresponding to transversally polarized photons was already in place.[11] In the following, I will nearly exclusively focus on the transversal part of the self-energy.

In most general terms, all the performed calculations contained mathematical structures of the form

$$E_2 = -\sum_n \frac{H_{mn} H_{nm}}{E_n - E_m} = -\sum_n \frac{|H_{mn}|^2}{E_n - E_m} \tag{10.1}$$

where H_{mn} refers to the matrix element of the interaction Hamiltonian between two unperturbed states of the system m and n, and E_i to the respective energy eigenvalue of the state m to ith order.[12] As the interaction Hamiltonian coupled states which differed by one photon and since the state m only consisted of one electron, there was one possible configuration for the intermediate state n: It consisted of the electron in a different state and an additional photon.

At the same time and in contrast to the general framing of time-dependent perturbation theory, this equation allows for another reading, which will become important in the following sections: The term after the second equal sign in Eq. (10.1) suggests that this phenomenon can be attributed to the probability (not only possibility) of a transition to state n, hence to the emission of a light quantum. It should also be noted that the probability to find the system in this state was according to time-dependent perturbation theory heavily fluctuating around a small but finite value, an identification we will also need later on.

---

[11] See Carson (1996b, Sect. 3.1) for a discussion of the replacement of the non-transversal part of the electromagnetic field by the Coulomb interaction. As a matter of fact, Oppenheimer presented both this procedure and the first calculation of the transversal self-energy in one and the same paper (Oppenheimer 1930a).

[12] At times, the interaction Hamiltonian was written out before the energy perturbation was calculated which rendered the equations rather complicated, as for example in Heisenberg and Pauli (1929, 50–51, Eq. (107/108)).

Diagnosis of the Infinite Self-Energy

The initial calculations of the self-energy were all carried out in close contact with Heisenberg and Pauli and were nearly exclusively performed in Zürich. The electrostatic self-energy was already perturbatively calculated by Heisenberg and Pauli (1929) but was, as the infinite zero point energy, only an additive constant and thus simply dropped.[13] Further, Heisenberg and Pauli did not describe any of their calculations in terms of successive transitions, although this would certainly have been possible according to the structure of their derivation.[14] They expressed the self-energy calculation, if not in terms of successive emission and absorption processes, at least verbally explicitly in terms of the occupation of the intermediate states only a few years later.[15]

The first calculation that showed that the self-energy problem was far worse than a simple additive constant was published by Robert Oppenheimer (1930a). In 1929, he spent part of his fellowship by the International Research Council in Zürich where he was involved in the further development of QED in the hands of Heisenberg and Pauli. Together with Pauli, he showed that the Coulomb interaction could be expressed non-perturbatively by the non-transversal part of the electromagnetic field. He further investigated the displacement of energy levels, given by an equation of the same structure as Eq. (10.1), as well as the energy difference between two such states and found a divergent result. Thus, the self-energy was state-dependent and could not simply be dropped.[16] In Oppenheimer's calculations, we find mathematical representations which we could directly identify with virtual transitions and particles.[17] But Oppenheimer did not describe the formulas in a verbally explicit manner.

The same holds for Léon Rosenfeld's work on QED and the self-energies from 1929 to 1930. After his dissertation, Léon Rosenfeld first spent one and a half years in Göttingen (1927–1929). At about the time Oppenheimer left for the US, Rosenfeld arrived in Zürich, where he worked for the academic year 1929/1930. In Rosenfeld's treatment of longitudinal solutions of Heisenberg's and Pauli's QED, contrasting these excitations to the "'normal' light quanta with transversal oscillation"

---

[13] As noted by the authors, relativistic invariance was spoiled thereby.

[14] The fact that Heisenberg and Pauli did not describe their calculations verbally was already noted by Cathryn Carson (1996b, 112) when she discussed their perturbative derivation of the Coulomb interaction.

[15] See Pauli (1933, 269–270) and Heisenberg to Pauli, February 5. 1934; as cited in Pauli (1985, 273).

[16] That the self-energy problem was far worse than expected was clear by mid July 1929 as we know from letters from Pauli to Bohr and Heisenberg to Pauli: "In particular, the self-energy of the electrons causes a lot more difficulties than Heisenberg initial thought." [*Original:* "Insbesondere macht die Eigenenergie der Elektronen viel größere Schwierigkeiten als Heisenberg anfangs gedacht hat."] Pauli to Bohr, July 17. 1929; as cited in Pauli (1979, 513). Heisenberg met with Oppenheimer to discuss these problems during his US visit at about the same time. Heisenberg to Pauli, July 20. 1929; as cited in Pauli (1979, 514).

[17] See, for example, Oppenheimer (1930a, 471, Eq. (30) through (32)).

(Rosenfeld 1929, 548), as well as in his calculation of the gravitational self-energy of a photon (Rosenfeld 1930) and his calculation of the self-energy of a harmonic oscillator according to Dirac's radiation theory (Rosenfeld 1931),[18] the perturbative evaluations were only represented in a mathematical format.

Finally, Ivar Waller, already in contact with Pauli during his calculations of the scattering of light from free as well as from bound electrons (see Sect. 7.3), spent a short period of time in Zürich in early 1930. There he finished his calculation of the self-energy of a free electron (Waller 1930a). Waller, as his Zürich colleagues, did not mention successive emission and absorption processes or transitions explicitly.

This might at first seem quite astonishing considering the modern description of these contributions as well as their presentation in the secondary literature. Yet, the most explicit verbal statements were

- that the displacement of energy levels "must be ascribed to the effect of the interaction of the particle with its own field" (Oppenheimer 1930a, 476)
- "the infinite interaction of the electron with itself" (Oppenheimer 1930a, 463)
- "the interaction of a particle with itself,"[19]
- the solutions "which corresponds to an electron [...] only in interaction with its self-field"[20]
- or "the interaction of the electron with its own electromagnetic field".[21]

Of course, according to the then current standards all of the actors included the occupation of the intermediate state correctly in their calculations but they did not discuss it in any temporal fashion or explicitly framed it in a verbal/diagrammatic representation in terms of successive emission and absorption processes. While they obviously conceived of and understood the theoretical restrictions, i.e. only one additional quantum in the intermediate state,[22] even the term "intermediate state" (or any synonym) was completely absent. Since Dirac presented his verbal model in 1927 and all actors were certainly aware of it, I am next to certain that a reading in terms of emission and absorption subprocesses must have occurred to the respective physicists in the course of their calculation. But then the question why neither of these actors provided any kind of non-mathematical expression becomes even more pressing.

[18] Rosenfeld was already employed in Liège and closely working with Bohr in Copenhagen when he finished this last paper.

[19] "die Wechselwirkung eines Elektrons mit sich selbst" Heisenberg and Pauli (1929, 3).

[20] "die [...] nur mit seinem Eigenfeld in Wechselwirkung befindlichen Elektron entspricht," Waller (1930a, 674).

[21] "die Wechselwirkung des Elektrons mit seinem eigenen elektromagnetischen Felde," Rosenfeld (1931, 456).

[22] For a very clear presentation, see e.g. Equation (12) and the following reflections on the calculational scheme in Rosenfeld (1931, 456–457).

Explanation of the Non-Invocation: Methodological and Sociological Reasons

I believe that this fact must be explained by taking into account three reasons, two of which are concerned with the methodological approach, while the other one has a stronger sociological component (although it is connected to the interpretation and methodology of quantum theory as well).[23] Arguing counterfactually, I would hold that if either of these reasons were not present an explication of the formulas in terms of successive transition processes would possibly have occurred earlier on.

Let's begin with the methodological reasons. First of all, there was only one intermediate state: The electron changed its state and a photon was present. This state had an infinity of instantiations according to the momentum transfer between photon and electron. As I argued previously, Dirac's verbal model was mostly used as a tool to arrive at all the possible ways which coupled the initial with the final state. The description of the self-energy problem was in these terms simple enough for rendering the invocation of Dirac's verbal model unnecessary.

My second methodological point is connected to the mode of perturbation theory applied and the nature of the phenomenon to be calculated. Oppenheimer, Rosenfeld, and Waller used time-independent perturbation theory. The problem was considered to be static, i.e. the second-order contribution to the energy of an electron. Although the way the matrix elements combined suggested a reading in terms of successive transitions, the difference in the phenomenon under study (energy correction vs. scattering phenomenon) and especially the mathematical modelling (time-independent vs. time-dependent perturbation theory) seemed to have rendered a verbal expression in terms of successive transitions inappropriate.

Both Oppenheimer and Waller used Dirac's verbal model to describe scattering phenomena and the application of time-dependent perturbation theory.[24] These contributions and Oppenheimer's as well as Waller's papers on the self-energy problem lay only a few months apart.[25] Furthermore, the physicists clearly differentiated between a time-dependent and a time-independent treatment. Already Heisenberg and Pauli remarked that they were "more interested in the perturbation of the eigenvalues".[26] In his self-energy paper, Waller pointed to the fact that "a difficulty corresponding to the infinite eigenvalue perturbation of second order occurs when using

[23] In Ehberger (2022, Sect. 2.2.3) I highlighted the sociological component.

[24] Oppenheimer (1930b), Oppenheimer (1930c), and Waller (1930b). See also my discussion in Sects. 7.3 and 7.6.2.

[25] Waller communicated his self-energy paper, in which he did not use Dirac's verbal model, only a month after his description of the scattering of light, where he did. Oppenheimer communicated his note on scattering problems only four months after his calculation of the shift of energy levels.

[26] "mehr für die Eigenwertstörung interessieren," Heisenberg and Pauli (1929, 51).

aperiodic solutions"[27] and cited Dirac's paper on dispersion. This differentiation was most clearly formulated by Léon Rosenfeld.[28]

Following a suggestion by Pauli,[29] Rosenfeld eventually discussed the self-energy problem in a "physically equivalent if not more instructive way".[30] Instead of calculating stationary solutions, Rosenfeld investigated the non-stationary case for which at $t = 0$ no light quanta and an electron in an arbitrary state n were present. He calculated the transition probability to a state in which the electron occupied a different state l and one light quantum was present. From the above probability, he investigated two different expectation values. He found that after a small time $t > 0$ the expectation value of the energy of an electron with a given light-quantum number of the radiation field remained finite, while the expectation value of the energy of the radiation field for a given value of the state of the electron diverged.[31] This treatment also explains the headline Rosenfeld chose for the respective paragraph: the "energy of the radiation field".[32]

Although the above two methodological points are important to acknowledge, there was yet another reason why especially this group of physicists did not include any kind of verbal model in their calculations. All of the papers on the self-energy were written either in Zürich or in close contact with Wolfgang Pauli and Werner Heisenberg. As Arthur Miller (1984, 154–155) noted, "from 1927 to 1932 Heisenberg resisted any imagery of atomic phenomena – that is, for Heisenberg and other quantum physicists such as Pauli, in the atomic domain visualizability did not yet posses a depictive or visual component." These physicist were still under a strong influence of the philosophical outlook of Niels Bohr and in general rather close to the philosophically motivated circles of Copenhagen.

While a deep analysis of Bohr's thinking goes beyond the scope of this investigation, already more general observations can elucidate the relevance of Bohr's influence.[33] As has often been pointed out, in Niels Bohr's writings the formalism of quantum theory, although most probably underlying Bohr's conclusions, was not

[27] "eine der unendlichen Eigenwertstörung zweiter Näherung entsprechende Schwierigkeit auch bei Benutzung unperiodischer Lösungen auftritt." Waller (1930a, 676).

[28] Besides the discussion in the main text, see also Rosenfeld (1931, 456, especially the last paragraph of Sect. 1 and the first paragraph of Sect. 2).

[29] Pauli to Rosenfeld, April 12. 1931; as cited in Pauli (1993, 746).

[30] "physikalische äquivalente oder gar instruktivere Weise," Rosenfeld (1931, 457).

[31] For clarity's sake: The transitions discussed by Rosenfeld were, as such, not possible, as a free electron cannot radiate a photon and simultaneously obey momentum and energy conservation. When Pauli proposed the method to Rosenfeld, he also remarked: "Of course this concerns entirely states, whose total energy is different from the one of the initial state [...]." [*Original*: "Natürlich handelt es sich hierbei durchweg um Zustände, deren Gesamtenergie von der des Anfangszustandes verschieden ist [...]."] Pauli to Rosenfeld, April 12. 1931; as cited in Pauli (1993, 746). Pauli did not comment why such an assumption was nonetheless suitable. But since the energy of the state was infinite, energy conservation was dragged into absurdity any way.

[32] "Energie des Strahungsfelds," Rosenfeld (1931, 456).

[33] Just to get a glimpse at Niels Bohr thinking, the reader is referred to Faye (2019).

discussed in any detail.[34] Niels Bohr put indeed great emphasis on language, but considered the concepts we use in physics as extensions of ordinary language, mostly a means to communicate empirical results, and regularly referred to the formalism as a "symbolic scheme". As far as I can judge, Bohr never used any verbal expression to refer to virtual processes during the investigated time period.[35]

The further development of thought of Rosenfeld and Oppenheimer constitutes a rather suggestive example for the connection between the (non-)usage of a physical language to describe the perturbative formalism and the influence of the Bohr-Heisenberg-Pauli philosophy. In 1930, Rosenfeld became lecturer for theoretical physics at the University in Liège, but also began closely collaborating with Niels Bohr.[36] In February 1931, Rosenfeld gave lectures on QED at the Institute Henri Poincaré in Paris. In the published version of these presentations, he introduced Dirac's verbal model for dispersion but, in reference to the relativistic description of scattering including negative energy intermediate states, warned his readers:

> First of all, the negative energy states play an important, although *purely formal*, role in the theory of dispersion. One knows that the dispersion formula [...] includes a summation over all energy states the diffusing body is capable of. *In a pictorial fashion, but which is not justified by the general precepts of the physical interpretation of quantum mechanics*, one can say that the phenomenon of diffusion consists of a double transition of the diffusing body.[37]

Robert Oppenheimer, quite on the contrary, extended the use of Dirac's verbal model shortly after he had left Zürich. In the same paper in which he gave the detailed calculation of the annihilation of electrons and protons according to Dirac's initial conception of the Dirac sea (Oppenheimer, 1930c), he also included a section on a problem priorly discussed by Wolfgang Pauli and Werner Heisenberg: radioactive decay accompanied by the emission of a photon.[38] Although quite cautiously, Oppenheimer noted:

---

[34] Although there is some rehabilitation of the role of formalism for Bohr's thinking on quantum theory (Diecks 2017), the fact remains that technical interpretations did not make it into Bohr's writings.

[35] On a side note, it was also Niels Bohr who criticised Richard Feynman for providing a too literal interpretation of trajectories in the quantum domain during Feynman's presentation at the Pocono Conference in 1948. See e.g. Schweber (1994, 444).

[36] Rosenfeld organized his lectures so that he could work several months of each year together with Niels Bohr in Copenhagen (Jacobsen 2012, 50).

[37] "Tout d'abord, les états d'énergie négative jouent un rôle important, *bien que purement formel*, dans la théorie de la dispersion. On sait que la formule de dispersion [...] comporte une sommation portant sur tous les états d'énergie dont est susceptible le corps diffusant. On peut dire *d'une manière imagée, mais qui n'est pas justifiée par les préceptes généraux de l'interprétation physique de la méchanique quantique*, que le phénomène de la diffusion consiste en une double transition du corps diffusant [...]." Rosenfeld (1932a, 56); emphasis added.

[38] It was treated in the last section of Heisenberg and Pauli (1929) and it goes without saying that Heisenberg and Pauli did not use any kind of verbal description for the virtual processes (modern terminology).

> Somewhat inaccurately we may say that the system can reach the same final state by a double jump, in which, e.g., the atom goes over into some arbitrary state in the continuum, and then – but there is no interval between the jumps, – jumps to the final state and emits a quantum. [...] This is the effect treated by H. P. [Heisenberg and Pauli]. (Oppenheimer1930c, 945)

While Rosenfeld, ever more strongly connected to the philosophically motivated circles of Central European quantum physicists, explicitly stated that Dirac's verbal model was not in accordance with the general interpretation of quantum theory even for the processes Dirac had first applied it to, Oppenheimer, no longer under the strong influence of the philosophical restrictions of continental theorists, extended its usage to a realm for which he himself considered it to be "inaccurate".[39]

Not to be misunderstood, this is not a criticism of Bohr, Heisenberg, or Pauli, but simply an explanation of the lack of an explicit non-mathematical reference to concepts connected to virtual particles. As already discussed, there are good reasons not to conceive of virtual particles or transitions as actually occurring or existing and this understanding was already in place in the early 1930s.[40] Considering the state of quantum theory and especially quantum electrodynamics at the time, an emerging theory the predictions and the conceptual framework of which were still explored, there were certainly good reasons not to spell out concepts which did not cohere with "the general precepts of the physical interpretation".

Weisskopf (1934), The Dirac Sea, and Vacuum Fluctuations

I discussed the reasons for the lack of a verbally explicit description of the calculations in some detail because it can partially be extended to Weisskopf's contribution on the self-energy of an electron in the Dirac sea. Quite notably, at the time both Heisenberg and Pauli already had explicitly referred to the concept of intermediate states in self-energy calculations. Pauli and Weisskopf (1934, 728) even included a temporal description of pair creation through intermediate states in their "anti-Dirac"-paper about the quantization of the scalar relativistic wave equation. But in Weisskopf's paper on the self-energy such a description was again lacking, which hence begs for an explanation.

In early 1933, the Dirac sea had been accepted; at least to such an extent that a closer theoretical investigation of it seemed reasonable also to opponents of this idea (most notably Wolfgang Pauli). The question whether the filling of the negative energy states had an impact on the self-energy problem seemingly became a natural question to ask, as it was independently investigated in the US, by the Berkeley group around Robert Oppenheimer on suggestion by Niels Bohr, as well as in Zürich and Leipzig. When Weisskopf became Pauli's assistant in October 1933, Pauli and

[39] Turned positively, Oppenheimer seemed to conceive of Dirac's model in connection with genuine scattering phenomena as accurate.

[40] Rosenfeld's refutation is an obvious example. But also Yakov Il'ich Frenkel (1929b, 758) noted hat "the usual assumption [...] does not regard the state n [intermediate state] as 'really' occurring [...]".

Heisenberg were engaged with the development of a consistent theoretical framework for hole theory. They had already roughly discussed the corresponding corrections to the self-energy[41] and Weisskopf became deeply involved in this programme. In early 1934, Pauli set Weisskopf to calculate the self-energy of the electron in a filled Dirac sea.[42]

Weisskopf's Calculation: "Anschauliche" QED and the Automatism of the Mathematical Representation

In most general terms, the self-energy of an electron e_0 was treated in hole theory in the following way: One had to calculate the energy of the filled vacuum plus e_0 and subtract the energy of the filled vacuum without the electron present: $E_{e_0} = E_{vac+e_0} - E_{vac}$. Since all negative energy states were assumed to be filled with electrons, only positive energy intermediate states of the electron needed to be considered. I will call this contribution $E_{e_0,pos}$. But there was another and important alteration.

Framed in terms of Dirac's verbal model,[43] E_{vac} consisted of transitions of vacuum electrons to any positive energy state and back, whilst emitting and reabsorbing a photon. When e_0 was present, one of the positive energy states was occupied and the respective transitions of negative energy electrons were forbidden due to the Pauli exclusion principle. Thus, the energy of the filled vacuum without e_0 consisted of more contributions than the energy of the vacuum with e_0 present. Following Nicholas Kemmer, I will call this contribution the "hole-theoretic supplement".[44] Abbreviating the contributions from the transitions of vacuum electrons when e_0 was present by E_{vac+e_0} and the forbidden ones by E_{p_0}, we have: $E_{e_0} = E_{e_0,pos} + E_{vac+e_0} - E_{vac} = E_{e_0,pos} + E_{vac+e_0} - (E_{vac+e_0} + E_{p_0}) = E_{e_0,pos} - E_{p_0}$. As should have been noticed, the initial occupation of the intermediate state of the transitions of the negative energy electrons was of particular relevance to arrive at the reduction of the self-energy.

But Weisskopf did not invoke the described physical picture of successive transitions in his paper on the self-energy. He applied a version of QED that was proposed by Werner Heisenberg (1931a).[45] From a modern point of view, it only differs from other formulations of QED by a contact transformation.[46] Heisenberg referred to it as a method "which follows the *intuitive* conceptions [*anschaulichen* Vorstellungen] of

---

[41] Initially, Pauli did not believe that the self-energy problem would be mitigated in hole theory. See, for example, Pauli to Heisenberg, June 16. 1933; as cited in Pauli (1985, 169).

[42] See the letter Pauli to Heisenberg, January 21. 1934; as cited in Pauli (1985, 254–257).

[43] I will follow Schweber (1994, 123) and Kemmer (1935, Paragraph 6). I still want to point out, as priorly, that I consider Schweber's use of Feynman diagrams problematic at this point.

[44] "löchertheoretische Ergänzung," Kemmer (1935, 697).

[45] For a historical contextualization of this approach as a reaction on the infinities of QED, see Darrigol (1986, 247–248) and Rueger (1992, 317–320).

[46] See Darrigol (1982, Chap. 2, especially Sect. 2.4) for a discussion of the different representations of QED at the beginning of the 1930s.

the classical theory and wave mechanics a lot more closely than the priorly known and which therefore delivers in most cases without detours the result expected on correspondence grounds [das korrespondenzmäßig zu erwartende Resultat]".[47] Weisskopf applied Heisenberg's version because "it follows the classical electrodynamics a lot more closely; the radiation field is calculated classically from the current and charge density of the atom, but the amplitudes are assumed to be non-commuting in the final result".[48]

Although formally equivalent to a regular perturbative evaluation of QED, the outlook on the theory in this approach was quite different, as the quotations above already suggest. It did not explicitly rely on a picture of successive transitions and showed rather striking resemblance to the classical theory of electrodynamics. The dynamical part of the self-energy was given by Weisskopf as

$$E^D = -\frac{1}{2}\int i_{tr}\cdot A_{tr}dr \tag{10.2}$$

where the index tr indicates that the transversal part of the respective quantity was to be taken and both for i, the current density, and for A, the vector potential, the values of the vacuum had to be subtracted.

Weisskopf first calculated a general expression for the second order transversal self-energy of a system with an arbitrary number of electrons (Weisskopf 1934, 37, Eq. (23)) by expanding both i_{tr} and A_{tr} in a power series in e. As he noted, the zeroth order approximation for the vector potential A_0 could not be set to zero as it corresponded to "the zero point oscillations of the field strength in empty space".[49] In the general result for the second order approximation, combinations of creation and annihilation operators of electrons occurred which rendered, through their connection to the occupation numbers,

$$E^D \propto \sum_{k,l} N^k(p)(1-N^l(p+k)) \tag{10.3}$$

In the verbal model of Dirac's radiation theory k corresponded to the initial state of the electron with momentum p. l corresponded to the intermediate state of the electron with momentum $p+k$. Thus, when the intermediate state was occupied initially, the second factor was zero and thus the respective term gave no contribution. All

---

47 "[...], die sich viel enger als die bisherigen an die *anschaulichen* Vorstellungen der klassischen Theorie und der Wellenmechanik anschließt und die deshalb in den meisten Fällen ohne Umwege das korrespondenzmäßig zu erwartende Resultat liefert." Heisenberg (1931a, 338); emphasis in the original.

48 "die sich viel enger an die klassische Elektrodynamik anschließt; das Strahlungsfeld wird klassisch aus Strom und Ladungsdichte des Atoms berechnet, die Amplituden der elektromagnetischen Potentiale werden aber im Endresultat als nicht vertauschbar angenommen." Weisskopf (1934, 27–28).

49 "die im leeren Raum vorhandenen Nullpunktsschwankungen der Feldstärke," Weisskopf (1934, 35).

Weisskopf had to do from this point onwards was to plug in the initial occupation of the states and the cancellation of the terms that were constructed above through Dirac's verbal model would occur automatically.

Hence, Weisskopf's approach took the occupation of the intermediate state into account through the mathematical representation. There was no necessity of introducing any kind of verbal expressions. Further, and as was also noted by Wendell H. Furry in a letter to Weisskopf, the physical picture connected to Heisenberg's intuitive or correspondence-like approach, as it was regularly called, and its perturbative evaluation did not rest on the conception of successive transitions through emission and absorption of photons.[50] Although essentially equivalent to a more regular perturbative evaluation, Weisskopf's specific theoretical framing rendered the usage of Dirac's verbal model unnecessary, if not unsuitable.

Did Weisskopf Think of Transitions through Intermediate States?

Furry's above mentioned letter is of further significance concerning the applications of such concepts as transitions through intermediate states. Weisskopf's paper is known for including a calculational mistake which rendered the divergence quadratic instead of logarithmic and Furry communicated the correct result to Weisskopf. But Weisskopf's mistake was not connected to the occupation of the intermediate state. It was a simple mathematical blunder. He changed the sign of an absolute value of a quantity while performing the replacement of an integration variable.[51] All the same, when Furry communicated the error to Weisskopf, he indicated that the mistake might have been prevented by a physical picture of successive transitions.[52]

I think that this is not the case. Weisskopf was an expert in the application of Dirac's radiation theory (see Sects. 8.1.2–8.1.4) and was aware that the different calculations of the self-energy led to the same result.[53] He actually used the concept of successive transitions explicitly a few months later in private communication with Werner Heisenberg. In his publication on the subtraction procedures in hole

[50] To be sure, Furry did not directly state that Heisenberg's version did not rely on the picture of successive transitions. But this was implicit in his statements as he contrasted Weisskopf's approach to a physical picture of successive emission and absorption subprocesses. Furry to Weisskopf, June 21. 1934; as cited in Schweber (1994, 124–125).

[51] See Blum (2015, 83–84) for a closer discussion of this mistake.

[52] "This [the correct result] is *physically evident* when we use the light quantum picture and treat the energy displacement as due to the transitions of the system from a given to other states and back with corresponding emission and reabsorption of light quanta." Furry to Weisskopf, June 21. 1934; as cited in Schweber (1994, 124–125); emphasis added.

[53] Weisskopf noted that his result was equivalent to Waller's if the problem was reduced to a single electron. He also noted that Waller's method, the more regular perturbative approach, also introduced the factor $N^k(p)(1 - N^l(p + k))$, the one responsible for the cancellation. See Weisskopf (1934, 38).

theory (see also Sect. 10.1.2), Heisenberg noted that the self-energy of a photon diverged and that "the analogy to the self-energy of the electrons is here complete".[54] But Heisenberg only provided a rough sketch of his calculations in the paper and Weisskopf and Pauli were unsure of how Heisenberg arrived at this result in detail. As Pauli suspected that also the self-energies might be subtracted by an extension of Heisenberg's subtraction scheme,[55] the relation between the self-energy of the photon and the electron as well as the exact way of calculating them was of great interest to the Zürich physicists. Although these discussions eventually faded without publishable results, the correspondence is important for the conception of the self-energy calculation of Weisskopf.

In a letter dated to October 24. 1934, Weisskopf explicitly compared the self-energy of the electron and the one of the photon in reference to the occupation of the intermediate state. He noted, that for the self-energy of an electron there are two kinds of processes through intermediate states. On the one hand, there were intermediate states

> which result from the transfer of the electron into a different state under emission of a quantum. [...] Further there are intermediate states, which stem from the creation of a pair (beside the considered electron). But these essentially represent the corresponding portion of the vacuum self-energy to be subtracted, except for those pairs, whose negative electron[56] has the same state as the considered one.[57]

As Weisskopf noted, it was the latter part of the energy which had to be subtracted from the energy of a free electron. Because of the Bose statistics of the photons, he argued, there was no part corresponding to this subtraction for the self-energy of a photon.[58]

The fact that Weisskopf did not use a verbal exposition of successive transitions in his self-energy paper, although he was aware of such a reading, can thus be explained in a similar way as for the contributions by Oppenheimer, Waller, and Rosenfeld. Weisskopf did not have to invoke it as his mathematical procedure already took care of the occupation of the intermediate states. The specific representation of QED he

---

[54] "Die Analogie zur Selbstenergie der Elektronen ist hier vollständig." Heisenberg (1934a, 229).

[55] See, for example, Pauli to Heisenberg, July 25. 1934; as cited in Pauli (1985, 338).

[56] *Comment by the author*: The term "negative electron" refers to the electron in the negative energy states.

[57] "[...] die von der Überführung des Elektrons in einen anderen Zustand unter Emission eines Quants herrühren. [...] Weiter gibt es Zwischenzustände, die von der Erzeugung eines Paares (neben dem betrachteten Elektron) herrühren. Diese stellen aber im wesentlichen den entsprechenden Anteil in der zu subtrahierenden Vakuumselbstenergie dar, bis auf jene Paare, deren negatives Elektron den gleichen Zustand hat, wie das betrachtete." Weisskopf to Heisenberg, October 24. 1934; as cited in Pauli (1985, 351).

[58] Weisskopf actually retracted this statement in the next letter to Heisenberg. But my point is not about the difference between photon self-energy and electron self-energy but that Weisskopf obviously could conceive of the self-energy in terms of successive transition processes. Weisskopf to Heisenberg, October 26. 1934; as cited in Pauli (1985, 353).

used in his paper even rendered its application unsuitable. Furthermore, Weisskopf's approach to quantum theory and QED changed appreciably shortly before he came to Zürich.

After receiving his PhD in Göttingen, he spent research fellowships in Leipzig, Berlin, and Kharkiv. In September 1932, he went to Copenhagen on a Rockefeller scholarship at a time when the question of the measurability of fields was discussed at Bohr's institute. Weisskopf helped Bohr in spelling out the ideas of the Bohr-Rosenfeld paper on this theme when Rosenfeld, Bohr's closest associate at the time, was not in Copenhagen. Later on, Weisskopf kept corresponding with Rosenfeld and Bohr about the publication.[59] Weisskopf himself remembered to have learned thinking about complementarity and correspondence during his stay in Copenhagen. Also his engagement with different formulations of QED started at Bohr's institute, while he learned the necessary mathematical techniques from Rudolf Peierls in Cambridge, where Weisskopf spent the shorter part of his yearlong Rockefeller scholarship.[60] It is rather suggestive that the specific approach taken by Weisskopf on the self-energy problem, often considered "correspondence-like", was due to the influence of the circles of Copenhagen, Leipzig, and Zürich.[61]

Outlook: Weisskopf and Vacuum Fluctuations

While the specific approach certainly contributed to Weisskopf not verbally explicating the formulas, it rendered another aspect connected to the self-energy and the virtual particle concept, which was also of relevance for the measurability paper of Bohr and Rosenfeld (1933),[62] a lot more transparent. Through the expansion of the vector potential the term A_0 occurred and was explicitly connected to vacuum fluctuations by Weisskopf. In the prior works on the self-energy this connection was not made explicit.[63]

---

[59] See the correspondence between Weisskopf and Rosenfeld/Bohr in October and November 1933 in BSC, MF 26, Section 2.

[60] See Weisskopf (1991, Chap. 4).

[61] The method had its predecessor in a paper by Klein (1927), at the time assistant to Bohr, was developed by Werner Heisenberg, and was especially supported by Bohr and Léon Rosenfeld (Rueger 1992, 317–320).

[62] See Darrigol (1991) for a historical discussion on the measurability of fields in QED.

[63] This does not mean that this connection was necessarily unknown. Just to name two examples: In a letter to Paul Ehrenfest, Wolfgang Pauli mentioned that "light-quantum number N and E and H [electric and magnetic field strength] do not commute". [*Original:* "Lichtquantenzahl N und E und H sind nicht vertauschbar"] Pauli to Ehrenfest, October 28. 1932; as cited in Pauli (1985, 132). This fact was heuristically used by Walter Heitler to connect vacuum fluctuations to the self-energy problem in his popular textbook (Heitler 1936, Chap. II, Sect. 7, Sect. 5). Gregory Breit (1933, 140) also traced the infinities "to the fact that the oscillators describing radiation may not be said to be in a state of absolute rest even when physically we like to deny the presence of light quanta".

A closer investigation of the connection between vacuum fluctuations and the self-energy of the electron, or more generally the virtual particle concept (which in the necessary detail goes beyond the scope of this book), will have to take Weisskopf's contributions into account. In 1935, Weisskopf provided the first argument why spontaneous emission was due to vacuum fluctuations (Weisskopf 1935, 653).[64] In his 1939 paper on the self-energy (Weisskopf 1939), he further decomposed the electrodynamic self-energy in a spin term and a fluctuation term, called W_{fluc}. After WWII, Theodor Welton, close friend of Richard Feynman and post-doctoral student working with Weisskopf at MIT, discussed an "intuitive explanation" of the Lamb shift and the self-energy by tracing it back to the coupling between the electron and electromagnetic field fluctuations (Welton 1948). Weisskopf, in turn, presented the most recent developments in QED according to this picture of vacuum and zero-point fluctuations, a presentation which should be "understandable to the physicist who is not specialized in this field" (Weisskopf 1949, 305). While the idea that the self-energy can be explained as a coupling of the electron to the quantum-theoretical vacuum fluctuations of the radiation field did not necessarily originate with Weisskopf, his further work was certainly very important for its popularization.[65]

Higher Orders: Late Application of Non-Mathematical Representations (1939)

While the prior subsections highlighted the lack of non-mathematical representations in the most popular publications on this problem, it is to note that other physicists, following regular quantum-theoretical perturbation theory more closely, explicitly used Dirac's verbal model for the description of the self-energy, both in an empty vacuum and in the Dirac sea.[66] But Weisskopf himself only applied it in the calculation of higher orders and as late as 1939. As Weisskopf (with the help of Furry) had shown, the second-order contribution to the self-energy diverged logarithmically in hole theory. Yet, it was still unclear whether higher order terms or even the whole series of approximations would strengthen the divergence or eventually lead to a convergent result. For these calculations, Victor Weisskopf was again the central actor.

---

[64] He already indicated that vacuum fluctuations were responsible for spontaneous emission in his self-energy paper but did not argue for this conjecture (Weisskopf 1934, 35).

[65] In the meantime, Wright (2024, Sects. 4.1–4.3) has provided such a historical account of vacuum fluctuations, in which Weisskopf is the main character and his influence up until the work of Julian Schwinger in the late 1940s is traced.

[66] See for example Kemmer (1935, especially Sect. 6) or, for a textbook presentation, Heitler (1936, Sect. 18, Sect. 2).

Mercier, Diagrams and the Reduction of the Contributing Terms

Weisskopf, since 1937 at Rochester University in the US, proposed the investigation of higher order terms to André Mercier,[67] who had studied physics in Geneva and Paris and spent a research trip in Copenhagen in 1939.[68] Mercier calculated with what he called "the method [...] of intermediate states",[69] i.e. with standard perturbation theory and the Born approximation. He identified the self-energy as being due to the capacity of the electron to "(virtually or really) emit or absorb light quanta".[70] Obviously following Weisskopf, Mercier noted that this constituted an interaction between the electron and the fluctuations of the electromagnetic field.[71]

Mercier's calculation included rough approximations, also for the next to leading order calculation, and he could only crudely and, as Weisskopf would prove just a few months later, wrongly estimate the degree of divergence for the complete perturbative series. But Mercier's approach is, from a methodological point of view, still interesting: He used the physical picture of successive transitions and non-mathematical representations to reduce the number of contributions he had to calculate in the fourth order approximation.

Mercier differentiated between terms which included the emission and absorption of a photon by different electrons (he called them "coupling terms"), i.e. the electron under consideration and the vacuum electrons, and terms for which each electron absorbed the photon it originally emitted. Mercier further differentiated between contributions which were composed of two successive second order processes, called B, and such contributions for which each of the three intermediate states was different from initial and final state, called A.

Besides verbally modelling the phenomenon, Mercier used diagrams as a representational means. Figure 10.1 displays Mercier's representation of a selection of the transitions of vacuum electrons actually forbidden because of the presence of the positive energy electron under consideration (e_0), i.e. the terms contributing to the "hole-theoretic supplement". Each diagram (I–VI) was read from left to right and represented the sequence from initial state over three intermediate states to the final state. The filled dots represented electrons, the empty dots holes, i.e. positrons. The horizontal lines represented the energies $-mc^2$ and $+mc^2$. e_0 was assumed to be at rest. It was not represented explicitly in the diagrams (as it would block the performed transitions) but can be thought of as lying on the $+mc^2$ line. All of the

---

[67] See Mercier (1939, 57).

[68] In Mercier and Gustafson (1938) the results for the self-energy of a free electron had already been published. But, as I am interested in the different representations the physicists applied, I will focus on the more comprehensive discussion in Mercier (1939).

[69] "méthode [...] des états intermédiares," Mercier (1939, 68).

[70] "[un électron est toujours susceptible] d'émettre et d'absorber (virtuellement ou réellement) des quanta de lumière," Mercier (1939, 56).

[71] Quite idiosyncratically, Mercier called the field fluctuations "virtual radiation field" ["champ de radiation virtuel"], Mercier (1939, 56). I would not overestimate this terminological variant and the connection to the BKS theory it suggests. When it comes to the history of QED, Mercier is a marginal figure and this terminological variant was not used by any other historical actor.

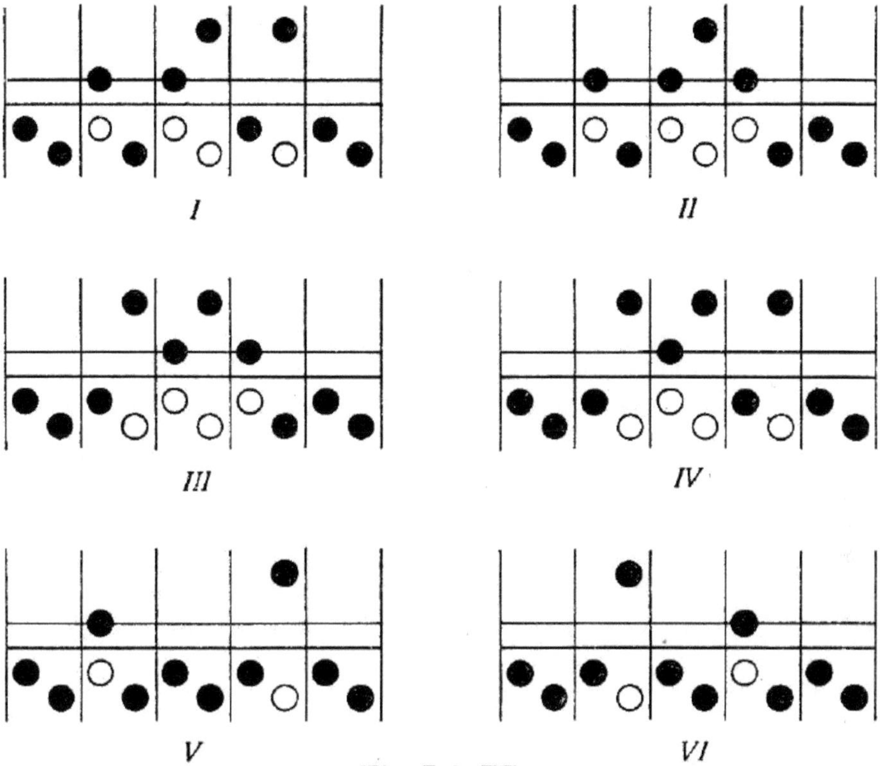

Fig. I à VI.
Schéma d'états intermédiaires (appartenant à J).

Fig. 10.1 Representation of some forbidden transitions to fourth order according to André Mercier. Figures I. to VI. in Mercier, André. 1939. "Sur l'énergie propre de l'électron." *Helvetica Physica Acta* 12, 55–75, here 64 (cited as Mercier (1939) in the bibliography). ETH-Bibliothek Zürich, E-periodica, https://www.e-periodica.ch. Reproduced with kind permission from the Swiss Physical Society

depicted processes could not occur as in each case one of the negative energy electrons would occupy the state of e_0 in an intermediate state. Thus, the mathematical terms corresponding to the diagrams had to be subtracted in the calculation. The diagrams I to IV represented a selection of terms of type A, diagrams V and VI of type B. Mercier provided similar diagrams for the coupling terms.

To exemplify the reading of the diagrams, let's take the third one (III) as an example. Initially, there was only the filled Dirac sea. The two electrons of the Dirac sea which were used to exemplify the process were represented as filled dots. In the first step, one of these electrons performed a transition to an unoccupied positive energy state, thus an electron-positron pair was created; the positron was represented by the empty circle, i.e. the hole in the Dirac sea. In the second step, the other electron performed a transition to the actually occupied state mc^2, leaving a hole in the negative energy contribution. Thus, another pair was created. This was the forbidden process. In the next two steps, each the electron of the first pair and of the second dropped to their initial states in the sea and thus two pair annihilation processes occurred. The final state was thus the same as the initial state: a filled Dirac sea.

Mercier could show, by transforming this representation into a mathematical one but without having to explicitly calculate the contributions, that the processes depicted in Fig. 10.1 cancelled each other.[72] Roughly stated, the argument went as follows: The integrand of the corresponding equation was a fraction, the numerator of which consisted of the matrix elements for the respective subprocesses and the denominator of the respective energy relations of the initial and intermediate states.[73] The numerator could be shown to be the same for all six processes. Thus, the comparison could be reduced to the energy relations expressed in the denominator. As Mercier showed, the terms could be grouped so that, in terms of the numbers of the processes represented in Fig. 10.1: $I + II - V = 0; \quad III + IV - VI = 0$.

Thus, by ordering the terms according to the above described system, heavily relying on a physical interpretation of the processes, Mercier reduced the contributions that needed detailed calculation to a small subset of the processes which one could imagine by invoking the verbal model. The difference to Feynman diagrams, i.e. the depiction of a succession of state transitions without any explicit depiction of photons (Mercier) vs. the depiction of a space-time scattering process in which also photons had a specific place (Feynman), is obvious and underscores the conceptual differences between QED during the 1930s and the late 1940s. But even more strikingly, one would not even consider such processes when using Feynman diagrams: Such contributions and the corresponding possible reduction of the self-energy are automatically taken into account in Feynman's way of doing QED.

Mercier's approach actually led him astray at some point, as he concluded that the reduction to a logarithmic divergence was only a phenomenon of the second order approximation.[74] For the estimation of the value of the complete perturbative series, Mercier framed the processes as a combinatorics problem of distributing the emitted and absorbed photons on the electrons taking part in each respective subprocess. Thereby he estimated the total number of terms contributing in the respective approximation. But Mercier's result was, at best, ambiguous. He concluded that the divergence was not as bad as quadratic but he was not able to give a reasonable estimation of its value. Thus, Mercier eventually concluded his paper in a rather pessimistic tone and argued that the divergence was due to the approximation scheme.[75] Just a few months later, Victor Weisskopf showed that matters were not as hopeless as Mercier had thought.

[72] The same held true for the so-called coupling terms.

[73] As always in such perturbative calculations.

[74] I have to admit that I could not locate the decisive mistake in his reasoning but I suspect that it occurs on page 68 to 69 of Mercier (1939). There he argued, again by using diagrams, that the compensation to second order was a lot stronger than to fourth order.

[75] Mercier (1939, 75).

Weisskopf, Emission and Absorption Subprocesses, and the Estimation of Higher Order Terms

In the larger part of his second paper on the self-energy problem, Weisskopf (1939) provided a closer investigation of the different contributions of the second order self-energy and its mitigation through the inclusion of the Dirac sea.[76] Again, he used an approach similar to the one of his first paper "in order to outline the physical significance of the different terms" (Weisskopf 1939, 79). He dissected the self-energy into three different contributions: the static Coulomb interaction, the contribution due to the electron's spin, and the contribution of vacuum fluctuations. He eventually showed how they combined to render a logarithmic divergence and interpreted them physically, mainly tracing the reduction of the divergence to the "broadening of the charge of the electron" (Weisskopf 1939, 79).

Only in the last section of his paper, where Weisskopf (1939, 83–85) discussed the higher order contributions to the self-energy problem, he turned to a verbally explicit description in terms of emission, absorption, and transition subprocesses. Weisskopf's argument consisted of two steps. First, he showed that in the limit $m \to 0$ the self-energy of an electron vanished to every order in perturbation theory. In a second step, he showed that this entailed that the highest degree of the self-energy divergence was logarithmic. Weisskopf's reasoning, especially in the first part of his proof, relied on the symmetries of the theory and the evaluation of the signs of the denominator and the numerator of each contribution in the perturbative series.

Without going into any detail, it is noteworthy that at this point, when Weisskopf considered higher order approximations and obviously used the "usual perturbation method" (Weisskopf 1939, 79), he started to talk about "consecutive transitions of the total system from one state to another" (Weisskopf 1939, 84). The central theme of the first part of Weisskopf's proof was a comparison of the self-energy of an electron and the self-energy of a positron.[77] This comparison of the two self-energies essentially rested on the picture of successive transitions and the respective general mathematical expression: For the n-th order approximation, Weisskopf grouped the possible transitions for the respective cases, either an additional electron or a positron was present, and showed that the energy contributions due to the forbidden transition possibilities exactly cancelled the ones due to the additional transition

[76] For a detailed discussion see Schweber (1994, 125–128).

[77] More explicitly, Weisskopf showed that in the limit $m \to 0$ the self-energy of the electron and the self-energy of the positron, each without the vacuum cancellation terms, combined to give twice the self-energy of the vacuum. Through the application of simple relations between the respective energy contributions, this directly entailed that the self-energy of the electron had to be equal to zero.

possibilities. The proof of $\lim_{m=0} W^n = 0$ finally allowed Weisskopf to show that the worst divergence one encounters to each order is logarithmic.[78]

The Tool Character of the Non-Mathematical Representations

Mercier's contribution, although incorrect, and Weisskopf's estimation, which eventually superseded it, nicely exemplify the tool character of the verbal model and especially its merits for higher order calculations. For Mercier, who tried to directly calculate the fourth order correction, the sheer amount of possible paths through intermediate states had to and could be reduced by grouping the terms according to verbal and diagrammatic representations. In contrast, Weisskopf did not aim at a brute force calculation. His proof that the self-energy vanished in the limit $m \to 0$ rested on the comparison between the perturbative contributions to the self-energy of the positron and the electron. And this comparison was based on Dirac's verbal model. In a letter to Weisskopf, Pauli highlighted this idea, "because one arrives at a reasonable assignment of 'paths' through intermediate states".[79] But these paths were not conceived of by Weisskopf as a physical picture of the self-energy. The physical reasons for the divergence were, according to his own assessment, better understood by the invocation of the calculation scheme he had already applied in his 1934 contribution.

The slow and gradual incorporation of non-mathematical representations of transitions to and from intermediate, i.e. virtual., states exemplifies the strong interrelation between the methods applied by the physicists and concepts connected to the historical development of the virtual particle. According to my argument, for the first contributions by Oppenheimer, Waller, and Rosenfeld, the difference between time-independent and time-dependent perturbation theory rendered the picture of successive transitions inadequate. The absence of the explicit mentioning of virtual entities in Weisskopf's 1934 paper can be traced back to the application of the correspondence-like picture of QED introduced by Heisenberg.

Of course, I cannot exclude the possibility that the actors simply regarded such an interpretation as obvious and, thus, did not include it. But given the historical material, I consider this scenario rather unlikely. I believe that the historical actors were capable of such an interpretation but considered it, in connection to the theoretical framework and the methods they applied, unsuitable. Especially since all of the above mentioned factors have to be understood against the background of the schools of theoretical

[78] To make this concrete: Weisskopf showed that the nth order self-energy contribution was, in the case of large momentum transfer in the intermediate steps, approximately given by $W^n = c\left(\frac{e^2}{hc}\right)^n \frac{P^N}{(mc)^N}\left(log\frac{P}{mc}\right)^t mc^2$. Since the energy vanishes in the limit $m \to 0$, the numerical factor N must be equal to zero or smaller (otherwise the expression diverged in this limit). Thus, the highest degree of divergence was logarithmic.

[79] "weil dadurch eine vernünfitge Zuordnung von 'Wegen' über Zwischenzustände erzielt wird." Pauli to Weisskopf, May 12. 1939; as cited in Pauli (1993, 820).

research the respective actors were in close contact with: Particularly physicists close to Copenhagen did not represent their calculations in an explicit verbal manner initially.

The (non-)application of the verbal model in self-energy calculations thus points to features of the conceptual development which we have encountered in prior chapters and will encounter in this chapter: the tool character of the concepts, the relevance of the respective theoretical and methodological framework, and finally a general trend, which took hold at about the mid 1930s, namely to connect static phenomena with a dynamical picture of successive transitions. All of these factors will contribute to the conceptual overlap between virtual particles and virtual transitions at the late 1930s.

10.1.2 Vacuum Polarization and the Electron Positron Cloud (1933–1935)

After the discovery of the positron, the idea of the Dirac sea became more popular. Vacuum polarization was amongst the first effects to be discussed. From a modern point of view, this effect is based on a cloud of virtual particles surrounding the particle under investigation. As we will see, the concept of such a charge cloud was soon generally accepted. But vacuum polarization is also of interest from a terminological point of view: Furry and Oppenheimer introduced the concept of "nascent pairs", at times used synonymously to "virtual pairs".

The history of the occurrence and handling of vacuum polarization is well-known and I will follow the general historical narrative.[80] But I will enrich it through a direct comparison of the concepts involved in the initial description of vacuum polarization in the Dirac sea picture and a second quantized version of the electron positron field. Further, I will particularly focus in on Furry's and Oppenheimer's "nascent pairs", a terminological variant which was proposed just as the term "virtual" became more popular.[81] Moreover, it has so far not been pointed out how naturally the investigation of vacuum polarization came to the historical actors when the Dirac sea was accepted more widely. The initial communication of the Delbrück effect was based on the polarization of the negative energy electrons in the vicinity of the nucleus (see Sect. 9.2.1). Guido Beck's theory of β-decay similarly rested, besides energy non-conservation, on the fact that pairs were created in the potential field of a

[80] See for example Darrigol (1982, Sect. 5.1), Schweber (1994, Sect. 2.5.2), or Pais (1986, Sects. 16.4–16.6).

[81] Wright (2024, Sect. 3.3) has recently provided a study of Furry and Oppenheimer's pair theory, focussing in on Furry's previous calculational exercises and the impact of Oppenheimer's pessimism on the paper. While Wright highlights that both from a Dirac Sea point of view and from a second quantized point of view physicists were led to the same results about vacuum phenomena, I will more closely consider which theoretical entities were made responsible. The question Wright tries to answer is "How did the vacuum become a theoretical resource?" while my question is "What kind of theoretical entities and concepts were made responsible in the respective conceptualizations of theory?"

nucleus (Beck 1933). According to Rudolph Peierls, Enrico Fermi immediately had the idea of vacuum polarization when he first heard of the Dirac sea.[82]

Peierls himself had already finished first calculations on vacuum polarization by mid July 1933 and he was in a perfect position to do so. He had investigated hole theory in early 1930 together with Wolfgang Pauli and, by mid 1933, worked in Cambridge with Dirac. Peierls found a divergent result "[which] can in no way be interpreted away, as it is connected to transitions, which necessarily have to occur in the theory, i.e. to the creation of particle pairs".[83] Wolfgang Pauli, before knowing any details about Peierls' calculation, was convinced "that [...] in a factually ["faktisch"] empty cavity ($N'_\alpha = 0$, $N_s = 0$) a charge contribution must be induced by an external (created for example by a proton) electrostatic potential field".[84] Weisskopf, at the time in Cambridge with Dirac and Peierls, also informed Pauli about the effect of vacuum polarization and the conjecture that the initial charge will be completely compensated by this effect.[85] By the end of July 1933, the main European actors involved in the investigation (Heisenberg, Pauli, Dirac, Peierls, and Weisskopf) were all informed about the divergence.[86] Independently and influenced by Niels Bohr, Robert Oppenheimer's group in Berkeley began investigating effects of electromagnetic fields on the Dirac sea in mid 1933.

The fact that the idea of vacuum polarization came so naturally to physicists at the time can be easily explained. The picture of scattering discussed by Dirac in his first communication of the Dirac sea (see Sect. 7.3.2) essentially rested on the capability of negative energy electrons to absorb and emit photons. Negative energy electrons interacted with the electromagnetic field. The delay of this discussion until 1933 can in turn be explained by a lack of faith in the Dirac sea picture until the discovery of the positron and its theoretical identification with a hole.

The second quantized formulation of the Dirac sea was immediately pursued. In correspondence between Heisenberg and Pauli, a quantized electron-positron field was on the table from mid 1933 onwards. Wendell H. Furry and Robert Oppenheimer

---

[82] Peierls to Pauli, July 17. 1933; as cited in Pauli (1985, 198). In the comment on the letter, the editors of Pauli's correspondence refer to Fermi and Uhlenbeck's calculation on the production of pairs in the vicinity of the nucleus (Fermi and Uhlenbeck 1933). I rather believe that Peierls is referring to an earlier period, as Fermi must have heard about the Dirac sea prior to 1933 and Peierls explicitly refers to "the alteration of the particle density due to the action of a potential field" [*Original*: "der Änderung der Teilchendichte durch die Wirkung des Potentialfeldes."].

[83] "Sie läßt sich auf keine Weise weginterpretieren, da sie mit Übergängen verknüpft ist, die in der Theorie unbedingt vorkommen müssen, nämlich mit der Erzeugung von Teilchenpaaren." Peierls to Pauli, July 17. 1933; as cited in Pauli (1985, 199).

[84] "daß [...] durch ein äußeres (etwa durch ein Proton erzeugtes) elektrostatisches Potentialfeld in einem faktisch leeren Hohlraum ($N'_\alpha = 0$, $N_s = 0$) eine Ladungsverteilung hervorgerufen werden muß." Pauli to Heisenberg, July 19. 1933; as cited in Pauli (1985, 202).

[85] Pauli to Heisenberg, July 19. 1933; as cited in Pauli (1985, 203).

[86] At the latest, Heisenberg knew about the divergence through a letter by Pauli in mid July. Pauli to Heisenberg, July 19. 1933; as cited in Pauli (1985, 202–203).

in Berkeley as well as Vladimir Fock (1933) in Leningrad turned to the investigation of this representation. In Cambridge, Dirac and Peierls kept up the Dirac sea formulation, which is rather unsurprising considering Dirac's disliking of the anti-commutation relations. Although the investigations in terms of the Dirac sea and a quantized electron-positron field were in their outcome essentially equivalent, the respective interpretation, most importantly the entities made responsible for vacuum polarization, differed appreciably. As it was conceptually more conservative and chronologically the first approach to be finished and publicly presented, I will begin with the Dirac sea picture.

The Dirac Sea and The Role of Negative Energy Electrons

Dirac presented his theory of vacuum polarization at three conferences in late 1933: at the yearly conference in Copenhagen (September 14–20), at the first All-Union Conference in Leningrad (September 24–30), and finally at the Solvay Conference in Brussels (October 22–29).[87] The vacuum in the Dirac sea picture was filled with an infinity of negative energy electrons and to arrive at a sensible description of the physical attributes of a state, one had to subtract a certain contribution of negative energy electrons. When there is no external field, the identification of the contribution to be subtracted is simply the whole Dirac sea. But as soon as an electromagnetic field was imposed on the vacuum, matters were not this simple: one had to specify which contribution of electrons produced a field and which contribution did not. Hence, Dirac's guiding question was which charge distribution should be considered as the source in the Maxwell equations, i.e. as ρ in $\nabla E = -4\pi\rho$.

For simplicity's sake, Dirac discussed the action of an electrostatic field, weak enough to allow for a perturbative evaluation. Dirac's calculation was based on a previously developed Hartree-Fock method. Its main theoretical entity was the density matrix defined as

$$(q'|R|q'') = \sum_r \psi_r^*(q')\psi_r(q'') \tag{10.4}$$

where the ψ_r were one-particle wave functions and the sum was extended over all occupied states. In the perturbative approach, the density matrix R was defined as $R_0 + R_1$. Dirac's task was to find the contribution R_1, so that $\dot{R} = 0 = HR - RH$, i.e. it did not change over time.

[87] See Roqué (1997, 105–106). The presentations at Leningrad (Dirac 1995) and Brussels (Dirac 1934b) seemed to have essentially the same content, while the published version in the Solvay proceedings is more detailed. I will only briefly sketch Dirac's approach, mostly following his Solvay presentation. For a more detailed account the reader is referred to Schweber (1994, 112–117).

The actual calculation is not of interest for us, but the outcome and its interpretation is. Dirac found that the induced charge density diverged logarithmically and took the form

$$-\frac{e^2}{\hbar c}\frac{2}{3\pi}\left(\log\frac{2P}{mc}-\frac{5}{6}\right)\rho-\frac{4}{15\pi}\frac{e^2}{\hbar c}\left(\frac{\hbar}{mc}\right)^2\nabla^2\rho \tag{10.5}$$

in which ρ was the charge density creating the static field and P was a momentum cut-off. Dirac then assumed, in harmony with the current trends in theoretical physics and especially Niels Bohr's views, that the theory would break down at energies approximately equal to $137mc^2$. With this assumption, the first term reduced to $-\frac{1}{137}\rho$, thus a small fraction of the electric density of the original distribution was shielded by the polarization of the vacuum.

At this point, Dirac's own interpretation of this effect is noteworthy. As already Sam Schweber (1994, 117) has pointed out, Dirac already described the physical effects connected to the "polarization of the vacuum" and introduced an important differentiation: The charges one actually observed in experiment were not the charges that figured in the basic equations because an additional charge was always induced in the vacuum. The different charge distributions are today called "dressed" and "bare", while Dirac called them "effective" and "real".

Yet, Dirac did not introduce any new concept such as virtual pairs in his consideration.[88] The phenomenon studied was considered as essentially static. In harmony with his modelling, Dirac explicitly considered vacuum polarization as an effect of the behaviour of negative energy electrons. For example, the cut-off was motivated by the failure of the relativistic wave function for energies bigger than 137 mc^2. Therefore, one should neglect "the disturbance that the field produces in negative-energy electrons with less energy than -137 mc^2". Dirac thus considered "the neutralization of charge [to be] produced by the other negative energy electrons".[89]

Although Dirac did not do so, the contributions might have as well been interpreted in terms of pairs as they coupled positive and negative energy states.[90] When working with a quantized electron-positron field instead of the Dirac sea, this identification was a lot more transparent. Although the procedure was also investigated in Leipzig and Zürich as well as in Leningrad, I will turn to the discussion of Furry and Oppenheimer (1934). It was most explicit about the new entities arising and they introduced a new terminological variant for entities we refer to as virtual: the "nascent pairs".

---

[88] Strictly adhering to the terminology at around 1933, one could argue that the only virtual entities in Dirac's derivation were the negative energy electrons. These were referred to as "virtual electrons" by Guido Beck (1933, e.g. p. 499) in his theory of the β-decay.

[89] Dirac to Bohr, September 10. 1933; as cited in Schweber (1994, 116).

[90] See for example the above quoted letter from Peierls to Pauli.

Furry and Oppenheimer's "MNtory" and "Nascent Pairs"

In Berkeley, the investigation of the interaction between the Dirac sea and electromagnetic radiation began when Niels Bohr was visiting the US in early to mid 1933,[91] exactly at the time when the interest in Dirac's hole theory started to increase.[92] In contrast to Dirac's hole theory, the main entities figuring in Oppenheimer and Furry's approach were electron-positron pairs. Oppenheimer referred to the theory as the "MNtory" in letters to his brother.[93] Since M referred to the number of positrons and N to the number of electrons, this term suggests an understanding of the theory as an inventory of electron-positron pairs.[94] Similarly, both Furry[95] and Oppenheimer[96] referred to their approach as a "pair theory" in letters. By October 1933, they were nearly finished with "cleaning up the formalism" and were also aware of "convergence difficulties, not unconnected with the breakdown of the theory at high energies, and similar to those of the radiation theory, which very seriously limit its applicability".[97] The polarization of the vacuum was thus independently of the European physicists diagnosed by Furry and Oppenheimer in a different conceptual framework.

For Oppenheimer, normally rather pragmatic, the paper was unusually convoluted in the physical description and placed enormous emphasis on questions of measurability, a topic that had been discussed with Bohr as his contribution with Rosenfeld (Bohr and Rosenfeld 1933) had just been finished.[98] In general, one senses the prior visit of Bohr all throughout the published paper. And this was not by chance. In June 1933, Oppenheimer gave a short outline of the reasoning that guided the introduction of his and Furry's paper in a letter to Bohr. He noted that "the situation is just what you [Bohr] had prepared us to expect" and thanked Bohr for "helping us to understand what we were talking about".[99]

---

[91] Starting in May 1933, Bohr was both lecturing at Caltech and the Summer Symposium at the University of Michigan. See e.g. Sopka (1988, 328).

[92] Prior to Furry and Oppenheimer's reformulation of the Dirac sea picture, the Berkeley group published short notes on pair production by electron impact, radiation and internal conversion (Furry and Carlson (1933), Oppenheimer and Plesset (1933), Nedelsky and Oppenheimer (1933)). Furry and Carlson also investigated the self-energy of an electron in the new picture of the vacuum, which allowed Furry to spot Weisskopf's mistake.

[93] Robert Oppenheimer to his brother, Frank Oppenheimer, October 1933; as cited in Smith and Weiner (1981, 164).

[94] See also Smith and Weiner (1981, 166).

[95] Furry to Weisskopf, June 21. 1934; as cited in Schweber (1994, 124).

[96] Robert Oppenheimer to Frank Oppenheimer, October 1933; as cited in Smith and Weiner (1981, 164).

[97] Robert Oppenheimer to his brother, Frank Oppenheimer, October 1933; as cited in Smith and Weiner (1981, 164).

[98] Furry and Oppenheimer (1934, 247, upper half of the right column), for example, is an obvious reference to Bohr and Rosenfeld's article. Yet, no citation was given and only Bohr's name was mentioned.

[99] Oppenheimer to Bohr, June 14. 1933; as cited in Smith and Weiner (1981, 162).

An important theoretical tool for Furry and Oppenheimer's discussion was the distinction between even and odd (parts of) operators.[100] In the language of a one-particle theory, odd operators coupled negative and positive energy states, i.e. possibly induced transitions between them, while even operators transformed a positive into a positive and a negative into a negative energy function. Furry and Oppenheimer already noted in their introduction:

> This distinction between even and odd operators is essential to the physical interpretation of the theory: for only those dynamical variables of a system which correspond to even operators may be determined without rendering the number of particles – electrons and positive[101] – in the system indeterminate. (Furry and Oppenheimer 1934, 245)

According to Furry and Oppenheimer, odd operators could surely be measured, "but this cannot be done without introducing an indeterminacy in the number of electrons and positives present". As they stated, "it is not, in general possible, to specify at the same time the number of particles in a system and the values of arbitrary dynamical variables of these particles".[102] In modern terms, as Sam Schweber has formulated it, this observation corresponded to the fact that "in QED the state vector describing a single charged particle had amplitudes for finding an arbitrary number of electron-positron pairs" (Schweber 1994, 88). Furry and Oppenheimer captured this phenomenon under the term "nascent pairs".

A specifically important odd operator was the charge density and its qualitative discussion is particularly clear about Furry and Oppenheimer's conception of the vacuum:

> As a matter of fact we shall see that in the present theory the vanishing of the mean charge density in a region of space does not necessarily mean that there are no particles present in the region; it may mean only that there are equal numbers of electrons and positives present and is no guarantee that the region is empty. For in the present theory, the charge density is not an even operator and can therefore be determined only at the expense of leaving indeterminate the number of particles present in the system. (Furry and Oppenheimer 1934, 247)

The formal development of their theory was closely modelled after already known procedures, but did no longer explicitly rest on a Dirac sea picture. Furry and Oppenheimer introduced operators $\sum_{r,r'} \Omega_{r',r} a^{\dagger}_{r'} a_r$ acting on the wave functions defined in occupation number space. The $a^{\dagger}_r$ and a_r were, respectively, the annihilation and creation operators of electrons fulfilling the regular anticommutation relations. Their connection to the number operator was given by $N_r = a^{\dagger}_r a_r$.

[100] This differentiation had been introduced by Erwin Schrödinger (1931) to avoid transitions to negative energy states in a one-particle theory of the Dirac equation. Wolfgang Pauli found a similar, slightly more general procedure already in 1930 but did not publish his results. Pauli to Schrödinger, March 11. 1931; as cited in Pauli (1985, 64–66). See also Pauli (1933, 229–231).

[101] *Note by the author*: At the time, Oppenheimer referred to the positron still as "positive".

[102] All quotes from Furry and Oppenheimer (1934, 246).

In a theory, for which the occupation of a negative energy state was considered as the presence of a particle, an operator as defined above and referring only to wholly positive (indicated by Latin indices) and wholly negative states (indicated by Greek indices) would be represented as

$$\sum_r \Omega_{rr} N_r + \sum_\rho \Omega_{\rho\rho} N_\rho \tag{10.6}$$

Simply equating $M_\rho = 1 - N_\rho$, which corresponded to considering the emptiness of a negative energy state as a particle, rendered

$$\sum_r \Omega_{rr} N_r - \sum_\rho \Omega_{\rho\rho} M_\rho + \sum_\rho \Omega_{\rho\rho} \tag{10.7}$$

Since a system of N electrons and M positrons, without any interactions between them, was expected to be of the form $\sum_r \Omega_{rr} N_r + \sum_\rho (-\Omega_{\rho\rho}) M_\rho$, Furry and Oppenheimer simply subtracted the last term $\sum_\rho \Omega_{\rho\rho}$ of Eq. (10.7).[103]

For an arbitrary operator $\widehat{\omega}_{(r)(s)}$,[104] this meant that

$$\widehat{\omega}_{(r)(s)} = \sum_{(r)(s)} \Omega_{(r)(s)} \omega_{(r)(s)} \tag{10.8}$$

with $\omega_{r(s)} = a_r^\dagger a_{(s)}$; $\omega_{(r)s} = a_{(r)}^\dagger a_s$; $\omega_{\rho\sigma} = a_\rho^\dagger a_\sigma - \delta_{\rho\sigma} = -a_\sigma a_\rho^\dagger$, where the last alteration was due to the subtraction in the transition from a filled to an empty vacuum; it effectively switched the role of the creation and annihilation operators for positrons in relation to electrons.[105] This last point was not very prominent in Furry and Oppenheimer's discussion, but it was important for the conception of their theory. It no longer depended on the Dirac sea, but on the electron-positron pairs.

The theory had some satisfactory features. The kinetic energy was always positive. The total charge of the system, i.e. the difference $e\nu = e(M - N)$ between the numbers of positrons and electrons, was conserved and through the action of the odd part of the Hamiltonian the number of electrons N and positrons M present in a system changed with time in equal amounts. Furry and Oppenheimer could establish the series of wave functions in configuration space, for which each subseries referred to a specific value of the total charge difference $e\nu$. They further brought their theory into contact with the one-particle theory of the Dirac equation and could

[103] As they themselves noted, this subtraction destroyed gauge invariance. A suitable subtraction was introduced by Peierls (1934).

[104] Indices in brackets can both refer to positive or negative states.

[105] See also Oppenheimer and Furry's description of pair annihilation and creation in reference to these operators in Furry and Oppenheimer (1934, 251, particularly the paragraph following their Eq. (2.25)).

show that its results were upheld in their formulation. They did so both in terms of a direct perturbative approach and by expressing the wave functions "by horrible determinants".[106]

In their perturbative approach, they subtracted the energy of a state (O) for which it was known that just as many electrons and positrons were present and which "corresponds to the normal state of the pairs in that field" (Furry and Oppenheimer 1934, 255) from the energy of the system for which it was known that either an additional electron (L) or positron (Λ) was present. The relevant contributions were, to second order, given by equations of the form (10.1), where the summation had to be taken both over positive and negative energy states. The respective interaction was the odd part of the Hamiltonian, i.e. the one responsible for pair creation. This procedure allowed them to get into contact with the one-particle theory of the Dirac equation. In general the systems which allowed for such an expression "are the systems in which, though pairs may be present, their presence may be unambiguously detected only by disturbing the system with high energy radiation" (Furry and Oppenheimer 1934, 258–259).

In the last section, in which Furry and Oppenheimer arrived through a direct perturbative evaluation at exactly the same numerical results for vacuum polarization as Dirac, they more explicitly discussed the state (O); or as they called it "a system in which directly observable pairs are not being created" (Furry and Oppenheimer 1934, 259). For such a system, there was some state of lowest energy $E^{(O)}$ which "is the energy of these 'nascent' pairs in the given field; when the field vanishes $E^{(O)} \to 0$; when the field does not vanish $E^{(O)}$ is in general negative" (Furry and Oppenheimer 1934, 259). As Dirac before them, Furry and Oppenheimer also introduced the basic ideas behind vacuum polarization and roots of charge renormalization, now formulated in their conceptual framework:

> Because of the polarizability of the nascent pairs, the dielectric constant of space into which no matter has been introduced differs from that of truly empty space. For fields which are neither too strong nor too rapidly varying the dielectric constant of the vacuum then has the constant value $\propto (1 + \kappa\alpha)$.[107] Because it is in practice impossible not to have pairs present, we may redefine all dielectric constants, as is customarily done, by taking that of a vacuum to be unity. (Furry and Oppenheimer 1934, 260)

Similar statements held for the "true and apparent charge", i.e. the regular one particle theory charge and the one including the action of the nascent pairs.

[106] Oppenheimer's words. Robert to Frank Oppenheimer, October 7. 1933; as cited Smith and Weiner (1981, 164).

[107] *Comment by the author*: α is the fine-structure constant; κ is a shorthand for the divergent integral Furry and Oppenheimer arrived at.

Furry and Oppenheimer's Theory: A Terminological and Theoretical Interlude

The way Furry and Oppenheimer connected the nascent pairs with the measurement process and the terminology they used can be drawn together to provide a conception that will also be relevant for the narrative discussed in Sect. 10.2.3. The mathematical formulation suggested that there was an actual probability of finding these pairs in the field. But Furry and Oppenheimer noted, that such pairs "may be unambiguously detected only by disturbing the system with high energy radiation", i.e. only in the case when the measurement itself provided the energy necessary for the particles to come into full existence. In this sense, the pairs were "nascent".

While the terminological variant of nascent or "in status nascendi" sounds rather strange to a modern reader, it can be understood from the terminology used at the time for pair production process: Physicists spoke of the "birth" of particle pairs.[108] The particle pairs were in the process of being born, i.e. created, but they were not created yet as to be observable.

The terminological variant of "nascent pairs" did not go unnoticed. During lectures in the US in the spring of 1935, Paul Dirac drew the analogy between the self-energy of the photon and electron and stated:

> Just as the self-energy of the electron can be regarded as due to many nascent light quanta surrounding it, so the theory gives around each photon many nascent electrons and positrons which give it self-energy; i.e. the Hamiltonian contains terms corresponding to such transitions as cause the creation of electrons and positrons.[109]

He also held a talk at the Kapitza Club entitled "Nascent electricity".[110]

George Gamow used the term "in status nascendi" in reference to Gian-Carlo Wick's explanation of the anomalous magnetic moments of the proton at the First Washington Conference on Theoretical Physics, held on April 19.-21. 1935 at the George Washington University.[111] In his textbook, Gamow (1937, 148) used the terms "in status nascendi" or "half-born" as synonyms for "virtual", which he also explicitly invoked.

The pairs were there and their presence had empirical consequences, yet they could not unambiguously be detected or, rather, still had to be brought into full existence. As late as 1949, Victor Weisskopf used a terminological variant with a similar connotation. He noted that "they [vacuum fluctuations] represent the *latent electron pairs* which, by means of light quanta, could be brought into real existence" (Weisskopf 1949, 311, emphasis added). While some physicists used the term, it still did not find its way into the standard vocabulary of quantum field theorists. In the mid 1930s, the term "virtual" started to spread (see Sect. 9.1.2).

---

[108] See, for example, the title of Heitler and Sauter (1933).

[109] Lectures on Quantum Electrodynamics 1934–35 (unpublished), as cited in Kragh (1990, 167–168).

[110] See Kragh (1990, 334, Note 8).

[111] See Sect. 10.2.2 for a historical discussion of this phenomenon and a full quote.

Also Furry and Oppenheimer's approach did not make a huge impact on the further development of the theory. Oppenheimer got only little response from European physicists directly,[112] but their paper was quite negatively received in Europe. All the same, the critique of European physicists was not directed towards the concepts Furry and Oppenheimer applied, but rather to the break with general invariance principles.[113] As such, their paper did not constitute any theoretical progress, neither in terms of formulating the theory invariantly, nor in the handling of the divergences.

Yet, shortly after these initial communications a consistent theoretical scheme was formulated which would eventually allow for a near to unambiguous singling out and subtracting of the infinities. Dirac (1934a) was the first to find a suitable solution based on a relativistic version of his density matrix. This procedure was subsequently generalized and also discussed for a quantized electron-positron field by Heisenberg (1934a). The guiding ideas of these papers, although eventually subtracting contributions which we would ascribe to virtual particles, did not essentially rest on this concept. Rather, the division of the density matrix into divergent and finite terms was led by general theoretical principles such as gauge and relativistic invariance.[114] Especially Heisenberg's formulation of the problem was subsequently applied and generalized to investigate the effects of vacuum polarization, also by the group around Robert Oppenheimer (Serber (1935), Uehling (1935), and Serber (1936)).

The "Electron Positron Cloud"

As natural as the evaluation of vacuum polarization came to the historical actors, as seemingly easily came its interpretation. Whether in the conception of the Dirac sea or in terms of Furry and Oppenheimer's nascent pairs, later to be called virtual, the idea that a particle did not occur on its own but was always accompanied by a charge cloud was a seemingly natural conclusion for the theorists. Whether against the background of nascent pairs or the Dirac sea: Empty space was no longer really empty. Such an interpretation was highly suggestive from the mathematical structures the theorists worked with. Also the number operators of electrons and positrons did not commute with the interaction Hamiltonian. Hence, only one of these characteristics of the

---

112 Oppenheimer to George Uhlenbeck, March 1934; as cited in Smith and Weiner (1981, 174–176).

113 Pauli noted in a letter to Heisenberg that Oppenheimer's paper "only treats the old non-gauge invariant form of hole theory [...]". [*Original*: "[...] nur die alte nicht eichinvariante Form der Löchertheorie behandelt [...]."] Pauli to Heisenberg, January 21. 1934; as cited in Pauli (1985, 222). According to Milton S. Plesset, "Bohr was pleased that this non-invariance [in this case relativistic] of Oppenheimer's theory could be demonstrated". Plesset to Guido Beck, December 21. 1933; GBA, 1121.1b.

114 These reasons were given by Dirac in correspondence with Pauli. Dirac to Pauli, November 11. 1933; as cited in Pauli (1993, 756); and Dirac to Pauli, December 3. 1933; as cited in Pauli (1993, 757).

system could be determined at each instant. When one was interested in the value of an odd operator, one could not know how many particles were actually present in the system.

It is important to note that the subtraction procedure devised during the 1930s by Dirac (1934a), Heisenberg (1934a) and then traced back to three simple assumptions by Weisskopf (1936)[115] could only handle additional charge contributions due to external electromagnetic fields which could as well be described classically.[116] The techniques could not simply be taken over into a complete quantum field theoretical description, as for example Heisenberg and Euler (1936, Paragraph 3) noted. Processes like the "proper energy of the vacuum", which was described by Milton S. Plesset in the following terms, still diverged:

> Transitions are considered from the initial state of the vacuum distribution to intermediate states in which an electron-positron pair and a quantum of radiation are present; then transitions are considered from these intermediate states back to the initial state. (Milton S. Plesset in Anonymous (1934, 109))

Heisenberg and Euler (1936, 730) framed this as "the virtual possibility of the creation and disappearance of a light quantum and a pair" while Plesset noted that "the intermediate states may be given the picturesque interpretation of representing fluctuations in the vacuum distribution" (Milton S. Plesset in Anonymous (1934, 109)).

Vacuum fluctuations of both charge and radiation as well as their connection to the divergences were clearly perceived of by the historical actors at the turn from 1933 to 1934. Parts of the contributions we ascribe to "virtual particles" today could already be divided into finite and infinite parts, where the latter were subtracted. One could name different references in which these phenomena and the picture of a charge cloud were made explicit and we will also encounter a prominent instance in the explanation of the anomalous magnetic momenta by Gian-Carlo Wick (see Sect. 10.2.2). Just to give one example, consider Wolfgang Pauli's assessment at a seminar held at the Institute for Advanced Studies in Princeton in 1935:

> The infinite polarization of the vacuum implies that no free charge can exist, for if we should place a charge in a vacuum it would surround itself with an equal and opposite charge. And, again, since the photon cannot exist without virtually generating pairs, it follows that its self-energy is also infinite in the same way as the particle self-energy. [...] We cannot even know how much of the energy is due to pairs since we cannot know how many pairs there are – only how much energy and charge. (Pauli 1936a, 4–5)

By the time, the concept of an "electron positron cloud"[117] or an additional induced charge and current in the vacuum had acquired a rather uncontroversial status. Just as a charged particle induced an electromagnetic field, an electromagnetic field induced charge and current densities in the vacuum.[118]

---

[115] See Wright (2024, Sect. 3.5) for Weisskopf's work.

[116] See Darrigol (1982, Sect. 5.1).

[117] See for example Lauritsen and Oppenheimer (1934, 80).

[118] See Heisenberg (1934a, 229) or Euler (1936, 400) who explicitly invoked this analogy for making the self-energy of a light quantum or light-by-light scattering accessible to their readers.

10.1.3 The Infrared Divergence (at Around 1937)

While the concepts of a bare and a dressed particle (modern terminology) were well established and theoretically evaluated to their empirical consequences for vacuum polarization, the corresponding conception for the inclusion of virtual photons into the concept of a particle itself showed up during the 1930s but soon faded into the background. To access this conception, the work on the infrared divergence is key. My assessment will mostly follow Alexander Blum (2015), who provided the most detailed and conceptually sensitive study of the infrared divergence and I refer the reader to his paper for the details of the historical development.[119]

The infrared divergence actually has a twofold relevance for the virtual particle concept. On the one hand, the above mentioned concept of a dressed electron, including its proper radiation field (in modern terminology: virtual photons), showed up in the initial solution to the infrared divergence. On the other hand, in the further evaluation of the infrared divergence physicists slowly realized that two contributions played a part in its theoretical description, one stemming from virtual processes and one from the emission of real photons. These contributions would exactly cancel (at least in the low energy spectrum).

The infrared divergence showed up multiple times in the theoretical evaluations of scattering problems, even before the advent of quantum electrodynamics, but gathered greater attention in the community at around 1935. It occurred most prominently in calculations of radiative corrections to the scattering of an electron by an external potential, most regularly the Coulomb field of an atom. While the self-energy as well as vacuum polarization diverged because of the contribution of high energy particles, the infrared divergence arose in the low frequency range of the emitted photons. When calculating the corrections to phenomena such as Bremsstrahlung, physicists arrived at a proportionality of the differential cross section to $\frac{d\nu}{\nu}$, where ν is the frequency of the emitted radiation. Integrating over this factor led to a logarithmic divergence for small values of ν, so far small energy values.

As was argued by Werner Heisenberg, in contrast to the ultraviolet divergences, QED itself was not to blame for the occurrence of the infrared divergence. Since for low frequencies the classical theory should be applicable, also QED, as a generalization of classical electrodynamics, should be correct. Actually, it was noted that the perturbation-theoretical approach to QED was the problem: According to classical theory, an electron passing a nucleus emitted a continuous spectrum, the intensity of which remained finite in the limit $\nu \to 0$. This in turn meant that in QED an infinity of photons would have to be emitted in this region to balance the tendency of the frequency towards zero. But the perturbative evaluation of QED was built around the

[119] Further relevant secondary literature includes Blum and Joas (2016, Sect. 2), Darrigol (1982, Sect. 5.2), and Aramaki (1987, Sect. 2.2).

emission (or absorption) of a single photon at each order. Thus, Heisenberg arrived at the conclusion that the problem was not one of QED proper but of its mode of evaluation.[120]

In the proposed solutions to the said problem by Bloch and Nordsieck (1937) as well as by Pauli and Fierz (1938), the perturbative approach to QED was consequently altered. In the standard approach to scattering problems in QED, the interaction of the electrons with the radiation field was treated as a perturbation while the Coulomb interaction could either be included as a perturbation or could be solved exactly. In the approaches to the infrared divergence, things were turned around: The interaction with the radiation field was solved non-perturbatively while the interaction with the Coulomb potential of the scatterer was treated as a perturbation.

In this approach, several approximations were applied, mainly based on the non-relativistic treatment and the focus on the low frequency spectrum, the interesting part for the infrared divergence. In the course of exactly solving the problem of the interaction of radiation field and electron, Pauli and Fierz as well as Bloch and Nordsieck arrived at a Hamiltonian which could be interpreted as a set of oscillators (the transverse part of the radiation field in the given approximation) with a perturbing force (the electron).[121] A simple transformation allowed them to rewrite this equation so that it contained (1) the kinetic energy of the electron, (2) the free modes of a radiation field, and (3) a term which was given by $\sum_s (ap)^2 \hbar\nu_s$, where ν_s referred to the frequency modes of the radiation field and p to the momentum of the electron. $a = \frac{2e}{m} \frac{1}{\hbar\nu_s} (\frac{\pi\hbar}{\nu_s V})^{1/2} \epsilon_s g(k)$, where ϵ_s was the polarization vector of the radiation and $g(k)$ a cut-off function introduced to focus on the low energy spectrum of the radiation field. In non-relativistic theory, the third term of the Hamiltonian could be interpreted as a electromagnetic mass of the electron, as Pauli and Fierz actually did.[122]

But Pauli and Fierz did not include this term in the mass of the electron.[123] Due to the cut-off they had introduced, they identified the term as small in comparison to the mechanical mass and simply dropped it. But Pauli and Fierz obviously were aware both of the interpretation in terms of an electromagnetic mass and the fact that they had subsumed all virtual effects by their transformation. As they noted:

> The quantum number n_s defined by $\frac{1}{2}(P_s'^2 + Q_s^2) = n_s + \frac{1}{2}$ [i.e. by the transformed oscillator variables] can be called the numbers of the "free light quanta" of frequency ω_s and they differ from the light-quantum numbers [...] which are usually used in radiation theory. But it follows from the Hamiltonian operator, that – in the approximation considered here – it is the free light quanta n_s, which can be directly observed with a spectrograph, as only these can, spatially separated from the charged body, propagate freely. In particular, for a

[120] This argument has been presented by Kockel (1937, 155–156) in print. In a draft for a possible solution to this "seeming difficulty" [scheinbare Schwierigkeit], Heisenberg presented this argument in full. AMPG, III. Abt., Rep. 93, V. 671.

[121] See Eq. (9) of Pauli and Fierz (1938).

[122] In the derivation by Bloch and Nordsieck which started with a relativistic formulation and introduced non-relativistic approximations in the course of the discussion, this identification was not possible. See Blum (2015, 77).

[123] See Pauli and Fierz (1938, 173, Eqs. (11) and (12)).

given particle momentum p the state of lowest energy, in which all quantum numbers n_s disappear, corresponds to the absence of radiation in the proper sense.[124]

The concept of a dressed electron, including photons which were attached to the electron, was present in Pauli and Fierz's approach.[125] But it was so only in a non-relativistic theory, which included several approximations due to the focus on soft photons. Actually, the concept existed only for a few lines in the respective paper, as it was nearly immediately dropped. And it was not dropped because of the inclusion of the term into the proper mass of the electron, but due to its smallness in comparison to the electron's mechanical mass. As was emphasized by Alexander Blum, the concept of renormalization, in a modern sense, was not yet developed.

The above described solution of the radiation problem was the first step in the solution of the infrared divergence. In the second step, a perturbing potential was introduced in the Hamiltonian of the system. Through the transformation that led to the decomposition of the radiation field into an electromagnetic mass term and "free light quanta" also the dependency of the potential was altered in such a way that the scattering process was necessarily accompanied by the emission of an infinity of photons. By neglecting energy conservation for the emission process, Bloch and Nordsieck were able to show, that the emission of infinitely many soft photons did not alter the scattering cross section. Hence, the initial problem of the infrared divergence was solved.

Although the infrared divergence could be taken care of with such a procedure, it was far from clear how this calculation was connected to intermediate or high energy phenomena. And in these investigations, the second aspect mentioned above, the interrelation between virtual and real processes for the cancellation of the infrared divergence, became manifest. Pauli and Fierz started to investigate the region of intermediate photon energies by imposing energy conservation for the emission process, thereby going beyond Bloch and Nordsieck's approach. But at this point something seemingly paradoxical happened: The scattering cross section vanished unless the cut-off frequency was introduced explicitly. But if the cut-off was introduced, the result depended critically on the cut-off frequency. Even worse, when the function depending on the energy cut-off was expanded in a power series in α, the second order term diverged when the cut-off was pushed to infinity.

Pauli and others did not understand this result at the time. It took Edgar Weinmann, a PhD student in Tübingen completely unknown until then, to make things apparent. In his dissertation, Weinmann made a similar expansion in a power series of the result

124 "Die durch $\frac{1}{2}(P_s'^2 + Q_s^2) = n_s + 1/2$ definierten Quantenzahlen n_s können als Anzahlen der 'freien Lichtquanten' der Frequenz ω_s bezeichnet werden und sind von den Lichquantenzahlen [...], die sonst in der Strahlungstheorie benützt werden, verschieden. Es folgt aber aus dem Hamilton-operator, dass – in der hier benützten Näherung – die freien Lichtquantenzahlen n_s es sind, die direkt in einem Spektrograph beobachtet werden können, da nur diese sich, vom geladenen Körper räumlich getrennt, frei fortpflanzen können. Insbesondere entspricht bei gegebenem Teilchenimpuls p der Zustand kleinster Energie, in welchem alle Quantenzahlen n_s verschwinden, der Abwesenheit von Strahlung im eigentlichen Sinn." Pauli and Fierz (1938, 174).

125 It also occurred in Bloch and Nordsieck's paper, but as already mentioned, could not be interpreted as an additional mass term.

as Pauli and Fierz did.[126] Weinmann and his supervisor Werner Braunbek (Braunbek and Weinmann 1938) developed the terms as follows: They assumed that there was an infinity of emitted light quanta which did not affect the cross section (which was the result of Bloch and Nordsieck) and, additionally, there was a finite amount of non-negligible photons in the final state as well. In their expansion and to the first non-trivial order the contributions of two terms, one with an observable final photon and one with no additional photon, combined to cancel the infrared divergence. Still, the term without photon emission diverged logarithmically for high energies.

Weinmann and Braunbek's calculation thus showed, although only implicitly and only to second order, that the two different contributions to the infrared divergence cancelled each other. The contributions of real photons and virtual ones exactly cancelled. Yet, whether this result would hold in a relativistic description was still unknown. This question was addressed in Sidney Dancoff's famously incorrect calculation (Dancoff 1939) of the radiative corrections to the scattering of electrons. As a matter of fact, Dancoff stumbled over the differentiation between Coulomb interaction and the radiation field, neglecting terms corresponding to Coulombian self-energy parts. As Alexander Blum (2015) has argued, if Dancoff would have included these contributions, he would have had all divergent integrals at hand which he could have combined to arrive at a term which he could have interpreted as an electromagnetic mass. But this would mean that Dancoff would have had to develop the modern understanding of renormalization, which was not yet in place. The attribution of the failure of Dancoff as "history changing wrong" can, as Blum argues, thus be attributed to an anachronistic reading of the usage of the term "renormalization".

Yet, the infrared divergence developed into an important methodological guideline and test case for the consistency of alterations to the practice of QED. Heitler and Peng's theory of radiation damping, for example, explicitly excluded higher order processes to avoid the ultraviolet divergence (see Sect. 9.3). But Heitler and Peng did not realize, that by throwing out these contributions, they also threw away the virtual contributions to the infrared divergence. Hence, the contributions from the real spectrum diverged and the infrared divergence was reintroduced through their procedure. This was realized by Pauli by November 1944[127] and proven in print by Bethe and Oppenheimer in 1946. As they noted: "What we have shown in this paper is that it is not possible to obtain a satisfactory answer either by including the effects of virtual quanta as they are given by present theory or by excluding them entirely." (Bethe and Oppenheimer 1946, 457–458).

---

[126] Weinmann was completely unaware of the Pauli-Fierz paper. But he had the same question in mind: How were regular perturbative treatments for high energy photons connected to the results obtained in the solution of the infrared divergence?

[127] Pauli to Wentzel, 27. November 1944; as cited in Pauli (1993, 249).

10.1.4 An Intermediate Summary

In the discussion so far, different aspects of the virtual particle concept took prominence that will be further developed and contextualized in the second part of this chapter. The discussion of the most prominent self-energy calculations from 1929 to 1939 shows that the invocation of non-mathematical representations of the processes, and hence also of virtual entities, was highly dependent on the mode of perturbation theory and the formulation of the theory itself. Only by the mid to late 1930s, so at least five years after Dirac's initial proposal, a processual description of the energy correction in terms of emission and absorption subprocesses took a hold in these calculations. But, and this is important to judge its relevance, these subprocesses were not necessarily conceived of as a physical picture of the phenomena. Weisskopf clearly invoked the correspondence-like method to spell out the physical reasons for the divergence and its mitigation in the Dirac sea. The picture of successive transitions, on the contrary, was used merely as a calculational aid in the evaluation of the higher order contributions.

Again by the mid 1930s, the conclusion that a particle is never on its own, the idea of a cloud of virtual particles, emerged in the discussion of vacuum polarization. This interpretation as its investigation seemed to have come quite naturally to the physicists and was rather suggestive from the mathematical representation of the theory and its perturbative evaluation. At the same time, the concept of vacuum fluctuations, both of the electromagnetic field and the electron-positron field, were connected to the ultraviolet divergences. And even though something along the lines of charge renormalization was developed in the subtraction procedures of Dirac and Heisenberg, the concept of a dressed electron, although partially available in the paper of Pauli and Fierz on the infrared divergence, was not investigated to its full potential. A point which is closely connected to Alexander Blum's observation that renormalization had not yet gathered its modern-day meaning.

The two different interpretations of the mathematical structure represented in Eq. (10.1) mentioned in the beginning of this chapter also came to the fore in the discussion of the divergences. A description in terms of successive transitions was invoked as a tool by Weisskopf and Mercier. But, as can be seen from the charge cloud picture and the comments on the "free" and "attached" photons in Pauli's and Fierz' paper on the infrared divergence, the mathematical structure could be interpreted differently. It represented a probability for the creation of photons or particles which were not there fully and on their own. They were "nascent", "half-born", or, as Weisskopf framed it, "latent" and, for example, "by means of light quanta, could be brought into real existence" (Weisskopf 1949, 311).

In the discussion of Furry and Oppenheimer's paper on vacuum polarization further traits of the conception, which will become important for nuclear physics, have been touched upon. On the one hand, and as will be discussed in Sect. 10.2.2, the explanation of the anomalous magnetic momenta of nucleons was (qualitatively) based on the charge cloud picture. On the other hand, already Furry and Oppenheimer (1934, 259) observed that the presence of the nascent pairs "may be unambiguously detected only by disturbing the system with high energy radiation", i.e. by performing a measurement that supplied the necessary energy for their coming into full existence. This influence of the measuring apparatus will eventually serve Gian-Carlo Wick as the entrance point to the most prominent argument connecting virtual particles with the energy-time uncertainty. This section has thus introduced important themes connected to the development of the virtual particle concept which we will as well encounter in the history of quantum field theoretical approaches to nuclear physics.

10.2 Nuclear Physics: Force Carriers, Qualitative Reasoning, and Feynman-Like Diagrams

The above observation, that specific themes and conceptions connected to the virtual particle concept will also be encountered in my historical investigation of quantum field theoretical approaches to nuclear theory, is partially due to my focus: I will exclusively discuss phenomena in which the particles were (according to a verbal model of successive transitions) only present in intermediate, i.e. virtual, states: the forces between nucleons and the most popular explanation of the anomalous magnetic moment of the proton and neutron. As such, the following section will serve to further contextualize the above described conceptual development from virtual transitions to virtual particles and provide for a historical explanation of their prominence. But each of these subsections also stands in its own right, not only in respect to its historical line of development but also due to its connection to the virtual particle concept.

The concept of a force carrier is a very popular instantiation of the virtual particle concept and its development will be discussed in Sect. 10.2.1.[128] The connected concept of exchange forces in quantum field theory has been studied by Cathryn Carson (1996b) and, even though I will mostly follow her assessment, some subtleties of the development connected to a dynamical conception of the interaction of particles deserve, as I will argue, a reassessment. Gian-Carlo Wick's explanation of the anomalous magnetic moments of the nuclei through "virtual decay/dissociation", discussed in Sect. 10.2.2, was, in the general structure of the argument, the most popular explanation of the said empirical observation until a completely new model of the proton in terms of quarks was proposed. Wick's argument, which at first completely transgressed the applicability of the theory from which it emerged and which afterwards only was upheld by neglecting divergence difficulties, is in its qualitative nature connected to the prominence of the narrative for virtual particles in nuclear physics, discussed in the Sect. 10.2.3. I will explain the fact that such an argumentative structure became prominent in quantum field theoretical approaches to nuclear physics by the uncertainties in the respective theoretical modelling. In this case, the

[128] I have already provided a first analysis of its (non-)occurrence in the paper by Bethe and Fermi (1932) in Sect. 8.2.1.

invocation of the narrative had an epistemic upshot which was independent of the specific mathematical description of the fields and interactions chosen by the actors.

In the last subsection, contrasting these mostly qualitative assessments by the historical actors, I will turn to a representational format strikingly similar to Feynman diagrams: Gregor Wentzel's "Schemata". I will discuss their development in the years 1937 to 1938, that is: their first instantiation shortly before Yukawa's meson theory was noticed in the West and their subsequent alteration in an article introducing the meson to the interested non-specialist. This will not only serve to highlight the conceptual and representational differences between QFT of the late 1930s and the late 1940s, but it will also serve as a prime example for the main theme of this chapter: the slow transition in the conception from virtual transitions to virtual particles that started to show in the mid to late 1930s.

But before going into the historical development itself I should point out that I will nearly exclusively discuss quantum field theoretical approaches to nuclear physics before Yukawa's theory was noticed in the West. Although Yukawa's proposal of the meson can hardly be overrated in its boldness (at least so in comparison to the more conservative approaches of European physicists) and in its impact on the further theoretical modelling of nuclear physics, the mechanisms for which virtual particles played a particularly important role were, on the one hand, discussed by European physicists before they became aware of Yukawa's proposal of the meson. The arguments I will revisit after Yukawa's theory became widely acknowledged in the community, on the other hand, were of a rather qualitative nature and there is no need to go into the details of the particular formulation of the respective theory.

10.2.1 Exchange Forces and the Quantum Electrodynamical Analogy (1932–1934)

The idea that nuclear interactions were due to an exchange force between neutron and proton was first expressed in Heisenberg's 1932 trilogy on the nucleus.[129] It was proposed nearly one and half years prior to Fermi's theory of β-decay, the first application of QFT in the realm of nuclei. To understand Heisenberg's conception of the nucleus, its constituents, and the roots of the particle exchange model in nuclear physics, we have to take a step back and shortly review both the concept of exchange forces as well as the conception of the nucleus prior to the discovery of the neutron in 1932.

The concept of exchange forces[130] had its roots in Heisenberg's 1926 works on many body problems, indistinguishability, and resonance phenomena, so it emerged

---

[129] Heisenberg (1932b), Heisenberg (1932c), and Heisenberg (1933). I will mostly follow Carson (1996b, Sect. 2). Further relevant secondary literature includes Bromberg (1971, 332–338), Miller (1984, 155–165), Darrigol (1988, Sect. I.2), Brown and Rechenberg (1996, Chap. 2), and Borrelli (2017, Sect. 2).

[130] The following paragraph is based on Carson (1996a).

shortly after quantum mechanics itself had been developed. It was subsequently used in the theory of atoms, molecules (most prominently by Heitler and London), ferromagnetism and electronic collisions, all with their own, overlapping but essentially different interpretations. This type of interaction is, from a mathematical point of view, due to the fact that the wave function describing a system of two identical particles can be (anti-)symmetrized, i.e. $\psi \propto \psi_a(1)\psi_b(2) \pm \psi_a(2)\psi_b(1)$, where the indices refer to the particles and the numbers to the variables describing them. Without going into details of its development and the interpretational issues at stake, it is to note that Heisenberg himself both represented exchange phenomena in terms of an intuitive ["anschauliches"] model of electrons oscillating back and forth between atoms of a molecule, changing their place in space and time, and without such intuitive pictures, taking the concept to correspond to a formal and only mathematical exchange of coordinates of the considered particles. By late 1928, although the intuitive interpretation was dominant in popular descriptions, in the diverse applications a concept emerged which described a quantum-mechanical effect due to the interchange of two identical systems, mostly electrons, which did not include their exchange as taking place in time and space, but simply as a mathematical replacement.

The conception of the nucleus was dominated by the electron-proton model up until 1932.[131] Yet, there were already different indications, both theoretical and empirical, that such a model was inadequate for the quantum-theoretical description of the nucleus. It was noted as early as 1929 that the nitrogen-14 nucleus behaved as if it consisted of an even number of fermions, although, according to an electron-proton model, it was composed of an odd number: fourteen protons and seven electrons, hence 21 fermions. Furthermore, the continuous spectrum of β-decay was at odds with energy conservation as long as electrons and protons were the only particles involved in this process.[132] In addition, electrons confined to a space of the radius of a nucleus would have to be treated relativistically due to their enormous kinetic energy. But problems of quantum theory at such length-scales or energies, besides the divergences of QED there was also the Klein paradox, seemed to restrict quantum theory to describe such phenomena correctly.[133] If electrons were constituents of nuclei, they had to, at least to some extent, loose their individuality as a particle and their physical characteristics such as spin.

---

[131] For more details on the electron-proton model of nuclear physics and, in general, the history of nuclear theory prior to the discovery of the neutron, see for example Darrigol (1988, 230–238 and 240–247), Brown and Rechenberg (1996, Chap. 1), or Stuewer (2018, Chaps. 5 and 6).

[132] Since the possibility of an accompanying γ-ray spectrum was empirically ruled out by the late 1920s, another explanation was needed. Energy non-conservation (or the breakdown of quantum theory) was rather popular for a time, followed for example by Bohr, Heisenberg, and Beck. Eventually and as is well-known, the neutrino hypothesis, first unofficially pronounced by Pauli in 1931, would govern the field theoretical explanations of the continuous β-spectrum.

[133] For the discussion of the problems of theoretical consistency, as most prominently voiced by Niels Bohr, I refer the reader to Bromberg (1971, Sect. 2) or Darrigol (1988, 242–245).

The Molecular Analogy and Heisenberg's Trilogy (1932)

After the discovery of the neutron in early 1932,[134] it turned out that the supposed failure of quantum theory on the length scales of nuclear constituents had an upside. Heisenberg could "blame all the difficulties of principle on the neutron and practise quantum mechanics in the nucleus".[135] Heisenberg's conception in his trilogy on the nucleus was accordingly the following: The neutron was built up of a proton and a (Bose) electron, but quantum theory could not describe how this bundle was constructed or how it gathered its stability. Heisenberg conceived of the neutron, not yet as a fundamental particle, but as a "fundamental building block"[136] of the nucleus and he could thus apply quantum mechanics to the behaviour of protons and neutrons. As he framed it, the empirical material "suggests the hypothesis, that the atomic nuclei are build up by protons and neutrons without the contributions of electrons".[137] Although Heisenberg considered the neutron as fundamental as far as the quantum-theoretical description of the constitution of nuclei was concerned, he still used the presence of electrons in the neutron to explain β-decay, the Meitner-Hupfeld effect, and, importantly for our purpose, to motivate his quantum-theoretical modelling of the forces between nucleons.

In the first paper of his trilogy, Heisenberg provided two representations of the interaction between proton and neutron, a verbal and a mathematical one. In the verbal description, Heisenberg drew the analogy to the binding of a hydrogen ion H_2^+:

> If one brings a neutron and a proton to within a distance comparable to the dimensions of the nucleus, then – in analogy to the H_2^+ ion – a change of place [Platzwechsel] of the negative charge will occur whose frequency is given by a function $(1/h)J(r)$ of the distance r between the two particles. The quantity $J(r)$ corresponds to the Austausch- or more correctly to the Platzwechsel-integral of molecular theory. One can illustrate this Platzwechsel again with the picture of electrons that have no spin and obey Bose statistics. But it is probably more correct to regard the exchange integral $J(r)$ as a fundamental property of the proton-neutron pair, without wanting to reduce it to motions of electrons.[138]

[134] For a detailed account of the discovery of the neutron, see for example Stuewer (2018, 155–166).

[135] Heisenberg to Bohr, June 20. 1932; as cited in Carson (1996b, 101).

[136] "Fundamentalbaustein," Heisenberg (1932b, 1).

[137] "[...] legt die Auffassung nahe, die Atomkerne seien aus Protonen und Neutronen ohne Mitwirkung von Elektronen aufgebaut." Heisenberg (1932b, 1).

[138] "Bringt man Neutron und Proton in einen mit Kerndimensionen vergleichbaren Abstand, so wird – in Analogie zum H_2^+ ion – ein Platzwechsel der negativen Ladung eintreten, dessen Frequenz durch eine Funktion $(1/h)J(r)$ des Abstandes r der beiden Teilchen gegeben ist. Die Größe $J(r)$ entspricht dem Austausch- oder richtiger Platzwechselintegral der Molekültheorie. Diesen Platzwechsel kann man wieder durch das Bild der Elektronen, die keinen Spin haben und der Bosestatistik folgen, anschaulich machen. Es ist aber wohl richtiger, das Platzwechselintegral $J(r)$ als eine fundamentale Eigenschaft des Paares Neutron und Proton anzusehen, ohne es auf Elektronenbewegungen reduzieren zu wollen." Heisenberg (1932b, 2). The translation follows Carson (1996b, 104), with a slight modification which includes Heisenberg's reference to the neutron as particle.

Similarly, Heisenberg invoked the H_2-molecule as an analogy for the neutron-neutron interaction.

At this point it is important to note that Heisenberg used the term "Platzwechsel" on purpose and that he favoured it over the term "Austausch" (i.e. exchange). As shortly outlined above, the exchange interaction was at the time connected to the exchange of two identical particles. But this was not the case for neutron and proton, at least so in the verbal description. Thus Heisenberg chose the "Platzwechsel" over the exchange terminology and drew the analogy to the H_2^+-ion.[139] Nevertheless, this terminological variant was soon to fade after Majorana's contribution to the theory of nuclear forces (see below). In the verbal description of Heisenberg, the Platzwechsel of an electron was certainly the most prominent feature of the interaction of neutron and proton but in its mathematical formulation this idea was next to absent.

In line with the conjecture that quantum theory was not applicable to the nuclear constituents as such and that one should "regard the exchange integral $J(r)$ as a fundamental property of the proton-neutron pair, without wanting to reduce it to motions of electrons", Heisenberg did not use variables of electrons in his mathematical formalism.[140] He described the interaction between neutron and proton as well as the wave function of the deuteron, the recently discovered bound state of proton and neutron and the only wave function Heisenberg represented explicitly in mathematical form, without reference to the variables of the electron. Rather, he introduced a new variable, later known as iso spin, to discern protons and neutrons. This allowed him to represent the verbal description in terms of mathematical formulas, including the exchange character of the forces, but without having to refer to any movement of the electron. As Arianna Borrelli has framed it: "The spin formalism was [...] used [...] as a tool to go from a verbal to a symbolic analogy between nuclei and other quantum systems." (Borrelli 2017, 85) As such, neutron and proton were, at least in the mathematical formalism, treated as different states of one particle.

In Heisenberg's trilogy we thus find an overlap of two representations, one verbal, the other mathematical, potentially guiding one another.[141] But in the mathematical representation specific aspects of the verbal description were absent. The Bose electron was no longer represented, any temporal dimension was lost. A purely formal conception of the exchange of certain variables used for the description of the particle either as proton or neutron was applied.

Ettore Majorana's study of nuclear forces put the formal character of the exchange in nuclear theory into sharper relief.[142] It was based on Heisenberg's papers but not

---

[139] For a full discussion of the usage of this terminology, see Carson (1996b, 103–105).

[140] See for example Darrigol (1988, 249–250), Carson (1996b, 106, Footnote 23), and especially Borrelli (2017, Sect. 2).

[141] Although it is generally assumed that Heisenberg used the heuristic concept of a Bose electron moving back and forth between two protons as the guiding idea behind his modelling of nuclear forces, in the most detailed study of the relation between mathematics and verbal representation, Borrelli (2017, 85) also suggested that matters may have occurred the other way around.

[142] Best see Carson (1996b, Sect. 2.2) for Majorana's background and conception of exchange forces.

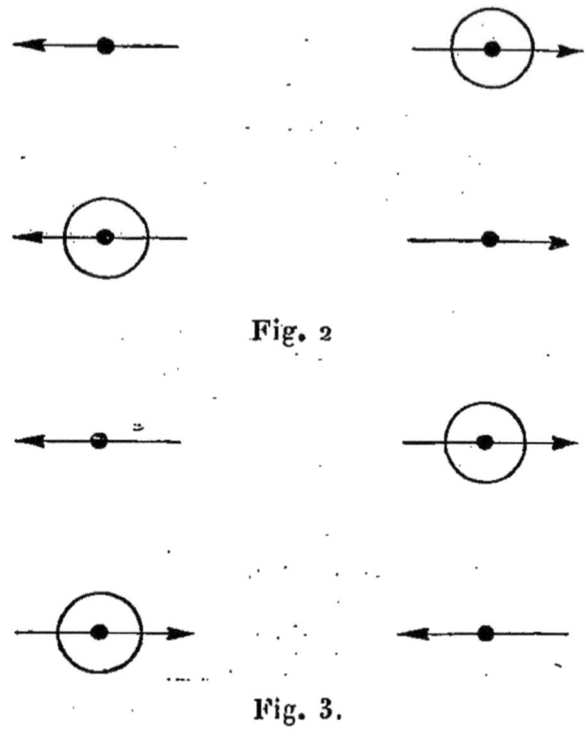

Fig. 10.2 Representation of Heisenberg (upper two lines) and Majorana (lower two lines) exchange forces. Taken from Heisenberg's contribution to the 1933 Solvay Conference: Heisenberg (1934b, 302), made available by the Bibliothèque nationale de France via Gallica

on the conception of an electron moving back and forth. Majorana (1933) did not provide any discussion of a compound neutron. He also focused on achieving the correct saturation behaviour of the forces, which was not taken account of by Heisenberg's consideration. The exchange of proton and neutron in Majorana's theory was therefore slightly different from Heisenberg's. According to Heisenberg, the charge of the two particles, proton and neutron, was exchanged. According to Majorana, the particles (neutron and proton) were exchanged, i.e. not only the charge but also their spin.

While Heisenberg initially gave a verbal representation,[143] the mathematical representation of exchange had taken over in nuclear theory in early 1933. In his contribution to the Solvay conference of 1933, Heisenberg (1934b) used diagrams to both represent his and Majorana's type of exchange forces (Fig. 10.2). The dot in the middle represented a positively charged particle, the arrow its spin direction. The circle around the dot represented the presence of an additional negative charge, i.e. the Bose

[143] There are several indications that Heisenberg took the concept of a neutron being composed of electron and proton rather seriously. See for example Heisenberg (1932c, Paragraph 3) or Heisenberg (1933, Paragraph 3) and the diagrams as reproduced in Fig. 10.2.

electron. Although Majorana did not describe the exchange of charge, but simply a spatial exchange between protons and neutrons, Heisenberg both represented his and Majorana's approach in such terms.[144]

Although these diagrams obviously referred to an exchange of charge (and spin) between two particles, they did so without an intermediary configuration. They followed the mathematical representation of exchange as put forward in Heisenberg's trilogy, rather than Heisenberg's original verbal heuristics in terms of a temporal process. A representation of such an intermediate state, both in a mathematical and diagrammatic form, only came with the application of field theories to the problem of nuclear forces and its evaluation through perturbation theory. And therefore we have to turn to Fermi's theory of β-decay.

The Interaction in Fermi's Theory of β-Decay (1933/34)

Directly after the Solvay conference of 1933, Fermi worked out a theory of β-decay which was closely modelled after QED.[145] The creation of an electron and a neutrino, combined with the transformation of the neutron into a proton, and its reversal, were the basic interactions of the theory. From a rather simple Hamiltonian and a few additional assumptions on the behaviour of the heavy particles, Fermi could reconstruct a number of observations relevant for understanding the empirical material on β-decay.

Heisenberg, Fermi,[146] as well as Tamm (1934) and Ivanenko (1934) soon realized that Fermi's theory of β-decay could incorporate nuclear forces if its perturbative evaluation was carried to second order.[147] They arrived at the following process: (1) the neutron transformed into a proton and an electron as well as a neutrino were created; (2) the proton with which the initial neutron was to interact transformed into a neutron and the electron as well as the neutrino were destroyed. In the intermediate state, two protons, one electron and a neutrino were present. Hence, there was also an intermediate configuration in terms of the mathematical representation. As is well-known,[148] in all of these calculations the interaction between neutron and proton according to Fermi's theory came out much too small.

---

[144] One could as well read the dot without a circle as a proton and the dot with a circle as a neutron and thus make the exchange of the particles neutron and proton more prominent. This would cohere with Majorana's point of view. But this was not how Heisenberg himself represented these ideas in his contribution to the Solvay conference.

[145] For the historical emergence and further development of Fermi's theory of β-decay, see Darrigol (1988, Sect. II.1), Brown and Rechenberg (1994), or Brown and Rechenberg (1996, Sect. 2.4 and Chap. 3).

[146] Heisenberg to Pauli, January 18. 1934; as cited in Pauli (1985, 250). Fermi to Heisenberg, January 30. 1934, AMPG, III. Abt., Rep. 93, Nachlaß Heisenberg, Nr. 1533.

[147] The most detailed discussion of these contributions was given by Carson (1996b, Sect. 3.3).

[148] See, for example, Brown and Rechenberg (1996, 56).

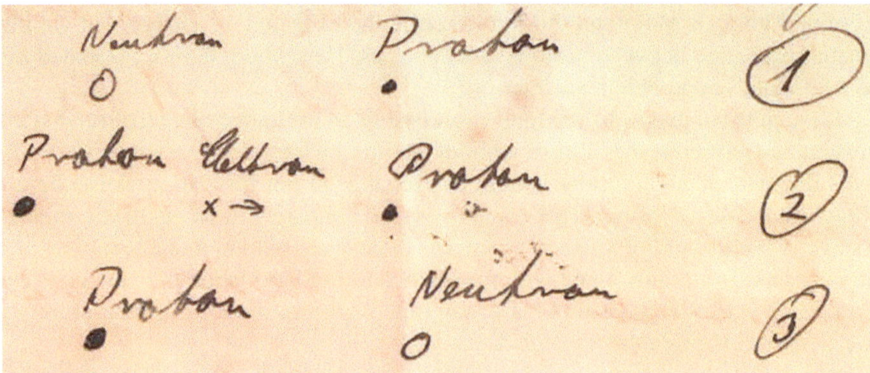

Fig. 10.3 Graphical representation of the interaction between neutron and proton according to Fermi's theory of β-decay. The exchanged neutrino is not represented. Taken from: Edward Teller to Guido Beck, undated [probably early to mid 1934]. Courtesy of and reproduced with kind permission from Guido Beck Archive

Nevertheless, this process obviously led to Heisenberg's exchange force between neutron and proton.[149] Thus, the verbal representation of Heisenberg found a firmer footing in the perturbative evaluation of Fermi's theory of β-decay. The important difference in the mathematical modelling, including an intermediate configuration, can also be exemplified by a comparison of Heisenberg's diagrammatic representation at the Solvay conference and the one given in Fig. 10.3. Edward Teller, who was at the time in Copenhagen and both working on molecular as well as nuclear theory, described the processes of interaction according to Fermi's theory in a letter to Guido Beck and depicted it in a diagrammatic format.[150] Figure 10.3 represents the sequence from (1) initial over (2) intermediate to (3) final state. Through the focus on the different states and the successive transitions, it certainly cohered with the perturbative methods of the time. But there was already a small but important alteration that would become more prominent in Gregor Wentzel's "Schemata" (to be discussed in Sect. 10.2.4): Teller represented an additional dynamical part, the arrow on the side of the electron in the intermediate state, which indicated its direction of movement from proton to proton.

[149] As Fermi's theory did not exchange the spin of the heavy particles (they were described non-relativistically), it reproduced Heisenberg- and not Majorana-type exchange forces.

[150] The historical context of this letter is the short-lived dispute between Fermi's theory of β-decay and a different proposal by Guido Beck. At the time, Beck communicated with diverse physicists to keep up with the current empirical and theoretical developments on the phenomenon of β-decay.

The Quantum-Electrodynamical Analogy and the Representations of QED

Thus far I have exclusively focused on nuclear theory and mainly followed Cathryn Carson's assessment of how the concept of an exchanged particle made its way into the neutron-proton interaction. But, as the standard story goes (the one Carson argued against, see below), a quantum-electrodynamical analogy already existed.[151] To some extent (I will comment on this in a second), it was already established that charged particles interacted with another through the emission and absorption of photons. In letters and in publications, physicists immediately drew the connection between the forces described by Fermi's theory of β-decay, Heisenberg's model of exchange, and the quantum-electrodynamical description of the Coulomb interaction being due to the emission and absorption of photons. And at this point an important question emerges: How was the idea that particles interact via force carriers, i.e. virtual particles, introduced into QFT?

Heisenberg's initial analogy to molecular theory is without doubt and is certainly important for connecting these kinds of interaction processes to the term "exchange". But Cathryn Carson provided an even stronger argument: She claimed that the concept of force transmission through the emission and absorption of a particle was first completely established in nuclear theory and then carried over to QED. How did she support this claim?

Her argument was based on the way QED was formulated and practised at the time,[152] namely as a quantized (transversal) radiation field and a separated Coulomb interaction. This treatment was upheld in the most influential papers and textbooks during the 1930s and early 1940s. The concept of a longitudinal photon was for the greater part of the 1930s certainly not practised, if it was developed at all. As already discussed in Sect. 8.2.1 and as Carson (1996b, 117–118) had already pointed out, Bethe and Fermi (1932) only described the relativistic corrections to the Coulomb interaction as being due to the emission and successive absorption of transversal photons. According to this argument, the concept of a virtual photon as a force carrier was only partially available for the description of QED processes, i.e. only for the transversal part and only for corrections to the Coulomb interaction, but not for the Coulomb interaction itself.

I share Carson's conviction that this differentiation is important to explain some of the ideas that were subsequently put forward. For example, shortly after finding that the exchange interaction between neutron and proton came out much too small according to his theory of β-decay, Fermi proposed that, besides the force between neutron and proton described by the emission and successive absorption of electron and neutrino, there might be an additional and much stronger force. He came up with

[151] This analogy is most explicitly discussed in Darrigol (1988, Sect. II).

[152] See Carson (1996b, especially Sects. 3.1–3.3) for Carson's far more detailed explication.

this conjecture in close analogy to the differentiation between Coulomb interaction and the transversal part in QED. Heisenberg was, although only for a short time, sympathetic to that idea.[153]

I also share Carson's assessment that Werner Heisenberg's 1934 invocation of the analogy between QED and Fermi's field theory did not include the idea of a particle being exchanged. Discussing his derivation of the nuclear forces according to Fermi's theory Heisenberg stated that "just as for atomic electrons the possibility that light quanta may come into existence leads to the Coulomb force"[154] a second order calculation should lead to nuclear forces. Carefully reading this passage, one notices, as Carson has argued, that an exchange of particles was not invoked at this point.[155] Similarly, the analogy between QED and nuclear physics which Werner Heisenberg opened up in letters to Bohr and which he presented in his article for the *Zeeman Festschrift* (Heisenberg 1935a) was of a more general nature and did not directly refer to the idea that forces are carried by intermediary particles.[156] For Heisenberg, field concepts were still primary, especially for the mediation of interactions. The analogy between QED and nuclear field theories as discussed by Heisenberg referred to the fields mediating the interaction and the particles which were associated to them, not to the interaction being mediated by the particles.[157]

I thus follow Carson's argument that Heisenberg and Fermi did not directly draw the analogy between nuclear forces and the Coulomb interaction as due to successive emission and absorption of photons. As I argued in Sect. 8.2.1 there is another, pragmatically informed reason why accrediting the concept of a force carrier, in the sense of an exchanged virtual particle, might, at least for Fermi's and Bethe's 1932 article, be a bit too hasty: Considering their methodological approach, the intermediate state was the "interlink", not the particle in it. But I also indicated that I consider the result of their specific modelling which showed that one electron absorbed a transversal photon and the other electron absorbed this photon definitely suggestive for the particle exchange picture.

[153] See Carson (1996b, 119–120).

[154] "ähnlich wie bei den Atomelektronen die Möglichkeit der Entstehung von Lichtquanten zur Coulombkraft führt [...]." Heisenberg to Pauli, January 18. 1934; as cited in Pauli (1985, 250). Translation according to Carson (1996b, 119).

[155] Especially for Heisenberg, there remains the question whether this was due to the lack of the concept of a longitudinal photon, as Carson argued, or rather due to his general reading of time-independent perturbation theory at the time. The wording Heisenberg used both for describing the perturbative evaluation of the exchange forces (connecting the Coulomb interaction to the possibility of light-quantum emission) and the self-energy problem (obviously referring to transversal photons at this point) were strikingly similar. See the two letters: Heisenberg to Pauli, January 18. 1934; as cited in Pauli (1985, 250–251); and Heisenberg to Pauli, February 5. 1934; as cited in Pauli (1985, 273).

[156] See Carson (1996b, 120–121).

[157] Heisenberg only slowly changed his interpretation during the years to come and spoke of successive acts of emission and absorption hesitantly. Carson (1996b, 121; Footnote 67) herself noted that, in 1937, Heisenberg would not talk about particle transmission but put the fields first.

In the following I will further argue for my reading of the development of the particle exchange picture as a mode of describing the interaction between two particles, i.e. that it emerged in an intricate interplay between heuristic ideas and suggestive mathematical models. Therefore, I will show that Carson underestimated the availability of a dynamical description of the Coulomb interaction between charged particles through a quantized field by downplaying a particular formulation of QED, the conception it entailed, and its impact on parts of the community.

Dirac's Reformulation of QED and the Coulomb Interaction

In 1932 and after nearly five years of silence on the formulation of QED, Dirac provided another approach to this theory (Dirac 1932).[158] Dirac's new formulation is most famous for introducing the many-time formalism, discussing ideas in line with S-matrix philosophy in a rather convoluted introduction, and leading to the explicit spelling out of the interaction picture formulation of QED. The initial fate of Dirac's formulation is also well-known: In April 1932, even before the publication of the paper, Dirac presented his new theory in Copenhagen. Léon Rosenfeld (1932b) immediately saw that Dirac's formulation was just Heisenberg and Pauli's QED in another form. The two theories were connected by a contact transformation, which was most clearly expressed afterwards by Dirac, Fock, and Podolsky (1932). The response from German and Copenhagen physicists was thus rather negative and Dirac's approach was not taken very seriously as progress in the formulation of the theory.[159]

Although Dirac's reformulation was thus essentially equivalent to the already existing versions of QED, the important point was Dirac's re-evaluation of the derivation of the Coulomb interaction. Dirac criticized Heisenberg and Pauli's QED in his 1932 paper and we also know from his notes on Heisenberg and Pauli's paper that he was rather unsatisfied with their applications. All the same, he considered one aspect of their contribution particularly interesting:

> They find the Coulomb law. [...] This is rather remarkable in view of the fact that Coulomb forces are nowhere directly assumed. They assumed only field equ[ation]s of Maxwell for continuous distribution of electricity and then applied quantum rules, and the appearance of particles with Coulomb force between them justifies this procedure.[160]

---

[158] For historical discussions of Dirac's approach, see e.g. Darrigol (1982, Sect. 2.4b), Kragh (1990, 132–138), Schweber (1994, 42–55), and Kojevnikov (2002, Sect. 8). Carson (1996b, 116) only shortly discussed this theory and did not consider its possible impact on the conception of physicists.

[159] Take, for example, Pauli's reaction, as always not mincing words: "After a confused introduction, that consisted of only half understandable, because only half understood, sentences you come finally to results in a simplified one dimensional example that are identical with those that the formalism of Heisenberg and I gives for that example." Pauli to Dirac, September 11. 1932; as cited in Pauli (1985, 115). Translation taken from Schweber (1994, 52–53).

[160] Dirac's "Manuscript notes for a critical presentation of Heisenberg-Pauli," AHQP, MF 36, Sect. 12. Dirac's notes are also discussed in Schweber (1994, 42–45).

For his treatment, Dirac considered a one-dimensional toy model of two electrons described non-relativistically. His starting point were the two Hamiltonian equations for the electrons, each with its specific time parameter

$$\left(i\hbar\frac{\partial}{\partial t_1} + \frac{\hbar^2}{2m_1}\frac{\partial^2}{\partial x_1^2} - e_1 V(x_1, t_1)\right)\Psi = 0 \tag{10.9}$$

$$\left(i\hbar\frac{\partial}{\partial t_2} + \frac{\hbar^2}{2m_2}\frac{\partial^2}{\partial x_2^2} - e_2 V(x_2, t_2)\right)\Psi = 0 \tag{10.10}$$

The potential obeyed the equation for a free field

$$\frac{\partial^2 V}{\partial x^2} - \frac{1}{c^2}\frac{\partial^2 V}{\partial t^2} = 0 \tag{10.11}$$

It was Fourier analysed and quantized by imposing commutation relations on the Fourier coefficients.

Dirac set all times equal to the field time t, and thus arrived at

$$\left(i\hbar\frac{\partial}{\partial t} + \frac{\hbar^2}{2m_1}\frac{\partial^2}{\partial x_1^2} + \frac{\hbar^2}{2m_2}\frac{\partial^2}{\partial x_2^2} - e_1 V(x_1, t) - e_2 V(x_2, t)\right)\Psi = 0 \tag{10.12}$$

where Ψ is the wave function for the whole system. Dirac then showed that a second order perturbative evaluation, expanding the wave function in a power series of e and treating the quantized potential as the perturbation, resulted in the Coulomb interaction of the two particles.[161]

To be sure, Dirac did not verbally describe his treatment in terms of particle exchange and throughout his paper the electromagnetic field enjoyed primacy over a photon picture. Still, this model was implicit in the mathematical calculation and as Dirac himself noted "Ψ_1 [the first approximation which was used for the second order evaluation] must consist of a sum of terms each referring to a state of the field in which just one oscillation is excited by one quantum" (Dirac 1932, 462). In order to arrive at a second approximation which corresponded to an interaction, this quantum subsequently had to be destroyed.[162]

The Impact of Dirac's Reformulation (in the Soviet Union)

While Central European field theorists reacted rather negatively to Dirac's proposal, physicists in the Soviet Union were far more sympathetic. Boris Podolsky, Ameri-

---

[161] Dirac initially supposed that an error in sign occurred as he arrived at an attractive force for particles with equal charges. In a note added in proof, he noted that there is no such error. As Werner Heisenberg had pointed out to him, also in classical theory the force is attractive for like particles in a one-dimensional model. See the "Note added" of Dirac (1932, 464).

[162] Dirac's discussion of the interaction can be found in Dirac (1932, 462–463).

can physicist spending a research stay at the Kharkiv institute, and Vladimir Fock, employed in Leningrad but visiting the Kharkiv institue at the time, developed Dirac's approach further, both by considering the three dimensional case with a scalar potential and non-relativistic electrons as well as the full Maxwellian case and the Dirac equation (Fock and Podolsky (1932b) and Fock and Podolsky (1932a)).[163] While their struggles with the quantization procedure is of lesser importance for us, it must be noted that the gist of Fock's and Podolsky's derivation of the Coulomb interaction or the relativistic interaction between two particles was the same as in Dirac's initial contribution, a perturbative one with intermediary quanta.

Why do I emphasize this point? Although Carson is certainly right that the splitting between Coulomb interaction and transversal radiation field was the most regularly practised version of QED during the 1930s, Dirac, Fock, and Podolsky's contributions in which the Coulomb interaction and the full relativistic description between two electrons was perturbatively traced back to emission and absorption processes of quanta had an impact. The difference in the formulation of the theory and the focus on the perturbative derivation of the Coulomb interaction from the basic equations was noted in the community.

For example, Dimitri Blokhintsev wrote in 1934 that

> today the interaction of charged particles (Coulomb's law) [reference to Dirac's article of 1932] is regarded *dynamically*, namely, as the result of continuous emission and absorption of light quanta by interacting particles.[164]

Blokhintsev was certainly not an immensely influential figure. But this quote is not about the influence of Blokhintsev. It is about the reference he chose for the dynamical picture of the interaction of particles, namely Dirac's 1932 article.[165] And it is about Blokhintsev's affiliation. He was a doctoral student of Igor Tamm at the time.[166]

---

[163] For a contextualization of Fock's and Podolsky's work in relation to Dirac's work, see e.g. Kojevnikov (2002, Sect. 8).

[164] The quote is taken from Frenkel and Gorelik (1994, 96).

[165] Matvei Bronstein's work on a quantized gravitational theory is also a point in case. According to Frenkel and Gorelik (1994, 96), Bronstein followed the ideas of Dirac, Fock, and Podolsky. In the same way they derived the Coulomb interaction he derived the Newtonian law of gravitation as a consequence of a quantized gravitation field. Although I consider this a strong argument in favour of the impact of Dirac, Fock, and Podolsky's work, I only included this as a footnote, since I am not capable of reading Russian and cannot wholly judge how deep the analogy was.

[166] On a side: Also in Central Europe the difference in the explanation of the Coulomb interaction was noted. Pascual Jordan (Jordan 1936b, 215) referred to Dirac and Fock's work when shortly discussing the derivation of the Coulomb interaction from near-distance effects. He also did so in his textbook (Jordan 1936a, 240). Yet, for Jordan the field aspect of the interaction was still primary.

Coming back to Fermi's theory of β-decay, the point is that Igor Tamm (1934) and Dimitri Ivanenko (1934) most probably applied methods along the lines of Dirac, Fock, and Podolsky.[167] As, Tamm noted "the role of the light particles (Ψ-field) [electron and neutrino] providing an interaction between heavy particles corresponds exactly to the role of the photons (electromagnetic field), providing an interaction between electrons" (Tamm 1934, 981). And comparing the derivation of the Coulomb interaction according to Fock and Podolsky (1932a), the correspondence is indeed exact. The interacting particles (neutron and proton/two electrons) were treated unrelativistically and according to a configuration space representation, the mediating entities were described relativistically and in terms of a quantized field theory (electron and neutrino/field quanta). Furthermore, Fermi's interaction Hamiltonian was "set up to be analogous to the scalar potential part of the QED interaction" (Carson 1996b, 118), exactly the part Fock and Podolsky used to reconstruct the Coulomb interaction.

Intermediate Analysis: Representations of QED, the Extent of the Force Carrier Analogy, and the Methodological Upshot

Following this line of argument and in analysis, the differentiation between the versions of QED that existed at the time and the mode of their evaluation are important to judge in how far the analogy between QED and Fermi's field theory also incorporated the force carrier analogy. Fermi, following his popular distinction between Coulomb interaction and radiation field, proposed to include another interaction besides the exchange force to explain the strength of nuclear forces. Heisenberg's discussion of the nuclear interaction as discussed in a letter to Pauli in early 1934 closely followed the (time-independent) perturbative evaluation of the Coulomb interaction in their first paper on QED.[168] Even though Pauli and Heisenberg did not give any verbal explanation in 1929, their derivation implicitly included intermediary longitudinal excitations. Igor Tamm and Dimitri Ivanenko, who most explicitly drew the force carrier analogy in their discussion of the nuclear interactions, followed Dirac's approach, which made the dynamical character of the Coulomb interaction as transmitted through a quantized field most transparent.

The connection between the respective derivations of nuclear forces and the version of QED that were practised has thus important consequences for the assessment of Cathryn Carson's discussion of exchange forces. I do not question the facts that

[167] This judgement is based, besides the points mentioned in the main text, on the expansion of the wave function, and Tamm's description of the second order approximation. The language as well as the symbol systems are strikingly similar to the ones given by Fock and Podolsky. According to Ivanenko (1934, 982), "the exact calculations were first carried out by Prof. Ig. Tamm, who also insisted on the development of this method". Ivanenko and Sokolov (1936), although pointing towards the difference between the transversal character of light quanta and the longitudinal character of the Coulomb law, also used Dirac's 1932 version of QED in a more detailed paper on Fermi's field theory and its connection to the neutrino theory of light.

[168] Heisenberg to Pauli, January 18. 1934; as cited in Pauli (1985, 250–251).

explicit verbal references to force carrying particles were only applied after the quantum field theoretical explanation of nuclear forces[169] and that in the historical path from Heisenberg's trilogy to the perturbative evaluation of the interaction according to Fermi's theory of β-decay the force carrier picture was certainly most prominent. But I think we can understand the diverse invocations of the force carrier analogy in QED and nuclear physics as well as the emphasis put on it a lot more coherently, if we acknowledge that the different representations of QED and specifically their evaluation provided for a more or less suggestive mathematical representation. Even if Fermi's version was the most common one practised, other versions were certainly known to the practitioners and could motivate the invocation of the analogy.

Accordingly, a perturbative evaluation of the Coulomb force with intermediary excitation was in place as early as 1932, and it were physicists who followed Dirac who most forcefully invoked the analogy between nuclear forces and electrodynamical forces as being transmitted through particle exchange. While I still hold that the method of calculation was based on virtual states and transitions to and from them, and to a lesser degree on virtual particles, we must take the results of these calculations and the configuration of the intermediate states into consideration when trying to understand how physicists developed a way to model and talk about quantum field theoretical interactions.

According to this argument, the idea that forces were transmitted via the exchange of virtual particles (modern terminology) emerged through the reinforcement of suggestive mathematical structures and explicitly invoked heuristic analogies (as in Heisenberg's 1932 treatment). The strength with which the force carrier analogy was put forward in the papers on the interaction of nucleons depended, on the one hand, on the version of QED applied, but also, on the other hand, on the respective conceptual outlook on the theory as such. For Heisenberg, the field concepts were primary in most of his verbal assessments of interactions. For Tamm, quite on the contrary, the explicit reference to verbal models including both successive emission/absorption acts and an emphasis on the particle aspects of the theory[170] was an important ingredient in the treatment of interaction processes in QED.

All the same, we have to acknowledge that the force carrier picture was, for the evaluation of the theory, quite probably not the most important part of the analogy between QED and Fermi's field theory.[171] It was not the possibly heuristically fruitful picture but the fact that the analogy between nuclear theory and QED directly suggested a mode of calculation (regardless of the specific interpretation of the intermediate states).

Heisenberg's evaluation was closely modelled after his and Pauli's discussion in their first paper on QED. Igor Tamm (1934, 981) argued that because of the analogy "we may adapt for our purposes the methods used in quantum electrodynamics to

---

[169] Nevertheless, I would not exclude the possibility that an investigation of the correspondence and notes particularly of Soviet physicists might reveal earlier invocations.

[170] See for example my discussion in Sects. 7.5 and 7.7.

[171] Of course, the analogy was highly important for the construction of Fermi's theory of β-decay, but here I focus on its practice.

deduce the expression for Coulomb forces". Ivanenko and Sokolov also noted that "this intuitive [anschauliche] viewpoint [...] *points immediately towards the method of calculation*".[172] According to Carl Friedrich von Weizsäcker "this derivation of the exchange forces [according to a second order calculation of Fermi's theory of β-decay] *is exactly analogous to the derivation of the Coulomb force* from the probability of the virtual emission and reabsorption of a (longitudinal) light quantum in quantum electrodynamics".[173] A didactic example of this methodological analogy can be found in the "Bethe Bible", a set of three massive articles on nuclear physics written by Hans Bethe and two co-authors in the mid 1930s. After describing the conception of the Coulomb interaction between two particles as "a second approximation of the interaction between field and individual particles", they noted: "*The same program may by carried out for Fermi's β-field and nuclear forces*" (Bethe and Bacher 1936, 201, emphasis added).

From Fermi and Bethe's discussion of the correction to the Coulomb interaction through the emission and absorption of a light quantum (see Sect. 8.2.1) over Heisenberg's heuristic picture of the *Platzwechsel* to the different derivation of the interaction between nucleons according to Fermi's theory of β-decay the idea that forces were transmitted through the exchange of virtual particles (modern terminology) became more and more common. This concept emerged through a mutual reinforcement of different heuristics and the evaluation of different formulations of the theory. But, regardless of the interpretation of the intermediate configuration of the mathematical structures in a quantum field theoretical evaluation, the main point of the analogy in terms of practice was the suggestion of a mathematical mode of evaluation: a perturbative one modelled closely after the formulation of QED the actors favoured at the time.

Outlook: Force Carriers and Yukawa's Meson Theory (1934)

Before concluding this subsection a look to Japan, Yukawa's development of meson theory and the role of the force carrier analogy is certainly necessary.[174] After translating Heisenberg's trilogy into Japanese, Yukawa (unsuccessfully) tried to develop a field theory of nuclear forces using the electron as the "mediator" of the nuclear interaction. In his manuscripts and talks, Yukawa also emphasized the methodological upshot: "On [sic] this idea, the problem of the interaction of neutrons and protons can be treated at the same time as the problem of β-ray emission by the [sic] method

[172] "Diese anschauliche Betrachtung [...] *weist sofort auf die Methode der Berechnung*." (Ivanenko and Sokolov 1936, 120; emphasis added).

[173] "Diese Ableitung des Austauschkraft *ist genau analog zur Ableitung der Coulomb-Kraft* aus der Wahrscheinlichkeit für die virtuelle Emission und Reabsorption eines (longitudinalen) Lichtquants in der Quantenelektrodynamik." (Von Weizsäcker 1936, 573; emphasis added).

[174] For more details on Yukawa's development of the meson theory in general, see Brown (1981), Darrigol (1988, Sect. II.2), or Brown and Rechenberg (1996, Chap. 5); and for the invocation of the force carrier analogy specifically see Carson (1996b, Sect. 3.4).

similar to the one used in the theory of interaction of matter and radiation."[175] Similarly he noted:

> The electron can be considered as the mediator of the interaction between proton and neutron or between one neutron and another, just as the radiation is the mediator of the interaction between one charged particle and another. We may go too far in thinking so. By this way of thinking, however we can determine not only the form of interaction between neutron and proton by method analogous to the case of radiation, but treat also the problem β-ray emission.[176]

But Yukawa stopped his field theoretical attempts even before he became aware of Fermi's theory of β-decay, somewhere in mid 1934. Although Yukawa first thought that Fermi had outrun him in developing a suitable theory of nuclear forces, the realization that the interaction according to this theory came out much too small led Yukawa, in contrast to the European physicists, to propose a new as yet unobserved heavy particle. But even though Yukawa performed perturbative calculations and previously referred to the electron as a mediator, in his 1935 paper (Yukawa 1935) the field conception is primary and a force carrier analogy is nowhere invoked. Only after the theory was noted in the West and its further investigation started in Japan, the idea of force carriers became more prominent. Even though the different formulations of the meson theory led to far more consistent theoretical results for the nuclear binding, the first publication on the meson did not essentially contribute to the development of the force carrier concept.

10.2.2 Anomalous Magnetic Momenta: Virtual Dissociation

In the prior section I argued that the specific way the theory of QED was mathematical represented and evaluated had decisive influence on the prominence of the force carrier analogy in the respective investigations. The extent to which virtual particle exchange (modern terminology) was made responsible for the interaction between particles depended directly on the practice and the favoured general conceptualization of the theory by the respective physicists. But in any case, the result of the calculation was finite and, according to Fermi's initial proposal, much too small.

In the following, I will turn to a phenomenon which was also theoretically described by particles in intermediate states but differed in two important aspects from the interaction between nucleons: First of all, the theoretical structures diverged but approximately cohered with empirically established values when seemingly sensible assumptions were made. Secondly, while the interaction between nucleons was conceived of as analogous to the interaction between two electrons, the conception and theoretical structures used in the evaluation of the so-called anomalous magnetic

[175] Unpublished Manuscript by Yukawa with the title "On the problem of nuclear electrons I." As cited in Darrigol (1988, 276).

[176] Yukawa's manuscript (probably April 1933) "A Comment on the Problem of Electrons in the Nucleus." As cited in Kawabe (1991, 248).

momenta of proton and neutron[177] show a striking similarity to the concept of a charge cloud, which I discussed in Sect. 10.1.2 in connection with vacuum polarization.

Shortly after its formulation, Fermi's theory of β-decay, or rather its extension, was used to explain the empirically established phenomenon of the anomalous magnetic moment of the nucleons. The problem was the following: The proton was known to be a spin $\frac{1}{2}$-particle and if one assumed that the Dirac equation held for the proton, this would imply that its magnetic moment should be $\mu_P = \frac{e\hbar}{2m_P}$, where e is the charge of the proton and m_P its mass. For the neutron, an uncharged particle, the magnetic moment should vanish. Both of these assumptions proved to be wrong.

By the turn of the year 1932/33, news was spreading about the measurement of the magnetic moment of the proton by Otto Stern and Otto Frisch in Hamburg.[178] By mid 1933, the results of Frisch and Stern (1933) and Estermann and Stern (1933) had been published. They indicated that the magnetic moment of the proton was about two to three times bigger than expected. A theoretical explanation was far from clear.[179]

Before discussing the proposed quantum field theoretical solution of the discrepancy between theory and experiment, it must be noted that the underlying assumption that the proton could be described by the Dirac equation was controversial, to say the least. Bohr, for example, took the anomalous magnetic moment of the proton as further proof that the Dirac equation could not be applied to nucleons.[180] In his first paper on the theory of β-decay, Fermi noted "that the relativistic wave equation for the heavy particle is unknown".[181] When Heisenberg reported the possible explanation by Wick to Pauli, the latter commented: "At this point, you introduce the new (i.e. otherwise not used in any way in the theory [Fermi's theory of β-decay]) assumption, that the Dirac equation should after all hold in some kind of approximation for the heavy particles."[182] But, as shall become clear, the applicability of the Dirac

---

[177] The anomalous magnetic momenta have, to the best of my knowledge, only found attention in overarching narratives but were not studied for their own sake. Brown and Rechenberg (1996, 54–57) discussed Wick's derivation and also observed that the general gist of the argument was taken over into meson physics (Brown and Rechenberg 1996, e.g. pp. 63, 154–157). Pais (1986, 427) only shortly mentioned "the fundamentally correct explanation" by Wick and the critique in its follow-up.

[178] By late December 1932, Pauli was aware of the results and communicated them to Bloch who in turn reported to Bohr (Pauli 1985, 146). At the time, Heisenberg has seemingly also been informed. On January 18. 1933, Pauli wrote to Heisenberg that "we are largely of the same opinion about Stern's proton". [*Original*: "Über das Sternsche Proton sind wir ja weitgehend derselben Ansicht."] As cited in Pauli (1985, 149).

[179] See Breit and Rabi (1934) for a discussion of the proposals up until mid 1934.

[180] See, for example, Darrigol (1988, 257).

[181] "[...], daß die relativistische Wellengleichung für die schweren Teilchen unbekannt ist." Fermi (1934, 166).

[182] "Du führst hier die neue (d.h. sonst in der Theorie nirgends benützte) Annahme ein, daß für die schweren Teilchen in irgend einer Näherung doch so etwas wie die Diracsche Gleichung gelten soll." Pauli to Heisenberg, November 7. 1934; as cited in Pauli (1985, 362).

equation was not the only problem the first proposition of a quantum field theoretical explanation of the anomalous magnetic moment by Gian-Carlo Wick would face.

Wick's Proposal (1934/35): A Different Concept of "Virtual Electrons" and the "Story of the Proton"

Gian-Carlo Wick had spent his postdoctoral years from 1930 to 1931 partially in Göttingen with Born and in Leipzig with Heisenberg. Since 1932, he was Fermi's research assistant in Rome and thus well informed about Fermi's theory of β-decay and deeply involved in spelling out its consequences, or rather its extensions. Only two month after Fermi's initial communication, Wick proposed to include a further process in the field theory: the decay of a proton into neutron, positron and neutrino,[183] i.e. the process of β^+-decay. It would turn out to be one of the decisive proposals for his explanation of the magnetic moment of the proton. On April 18. 1934, Wick wrote a letter to Heisenberg proposing "a rather crazy consideration" on how to explain Stern's measurement, closing with the words: "But please do not think that I believe in all this!"[184]

I will turn to the more detailed published paper Wick (1935) for the discussion of Wick's suggestion. Yet, another part of Wick's letter is of terminological interest. Although he applied perturbation theory to explain the magnetic moment of the proton in the last part of the letter to Heisenberg, he used the term "virtual" just once and in a different context: He referred to the Bose electrons of Heisenberg's 1932 neutron model as "the electrons virtually present in the neutrons".[185] Although these electrons, to some extent (see above), were used to explain the nuclear exchange interactions, it is clear from context that Wick did not directly connect the term to the exchange forces.[186] Rather, he used it to point towards the compound structure of a neutron in Heisenberg's model, i.e. Heisenberg's shoving of the nuclear electrons, and thus all the problems connected to them, inside the neutrons. Wick's invocation was consequently essentially unconnected to the modern concept of a virtual particle.

As Guido Beck's "virtual electrons", a shorthand for negative energy electrons in the Dirac sea (Beck 1933), the "virtual photons" of the Weizsäcker-Williams method (Williams 1935), or the concept of a "virtual state" (Beck 1930) as an accumulation

---

[183] For the empirical context of Wick's proposal of β^+-decay, see Stuewer (2018, 298–299).

[184] "eine ziemlich verrückte Betrachtung;" "Aber bitte denken Sie nicht[,] dass ich an all dies glaube!" Wick to Heisenberg, April 16. 1934; AMPG, III. Abt., Rep. 93, Nachlaß Heisenberg, Nr. 1533.

[185] "die in den Neutronen virtuell vorhandenen Elektronen," Wick to Heisenberg, April 16. 1934; AMPG, III. Abt., Rep. 93, Nachlaß Heisenberg, Nr. 1533.

[186] Wick invoked Heisenberg's model and the term "virtuell" in reference to the Meitner-Hupfeld effect, an anomaly in the scattering cross section, which Heisenberg tried to explain through the presence of the electrons inside the neutron.

point in the continuous spectrum of a potential well[187] it shows that by the mid 1930s the term "virtual" was not yet exclusively used to refer to concepts connected to the virtual particle concept as understood today. At the same time, concepts and processes we connect with the terminology, as the nascent pairs of Furry and Oppenheimer, were not yet necessarily referred to as virtual. In Wick's explanation of the anomalous magnetic moment, although we would certainly say that it invokes virtual processes, maybe even virtual particles, the term was completely absent, both in the letter to Heisenberg and in the published paper.

Wick's explanation[188] was the first one in terms of a field theory and what later was called "virtual decay/dissociation".[189] According to Wick (1935), the reason for the discrepancy between the theoretical and the observed value of the magnetic moment of the proton can be explained by the

> possibility that the proton momentarily splits up into a neutron + a positron + a neutrino and then immediately falls back into the state of the proton (a permanent transition cannot occur, because it is contrary to energy conservation [...]).[190]

Wick expressed the empirically observed magnetic moment as a combination of the magnetic moment of the proton "in the non-dissociated state"[191] μ_P, and the magnetic moment of the "dissociated state", μ:[192]

$$\frac{\mu_P + \tau\mu}{1+\tau} \tag{10.13}$$

where τ was the fraction of the time the proton spends in the dissociated state. Since the mass of the electron or positron is much smaller than the proton's ($\frac{m_e}{m_P} \propto 1/1860$) and the magnetic moment of a Dirac particle is indirectly proportional to its mass, μ is much larger than μ_P. Thus, if τ, the fraction of time the proton was dissolved into neutron, neutrino, and positron, was not very small, the corresponding conception and Formula (10.13) could, according to Wick, reproduce the empirically observed value.

Wick's aim was not a completely rigorous quantitative evaluation. He was well aware that the interaction between neutron and proton came out much too small according to Fermi's field theory. He also noted that Fermi's theory reproduced Heisenberg exchange forces, while Majorana exchange forces were much more likely. Thus Wick abstracted from the specific form and strength of the interaction

---

187 This concept was recently discussed by Jean-Philippe Martinez (2024, Sect. 2).

188 Wick obviously hesitated to publish his consideration. The principle idea and the estimations performed were already laid out in the letter to Heisenberg, while the paper was communicated as late as February 2. 1935, nearly a year later.

189 "virtueller Zerfall," e.g. von Weizsäcker (1936, 591).

190 "possibilità che ha il protone di scindersi momentaneamente in un neutrone + un positrone + un neutrino, per poi ricadere immediatamente all stato di protone (una transizione permanente non puó avvenire, poiché contraria alla conservazione dell'energia [...])." Wick (1935, 171).

191 "allo stato non dissociato," Wick (1935, 171).

192 "stato dissociato," Wick (1935, 171).

but still kept the basic processes of $\beta^{(+)}$-decay. The mathematical representations of the interaction terms were consequently no longer concretely defined but simply stood as placeholders for the process $P \rightarrow N + e^+ + \nu$ and its reversal.

As a concrete mathematical evaluation of the interaction terms was unreasonable, the estimation of τ had to be performed otherwise. Therefore, Wick established a connection between τ and the exchange interaction resulting from Fermi's theory. The latter could be expressed as

$$J(r) = \sum_n \frac{V_{0n}(P)V_{n0}(P')}{E_n - E_0} \tag{10.14}$$

where V_{n0} was the matrix element for a transition from the non-dissociated to the dissociated state n, V_{0n} the respective reverse process. P and P' were the locations of the nucleons. This was essentially an abstract expression for the second order perturbative calculation as performed by Heisenberg, Fermi, Tamm, and Ivanenko.

Interestingly, when establishing the connection between the interaction and τ, Wick actually corrected his description of the proton's dissociation:

> We have previously described the story [la storia] of the proton as a succession of transitions from the state 0 to the state n and vice versa. More correctly, what we observe is a stationary state, which is a mixture of 0 and n[...].[193]

τ was, accordingly, the probability of finding the system in any of such states n within this mixture and thus given by

$$\tau = \sum_n \left| \frac{V_{n0}(P)}{E_n - E_0} \right|^2 \tag{10.15}$$

.

By some further approximations,[194] Wick derived the relation

$$\tau \propto \frac{1}{137\ \mathrm{mc}^2} J(0) \tag{10.16}$$

The exact value of $J(0)$ was unknown. But using the most probable values gathered from empirical findings available at the time,[195] τ would turn out to be of the order 1 (or a smaller fraction, but essentially not very small), which would thus be sufficient to explain the anomalous magnetic moment of the proton.

---

[193] "Abbiamo descritto precedentemente la storia di un protone come una succcessione di transizioni dallo stato 0 agli stati n e viceversa. Più corretamente, ciò che noi osserviamo è uno stato stazionario che è una miscela di 0 con gli n [...]." Wick (1935, 172).

[194] Wick neglected the kinetic energy of the heavy particles, the mass difference between proton and neutron, and the masses of the light particles in comparison to their kinetic energy. He estimated the kinetic energy of the light particles to be of the order 137 mc^2, as the forces were only appreciable for distances of the electron radius or smaller.

[195] Wick cited a paper by himself and another one by Wigner (1933). In Wick's nomenclature, Wigner estimated $J(0) \approx 140$ mc^2.

Wick's argument was certainly quite bold. But not because he attributed the additional magnetic moment to the dissociated state. In this respect Wick's argument was conceptually (and also in its mathematical representation) very close to the charge cloud of vacuum polarization, which was seemingly uncontroversially accepted. Arnold Nordsieck (1934, 234), when investigating (on suggestion of Oppenheimer) the interaction of neutrons with charged particles according to Fermi's theory of β-decay, also explained that "any charged particle may interact with the neutron by polarizing the charge cloud produced by oscillation ($N \rightleftarrows P + e^- + n$) between the two states for heavy particles by Fermi". As already noted, George Gamow used the terminology of the "nascent", which Furry and Oppenheimer introduced for their description of vacuum polarization, to describe Wick's proposal. All the same, Wick's argument was bold because he assumed that the proton could be described by the Dirac equation and because he kept the basic processes of Fermi's theory while completely abstracting from their strength and their specific mathematical form. He knowingly transgressed the limits of the applicability of the theory.

In respect to the modelling of the phenomena, Wick's presentation also exemplifies a trend in the theoretical reasoning of the time. He used the description in terms of a dynamical process, i.e. in terms of Dirac's verbal model, to introduce his readers to the general idea: Namely, that the proton spent some fraction of time in the dissociated state. But when he turned to the time-independent perturbative evaluation, he retracted this "story of the proton" and referred to the mixture of states in which the proton was actually represented. The dynamical picture was used to make this probability accessible and thus was also applied to an actually static phenomenon.

Popularization and Evaluation of Wick's Idea (1934–1937)

Werner Heisenberg was very sympathetic to Wick's idea and worked on its popularization, both before and after its publication. On October 28. 1934 he wrote to Pauli that "the connection between the exchange forces (see Ivanenko and Tamm) and Stern's moment of the proton"[196] has always been a side of Fermi's theory that appealed to him. Pauli was obviously not aware of Wick's argument at this point.[197] Heisenberg replied on November 4. by taking up Wick's idea but altering it slightly.

---

[196] "der Zusammenhang mit den Austauschkräften (siehe Iwanenko und Tamm) und mit dem Sternschen Moment des Protons," Heisenberg to Pauli, October 28. 1934; as cited in Pauli (1985, 354).

[197] "You mention a possible connection between the magnetic moment of the proton and the Fermi theory. On what is the hypothesis of this connection based? (I only heard rumours, that Wigner might have thought something about that.)" [*Original:* "Du erwähnst noch einen möglichen Zusammenhang zwischen dem magnetischen Moment des Protons und der Fermischen Theorie. Worauf stützt sich die Hypothese eines solchen Zusammenhangs? (Ich hörte nur gerüchteweise, daß Wigner etwas darüber überlegt haben soll.)"] Pauli to Heisenberg, November 1. 1934; as cited in Pauli (1985, 357).

Heisenberg described the process perturbatively a little bit differently and further included an explanation of the magnetic moment of the neutron.[198] He considered a heavy particle in an external magnetic field. The "path" ["Pfad"; Heisenberg's wording] one had to choose to get an additional moment for the neutron was according to Heisenberg the following: "(1) emergence of electron + neutrino, (2) influencing of the electron by the external field, (3) disappearance of the electron + neutrino [...]."[199] The result, again estimating that the main contributions came from electron and neutrino energies $\approx$ 137 mc^2, was according to Heisenberg in general agreement with experiment.

Pauli was sceptical, to say the least. He had three objections. First of all, if one assumed, as Pauli did at the time, that negatively charged anti-protons existed, an intermediate state of the form negative proton + positron + neutrino should exist. Hence, either the magnetic moment of the neutron was zero, as the contributions of the intermediate states cancelled, or one had to assume "two sorts of neutrons, which would differ by an opposite direction of the magnetic moment [...]".[200] Secondly, Heisenberg and Wick's treatment rested on the assumption that one could apply the Dirac equation to the proton, which Pauli doubted and which was not part of the theoretical modelling of Fermi's original theory of β-decay. Thirdly, Pauli did not believe that the magnetic momenta of two nucleons could simply be added, a consideration which grounded the idea that the neutron also had a magnetic moment.[201]

Heisenberg did not directly respond to Pauli's critique at first but kept his faith in Wick's explanation and included it in his contribution to the *Zeeman Festschrift* (Heisenberg 1935a). Therein, Heisenberg held that the additional terms of second order corresponded to the "virtual emergence [Entstehen] and disappearance of positron-neutrino pair for the proton and an electron-neutrino pair for the neutron".[202] In contrast to Wick, Heisenberg also noted that these additional terms were structurally equivalent to the self-energy terms and diverged. Still, suitable subtractions corresponding to reasonably chosen length and energy scales would render the

[198] Heisenberg derived the conclusion that also the neutron had to have a magnetic moment from the measured values of the magnetic moment of the proton and the deuteron, composed of neutron and proton, and the assumption that the magnetic momenta of nucleons could be added vectorially.

[199] "(1) Entstehen von Elektron + Neutrino, (2) Beeinflussung des Elektrons durch das äußere Feld, (3) Verschwinden von Elektron + Neutrino [...]," Heisenberg to Pauli, November 4. 1934; as cited in Pauli (1985, 359).

[200] "zwei Sorten von Neutronen, die sich durch entgegengesetzte Lage des magnetischen Momentes [...] unterscheiden." Pauli to Heisenberg, November 7. 1934; as cited in Pauli (1985, 362). For a short while, Pauli actually made some peace with the idea that there are "two neutron types, one of which results only from positive, the other only from negative protons". [*Original:* "zwei Neutronensorten, von denen die eine nur aus positiven, die andere nur aus negativen Protonen entstehen [...]."] Pauli to Heisenberg, December 16. 1934; as cited in Pauli (1985, 373).

[201] Pauli stressed the application of the Dirac equation to the proton, the consequent transgression of Fermi's Ansatz, and finally added the divergence difficulties occurring in the calculation to the list of objections in a letter to Heisenberg after Wick's paper was published. Pauli to Heisenberg, June 15. 1935; as cited in Pauli (1985, 404–405).

[202] "virtuellen Enstehen und Wiederverschwinden eines Paares Positron-Neutrino beim Proton bzw. Elektron-Neutrino beim Neutron," Heisenberg (1935a, 113).

contribution of the "virtual electron-neutrino pair"[203] in harmony with experiment. Moreover, the additional magnetic moment of the proton and the neutron were of equal strength but of opposite direction, as was expected from experiment.

By the time, news about Wick's treatment had also made it to the US. The topic of the first Washington Conference on Theoretical Physics, held on April 19 to 21 1935, was "Nuclear Physics". The first session was devoted to the magnetic moment of the proton. Gamow reported on Bohr's conviction that the Dirac equation could not be used for the proton and that thus "one must [...] expect its magnetic moment to be different from 1/1840 of Bohr's magneton".[204] Bethe's reaction to this proposal, as reported by Gamow, wrongly ascribed the idea of Wick to Heisenberg and von Weizsäcker but is nevertheless quoted here in full:

> In this connection Bethe told about an as yet unpublished theory of Heisenberg and Wiezecker [sic!] according to which the large value of magnetic momentum of a proton, as observed by Stern, can be explained as due to the *virtual disintegration* of a free proton into a neutron and a positive electron. The electron *existing for a very short time "in status nascendi"* contributes to the total magnetism momentum and this makes it as observed about 2.5 times larger than would be expected from Dirac's theory.[205]

The false accreditation to Heisenberg and von Weizsäcker is probably due to the fact, that Heisenberg popularized the idea and that by mid to early 1935 the problem was already investigated in more detail in Leipzig, mainly by von Weizsäcker.[206]

But even before von Weizsäcker could finish his publication, one of Pauli's objections was also noticed by George Uhlenbeck and his student Boris Kahn. Uhlenbeck was among the participants of the Washington conference and himself working on a possible modification of Fermi's theory of β-decay. Through direct calculation Kahn (1936) eventually proved Pauli's objection that according to Fermi's theory no additional magnetic moment could be induced. But he also acknowledged that Wick had already pointed out that Fermi's theory led to the wrong kind of exchange force (Heisenberg instead of Majorana). Thus, Fermi's theory had to be modified to include the exchange of spin for two reasons: to arrive at the correct kind of exchange forces and to include the possibility of explaining the anomalous magnetic moment according to Wick's proposition.

---

[203] "virtuellen Elektron-Neutrino-paares [sic!]," Heisenberg (1935a, 115).

[204] Gamow referred to the magnetic moment of the electron as a Bohr magneton. The ratio 1/1840 derives from the mass difference between electron and proton: $\frac{m_e}{m_P} \approx 1/1840$.

[205] Taken from Gamow's report on the *First Washington Conference on Theoretical Physics*, as reproduced in The GW Astrophysics Group (2022).

[206] Becoming aware of Wick's publication Pauli wrote to Heisenberg "that he [Wick] spoiled all punchlines" in his article. [*Original*: "daß er sämtliche Pointen verpatzt hat,"] Pauli to Heisenberg, June 15. 1935; as cited in Pauli (1985, 404). Heisenberg agreed with Pauli that "Wick's work is certainly very incomplete [lückenhaft]". [*Original*: "Die Arbeit von Wick ist zweifellos sehr lückenhaft."] Heisenberg to Pauli, July 2. 1935; as cited in Pauli (1985, 408). Heisenberg referred Pauli to the forthcoming Habilitation of von Weizsäcker.

This modification of the Fermi Hamiltonian was introduced by von Weizsäcker in his Habilitation. Von Weizsäcker investigated four relativistically invariant formulations leading to a Fermi type of β-decay.[207] The general perturbative calculation closely followed the conception outlined by Heisenberg in the letter to Pauli. Von Weizsäcker again noted that the "process can thus be called a correction to the self-energy by an external field".[208] Unsurprisingly, and as for all such calculations, "the possibility of the virtual decay/dissociation"[209] diverged but was made finite by suitably chosen[210] cut-off procedures. For the respective values, the magnetic moment (for specific modifications of the Hamiltonian) had at least the right order of magnitude and the additional moment of neutron and proton were of opposite direction, as was necessary to explain the magnetic moment of the deuteron.[211]

But the explanation of the additional magnetic moment through virtual dissociation had a further function for the modifications of the Hamiltonian and approaches to Fermi's theory of β-decay in general. As von Weizsäcker explicitly stated, "despite the results of Stern it is thus [due to explanation through virtual dissociation] justified to use the Dirac equation as the unperturbed wave equations of the heavy particles also for the derivation of the exchange forces".[212] This was the starting point of von Weizsäcker's investigation. One could thus either argue that the anomalous magnetic moment of the proton indicated that relativistic quantum theory broke down for the length scales of nuclei, as Bohr would. Or one could accept Wick's mechanism and build a relativistic theory which at least would qualitatively cohere with experiment. The latter option was far more popular.

Nevertheless, Pauli severely criticized von Weizsäcker's paper in a letter to Werner Heisenberg. But not due to the conceptualization of the processes, but due to diverse calculational mistakes.[213] Pauli also set his assistant Markus Fierz to re-evaluate von Weizsäcker's calculations and to estimate the magnetic momenta. Fierz and Pauli expressed their doubts in the publication:

---

[207] As was shown by Bethe and Bacher (1936, 190–191) in total five relativistic invariant formulations of such an approach were possible.

[208] "unser Prozeß daher als die Störung der Selbstenergie durch ein äußeres Feld bezeichnet werden kann." von Weizsäcker (1936, 592).

[209] "die Möglichkeit des virtuellen Zerfalls," von Weizsäcker (1936, 591).

[210] The cut-offs were chosen so that the exchange forces were of the order of the values expected from experiment.

[211] Nevertheless, von Weizsäcker's conclusions were still somewhat ambiguous. The one modification which led to pure Majorana-type forces, as was expected from experiment, would not lead to an additional moment of the nucleons.

[212] "Es ist demnach trotz der Ergebnisse von Stern berechtigt, auch für die Ableitung der Austauschkraft als ungestörte Wellengleichung der schweren Teilchen die Dirac-Gleichung zu verwenden." von Weizsäcker (1936, 574).

[213] "Closer inspection has shown that the amount of correct formulas in this publication can be neglected." [*Original*: "Es hat sich nämlich bei näherem Studium gezeigt, daß die Anzahl der richtigen Formeln in dieser Schrift vernachlässigt werden kann."] Pauli to Heisenberg, November 24. 1936; as cited in Pauli (1985, 479).

> The calculations of the additional magnetic moment of the heavy particle is included only for comprehensiveness [...], because it seems very doubtful to us, whether a physical sense can be connected with it.[214]

It is important to note that Fierz (and Pauli) did not mention the conception of the process as a reason for the lack of physical sense. Rather, they pointed towards the divergence of the formulas.

Implementing the Mechanism in Meson Theory (1937 Onwards)

This is where things stood when Yukawa's theory was noticed in the West. The conceptual structure of the explanation of the additional magnetic momenta in terms of a virtual dissociation process was, although the formalism and the quantitative results were on entirely shaky grounds, considered suitable. It was, for example, included in the "Bethe Bible" (Bethe and Bacher 1936, Sect. 45), while pointing to the problems of quantitative evaluation within Fermi's field theory. Even the ever sceptical Wolfgang Pauli seemed to have accepted it.[215] Thus, it is unsurprising that among the first papers on Yukawa's meson theory were discussions of the additional magnetic momenta.

In Japan, Wick's explanation was simply taken over for a rough estimation. Yukawa, Sakata, and Taketani (1938, 34) noted that "the probability of the virtual presence of the heavy quanta" led to an additional magnetic moment. "The fraction of time, during which the neutron is splitting up into a proton and a heavy quantum with the negative charge virtually" was estimated to be of the order $g^2/\hbar c$, where g is the coupling constant.[216] "The virtual presence of the heavy quantum" (Yukawa et al.1938, 35) resulted, according to the authors, in a reasonable agreement with empirical values.

---

[214] "Die Rechnungen über das magnetische Zusatzmoment der schweren Teilchen sind hier nur der Vollständigkeit halber aufgenommen [...]; denn es scheint uns sehr zweifelhaft, ob ein physikalischer Sinn damit verbunden werden kann." Fierz (1937, 554).

[215] That Pauli's assistant Fierz was set to perform the respective calculation is rather indicative. In letters to Heisenberg (Pauli to Heisenberg, November 24. 1936; as cited in Pauli (1985, 479)) and Weisskopf, Pauli directly used the term "magnetic moments" of the nucleons when obviously referring to the mechanism Wick had introduced. Discussing the first results of Yukawa's theory, Pauli mentioned to Weisskopf that "the self-energies and the magnetic momenta of the particles become infinite". [*Original:* "die Selbstenergien und magnetischen Momente der Teilchen werden unendlich [...]"] Pauli to Weisskopf, January 13. 1938; as cited in Pauli (1985, 549). Pauli certainly questioned details of the performed calculations or of the respective theories in these letters, but not the general mechanism.

[216] As I am not capable of reading Japanese I was not able to access the prior publication by Taketani on this topic and thus cannot say for sure how Yukawa, Sakata, and Taketani arrived at this estimation. Considering the given value, the scheme proposed by Wick, and the assessment given in a footnote of Fröhlich, Heitler, and Kahn (1939, 274), the value was due to a self-energy calculation for which the cut-off was chosen in such a way as to cohere with the range of nuclear forces. This would result in the divergent integral approximately taking the value 1 and thus in total to the value $g^2/\hbar c$ for the additional magnetic moment.

In Britain, which became with the growing threat of the Nazis and the looming outbreak of WWII one of the centres for theoretical research on nuclear physics, Herbert Fröhlich, Walter Heitler, and Nicholas Kemmer collaborated on the further development of Yukawa's theory.[217] The first application of Wick's idea in connection with meson theory given by Fröhlich and Heitler (1938) nicely exemplifies how well established the corresponding explanation had become. Fröhlich and Heitler did not actually calculate the value of the magnetic moment of the neutron or proton. Rather, they used "the fraction of time which a proton (neutron) spends in a virtual state as a neutron (proton) + heavy electron" (Fröhlich and Heitler 1938, 37) as a placeholder variable to estimate the mass of the Yukawa particle.

Assuming that the proton fulfilled the Dirac equation, the magnetic moment of the proton and the neutron were given by

$$\mu_P = \mu \left(1 - \alpha + \alpha \frac{M}{m}\right) \tag{10.17}$$

$$\mu_N = \mu \left(-\alpha - \alpha \frac{M}{m}\right) \tag{10.18}$$

where μ stood in for the magnetic moment of the unperturbed proton. M was the mass of the proton and m the mass of the Yukawa particle. Combining the equations, Fröhlich and Heitler obtained

$$\frac{M}{m} = \frac{\mu_P + |\mu_N| - 1}{1 - \mu_P + |\mu_N|} \tag{10.19}$$

By plugging in the empirically known (or estimated) values of the magnetic momenta they obtained a mass of about 80 electron masses, which they considered in good agreement with experiment. In their subsequent more detailed article they considered it "strong support for such an exchange theory of the nuclear properties" (Fröhlich, Heitler, and Kemmer 1938, 155) that the mass of the heavy electron as estimated both from the value of the magnetic momenta of the nucleons and by the range of the nuclear forces, as initially given by Yukawa (1935), led approximately to the same value.

Of course, in every formulation of the meson theory the magnetic momenta diverged. But the estimations and the qualitative consequences seemed to cohere with experiment and were conceived of as signs that the theorists were on the right track.[218] Thus, the explanation was upheld. When renormalization schemes for QED were worked out in the late 1940s, diverse (unsuccessful) attempts at calculating the magnetic momenta according to a mechanism resembling Wick's explanation were put forward.[219] But a consistent theoretical explanation of the magnetic moment of

[217] See Brown and Rechenberg (1996, Sects. 7.4–7.7) for the initial collaboration on nuclear theory by the German refugees in Britain, at the time mainly based in Bristol.

[218] See also the following subsection.

[219] See Pais (1986, 483).

nucleons was only provided with a completely different picture of proton and neutron, namely as bound states of quarks. But for the time being, the possibility of invoking such virtual dissociation processes had a theoretical upshot: It ensured that the Dirac equation could be applied to the heavy particle. At least there was an explanation of the additional magnetic moment, that allowed for such an application. The arguments were shaky, of course, but not on conceptual, but on theoretical grounds.

In analysis, it might seem surprising that a mechanism, closely connected to the self-energy and the modelling of vacuum polarization and as these effects divergent, was upheld as a suitable explanation of an empirically measured effect. Yet, it is to note that at around the time concepts as the virtual charge cloud were rather uncontroversial. The calculations might have been divergent, but the conceptualization seemed reasonable enough. Moreover, the realm of nuclear and high energy physics was in an epistemically uncertain state. Uncertainties in the concrete quantum field theoretical description, problems of inconsistencies internal to theory (the divergences), but also problems with the coherence of experiment and theory begged for an explanation.[220] Qualitative arguments based on the mean lifetime of the proton/neutron in the virtually dissociated state could provide for signs that the theory was on the right track as long as it could qualitatively reproduce the empirical data. And, as I will argue in the following section, exactly in this epistemically uncertain constellation the narrative, i.e. a way of arguing for the inclusion of a concept into the conceptual framework, became important: It allowed for qualitative estimations when quantitative ones were out of the question.

10.2.3 Arguing for Virtual Transitions and Particles: The Slow Emergence of a Narrative

There is an important part of the story of virtual transitions and particles which I have only touched upon: the connection between energy non-conservation in these processes and the uncertainty relation[221] for energy and time. This is partially due

---

[220] See for example Cassidy (1981) for the problems of reconciling experiment with theory in the second half of the 1930s. Brown and Rechenberg (1996, 254–291) described the diverse attempts at modelling quantum field theories of nuclear forces from the late 1930s to the mid 1940s. See also Mukherji (1974) for a grand tour through the theoretical attempts from 1935 to 1952. Monaldi (2005) discussed the same period with a specific focus on the decay of the mesotron. As already noted in Sect. 9.3.2, the theory of radiation damping had its origins in the discrepancies between theoretical modelling of scattering and the observed cross sections. These difficulties could only be settled when the two meson hypothesis, i.e. the identification of the pion with the meson responsible for nuclear interactions and the identification of the only weakly interacting muon with the particles observed in the cosmic rays, became widely accepted.

[221] I will exclusively use the term "uncertainty" in reference to Heisenberg's principle, as it is today the most common one, and I will not discern between the term "uncertainty relation" or "uncertainty principle". This means as well that I will translate terms like "Ungenauigkeit", which is closer to the term "inaccuracy", with "uncertainty", as long as it is clear that the historical actors referred to Heisenberg's uncertainty relations.

to the fact that this connection was not as prominent as one would expect. In most articles explicitly using the concept of energy non-conserving transitions, either in time-dependent or time-independent perturbation theory, the uncertainty relations were simply not invoked for their explanation. Even didactic discussions did not necessarily introduce this connection.

Walter Heitler's textbook, probably the most popular one on radiation theory of the time, can serve as an example. Discussing perturbation theory Heitler noted that energy conservation held for the entire process while "for the transitions from or to the intermediate states n' the energy is, of course, in general not conserved" (Heitler 1936, 90) without giving any further argument. Also in the introductory text by Fermi (1932, 120–123) and in the textbook by Gregor Wentzel (1943, 41–42) energy non-conservation was noted but not discussed or justified. Gregory Breit (1932, 530), in an overview article written mainly for didactic purposes, also stated that "the picture of the absorption of $h\nu$ via the intermediate state J'' has a certain degree of reality even though in the temporary condition J'' the conservation of energy is violated". He went on: "Considerable light has been thrown since on the reason for this temporary lack of conservation of energy by the more modern quantum theory."[222] But he did not mention the uncertainty relations. Of course, there is the possibility that the connection between the energy-time uncertainty and the violation of energy conservation by the only shortly occupied intermediate states was simply obvious to the practitioners and they felt no need to point out this connection explicitly. But, still, one would expect this connection to be spelled out in the didactic literature.

First Signs of the Narrative (1929–1936)

This is not to mean that a connection between the (apparent) energy non-conservation and the duration of the process through the uncertainty principle was completely absent. As I have discussed in Chap. 5, the mathematical representation, and especially time-dependent perturbation theory, made the development of a verbal model as well as the short duration of the lifetime of the intermediate states highly suggestive. Dirac, for example, did not invoke the uncertainty relation explicitly, but upheld that "the intermediate state [...] lasts only a very short time" (Dirac 1930f, 365).

There were some explicit invocations of the connection between energy-time uncertainty and the intermediate states or other versions of arguing for the inclusion. As discussed in Sect. 6.3, Yakov Frenkel explicitly invoked the role played by the perturbation, i.e. as inducing transitions and not explicitly being part of the energy balance, to, at first, realistically interpret such energy non-conserving transitions and, after this conceptualization could not be upheld, to make the processes accessible in his textbook. When disproving Frenkel's conception of independent absorption and emission process, also Gregor Placzek referred to the energy-time uncertainty.

---

[222] "More modern" refers to the post-1925 quantum theory, as Breit discussed the approach of Kramers and Heisenberg to dispersion at this stage.

Some of the further, but few explicit references to the uncertainty relations were given by close associates of Werner Heisenberg. In his dissertation on light-by-light scattering, Hans Euler (see Sect. 9.1.3) explicitly stated that the time-dependent term of the first approximation is "only considerable in the region of uncertainty ["Ungenauigkeitsbereich"] $|E_k - E_i| < \frac{h}{t}$".[223] Carl Friedrich von Weizsäcker noted that to derive the exchange forces between nucleons in field theory one has to consider an energy non-conserving "intermediate state, which is only occupied during the time $\frac{h}{E}$".[224] When deriving Dirac's trick (see Sect. 5.1.3 for its explanation) von Weizsäcker explicitly referred to Heisenberg's uncertainty relations and the possibility of measurement during the respective interval of time (von Weizsäcker 1937, 201). Bernhard Kockel obviously did not consider Dirac's verbal model to cohere with reality when he called it the "picture of 'successive' transitions".[225] Yet, in his discussion of the *Schüttelwirkung*, he noted that besides transitions induced by the sudden switching on of the perturbation, there existed "real/actual [wirklich]" ones, "which obey the energy law",[226] even though the energy difference $|E_i - E_k| < \frac{\hbar}{t_0}$ between initial and final state was finite (Kockel (1937, 164); see also Sect. 9.1.4).

In the contributions discussed in the prior subsections (especially 10.2.2 and 10.1.2), similar connections between the energy non-conserving transitions and a temporal dimension of the processes were invoked. Although Wick retracted this picture initially, the temporal dimension of the dissociated state played an important role in introducing his idea and its further development. As noted, Arnold Nordsieck spoke of "fluctuations" between the neutron and the proton-electron-neutrino state. In the "Bethe Bible" the temporality of the phenomenon is discussed explicitly (Bethe and Bacher 1936, 205). Fröhlich and Heitler (see above) even used the time spent in the dissociated state as a place-holder to estimate the mass of the meson. Finally, in the conceptually close concept of vacuum polarization another related idea was invoked by Furry and Oppenheimer: They noted that the nascent pairs could only be observed by considerably perturbing the system through the measurement, i.e. by providing enough energy for the "nascent pairs" to be born.

Thus, I would hold that physicists were certainly aware of the possibility of connecting the temporality of virtual processes with the time-energy uncertainty and the possibility of the measurement of the respective states.[227] But still, the question remains: Why wasn't this connection made more prominent? Why didn't physicists

223 "nur beträchtlich innerhalb des Ungenauigkeitsbereichs $|E_k - E_i| < \frac{h}{t}$." Euler (1936, 412).

224 "[...] Zwischenzustand, der jeweils nur während der Zeit $\frac{h}{E}$ angenommen wird." von Weizsäcker (1937, 189).

225 "Bild 'aufeinander folgender' Übergänge," Kockel (1937, 162).

226 "dem Energiesatz gehorchenden," Kockel (1937, 164).

227 How seriously they took this possibility is another question and, considering the sources I have evaluated, not only a hard one to answer, but highly dependent on the views of the respective historical actors.

refer to the energy-time uncertainty when this would have made the inclusion of virtual transitions far more suggestive? Why didn't they argue for the concept of virtual transitions based on basic principles of the theoretical description, i.e. why didn't they stress the role of the narrative in the sense I defined in my methodological introduction (see Sect. 2.3.2)? And why would the narrative become prominent, as we will see in a moment, in nuclear physics in the later half of the 1930s?

Landau and Peierls (1931): Uncertainty in the Interaction Energy

To partially answer this question,[228] let me first turn to a paper by Landau and Peierls (1931) on the measurement in field theory. Peierls and Landau were both in Zürich at the time and their paper falls in one historical line with the Bohr-Rosenfeld cooperation on the measurability of fields in QED.[229] But I include Peierls and Landau's contribution at this point because of their discussion of the connection between perturbation theory, the uncertainty relations, and the measurement process. Importantly, their discussion was seemingly in line both with Werner Heisenberg and Niels Bohr's thinking about the uncertainty relations at the time: Landau and Peierls explicitly based their reasoning on Bohr's Como lecture and, while Heisenberg was critical of Peierls and Landau's assessment of the measurability of fields, he did not criticize the general interpretation of the uncertainty relations I will present.[230]

Landau and Peierls noted, that the energy-time uncertainty did neither mean that energy conservation was violated nor that energy could not be measured with arbitrary accuracy in a short amount of time. Rather, it referred to the change of the system due to the measuring process. I quote the connection they thereby established in full:

> The method of the variation of constants [time-dependent perturbation theory] shows, that the transitions during short times by no means only occur between such states, which obey the condition $E + \epsilon = E' + \epsilon'$. ($E$, E' energy of the system before and after the transition, ϵ, ϵ' the ones of the measuring apparatus) Only after longer times these states are favoured by resonance, as the corresponding transition probability considerably grows with time. After the time Δt only such transitions practically play a role, for which $|E + \epsilon - E' - \epsilon'| \propto \frac{h}{\Delta t}$. Of course, this fact in no way contradicts the strict validity of the energy law in wave

[228] A concise answer would have to go into the details about the different interpretations and the philosophy of the uncertainty relation, its connection with the measurement process and vacuum fluctuations. But this certainly exceeds the scope of this investigation.

[229] See, for example, Darrigol (1991, especially pp. 159–163).

[230] See Heisenberg to Peierls and Landau, January 26. 1931; as cited in Pauli (1985, 53–54).

mechanics, *but the interaction energy between system and apparatus is undefined by the amount in question.* In the most favourable case, where ϵ and ϵ' are exactly known, the uncertainty has to be $\Delta(E - E') > \frac{h}{\Delta t}$.[231]

First of all, it should be noted that Peierls and Landau obviously spoke about actual transitions as they considered a measurement process. Therefore, Peierls and Landau did not talk about virtual transitions as such, but about the probability of the system performing a transition to a state with a different energy than the initial state. Yet, in multiple processes, the respective states corresponded to the intermediate or virtual states. Secondly, they brought in the interaction energy between measuring apparatus and system. As they stated, the energy is not not-conserved, but the apparent energy non-conservation was due to the fact that the interaction energy between measuring device and system was undetermined by exactly this amount.

Reasoning with the Narrative: Qualitative Conclusions in Nuclear Theory (1938 Onwards)

Landau and Peierl's kind of reasoning was also underlying the most prominent connection between virtual processes and the uncertainty relations during the 1930s. It was developed by Gian-Carlo Wick (1938) during a research stay in Copenhagen. Its aim was "a derivation as elementary as possible" of the range of nuclear forces $r = \frac{h}{mc}$, where m is the mass of the force carrying particle.[232] According to Wick, the interaction between two nucleons was caused by successive emission and absorption processes. Yet:

> These are not, of course, actual emission and absorption processes, which would be contrary to the energy principle; they are called, therefore, virtual transitions. Let us see, however, a little closer how it comes about that the energy principle is respected. (Wick 1938, 994)

For his argument, Wick brought in a device which was supposed to observe the emitted particle. "In this case the energy principle can only be saved, as usual, if the uncontrollable energy exchange involved in the operation of the device is so large as to cover the energy excess actually observed, which is at least mc^2." To detect the particle, the observation had to be performed during the passage of the particle from one nucleon to the other. The duration t of this process was at least $\frac{r}{c}$, where

[231] "Die Methode der Variation der Konstanten [zeitabhängige Störungstheorie] zeigt, daß die Übergänge innerhalb kurzer Zeiten keineswegs nur zwischen solchen Zuständen vor sich gehen, die der Bedingung $E + \epsilon = E' + \epsilon'$ genügen. (E, E' Energie des System vor und nach dem Übergang, ϵ, ϵ' diejenigen des Apparats.) Erst nach längerer Zeit werden diese Zustände durch Resonanz bevorzugt, indem die entsprechenden Übergangswahrscheinlichkeiten mit der Zeit stark anwachsen. Praktisch spielen nach der Zeit Δt nur solche Übergänge eine Rolle, für welche $|E + \epsilon - E' - \epsilon'| \propto \frac{h}{\Delta t}$ ist. Diese Tatsache widerspricht natürlich keineswegs der strengen Gültigkeit des Energiesatzes in der Wellenmechanik, *sondern die Wechselwirkungsenergie zwischen System und Apparat ist eben um den betreffenden Betrag unbestimmt.* Im günstigsten Falle, wo ϵ und ϵ' exakt bekannt sind, muß dann die Ungenauigkeit $\Delta(E - E') > \frac{h}{\Delta t}$ sein." (Landau and Peierls 1931, 59; emphasis added).

[232] This connection was known (at least) since Yukawa's initial paper on the meson theory: Yukawa (1935). See Darrigol (1988, 279–282) for a historical contextualization of Yukawa's derivation.

r denoted the distance between the two particles. Since $\Delta E \Delta t \approx h$ it followed that $\Delta E \approx \frac{h}{\Delta t} = \frac{hc}{r}$. And as ΔE was at least mc^2, it followed that $r \approx \frac{h}{mc}$. As Wick noted, "the condition [$\Delta E > mc^2$] actually gives the distance r as the limit up to which virtual transitions can make themselves felt without contradiction of the energy principle".[233]

The difference to the modern and often hand-waving argument should be noted. Neither was energy conservation violated, nor was the vacuum or any other kind of medium assumed to provide for the necessary energy. Rather, Wick argued counterfactually that if we were to measure we would find the particle, as long as the indeterminacy in the interaction energy between measuring device and system was large enough and the time short enough to allow for such a measurement. The narrative, the connection between theoretical principles and the concepts it was supposed to support, had a much more involved character than today.

Moreover, the energy uncertainty was not only invoked to explain that such processes "can make themselves felt without contradiction of the energy principle". Rather, the uncertainty relation allowed for the derivation of an important quantity, the range of nuclear forces. At this point, what I called a narrative had an epistemic upshot. The sole assumptions Wick had to introduce were (1) the energy-time uncertainty and (2) the interaction of nucleons via the virtual transmission of another massive particle. Wick could thereby, although essentially qualitatively, establish an important empirically testable parameter of the interaction, i.e. its range.

It is rather unsurprising that it was Gian-Carlo Wick who proposed this kind of reasoning. As discussed in the previous subsection, Wick established the derivation of the magnetic moment of nucleons by taking, on the one hand, the concept of virtual transitions rather seriously, while, on the other hand, nearly exclusively relying on qualitative reasoning, i.e. abstracting from the specific form of the interaction Hamiltonian. The same holds for Heitler and Fröhlich's estimation of the mass of the Yukawa particle. The only assumptions they made were that (1) the empirically established values of the magnetic momenta were approximately correct and (2) that Wick's 1934/35 conceptualization of the anomalous magnetic momenta was adequate.

In an article, obviously written for interested physicists who were not specialized in quantum field theory, Heitler (1941b) nearly exclusively invoked such qualitative reasoning.[234] As Heitler noted, this approach was not only chosen for didactic purposes, but because of "the very preliminary character of the theory" and because of the fact that the fundamental properties of elementary particles "turn out to be largely independent of the particular form we choose for the theory and can be explained by using general arguments only" (Heitler 1941b, 680). In the same vein as Wick derived the range of nuclear forces with the help of the energy-time uncertainty, Heitler provided an argument based on the position-momentum uncertainty. The "'virtual emission and reabsorption' of a meson" has according to Heitler "a very

[233] All quotes taken from Wick (1938, 994).

[234] Heitler used both field and particle concepts, while I will only shortly consider the particle side of his arguments.

literary [sic!] meaning",[235] because "if we would carry out an experiment to find the charge at a distance r, say, from the proton (or neutron), we shall, indeed, with a certain probability, find a charged particle at that point". Yet, "the word 'virtual' means that, in contrast to a real emission, energy need not be conserved" and the "experiment will only give a positive result if the necessary energy μc^2 is *supplied by the measuring process itself*".[236]

While the exact quantum field theoretical approach to nuclear physics, i.e. the modelling of the interaction and the wave functions, was unclear, each of its possible formulation had to face the divergences that came with every quantum field theoretical approach at the time. Although there is certainly room for further investigation at this point, I do not consider it a coincidence that the uncertainty relations and the concept of virtual transitions were so prominently connected at this point of epistemic uncertainty. Establishing a connection to one of the grounding pillars of quantum theory allowed for qualitative reasoning with partial but coherent results. Drawing together his argument about the range of nuclear forces (based on Wick's argument) with his argument about the mass of the heavy particle through the invocation of virtual dissociation (see the previous section), Heitler stated in his 1941 article: "The consistency of these results suggests that the fundamental concepts of this theory are correct." (Heitler 1941b, 683)

In analysis, it is noteworthy that the narrative for the inclusion of virtual transitions in the conceptual framework of quantum theory, which will later on be used for the virtual particle concept, became particularly prominent when the exact theoretical modelling of nuclear interactions was on particularly shaky grounds. Andrea Monaldi has already observed that Wick's "trans-theoretical justification of Yukawa's mass-range relation became a recurrent argument in particle physics" (Monaldi 2005, 429). Although the exact theoretical modelling was unknown and each attempt ran into inner-theoretical difficulties, the coherence of such arguments based on the narrative of the virtual particle reassured that one was, in principle, on the right track.

Again, I want to stress that this is not to mean that the narrative was not known (or accepted, while those are certainly two different things) but that it was not prominent priorly. Even though these arguments might have strengthened the concept's incorporation into the conceptual framework of quantum field theory, the connection between the energy-time uncertainty, the measuring process, and the virtual processes were mainly used for estimations that aimed at exemplifying the consistency of the general assumptions. And, importantly, they did so without having to go into the details of the formalism but still allowed "to obtain a large number of qualitative results and order of magnitude relations between the fundamental properties of the elementary particles" (Heitler 1941b, 680).

[235] From context it is clear that Heitler meant "literal".

[236] All quotes taken from Heitler (1941b, 681, emphasis in the original).

10.2.4 Wentzel's Schemata *(1937/38): Feynman-Like Diagrams*

The mainly qualitative estimations described in the last two subsections elucidate the prominence the reasoning with theoretical entities connected to the development of the virtual particle concept had acquired in quantum field theory during the second half of the 1930s. In this final subsection, I will turn to the application of a representational system, the "Schemata" introduced by Gregor Wentzel, which are more closely related to the quantitative evaluation of the theory. According to the following analysis, the diagrams introduced by Wentzel (especially his schemata i, see below) come closest to a paper tool during the 1930s, at least so as long as only diagrammatic representations are concerned. At the same time, Wentzel's usage of representations strikingly exemplifies the conceptual overlap of *virtual particles* and *virtual transitions* which formed in the second half of the 1930s.

The background of the introduction of Wentzel's schemata was an important quantitative drawback of Fermi's theory of β-decay. As previously noted, this theory could not even approximately reproduce the strength of attraction between nucleons.[237] But even if cut-offs were appropriately chosen or other modifications were applied, there was another problem directly connected to the modelling of the interaction: The force between neutron and proton was supposed to be particularly strong in comparison to the proton-proton or the neutron-neutron forces. In Fermi's theory, nuclear forces were charge-dependent.

In late 1935, empirical evidence from proton-proton scattering experiments was gathering which indicated that nuclear forces were equally strong amongst all nuclear constituents, an observation known as the charge-independence of nuclear forces.[238] But this fact was hard to incorporate into the original form of Fermi's field theory: Forces between unlike particles resulted from the successive emission and absorption of an electron/positron and an (anti-)neutrino. They were thus a second order effect. But forces between like particles required two of such processes, hence they constituted a fourth order effect.

Wolfgang Pauli, at the time at the Institute for Advanced Study, reported the empirical results obtained in the US to Gregor Wentzel in a letter dating to February 24. 1936,[239] and Wentzel set himself to solve the respective problem. He had contributed to quantum electrodynamics, particularly the investigation of the self-energy problem (Wentzel 1933b, Wentzel 1933a, and Wentzel 1934a), the neutrino theory of light (Wentzel 1934b), and to the development and understanding of Fermi's theory of β-decay (Wentzel 1935). Although Wentzel was closely associated with

[237] There were further problems of empirical validity but I will not go into any details. Also alterations of the theory as proposed by Bethe and Peierls or Konopinski and Uhlenbeck, introducing derivatives in the Hamiltonian, could not undo the discrepancy between theory and experiment. See Brown and Rechenberg (1996, Sect. 3.4).

[238] For a short introduction to the theme, see Brown and Rechenberg (1996, 60–62).

[239] Pauli to Wentzel, February 24. 1936; as cited in Pauli (1985, 440–441).

Pauli,[240] he, quite contrary to Pauli, used non-mathematical representations to refer to perturbative evaluations regularly.[241]

In late 1936 and early 1937, Wentzel proposed two quantum field theoretical models accompanying or replacing Fermi's original formulation of β-decay. Wentzel's first proposal introduced processes that complemented Fermi's theory. Wentzel (1937a) assumed that neutrons and protons could as well emit electron/positron or neutrino/anti-neutrino pairs.[242] The second proposal included hypothetical states of proton and neutron in which they would acquire integer spin values (Wentzel 1937b; Wentzel 1937c). Wentzel called them "Singulett-Neutron/Proton" (spin 0) and "Triplett-Neutron/Proton" (spin 1). In this theory, β-decay was a second order process: A neutron could decay into a proton, a neutrino, and an electron "via a 'virtual' intermediate state"[243] of a singlet neutron and a neutrino or a singlet proton and an electron (see left side of Fig. 10.4).

Both of these propositions were rather short-lived as news about Yukawa's meson theory arrived in the US and Europe at about the same time and attracted the attention of theoretical physicists. Also Wentzel would turn to Yukawa's theory in a popular exposition only a year later (Wentzel 1938). But I will not discuss Wentzel's propositions due to their relevance for the general modelling of nuclear forces, but because Wentzel is known for the introduction of diagrams which on first sight show striking resemblance to Feynman diagrams. A resemblance, that has been, according to the following argument, been overstated.[244]

The terminology Wentzel used gives first indications: Wentzel nearly exclusively used the term "virtual" to refer to transitions, processes or states. He never designated a particle with the term. This was in harmony with the then current method of perturbation theory in which energy was not conserved in transitions. The closest Wentzel came to attribute a particle with the term "virtual" was in Wentzel (1937a).

[240] Wentzel and Pauli were employed at universities in the same city (Zürich) and institutionalized a shared weekly theoretical seminar (Enz 2002, 207). Some of Wentzel's doctoral students (Nicholas Kemmer, Markus Fierz) became Pauli's assistants.

[241] Nevertheless, Wentzel would often set the word "intermediate state" in quotation marks. This is for example the case for his proof that the mass of the neutron must, according to an extended version of Fermi's theory of β-decay, be larger than the mass of the electron and proton combined. Thus a free neutron would be subject to β-decay (Wentzel 1935).

[242] Gamow and Teller (1937) independently proposed the same model to take account of the charge independence of nuclear forces.

[243] "über einen 'virtuellen' Zwischenzustand," Wentzel (1937b, 34).

[244] See Miller (1984, 165–167). Miller seems to suggest a historical connection between Wentzel's and Feynman's diagrams, but does not follow the historical route due to its complexity. The historical reconstruction of Feynman diagrams by Wüthrich (2010) does not discuss Wentzel's schemata in detail. But Wüthrich's reconstruction already indicates that the schemata did not play a prominent role in Feynman's reasoning process.

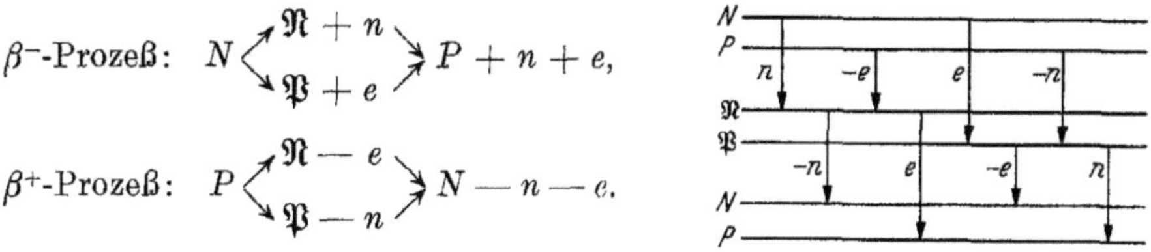

Fig. 10.4 Diagrammatic representations of β-decay in Wentzel's singlet-theory. The left diagram, an example of his schemata i, was used by Wentzel for virtual processes, while the right represented successions through "real" states. Both figures in Wentzel, Gregor. 1937b. "Zur Theorie der Beta-Umwandlung und der Kernkräfte. I." *Zeitschrift für Physik* 104, 34–47, here 36 (left) and 37 (right). https://doi.org/10.1007/BF01337892 (cited as Wentzel (1937b) in the bibliography). Reproduced with permission from Springer Nature

There, he described the β-field as "the virtually present light particles in the vicinity of protons and neutrons".[245] This, in turn, was in line with the particle-cloud picture which was already known and practised at the time.

My discussion of the diagrammatic representation will focus on Wentzel (1937b) (singlet-theory) and Wentzel (1938) (popular article already including Yukawa's theory). Since Wentzel presented two different versions of "schemata", as he called them, in the two papers, I will refer to them as *schemata i* (Wentzel 1937b) and *schemata ii* (Wentzel 1938). But before going into its details, it must be noted that my historical reconstruction of Wentzel's schemata will differ appreciably from the only account that I am aware of. Miller (1984, 165–167) identified Feynman-like diagrams, the *schemata ii* as depicted in Fig. 10.7, in Wentzel's textbook of 1943 and conceived of them as an alteration of energy level diagrams, such as the ones depicted on the right hand side of Fig. 10.4. Miller further suggested a historical connection to Feynman's work but did not substantiate this claim.

As I will argue in the following, the closest one got to Feynman diagrams during the 1930s, at least so in terms of functionality, were the diagrams depicted on the left of Fig. 10.4, the *schemata i*. Theses diagrams did not emerge from the pictorial tradition of energy-level diagrams but rather resembled reaction schemes as they were used in the theory of radioactivity at the time. Finally, the Feyman-like diagrams, which I call *schemata ii*, were historically far more likely connected to the diagrams on the left of Fig. 10.4 and not, as Miller suggested, to energy level diagrams. As I will argue there is a deeply rooted difference between the Feynman-like *schemata ii* of Wentzel and the diagrams actually given by Feynman himself, as they emerged from different pictorial traditions.

245 "in der Umgebung von Protonen und Neutronen virtuell vorhandenen leichten Teilchen", Wentzel (1937a, 109).

$$N \longleftrightarrow \mathfrak{N} + n, \qquad P + e \longleftrightarrow \mathfrak{N},$$
$$N \longleftrightarrow \mathfrak{P} + e, \qquad P + n \longleftrightarrow \mathfrak{P}.$$

Fig. 10.5 The four basic transition processes of Wentzel's singlet/triplet-theory. Figure in Wentzel, Gregor. 1937b. "Zur Theorie der Beta-Umwandlung und der Kernkräfte. I." *Zeitschrift für Physik* 104, 34–47, here 35. https://doi.org/10.1007/BF01337892 (cited as Wentzel (1937b) in the bibliography). Reproduced with permission from Springer Nature

The Theory of Singlet and Triplet Nucleons: The Diagrammatic Representation, Its Tool Character, and the Pictorial Tradition from Which It Emerged

To understand the usage of the *schemata i*, we have to turn to Wentzel's singlet-theory. It was build on four basic transition processes, depicted in Fig. 10.5. For their identification Wentzel explicitly disregarded energy conservation. Neutrons and protons (represented by capitalized letters) could transform into singlet protons or singlet neutrons (fractured capitalized letters) by emitting or absorbing electrons, positrons, and (anti-)neutrinos (small letters; the anti-particles are represented by a minus sign in front). To all of these and only to these processes corresponded non-zero matrix elements.[246] Wentzel could thus construct every possible process in his theory from these basic building blocks and their reversals. In each subprocess one of these basic transmutations occurred. $\beta^{(\pm)}$-decay was constructed as a second order process (see Fig. 10.4, left side). Six of the in total 24 "paths"[247] which contributed to the interaction between neutron and proton are represented in Fig. 10.6. After constructing these schemata, all Wentzel had to do to translate the respective processes into a mathematical representation was to identify the respective matrix elements with the sequence of basic processes in the general perturbative expansion.[248]

But Wentzel actually used two different kinds of diagrams already in Wentzel (1937b). He referred to them differently and they performed different functions. Wentzel called the kind of diagrams on the left of Fig. 10.4 "schemata" (Wentzel 1937b, e.g. p. 36) while the diagram on the right, resembling an energy level diagram, did not have a particular name. Wentzel referred to such diagrams as figures. In both of the diagrams in Fig. 10.4 β-decay processes are depicted. Just to exemplify the reading, lets focus on the β^--decay.

[246] This is not exactly true. To each of these processes existed a corresponding process including the respective anti-particle of the light particles. These processes can be easily constructed by subtracting the respective particle from both sides of the diagrams. For example, $N \leftrightarrow \mathfrak{P} + e$ becomes $N - e \leftrightarrow \mathfrak{P}$.

[247] "Wege," Wentzel (1937b, 38).

[248] The translation into mathematical formulas and their calculation was performed by Wentzel in part B of Wentzel (1937b) and his second paper on his singlet-theory: Wentzel (1937c).

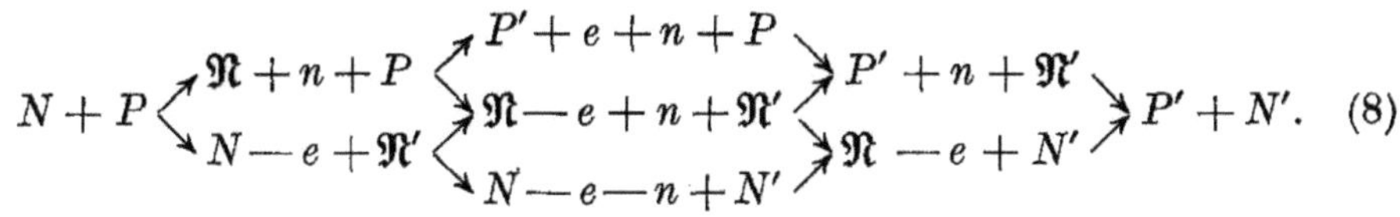

Fig. 10.6 Six of the in total 24 "paths" contributing to the proton neutron interaction in Wentzel's singlet theory. Figure in Wentzel, Gregor. 1937b. "Zur Theorie der Beta-Umwandlung und der Kernkräfte. I." *Zeitschrift für Physik* 104, 34–47, here 38. https://doi.org/10.1007/BF01337892 (cited as Wentzel (1937b) in the bibliography). Reproduced with permission from Springer Nature

Wentzel's *schemta i* were read from left to right. The upper schemata in Fig. 10.4 (β^--Prozeß) described the process of a neutron decaying into a singlet neutron and a neutrino or into a singlet proton and an electron (respectively represented as the upper or lower intermediate state) followed by a decay of the singlet neutron/proton into a neutron and a neutrino/a proton and an electron. Thus, one arrived at the final state in which a proton, an electron, and a neutrino were present. In the energy level diagram, diverse processes were depicted, among them β^--decay according to Wentzel's theory. To get the same subprocesses as in the first possibility of the β^--decay of the schemata, one had to start from the horizontal line denoted by N and follow the first line on the left, the neutrino emission process, down to the line denoted by the fractured N. Thus a state of singlet neutron and neutrino was reached. From this point, one had to follow the second vertical line downwards, denoted with an e, arriving at the horizontal line P. Since the downward lines referred to the emission of particles, one had finally reached a state in which a proton, a neutrino and an electron was present, i.e. β-decay has occurred.

So, why did Wentzel represent the same processes in two different diagrams? Because Wentzel did not use them to represent exactly the same processes. The diagram on the left, Wentzel's schemata, represented virtual (sub-)processes. In contrast, Wentzel used the diagram on the right to indicate "cascadelike" processes through "'real' states".[249] In each step in the energy level diagram on the right hand side of Fig. 10.4 energy was conserved. The analogy of these diagrams to the ones used by Weisskopf and Wigner for the composition of real and successive emission processes (see Fig. 8.2) is obvious. But this representation had important drawbacks as a calculational tool for virtual processes: The occupation of the intermediate states was not directly represented and could only be reconstructed by following the processes from the upper line to the end point. Moreover, one could not directly see which different paths over intermediate states led to the same final state.[250]

Both aspects were far more obvious in the *schemata i*. Similar diagrams had been used by Amaldi (1929) (compare Fig. 6.1) to represent the Raman effect. The dynamic character of the diagram, representing a process which was read from left to right, was as obvious as in the earlier example, but the things that were represented

---

[249] "kaskadenartig" and "'reelle' Zustände," Wentzel (1937b, 36–37).

[250] To be sure, Wentzel did not make the above drawbacks explicit. They are part of my reconstruction.

changed. Amaldi represented probability amplitudes, which carried the occupation of the respective states as indices. Wentzel, on the other hand, directly represented the particles (although in an abbreviated form) present in the respective (intermediate) states[251]: To some extent, Wentzel represented *virtual particles*, although he himself never used this kind of terminology. As a common feature, all of these diagrams allowed for reading out the occupation of the respective states directly and, thus, also for reconstructing the sequence of transition processes quite easily. Although I consider it likely that Wentzel was aware of Amaldi's or other diagrams that shared this feature of representing a stepwise process while explicitly representing (aspects of) the particles present in the respective states, I do not think that this was necessary for Wentzel to come up with his *schemata i*.

Actually, Wentzel followed a representational tradition which was prominently used in nuclear physics at the time, especially for nuclear decay[252]: chemical or reaction equations, such as $N \rightarrow P + e^{-} + n$ to represent β-decay.[253] Obviously, there were a few simplifications but also extensions in respect to this tradition. On the one hand, Wentzel discussed particles and not molecules, so no atomic weights or anything similar needed to be indicated. Wentzel's schemata were further not used to balance out the respective process in terms of the involved numbers of atoms or molecules. On the other hand, Wentzel described processes composed of the successive occurrence of the four basic processes of his theory. Nearly always, more than one possibility of successions led to the same final result. As they were virtual processes, they all needed to be included at the same time in the calculation and thus the different paths were also represented in one and the same diagram (as for example in Fig. 10.6).

Even though I did not consult manuscript notes on Wentzel's papers on the singlet-theory, I consider it rather likely that he effectively used the *schemata i* as paper tools. The construction of all paths from initial to final state through successive transitions can easily be undertaken with this method at hand. That Wentzel considered the schemata as more than a short-hand notation is also suggested from the way they were included in the papers: The energy-level diagrams were referred to as figures, but the schemata were included in the numbering of the equations, as one can see on the outer most right of Fig. 10.6.[254] The whole conception of processes they derived from was obviously equivalent to Dirac's verbal model of QED, while more complicated through the inclusion of four elementary processes. Essentially, one can

---

[251] This should not be misunderstood as a statement about the corresponding mathematical representation, which obviously took place on the quantum-amplitude level of the theory. The above is an observation exclusively concerning the diagrammatic representation.

[252] I did not study the history of the usage of this representational format in nuclear theory. For the point I am about to make, it suffices to say that it was available and prominently applied in nuclear physics when Wentzel proposed his theory.

[253] Wentzel himself denoted Fermi's process by $N \rightarrow P + e^{-} + n$ in a short note communicated in December 1934 (Wentzel 1935, 35).

[254] This holds for each of his papers discussed here: Wentzel (1937a, b, c).

conceive of Wentzel's *schemata i* as a diagrammatic representation of the verbal model with an extended set of transition possibilities.

Exclusively using these schemata, Wentzel could even draw some conclusions about the nature of the forces between neutron and proton, i.e. their spin dependence and the fact, that neutron-proton forces were of the same order as neutron-neutron and proton-proton forces (Wentzel 1937b, 39–40), the empirical observation initially motivating Wentzel's approach. But the main reason why Wentzel included the schemata in his papers was the possibility of constructing all the different paths leading from initial to final state through successive transitions. These paths were build up from the four building blocks he had introduced for his theory and were subsequently translated into a mathematical representation for quantitative evaluation. Concerning the tool character of diagrammatic representations, I would hold that, during the 1930s, Wentzel's *schemata i* are the closest one gets to Feynman diagrams.

Of course, Wentzel's *schemata i* appreciably differed from Feynman diagrams in their specific function and conceptualization. The schemata represented a stepwise process of emission and absorption in occupation number space and not the interrupted free movement of particles in space and time. From a methodological point of view, there was also a huge difference, which has its roots in the mathematical representation of the perturbative evaluation. In the perturbative expression of the 1930s, the matrix elements of the respective subprocesses entered. But these were not represented by Wentzel in the schemata. He only represented the paths and the corresponding intermediate configurations leading from initial to final state. Following the paths one could easily reconstruct which processes and thus which matrix elements were involved and had to be included at which place in the mathematical expression. But they were not directly represented. In Feynman diagrams, each line and interaction point represents a mathematical expression directly. The translation procedure is, also due to the mathematical concepts of propagators, direct, without any detour.

While Wentzel's singlet-theory faded into the background quite quickly after Yukawa's meson theory was taken up in Europe, other physicists still used diagrams like Wentzel's *schemata i*. Fröhlich, Heitler, and Kemmer (1938) or Peierls (1939) represented the basic interactions of meson theory in such schemes.[255] Yukawa and Sakata (1937, 1089) described the six paths through which two neutrons can interact according to a charged meson theory by diagrams of exactly the same structure as Wentzel's *schemata i*. They also called them "schemes".[256] But also Wentzel himself reiterated on his schemata. In a popular account on the recent developments of nuclear theory, a quick tour starting with Fermi's theory of β-decay and already including some information about the first works on meson theory in Europe (Wentzel 1938),

[255] In order to do so, they did not necessarily have to rely on Wentzel's schemata. The representation of such basic processes in terms of reaction equations was rather usual.

[256] Although they did not cite Wentzel's paper, I consider it rather likely that they knew of Wentzel's proposal. The usage of the same term to refer to the diagrams as well as the structural similarity to Wentzel's *schemata i* are strong indications.

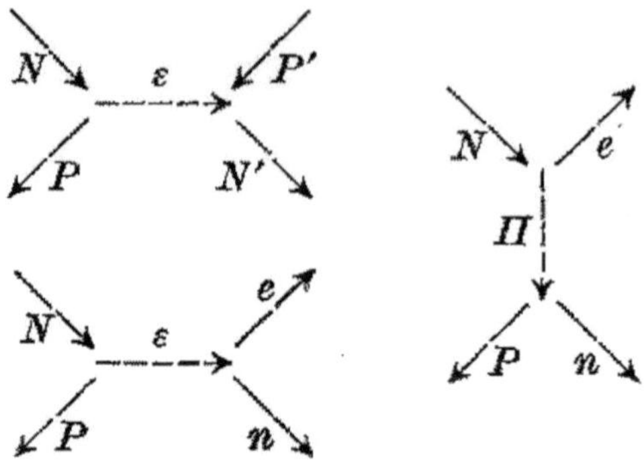

Fig. 10.7 Three different processes as represented by Wentzel (1938) (*schemata ii*). Upper left: Exchange interaction of neutron and proton. Bottom left: β-decay according to Yukawa's meson theory. Right side: β-decay according to Wentzel's singlet-theory (N=Neutron, P=Proton, ϵ=Yukawa meson, e=electron, n=neutrino, Π=Singlet-Proton). Figures in Wentzel, Gregor. 1938. "Schwere Elektronen und Theorien der Kernvorgänge." *Die Naturwissenschaften* 26, 273–279, here 274, 276, and 277. https://doi.org/10.1007/BF01590308 (cited as Wentzel (1938) in the bibliography). All reproduced with permission from Springer Nature

as well as in his well-known textbook (Wentzel 1943) Wentzel provided diagrams, again referred to as "schemata", but in a considerably modified form.

Schemata ii: Feynman-Like Diagrams But Without Calculational Application

Three examples of the *schemata ii*, as I will call them, are depicted in Fig. 10.7. As one easily notices, some rather important changes have occurred. In the *schemata ii*, arrows did no longer refer to processes but to particles. Each connecting point represented an interaction. The direction of the arrows indicated whether the particle entered the interaction as an already produced particle (arrow pointing towards the point) or whether it was produced at the interaction point (arrow pointing away from it). Reading the depicted diagrams, one should always start with the upper left arrow.[257] The diagram on the upper left represented the interaction between a neutron and a proton according to Yukawa's meson theory, the diagram below depicted β-decay according to this theory, and the diagram on the right represented one of the possibilities of how β-decay could occur according to Wentzel's singlet-theory.

For the diagram on the upper left, time seems to be flowing from top to bottom. But, in general, this was not the case. The electron lines in the other two diagrams in Fig. 10.7 represented particles in the final state even though the arrows were pointing upwards. Time was also not running from left to right, as was clear from the direction of the proton arrows in each diagram. In general, the sequence of events in this kind of

[257] But there were also diagrams for which one should start on the upper right hand side, as for example Wentzel (1938, 276, Fig. 4).

schemata had to be reconstructed by finding the first decay process, i.e. an interaction point for which one particle entered and two particles emerged.[258] In most general terms, there was no time dimension in which all diagrams could be aligned.

The *schemata ii* of Wentzel (1938) actually had a little drawback. They did not directly indicate which particles were present in intermediate states. Wentzel, for example, stated in reference to the diagram on the right of Fig. 10.7: "The fact, that the emitted electron is already present in the virtual intermediate state, actually causes a specific asymmetry of the β-spectrum in the wished for sense."[259] That the electron was already present in the intermediate state was, just by looking at Fig. 10.7, not obvious and could only be concluded by knowing that the upper interaction point was the first one.[260]

Furthermore, Wentzel seemed not to have used the *schemata ii* for calculational purposes, in contrast to his *schemata i*. Wentzel would reproduce them to describe simple interaction phenomena in his 1943 textbook, but did not do so in research papers. While second order processes could nicely be depicted with Wentzel's second version of schemata, higher order processes as the ones given in Fig. 10.6 were neither easy to construct nor, when constructed, easy to read.[261]

At first, one might think that the *schemata ii* had an advantage over the *schemata i* as the interaction points are represented, thus one could in principle read out the respective matrix elements directly. But the conceptualization of QED and QFT during the 1930s focused on the occupation of the intermediate states and the constructions of paths from initial to final state. Furthermore, in contrast to the *schemata i*, the *schemata ii* were only representatives of one specific sequence. This is essentially true for Feynman diagrams as well, but physicists during the 1930s had to imagine a lot more processes than physicists working with the version of QED and QFT after the late 1940s. In Wentzel's singlet-theory, for example, there were 24 paths which contributed to the proton-neutron interaction: Wentzel would have had to draw 24 *schemata ii* to represent them. Furthermore, as the *schemata i* were built up of the successive application of the basic building blocks representing the basic processes of the theory and the respective paths were represented within one and the same schematic display, it was a lot more apparent if actually every possible path had been found.

---

[258] If the first process was meant to be an absorption, like for example in Fig. 8 of Wentzel (1938, 277), the interaction point at which two incoming particles produced a third particle had to be located.

[259] "Der Umstand, daß hier das emittierte Elektron im virtuellen Zwischenzustand bereits anwesend ist, bedingt in der Tat eine Unsymmetrie des β-Spektrums in dem gewünschten Sinne." Wentzel (1938, 277).

[260] The diagram also represented only one of the possible processes leading to β-decay in Wentzel's singlet-theory. But this would also be true for Feynman diagrams.

[261] I did not find any historical example of such a construction. But the reader can try, as I did, to translate the processes of Fig. 10.6 into the second version of Wentzel's schemata. Even after figuring out how to incorporate the respective particles, reading out the initial and final state, let alone the respective intermediate states, is far from trivial.

Comparing the *schemata ii* to Feynman diagrams, it is to note that the former were not represented in space and time. As discussed, time runs through the diagram in terms of the direction of the arrows and not from top to bottom or bottom to top. Methodologically, there is another important difference, again connected to the mathematical representation of the theory: The concept of a propagator was not used in perturbative evaluations of the 1930s. Thus, the connecting central lines, the ones we might read as virtual particles (the ϵ's and the Σ in Fig. 10.7)[262] had no direct mathematical counterpart. They would acquire such only through the work of Richard Feynman in the late 1940s. If there was a translational procedure, which was, to the best of my knowledge, never used for the *schemata ii*, it was at best indirect.

Nevertheless, the fact that Wentzel referred to both kinds of diagrams as schemata and that he published them in short succession suggests a historical connection. Consequently, there is still the obvious and big question: Why did Wentzel change the representation in his 1938 article?

A Tentative Explanation: Pictorial Traditions of Theory and Experiment

Although I cannot proof this conjecture at the moment I would suggest two reasons for the change from *schemata i* to *schemata ii*. On the one hand, Wentzel (1938) was a purely popular article and only the general conceptions of the theories were expounded by describing second order processes. For such simple cases, the *schemata ii* provided a rather good intuitive picture. On the other hand, photographs of cloud chamber tracks surely enjoyed prominence in the experimental literature for some time before 1938 and obviously theorists as Wentzel were well aware of these pictures.[263] But in the years shortly before Wentzel published his schemata, theorists started to investigate high energy phenomena, especially from cosmic radiation, ever more closely and more regularly referred to empirical articles using such experimental means.[264] The Yukawa particle, the experimental discovery of which was the main motivation for Wentzel's popular article, had been identified in cloud chamber pictures of cosmic radiation. Wentzel cited empirical papers in which exactly such cloud chamber pictures had been reproduced. A connection between the change in representation and cloud chamber pictures would also explain the orientation of Wentzel's *schemata ii*: As for the respective cloud chamber tracks, the starting point in reading them was at the top of the diagrams.[265]

If we accept a connection between the change in diagrammatic representation and cloud chamber pictures as well as a historical connection between Wentzel's *schemata i* and *schemata ii*, which I consider due to the name and the short interval of time between the two versions rather likely, the above noted differences to

[262] Again, Wentzel only spoke of virtual processes or states.

[263] For a historical account of experimentation with cloud chambers, see Galison (1997, Chap. 2).

[264] See for example Cassidy (1981) and Galison (2008, Sect. V).

[265] But, as already noted, this did not exclude the possibility that also the arrows of particles in the final state could be pointing upwards.

Feynman diagrams (but also their likeness) can be explained by the pictorial traditions from which they emerged. As David Kaiser has argued,[266] Feynman diagrams fall into the pictorial tradition of Minkowski diagrams (theory) and show striking resemblance to cloud chamber and the later bubble chamber tracks (experiment). Wentzel's schemata combine the traditions of reaction schemes (theory) and cloud chamber tracks (experiment). While the pictorial tradition of the experiment is the same for both types of diagrams, the theoretical tradition of the schematic or diagrammatic displays differed.

That Wentzel's *schemata ii*, although different in form, were still connected to the pictorial tradition of reaction schemes is further suggested from the way they were embedded in his textbook *Einführung in die Quantentheorie der Wellenfelder*: Wentzel used both reaction schemes and his *schemata ii* to represent second order processes (Wentzel 1943, 56, 59). He further embedded the second usage of his *schemata ii* for second order processes directly in the reaction schematic display of the contributing first order process (Wentzel 1943, 103–104).

According to this reconstruction, the resemblances of Feynman diagrams and the *schemata ii*, particles as tracks or arrows and intersections as interaction points, result from the shared visual tradition of cloud chamber tracks. The coherent alignment of Feynman diagrams according to a temporal axis, which connects them to Minkowski diagrams, was absent in Wentzel's *schemata ii*. Rather, the latter were still more strongly rooted in the tradition of reaction schemes and a conception of successive transformation processes. Needless to say, the mathematical representations which Feynman diagrams, on the one hand, and Wentzel's schemata, on the other, were connected to and embedded in differed considerably. All the same, it is noteworthy that the diagrammatic techniques which function(ed) as paper tools for quantum field theorists, Feynman diagrams and Wentzel's *schemata i*, do not fall into the pictorial tradition of energy level diagrams. And, according to my reconstruction above and Adrian Wüthrich's reconstruction (Wüthrich 2010), both of these diagrammatic or schematic representational formats could circumvent drawbacks of energy level diagrams in the respective theoretical framework.

10.3 Summary and Analysis: Conceptual Overlap Through Different Representational Formats

In the following, mostly chronologically ordered summary and analysis I will draw the different observations on concepts and practices connected to the virtual particle together. During the 1930s, vacuum polarization, the charge cloud picture, and the concept of force carriers, which are today closely connected to the virtual particle concept, were established. The terminology of the "virtual" became more common in the later half of the 1930s. According to my reconstruction, this was due to the

[266] See Kaiser (2005, 185–188; 368–373).

usage of the terminology by Werner Heisenberg and his student Hans Euler (see Sect. 9.1.2). As I will argue in the following, by the end of the 1930s there was a considerable overlap of the concepts of *virtual processes* and *virtual particles* which can be accessed through the different formats of representations the actors themselves provided.

As discussed in Sect. 10.1.1, at the turn from the 1920s to the 1930s physicists did not explicitly apply non-mathematical representations of virtual processes in their evaluation of the self-energy problem. I argued that this fact should be explained by (1) the simplicity of the phenomenon in terms of a model of successive transitions, (2) the close relation of these works to a group of physicists who tended not to give pictorial explanations of quantum phenomena, and, importantly, (3) the application of time-independent perturbation theory to a phenomenon which was essentially conceived of as static. All of these reasons tended to loosen their grip in the course of the 1930s and we encounter a particularly dense period for the development of concepts in one historical line with the virtual particle at the turn from the year 1933 to 1934. This fact is connected to two developments in theory. On the one hand, the concept of the Dirac sea was taken seriously after the discovery of the positron; on the other hand, the first extension of field theoretical methods to nuclear phenomena was proposed.

Let's first turn to quantum electrodynamics. In Weisskopf's 1934 contribution to the self-energy of the electron in the Dirac sea, as far as I know, the first explicit and published connection between vacuum fluctuations of the radiation field and the self-energy calculations was provided. At the same time, although only implicit in Weisskopf's paper, the fluctuations of the electron-positron distribution in the vacuum were included and mitigated the self-energy problem: The "hole-theoretic supplement" essentially rested on these contributions.[267] The connection between virtual transitions and vacuum fluctuations was established.

Also in quantum electrodynamical theory and not unconnected to the fluctuations of the electron-positron field, the effect of vacuum polarization was investigated from the second half of 1933 onwards (Sect. 10.1.2). With its evaluation in terms of a quantized electron-positron field, the idea that empty space was no longer really empty but included "nascent pairs", to use Furry and Oppenheimer's terminology, became prominent. By mid 1934, the concept of a particle cloud formed around other particles had been established and could, through the subtraction procedures of Dirac and Heisenberg, (partially) be evaluated to its empirical consequences. Yet, as the state of the infrared divergence in the second half of the 1930s shows (Sect. 10.1.3), the concept of a bare and dressed particle (modern terminology) was not yet fully formed. The concept of renormalization, in a modern sense, had not been established.

In nuclear physics, also starting at the turn from 1933 to 1934 and according to my reconstruction (Sect. 10.2.1) through a mutual reinforcement of heuristic verbal

---

[267] As already indicated, there is still room for further investigation of the connection between vacuum fluctuations, the self-energy calculations, and concepts connected to the virtual particle. Yet, Weisskopf's contributions and his previous stay in Copenhagen will surely play an essential role therein.

invocations and suggestive mathematical (but previously only partially interpreted) structures, the concept of a force carrier played an ever more prominent role in the historical actors' assessments. A dynamical description of phenomena previously considered to be static, as the exchange and the Coulomb force, was reinforced by the incorporation of the verbal model into time-independent perturbation theory. Wick's explanation of the anomalous magnetic moment (Sect. 10.2.2), conceptually close to the effect of vacuum polarization and the self-energy problem, is a case in point. Although Wick eventually described the phenomenon as a mixture of (non-) dissociated states of the proton, he introduced his readers to his considerations by invoking a dynamical picture of successive transitions.

The further development of nuclear theory and the self-energy problem exemplifies the tool character of both the verbal model of successive transitions and the temporality of the descriptions. Weisskopf's proof of the logarithmic nature of higher order divergences (Sect. 10.1.1) essentially rested on the picture of successive transitions. And he obviously conceived of it merely as a tool, while he considered the physical reasons for the mitigation to be best explained through the invocation of the correspondence-like approach to QED. In the qualitative arguments in nuclear theory, concerning both the epistemic exploitation of the narrative and the explanation of the anomalous magnetic momenta (Sects. 10.2.2–10.2.3), the temporal description of the processes was the key element. The time spent in the dissociated state or the time between emission and absorption in interaction processes allowed for qualitative estimation of empirically established values. The fact, that such qualitative arguments were put forward in the late 1930s, can be understood by the wide acceptance of the respective conceptual structures purportedly describing the effects, i.e. virtual particles/transitions, while the concrete theoretical description (in terms of the formulation of the interactions and the quantized fields) was disputed and, in general, led to ambiguous results.

All the same, the invocation of a dynamical description (or concepts connected to the virtual particle in more general terms) depended heavily on the respective formulation of the theory and the way of its evaluation. According to my reconstruction, the lack of a picture of successive transitions in the first contributions to the self-energy problem should be explained by a difference in the conceptions of time-independent and time-dependent perturbation theory. The application of the correspondence-like formulation of QED by Weisskopf in his 1934 and the major part of his 1939 paper on the self-energy of the electron made the explicit usage of non-mathematical representations of virtual processes unsuitable, although Weisskopf was most certainly aware of such a reading. Similarly, and according to my argument in Sect. 10.2.1, the strength with which the force carrier analogy was put forward in the respective evaluations of Fermi's theory of β-decay should be understood against the background of the different formulations of QED available and practised at the time: Soviet physicist applying methods along the lines of Dirac's 1932 interaction picture formulation most explicitly referred to a dynamical description of the Coulomb and exchange interaction in terms of successive emission and absorption processes.

The mathematical formulation of the theory thus played a crucial role in setting the frame for the possible invocation of non-mathematical formats of concepts connected to the virtual particle. At the same time, we can identify a shift in the historical actors' conceptions through the role these theoretical entities played in the respective effects. According to the verbal model of radiation theory, the particles were only present in an intermediate state. According to the charge cloud picture, the particles were present but could, due to the lack of sufficient energy, not escape into free space. But these two pictures and even their connection to a field description of interactions were not mutually exclusive. Just consider Rudolph Peierls' comments (1939) on the interaction of neutron and proton according to Yukawa's meson theory:

> In this theory we obtain an interaction between proton and neutron owing to the interaction of the one with the *meson field* of the other. Indeed, the proton has a tendency to become a neutron, with the emission of a positive meson. This process cannot actually take place in the sense that the meson can be set free, because this would require at least the energy μc^2, which is not available. However, as long as *the meson is in the immediate neighbourhood of the heavy particle*, the interaction energy associated with the *the process of emission and re-absorption* may make up the balance. Hence *an isolated meson carries with it a field of mesons which are continuously being absorbed and re-emitted, but which cannot escape.* This situation, which is familiar in quantum mechanics, is customarily described as the 'virtual' existence of a meson. (Peierls 1939, 85–86, emphasis added)

A temporal description, a continuous absorption and re-emission, of effects which were considered to be static in nature (a charge cloud, the self-energy of a particle, the Coulomb interaction or the interaction energy between particles in more general terms) was ever more strongly pronounced in the description of the perturbative evaluation.

Peierls' above reference to a "virtual existence" or "presence" of a particle, which was similarly invoked by Wentzel (1937a, 109) or Yukawa, Sakata, and Taketani (1938, 34), further points towards the trend of connecting the particles, and not the processes, with the term "virtual". And indeed, although the terminology of "virtual processes" was still common, physicists started to apply the term virtual, or synonyms of it,[268] directly to particles. Furry and Oppenheimer spoke of "nascent pairs", Werner Heisenberg of the "virtual electron-neutrino pair",[269] Evan J. Williams (1935, 48) of "virtual pairs",[270] Wendell H. Furry (1937, 125) of "a (virtual) pair",

[268] As pointed out, the fixation of the term virtual as referring to theoretical entities we connect to the virtual particle occurred in the second half of the 1930s. The term "synonym" above, thus, should not point to an exact overlap of meaning in the everyday usage of the words, but rather to the fact that physicists used the terms to refer to one and the same theoretical entity.

[269] "des virtuellen Elektron-Neutrino-paares [sic!]," Heisenberg (1935a, 115).

[270] It must be noted that Williams both referred to the equivalent photons of the Weizsäcker-Williams method and to concepts we can directly connect with virtual particles as "virtual". Seemingly, Williams saw some connection, possibly in them being both due to specific approximation procedures.

Hans Euler of "the virtual light particles",[271] and Akhiezer and Pomeranchuck of "a virtual pair".[272] Havas (1944, 69) or Bethe and Oppenheimer (1946, 457) would talk about "virtual quanta". All of these (nearly) direct connections between the term "virtual" and the term "particle" were put forward by the respective actors, when discussing phenomena for which the particles were exclusively present, according to the verbal model, in the intermediate states.

The mathematical representation obviously worked with energy non-conserving transitions, i.e. with virtual (sub-)processes, and some actors explicitly stated that the meaning of the term "virtual" was the energy non-conservation of these transitions.[273] All the same, the fact that some phenomena were theoretically described by particles exclusively present in the intermediate states (or intimately connected to other particles due to the lack of energy for their escape into free space) led physicists to apply the terminology of a "virtual particle".

While the above described tendency was most evidently expressed in the terminology the actors used, also in diagrammatic representations "virtual particles" became more pronounced. The diagrams of Mercier (see Fig. 10.1), the representation of nuclear forces by Teller (see Fig. 10.3) and Wentzel's *schemata i* (see Fig. 10.6) can serve as examples. Of course, these representations rested on a model of successive transitions: Mercier and Wentzel (*schemata i*) ordered the respective states from left to right, while Wentzel additionally indicated transition processes by an arrow; Teller ordered the states from top to bottom. Nevertheless, these diagrams represented particles in intermediate states directly. They did not use their energy levels as Maria Göppert-Mayer (see Fig. 8.1) or Hans Euler (see Fig. 9.1) did, nor their kinematic properties (as for example Walther Heitler, see Fig. 8.3), nor the amplitudes of states (as Amaldi, see Fig. 6.1). Teller even indicated a direction of movement of the intermediary electron, a dynamical aspect which was even more prominent in Wentzel's *schemata ii* (see Fig. 10.7).

As said, Wentzel nearly exclusively referred to virtual processes and explicitly stated that the term "virtual" referred to energy non-conservation during transitions. But his *schemata ii* already speak, so to say, a different language. The concept of an intermediary particle, represented by an arrow moving from one nucleon to the other, being created at one interaction point and destroyed at the other, is strikingly similar to a depiction in Feynman diagrams. All the differences to Feynman diagrams set aside for the moment (see Sect. 10.2.4), one might say that in Wentzel's schemata a dynamic conception of a virtual particle was already represented diagrammatically.

In analysis and according to my analytical framework, there existed a considerable conceptual overlap of, maybe even a tension between *virtual transitions* and *virtual particles* in the later half of the 1930s. As I argued in Sect. 2.3, we should access the

[271] Hans Euler to Werner Heisenberg, March 13. 1937; AMPG, III. Abt. Rep. 93, Nachlaß Heisenberg, Nr. 1733.

[272] "ein virtuelles Paar," Akhiezer and Pomeranchuck (1937, 478).

[273] See e.g. Wentzel (1938, 275), Wick (1938, 994), or Heitler (1941b, 681).

historical development of a concept through the diverse means of representation the actors used to depict and apply it, of course always located in the respective theoretical framework. Up until the late 1940s, the mathematical representation was still firmly rooted in energy non-conserving transitions, i.e. "virtual transitions". Actors used the energy non-conservation of these transitions to explain the terminology of the "virtual". But the evaluation of phenomena and processes, in which the particles were only present in intermediate states (or could not escape into free space) led to the terminological combination of "virtual" and the respective "particles". As discussed, also in diagrams or schematic displays the particles became more prominent, although these representational formats still relied on a picture of successive transitions. Physicists started to think with and about particles in intermediate states and represented them accordingly.

This overlap can not only be exemplified by the references of different historical actors to the said entities or by one and the same actor's use of different representational formats. To some extent, it can be located within the representational format of Wentzel's *schemata ii* itself: According to my argument, this mode of representation emerged in the combination of the pictorial tradition of reaction schemes, i.e. a conception of successive transitions, with cosmic ray photographs, i.e. a depiction of a dynamical scattering processes in which particles are represented.

This trend from *virtual transitions* to *virtual particles* and their overlap/tension in the representational formats cohered with another shift in perception: Different types of theoretical entities were made responsible for the respective effects; what was conceived of as the connecting piece for the occurrence of the process changed from states to particles. While Walter Heitler stated in his highly influential textbook that processes occurred "through the *agency of intermediate states*",[274] Wendell H. Furry noted that for effects such as light-by-light scattering or the Delbrück effect "electrons and positrons, though they are not actually created so as to be experimentally observable, are able to play the part of *catalytic agents* producing the effects".[275] "Virtual particles" slowly emerged as the effective entities in the theoretical description of quantum field theoretical processes during the later half of the 1930s. But for the full establishment and a coherent application of them it needed a reformulation of QED and its practice: The following outlook on the work of Richard Feynman and its relevance for the virtual particle concept will constitute the endpoint of my historical investigation.

[274] Heitler (1936, 131, emphasis extended to include "agency of").

[275] Furry (1937, 125, emphasis added).

Chapter 11
Outlook: Feynman, Diagrams, and Virtual Particles (1948–1949)

In the previous chapter I argued that the concept of a *virtual particle*, in contrast to the concept of a *virtual transition*, gradually emerged in the second half of the 1930s. According to my reconstruction, physicists started to connect the term "virtual" directly with particles or pairs because some effects were theoretically described with particles exclusively present in the intermediate states. All the same, the perturbative evaluation was still based on energy non-conserving transitions and not on off-shell particles, i.e. particles which did not obey the relativistic energy-momentum relation.[1] And even the most dynamical diagrammatic representations of the 1930s, i.e. Gregor Wentzel's schemata, were still (partially) rooted in a model of successive transitions.[2] Accordingly, although the terminology of *virtual particles* was used, the reference to *virtual processes* was common. In this final chapter I will provide an outlook on how the virtual particle became prominent in the late 1940s mainly by focusing on Richard Feynman's space-time approach, a manifestly covariant formulation of QED, and its perturbative evaluation.

The landscape of QFT changed considerably during and after WWII.[3] This holds for theory, experiment, and the centres of theoretical research. The rise to power of the NS regime and the outbreak of WWII led to an enormous brain-drain from Central Europe, particularly from Germany. Only a small number of quantum field theorists remained in Germany and German speaking countries, as for example Werner Heisenberg (Leipzig) and Gregor Wentzel (Zürich). Only few of the refugees and emigrants could secure positions in Europe, as for example Max Born (Edinburgh)

[1] In how far the term "off shell" is an anachronism at this point would need further investigation. It did not occur in the papers by Feynman or Dyson during the late 1940s, while the idea that the particles did not obey the relativistic energy-momentum relation and that energy and momentum were conserved at the interaction points was surely expressed in mathematical terms.

[2] See my reconstruction in Sect. 10.2.4.

[3] For the following two paragraphs see also my short outline in Sects. 3.3–3.4 and the literature cited therein.

M. Ehberger, *Representing the Unobservable*, Science Networks. Historical Studies 68,
https://doi.org/10.1007/978-3-032-09188-8_11

or Walter Heitler (Dublin). The greater part of the scientific personnel emigrated to the US, where theoretical physics flourished after WWII.

There were diverse reasons for the prospering of theory in the US directly after the war. The European emigrants assimilated in an already developed community of theoretical physicists; the advances in experimental technologies led to new input for theory which eventually contributed to the development of the renormalization techniques; the greater esteem of theoretical physics in the after-war period allowed young theorists to secure professor positions early in their career; and, hand in hand with the reorganization of theoretical physics after the war, opportunities of direct idea exchange were established, as epitomized by the Shelter Island (June 1947), Pocono (April 1948), and Old Stone (April 1949) conferences.[4] It is quite unsurprising that the further development of theory was most importantly an American achievement.

To be sure, I do not want to downplay the contributions by Japanese or Swiss physicists, especially by Tomonaga and Stueckelberg.[5] But as these contributions were noticed by US physicists, the development in the US was already in progress and had taken on its own pace. A rather striking example and especially important for the *virtual particle* concept is Stueckelberg's covariant perturbation theory, which was developed as early as 1934, but found next to no response in the community.[6] As Lacki, Ruegg, and Telegdi (1999, 496) pointed out, in Stueckelberg's perturbation theory energy and momentum were conserved at the interaction points and the particles in intermediate states were off shell. But it had no impact on the further development as next to no physicist seemed to have noticed it. Concepts, representations, as well as theoretical techniques do not only have to be communicated, but taken up by the community; they have to be communicated in such a way that the community is able to grasp its merits for the further development of the theory. For Stueckelberg's covariant perturbation theory this was not the case.

Considering the contents of theoretical research in the US after WWII, the development of consistent renormalization techniques was certainly the outstanding achievement. Just by throwing a short glimpse at these techniques and its most important concepts, the relevance for the virtual particle concept becomes apparent.[7]

---

[4] See (Schweber 1994, Chap. 4) for the organization and realization of these conferences.

[5] See Schweber (1994) Chap. 6 (for Tomonaga) and Chap. 10 (for the Swiss context and Stueckelberg). See also Blum (2017, Sect. 2.5.2) for a discussion of Stueckelberg's work in relation to the S-matrix program.

[6] See Lacki, Ruegg, and Telegdi (1999) for a historical reconstruction of Stueckelberg's covariant perturbation theory and its relation to modern methods. See Sect. 7 of this paper for the lack of response by the community.

[7] For historical outlines, see Darrigol (1982, Chap. 8), Pais (1986, Chap. 18), Aramaki (1987), Brown (1993), Dresden (1993), or Blum and Joas (2016, Sect. 2). Schweber (1994, Chap. 5) provided a detailed discussion of the empirical background and the first, still old-fashioned calculations of the Lamb Shift and the anomalous magnetic moment of the electron. For a technically detailed discussion embedded in the biographies of the main protagonists in the reformulation of theory, see Schweber (1994, especially Chaps. 6–9). Wüthrich (2010, Chap. 6) provided a detailed reconstruction of the role of Feynman diagrams for Dyson's formulation of renormalization. For a very accessible modern discussion on the technical implementation of renormalization, see Mills (1993).

During the late 1940s, the concepts of a "dressed" and a "bare" particle became firmly established. The "bare" particle, figuring in the basic equations of the theory, priorly enjoying ontological primacy and defined without the contributions of virtual particles, was now conceived of as a theoretical idealization. The "dressed" particle, including the presence of virtual particles, was now and in turn conceived of as the real particle, i.e. the one observed in experiments. As Adrian Wüthrich (2010, 186) has put it:

> Through mass and charge renormalization, the inert, that is force-free, state of a particle is redefined. For instance, one defines the energy of a free electron as the observed (finite) energy of an electron, which necessarily includes the virtual processes, such that the electron is not free in the strict sense of the word.

Thus, basic concepts of the theory were altered in direct relation to the virtual particle concept through renormalization. Nevertheless, the relevance of the different representations of the virtual particle for the necessary theoretical and mathematical manipulations is a different question. It is certainly worthy of pursuit but will not be followed in this outlook.[8] But renormalization was not the only major alteration of the practice of QED in the late 1940s.

As has been discussed in detail by Alexander Blum (2017), while QED remained stable in its conceptual basis, the exemplar, i.e. the paradigmatic problem (Kuhnian terminology), of the theory changed. The focus of quantum mechanics on the properties of states was replaced by a formulation in which only asymptotically free states entered and the formalization of scattering processes took centre stage. As the subtitle of Blum's paper goes, "quantum field theory became a theory of scattering". Adrian Wüthrich (2010) discussed how the change in theoretical reasoning was reflected in the conception of processes and their representation through diagrams. I, in turn, want to emphasize the respective alterations to the virtual particle concept.

In the following, I will focus on the work of one of the major actors of the above mentioned paradigmatic shift: Richard Feynman. I will discuss how the concept of a virtual particle changed due to his reformulation of QED. For this outlook, it shall suffice to discuss the characteristics and the usage of the concept in the form of the theory Feynman eventually presented in his two major publications, *The Theory of Positrons* (Feynman 1949b) and *Space-Time Approach to Quantum Electrodynamics* (Feynman 1949a). With Feynman's reformulation and, eventually, Dyson's incorporation of it in regular QFT, the shift from the concepts of virtual transitions to virtual particles was finally and completely realized.

In the Sect. 11.1, I will describe Feynman's path towards his "space-time approach" in order to make this new formulation and its relevant features accessible to the reader. In Sect. 11.2, I will discuss the different alterations to the virtual particle concept, its characteristics and its representational formats, which came with Feynman's

[8] According to Adrian Wüthrich (2010), Feynman diagrams functioned as a particularly adequate representational system for the technical implementation of renormalization. Thus, the impact of the new representation of the virtual particle, giving rise to the divergences, suggests investigation. Yet, according to historical reconstructions, the concept of the virtual particle itself seems not to have been an important driving force behind Feynman's reformulation.

approach. In Sect. 11.3, I will shortly revisit Dyson's systematization of Feynman diagrams, his equivalence proof of Feynman's QED to regular quantum field theoretical approaches, and point towards the merits but also one of the short-comings of the new means of representation.

11.1 Feynman's Path to the Path Integral Formulation (1939–1948)

Feynman's work on and his path towards his reformulation of QED have found a lot of attention in the secondary literature. Biographical details on Feynman can be found in Mehra (1994) and Schweber (1994, Chap. 8), while both also included technically detailed descriptions of Feynman's research. Christian Forstner (2007, Chap. 4) discussed Feynman's portrayal as a pragmatist in connection to the thought collective[9] of American physicists.[10] Adrian Wüthrich (2010) provided a detailed reconstruction of the emergence of the representational system of diagrams in Feynman's research. Alexander Blum (2017, Sect. 3) focused on Feynman's work in connection to what Blum identified as a paradigmatic shift of the exemplar (Kuhnian terminology) of the theory in the late 1940s: the shift of QED from a theory which focused on states to a theory which had scattering as its exemplar. Peter Galison (1998) discussed the impact on Feynman's war related work in Los Alamos to his modelling of physical process and the corresponding techniques after the war. Finally, Olivier Darrigol (2019) provided a concise and very accessible reconstruction of Feynman's work on QED. The following reconstruction of Feynman's path to the path integral formulation is based on these contributions.

The story of Feynman's reformulation of QED is primarily an American one. He portrayed himself, and was portrayed by others, as a practical man with a drive towards getting the right numbers out. Feynman studied physics in the US. During his graduate days in Princeton John Wheeler became his mentor, who himself had only spent half a year abroad in Copenhagen. During the war, Feynman worked in Los Alamos on the Manhattan Project, where his ability to translate between theorists and experimentalists was recognized and highly appreciated. As discussed in detail by Galison (1998), the specific approaches taken towards diffusion problems and the whole of the rather pragmatic theoretical culture prevailing at Los Alamos deeply influenced Feynman's further style of research. After the war, Feynman stayed in the US, now working with Hans Bethe at Cornell. Bethe had received his education in physics and spent his post-doctoral years in Europe, but was removed from his assistant professor position in Tübingen under the racist Nürnberg laws of the NS

---

[9] Forstner worked with the analytical framework of Ludwik Fleck (1980, originally published 1935) in which the connection of scientific research to a "thought style" ("Denkstil") and the respective integration of the scientists in a "thought collective" ("Denkkollektiv") is stressed.

[10] For the prevailing pragmatist attitude of American physics from 1920 to 1950, see Schweber (1986).

regime. Since 1935 he was Professor at Cornell. Sam Schweber described Bethe as "a pragmatic, highly mathematically skilled problem solver" (Schweber 2012, 389), who had assimilated well into the American style of physical research, emphasizing the close connection between experiment and theory.

Feynman's path to his reformulation of QED started when he arrived as a graduate student in Princeton in 1939. He had clearly noticed the problems quantum field theorists were struggling with and came up with an idea of how to avoid them. If the divergences were due to the infinite degrees of freedom of the electromagnetic field and the self-interaction of the particles, as Feynman believed, then one should simply avoid both of these concepts. Thus, Feynman aimed at an action-at-a-distance electrodynamics with retarded interactions, in which the self-interaction was simply excluded.

Feynman himself came to the conclusion that this would not work. By throwing out the self-energy one would also omit effects as radiation resistance.[11] John Wheeler, well aware of possible modifications of classical electrodynamics, suggested the following workaround to Feynman: Instead of only considering retarded interactions, one could work with both retarded and advanced interactions. The backreaction of the electron on the emitted radiation would accordingly be modelled as an advanced interaction with the other particles. Feynman indeed found that radiation resistance was properly taken into account if one incorporated half advanced and half retarded interactions while assuming that all the emitted radiation was eventually completely absorbed. The classical theory which Feynman and Wheeler worked out is therefore also known as Wheeler-Feynman absorber theory.

Since this classical formulation of electrodynamics worked with both advanced and retarded interactions, i.e. the behaviour of a particle depended on the behaviour of other particles in both past and future, the theory could not be expressed in a Hamiltonian form. Nevertheless and as Feynman became aware through Wheeler, Adrian Fokker had already shown that such a theory could be expressed in terms of an action. The classical trajectories over all space and time could be calculated by minimizing the action along the paths of the particles. In the nomenclature of Wheeler and Feynman, the respective action read:

$$S = \sum_i \int m_i \sqrt{dx^{\mu(i)} dx_\mu^{(i)}} + \sum_{i \neq j} \int \int e_i e_j \delta((x^{(i)} - x^{(j)})^2) dx^{\mu(i)} dx_\mu^{(j)} \quad (11.1)$$

The second term described the interaction of the particles, both forward and backwards in time. It was restricted to particles on each others light-cone by the δ-function. Self-interaction was excluded by the omission of the terms of the sum which referred to the same particle.

Wheeler and Feynman had thus formulated a theory which only involved the particle trajectories as variables, which could be expressed in terms of an action

[11] An accelerated particle emits radiation. Thus, more energy is needed to accelerate a charged than a neutral particle. By simply throwing out the self-interaction, this effect could not be taken into account. See Schweber (1994, 380).

and which allowed to calculate the entire trajectories of the interacting particles through space and time, but which could not be brought into a Hamiltonian form. As quantization normally worked with a Hamiltonian formulation of the system, the immediate question was how to quantize such a theory. This would be the concern of Feynman's dissertation (Feynman 1942), finished in 1942 shortly before leaving for Los Alamos and inspired by a paper by Paul Dirac on *The Lagrangian in Quantum Mechanics* (Dirac 1933).

To be sure, Feynman did not quantize the Wheeler-Feynman theory in his dissertation, nor any other relativistic theory of the interaction of particles.[12] But he came up with a quantization method that worked for systems that were only described by an action. Today his method is known as path integral quantization. The amplitude for a particle initially at q' at time t' to be at q'' at time t'', using modern and far less unwieldy notation than Feynman himself, can be expressed as:

$$\langle q'', t''|q', t'\rangle = \int e^{iS[q(t)]/\hbar}\mathfrak{D}q \tag{11.2}$$

where $\mathfrak{D}q$ refers to the integration over the paths of the particles and $S[q(t)]$ to the action along each path. Thus the quantum-mechanical amplitude was an integral, or sum, over all the possible paths, not only the classical, stationary one, a particle might take, each contributing with a phase $e^{iS[q(t)]/\hbar}$.[13]

Feynman started to visualize his theory in terms of space-time paths when writing down the results of his dissertation for an article in late 1947 (Feynman 1948b). But even though the strong visual component, that would also play an important role for Feynman's diagrammatic technique, was not part of his dissertation, this work brought important advances. Feynman had found a new of quantization. It could be applied to systems which were, for an interval of time, not describable through an Hamiltonian but only through an action. The important point here was the interval of time. Feynman still needed the concept of a wave function, derivable from a Hamiltonian equation, at least for asymptotic states.[14]

His path integral formulation was more powerful than canonical quantization and essentially equivalent to it if the system could be described by a Hamiltonian. But importantly, it further allowed Feynman to integrate out some of the dynamical

[12] Applying his approach to the Dirac equation would become the main obstacle Feynman had to face after the war (see below).

[13] Since one integrated over all possible paths, also the ones due to "virtual displacements", often called "virtual paths", and the Feynman propagators, describing the virtual particles in his approach, essentially derived from these mathematical entities, this might be a reason why Feynman referred to the particles transmitting the forces as virtual particles. But since the concept of a virtual particle had already been around for some time and Feynman's approach to perturbation theory made the connection between the term virtual and the term particle far more suggestive, I do not think that the idea of virtual paths played an important role for Feynman finding a suitable terminology.

[14] Alexander Blum (2017, 69–71) has argued that the unwillingness of Feynman to apply quantum boundary conditions even for asymptotic states of the systems might have contributed to a delay in the application of the path integral methods to the electromagnetic field.

variables of the system under study. In his dissertation, Feynman provided the example of two particles coupled by an oscillator. The choice of this system was due to the fact that in regular QED the electromagnetic field could be described as a set of harmonic oscillators. For suitable boundary conditions, the oscillator variables could be integrated out, leaving an effective action which only depended on the particle variables.

After his war related work,[15] Feynman worked out a way of applying his path integral methods to QED proper. In late 1946 and early 1947, Feynman worked on ways of formalizing the behaviour of a relativistic electron in terms of his integral-over-paths approach. These attempts were successful for one dimension but Feynman failed to generalize them to three dimensions and interacting particles. Yet, he extensively used diagrammatic methods to represent the respective contributions to the paths of the electrons. He even could extract physical conclusions directly from these diagrams.[16]

Although Feynman failed to describe the relativistic electron with his path-integral methods, he succeeded in eliminating all the oscillators of the electromagnetic field. This resulted in an effective action, which included what we today know as the Feynman propagator for the photon and which Feynman referred to as $\delta_+(x^2)$. The nomenclature should emphasize the connection to the delta-function in the Wheeler-Feynman absorber theory.[17] Furthermore, Feynman had figured out a way of bringing the path integral formalism into contact with the solutions of the Hamiltonian equations, i.e. the Green's functions or kernels,[18] which was rather suggestive from the perturbative evaluation of his theory.[19] I will discuss the interpretation of this expansion in connection with the published articles.[20]

---

[15] For the relevance of Feynman's war time experiences for taking up the kernel approaches as discussed in his two major articles on quantum electrodynamics, see Galison (1998).

[16] See Wüthrich (2010, Chap. 4) for a detailed reconstruction of the archival material.

[17] I will discuss Feynman's interpretation of the δ_+ function in connection with the published paper (see below).

[18] I will use these two terms interchangeably in the following. For their definition, see my discussion of Feynman's published papers.

[19] See Darrigol 2019, 359, 361–362) and Schweber (1994, 409–411) for the identification of Green's functions with path integrals in the perturbative expansion. See Wüthrich (2018, 511–513), Blum (2017, 74–76) and Galison (1998, Especially Sect. 4) for similar arguments and their application to the Dirac equation.

[20] At this point, it must be noted, that there is some disagreement in the literature how Feynman diagrams exactly emerged and if they were read by Feynman as directly modelling the physical process. According to Schweber (1994, 434) and Darrigol 2019, 359–360), the diagrammatic method emerged as a representation of the perturbative expansion and is directly connected with Feynman's integral over path methods. Wüthrich (2010), on the other hand, argued that Feynman conceived of the diagrams to represent physical processes and that their emergence was closely related to Feynman taking up the Green's function approach which he applied in his war related work. Although, the representational format of Feynman diagrams is of great importance for this historical investigation, the exact way of how it emerged and whether Feynman conceived of his diagrams to cohere with physical reality is unimportant for the following discussion. All the same, that Feynman interpreted the diagrams using a language resembling physical processes and what this fact allowed him to do is of importance and will be discussed.

As Feynman had failed to find an action for the Dirac equation, he shifted gears in late 1947 and directly worked with the perturbative expansion of the non-relativistic theory and simply plugged in solutions of the Dirac equation. By "tweaking the kernel", as Alexander Blum (2017, 76) has expressed it, of the one particle Dirac equation, including the suggestion of John Wheeler to conceive of positrons as electrons travelling backwards in time, Feynman could finally derive a completely new description of QED.

Feynman's version differed appreciably from quantum field theoretical formulations. It was conceptually embedded in his space-time approach: Instead of conceiving of systems through their temporal development from past to future, as a Hamiltonian method would do, the whole of the space-time development was taken into account. Feynman pictured the behaviour of systems, not through temporal snapshots, but as laid out in front of him over all space and time. Even if the problem could be expressed in a Hamiltonian form, Feynman preferred to work, not with the dynamical equations themselves, but with their solutions, i.e. their Green's functions. These in turn where interpreted as sums over paths or amplitudes to go from one space-time point to another. Feynman's theory was furthermore not based on field quantization, neither for the electromagnetic nor for the electron-positron field. Quite on the contrary, the Dirac equation was treated as a one particle equation and all of the degrees of freedom of the electromagnetic field were integrated out to give a delayed action-at-a-distance formulation in terms of the δ_+-function.

Feynman's approach was thus so different from field theoretical conceptions of QED that most physicists could at first neither appreciate the entire approach nor the calculational ease it came with. The first presentation of his QED at the Pocono Conference in April 1948 famously flopped. But it convinced Feynman of the need to write up his results and publish them in a set of two papers,[21] which we shall turn to now.

11.2 Feynman's Space-Time Approach to QED (1948/49) and Its Influence on the Virtual Particle Concept

Although the incorporation of a relativistic description of electrons into his theory was the last obstacle Feynman took, I will follow the chronology of his publications, start with his theory of positrons and then go over to the full quantum-electrodynamical description of the interaction of charged particles.[22] This makes the alterations of the

[21] The emphasis is on results at this point. More or less rigorous derivations of his methods were only published in 1950 and 1951. The proof of the equivalence between regular field theoretical methods and Feynman's approach was also put forward by Freeman Dyson and not by Feynman himself.

[22] The following will, besides the papers published by Feynman himself, draw especially from the reconstruction by Wüthrich (2010, Chap. 5). I will particularly focus on the alterations to the processes and conceptions connected to the virtual particle concept.

conceptions in connection with the virtual particle as well as Feynman's approach in total a lot more comprehensible and is actually how Feynman himself advised his colleague and friend Theodore Welton to approach his theory.[23]

11.2.1 *The Basis: Scattering as Interrupted Free Movement of Particles*

As Feynman noted in the very beginning of his first article, "the main principle is to deal directly with the solutions to the Hamiltonian differential equations rather than with these equations themselves." (Feynman 1949b, 749) No attention was paid to the continuous time evolution from past to present to future, but "we imagine the entire space-time history laid out, and that we just become aware of increasing portions of it successively" (Feynman 1949b, 749). Feynman introduced this approach to his readers by the example of a non-relativistic particle described by the Schrödinger equation.

According to the above statements, Feynman did not work with the Schrödinger equation directly but with its Green's functions, i.e. with the solutions $K(2, 1)$ of the equation:

$$\left(i\frac{\partial}{\partial t_2} - H_2\right) K(2, 1) = i\delta(2, 1) \tag{11.3}$$

where H referred to the Hamiltonian operator, δ to the Dirac delta function, and the numbers 2 and 1 in brackets of the functions K and δ to the space-time points x_2, t_2 and x_1, t_1.

As Feynman noted, the wave function of a particle governed by the Schrödinger equation at some space-time point x_2, t_2 was related to the wave function at a different space-time point x_1, t_1 by

$$\psi(x_2, t_2) = \int K(2, 1)\psi(x_1, t_1)d^3x_1 \tag{11.4}$$

where $K(2, 1) = \sum_n \Phi_n^*(x_1)\Phi_n(x_2)e^{-iE_n(t_2-t_1)}$ for $t_2 > t_1$ and $K(2, 1) = 0$ for $t_2 < t_1$.[24] $K(2, 1)$, the Green's function or kernel, could thus be called "the total amplitude for arrival at x_2, t_2 starting from x_1, t_1" (Feynman 1949b, 750). The amplitude for finding the particle in some state $\chi(x_2, t_2)$, if the particle was priorly in the state $\psi(x_1, t_1)$, was given by

$$\int \chi^*(2)K(2, 1)\psi(1)d^3x_1d^3x_2 \tag{11.5}$$

[23] This advice can be found in Kaiser (2005, 178).

[24] See Feynman (1949b, 749–750) or Wüthrich (2010, 115–116) for the derivation.

Even though the descriptions in terms of the Hamiltonian equation and in terms of the kernel were essentially equivalent, "for some purposes the specification in terms of K is easier to use and visualize" (Feynman 1949b, 750).

What Feynman meant by "easier to use and visualize" was exemplified by the treatment of the interaction of the non-relativistic particle with a weak external potential $U(x, t)$. Feynman expanded K in a power series:

$$K(2, 1) = K_0(2, 1) + K^{(1)}(2, 1) + K^{(2)}(2, 1) + ... \tag{11.6}$$

For the first order correction, Feynman derived

$$K^{(1)} = -i \int K_0(2, 3)U(3)K_0(3, 1)d\tau_3 \tag{11.7}$$

where $d\tau_3 = d^3x_3dt_3$, so that the integral extended over all space and time. Due to the temporal restrictions on the kernels the contributions were restricted to the interval between t_1 and t_2.[25]

Feynman provided the following interpretation of Eq. (11.7):

> We can imagine that a particle travels as a free particle from point to point, but is scattered by the potential U. Thus the total amplitude for arrival at 2 from 1 can be considered as the sum of the amplitudes for various alternative routes. It may go directly from 1 to 2 (amplitude $K_0(2, 1)$, giving the zero order term [...]). Or [...] it may go from 1 to 3 (amplitude $K_0(3, 1)$), get scattered there by the potential (scattering amplitude $-iU(3)$ per unit volume and time) and then go from 3 to 2 (amplitude $K_0(3, 2)$). This may occur for any point 3 so that summing over these alternatives gives [11.7]. (Feynman 1949b, 751)

Feynman used this verbal representation to immediately construct the second order contribution to $K(2, 1)$. By incorporating two successive scattering events at space-time points τ_4 and τ_3 he stated that the second approximation should be:

$$(-i)^2 \int\int K_0(2, 4)U(4)K_0(4, 3)U(3)K_0(3, 1)d\tau_3 d\tau_4 \tag{11.8}$$

Instead of deriving this result from the underlying equation, Feynman only noted that this was possible. He chose the physical but highly abstracted conception of free particle movement interrupted by the scattering by a potential to write down the respective mathematical expression. As he noted: "One can in this way [through his verbal modelling] obviously write down any of the terms of the expansion." (Feynman 1949b, 751)

[25] See Feynman (1949b, 750–751) or Wüthrich (2010, 117–118).

11.2.2 Inclusion of Scattering Via Virtual Pair Production

Having described the perturbative expansion as an interrupted movement of a free particle, Feynman turned to the relativistic description of the electron. At this point, Wheeler's suggestion that positrons could be perceived of as electrons travelling backwards in time became important.[26] With definitions that are today known as Feynman's slash notation, he presented the Dirac equation as

$$(i\not{\partial} - m)\psi = \not{A}\psi \tag{11.9}$$

where the gamma matrices and the charge of the electron have been absorbed into the operators. The equation "determining the propagation of a free particle" (Feynman 1949b, 752), i.e. the one solved by the kernel $K_+(2, 1)$, was thus given by

$$(i\not{\partial}_2 - m)K_+(2, 1) = i\delta(2, 1) \tag{11.10}$$

With this reformulation at hand Feynman provided exactly the same interpretation of successive scattering processes as he did for the Schrödinger equation. He simply replaced the K's and U's in the above expansion (11.7) and (11.8) by K_+ and A.

But simply replacing the eigenfunctions and energy eigenvalues in K by the solutions of the Dirac equation and setting $K_+(2, 1)$ equal to zero for $t_1 > t_2$, as he had down for the non-relativistic case, led, as Feynman noticed, to the one particle theory of the Dirac equation. Hence, negative energies would be allowed after a scattering process, which was contrary to any expectation and to hole theory as it was practised.

But Eq. (11.10) also allowed for different solutions. Feynman restricted $K_+(2, 1)$ to the sum over positive energies for $t_2 > t_1$ and minus the sum over negative energy contributions for $t_2 < t_1$. Expressed in formulas this meant:

$$\begin{aligned} K_+(2, 1) = &\sum_{Pos\ E_n} \Phi_n(2)\bar{\Phi}^{(}_n 1)e^{-iE_n(t_2-t_1)}\theta(t_2 - t_1) \\ &- \sum_{Neg\ E_n} \Phi_n(2)\bar{\Phi}_n(1)e^{-iE_n(t_2-t_1)}\theta(t_1 - t_2) \end{aligned} \tag{11.11}$$

where the Φ_n referred to the unperturbed solutions of the Dirac equation, respectively for positive (first summation) and negative (second summation) energies. The function $\theta(t_x - t_y)$ is equal to 1 as long as $t_x > t_y$ and zero otherwise. Thus, negative energy solutions would propagate backwards and positive energy solutions forwards in time.

---

[26] Wheeler first suggested this idea in a phone call to Feynman in 1941. This conception was independently put forward by Ernst Stückelberg in 1941. Feynman himself had already discussed this idea on a classical level in connection to the Klein paradox (Feynman 1948a, 943–944). In his paper on the positron theory and in the corresponding notes of 1947 (see the discussion in Schweber (1994, 428–434)) Feynman incorporated this conception in his kernel methods.

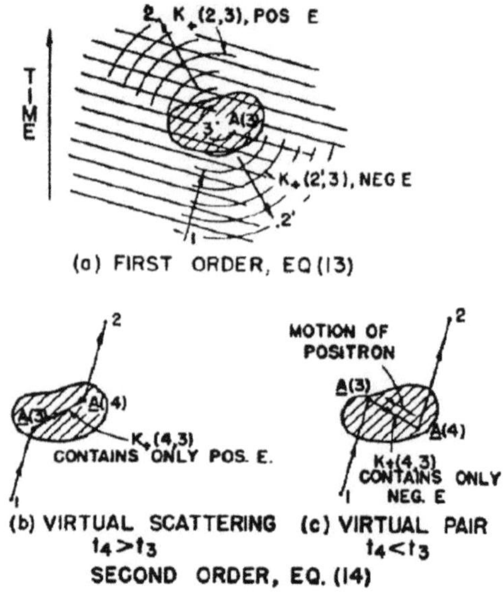

Fig. 11.1 Graphical representation of the scattering of a relativistic particle from an external potential according to Feynman. Reprinted figure with permission from Feynman, Richard, "The theory of positrons", *Physical Review* 76, 749–759, here 752, 1949, https://doi.org/10.1103/PhysRev.76.749 (cited as Feynman (1949b) in the bibliography). Copyright 2025 by the American Physical Society

Feynman most probably arrived at this equation by comparing his methods with the perturbative treatments of hole theory[27] and incorporated the priorly known conception of the positron being an electron moving backwards in time (see also the waves and the arrows pointing to $2'$ in the diagram (a) in Fig. 11.1). With this interpretation and the corresponding kernel (11.11), Feynman could provide for a simplified representation of successive scattering events equivalent to those of hole theory.

To exemplify how the simplification came about, Feynman discussed the second order equation

$$K_+^{(2)} = -\int\int K_+(2,4)A(4)K_+(4,3)A(3)K_+(3,1)d\tau_4 d\tau_3 \tag{11.12}$$

with the help of the diagrams (b) and (c) of Fig. 11.1. The diagram (b) designated with "virtual scattering" can simply be read as an electron moving freely towards the space-time point 3, where it was scattered by the potential $A(3)$, then proceeded with only positive energy amplitudes to space-time point 4, where it was scattered again ($A(4)$), and then moved towards the future to space-time point 2. Since time increased both from the bottom to the top of the diagram and along the world line of the electron, there was essentially one way of reading this diagram.

But in hole theory there was an additional process that led from the initial to the final state. Following the direction of time upwards in diagram (c) "Virtual Pair", the following possibility could be conceived: An electron-positron pair was created by the potential at 4. The initial electron annihilated the positron of this pair at point 3 under the influence of the potential. The electron of the pair went off to point 2. This

[27] See, for example, Blum (2017, 75) and Schweber (1994, 433).

possibility was also included through the definition of Feynman's kernel and was, as Feynman noted, equivalent to hole-theoretical considerations.

Following the world line of the electron, one could read this diagram in another way. The electron arriving from 1 was scattered backwards in time at point 3. It arrived at space-time point 4 where its time direction was again reversed so that it propagated towards the future to point 2. According to Feynman's definition of K_+, the contributions between t_3 and t_4 only included negative energy functions.

The second interpretation, the one with an electron moving backwards in time being a positron, was motivated by Feynman in the introduction of the paper. As he noted, although the number of particles was in general not conserved, charge was conserved. This "suggests that if we follow the charge, not the particle, the results can be simplified" (Feynman 1949b, 749). He went on:

> In the approximation of classical relativistic theory the creation of an electron pair (electron A, positron B) might be represented by the start of two world lines from the point of creation, 1. The world lines of the positron will then continue until it annihilates another electron, C, at a world point 2. Between the times t_1 and t_2 there are then three world lines, before and after only one. However, the world lines of C, B, and A together form one continuous line albeit the 'positron part' B of this continuous line is directed backwards in time. Following the charge rather than the particles corresponds to considering this continuous world line as a whole rather than breaking it up into its pieces. It is as though a bombardier flying low over a road suddenly sees three roads and it is only when two of them come together and disappear again that he realizes that he has simply passed over a long switchback in a single road.[28] (Feynman 1949b, 749)

In Feynman's space-time point of view, "one automatically takes into account the effect of virtual pair productions" (Feynman 1949b, 749). At this point, the overall space-time conception and the idea that a positron can be conceived of as an electron travelling backwards in time led to a first simplification: The combination of the second order perturbative expansion (Eq. (11.12)) with Feynman's kernel for the Dirac equation (Eq. (11.11)) included both the processes of regular scattering and of virtual pair production[29] as "the temporal order of events during the scattering, which is analyzed in such detail by the Hamiltonian differential equation, is irrelevant" (Feynman 1949b, 749).

11.2.3 Disregarding the Exclusion Principle

Feynman's formulation of positron theory introduced a further simplification in the treatment of perturbations, namely that "no account need be taken of the

---

[28] *Comment by the author*: See also Schweber (1994, 431–433) for a discussion of this bombardier and a similar metaphor involving a thread laid out as the letter N.

[29] Incidentally, this fact is the basis of the argument of topological equivalence discussed in Sect. 2.2.1.

exclusion principle in intermediate states"[30] (Feynman 1949b, 755). As discussed in Sect. 10.1.1, the exclusion principle was of prime importance for the mitigation of the self-energy divergence in hole theory but could be neglected for scattering processes.[31] Thus, it is quite understandable that Feynman's observation caused some puzzled reactions during his presentation at the Pocono conference, especially since Feynman had not derived this conclusion rigorously but only had found it as what he called an "empirical rule": by working out concrete problems.[32]

In his published account, Feynman exemplified this "empirical rule" by invoking diagram (c) of Fig. 11.1. In between the times t_4 and t_3 two electrons and a positron exist. The exclusion principle would now state that, if the incoming electron and the one created at space-time point 4 are in the same state, this process is forbidden, i.e. the respective amplitude for exactly this case has to be subtracted from the entire amplitude. In manuscript notes and the published paper, Feynman provided two arguments why the exclusion principle can and should be disregarded in the intermediate state.

The first reason directly derived from his over all space-time conception and the idea of a positron being an electron travelling backwards in time. In his notes, Feynman conjectured that according to regular formulations of quantum electrodynamics two electrons were present and therefore the exclusion principle had to be invoked. But in his formulation, it is one and the same electron so that one could ignore the exclusion principle. Thus, there was an intuitive reason why one can disregard the exclusion principle.[33]

The second reason was also discussed in Feynman's manuscript notes, but I will turn to the more comprehensible presentation in the published paper. Feynman argued that if we were to take the exclusion principle into account, another alteration would occur to the calculation which "reinstates the [above described] quantity":

> For we are computing amplitudes relative to the amplitude that a vacuum at t_1 will still be a vacuum at t_2. We are interested in the alteration in this amplitude due to the presence of an electron at 1. Now one process that can be visualized as occurring in the vacuum is the creation of a pair at 4 followed by a re-annihilation of the *same* pair at 3 (a process which we shall call a closed loop path). But if a real electron is present in a certain state 1, those pairs for which the electron was created in state 1 in the vacuum must now be excluded. We must therefore subtract from our relative amplitude the term corresponding to this process. But this just reinstates the quantity which it was argued should not have been included in [Eq. (11.12)], the necessary minus sign coming automatically from the definition of K_+. (Feynman 1949b, 755)

---

[30] For real electrons and positrons the Pauli exclusion principle could simply be taken into account by anti-symmetrizing the respective kernel. See Feynman (1949b, 755) or Wüthrich (2010, 124–125).

[31] See also Sects. 7.7.2 and 9.1.4 for Tamm's and Kockel's comments on the exclusion principle for hole-theoretical calculations.

[32] See, for example, Feynman's recollections on the Pocono conference as cited in Mehra (1994, 247) for him framing the rules for the exclusion principle as "empirical rules" and also for some of the puzzled to hostile reactions on this "empirical rule".

[33] See the manuscript notes of Feynman on the theory of positrons as cited in Wüthrich (2010, 128).

Thus, Feynman's argument went, if we were to apply the Pauli exclusion principle in intermediate states we would reinstate the contribution of the term we had neglected because of the application of the Pauli exclusion principle. This was due to the effects of vacuum fluctuations, or more precisely: the exclusion of some of the terms due to vacuum fluctuations.[34] Feynman concluded that "it is obviously simpler to disregard the exclusion principle completely in the intermediate states" (Feynman 1949b, 755).

Feynman did not give a rigorous proof of the possibility of disregarding the exclusion principle for intermediate states in his published paper. It is even possible that Feynman himself had not worked out such a proof at the time. As said, when he first presented the argument it was an "empirical rule". All the same, the conclusion was important. It considerably simplified the calculations and entailed that one could disregard (diagrammatically unconnected) vacuum fluctuations.[35]

In his follow-up paper on quantum electrodynamics Feynman only devoted a footnote to unconnected virtual pairs and the Pauli principle.[36] Freeman Dyson proved that disconnected graphs could be disregarded in his paper on the equivalence of the Schwinger-Tomonaga approach with Feynman's theory. He noted that this fact was, "from a physical point of view, somewhat trivial, since these graphs arise merely from the fact that meaningful physical processes proceed simultaneously with totally irrelevant fluctuations of fields in the vacuum" (Dyson 1949a, 496).

11.2.4 The Exchange of Virtual Photons

Turning from the interaction with external potentials to quantum electrodynamics, Feynman highlighted two points in the introduction of his follow-up article: "It appears that a considerable simplification can be achieved in writing down these matrix elements [of the perturbative expansion] for complex processes. Furthermore, each term in the expansion can be written down and understood directly from a physical point of view [...]." (Feynman 1949a, 769) The simplifications mentioned in the beginning of his paper, making his methods superior to prior calculations, were all (either directly or indirectly) connected to virtual processes:

[34] As Adrian Wüthrich, I also consider it quite hard to exactly follow Feynman's prescription at this point mathematically, but want to point the reader's attention to Wüthrich's "graphical proof" that both terms amount to the same contribution: see Fig. 5.6 in Wüthrich (2010, 127).

[35] "Real processes are completely independent of what goes on in the vacuum." Feynamn (1949b, 756).

[36] He noted: "There are loops completely without external interactions. For example, a pair is created virtually along with a photon. Next they annihilate, absorbing this photon. Such loops are disregarded on the grounds that they do not interact with anything and are thereby completely unobservable. Any indirect effects they may have via the exclusion principle have already been included." Feynman (1949a, 780, Footnote 22).

> The simplification in the formulae results mainly from the fact that previous methods unnecessarily separated into individual terms processes that were closely related physically. For example, in the exchange of a quantum between two electrons there were two terms depending on which electron emitted and which absorbed the quantum. Yet, in the virtual states considered, timing relations are not significant. Only the order of operators in the matrix must be maintained. We have seen (I), that in addition, processes in which virtual pairs are produced can be combined with others in which only positive energy electrons are involved. Further, the effects of longitudinal and transverse waves can be combined together. The separations previously made were on an unrelativistic basis (reflected in the circumstance that apparently momentum but not energy is conserved in intermediate states). When the terms are combined and simplified, the relativistic invariance of the result is self-evident. (Feynman 1949a, 769)

As in his paper on positron theory, Feynman started his presentation with a non-relativistic example, in this case the Coulomb interaction of two Schrödinger electrons. The details of Feynman's derivation of the relativistic interaction, neither the original one in Feynman's notes nor the one provided in the published paper,[37] are of particular importance for us. Nevertheless, it must be noted that the emergence of the δ_+-function, which described the relativistic generalization of the interaction and is today known as the Feynman propagator, was in either case not directly related to the concept of a photon.

In the published paper, Feynman derived it as a relativistic generalization of the Coulomb interaction, which was restricted to positive frequencies and by construction took both sequences of interactions into account. In the manuscript notes, it emerged as part of the effective action through the elimination of the transverse field oscillators within Feynman's path integral formalism. All the same, Feynman immediately interpreted the δ_+-function as describing the emission, absorption, and propagation of positive frequency photons of all four polarizations. Feynman must have been aware of such a reading through his acquaintance with regular QED, which was furthermore suggested by his interpretation of his formalism as the interrupted propagation of freely moving particles, the description of Wheeler-Feynman theory, and finally his use of diagrammatic representations.[38] This interpretation incorporated the verbal model of the 1930s into Feynman's approach to QED, altering it at the same time. In Feynman's interpretation, it was unimportant which electron emitted the photon and which absorbed it (both processes were taken into account by construction) and there was no splitting into transversal and non-transversal polarizations. They were treated on an equal footing.

[37] See Schweber (1994, 419–421) for a discussion of how the δ_+-function entered Feynman's QED through the elimination of the field degrees of freedom and of how Feynman used it in his calculation of the self-energy in late 1947. Blum (2017, 72–73) highlighted the role of the suitable choice of boundary conditions for its derivation. Darrigol (2019, 357–359) provided a simplified derivation for a scalar field. Wüthrich (2010, Sect. 5.6) reconstructed the published derivation by Feynman in detail.

[38] For the most detailed discussion of the incorporation of the concept of photon exchange into Feynman's theory, see Wüthrich (2010, Sect. 5.6.3).

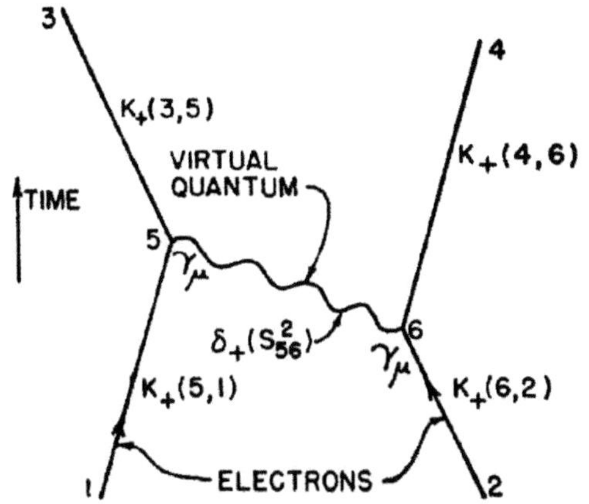

Fig. 11.2 Diagrammatic representation of the interaction of two electrons in terms of the exchange of a virtual quantum. Reprinted figure with permission from Feynman, Richard, "Space-time approach to quantum electrodynamics", *Physical Review* 76, 769–789, here 772, 1949, https://doi.org/10.1103/PhysRev.76.769 (cited as Feynman (1949a) in the bibliography). Copyright 2025 by the American Physical Society

The mathematical representation of the interaction of two charged particles which Feynman derived according to his construction and interpretation of the δ_+-function[39] was given by

$$K^{(1)}(3,4;1,2) = -ie^2 \int\int K_{+a}(3,5)K_{+b}(4,6)\gamma_{a\mu}\gamma_{b\mu}\times \\ \times\delta_+(s_{56}^2)K_{+a}(5,1)K_{+b}(6,2)d\tau_5 d\tau_6 \tag{11.13}$$

Feynman referred to Eq. (11.13) as the "fundamental equation for electrodynamics" (Feynman 1949a, 772). Based on his (highly abstracted) conception of the interrupted propagation of freely moving particles and on the incorporation of the verbal model of the 1930s, Feynman provided an immensely useful interpretation which was graphically represented in Fig. 11.2:

> We shall now interpret Eq. [11.13] in a manner which will permit us to write down the higher order terms. It can be understood (see Fig. 11.2) as saying that the amplitude for "a" to go from 1 to 3 and "b" to go from 2 to 4 is altered to first order because they can exchange a quantum. Thus, "a" can go to 5 (amplitude $K_+(5,1)$) emit a quantum (longitudinal, transverse, or scalar $\gamma_{a\mu}$) and then proceed to 3 ($K_+(3,5)$). Meantime "b" goes to 6 ($K_+(6,2)$), absorbs the quantum ($\gamma_{b\mu}$) and proceeds to 4 ($K_+(4,6)$). The quantum proceeds from 5 to 6, which it does with amplitude $\delta_+(s_{56}^2)$. We must sum over all the possible quantum polarizations μ and positions and time of emission 5, and of absorption 6. Actually, if $t_5 > t_6$ it would be better to say that "a" absorbs and "b" emits but no attention need be paid to these matters, as all such alternatives are automatically contained in [11.13].

[39] In the following equation, the factorization of the kernels for two independent particles, which Feynman already discussed in Feynman (1949b), was also taken into account.

> The correct terms of higher order in e^2 or involving larger numbers of electrons (interacting with themselves or in pairs) can be written down by the same kind of reasoning. (Feynman 1949a, 772–773)

As already discussed in Sect. 2.4, this verbal model, the corresponding diagram in Fig. 11.2, and the mathematical representation now functioned as a tool kit for the derivation of the concrete quantitative results. Translating the physical conception into diagrams allowed Feynman to directly write down the respective mathematical expression. And especially for Feynman, who had not yet systematized the use of his diagrams, the physical interpretation was highly relevant in his handling of the theory. As he both stated at the beginning and the end of the above quote, it provided him with the means to construct higher order processes.

Concerning the diagrammatic representation of the virtual quantum exchange, it must be noted that it most probably did not rely on the seemingly similar diagrams of Wentzel (see Sect. 10.2.4). As discussed, the latter were rooted in chemical reaction schemes as well as cloud chamber pictures and did not have a universal time direction. Feynman's diagrams, on the contrary, emerged from a conception that was rooted in space-time representations. As David Kaiser (2005, Chap. 5) has most explicitly discussed, Feynman diagrams resemble Minkowski diagrams. Feynman himself, together with Wheeler, provided similar diagrams for the description of the classical Wheeler-Feynman theory.[40]

From a practical point of view, there was a further difference. Wentzel's diagrams described the behaviour of the systems, but the respective arrows did not directly stand in for a mathematical term. In contrast, Feynman diagrams allowed for a direct translation into mathematical expressions. But the systematization was not undertaken by Feynman himself. It was Freeman Dyson, the work of which we shall turn to in a moment, who performed this task. Yet, there is one further, and with respect to the virtual particle very important point, I want to discuss first: the conservation of energy and momentum at the interaction points.

11.2.5 Momentum and Energy Conservation at the Interaction Points

In relation to the question what kind of theoretical entities the term "virtual" was and is connected to, either particles or processes, Feynman's mathematical representation directly implied an important alteration. Both energy and momentum were conserved at the interaction points, while the momentum of the particles in intermediate states figured as an integration variable. Feynman mentioned this fact in his introduction (see the quote opening the previous subsection) but did not explicitly discuss how it arose. Most probably because it was rather easy to see.

---

[40] See for example Wüthrich (2010, 184, Fig. 7.5) for a direct comparison. I reproduced this comparison in Fig. 12.9.

Most of the concrete calculations Feynman performed in the papers were carried out in momentum space. And in the transition from the space-time representation to the energy-momentum representation the above mentioned fact becomes apparent.[41] To see how, we need the four dimensional Fourier representation of the propagators δ_+ and K_+. As given by Feynman,[42] they are proportional to

$$\delta_+(s_{21}) \propto \int \frac{exp(-ikx_{21})}{k^2} d^4k \tag{11.14}$$

and

$$K_+(2,1) \propto \int \frac{exp(-ipx_{21})}{p-m} d^4p \tag{11.15}$$

where p refers to the momentum of the electron/positron and k to the wave vector of the photon.

First, it is to note that the space and time variables only occur in the exponentials. These exponentials can be combined for the respective interaction points, so that they depend on the sum over the momenta of the in- and outgoing particles. The integration over the variables in Eq. (11.13) stretches over all space and time. Thus, one arrives for each interaction point x at an expression which is proportional to

$$\propto \int e^{ix(p_1 \pm p_2 \pm k)} dx \tag{11.16}$$

where the p_i and k belong to the respective in- and outgoing particles.

Expression (11.16) is a representation of the Dirac δ-function. Thus, at each interaction point a δ-function depending on the sum of the four-momenta of the ingoing and outgoing particles occurred. Hence, energy and momentum was conserved at each interaction point. Now, one might wonder, how this is possible. Wasn't it common knowledge, at least since the first discussions on the Compton effect at around 1923, that energy and momentum cannot be conserved simultaneously when an electron emits or absorbs a photon?

This is true. But there is another alteration which concerns the denominators of, for example, Eq. (11.15). First of all, one should note that these denominators are now relativistically invariant expressions. They refer to the four-momenta and not, as in previous perturbative accounts, to the energy of the system. Secondly, one can carry out (some of) the momentum integrations by using the Dirac δ's.[43] If for example,

[41] While the Feynman rules, i. e. the correspondence between mathematical terms and the parts of the diagrams, are altered, the form of the diagrammatical representation essentially remains stable.

[42] The Fourier transform of $K_+(2,1)$ was presented in Feynman (1949b, 757), the one of δ_+ in Feynamn (1949a, 774). For this qualitative argument, all numerical factors as well as the infinitesimal imaginary displacement are neglected.

[43] When there is a self-energy part or electron-positron loop, there remains an unrestricted integration which gives rise to the divergences. For processes, which are finite without renormalization, as the first orders of Compton scattering and electron-electron scattering, the momentum of the virtual particle is fixed through the momenta of the in- and outgoing particles.

an electron of (momentum p) emits or absorbs a photon (k) in a subprocesses, the denominator is changed to $\frac{1}{p \pm k - m}$. Thus, the relativistic energy momentum relation of the electron $p^2 = m^2$ is no longer fulfilled in the intermediate state. The particle is off shell.

Thus, the relativistically invariant formulation, based on the four-momenta of the particles and the four-dimensional Fourier transformation, led to an important alteration concerning the characteristics of the processes and particles. In contrast to the prior calculational schemes, energy and momentum were conserved at the interaction points in Feynman's calculations, but the particles in intermediate states did no longer fulfil the relativistic energy-momentum relation. It were the particles, not the (sub-)processes, which broke with known and accepted properties of the corresponding real entities. Hence, from this point of view, the particles were virtual and not the processes.

11.2.6 A Short Summary

With Feynman's new outlook on QED the concept of a "virtual particle" became firmly established. But Feynman's version of QED also considerably changed the characteristics of these theoretical entities and the description of the processes they were involved in. Feynman's formulation "automatically"[44] took regular scattering and scattering with virtual pair production into account, in one and the same mathematical term. During the 1930s, these two types of processes had to be differentiated. The Pauli exclusion principle could be ignored entirely for the intermediate states. The most detailed evaluation of the Pauli exclusion principle during the 1930s by Bernhard Kockel came to a different conclusion (see Sect. 9.1.4). For some effects, such as the self-energy, the exclusion principle had to be taken into account. This last fact also entailed that the vacuum fluctuations, which were so important for the mitigation of the self-energy of the electron in the calculations of the 1930s, could be disregarded in Feynman's version of QED.

Through the symmetric treatment of the (non-)transversal parts of the electromagnetic field, the expression referring to virtual photons contained all four polarizations. During the 1930s the differentiation between Coulomb interaction and radiation field was customary and contributed, for example, to Dancoff's famous mistake in the calculation of the radiative corrections to the relativistic scattering process.[45] The symmetric treatment in Feynman's version also entailed that the idea of a virtual particle as a force carrier was now fully applicable in QED.[46] The virtual particles had mathematical representations: the δ_+- and K_+-functions. They also had their

[44] This is the term Feynman actually used himself. I quote in full: "For example, when considering the scattering of an electron by a potential one automatically takes into account the effects of virtual pair production." Feynman (1949b, 749).

[45] It was shortly mentioned in Sect. 10.1.3. For a full discussion, see Blum (2015).

[46] See Carson (1996b), but also my discussion in Sects. 8.2.1 and 10.2.1.

own, distinct diagrammatic representation as internal lines in Feynman diagrams. Importantly, the theoretical entities which broke with known physical rules changed. During the 1930s, energy was not conserved at the interaction points but the particles in intermediate states behaved regularly. In Feynman's version, the particles in intermediate states broke with the relativistic energy-momentum relation and energy-momentum conservation was imposed at the interaction points. The particles were virtual, not the subprocesses.

The verbal model Dirac had introduced was used by Feynman, but was altered as well. Besides the above mentioned observations (which are partly based on the following point), timing relations for intermediate states could be ignored. It was regardless whether a virtual electron propagated forwards or backwards in time and which of the two interacting electrons emitted and which absorbed the photon. It was the causal connection which the photon established that was important. All the same, the verbal model had, also in this modified form, an important function: It formed the basis of Feynman's physical interpretation which was enriched with a corresponding diagrammatical representation.[47] Each diagram, constructed by Feynman through physical reasoning, could be directly translated into a mathematical equation which could then be evaluated.

Thus, Feynman certainly did not introduce the concept of a virtual particle from scratch. Rather, his version of QED led to diverse alterations in the practice with and the characteristics of such theoretical entities, which, as often stressed, made the theory easier to handle but also shifted focus on the virtual aspect of the particles themselves. But Feynman himself worked with rather idiosyncratic methods and he used his diagrams relying on physical reasoning. It took another physicists to bring the new means of representation, and with it the virtual particle, to the masses. Freeman Dyson, the work of whom we shall turn to in the following, established the contact of Feynman's work with more regular quantum field theory and systematized the application of Feynman diagrams.

11.3 Dyson's Intervention (1949): Equivalence, Systematization, and Furry's Theorem

In late 1948, when Dyson finished his proof of the equivalence of the more regular quantum field theoretical approaches to QED of Schwinger and Tomonaga with Feynman's theory, he was in a perfect position to do so.[48] He had studied mathematics in Cambridge, where he received his Bachelor degree in 1945. After a short occupation as mathematician at the Imperial College, he turned his attention to theoretical

[47] See my discussion in Sect. 2.4 for the interrelations between the verbal, mathematical, and diagrammatic representations in Feynman's original paper.

[48] For Dyson's background and his work on quantum electrodynamics, best see Schweber (1994, Chap. 9). The following short outlook on Dyson's work is particularly informed by Schweber's contribution and the discussion in Wüthrich (2010, Chap. 6).

physics in 1946 and moved to Cambridge. Nicholas Kemmer, student of Gregor Wentzel and Wolfgang Pauli, who had made important contributions to the quantum field theoretical formulation of nuclear physics in the mid to late 1930s, was lecturing at Cambridge at the time. Dyson considered Kemmer his first real physics teacher. He learned quantum field theory from Kemmer and Wentzel's textbook *Einführung in die Quantentheorie der Wellenfelder*. While in Cambridge, Dyson had already become aware of Heisenberg's S-matrix but initially thought of it as an attempt of altering the basis of physics, not a reformulation of it.

In September 1947, Dyson joined Hans Bethe and Richard Feynman in Cornell. There he worked on a relativistic calculation of the Lamb shift for a scalar electron while applying the methods of the 1930s. In Cornell, Dyson became acquainted with Feynman's methods through personal contact (Feynman had not yet published anything on his formulation of QED). When Dyson left for Princeton in September 1948, he had realized that the incorporation of Feynman's methods into the new formulations of quantum field theory by Tomonaga and Schwinger was of prime importance. And at this point, it became apparent to Dyson that Feynman's formalism was essentially a way of calculating elements of the S-matrix in quantum field theory which finally allowed him to establish the connection between the Schwinger-Tomonaga and the Feynman formalism.

11.3.1 Equivalence and Systematization

In his first paper on the equivalence of the different radiation theories, Dyson derived a perturbative representation of the S-matrix in the form of

$$H_F(x_0) = \sum_{n=0}^{\infty} \left(\frac{-i}{\hbar c}\right)^n \int_{-\infty}^{\infty} dx_1 \ldots \int_{-\infty}^{\infty} dx_n P(H^e(x_0), H^I(x_1), \ldots, H^I(x_n)) \quad (11.17)$$

where the function P refers to the time-ordered product of the operators in brackets. H^e is an external perturbation and H^I the interaction between the positron-electron field and the electromagnetic field. The above statement is not entirely correct, as only when $H^e = 1$, $H_F = S$, i.e. the S-matrix of Heisenberg.

Dyson derived the above expression from a perturbative expansion of quantum field theory in the interaction picture, this is from Tomonaga's and Schwinger's formulation of it. He consequently had to show that the above expansion was equivalent to the procedures and functions Feynman used. This would establish the connection. Plugging in the concrete form of the interaction Hamiltonian in terms of field operators, Dyson showed that the integrand of Eq. (11.17) could be expressed as the product of the vacuum expectation values of time-ordered products of each two field operators of the electromagnetic field or the electron-positron field:

$$\langle P(A_\mu(x), A_\nu(y))\rangle_0 ; \quad \langle P(\bar{\Psi}_\alpha(x), \Psi_\beta(y))\rangle_0 \quad (11.18)$$

These in turn were just Feynman's kernels δ_+ and K_+, which Dyson renamed as D_F and S_F.

This finally meant that any matrix element occurring in this formulation of the theory might be written as

$$M = \epsilon' \prod_{i \neq k} (\frac{1}{2} S_F(x_i - x_{r_i})) \prod_j (\frac{1}{2} \hbar c D_F(x_{s_j} - x_{t_j})) \bar{\Psi}(x_k) \Psi(x_{r_k}) \tag{11.19}$$

where ϵ' depended on the permutations one had to perform to order the creation and annihilation operators of the electron-positron field to arrive at the given expression only containing vacuum expectation values. In his first paper Dyson exclusively considered initial and final states with one electron present. The last two terms, $\bar{\Psi}$ and Ψ, respectively created the electron in the final state and annihilated the electron in the initial state. Substituting all the possible $M's$ to a respective order into Eq. (11.17), Dyson found a way to express the perturbative evaluation of a quantum field theoretical approach to QED in terms of the functions D_F and S_F.

Although the problem was thus essentially solved, Dyson went on and introduced the diagrams as "a much simpler and more practical procedure" (Dyson 1949a, 495). As David Kaiser (2005, Chap. 5) has stressed, Dyson's introduction and explanation of the diagrams resembled an instruction in terms of a "paint by numbers"-pattern. In contrast to Feynman, Dyson did not interpret the diagrams as a picture of a physical process[49] but simply as topological structures, a representational and rule governed means for writing down matrix elements. This difference in conception is what David Kaiser has called the "Feynman-Dyson-split".

As Dyson himself noted:

> In Feynman's theory the graph corresponding to a particular matrix element is regarded, not merely as an aid to calculation, but as a picture of the physical process which gives rise to that matrix element. For example, an electron line joining x_1 and x_2 represents the possible creation of an electron at x_1 and its annihilation at x_2, together with the possible creation of a positron at x_2 and its annihilation at x_1. This interpretation of a graph is obviously consistent with the methods, and in Feynman's hands has been used as the basis for the derivation of most of the results of the present paper. For reasons of space, these ideas of Feynman will not be discussed in further detail here. (Dyson, 1949a, 496)

In the new field theoretical context and after Dyson's systematization, the diagrams did no longer depend on the imagination of a physical process taking place in space and time. The topological features of this representational system were pushed to the fore and allowed for the derivation of the non-zero matrix elements contributing to the perturbative expansion. In this context of systematization, Dyson (1949a, 500) was the first to state that at each vertex two electron and one photon line had to

[49] Again, for my approach it is irrelevant if Feynman considered his physical interpretation ontologically, i.e. his diagrams as (a highly abstracted) representation of physical processes actually taking place in space and time. But it is important that he used a physical language and how its usage governed his reasoning processes.

meet.[50] Dyson (1949a, 496) further provided a rigorous argument for the omission of the unconnected graphs within the quantum field theoretical framework.

In his follow-up paper (Dyson 1949b), Dyson introduced Feynman rules for processes involving more than one electron in initial and final state, external photons and the momentum space representation. As he noted, each vertex stood in for a δ-function which depended on the "algebraic sum of the k^i and p^i 4 vectors corresponding to those lines of [the graph] which meet at x_j. [This factor] in the integrand of M [the matrix element] expresses the conservation of energy and momentum in the interaction occurring at the point x_j." (Dyson 1949b, 1740) Integration had to be performed over the four-momenta of the involved particles.

Dyson used the graphs[51] extensively to discuss the divergences and their cancellation through renormalization. As Wüthrich (2010, Chap. 6) argued, Feynman diagrams constituted a particularly suitable means of representation for exactly this purpose. All the same, there was a little short-coming of the diagrams. It is often stated that Feynman diagrams allow to derive all the non-zero matrix elements of the theory. This might be true. But still, since one effect can occur through different sequences involving virtual particles, this does not mean that every effect one can imagine with Feynman diagrams is non-zero: Furry's theorem, for example, is not taken into account by the diagrammatic system itself.

11.3.2 The Incorporation of Furry's Theorem

The first troubles with Furry's theorem occurred in October 1948, when Dyson was visiting Feynman in Cornell.[52] Feynman and Dyson discussed exactly the problem context in which Furry's theorem was initially introduced: light-by-light scattering and the Delbrück effect (see Sect. 9.2.4). Dyson, who had just finished his paper on the equivalence of Schwinger, Tomonaga, and Feynman's theories, was still not sure whether or not these effects remained finite in the new formulation. According to Dyson (as he reported in a letter to his parents), Feynman sat down and arrived at sensible and finite results within two hours.

But, as Feynman communicated to Dyson only a week later, he had discovered that the matrix elements are equal to zero. Feynman had calculated the effects at an odd order in perturbation theory, most probably the third. In the letter to Dyson, Feynman already concluded that any loop including an odd number of quanta is equal to zero. For any loop, there are two directions in which the electron might go around it. Changing the direction resulted in a reversal of the sign of the charge. This

[50] In Feynman's 1949 lecture notes there is a famously "wrong" Feynman diagram for light-by-light scattering. It includes only one electron-positron bubble, so that two photon lines meet at the interaction points. It is a striking proof of Feynman's intuitive usage of his physical reasoning and the diagrams. See for example Kaiser (2005, 187–188).

[51] More precisely, he mostly used references to the graphs. The articles of Dyson introducing the Feynman graphs and rules included surprisingly few diagrams.

[52] The following is based on Schweber (1994, 449–451).

meant that all interactions also changed signs and thus combining the two directions resulted in a cancellation.[53] Feynman incorporated this observation only in a footnote in his article on the *Theory of Positrons* (Feynman 1949b, 756, Footnote 9).

Now, it is important to note that Feynman's inclusion of graphs only involving virtual electron-positron pairs and having an odd number of interaction points was not due to Feynman's intuitive usage of the diagrams. Such diagrams could still be drawn after Dyson's systematization. Dyson (1949b, 1743) himself introduced Furry's theorem as an "additional reduction in the number of graphs". Still, while it took Furry a whole paper to prove that such terms were equal to zero, Feynman reduced the argument to a footnote and it took Dyson one paragraph (Dyson 1949b, 1743–1744) to prove Furry's theorem in the new formulation of QED. As he noted: "The theorem was shown by Feynman to be an elegant consequence of his theory." (Dyson 1949b, 1743)

The simplification, not only of calculations but also in the investigation of the structure of the theory, was also noted in textbook presentations. After introducing their readers to the S-matrix, Feynman diagrams, and Dyson's systematization, also in momentum space, Jauch and Rohrlich (1959, 159; originally published 1955) noted:

> The graphical representation of the covariant perturbation calculation has proved not only an aid in computation but also a powerful tool in the investigation of general problems. An example will be found in the next section, where Furry's theorem is proved. This theorem is much more difficult to prove in the older formalism.

Within three pages (Jauch and Rohrlich 1959, 159–161) they provided a proof of Furry's theorem on a textbook level.

11.4 Conclusion: The *Virtual Particle* Takes Centre Stage

With Feynman's space-time approach and Dyson's systematization of Feynman's theory the practice of QED was appreciably altered. Feynman's approach and his diagrams did not only bring quantum electrodynamical calculations "to the masses", as Julian Schwinger once noted. They also came with a completely new conceptualization of scattering processes as an interruption of the space-time propagation of freely moving particles.[54] Furthermore, the concept of a stationary state was pushed to the fringes of the theoretical description. Since the integrations involved in the perturbative calculations stretched over all space and time, only asymptotically such a concept as the stationary state prevailed. In Feynman's formulation, QED was a theory of scattering.[55] And even though Feynman's approach initially did not rest on the quantization of fields and also after Dyson's intervention the distinctive change

[53] See Feynman to Dyson, October 29. 1948, as cited in Schweber (1994, 450).

[54] As said, the most detailed discussion of this shift of conception was given by Adrian Wüthrich (2010).

[55] As most explicitly discussed by Alexander Blum (2017).

in the conception of scattering processes prevailed, Feynman had incorporated the verbal model of the 1930s, i.e. the emission and absorption of photons describing the interaction, into his theory.

Particularly for Feynman, the incorporation of the verbal model was important since he used his diagrams more or less intuitively. The physical interpretation, regardless of any ontological connotation, allowed him to write down a respective diagram and to translate it into a mathematical format. As noted, according to Feynman, the interpretation of the formula, also represented in a diagrammatic format, allowed for the construction of the higher order terms. When Dyson systematized the usage of Feynman diagrams, giving rules of how and what to draw, the physical interpretation performing the function of a verbal model, faded into the background and was only shortly mentioned. It was simply no longer necessary.

Feynman's reformulation of theory had several and direct impacts on the concepts under study. His method automatically incorporated scattering via virtual pair creation; it did not temporally resolve which electron emitted and which absorbed a photon, nevertheless showed their connection; the Pauli exclusion principle could be entirely neglected in intermediate states; and the effects of unconnected vacuum fluctuations could simply be ignored.[56] The description of virtual photons contained all four directions of polarization, thus making the concept of photon exchange even more suggestive. All the same, the reformulation of perturbative QED and the new means of representation could not take all aspects of theory into account: Furry's theorem had to be incorporated to reduce the number of diagrams. But its proof was appreciably simplified.

From a terminological point of view, it is particularly striking that Feynman spoke of *virtual particles*, in contrast to virtual processes, far more frequently than his contemporaries. As said and indicated throughout this chapter, with Feynman's work (and eventually Dyson's systematization) the transformation from virtual processes to virtual particles, which started to show in the mid to late 1930s, came to a first finalization. Let me draw together the reasons for this conclusion, which were (implicitly) made in my historical discussion.

First of all and even though this holds for the entire field, I consider it important to note that with the development of renormalization physicists focused on processes which included particles exclusively present in the intermediate states. As argued priorly, this was one of the main reason why physicists started to talk about "virtual particles" in the later half of the 1930s. Secondly, through the manifest covariance of the techniques and in contrast to prior methods, energy and momentum were conserved at each interaction point. The four-momenta of the particles in intermediate states were not equivalent to the ones of free particles; they were off shell. Thus, from this point of view, the subprocesses were no longer virtual (i.e. energy non-conserving), but the particles were (as they laid off the mass-shell).

Thirdly, and this became most apparent in Dyson's systematization in a quantum field theoretical context, virtual particles acquired two additional forms of representation which were (at least so at first and in QED) exclusively reserved for them:

[56] The last two points are, as discussed, interrelated.

(1) the propagators as a mathematical one, and (2) the internal lines in Feynman diagrams as a diagrammatical one. Dyson exclusively used propagators for virtual particles while single field operators represented the real particles (external lines).[57] This differentiation between internal and external lines, virtual and real particles, in mathematical terms was also immediately taken up in textbooks and can, for example, be found in Jauch and Rohrlich (1959, originally published 1955).[58]

Taking these points together I argue, that with Feynman's approach to QED the virtual particle concept became firmly established. But within Feynman's approach, and let me conclude by shortly reflecting on this observation, the virtual particle did not only become firmly established, it took centre stage. Looking at Fig. 11.2, the prominence of the virtual particle is manifest. It was the central piece of what Feynman himself called the "fundamental interaction" of QED. And the centrality of the virtual particle was even more broadly reflected in Feynman's papers. He stressed that the application of his theory was far more suitable if the quanta were virtual.[59] In effect, Feynman did not construct processes involving virtual quanta from real ones but went the other way around. As he mentioned in a footnote in his *Space-Time Approach to Quantum Electrodynamics*:

> Although in the expression stemming from [Eq. (11.13)] the quanta are virtual, this is not actually a theoretical limitation. One way to deduce the correct rules for real quanta from [Eq. (11.13)] is to note that in a closed system all quanta can be considered virtual (i.e. they have a source and are eventually absorbed) so that in such a system the present description is complete and equivalent to the conventional one. (Feynman 1949a, 773)

When Feynman himself presented not only the results of his theory, but its derivation in Feynman (1950), he also wrote a section on the incorporation of real quanta into his methods. All the same, he provided an appendix (Feynman 1950, 455–456: B. "The relation between real and virtual processes") where he directly derived expressions for processes involving real quanta from those describing virtual ones. The virtual particle had become *primary*, at least for Feynman.

When the stationary states where pushed to the theoretical fringes, their place was taken by the virtual particles. Making QED a theory of scattering, in this respect, also meant to make QED a theory of virtual particles. As already noted, Alexander Blum (2017) has argued, that the exemplar (Kuhnian terminology) of QED changed in the late 1940s: a paradigmatic shift of the practice of theory, and not of its fundamentals, occurred. As argued specifically in Chap. 2, but essentially throughout this whole

[57] Compare Dyson (1949a, 492–495) for external electrons/positrons and Dyson (1949b, 1738) for both external electrons/positrons and photons.

[58] See Jauch and Rohrlich (1959, 146, 154) for the Feynman rules and the corresponding differentiation, also in mathematical terms, of virtual and real particles in position and momentum space.

[59] As for example in Feynman (1949a, 770): "Roughly, the field point of view is most practical for problems involving real quanta, while the interaction point of view [i.e. Feynman's] is best for the discussion of the virtual quanta involved." Similarly, Dyson (1949b, 1736) observed, that Schwinger's method should be applied when bound state problems are treated while Feynman's methods are handy for scattering phenomena.

book, the concepts connected to the *virtual particle*'s historical development should be conceived of as *concepts arising in the practice of theory*. Thus, the radical shift both in the concept's characteristics and its status, can be understood from a historio-philosophical point of view: *Paradigmatic change* in *theoretical practice* caused *fundamental change* for one of the most prominent *concepts of practice*; it altered both the characteristics and the status of the virtual particle concept.

Part IV
Analysis, Summary, and Conclusion

Chapter 12
Representations and Practices in the Formation of the Virtual Particle Concept

The question guiding my research can be stated quite simply: When, how, and why did the concept of the virtual particle emerge? But before being able to engage with this question, another one had to be answered: How can we study concepts and their development historically? Methodological reflections are certainly a necessary prerequisite for any historical investigation. But the problem is rather pressing when it comes to the analysis of concept formation and development. Even the more basic question "*what* do we study when we study concepts historically?" unfolds diverse nuances since the object of investigation, the concept(s) of the historical actors, cannot be conceived of as disembodied, and therefore inaccessible, elements of thought. For the virtual particle the problem might seem even worse: One of its defining features is its inaccessibility by empirical means.

To encounter this problem, I proposed to conceive of the virtual particle as a concept of practice. I took a pragmatically informed perspective on concepts as context dependent tools. They should be studied through their specific representations and the reasoning they allow for.[1] These representations and the actors who introduce, alter, and use them should in turn always be contextualized in the specific historically situated epistemic constellation.[2] Concepts do not come on their own but are always embedded in a larger conceptual framework. Therein, they serve a specific purpose in explaining, interpreting, or working with the theory. The same holds for their representations. Most often, physicists do not represent a single concept but a model of a physical effect. At the same time, physicists use diverse means of representing one and the same concept in their aim to answer the research questions.

In this concluding chapter, I will first turn to the historical development of specific representational formats, namely the terminological variants (Sect. 12.1.1) as well as the schematic, tabularic, and diagrammatic representations (Sect. 12.1.2). These representational formats deserve special attention: the studied concept carries the

[1] See Sect. 2.3.

[2] For a broad look on the development of the community of practitioners, see Chap. 3.

M. Ehberger, *Representing the Unobservable*, Science Networks. Historical Studies 68,
https://doi.org/10.1007/978-3-032-09188-8_12

(to me at least) enigmatic epithet "virtual" and its most prominent representation are Feynman diagrams. In Sect. 12.2, I will discuss occurrences of what I called "false predictions".[3] Thereby, I will also contribute to the debate about the nature of QED's transformation in the late 1940s: I will show that not only the way *how* theoretical processes were perceived and investigated changed.[4] *What* was conceivable had changed:[5] Physicists during the 1930s and physicists using Feynman diagrams could imagine different *actual* processes.

The second half of this chapter constitutes a concise summary of the historical development. Together with the first part it will form the basis for my answer to the guiding question of this historical investigation: When, how, and why did the virtual particle emerge?

12.1 The Evolution of Specific Representations

In my historical reconstruction I argued that the verbal representation of the model of QED was specifically important for the practice of theory. Thus, one might wonder: Why discuss the terminology and the diagrams specifically? I want to provide two reasons for my focus in this section.

First of all, the verbal model of QED, namely the storyline of successive (energy non-conserving) emission and absorption subprocesses, as introduced by Dirac (1927b) (see Chap. 5), remained relatively stable throughout the 1930s and, in a modified form, is still underlying the representation of Feynman diagrams. The same holds for the narrative, i.e. the way of arguing for the inclusion of the virtual transitions and particles in the conceptual framework: As late as 1970, Mario Bunge (1970) identified the energy-time uncertainty combined with the short lifetime of virtual particles as the premise for a realistic reading of the concept.[6]

Secondly, a concise discussion of the terminological development is certainly expedient. And the following will go beyond an evaluation of the term "virtual". I will take into account all the different variants of terminology I identified in the course of my historical investigation. This reconstruction will make their historical

[3] See Sect. 2.3.3 for the introduction of this analytical term.

[4] The alteration of the theoretical framework of QED in the late 1940s has most coherently been discussed by Alexander Blum (2017). He argued that a genuine paradigm shift occurred, but not to the basis of the theory but in terms of its exemplar. Thus, Blum's observation is already deeply connected to the practice of theoretical physicists.

[5] Wüthrich (2010) has not only observed the shift from the theoretical modelling in terms of transitions between stationary states to a conception which takes the interrupted free movement of particles as its basis. He also indicated that the representational formats of the 1930s (and their further development) was not as well suited as Feynman diagrams to single out the divergences of QED. My point has a different focus: I will argue that Feynman diagrams restrict the thinking possibilities of physicists and thereby avoid some of the false predictions which occurred during the 1930s.

[6] See also Sect. 2.2.1.

(dis-)connections apparent. It will also show how the focus on terminology can inform a historical investigations, how names are more than mere referential signs.

The same holds for the diagrammatic/tabularic/schematic representations. They might have served a purely didactic purpose or were only used as book-keeping devices. Yet, the historical development from an atemporal understanding over a temporally ordered sequence of state transitions to a conception of the interrupted movement of freely moving particles is mirrored in these representations. They also show how the concept of the virtual particle slowly emerged.

12.1.1 The Plethora of Terminological Variants

According to my reconstruction in Sect. 4.1.3, the concept of "virtuality" in the BKS theory, named after their inventors, Niels Bohr, Hendrik Anthony Kramers and John Clarke Slater, had two interrelated connotations. (1) The terminology indicated that a "virtual", unobservable world of virtual oscillators and virtual radiation field could be used to describe the behaviour of the real and observable quantum atoms and transitions. (2) The interaction between virtual radiation field and real atoms was only statistical. According to my argument, only through this non-classical interaction the virtual radiation field, although carrying energy in a formal way, became unobservable and thus inhabited the same plane of reality (or physical modelling) as the virtual oscillators.

Energy non-conservation was not a connotation of the term virtual for BKS. It was real and observable. Furthermore, BKS did not connect the term virtual to quantum concepts, as quantum states and the transitions between them. Only classical concepts carried the epithet. These concepts emerged, according to the reconstruction of Blum and Jähnert (2022), by merging the *Ersatzstrahler* (Kramers and others) and the *Ersatzoszillatoren* (first explicitly referred to in print by Ladenburg and Reiche (1923)). And this "Ersatz-", i.e. substitute character did survive the demise of the BKS theory and was partly underlying the reference to virtual oscillators after the advent of matrix and wave mechanics.

The terminology used by the historical actors after the demise of BKS actually exemplifies the conceptual break between virtual oscillators and (later to be called) virtual transitions (Sect. 5.2). Dirac did not use the term when introducing the verbal model of QED (see Chap. 5). Up until the mid 1930s, the term "virtual" was nearly absent in theoretical papers in QED. Maria Göppert-Mayer (Sect. 8.1.1), for example, applied both Dirac's verbal model of successive transitions without invoking the term "virtual". But in the same paper, she used the concept as well as the term "virtual oscillator" (Göppert 1929 and also Göppert-Mayer 1931). Dirac's intermediate states and the corresponding verbal model were a means to find out which matrix elements had to be calculated; but Göppert-Mayer referred to the "virtual oscillator" only when she calculated the radiation expected to be emitted by a corresponding classical dipole (i.e. the classical substitute for a quantum system). All the same, she located Dirac's verbal model on the same plane of reality (or level of physical modelling) as the virtual oscillators: She connected both to the terminology of the "as if".

That Göppert-Mayer highlighted this ontological demarcation line of the "as if", was certainly not by chance. Also the term "intermediate state" had a prehistory. Intermediate states, as classical, only partially or weakly quantized orbits, played an important role in the theoretical description of dispersion in a light-quantum framework prior to Heisenberg's *Umdeutung* (see Sect. 4.3). The transitions to and from the intermediate states were actual, described by Einstein coefficients, and short-lived. They were on the same level of reality as the transitions between regular quantized states. Through Yakov Frenkel, this conception had a short revival in the late 1920s. But Frenkel altered the meaning of the term again. The transitions were now energy non-conserving (in contrast to light-quantum explanations before the advent of quantum mechanics) but actual (in contrast to Dirac's description).

The terminology used before the term "virtual" became more regular for the transitions to and from intermediate states can inform us on the ontological status that the actors assigned to the respective concepts. Göppert-Mayer's "as if" is a case in point. "The picture of 'successive' transitions", which Bernhard Kockel (1937) used, clearly indicated the fictional character of the involved concepts. Evan James Williams (1935) would baptise the obvious theoretical fiction of the photons in the Weizsäcker-Williams method "virtual photons". While there is little to no conceptual and methodological overlap with the concepts we refer to today as virtual, Williams would use the term "virtual pairs" for the perturbative evaluation of vacuum polarization in the same paper. For Williams, the virtual entities of perturbation theory were seemingly theoretical fictions as well.[7]

Quite on the contrary, Igor Tamm used the term "elementary process", normally reserved for actual transitions, to speak about processes involving intermediate states. As discussed in Sect. 7.5, Tamm and Dirac had developed quite a literal reading of these transitions, while being well aware that they played their essential role on the probability-/quantum-amplitude level of the theory.

Terminology is even indicative of the role the concepts played in the application of perturbation theory. Walter Heitler referred to the "intermediärer Zwischenzustand" [intermediary intermediate state] (Heitler 1933b, 152). At first sight this might seem like a duplication, yet it indicates the function of the intermediate state: "Intermediary" goes conceptually well beyond the term "intermediate": It includes a mediating function. According to Heitler, processes were carried out "through the agency of the *intermediate states*" (Heitler 1936, 131, emphasis in the original). Bethe and Fermi

[7] This point might further be interesting as Williams was in close contact with the Copenhagen physicists at the time. The specific terminology of "virtual radiation of the nucleus" showed up first in correspondence with Bohr; Williams to Bohr, July 25. 1934; BSC, MF 26, Section 4. Although an extrapolation is suggestive, especially considering the lack of objection by Bohr, it would need further historical research for definite conclusions.

similarly called the intermediate state an "interlink".[8] The intermediate state did not simply lie in between initial and final state. It connected the two states and therefore allowed for the specific processes to occur.

The connection between these intermediate states, or rather the energy non-conserving transitions to and from them, and the term "virtual" was established in two ways. When Chandrasekhara Venkata Raman used the term "virtual" to refer to such transitions (see Sect. 6.2.2) he wanted to underline their purely theoretical nature and the breach with energy conservation. But energy non-conservation was none of the connotations in BKS. Rather, I argued, Raman blended the two prominent aspects of this framework: energy non-conservation and the terminology of the "virtual".

Raman's talk led to the spread of the term only in a small circle of physicists. The breakthrough of the connection between Dirac's verbal model and the term "virtual" came with Werner Heisenberg's "virtual possibilities" and the widely received work of Hans Euler on light-by-light scattering (see Sect. 9.1.2). As I argued, Heisenberg's invocation of the term can more closely be connected to the BKS paper, namely by reference to an unobservable/unobserved virtual world. The measurement process always played an important role in Heisenberg's thinking (at least so since his uncertainty paper) and, according to Heisenberg, implied a strict division of the physical world: an unobserved one, governed by quantum amplitudes, and the observed one, which can be described classically. This division is generated through the measurement and the boundary is known as the "Heisenberg cut". That Heisenberg used the terminological combination of "virtual possibilities", where the term "possible" also traces back to the (pre-)BKS descriptions of dispersion, is another indication of the connection. All the same, the "possibilities" were soon dropped by Heisenberg himself, by his students and the physicists applying the term in connection to Dirac's verbal model.

In this development, the meaning of the term "virtual" changed. It was not necessarily an unobservable, potentially hypothetical virtual world that was referred to and discussed by the historical actors. Rather, the physical aspects of non-observation and the energy non-conservation were used to determine the term "virtual".[9]

All the same, the term had not been used exclusively to refer to concepts we can put in one line with the virtual particle. Williams inclusion of "virtual photons" is an instance amongst many, where historical actors named concepts "virtual" which have no direct connection to the virtual particle concept today. In X-ray spectroscopy, optically active states in the fine-structure of X-ray absorption edges were referred to as "virtual states".[10] Guido Beck (1930) used the same term to refer to accumulation points in the continuous spectrum of a potential-well model.[11] The "virtual electron" as a shorthand for negative energy electrons was as well introduced by Guido Beck[12]

---

[8] "Zwischenglied," Bethe and Fermi (1932, 301).

[9] See also the discussion by Martinez (2024, especially pp. 363–364).

[10] See, for example, Stoner (1926, 103).

[11] For a discussion of the development of this concept, see Martinez (2024, Sect. 2).

[12] "virtuelle Elektronen," Beck (1933, e.g. p. 499).

and used, for example, by Bhabha (1936, 196). Gian-Carlo Wick referred to the electrons inside neutrons of Heisenberg's 1932 nuclear model as "virtual".[13]

Vice versa, even after Heisenberg had introduced the term "virtual" in connection with transitions to intermediate states other terminological variants were used. Some actors simply did not use the term virtual and referred to intermediate states and transition to and from them. But also phrases such as "nascent pairs" or "in status nascendi" were introduced by Furry and Oppenheimer (1934) and used as a synonym for "virtual" by Dirac or Gamow (1937). They point to another layer of meaning of the concept of virtual particles which we find to this day (see Sect. 10.1.2, especially 10.1.2). Weisskopf called the pairs of vacuum fluctuations "latent pairs which, by means of light quanta, could be brought into real existence" (Weisskopf 1949, 311). The pairs and particles were ready to be "born",[14] i.e. really created, but were not there yet. But if we were to apply the amount of energy necessary, we could actually observe these particles. This kind of reasoning was also the basis of the derivation of the range of nuclear forces from the Heisenberg uncertainties by Gian-Carlo Wick (1938) (see Sect. 10.2.3).

Besides the mixture of adjectives and adverbs used as synonyms for virtual during the 1930s, there was also an overlap of the theoretical entities carrying the epithet, either being processes or particles. As discussed in Chap. 10, in the second half of the 1930s the term virtual was at times also directly attached to the particles. Wendell Furry, for example, used the term "virtual pair". At the same time, he pointed out that "though they [the virtual pairs] are not actually created so as to be experimentally observable, are able to play the part of catalytic agents producing effects which could otherwise not appear" (Furry 1937, 125). The shift of the referent of the term virtual was connected to the observation that these pairs formed—to use Heitler's term from above—the intermediaries. For Heitler the intermediate states had "agency", for Furry the particles were the "catalytic agents". This conception and corresponding terminological combinations emerged when more and more effects, static or dynamic in nature, were describe by additional, unobservable pairs and particles.

Finally, in the late 1940s the term virtual was attached more frequently to the concept of a particle, most prominently by Richard Feynman. Feynman's source for the term is not easy to reconstruct. He seems to have assumed his readers to be familiar with the term: He did not waste a line to properly introduce it. Feynman's methods and his use of the term surely fostered the attachment of the term virtual to a particle. While far-reaching alterations to the concept of the virtual particle occurred in Feynman's work, but also in the more usual approaches to quantum field theory in the late 1940s (see Chap. 11), Feynman himself seemed to have considered the "virtual quantum" a rather uncontroversial term.

[13] "in den Neutronen virtuell vorhandenen Elektronen," Wick to Heisenberg, April 16. 1934; AMPG, III. Abt., Rep. 93, Nachlaß Heisenberg, Nr. 1533.

[14] Note that pair production was also described anthropomorphically by the term "birth". See, for example, the title of Heitler and Sauter (1933).

Definitions of the term "virtual" in a more general philosophical sense, abstracting from the concrete physical attributes of the virtual entities, have been given by Blum and Jähnert (2024) (for the BKS theory) and Martinez (2024) (aiming to capture both the BKS terminology but also the varied uses throughout the 1930s). In both papers, (a modification of) Charles Sanders Peirce's definition of the virtual as "something, not an X, which has the efficiency (virtus) of an X" (Peirce 1902, 763) was invoked. I have hinted at my own take on this question in Ehberger (2024). The concepts we can reconstruct as lying in one historical line with the virtual particle and which sooner or later carried the epithet "virtual" had the common feature that they served as bridging elements. I also indicated that these concepts performed this function in a very specific way: They extended the space of possible processes and made this set of possibilities, although not actualized, effective. I will argue for this definition in connection with my analysis given in the final sections of this chapter.

In analysis, the plethora of terminological variants and the occurrence of "virtual" entities which are not connected to the virtual particle lead to an important conclusion: The coherence of the concept of the virtual particle and the concepts in its historical trajectory was not (exclusively) warranted by a strict definition or a stable terminological reference point. Terminological variants are telling in reference to the ontological status or to the role of the concepts in theoretical explanations. And, of course, the (non-)usage and (re-)occurrence of the term "virtual" needs explanation, as well as it gives insights into the actors' conceptual framework. Yet, the virtual particle and the concepts connected to its emergence are striking examples of concepts the identification and use of which did not need the backdrop of a stable terminological reference point. Rather, the terminology altered meanings, had competitors or was simply not applied. Still, the actors did not misunderstand each other. Thus, a broader look taking other representational formats and their role in practice into account is necessary. I will turn to this reconstruction in Chap. 12.3. For the moment, I want to recapitulate the development of the diagrammatic displays in which such virtual entities played a role.

12.1.2 Diagrammatical Representations: From Formal Structures to Processes in Space-Time

Let's start with Kramers' and Heisenberg's diagrams (1925) describing the incoherent scattering of light, later to be known as the Raman effect (Sect. 4.2). The diagrams did not include a distinct temporal order of events. This is in harmony with Kramers and Heisenberg not providing any kind of temporal description in written words. The diagrams had no temporal dimension; they were drawn in the action-variable space of the corresponding motion. One might be tempted to read an ordering into the numbering of the different transitions in the two diagrams on the right. Yet, these were not used for the translation of the classical formulas into quantum theory:

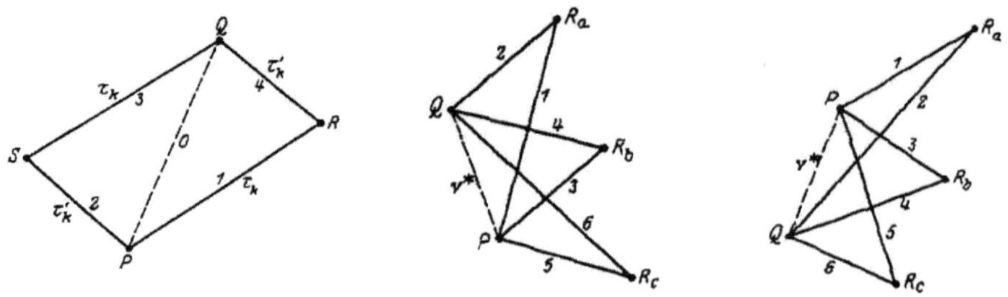

Fig. 12.1 Diagrams in the dispersion theoretical treatment by Heisenberg and Kramers. The left diagram was used for the translation of the formulas into quantum theory, while the two diagrams on the right were provided for the discussion of the relative energy position of the states R. Figures in Kramers, Hendrik, and Werner Heisenberg. 1925. "Über die Streuung von Strahlung durch Atome." *Zeitschrift für Physik* 31, 681–708, here 694 (left) and 699 (right). https://doi.org/10.1007/BF02980624 (cited as Kramers and Heisenberg (1925) in the bibliography). Both reproduced with permission from Springer Nature

Kramers and Heisenberg used the outer most left of the three diagrams for this purpose. Therein, the processes we would want to interpret as successive transitions were given by the numbers 1 and 4 (Fig. 12.1).[15]

As soon as Dirac (1927b) introduced a temporal language in respect to what we call virtual processes, diagrams reflected this temporal dimension.[16] Figures 12.2 and 12.3 are two examples. The diagrams in Fig. 12.2 were given by Edoardo Amaldi (1929).[17] They read from left to right in the sequence: initial-intermediate-final state. By ordering the processes in such a way and by representing the probability amplitudes Amaldi broke with the pictorial tradition of energy-level diagrams.[18] But as can be seen from Fig. 12.3 provided first by Maria Göppert-Mayer in 1929 (see Sect. 8.1.1), also in energy-level diagrams we can find an indication of temporality (note the arrow on the dotted line referring to the behaviour of the material part of the system).[19]

These diagrams were explicitly used for illustration. But actors also provided representations that served a methodological, and not mostly didactic, purpose. The

[15] Again, I want to stress here that this is not an in principle argument against the idea that actors might have thought about these processes in terms of successive transitions. Yet, since this conception is nowhere spelled out, I believe that the actors did not consider it an important ingredient in their methodological tool kit.

[16] See also the discussion in Enberger (2022, Sect. 2.1).

[17] See Sect. 6.2.1 for more details and the contextualization of Amaldi's work in the discussion around the Raman effect in the late 1920s.

[18] Amaldi was not alone in his introduction of new forms of diagrams. One finds similar examples in Heitler's and Nordheim's diagrams (see Fig. 8.3) and, of course, in Wentzel's schemata, discussed fully in Sect. 10.2.4, but also below.

[19] The arrows of the light quanta (solid lines) indicated whether a photon was absorbed (pointing upwards) or emitted (pointing downwards). No order was implied. This was in harmony with the conceptualization as the involvement of the photons in either processes had to be permuted.

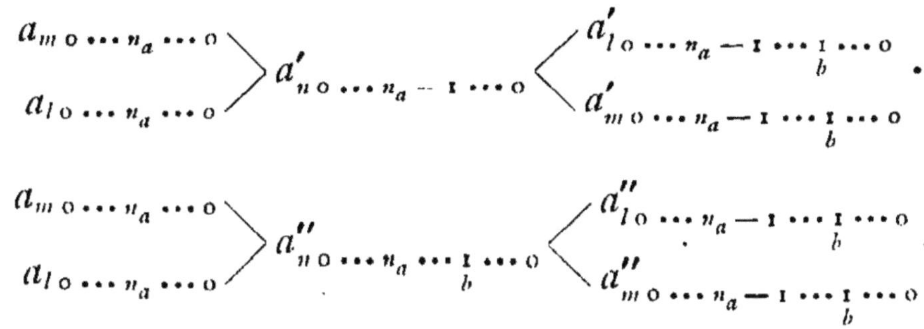

Fig. 12.2 Representation of the two possible sequences of events of Raman scattering. Figures in Amaldi, Edoardo. 1929. "Sulla teoria quantistica dell'effetto Raman." *Rendiconti Accademia dei Lincei. Classe di Scienze Fisiche, Mathematiche e Naturali* 9, 876–881, here 878–879 (cited as Amaldi (1929) in the bibliography). Reprinted with kind permission by Accademia Nazionale dei Lincei

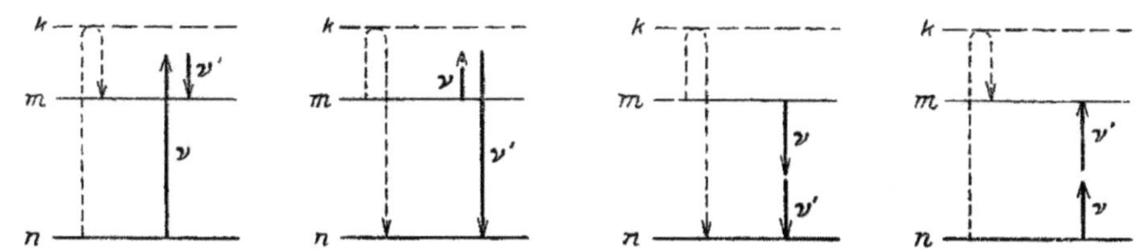

Fig. 12.3 Representation of second order processes. From left to right they represent: Stokes Raman scattering, Anti-Stokes Raman scattering, double absorption and double emission. Diagrams in Göppert, Maria. 1929. "Über die Wahrscheinlichkeit des Zusammenwirkens zweier Lichtquanten in einem Elementarakt." *Naturwissenschaften* 17, 932, here 932. https://doi.org/10.1007/BF01506585 (cited as Göppert (1929) in the bibliography). Reprinted with permission from Springer Nature

best known are certainly Hans Euler's diagrams (shown in Fig. 12.4, lower part).[20] But at the same time, Euler provided a table that listed the possible sequences of events (Fig. 12.4, upper part). Bernhard Kockel, close associate of Euler and working with Euler on the perturbative evaluation of light-by-light scattering, exclusively used tables (Fig. 12.5)[21] to list the possible sequences of subprocesses in his dissertation. These tables and diagrams seemed to have served the same purpose as the verbal model or short-hand notations: As abstracted representations they were used in the construction of the specific matrix elements to be calculated and served as book-keeping devices for the diverse terms. But I consider them mostly after the fact representations. They were a means of bringing the physical reasoning with the verbal model in a concise form.

Kockel's tables bring the role of the photons in the corresponding theoretical modelling to the fore. They are not represented. The radiation field served as taking up and providing energy and momentum through quanta, but the quanta could be

[20] See Sect. 9.1.3 and the literature cited therein for an analysis.

[21] See Sect. 9.1.4 for a discussion of Kockel's dissertation.

Licht-quanten / Elek-tronen	Absorption von g^1	Absorption von g^2	Emission von $-g^3$	Emission von $-g^4$
$\mu = 1$	Paarerzeugung	Elektronensprung	Elektronensprung	Paarvernichtung
$\mu = 2$	Paarerzeugung	Elektronensprung	Positronensprung	Paarvernichtung
$\mu = 3$	Paarerzeugung	Positronensprung	Elektronensprung	Paarvernichtung
$\mu = 4$	Paarerzeugung	Positronensprung	Positronensprung	Paarvernichtung
$\mu = 5$	Paarerzeugung	Paarerzeugung	Zerstrahlung des Elektrons vom 1. mit dem Positron vom 2. Paar	Zerstrahlung des Positrons vom 1. mit dem Elektron vom 2. Paar
$\mu = 6$	Paarerzeugung	Paarerzeugung	Zerstrahlung des Positrons vom 1. mit dem Elektron vom 2. Paar	Zerstrahlung des Elektrons vom 1. mit dem Positron vom 2. Paar

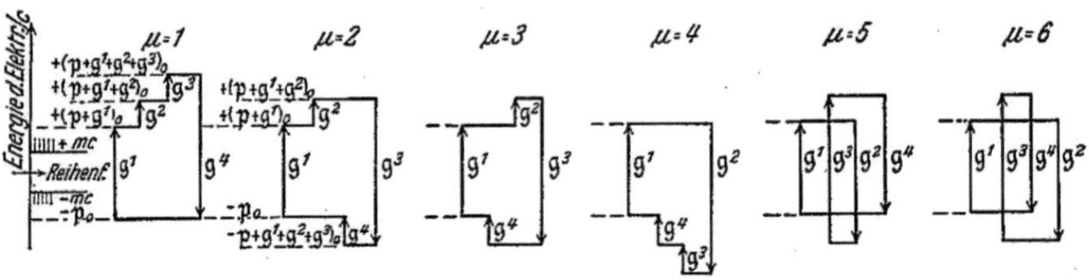

Fig. 12.4 Table and diagrams provided by Hans Euler to represent the process of light-by-light scattering. Figure and Table in taken from Euler (1936, 418)

	1. Übergang	2. Übergang	3. Übergang
I	Paarerzeugung ($\mathfrak{p}_0^- \to \mathfrak{p}_1^+$)	Elektronensprung ($\mathfrak{p}_1^+ \to \mathfrak{p}_2^+$)	Elektronensprung ($\mathfrak{p}_2^+ \to \mathfrak{p}^+$)
II	„ ($\mathfrak{p}_1^- \to \mathfrak{p}_2^+$)	„ ($\mathfrak{p}_2^+ \to \mathfrak{p}^+$)	Positronensprung ($\mathfrak{p}_0^- \to \mathfrak{p}_1^-$)
III	„ ($\mathfrak{p}_1^- \to \mathfrak{p}_2^+$)	Positronensprung ($\mathfrak{p}_0^- \to \mathfrak{p}_1^-$)	Elektronensprung ($\mathfrak{p}_2^+ \to \mathfrak{p}^+$)
IV	„ ($\mathfrak{p}_2^- \to \mathfrak{p}^+$)	„ ($\mathfrak{p}_1^- \to \mathfrak{p}_2^-$)	Positronensprung ($\mathfrak{p}_0^- \to \mathfrak{p}_1^-$)
V	„ ($\mathfrak{p}_0^- \to \mathfrak{p}_1^+$)	Paarerzeugung ($\mathfrak{p}_2^- \to \mathfrak{p}^+$)	Paarvernichtung ($\mathfrak{p}_1^+ \to \mathfrak{p}_2^-$)
VI	„ ($\mathfrak{p}_2^- \to \mathfrak{p}^+$)	„ ($\mathfrak{p}_0^- \to \mathfrak{p}_1^+$)	„ ($\mathfrak{p}_1^+ \to \mathfrak{p}_2^-$)

Fig. 12.5 Table representing the different sequences of events leading to pair creation and the emission of a photon by two photons as given by Bernhard Kockel. Table in Kockel, Bernhard. 1937. "Über einige Mehrfachprozesse zwischen Elektronen, Positronen und Lichtquanten." *Zeitschrift für Physik* 107, 153–176, here 167. https://doi.org/10.1007/BF01330357 (cited as Kockel (1937) in the bibliography). Reproduced with permission from Springer Nature

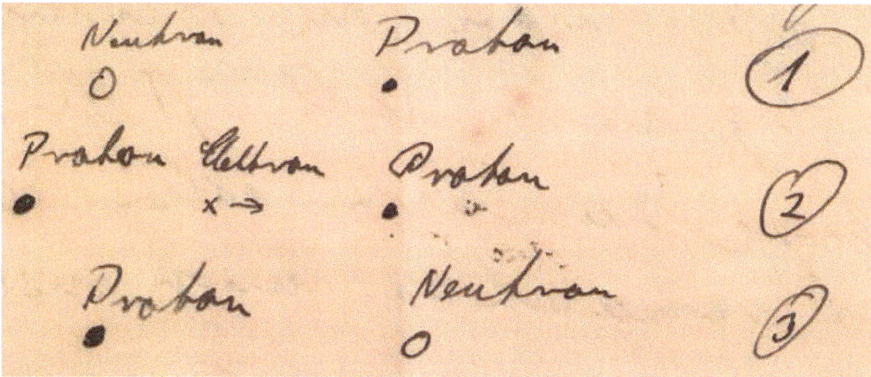

Fig. 12.6 Representation of nuclear forces according to Fermi's theory of β-decay. Communicated by Edward Teller in a letter to Guido Beck dating to early 1934. Courtesy of and reproduced with kind permission from Guido Beck Archives

permuted and could participate in any of the processes.[22] The important part to represent was the material part: Its behaviour would allow for the construction of the possible sequences of events. As a matter of fact, Kockel did not represent the material system itself, but the different virtual subprocesses, i.e. exactly these possible sequences of events.

Comparable to the terminological shift from "virtual transitions" to "virtual particles", at around the mid 1930s physicists started to represent the *particles* present in intermediate states, most prominently in the realm of nuclear physics. The first graphical representation of the interaction of nucleons according to Fermi's theory of β-decay I know of was an intermediary between a state-transition model and a conception of the interaction as scattering (see Fig. 12.6).[23] On the one hand, the initial (1), intermediate (2) and final (3) states were clearly indicated. Therefore, the focus on the transitions was to some extent still upheld. On the other hand, in the intermediate state (2) we can see an electron propagating (note the arrow) from proton to proton.[24]

The chronologically next diagrams which represented the particles in the intermediate states and provided a more dynamical picture of the mediation of forces were published by Gregor Wentzel in 1938 (Sect. 10.2.4). These diagrams have to be contextualized in Wentzel's prior work on a quantum field theoretical description of nuclear physics. Diagrams like the ones shown on the left of Fig. 12.7, Wentzel's *schemata i*, as I called them, clearly fall into the tradition of reaction

[22] Adrian Wüthrich (2010, Especially Chap. 2) has already pointed out that this fact is a decisive difference to Feynman diagrams.

[23] See Sect. 10.2.1 for the context of this diagram.

[24] Similar representations were used by Wolfgang Pauli in late 1939 when discussing a possible modification of meson theory proposed by Bhabha. Pauli to Heisenberg, April 27. 1939, and Pauli to Bhabha, May 4. 1939; as cited in Pauli (1985, 637, 643).

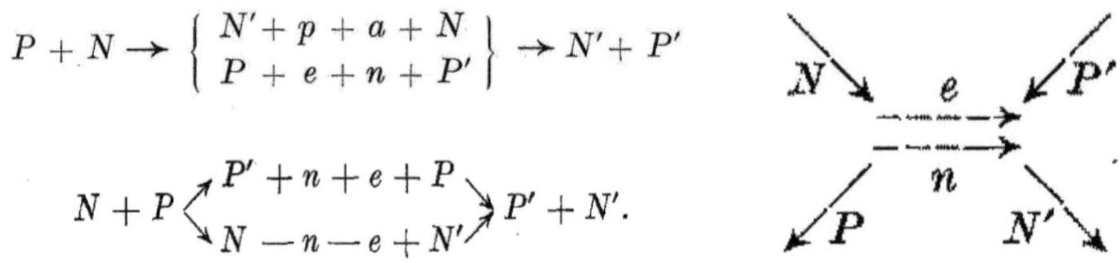

Fig. 12.7 Representations of nuclear interaction in terms of Fermi's theory of β-decay, all given by Gregor Wentzel. Figure on upper left side in Wentzel, Gregor 1937a. "Zur Frage der Beta-Wechselwirkung." *Helvetica Physica Acta* 10, 107–111, here 109 (cited as Wentzel (1937a) in the bibliography). ETH-Bibliothek Zürich, E-periodica, https://www.e-periodica.ch. Reproduced with kind permission from the Swiss Physical Society. Figure on bottom left side in Wentzel, Gregor. 1937b. "Zur Theorie der Beta-Umwandlung und der Kernkräfte. I." *Zeitschrift für Physik* 104, 34–47, here 38. https://doi.org/10.1007/BF01337892 (cited as Wentzel (1937b) in the bibliography). Figure on right side in Wentzel, Gregor. 1938. "Schwere Elektronen und Theorien der Kernvorgänge." *Die Naturwissenschaften* 26, 273–279, here 274. https://doi.org/10.1007/BF01590308 (cited as Wentzel (1938) in the bibliography). Bottom left and right figure reproduced with permission from Springer Nature

schemes, known from chemistry and extensively used in the context of β-decay. Yet, while Amaldi still represented the probability amplitudes in his diagrammatic display, Wentzel represented the particles present in the intermediate state.[25] In a popular article published in 1938, Wentzel eventually provided the diagrammatic display shown on the right of Fig. 12.7, which strikingly resembles a Feynman diagram.

Such diagrams clearly show that the idea, that processes were mediated by particles not present in the initial or final state, gathered more and more prominence in the community. But the *schemata ii* actually served the purpose of illustration. Wentzel himself did not use the diagrams on the right in his theoretical evaluation; the diagrams on the left served the purpose of book-keeping devices in his research papers. As discussed in Sect. 10.2.4, diagrams like the one on the right-hand side of Fig. 12.7 had drawbacks for Wentzel's purposes. Although they might resemble Feynman diagrams, they emerged from a different pictorial tradition: Wentzel's *schemata ii* developed as an overlap of reaction schemes (theoretical tradition) and cloud chamber photographs (empirical tradition).

Diagrams used by André Mercier (see Fig. 12.8) in his calculation of higher order contributions to the self-energy of the electron show that the "picture of 'successive' transitions", as Bernhard Kockel has called it, was still underlying the physical interpretation of the theory in the late 1930s. They nicely exemplify the tool character of the respective representational format: Without having to go through detailed calculations, Mercier employed the diagrams and the underlying physical picture of successive virtual transitions to group certain terms and to show that their contributions effectively cancelled. Thus, the calculation effort was considerably reduced (Sect. 10.1.1).

[25] N = neutron, P = proton, e = electron, –e/p = positron, n = neutrino, and –n/a = anti-neutrino.

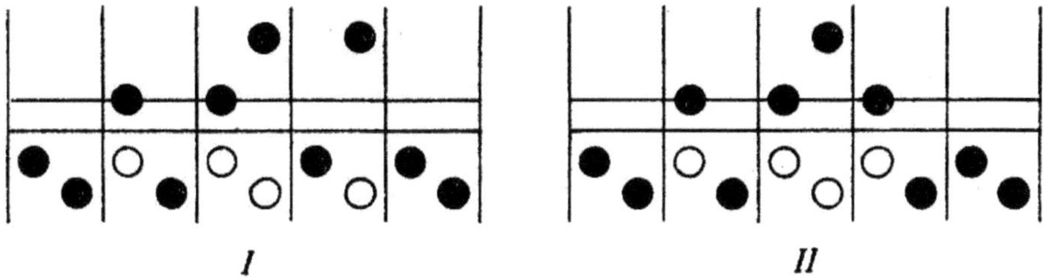

Fig. 12.8 Two examples of the diagrams André Mercier used to reduce the terms he had to calculate in the fourth order contribution to the self-energy of the electron. Figures in Mercier, André. 1939. "Sur l'énergie propre de l'électron." *Helvetica Physica Acta* 12, 55–75, here 64 (cited as Mercier (1939) in the bibliography). ETH-Bibliothek Zürich, E-periodica, https://www.e-periodica.ch. Reproduced with kind permission from the Swiss Physical Society

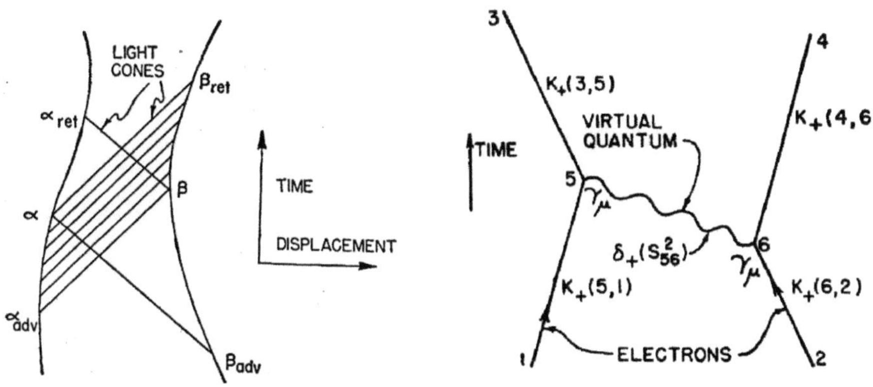

Fig. 12.9 Representation of the classical retarded interaction between two particles in Wheeler-Feynman absorber theory (left side) and the first Feynman diagram presented by Richard Feynman himself. This combination of diagrams was also given by Wüthrich (2010, 184). Left reprinted figure with permission from Wheeler, John, and Richard Feynman, "Classical electrodynamics in terms of direct interparticle action", *Review of Modern Physics* 21, 425–433, here 431, 1949, https://doi.org/10.1103/RevModPhys.21.425 (cited as Wheeler and Feynman (1949) in the bibliography). Right reprinted figure with permission from Feynman, Richard, "Space-time approach to quantum electrodynamics", *Physical Review* 76, 769–789, here 772, 1949, https://doi.org/10.1103/PhysRev.76.769 (cited as Feynman (1949a) in the bibliography). Both: Copyright 2025 by the American Physical Society

For the coherent use and representation of scattering diagrams as (pseudo-)causal processes, a change in the theoretical and physical picture was still needed. This transformation was most closely studied by Wüthrich (2010) and Blum (2017). The development of Feynman diagrams (see right of Fig. 12.9 for the first one presented by Richard Feynman himself) was connected to a shift of the conception of scattering processes, its mathematical display and the corresponding theoretical modelling (see Chap. 11). Whether or not these diagrams emerged as a reaction to a concrete theoretical problem and through a transformation of the classical picture of the electrodynamical interaction (see the left of Fig. 12.9) (Wüthrich 2010 and Wüthrich 2018), whether they derived from Feynman's war-time work (Galison 1998), are directly and genealogically related to the perturbative evaluation of his space-time approach (Schweber 1994 and Darrigol 2019), or whether a combination of all of these factors finally led Feynman to represent his theory in terms of the diagrams: They served a calculational purpose, restricted the prior thinking possibilities, and finally put (as in Feynman's whole approach) the virtual particle in the centre of the evaluation.

All of the above diagrams and tables were representations of a physical modelling. This modelling was used in the search for the matrix elements to be calculated in perturbative evaluations. Most often, the diagrams were accompanied by the description of the processes in terms of the verbal model which Dirac had introduced in 1927. And even in Feynman's papers the corresponding (pseudo-)physical reasoning was a necessary part for the coherent evaluation and construction of the diagrams.[26] While the graphical representations of the 1930s mostly served a didactic purpose or were book-keeping devices, the Feynman diagrams were eventually connected to specific rules of how to draw them and how to translate them directly into mathematical objects. Through this systematization by Dyson, the physical interpretation receded into the background and focus was laid on the topological features of the diagrams.[27]

But Feynman diagrams were not only particularly suited for a systematization and, as Adrian Wüthrich (2010) has argued, the clear identification of the divergent terms of the perturbative series. The representational format of Feynman diagrams restricted the thinking possibilities of the historical actors. As I will argue in the following section, the real processes conceivable during the 1930s were different from the ones in the late 1940s. Thus, let me turn to the evaluation of what I called "false predictions".

12.2 Reasoning with Representations: False Predictions

During the 1930s the verbal representation of QED as introduced by Paul Dirac in 1927 guided theoretical evaluations. It served as a construction method, as a tool, in the search for the specific matrix elements in the application of perturbation theory.

In its emergence, the verbal model of QED (Dirac 1927b) was closely connected to specific theoretical and mathematical techniques (see Chap. 5). Also in its further elaboration, this verbal representation of successive acts of emission and absorptions was structured and always backed up by the mathematical and theoretical machinery the actors had at their disposal. In the following, in order to show that the verbal model is still not reducible to the mathematical formulation of the theory, I will draw

[26] See also the discussion of the interrelation between the different representational formats used in Feynman's space-time approach to quantum electrodynamics in Sect. 2.4.

[27] See also David Kasier (2005, especially Chap. 5) on what he called the "Feynman-Dyson-split".

on instances of what I called "false predictions". Thereby, I will also contribute to the questions of conceptual differences between QED of the 1930s and the late 1940s,

According to my definition, false predictions are instances where the historical actors were, at first, led astray by the representations they had at their disposal.[28] Processes they had envisioned through one means of representation turned out to be non-existent when transformed into another. Importantly, the envisioning of the possible process and the realization that this process was not part of the theory the historical actors studied occurred in one and the same conceptual and theoretical framework. Often, the same historical actor came to this realization, often within a rather short amount of time (weeks to months). This does not mean that the representation should not be applied in the evaluation of the theory. Often, such representations were indispensable tools, as for example the verbal model. On the contrary, false predictions might spur the further evaluation of this representation by the introduction of new rules for its application.

There are two things to mention at this point. First of all, the term and the concept of a false prediction is not an actors category but an analytical tool I propose.[29] Secondly, the instances discussed below have a common feature: The effects were predicted by non-mathematical forms of representation and were found to be non-existent when transformed into the mathematical representation. This should not be misunderstood as indicating an epistemological superior status of the mathematical language. While the verbal, tabularic, or diagrammatic representations might have led astray, they were and are nonetheless an indispensable part of the theoretical tool kit of practising quantum field theorists.

The instances of false predictions[30] I identified are the scattering of light by two electrons in second order (Tamm 1930a, see Sect. 7.7),[31] the idea of light-quantum splitting in the vacuum (Halpern 1933, see Sect. 9.1.1), the creation of two pairs from three photons (Kockel 1937, see Sect. 9.1.4) and the idea that Delbrück scattering could occur in third order (Max Delbrück, as well as groups in Berkeley, Zürich, and Kharkiv; see Sect. 9.2). In the following, I will use the comparison to modern methods to group and discuss these instances.

---

[28] See Sect. 2.3.3 for the introduction of this analytical term.

[29] The one time I encountered the term in the historical literature was Robert Oppenheimer referring to the infinite displacement of spectral lines as a "false prediction" (Oppenheimer 1930a, 477). In my definition, this is not a false prediction. QED as practised during the 1930s included these divergences. Theoretical and conceptual progress was necessary to eventually handle the occurrence of the infinities.

[30] For the details of the calculations and their historical background I refer the reader to the respective sections.

[31] Bethe and Fermi (1932) identified a similar process of the emission of two light quanta by two electrons.

12.2.1 *False Predictions that Still Occur when Using Feynman Diagrams*

Light-Quantum Splitting

On the basis of "the language of Dirac's theory of radiation" (Halpern 1933, 856) Otto Halpern predicted that light quanta might split up into several others in the vacuum. During the 1930s, this was shown to be a false prediction (although it seems not to have been recognized widely). Based on arguments from special relativity, Bronstein (1936) showed that the hypothesis of the redshift being due to light-quantum splitting was untenable. Further, he stated that a light quantum could spontaneously decay into three light quanta if calculated directly from the interaction term. But, when including the subtraction terms of Heisenberg (1934a), the probability of such a process vanished.

Halpern's inference and its refutation can be understood in two ways. On the one hand, his conclusion was actually suitable for QED without the subtraction terms which had not yet been obtained as he published his paper. Thus, Halpern's inference was not a false prediction in the sense I defined it. On the other hand, the possibility of light-quantum splitting was discussed, for example, by Walter Heitler (1936) even after the modifications by Heisenberg had been proposed.[32] Simply applying Dirac's verbal model, which was surely guiding Heitler's discussion, we can imagine that one photon creates a pair and after two emission processes coupled to state transitions the pair annihilates and sends out the third photon. Even more strikingly, the process can be depicted in terms of Feynman diagrams[33] and it is not ruled out by purely kinematic dependencies. Thus, the verbal model of the 1930s as well as the diagrammatic format of Feynman diagrams lead to the false prediction of light-quantum splitting in the vacuum, which was known (by 1936) not to be part of the theory these formats represented.

Delbrück Scattering as a Third Order Effect

The second, and actually similar, case is the prediction of Delbrück scattering as a third order process. Delbrück himself could exclude this possibility only through the help of Hans Bethe, but did not publish his reasoning. The groups investigating the process in Zürich and Kharkiv independently assumed a third order Delbrück scattering at first. Robert Oppenheimer and Charles Lauritsen (Lauritsen and Oppenheimer 1934) were the first to publicly state that such a process could not occur to third order, but without going into any details of the explanation. In Berkeley, the

---

[32] Halpern did not restrict the possibility in any sense. But Heitler, for example, already noted that to lowest order the photon must split into three photons. Today this is easily proved through Furry's theorem.

[33] It is simply the Feynman diagram for light-by-light scattering with one incoming and three outgoing photons.

Delbrück effect had been investigated as well by one of Oppenheimer's PhD students, Arnold Nordsieck. In this case, the reoccurrence of the false prediction would (partially) motivate Wendell H. Furry to investigate the rules for such processes. He gave a general proof that odd order processes vanished identically if "the appearance of electrons and positrons is transitory only" (Furry 1937, 125). Today this statement is known as Furry's theorem.

In the modern representational format of Feynman diagrams such processes are in principle imaginable. Feynman himself stumbled over this problem in late 1948 (see Sect. 11.3.2). Judging from Feynman's correspondence with Dyson, he had gone through the calculations only to find that the result was zero. But in relation to Furry's theorem, Feynman's methods also provided considerable improvement. While Furry's original proof took about three and half double-columned pages, Feynman's way of visualizing showed the "physical" reason for the cancellation without any further calculational efforts. It took Feynman only a view lines to provide an argument for the vanishing of such processes. Dyson considered this "an elegant consequence of his [Feynman's] theory" (Dyson 1949b, 1743).[34]

Although Feynman's interpretation and representation would allow for a simplified and initially only verbally communicated proof based on physical reasoning, the problem was not resolved by the representation in Feynman diagrams itself. This case nonetheless exemplifies a possible consequence of a false prediction: the further evaluation of the theoretical structures in order to find supplementary rules for the application of the representations, in this case Furry's theorem.

12.2.2 False Predictions Ruled out by Feynman Diagrams

The Scattering of Light from a Pair of Electrons

Igor Tamm envisioned the scattering of light from a pair of electrons, where one electron absorbed the incident photon, the other one emitted the final photon. Tamm's reasoning is historically connected to Dirac's reconceptualization of scattering in a filled Dirac sea. According to Dirac's model, two electrons were involved: One electron in the sea which jumped to the final state, and the initial electron which jumped into the hole created in the first step (both processes were either connected to emission or absorption of a light quantum). Thus, Tamm's idea is understandable from the context of the theoretical research of the time.

From a modern vantage point, Tamm's suggestion is rather curious. It would correspond to putting two basic and unphysical building blocks of Feynman diagrams side by side. Thus, the representational format of Feynman diagrams allows for the direct exclusion of this false prediction. When Tamm proposed the effect, neither he nor Dirac, both at the top of their game and both experts in QED, could exclude

[34] The specific simplification of the reasoning can also be judged by having a look at the proof of an "extended Furry's theorem" in the textbook by Jauch and Rohrlich (1959, 159–161).

the possibility of this process without invoking a mathematical representation.[35] Moreover and again from a modern vantage point, one might not even be led to the idea that such processes should occur. In Feynman's overall space-time interpretation of what had priorly been conceptualized as scattering with an intermediate pair, there was actually only one electron involved, although travelling backwards in time for a short while. Although this reasoning is surely counterfactual, the basic idea that had led Tamm to conceive of the scattering from a pair of electrons was no longer present in Feynman's interpretation.

As in the case of Furry's theorem, Tamm's elaboration on the scattering process would result in additional rules for the application of the verbal model. Tamm (1930a) engaged with the problem more generally and found that the scattering from a pair of electrons was only possible when the initial state of one electron was equivalent to the final state of the other, and for this case the exclusion principle could be neglected for intermediate states.

The Creation of Two Pairs from Three Photons

The problem discussed by Bernhard Kockel (1937), namely the creation of two pairs from three photons, should have been possible, as Kockel explicitly stated, within the restrictions posed on energy and momentum conservation for the sequences of subprocesses. From a modern vantage point, this process can also be excluded by simply trying to construct a corresponding matrix element to third order through Feynman diagrams. Such a diagram does not exist, at least not a connected one. By looking at the possible sequences of subprocesses Kockel provided in his tables (Fig. 9.4), we immediately see why such processes should not occur (again from a modern vantage point). In each sequence the creation of a pair from two photons (possible) was described, while one subprocess corresponded to the creation of a pair from one photon (impossible).

Although Kockel reflected on the relevance of the Pauli principle for the intermediate states in his dissertation, this false prediction seemed not to have had any direct impact on the evaluation of the theory. Yet, as late as 1949, Kockel (1949) turned again to the possibility of processes in radiation theory for his Habilitation. In the published paper, Kockel proved a more general theorem that would exclude the process he had initially envisioned: As long as there was no interaction between electrons (either through photons or the Coulomb interaction) no higher order processes existed in which more than one electron was involved.[36] Kockel's rather lengthy and complicated proof in radiation theory was obsolete as the new physical picture

[35] Also Enrico Fermi and Hans Bethe would shortly indicate that a similar process, the emission of two photons by two electrons, was conceivable. Yet, they immediately stated that this process was forbidden due to the impossibility of fulfilling momentum conservation twice and energy conservation once as long as all electrons should have positive energy in the final state (Bethe and Fermi 1932, 306).

[36] Note that Kockel was working in the Dirac sea picture and conceptualized pair creation accordingly.

and methods of Feynman were already published (or in the course of publication). Applying the graphs in this framework, it took Dyson half a page to proof "the elimination of disconnected graphs [which] is, from a physical point of view, somewhat trivial" (Dyson 1949a, 496).

12.2.3 *False Predictions as Indicators of Methodological and Conceptual Differences*

The historiographical concept of a false prediction is directly connected to the cornerstones of my analytical framework: Physicists use different forms of representations as tools in their reasoning processes. Such representations might be more or less efficient, they might widen or restrict the scope of the thinking possibilities, and they are in contact with each other, as physicists assign them to the same theoretical entity or process, but they are not necessarily reducible to each other.

This last statement is strikingly exemplified by the occurrence of false predictions. The verbal model of Dirac (and partially the diagrammatic representation of Feynamn diagrams) allowed to conceive of processes which were shown to be non-existent when put into a mathematical form. Although the verbal model was motivated by the mathematical structures of the theory, it was not equivalent to it. The scope of the predictions drawn by using the verbal model or Feynman diagrams was actually wider than the scope of the effects which were included in the theory. In this sense, the verbal model had some independence from the underlying theory.

Nonetheless, the verbal model was an integral part of the theoretical investigation. None of the actors who hit upon a false prediction actually questioned the application of the verbal model or other formats of representation. Rather, they turned to the investigation of their application: Tamm searched for the processes which might be possible when two electrons are involved and eventually also investigated the role of the Pauli exclusion principle for intermediate states. Motivated by the diverse predictions of the Delbrück effect as a third order process, Furry came up with a general proof that odd order process vanished if only virtual electron-positron pairs were involved. Tamm and Furry thus came up with additional rules which reduced the calculational efforts and thereby unravelled characteristics of the concepts involved.[37] The physicists sharpened their tools and learned how to handle them.

The comparison with Feynman diagrams allows to gain further insight into the refinement of the methods this new tool came with. Although we can draw a connected Feynman diagram which would correspond to a third order Delbrück effect (and Feynman and Dyson themselves obviously did so), the proof of Furry's theorem was considerably simplified by the more modern set of methods. Even more strikingly,

[37] In Tamm's case, the calculation without having to take the Pauli exclusion principle into account for intermediate states reduced the matrix elements to be calculated (this can also clearly be seen from Bernhard Kockel's dissertation). In Furry's case, a whole class of possibly conceivable processes was excluded.

Tamm's and Kockel's false predictions are immediately ruled out. While Tamm and Kockel still had to transform their verbal or tabularic display into matrix elements, we can directly exclude the corresponding processes by using Feynman diagrams and the general rules of how to draw them. And we do not even need Feynman diagrams. Only by invoking energy and momentum conservation at the interaction points, we cannot think of Tamm's or Kockel's false prediction. The set of real processes physicists could conceive of by patching together virtual processes before going into a detailed evaluation had changed. Thus, as a historiographic tool, the false prediction allows us to get a more nuanced picture of the conceptual differences between theories which purportedly, to quote Steven Weinberg (1977, 18), "changed only just enough so that they could stay the same".

Kockel's and Tamm's contribution point to a further, deeper conceptual gap between QED as practised during the 1930s and the late 1940s. Kockel and Tamm focused on the transitions of the material part of the system. The radiation field was a reservoir of energy and momentum, exchanged through the emission or absorption of photons. Feynman diagrams, on the contrary, explicitly represent which processes are connected and which processes are to be considered as separate ones.[38] The line drawn between different physical systems has shifted. In Feynman diagrams only particles actually in some way in interaction with each other are part of the same physical system. At least Tamm and Kockel included non-interacting particles in one and the same system, as long as the energy and momentum restrictions for the whole process could be upheld. How close the connection between this conceptual difference and the difference between virtual particles (particles being off shell but ensuring energy and momentum conservation at the interaction points) and virtual transitions (breach with energy conservation at the interaction points) actually is, remains to be investigated more fully.

But just judging from Tamm's and Kockel's use of virtual transitions, there seems to be a connection. Tamm's processes cannot be conceived of if we impose energy and momentum conservation at the interaction points. Also Kockel stated that the creation of two pairs from three light quanta should be possible according to the restrictions imposed on momentum and energy conservation. On the contrary, in terms of energy and momentum conservation at the interaction points, each of the sequences of suprocesses Kockel discussed is obviously impossible. Thus, Feynman diagrams do not only show which particles should be considered to be part of the same physical system. They also seem to codify the concept of the conservation of energy and momentum at the interaction points in a diagrammatic form.

[38] This fact was most prominently argued for by Wüthrich (2010).

12.3 The Slow Formation of the Virtual Particle Concept

Having discussed specific representations and parts of my analytical framework, it is time to present my final reflections on the development of the virtual particle concept and turn to the question: "When, how and why did the virtual particle emerge?" My historiographical approach necessarily shifted focus to the practices and the explanatory structures of the historical actors. Consequently and in a first approximation, the attention did not lie on the concepts to be studied. In the historical analysis, a second step was necessary in each of the chapters to concentrate on the development of the concept itself.

In the following, I will take both steps of the analysis in a highly condensed but also focused form. I will take up the three part structure of my historical analysis. In each of these parts a specific focus was taken. The first part had the concept of virtual transitions as its endpoint. In the second part, I followed the practice of Dirac's radiation theory and its representations, including the concepts of virtual transitions and states, during the 1930s. Finally, in the third part the shift from virtual transitions (or processes in more general terms) to virtual particles was investigated. From a broader perspective, I reconstructed how specific parts of the conceptual framework—connected to the virtual particle concept—emerged during the 1930s. And finally I provided an outlook on the alterations of the investigated concepts in Feynman's version of QED.

12.3.1 From Virtual Oscillators to Virtual Transitions

It is all but trivial to fix a specific date for the emergence of the concept of virtual transitions and the related concept of an intermediate state. Similarly, the historical connection between virtual oscillators and virtual transitions is so far not answered sufficiently in terms of generative relations. I tried to engage with these ambiguities by looking closely at the representations and their usage and by distinguishing different characteristics of the virtual transitions: the probability-amplitude level of the theory, energy non-conserving transitions, and a temporal storyline of successive transitions to intermediate states.

Virtual Oscillators and "Intermediate States" Before Quantum Mechanics (1923–1925)

In the BKS theory the transitions of the atom and its reaction to the radiation field were disconnected so that energy conservation would only hold statistically. The atom stayed in one stationary state during the interaction process. The virtual oscillators were a means for including all *possible transitions* in this conceptual framework, but—and this is important—they did not do so in the form of transition processes but rather in the form of "grafted on" (Duncan and Janssen 2007a, 580) entities.

The stationary states of quantized systems were the only real, because observable, entities. The virtual radiation field and the virtual oscillators constituted an unobservable, mostly classically governed virtual world. It was only statistically connected to observation. According to my reconstruction, the terminology of virtuality in BKS referred to two interrelated pillars of the framework: (1) to the differentiation between a real world of quantised atoms and an unobservable world of virtual oscillators and virtual radiation field;[39] and (2) to the non-classical, probabilistic interaction between the radiation field and matter, i.e. between these two worlds. Energy non-conservation was not the reason for the terminology. And only the radiation field and the oscillators were virtual. Quantum concepts, such as states or transitions, were actual.

The mathematical and diagrammatic representations in the work on dispersion theory by Werner Heisenberg and Hendrik Anthony Kramers (Kramers and Heisenberg 1925) seem to suggest successive acts of transitions of the atom. But such a mental imagery was neither spelled out nor was it necessary for—at times contradictory to—their evaluation (Sect. 4.2). Up until Dirac's work in 1927, only Wolfgang Pauli described the formulas as a "detour" of the atom through a third state.[40]

In another, mostly independent approach (see Sect. 4.3), the light quantum was taken more seriously or, more accurately and correct for all involved actors, energy conservation was upheld. From a purely verbal point of view, the description of the light-matter interaction was the same as in modern parlance: An atom absorbed a photon, went to an intermediate state, only existing for a short amount of time, and then re-emitted a photon. But emission and absorption processes were considered to be actual and energy conserving, the intermediate state was energywise and temporarily in between initial and final state. While it was not followed, this line of research still exemplifies the close connection between the concept of a light quantum and the description of scattering as a two-step process.

Dirac and the Introduction of a Verbal Model (1927)

In 1927, Paul Dirac explicitly framed dispersion as a two-step process of energy non-conserving transitions connected to emission and absorption subprocesses on a probability-amplitude level. This is what I called Dirac's verbal model. Dirac was neither influenced by the BKS theory nor by the description of the light-quantum framework of the old quantum theory. Rather, the introduction of the verbal model has to be understood against the background of Dirac's research and philosophical stance at the time (see Sect. 5.1).

---

[39] Today we would most probably connect the second of these worlds to the probability-amplitude, i.e. quantum-amplitude, level of a semi-classical description.

[40] It should be noted that Pauli would not use this kind of framing in his 1926 article in the *Handbuch der Physik*.

The verbal representation had a direct connection to practice. Dirac used it to construct the different orders of events necessary to get into contact with the priorly known Kramers-Heisenberg formula. The verbal model was not only an interpretation of the formulas, it served as a tool.

Virtual oscillators and virtual transitions share some features, besides the mathematical representation their location on a quantum-amplitude level of theory and their function to incorporate all *possible* (unbound by energy conservation), yet not actualized transitions of the system under study. But they did so in different conceptual frameworks and in very different ways. The concepts were not conceived of as identical and thus the non-invocation of the term "virtual" is explained (see Sect. 5.2).

Finally—and this point seems to have been overlooked so far—Dirac redefined one of the basic concepts of the broader framework of quantum theory. Up until Dirac's work only actual transitions were referred to as transitions. Dirac opened the doors to speak about transitions on the probability-amplitude, i.e. quantum-amplitude, level of theory, suggesting temporality and causality. This provided an immensely important tool in the conceptualization of scattering processes.

Connecting Terminology and Concepts: The Raman Effect (1929) and Heisenberg's "Virtual Possibilities" (1934)

How, then, was the "virtual" connected to the transitions? Chandrasekhara Venkata Raman was the first to do so explicitly in 1929. Raman's initial interpretation of the effect named after him (see Sect. 6.1) was called into question by more refined empirical investigations (see Sect. 6.2.1). Raman therefore pushed the virtual transitions in the purely mathematical terrain; for Raman they were theoretical fictions defined by energy non-conservation (see Sect. 6.2.2). In comparison to BKS, Raman's use of the term virtual constituted a break in meaning,[41] but also in its referents: Raman connected the term to a quantum concept, a transition.

At around the same time and independently, Yakov Il'ich Frenkel went to the other extreme: He conceived of the subprocesses as actually occurring and thereby connected Dirac's time-dependent perturbation theory with the light-quantum interpretation of the old quantum theory. All the same, this third interpretation of the term "intermediate state" (see Sect. 6.3) was quickly refuted.

To explain the wider spread of the terminology—starting at around the mid 1930s—one has to take into account Werner Heisenberg's invocation of "virtual possibilities" (see Sect. 9.1.2). According to my reconstruction in Sect. 9.1.2, Heisenberg had a connotation of the BKS theory in mind: I argued that Heisenberg referred to the unobserved virtual (quantum) world, when he used the terminology. But regardless of Heisenberg's reasons, by the mid 1930s the connection of the term "virtual" with the transitions to and from intermediate states was firmly established (the possibilities were dropped pretty quickly). Through the status of Heisenberg and the widely

[41] Still, I believe that BKS was the most probable source for the terminology.

received work of his students the term spread throughout the community. And corresponding to the then current standards of perturbative evaluations and the involved concepts, it were the transitions that were virtual: They broke with known physical laws, i.e. energy conservation.

12.3.2 The Practice of Radiation Theory

Throughout the 1930s, the general conceptualization of scattering in radiation theory, i.e. in terms of energy non-conserving transitions between stationary states of the system, did not change. The concept of virtual transitions would remain intact in its basic characteristics. But it had important influence on the conceptual framework of quantum electrodynamics and physicists explored new means of representing the verbal model.

Scattering and the Sea (1928–1931)

In late 1929, Ivar Waller made Dirac aware that states of negative energy were necessary as intermediate states for the relativistic description of scattering. This was the motivation for Dirac to think about the negative energy states, which were part of his relativistic wave equation, again (Sects. 7.1–7.4). The motivation to consider the negative energy states as "physical"[42] was their occurrence as virtual states (modern terminology), and it led to the most practised conceptualization of the vacuum during the 1930s: the Dirac sea.

Dirac's move from virtual negative energy states to considering negative energy states physically was the first instantiation of a more general argumentative structure: From the effects hypothetical particles and states might have as virtual ones, the inclusion of such particles and states in the conceptual framework of QFT is suggested or ruled out. In the late 1930s, physicists applied this kind of reasoning to explain discrepancies between theory and experiment, and today it is used to probe the realm of beyond standard model physics. In these cases, new theoretical entities are introduced to the conceptual framework (at least at first) as virtual ones.

In the further evaluation of the Dirac sea the scattering mechanism and virtual transitions played an important role (Sect. 7.6). Dirac's reconceptualization of scattering in a filled vacuum, the analogous structure of the formulas of scattering and pair annihilation as well as their physical interpretation did not only indicate numerical but conceptual inadequacies. Physicist clearly perceived of these problems in such a manner and the situation paved the way for new research questions.

[42] Waller (at least initially) explicitly considered the intermediate states in semi-classical methods as "more formal" than in radiation theory. Also Dirac would speak of the scattering process "really" being a double transition. Igor Tamm shared Dirac's quite physical reading of the intermediate state (see Sect. 7.5).

General Observations and the Practice of Radiation Theory During the 1930s

Chapters 8 and 9 engaged with a (necessarily selective) review of the practice of time-dependent perturbation theory in the realm of radiation theory during the 1930s. For a mostly chronologically ordered discussion, the reader is referred to these chapters. Here, I will provide some of the more general findings.

Throughout the 1930s, theoretical practice was firmly based on energy non-conserving transitions through intermediate states.[43] Correspondingly, the term "virtual" often referred to transitions or subprocesses, and rather seldom to the particles themselves.

Although the methods were surely cumbersome, at the latest by the mid 1930s a clear algorithm had been established for perturbative calculations in radiation theory (see Sect. 8.3 for my reconstruction). The virtual transitions and the connected "picture of 'successive' transitions",[44] their verbal, diagrammatic, tabularic, or schematic representations, played an important role in this theoretical practice: They were used for the construction of the specific matrix elements that had to be calculated for a given process. They took the same place as Feynman diagrams do today.

Physicists also started to study the rules for virtual transitions, virtual states and their combinations: Furry's theorem (1937) is a statement about which sets of intermediate states have to be considered in a perturbative evaluation if the existence of electron-positron pairs is transitory only (see Sect. 9.2.4). The validity of the Pauli exclusion principle for intermediate states was discussed by Igor Tamm (1930a) and Bernhard Kockel (1937) (see Sect. 7.7 and Sect. 9.1.4, respectively). Furry's theorem is still valid for modern QED, but especially Kockel's research indicates an important difference between QED in the 1930s and the late 1940s: Feynman could ignore the Pauli principle in all cases. According to Kockel this was only the case for a specific class of processes.

Actors also extended the conceptual framework of perturbation theory. The *Schüttelwirkung* was (re-)introduced by Maria Göppert-Mayer in 1931 (see Sect. 8.1.1). This concept was used in the first general perturbative expansion to an arbitrary order by Victor Weisskopf (1933) (see Sect. 8.1.4). Bernhard Kockel's discussion of the *Schüttelwirkung* led Edwin Góra in Poland to propose a version of radiation damping, which was generalized by Walter Heitler. For some time, this theory was considered as a possible candidate for a divergence-free formulation of QED (see Sect. 9.3).

---

[43] This is in contrast to the invariant perturbation theory, for which energy and momentum are conserved at the interaction points but the particles do not obey the relativistic energy-momentum relation. An invariant form of perturbation theory had been proposed by Ernst Carl Gerlach Stueckelberg (1934) but remained without any impact on the modelling of the phenomena up until the late 1940s (Lacki et al. 1999).

[44] "Bild 'aufeinander folgender' Übergänge", Kockel (1937, 162).

In this development, the intermediate states were directly manipulated: They co-existed with the initial state. But the general procedure Heitler envisioned eventually failed; it reintroduced the infrared divergence. The intermediate states and the perturbative series could be manipulated, but not at will.

12.3.3 From Virtual Transitions to Virtual Particles

Conceptual Overlap and an Emerging Narrative During the 1930s

When physicist applied time-independent perturbation theory and discussed static phenomena, such as self-energies, they did not verbalize the mathematical structures in terms of successive emission and absorption subprocesses (or virtual particles) initially. The incorporation of the verbal model into static phenomena was a slow process, but as higher orders of the self-energy were approached in 1939, representations in terms of successive transitions were introduced as tools in the evaluation (see Sect. 10.1.1).

In 1934, vacuum polarization (see Sect. 10.1.2) and the electron self-energy were connected with vacuum fluctuation. From that point onwards these effects of virtual transitions or particles were closely connected to the redefinition of the vacuum in QED and QFT: It was no longer empty but an undetermined number of virtual or nascent particles was present. At the same time, it was realised that a particle was always surrounded by an "electron positron cloud" (Lauritsen and Oppenheimer 1934, 80). The modern concept of a "dressed electron" and germs of renormalization were developing, but had to await the late 1940s to be spelled out explicitly. This is most strikingly underlined by the status of the infrared divergence in the 1930s (Sect. 10.1.3).

In the realm of nuclear physics and in a rather entangled mixture of intuitive ideas and more or less suggestive mathematical representations, the concept of virtual particles as force carriers developed in the early to mid 1930s (Sect. 10.2.1). The so-called anomalous magnetic momenta of neutron and proton were explained by the "virtual decay/dissociation" of the nucleons, a conceptualization similar to the electron positron cloud (Sect. 10.2.2). Although a quantitative evaluation led to divergent results, this mechanism was upheld. Qualitative estimations in approximate agreement with observation served as reassurance for the theoretical models of QFT in more general terms.

In this epistemologically uncertain terrain, the narrative, the connection between the time-energy uncertainty, the measurement process, and the concept of the virtual particle came to the fore (Sect. 10.2.3). This connection was certainly known,[45] but not very prominent up until the late 1930s. But as the reliability of concrete

[45] How seriously it was taken is a different question and the answer would most probably differ from physicist to physicist. Frenkel, for example, took it so serious as to propose that virtual transitions are actual ones.

quantitative evaluations became under pressure, it was used for qualitative estimations, which should back up the coherence of the theoretical modelling in general terms.

From the mid to the late 1930s a conceptual overlap of virtual particles and virtual transitions had emerged.[46] In the theoretical description, energy was not conserved in the transition processes and actors would often refer to virtual processes. But the concept of particles only present in intermediate states became ever more prominent. Particles served as the intermediaries, as the "catalytic agents" (Furry 1937, 125). Both in the conceived of function and in their diverse representations, we can find a considerable overlap of the concepts of virtual transitions and virtual particles at around the late 1930s.

Feynman and His Diagrams (1949): The Virtual Particle Takes Centre Stage

The developments of the late 1940s had a considerable impact on the concept of the virtual particle. Renormalization procedures directly connected the virtual particle concept to the concept of a particle itself. Only the "dressed particle", i.e. including the virtual particles surrounding it, was conceived of as the real description of a particle.[47] In Richard Feynman's theoretical approach (Chap. 11) diverse alterations to the virtual particle concept occurred: the scattering through intermediate pair production was no separate process; the Pauli exclusion principle could be completely ignored for particles in intermediate states; timing relations for "virtual quanta" were irrelevant; and all polarizations (longitudinal, scalar, and transversal) of virtual quanta were taken into account. The verbal model of QED was altered, but still played an important role for Feynman's application of his diagrams.

In Feynman's work, the concept of the *virtual particle* was fully established. He spoke rather consistently of *virtual particles* or *virtual quanta*. Particles broke with known laws of dynamical behaviour, they did not have to obey the relativistic energy-momentum relation while energy and momentum was conserved at the interaction points. The virtual particles acquired two new forms of representation which were exclusively, at least so in Dyson's systematization, used for them: internal lines in Feynman diagrams and the corresponding mathematical structure of a propagator. For Feynman himself, the virtual particle had become the primary entity of his theory. When QED became a theory of scattering (Blum 2017), it also became a theory of the virtual particle.

[46] See Sect. 10.3 as well as my reflections on the terminological and diagrammatic representations in Sects. 12.1.1 and 12.1.2.

[47] This important role of the virtual particle concept for the fundamentals of the theory has already been observed and has very prominently been discussed by Wüthrich (2010).

12.4 So, When, How, and Why Did the Virtual Particle Emerge?

Finally, let's turn to the question motivating this research by taking into account the arguments of this chapter. I will not give an unequivocal answer to the question: When, how and why did the concept of the virtual particle emerge?[48] Since the how and why partially depend on when the concept supposedly emerged, I will discuss this question first.

One might argue that the first instance of a virtual particle were the virtual oscillators of the BKS theory and, later on, of semi-classical evaluations. Although the virtual oscillators in BKS did not (necessarily) behave classically and were potentially motivated by an only partial, non-dynamical analogy, they were radiating entities, i.e. charged particles, with the eigen-frequencies of the possible transitions of the atom.[49] All the same, there is a deeper conceptual gap between virtual oscillators and virtual transitions.

Dirac, the inventor of the verbal model of "virtual transitions", did not apply the terminology. Yet, this seems like a rather small issue. More importantly, in the processes which Dirac discussed all the particles were real particles, although inhabiting a virtual state (modern terminology) for a little while. Creation and annihilation operators for electrons had not yet been developed and, in either case, Dirac would most probably not have applied them. Furthermore, in Dirac's perturbation theory the particles were on shell while energy was not conserved at the interaction points.

Throughout the 1930s, we have diverse further possibilities. Maybe the concept emerged with the idea that interactions between charged particles is due to particles in intermediate states. But then we would have the problem off assigning this idea to a particular actor (which, as my discussion and Cathryn Carson (1996b) have shown, is hardly possible). The same holds for self-energy calculations, vacuum polarization, higher order scattering processes, or light-by-light scattering. All of these identifications would cause the problem that the methods still rested on energy non-conserving transitions instead of off shell particles.

Rather, through the combinations of all these effects and the connected insight that particles only present in intermediate states were responsible for specific processes to occur, by the mid to late 1930s physicists had gathered an understanding of the virtual particle concept which has a strong conceptual overlap with the modern term. Yet, in the construction of the matrix elements the virtual states were the connecting pieces, the "interlink" (Fermi and Bethe) or the "intermediaries" (Heitler). While the vacuum was surely no longer empty and no longer passive, but fluctuating due to the virtual pair production, the concept of a bare particle and the idea of renormalization

[48] See also my discussion of the "birth" of the virtual particle in Ehberger (2020).

[49] As outlined in Sect. 4.1.3, there is discussion whether or not the virtual oscillators can be conceived of as a substitute model. If one follows Blum and Jähnert (2024) and conceives of virtual oscillators as classically behaving entities, the above argument is much stronger.

had not been fully developed. Furthermore, the discussion of the false predictions (Sect. 12.2) above indicates a striking contrast between the conceptual resources of QED during the 1930s and the late 1940s. A decisive alteration of QED and, with it, of the virtual entities of QED was still to come.

Lastly, one might go for the late 1940s, the development of renormalization and invariant calculation techniques, in which the particles were actually off shell and the energy was conserved at the interaction points. As already indicated (Ehberger 2020), if we would restrict the concept to this technical issue the question remains how to interpret the invariant perturbation theory of Stueckelberg which was largely ignored by the scientific community.

Still we should not underestimate the changes in the physical picture of processes and the corresponding representations which occurred in the late 1940s and which also derived from Feynman's intuitive approach. But still, and finally, Feynman himself did not apply creation and annihilation operators. So would we have to wait for Dyson's amelioration of Feynman's approach with quantum field theory. But then the physical interpretation of Feynman diagrams (and with it the explicit reference to virtual particles) receded into the background and the topological features of the same were stressed.

To summarize, I cannot find a way to distinguish one particular moment in time or one particular actor who introduced the concept of the virtual particle. Rather, I see the emergence of this concept as a gradual process that combines continuities and breaks in the technical and conceptual apparatus of the theory. There were certain actors, like Dirac and Feynman, who more radically changed the representations which the historical actors had at their disposal for developing and applying their theory. But some changes in the practical and conceptual framework took some time, they emerged slowly and led, for example, to historical actors talking about virtual particles or pairs during the 1930s.

Altogether, the concept of the virtual particle emerged gradually, by the transformations of the representational formats, of the reasoning they allowed for, and of the underlying conceptual and theoretical framework. These transformations were not necessarily synchronized. In the case of vacuum polarization and vacuum fluctuation, for example, there was a considerable alteration of the conceptual framework through the necessity of including nascent or virtual pairs. At the same time, the methods to evaluate the frameworks did not change considerably. Similar comments could be made in the case of exchange forces and in the case of the magnetic moment of the nucleons, at least during the 1930s.

Finally we come to the question: Why did physicists introduce the concept of a virtual particle? If we make the (hopefully uncontroversial) assumption, that new concepts are not introduced light-heartedly by the historical actors, we are led to the related question: What function did the concepts perform for the historical actors? Instead of going through the successive instances of virtual oscillator, virtual transition and state, and virtual particle, I will look at the features they shared in their respective frameworks and practical functions.

First of all a common feature of the "virtual" concepts was the mathematical machinery of perturbation theory to which all of them were connected. In each case, they gave a suitable interpretation of the involved mathematical structures. They allowed the actors to work on or with the mathematical apparatus and the conceptual framework they had at their disposal. The concepts provided a handle on inaccessible processes, most importantly of the light-matter interaction.

Secondly, the concepts we met served a specific function in the respective theories and frameworks. Interpreted physically, virtual oscillators, virtual transitions and states, as well as virtual particles functioned as intermediaries in the theoretical modelling of the interaction between matter and radiation: They served as the bridging elements in the theoretical description of a physical process.[50] They did so in different manners in their specific conceptual frameworks: Virtual oscillators bridged the black-boxing of the light-matter interaction in Bohr's atomic model, i.e. the conceptual gap between between the incoming radiation and the one emitted by the atoms. Virtual transitions would allow for emission and absorption of non-fitting quanta of radiation and thereby the virtual states could function as the interlinks between the initial and final states. Lastly, the mediating function of virtual particles is best displayed by its most prominent graphical representation: They are the particles which do not get out of the diagrams, their function is to allow for the full variety of the processes. Hence, comes the ascribed characteristic of force carriers.

This bridging function has another connotation which came to the fore in Dirac's 1927 article on dispersion: The theoretical entities sooner or later carrying the epithet "virtual" bridged the gap between the abstract mathematical representation of a perturbative expansion and the concrete problem under study. The verbal model, the diagrams and tables of the 1930s, or Feynman's diagrams filled the perturbative series with physical content. They bridged the gap between abstract mathematics and physical instantiation.

In a more philosophical vein, all virtual entities fulfilled this intermediary function in a similar manner. Virtual oscillators stood in for every possible transition from one specific state to any other, without any restrictions on the energy difference between these states. These transition possibilities, although not actualized, became effective, the atom would send out radiation. In Dirac's verbal model, the atom carried out transitions, but the processes themselves were not bound by energy conservation; they could not actually occur. The virtual particle allows for the construction of physical processes from unphysical building blocks.[51] But since energy and momentum are now conserved at the interaction points, the particles themselves do not behave as they should: They do not obey the relativistic energy-momentum relation. All of these concepts, thus, extend the space of possible processes, and the possibilities of these processes become effective agents in the theoretical description. An extended set of possibilities became effective without actualization.

[50] I already referred to this bridging function in Ehberger (2022) in connection with the verbal model of Dirac. Here, I abstract to the different virtual entities discussed in this investigation.

[51] Emission, absorption, pair creation and pair annihilation are physically impossible as long as only one photon and no other potential is involved.

Despite the intricacies of the historical development, the discussed concepts have common features. Virtual oscillators, virtual states, virtual transitions, and virtual particles are concepts belonging to the probability-amplitude level, unobservable by definition. They fulfil their function as intermediaries in the (perturbative) application of the theory, and they did so by extending the set of physically effective entities and processes. The concepts discussed were all central in their conceptual environments as well as in the practical application and evaluation of the theories. When QED became a theory of scattering the virtual particle took centre stage in the practice of theory: In Feynman's hands, QED became a theory of virtual particles.

Archives

AHQP	Archive for the History of Quantum Physics Microfiches accessed at the Library of the Max Planck Institute for the History of Science, Berlin
AHUB	Universitätsarchiv der Hulmboldt-Universität zu Berlin, Berlin
AMPG	Archiv der Max-Planck-Gesellschaft, Berlin
BSC	Bohr Scientific Correspondence Microfiches accessed at the Library of the Max Planck Institute for the History of Science, Berlin
GBA	Guido Beck Archives, Universidade do Estado do Rio de Janeiro
IWA	Ivar Waller Archives, Center for History of Science, Royal Swedish Academy of Science, Stockholm
PAMDP	Paul A.M. Dirac Papers, Florida State University, Special Collections and Archives

M. Ehberger, *Representing the Unobservable*, Science Networks. Historical Studies 68,
https://doi.org/10.1007/978-3-032-09188-8

References

Aaserud, Finn. 1990. *Redirecting science. Niels Bohr, philanthropy, and the rise of nuclear physics*. Cambridge: Cambridge University Press.

Akhiezer, Aleksandr. 1937. Ueber die Streuung von Licht an Licht. *Physikalische Zeitschrift der Sowjetunion* 11: 263–276.

Akhiezer, Aleksandr. 1989. Teacher and friend. In *Landau. The physicist and the man. Recollections of L. D. Landau*, ed. by Isaak Markovich Khalatnikov, 36–56. Oxford: Pergamon Press.

Akhiezer, Aleksandr, Lev Landau, and Isaak Pomeranchuck. 1936. Scattering of light by light. *Nature* 138: 206.

Akhiezer, Aleksandr, and Isaak Pomeranchuck. 1937. Kohaerente Streuung von Gamma-Strahlen durch Kerne. *Physikalische Zeitschrift der Sowjetunion* 11: 478–498.

Alvargonzález, David. 2013. Is the history of science essentially whiggish? *History of Science* 51: 85–99.

Amaldi, Eduardo. 1929. Sulla teoria quantistica dell'effetto Raman. Rendiconti Accademia Nazionale dei Lincei. *Classe di Scienze Fisiche, Matematiche e Naturali* 9: 876–881.

Amaldi, Eduardo, and Emilio Segré. 1929. Sulla teoria dell'effetto Raman. Rendiconti Accademia Nazionale dei Lincei. *Classe di Scienze Fisiche, Matematiche e Naturali* 9: 407–409.

Anonymous. 1925. Minutes of the Washington Meeting, December 29, 30 and 31, 1924. *Physical Review* 25: 233–258.

Anonymous. 1934. International congress on theoretical physics at Kharkov. *Nature* 134: 109–110.

Antoci, Salvatore, and Dierck-E. Liebscher. 1996. The third way to quantum mechanics is the forgotten first. *Annales de la Foundation Louis de Broglie* 21: 349–368.

Arabatzis, Theodore. 2006. *Representing electrons. A biographical approach to theoretical entities*. Chicago, London: The University of Chicago Press.

Aramaki, Seiya. 1987. Formation of the renormalization theory in quantum electrodynamics. *Historia Scientiarum* 32: 1–42.

Arthur, Richard T. W. 2012. Virtual processes and quantum tunnelling as fictions. *Science & Education* 21: 1461–1473.

Bacelar Valente, Mario. 2011a. Are virtual quanta nothing but formal tools? *International Studies in the Philosophy of Science* 25: 39–53.

Bacelar Valente, Mario. 2011b. *Conceptual problems in quantum electrodynamics: a contemporary historical-philosophical approach*. Thesis: Universidad de Sevilla, Sevilla.

Banerjee, Somaditya. 2014. C. V. Raman and colonial physics: acoustics and the quantum. *Physics in Perspective* 16: 146–178.

Beck, Guido. 1930. Über die Streuung von Teilchen durch Kraftfelder. *Zeitschrift für Physik* 62: 331–351.

M. Ehberger, *Representing the Unobservable*, Science Networks. Historical Studies 68,
https://doi.org/10.1007/978-3-032-09188-8

Beck, Guido. 1933. Hat das negative Energiespektrum einen Einfluß auf Kernphänomene? *Zeitschrift für Physik* 83: 498–511.

Becker, Richard. 1924. Über Absorption und Dispersion in Bohrs Quantentheorie. *Zeitschrift für Physik* 27: 173–188.

Beller, Mara. 1999. *Quantum Dialogue. The making of a revolution.* Chicago: The University of Chicago Press.

Bethe, Hans. 1935. On the annihilation radiation of positrons. *Proceedings of the Royal Society of London A: Mathematical, Physical and Engineering Sciences* 150: 129–141.

Bethe, Hans, and Robert Fox Bacher. 1936. Nuclear physics A. Stationary states of nuclei. *Reviews of Modern Physics* 8: 82–229.

Bethe, Hans, and Enrico Fermi. 1932. Über die Wechselwirkung von zwei Elektronen. *Zeitschrift für Physik* 77: 296–306.

Bethe, Hans, and Walter Heitler. 1934. On the stopping of fast particles and on the creation of positive electrons. *Proceedings of the Royal Society of London A: Mathematical, Physical and Engineering Sciences* 146: 83–112.

Bethe, Hans, and Julius Robert Oppenheimer. 1946. Reaction of radiation on electron scattering and Heitler's theory of radiation damping. *Physical Review* 70: 451–458.

Bhabha, Homi Jehangir. 1936. The scattering of positrons by electrons with exchange on Dirac's theory of the positron. *Proceedings of the Royal Society of London A: Mathematical, Physical and Engineering Sciences* 154: 195–206.

Bhabha, Homi Jehangir. 1939. Classical theory of mesons. *Proceedings of the Royal Society A: Mathematical, Physical and Engineering Sciences* 172: 384–409.

Bhabha, Homi Jehangir. 1940. Classical theory of point dipoles. *Nature* 145: 819–820.

Bhabha, Homi Jehangir. 1940. On elementary heavy particles with any integral charge. *Proceedings of the Indian Academy of Sciences-Section A* 11: 347–368.

Bhagavantam, Suri. 1935. Light scattering and Raman effect. *Current Science* 3: 526–531.

Bloch, Felix, and Arnold Nordsieck. 1937. Note on the radiation field of the electron. *Physical Review* 52: 54–59.

Blum, Alexander. 2015. QED and the man who didn't make it: Sidney Dancoff and the infrared divergence. *Studies in History and Philosophy of Science Part B: Studies in History and Philosophy of Modern Physics* 50: 70–94.

Blum, Alexander. 2017. The state is not abolished, it withers away: How quantum field theory became a theory of scattering. *Studies in History and Philosophy of Science Part B: Studies in History and Philosophy of Modern Physics* 60: 46–80.

Blum, Alexander, and Martin Jähnert. 2022. The birth of quantum mechanics from the spirit of radiation theory. *Studies in History and Philosophy of Science* 91: 125–147.

Blum, Alexander, and Martin Jähnert. 2024. Real virtuality and actual transitions: Historical reflections on virtual entities before quantum field theory. *Perspectives on Science* 32: 329–349.

Blum, Alexander, Martin Jähnert, Christoph Lehner, and Jürgen. Renn. 2017. Translation as heuristics: Heisenberg's turn to matrix mechanics. *Studies in History and Philosophy of Science Part B: Studies in History and Philosophy of Modern Physics* 60: 3–22.

Blum, Alexander, and Christian Joas. 2016. From dressed electrons to quasiparticles: The emergence of emergent entities in quantum field theory. *Studies in History and Philosophy of Science Part B: Studies in History and Philosophy of Modern Physics* 53: 1–8.

Bohr, Niels. 1918. On the quantum theory of line-spectra. Part I: On the general theory. Det Kgl. Danske Videnskabernes Selskab. *Skrifter. Naturvidenskabelig of Matematisk Afdeling* 4: 1–36.

Bohr, Niels. 1921. Die Anwendung der Quantentheorie auf periodische Systeme (unpublished 1916). In *Abhandlungen über den Atombau aus den Jahren 1913–196*, ed. by Niels Bohr, 123–151. Braunschweig: Friedrich Vieweg und Sohn.

Bohr, Niels. 1923. Über die Anwendung der Quantentheorie auf den Atombau. *Zeitschrift für Physik* 13: 117–165.

Bohr, Niels, Hendrik Anthony Kramers, and John Clarke Slater. 1924a. LXXVI. The quantum theory of radiation. *The London, Edinburgh, and Dublin Philosophical Magazine and Journal of Science* 47: 785–802.

Bohr, Niels, Hendrik Anthony Kramers, and John Clarke Slater. 1924b. Über die Quantentheorie der Strahlung. *Zeitschrift für Physik* 24: 69–87.

Bohr, Niels, and Léon. Rosenfeld. 1933. Zur Frage der Messbarkeit der Elektromagnetischen Feldgrössen. Det Kgl. *Danske Videnskabernes Selskab. Mathematisk-fysiske Meddelelser* 12: 57–121.

Bokulich, Alisa. 2009. Explanatory fictions. In *Fictions in Science. Philosophical Essays on Modeling and Idealization*, ed. by Mauricio Suárez, 91–109. New York, London: Routledge.

Bokulich, Alisa. 2012. Distinguishing explanatory from nonexplanatory fictions. *Philosophy of Science* 79: 725–737.

Born, Max. 1923. Quantentheorie und Störungsrechnung. *Die Naturwissenschaften* 11: 537–542.

Born, Max. 1924. Über Quantenmechanik. *Zeitschrift für Physik* 26: 379–395.

Born, Max. 1925. *Vorlesungen über Atommechanik*, Vol. 1. Berlin: Verlag von Julius Springer.

Born, Max. 1927. Das Adiabatenprinzip in der Quantenmechanik. *Zeitschrift für Physik* 40: 167–192.

Born, Max. 1928. Zur Theorie des Ramaneffekts. *Naturwissenschaften* 16: 673.

Born, Max. 1938. Some philosophical aspects of modern physics. *Proceedings of the Royal Society of Edinburgh* 57: 1–18.

Born, Max, and James Franck. 1925. Quantentheorie und Molekelbildung. *Zeitschrift für Physik* 31: 411–429.

Born, Max, and Werner Heisenberg. 1923. Über Phasenbeziehungen bei den Bohrschen Modellen von Atomen und Molekeln. *Zeitschrift für Physik* 14: 44–55.

Born, Max, Werner Heisenberg, and Pascual Jordan. 1926. Zur Quantenmechanik. *II. Zeitschrift für Physik* 35: 557–615.

Born, Max, and Pascual Jordan. 1925. Zur Quantenmechanik. *Zeitschrift für Physik* 34: 858–888.

Born, Max, and Pascual Jordan. 1925. Zur Quantentheorie aperiodischer Vorgänge. *Zeitschrift für Physik* 33: 479–505.

Born, Max, and Pascual Jordan. 1930. *Elementare Quantenmechanik.* Berlin: Springer.

Born, Max, and Wolfgang Pauli. 1922. Über die Quantelung gestörter mechanischer Systeme. *Zeitschrift für Physik* 10: 137–158.

Borrelli, Arianna. 2015. Between logos and mythos: Narratives of "Naturalness" in today's particle physics community. In *Narrated communities–narrated realities. Narration as cognitive processing and cultural practice*, ed. by Hermann Blume, Christoph Leitgeb, and Michael Rössner, 69–83. Leiden, Boston: Brill Rodopi.

Borrelli, Arianna. 2017. The uses of isospin in early nuclear and particle physics. *Studies in History and Philosophy of Science Part B: Studies in History and Philosophy of Modern Physics* 60: 81–94.

Borrelli, Arianna. 2021. Between symmetry and asymmetry: spontaneous symmetry breaking as narrative knowing. *Synthese* 198: 3919–3948.

Bothe, Walther, and Hans Geiger. 1925. Über das Wesen des Comptoneffektes; ein experimenteller Beitrag zur Theorie der Strahlung. *Zeitschrift für Physik* 32: 639–663.

Brand, John C. D. 1989. The discovery of the Raman effect. *Notes and Records of the Royal Society of London* 43: 1–23.

Braunbek, Werner, and Edgar Weinmann. 1938. Die Rutherford-Streuung mit Berücksichtigung der Ausstrahlung. *Zeitschrift für Physik* 110: 360–372.

Breit, Gregory. 1924. The quantum theory of dispersion. *Nature* 114: 310.

Breit, Gregory. 1932. Quantum theory of dispersion (I). *Reviews of Modern Physics* 4: 504–576.

Breit, Gregory. 1933. Quantum theory of dispersion (continued). Parts VI and VII. *Reviews of Modern Physics* 5: 91–140.

Breit, Gregory, and Isidor Isaac Rabi. 1934. On the interpretation of present values of nuclear moments. *Physical Review* 46: 230–231.

Breit, Gregory, and John Archibald Wheeler. 1934. Collision of two light quanta. *Physical Review* 46: 1087–1091.

Bretscher, Egon. 1933. Bericht über die physikalische Vortragswoche der Eidg. *Technischen Hochschule Zürich. Helvetica Physica Acta* 6: 411–456.

Bromberg, Joan. 1971. The impact of the neutron: Bohr and Heisenberg. *Historical Studies in the Physical Sciences* 3: 307–341.

Bromberg, Joan. 1977. Dirac's quantum electrodynamics and the wave-particle duality. In *History of Twentieth Century Physics*, ed. by Charles Weiner, 147–157. New York, London: Academic Press.

Bronstein, Matvei. 1936. Über den spontanen Zerfall der Photonen. *Physikalische Zeitschrift der Sowjetunion* 10: 686–688.

Bronstein, Matvei. 1937. O vozmozhnosti spontannogo rachsheplenia fotonov [On the possibility of spontaneous splitting of photons]. *Zhurnal Eksperimentalnoy i Toreticheskoy Fiziki [Journal of Experimental and Theoretical Physics]* 7: 335–358.

Brown, James Robert. 2018. How do Feynman diagrams work? *Perspectives on Science* 26: 423–442.

Brown, Laurie Mark. 1981. Yukawa's prediction of the meson. *Centaurus* 25: 71–132.

Brown, Laurie Mark. 1993. Introduction: Renormalization. 1930–1950. In *Renormalization. From Lorentz to Landau (and Beyond)*, ed. by Laurie Mark Brown, 1–27. New York: Springer.

Brown, Laurie Mark. 2002. The Compton effect as one path to QED. *Studies in History and Philosophy of Science Part B: Studies in History and Philosophy of Modern Physics* 33: 211–249.

Brown, Laurie Mark, and Lillian Hoddeson, eds. 1983. *The birth of particle physics*. Cambridge: Cambridge University Press.

Brown, Laurie Mark, and Donald F. Moyer. 1984. Lady or Tiger? The Meitner-Hupfeld effect and Heisenberg's neutron theory. *American Journal of Physics* 52: 130–136.

Brown, Laurie Mark, and Helmut Rechenberg. 1994. Field theories of nuclear forces in the 1930s: The Fermi-Field theory. *Historical Studies in the Physical and Biological Sciences* 25: 1–24.

Brown, Laurie Mark, and Helmut Rechenberg. 1996. *The origin of the concept of nuclear forces*. Bristol: Institute of Physics Publishing.

Brush, Stephen G. 2007. How ideas became knowledge: The light-quantum hypothesis 1905–1935. *Historical Studies in the Physical and Biological Sciences* 37: 205–246.

Bunge, Mario. 1970. Virtual processes and virtual particles: real or fictitious? *International Journal of Theoretical Physics* 3: 507–508.

Camilleri, Kristian. 2006. Heisenberg and the wave-particle duality. *Studies in History and Philosophy of Science Part B: Studies in History and Philosophy of Modern Physics* 37: 298–315.

Cao, Tian Yu. 1997. *Conceptual development of 20th century field theories*. Cambridge: Cambridge University Press.

Carrelli, Antonio, Peter Pringsheim, and B. Rosen. 1928. Über den Ramaneffekt an wässerigen Lösungen und über den Polarisationszustand der Linien des Ramaneffekts. *Zeitschrift für Physik* 51: 511–519.

Carson, Cathryn. 1996a. The peculiar notion of exchange forces-I: Origins in quantum mechanics, 1926–1928. *Studies in History and Philosophy of Science Part B: Studies in History and Philosophy of Modern Physics* 27: 23–45.

Carson, Cathryn. 1996b. The peculiar notion of exchange forces-II: From nuclear forces to QED, 1929–1950. *Studies in History and Philosophy of Science Part B: Studies in History and Philosophy of Modern Physics* 27: 99–131.

Casimir, Hendrik. 1933. Über die Intensität der Streustrahlung gebundener Elektronen. *Helvetica Physica Acta* 6: 287–304.

Cassidy, David C. 1981. Cosmic ray showers, high energy physics, and quantum field theories: Programmatic Interactions in the 1930s. *Historical Studies in the Physical Sciences* 12: 1–39.

Cassidy, David C. 1995. *Werner Heisenberg. Leben und Werk*. Heidelberg, Berlin, Oxford: Spektrum Akademischer Verlag.

Castagnetti, Giuseppe, and Hubert Goenner. 2020. *Establishing quantum physics in Berlin. Einstein and the Kaiser Wilhelm Institute for Physics, 1917–1922*. Cham: Springer Nature Switzerland.

Chang, Hasok. 2012. Beyond case-studies: History as philosophy. In *Integrating history and philosophy of science. Problems and prospects*, ed. by Seymour Mauskopf and Tad Schmaltz, 109–124. Dodrecht, Heidelberg, London, New York: Springer.

Cini, Marcello. 2001. Fermi and quantum electrodynamics. In *Enrico Fermi: His Work an Legacy*, ed. by Carlo Bernardini and Luisa Bonolis, 126–137. Berlin, Heidelberg: Springer.

Compton, Arthur Holly. 1923. A quantum theory of the scattering of X-rays by light elements. *Physical Review* 21: 483–502.

Compton, Arthur Holly. 1924. The scattering of X-rays. *Journal of the Franklin Institute* 198: 57–72.

Compton, Arthur Holly, and Alfred W. Simon. 1925. Directed quanta of scattered X-rays. *Physical Review* 26: 289–299.

Crull, Elise, and Guido Baccuagaluppi, eds. 2016. *Grete Hermann-between physics and philosophy*. Dodrecht: Springer.

Czerny, Marianus. 1929. Zum Raman-Effekt des Quarzes. *Zeitschrift für Physik* 53: 317–325.

Dancoff, Sidney M. 1939. On radiative corrections for electron scattering. *Physical Review* 55: 959–963.

Darrigol, Olivier. 1982. *Les debuts de la théorie quantique des champs (1925–1948). Thèse pour le Doctorat de Troisième Cycle présentée à l'Université de Paris I (Panthéon-Sorbonne)*. Doctoral thesis, Université de Paris I, Paris.

Darrigol, Olivier. 1986. The origin of quantized matter waves. *Historical Studies in the Physical and Biological Sciences* 16: 197–253.

Darrigol, Olivier. 1988. The quantum electrodynamical analogy in early nuclear theory or the roots of Yukawa's theory. *Revue d'histoire des sciences* 41: 225–297.

Darrigol, Olivier. 1991. Cohérence et complétude de la mécanique quantique: l'exemple de "Bohr-Rosenfeld.âŁž *Revue d'histoire des sciences* 44: 137–179.

Darrigol, Olivier. 1992. *From c-numbers to q-numbers. The classical analogy in the history of quantum theory*. Berkeley: University of California Press.

Darrigol, Olivier. 2008. The modular structure of physical theories. *Synthese* 162: 195–223.

Darrigol, Olivier. 2019. The magic of Feynman's QED: from field-less electrodynamics to Feynman diagrams. *European Physical Journal H* 44: 349–369.

Darwin, Charles Galton. 1928. The Sixth Congress of Russian Physicists. *Nature* 122: 630.

Debye, Peter. 1923. Zerstreuung von Röntgenstrahlen und Quantentheorie. *Physikalische Zeitschrift* 24: 161–166.

Delbrück, Max. 1979. Interview with Max Delbrück by Carolyn Harding. Passadena, California, July 14-September 11, 1978. Oral History Project, California Institute of Technology Archives. Caltech Oral Histories. https://resolver.caltech.edu/CaltechOH:OH_Delbruck_M.

Dieckş, Dennis. 2017. Niels Bohr and the formalism of quantum mechanics. In *Niels Bohr and the philosophy of physics. Twenty-first-century perspectives*, ed. by Jan Faye and Henry Folse, 303–334. London, New York: Bloomsbury Academic.

Dieke, Gerhard Heinrich. 1929. Difference between the absorption and the Raman spectrum. *Nature* 123: 564.

Dirac, Paul Adrien, and Maurice. 1924. The conditions for statistical equilibrium between atoms, electrons and radiation. *Proceedings of the Royal Society of London Series A: Mathematical, Physical and Engineering Sciences* 106: 581–596.

Dirac, Paul Adrien, and Maurice. 1926. On the theory of quantum mechanics. *Proceedings of the Royal Society of London A: Mathematical, Physical and Engineering Sciences* 112: 661–677.

Dirac, Paul Adrien, and Maurice. 1927a. The physical interpretation of the quantum dynamics. *Proceedings of the Royal Society of London A: Mathematical, Physical and Engineering Sciences* 113: 621–641.

Dirac, Paul Adrien, and Maurice. 1927b. The quantum theory of dispersion. *Proceedings of the Royal Society of London A: Mathematical, Physical and Engineering Sciences* 114: 710–728.

Dirac, Paul Adrien, and Maurice. 1927c. The quantum theory of the emission and absorption of radiation. *Proceedings of the Royal Society of London A: Mathematical, Physical and Engineering Sciences* 114: 243–265.

Dirac, Paul Adrien, and Maurice. 1927d. Über die Quantenmechanik der Stoßvorgänge. *Zeitschrift für Physik* 44: 585–595.

Dirac, Paul Adrien, and Maurice. 1928a. The quantum theory of the electron. *Proceedings of the Royal Society of London A: Mathematical, Physical and Engineering Sciences* 117: 610–624.

Dirac, Paul Adrien, and Maurice. 1928b. Zur Quantentheorie des Elektrons. In Leipziger Vorträge,. 1928. Quantentheorie und Chemie. In *85–94*, ed. by Hans Falkenhagen. Leipzig: S. Hirzel.

Dirac, Paul Adrien, and Maurice. 1930a. Note on exchange phenomena in the Thomas atom. *Mathematical Proceedings of the Cambridge Philosophical Society* 26: 376–385.

Dirac, Paul Adrien, and Maurice. 1930b. On the annihilation of electrons and protons. *Proceedings of the Cambridge Philosophical Society* 26: 361–375.

Dirac, Paul Adrien, and Maurice. 1930c. *The principles of quantum mechanics*. Oxford: Clarendon Press.

Dirac, Paul Adrien, and Maurice. 1930d. The proton. *Nature* 126: 605–606.

Dirac, Paul Adrien, and Maurice. 1930e. Quelques problèmes de mécanique quantique. *Annales de l'institut Henri Poincaré* 1: 357–400.

Dirac, Paul Adrien, and Maurice. 1930f. A theory of electrons and protons. *Proceedings of the Royal Society of London A: Mathematical, Physical and Engineering Sciences* 126: 360–365.

Dirac, Paul Adrien, and Maurice. 1931. Quantised singularities in the electromagnetic field. *Proceedings of the Royal Society A: Mathematical, Physical and Engineering Sciences* 133: 60–72.

Dirac, Paul Adrien, and Maurice. 1932. Relativistic quantum mechanics. *Proceedings of the Royal Society of London A: Mathematical, Physical and Engineering Sciences* 136: 453–464.

Dirac, Paul Adrien, and Maurice. 1933. The Lagrangian in quantum mechanics. *Physikalische Zeitschrift der Sowjetunion* 3: 64–72.

Dirac, Paul Adrien, and Maurice. 1934. Discussion of the infinite distribution of electrons in the theory of the positron. *Proceedings of the Cambridge Philosophical Society* 30: 150–163.

Dirac, Paul Adrien Maurice. 1934b. Théorie du positron. In *Structure et propriétés des nouyaux atomiques. Rapports et discussions du septième conseil de physique tenu à Bruxelles du 22 au 29 Octobre 1933*, ed. by Institut International de Physique Solvay, 203–212. Paris: Gauthier Villars.

Dirac, Paul Adrien, and Maurice. 1936. Does conservation of energy hold in atomic processes? *Nature* 137: 298–299.

Dirac, Paul Adrien Maurice. 1942. Bakerian Lecture–The physical interpretation of quantum mechanics. *Proceedings of the Royal Society of London. Series A: Mathematical, Physical and Engineering Sciences* 180: 1–40.

Dirac, Paul Adrien, and Maurice. 1947. *The principles of quantum mechanics*, 3rd ed. Oxford: Clarendon Press.

Dirac, Paul Adrien Maurice. 1995. Teoriya pozitrona [Theory of the positron] (originally published: 1934). In *The Collected Works of P.A.M. Dirac*, ed. by Richard Henry Dalitz, Vol. 1, 720–759. Cambridge, New York: Cambridge University Press.

Dirac, Paul Adrien Maurice, Vladimir Fock, and Boris Podolsky. 1932. On quantum electrodynamics. *Physikalische Zeitschrift der Sowjetunion* 2: 468–479.

Dresden, Max. 1987. *H. A. Kramers. Between tradition and revolution*. New York: Springer.

Dresden, Max. 1993. Renormalization in historical perspective–The first stage. In *Renormalization. From Lorentz to Landau (and beyond)*, ed. by Laurie Mark Brown, 29–56. New York: Springer.

Duncan, Anthony. 2012. *The conceptual framework of quantum field theory*. Oxford: Oxford University Press.

Duncan, Anthony, and Michel Janssen. 2007. On the verge of Umdeutung in Minnesota: Van Vleck and the correspondence principle. Part one. *Archive for History of Exact Sciences* 61: 553–624.

Duncan, Anthony, and Michel Janssen. 2007. On the verge of Umdeutung in Minnesota: Van Vleck and the correspondence principle. Part two. *Archive for History of Exact Sciences* 61: 625–671.

Duncan, Anthony, and Michel Janssen. 2009. From canonical transformations to transformation theory, 1926–1927: The road to Jordan's Neue Begründung. *Studies in History and Philosophy of Science Part B: Studies in History and Philosophy of Modern Physics* 40: 352–362.

Dyson, Freeman John. 1949. The radiation theories of Tomonaga, Schwinger, and Feynman. *Physical Review* 75: 486–502.

Dyson, Freeman John. 1949. The S Matrix in quantum electrodynamics. *Physical Review* 75: 1736–1755.

Dyson, Freeman John. 1952. Divergence of perturbation theory in quantum electrodynamics. *Physical Review* 85: 631–632.

Dyson, Freeman John. 2007. Hans Bethe and quantum electrodynamics. *Physics Today* 58: 48.

Eckert, Michael. 2001. The emergence of Quantum Schools: Munich, Göttingen and Copenhagen as new centers of atomic theory. *Annalen der Physik* 10: 151–162.

Eckert, Michael. 2013. *Arnold Sommerfeld. Science, life and turbulent times. 1868–1951*. New York: Springer.

Eckert, Michael. 2020. *Establishing quantum physics in Munich: Emergence of Arnold Sommerfeld's quantum school*. Cham: Springer Nature Switzerland.

Eckert, Michael, and Karl Märker, eds. 2004. *Arnold Sommerfeld. Wissenschaftlicher Briefwechsel. Band 2: 1919–1951*. Berlin, Diepholz, München: Verlag für Geschichte der Naturwissenschaften und der Technik.

Ehberger, Markus. 2020. I'm not there. Or: Was the virtual particle ever born? In *Biographies in the history of physics. Actors, objects, institutions*, ed. by Christian Forstner and Mark Walker, 261–280. Cham: Springer Nature Switzerland.

Ehberger, Markus. 2022. "The language of Dirac's theory of radiation": The inception and initial reception of a tool for the quantum field theorist. *Archive for History of Exact Sciences* 76: 531–571.

Ehberger, Markus. 2024. How to study virtual entities historically? A proposal. *Perspectives on Science* 32: 278–299.

Ehrenfest, Paul, and Richard Chace Tolman. 1924. Weak quantization. *Physical Review* 24: 287–295.

Einstein, Albert. 1905. Über einen die Erzeugung und Verwandlung des Lichtes betreffenden heuristischen Gesichtspunkt. *Annalen der Physik* 322: 132–148.

Einstein, Albert. 1916a. Quantentheorie der Strahlung. *Mitteilungen der Physikalischen Gesellschaft, Zürich* 18: 47–62.

Einstein, Albert. 1916b. Strahlungs-Emission und -Absorption nach der Quantentheorie. *Verhandlungen der Deutschen Physikalischen Gesellschaft* 18: 31–323.

Einstein, Albert, and Paul Ehrenfest. 1923. Zur Quantentheorie des Strahlungsgleichgewichts. *Zeitschrift für Physik* 19: 301–306.

Ellis, J. W. 1929. An apparently anomalous Raman effect in water. *Nature* 123: 205–206.

Enz, Charles P. 2002. *No time to be brief. A scientific biography of Wolfgang Pauli*. Oxford: Oxford University Press.

Estermann, Immanuel, and Otto Stern. 1933. Über die magnetische Ablenkung von Wasserstoffmolekülen und das magnetische Moment des Protons. *II. Zeitschrift für Physik* 85: 17–24.

Euler, Hans. 1936. Über die Streuung von Licht an Licht nach der Diracschen Theorie. *Annalen der Physik* 418: 398–448.

Euler, Hans, and Bernhard Kockel. 1935. Über die Streuung von Licht an Licht nach der Diracschen Theorie. *Naturwissenschaften* 23: 246–247.

Fabelinskii, Immanuil L. 1978. The discovery of combinational scattering of light (the Raman effect). *Soviet Physics Uspekhi* 21: 780–797.

Farmelo, Graham. 2009. *The strangest man. The hidden life of Paul Dirac, quantum genius*. London: Faber and Faber.

Faye, Jan. 2019. Copenhagen interpretation of quantum mechanics. The Stanford Encyclopedia of Philosophy (Winter 2019 Edition). https://plato.stanford.edu/archives/win2019/entries/qm-copenhagen/.

Feinberg, Gerald. 1972. Philosophical implications of contemporary particle physics. In *Paradigms and paradoxes. The philosophical challenge of the quantum domain*, ed. by Robert G. Colodny, 33–46. Pittsburgh: University of Pittsburgh Press.

Fermi, Enrico. 1929. Sulla teoria quantistica delle frange di interferenza. Atti della Reale Accademia dei Lincei. *Rendiconti* 10: 72–77.

Fermi, Enrico. 1932. Quantum theory of radiation. *Reviews of Modern Physics* 4: 87–132.

Fermi, Enrico. 1934. Versuch einer Theorie der Beta-Strahlen. *I. Zeitschrift für Physik* 88: 161–177.

Fermi, Enrico. 1950. *Nuclear physics: A course given by Enrico Fermi at the University of Chicago*. Chicago: University of Chicago Press.

Fermi, Enrico, and George Eugene Uhlenbeck. 1933. On the recombination of electrons and positrons. *Physical Review* 44: 510–511.

Feynman, Richard Phillips. 1942. *The principle of least action in quantum mechanics*. Thesis: Princeton University, Princeton.

Feynman, Richard Phillips. 1948. A relativistic cut-off for classical electrodynamics. *Physical Review* 74: 939–946.

Feynman, Richard Phillips. 1948b. Space-time approach to non-relativistic quantum mechanics. *Reviews of Modern Physics* 20.

Feynman, Richard Phillips. 1949. Space-time approach to quantum electrodynamics. *Physical Review* 76: 769–789.

Feynman, Richard Phillips. 1949. The theory of positrons. *Physical Review* 76: 749–759.

Feynman, Richard Phillips. 1950. Mathematical formulation of the quantum theory of electromagnetic interaction. *Physical Review* 80: 440–457.

Fick, Dieter, and Horst Kant. 2013. The concepts of light atoms and light molecules and their final interpretation. In *Traditions and transformations in the history of quantum physics: Third International Coference on the History of Quantum Physics, Berlin, June 28- July 2, 2010*, ed. by Shaul Katzir, Christoph Lehner, and Jürgen. Renn. Berlin: Edition Open Access.

Fierz, Markus. 1937. Zur Fermischen Theorie des Beta-Zerfalls. *Zeitschrift für Physik* 104: 553–565.

Fierz, Markus. 1974. Wolfgang Pauli. In *Dictionary of scientific biography*, ed. by Ch. C. Gillispie, Vol. 10, 422–425. New York: Charles Scribner and Sons.

Fierz, Markus, and Victor Weisskopf, eds. 1960. *Theoretical physics in the twentieth century. A memorial volume to Wolfgang Pauli*. New York: Interscience Publishers.

Fleck, Ludwik. 1980. *Entstehung und Entwicklung einer wissenschaftlichen Tatsache. Einführung in die Lehre vom Denkstil und Denkkollektiv* (originally published 1935), 9th ed. Frankfurt a. M.: Suhrkamp.

Fock, Vladimir. 1933. Zur Theorie der Positronen. *Doklady Akademii nauk SSSR* 1: 267–271.

Fock, Vladimir, and Boris Podolsky. 1932. On the quantization of electromagnetic waves and the interaction of charges in Dirac's theory. *Physikalische Zeitschrift der Sowjetunion* 1: 801–817.

Fock, Vladimir, and Boris Podolsky. 1932. Zur Diracschen Quantenelektrodynamik. *Physikalische Zeitschrift der Sowjetunion* 1: 798–800.

Forstner, Christian. 2007. *Quantenmechanik im Kalten Krieg. David Bohm und Richard Feynman*. Diepholz, Stuttgart, Berlin: Verlag für Geschichte der Naturwissenschaften und der Technik.

Fox, Tobias. 2008. Haunted by the spectre of virtual particles: A philosophical reconsideration. *Journal for General Philosophy of Science* 39: 35–51.

Franck, James. 1928. Beitrag zum Problem der Wiedervereinigung von Ionen und Elektronen. *Zeitschrift für Physik* 47: 509–516.

Frenkel, Jakow. 1926. Zur Theorie des Faradayeffektes. *Zeitschrift für Physik* 36: 215–250.

Frenkel, Jakow. 1929a. *Einführung in die Wellenmechanik*. Berlin: Springer.

Frenkel, Jakow. 1929b. The quantum theory of the absorption of light. *Nature* 124: 758–759.

Frenkel, Jakow. 1929c. Über quantenmechanische Energieübertragung zwischen atomaren Systemen. *Zeitschrift für Physik* 58: 794–804.

Frenkel, Jakow. 1934. *Wave mechanics. Advanced general theory*. Oxford: Clarendon Press.

Frenkel, Victor. 1974. Yakov (James) Il'ich Frenkel (1894–1952): Materials for his scientific biography. *Archive for History of Exact Sciences* 13: 1–26.

Frenkel, Victor. 1996. *Yakov Ilich Frenkel. His work, life and letters.* Basel, Boston, Berlin: Birkhäuser Verlag.

Frenkel, Victor, and Gennady E. Gorelik. 1994. *Matvei Petrovich Bronstein and Soviet theoretical physics in the thirties.* Basel: Birkhäuser.

Frisch, Otto Robert, and Otto Stern. 1933. Über die magnetische Ablenkung von Wasserstoffmolekülen und das magnetische Moment des Protons. *I. Zeitschrift für Physik* 85: 4–16.

Fröhlich, Herbert, and Walter Heitler. 1938. Magnetic moments of the proton and the neutron. *Nature* 141: 37–38.

Fröhlich, Herbert, Walter Heitler, and Boris Kahn. 1939. Deviation from the Coulomb law for the proton. *Proceedings of the Royal Society A: Mathematical, Physical and Engineering Sciences* 171: 269–280.

Fröhlich, Herbert, Walter Heitler, and Nicholas Kemmer. 1938. On the nuclear forces and the magnetic moments of the neutron and the proton. *Proceedings of the Royal Society of London A: Mathematical, Physical and Engineering Sciences* 166: 154–177.

Furry, Wendell Hinkle. 1937. A symmetry theorem in the positron theory. *Physical Review* 51: 125–129.

Furry, Wendell Hinkle, and J. F. Carlson. 1933. On the production of positive electrons by electrons. *Physical Review* 44: 237–238.

Furry, Wendell Hinkle, and Julius Robert Oppenheimer. 1934. On the theory of the electron and positive. *Physical Review* 45: 245–262.

Galison, Peter. 1997. *Image and logic. A material culture of microphysics.* Chicago: The University of Chicago Press.

Galison, Peter. 1998. Feynman's war: Modelling weapons, modelling nature. *Studies in History and Philosophy of Science Part B: Studies in History and Philosophy of Modern Physics* 29: 391–434.

Galison, Peter. 2008. The discovery of the muon and the failed revolution against quantum electrodynamics (originally published 1982). *Centaurus* 50: 105–159.

Galison, Peter, and Andrew Warwick. 1998. Introduction: Cultures of theory. *Studies in History and Philosophy of Science Part B: Studies in History and Philosophy of Modern Physics* 29: 287–294.

Gamow, George. 1937. *Structure of atomic nuclei and nuclear transformation.* Oxford: Clarendon Press.

Gamow, George, and Edward Teller. 1937. Some generalizations of the beta transformation theory. *Physical Review* 51: 289–289.

Ganesan, A. S. 1929. Bibliography of 150 papers on the Raman effect. *Indian Journal of Physics* 4: 281–346.

Giulini, Domenico. 2013. Max Born's Vorlesungen über Atommechanik, Erster Band. In *Research and pedagogy: A history of quantum physics through its textbooks*, ed. by Massimiliano Badino and Jaume Navarro, 203–226. Berlin: Edition Open Access.

Göppert, Maria. 1929. Über die Wahrscheinlichkeit des Zusammenwirkens zweier Lichtquanten in einem Elementarakt. *Naturwissenschaften* 17: 932–932.

Göppert-Mayer, Maria. 1931. Über Elementarakte mit zwei Quantensprüngen. *Annalen der Physik* 401: 273–294.

Gordon, Walter. 1926. Der Comptoneffekt nach der Schrödingerschen Theorie. *Zeitschrift für Physik* 40: 117–133.

Grandin, Karl. 2008. Intermediate theoretical physics. In *Aurora Torealis: Studies in the history of science and ideas in honor of Tore Frängsmyr*, ed. by Marco Beretta, Karl Grandin, and Svante Lindqvist, 193–214. Sagamore Beach: Science History Pbulications.

Greenspan, Nancy Thorndike. 2005. *The end of a certain world. The life and science of Max Born. The Nobel physicist who ignited the quantum revolution.* New York: Basic Books.

Griffiths, David. 2004. *Introduction to elementary particles.* Weinheim: Wiley-VCH Verlag.

Güttinger, Paul. 1932. Über Streuprozesse höherer Ordnung. *Helvetica Physica Acta* 5: 238–261.

Góra, Edwin. 1938. Zur Theorie des Comptoneffekts. *Acta Physica Polonica* 7: 159–176.

Góra, Edwin. 1943. Quantentheorie der Strahlungsdämpfung. *Zeitschrift für Physik* 120: 121–147.

Hall, Karl. 2008. The schooling of Lev Landau: The European context of postrevolutionary Soviet theoretical physics. *Osiris* 23: 230–259.

Halpern, Otto. 1933. Scattering processes produced by electrons in negative energy states. *Physical Review* 44: 855–856.

Hartmann, Stephen. 1999. Models and stories in hadron physics. In *Models as mediators. Perspectives on natural and social science*, ed. by Mary S. Morgan and Margaret Morrison, 236–246. Cambridge: Cambridge University Press.

Havas, Paul. 1944. On the interaction of radiation and two electrons. *Physical Review* 66: 69–76.

Heisenberg, Werner. 1925. Über quantentheoretische Umdeutung kinematischer und mechanischer Beziehungen. *Zeitschrift für Physik* 33: 879–893.

Heisenberg, Werner. 1927. Über den anschaulichen Inhalt der quantentheoretischen Kinematik und Mechanik. *Zeitschrift für Physik* 43: 172–198.

Heisenberg, Werner. 1930a. *The physical principles of the quantum theory*. Translated by C. Eckart and F. C. Hoyt. Chicago: Dover Publications.

Heisenberg, Werner. 1930b. *Die physikalischen Prinzipien der Quantentheorie*. Leipzig: Verlag von S. Hirzel.

Heisenberg, Werner. 1931a. Bemerkungen zur Strahlungstheorie. *Annalen der Physik* 401: 338–346.

Heisenberg, Werner. 1931b. Über Energieschwankungen in einem Strahlungsfeld. *Berichte über die Verhandlungen der sächsischen Akademie der Wissenschaften zu Leipzig. Mathematisch-Physische Klasse* 23: 3–9.

Heisenberg, Werner. 1932. Theoretische Überlegungen zur Höhenstrahlung. *Annalen der Physik* 405: 430–452.

Heisenberg, Werner. 1932. Über den Bau der Atomkerne. *I. Zeitschrift für Physik* 77: 1–11.

Heisenberg, Werner. 1932. Über den Bau der Atomkerne. *II. Zeitschrift für Physik* 78: 156–164.

Heisenberg, Werner. 1933. Über den Bau der Atomkerne. *III. Zeitschrift für Physik* 80: 587–596.

Heisenberg, Werner. 1934. Bemerkungen zur Diracschen Theorie des Positrons. *Zeitschrift für Physik* 90: 209–231.

Heisenberg, Werner. 1934b. Considérations théoriques générales sur la structure du noyau. In *Structure et propriétés des nouyaux atomiques. Rapports et discussions du septième conseil de physique tenu à Bruxelles du 22 au 29 Octobre 1933*, ed. by Institut International de Physique Solvay, 289–323. Paris: Gauthier Villars.

Heisenberg, Werner. 1934. Über die mit der Entstehung von Materie aus Strahlung verknüpften Ladungsschwankungen. *Verhandlungen der Sächsischen Akademie der Wissenschaften* 86: 317–322.

Heisenberg, Werner. 1934. Wandlungen der Grundlagen der exakten Naturwissenschaften in jüngster Zeit. *Die Naturwissenschaften* 22: 669–675.

Heisenberg, Werner. 1935a. Bemerkungen zur Theorie des Atomkerns. In *Pieter Zeeman 1865–25 Mei-1935. Verhandelingen Op 25 Mei 1935 aangeboden aan Prof. Dr. P. Zeeman*, 108–116. 's-Gravenhage: Martinus Nijhoff.

Heisenberg, Werner. 1935. Die Struktur der leichten Atomkerne. *Zeitschrift für Physik* 96: 473–484.

Heisenberg, Werner. 1937. Gedanken der antiken Naturphilosophie in der modernen Physik. Die Antike. *Zeitschrift für Kunst und Kultur des klassischen Altertums* 13: 118–124.

Heisenberg, Werner. 1946. Der unanschauliche Quantensprung. *Physikalische Blätter* 2: 4–6.

Heisenberg, Werner. 1949. *Wandlungen in den Grundlagen der Naturwissenschaften (8.* Enlarged. Stuttgart: S. Hirzel Verlag.

Heisenberg, Werner. 1958. *Physics and philosophy. The revolution of modern science*. New York: Harper and Brothers Publishers.

Heisenberg, Werner. 1972. *Der Teil und das Ganze. Gespräche im Umkreis der Atomphysik*, 4th ed. München: Piper und co.

Heisenberg, Werner, and Hans Euler. 1936. Folgerungen aus der Diracschen Theorie des Positrons. *Zeitschrift für Physik* 98: 714–732.

Heisenberg, Werner, and Wolfgang Pauli. 1929. Zur Quantendynamik der Wellenfelder. *Zeitschrift für Physik* 56: 1–61.

Heitler, Walter. 1933. Bemerkung über die Linienform bei Anregung durch Elektronenstoß. *Zeitschrift für Physik* 82: 146–150.

Heitler, Walter. 1933. Über die bei sehr schnellen Stößen emittierte Strahlung. *Zeitschrift für Physik* 84: 145–167.

Heitler, Walter. 1936. *The quantum theory of radiation*. Oxford: Clarendon Press.

Heitler, Walter. 1940. Scattering of mesons and the magnetic moments of proton and neutron. *Nature* 145: 29–30.

Heitler, Walter. 1941. The influence of radiation damping on the scattering of light and mesons by free particles I. *Proceedings of the Cambridge Philosophical Society* 37: 291–300.

Heitler, Walter. 1941. Physical concepts of the meson theory of the atomic nucleus. *Nature* 148: 680–683.

Heitler, Walter. 1944. *The quantum theory of radiation*, 2nd ed. London: Oxford University Press.

Heitler, Walter, and Gerhard Herzberg. 1929. Gehorchen die Stickstoffkerne der Boseschen Statistik? *Naturwissenschaften* 17: 673–674.

Heitler, Walter, and S. T. Ma. 1940. Inner excited states of the proton and neutron. *Proceedings of the Royal Society of London A: Mathematical, Physical and Engineering Sciences* 176: 368–397.

Heitler, Walter, and Lothar Nordheim. 1934. Sur la production des paires par des chocs de particules lourdes. *Journal de Physique et le Radium* 5: 449–454.

Heitler, Walter, and Lothar Nordheim. 1934. Über die Wahrscheinlichkeit von Mehrfachprozessen bei sehr hohen Energieen. *Physica* 1: 1059–1072.

Heitler, Walter, and Huanwu Peng. 1942. The influence of radiation damping on the scattering of mesons: II. Multiple Processes. *Proceedings of the Cambridge Philosophical Society* 38: 296–312.

Heitler, Walter, and Fritz Sauter. 1933. Stopping of fast particles with emission of radiation and the birth of positive electrons. *Nature* 132: 892.

Hendry, John. 1981. Bohr-Kramers-Slater: A virtual theory of virtual oscillators and its role in the history of quantum mechanics. *Centaurus* 25: 189–221.

Hentschel, Klaus. 2014. Zur Rezeption von Vaihingers Philosophie des Als Ob in der Physik. In *Fiktion und Fiktionalismus. Beiträge zu Hans Vaihingers Philosophie des Als Ob*, ed. by Matthias Neuber, 161–186. Würzburg: Königshausen und Neumann.

Hentschel, Klaus. 2017. *Lichtquanten. Die Geschichte des komplexen Konzepts und mentalen Modells von Photonen*. Berlin: Springer Spektrum.

Herzfeld, Karl Fedinand. 1924. Versuch einer quantenhaften Deutung der Dispersion. *Zeitschrift für Physik* 23: 341–360.

Hill, E. L., and Edwin C. Kemble. 1929. On the Raman effect in gases. *Proceedings of the National Academy of Sciences of the United States of America* 15: 387–392.

Hoch, Paul K. 1983. The reception of Central European refugee physicists of the 1930s: U.S.S.R., U.K., U.S.A. *Annals of Science* 40: 217–246.

Hulme, Henry R. 1932. The Faraday effect in ferromagnetics. *Proceedings of the Royal Society of London A: Mathematical, Physical and Engineering Sciences* 135: 237–257.

Hund, Friedrich. 1967. *Geschichte der Quantentheorie*. Mannheim: Bibliographisches Institut.

Im, Gyeong Soon. 1996. Experimental Constraints on Formal Quantum Mechanics: The Emergence of Born's Quantum Theory of Collision Processes in Göttingen, 1924–1927. *Archive for History of Exact Sciences* 50: 73–101.

Institut International de Physique Solvay, eds. 1950. *Les Particules élémentaires: Huitième Conseil de Physique, tenu à l'Uinversité de Bruxelles du 27 Septembre au 2 Octobre 1948. Rapports et Discussions*. Bruxelles: R. Stoops.

Ito, Kenji. 2002. *Making sense of Ryoshiron (Quantum Theory): Introduction of quantum mechanics into Japan, 1920–1940*. Thesis: Harvard University, Cambridge, Massachusetts.

Ivanenko, Dimitri. 1934. Interaction of neutrons and protons. *Nature* 133: 981–982.

Ivanenko, Dimitri, and Vladimir Fock. 1929. Vorträge und Diskussionen der theoretisch-physikalischen Konferenz in Charkow (19.-25. Mai 1929). *Physikalische Zeitschrift* 30: 700–720.

Ivanenko, Dimitri, and Arseni Sokolov. 1936. Zur Wechselwirkung der schweren Teilchen. *Zeitschrift für Physik* 102: 119–131.

Jackson, J. David., and Kurt Gottfried. 2004. Victor Frederick Weisskopf. 1908–2002. *Biographical Memoirs of the National Academy of Sciences* 84: 373–397.

Jacobsen, Anja Skaar. 2012. *Léon Rosenfeld. Physics, philosophy, and politics in the twentieth century*. Singapore: Wold Scientific Publishing Co. Pte. Ltd.

Jaeger, Gregg. 2019. Are Virtual Particles Less Real? *Entropy* 21: 141.

Jähnert, Martin. 2019. *Practicing the correspondence principle in the old quantum theory. A transformation through implementation*. Cham: Springer Nature Switzerland.

James, Jeremiah, and Christian Joas. 2015. Subsequent and subsidiary? Rethinking the role of applications in establishing quantum mechanics. *Historical Studies in the Natural Sciences* 45: 641–702.

Jammer, Max. 1966. *The conceptual development of quantum mechanics*. New York: McGraw-Hill.

Jauch, Josef M., and Fritz Rohrlich. 1959. *The theory of positrons and electrons. The relativistic quantum field theory of charged particles with spin one-half*, 2nd ed. Reading: Addison-Wesley Publishing.

Joas, Christian, and Thiago Hartz. 2019. Quantum cultures: Historical perspectives on the practices of quantum physics. *Berichte zur Wissenschaftsgeschichte* 42: 281–289.

Jordan, Pascual. 1924. Zur Theorie der Quantenstrahlung. *Zeitschrift für Physik* 30: 297–319.

Jordan, Pascual. 1927. Über eine neue Begründung der Quantenmechanik. *Zeitschrift für Physik* 40: 809–838.

Jordan, Pascual. 1936a. *Anschauliche Quantentheorie. Eine Einführung in die moderne Auffassung der Quantenerscheinungen*. Berlin: Verlag von Julius Springer.

Jordan, Pascual. 1936. Fortschritte der Theorie der Atomkerne. *Die Naturwissenschaften* 24: 209–216.

Jordan, Pascual, and Oskar Klein. 1927. Zum Mehrkörperproblem der Quantentheorie. *Zeitschrift für Physik* 45: 751–765.

Jordan, Pascual, and Wolfgang Pauli. 1928. Zur Quantenelektrodynamik ladungsfreier Felder. *Zeitschrift für Physik* 47: 151–173.

Jordan, Pascual, and Eugene Wigner. 1928. Über das Paulische Äquivalenzverbot. *Zeitschrift für Physik* 47: 631–651.

Jordi Taltavull, Marta. 2013. Challenging the boundaries between classical and quantum physics: The case of optical dispersion. In *Traditions and transformations in the history of quantum physics. HQ-3: Third International Conference on the History of Quantum Physics, Berlin, June 28- July 2, 2010.*, ed. by Jürgen Renn, Shaul Katzir, and Christoph Lehner, 29–59. Berlin: Edition Open Access.

Jordi Taltavull, Marta. 2016. Transmitting knowledge across divides: Optical dispersion from classical to quantum physics. *Historical Studies in the Natural Sciences* 46: 315–359.

Jordi Taltavull, Marta. 2017. *Transformation of optical knowledge from 1870 to 1925: Optical dispersion between classical and quantum physics*. Doctoral thesis, Humboldt Universität, Berlin.

Jordi Taltavull, Marta. 2020. Rudolf Ladenburg and the first quantum interpretation of optical dispersion. *European Physical Journal H* 45: 123–173.

Josephson, Paul, Yury Ranyuk, Ivan Tsekhmistro, and Karl Hall. 2010. Science and the periphery under Stalin: Physics in Ukraine. In *Physics and politics. Research and research support in twentieth century Germany in international perspective*, ed. by Helmuth Trischler and Mark Walker, 197–226. Stuttgart: Franz Steiner Verlag.

Jost, Res. 1972. Foundation of quantum field theory. In *Aspects of quantum theory*, ed. by Abdus Salam and Eugene Wigner, 61–78. Cambridge: Cambridge University Press.

Kahn, Boris. 1936. On some further consequences of Fermi's theory of the Beta-radioactivity. *Physica* 3: 495–502.

Kaiser, David. 2005. *Drawing theories apart: The dispersion of Feynman diagrams in postwar physics*. Chicago, London: The University of Chicago Press.

Kane, Gordon. 2006. Are virtual particles really constantly popping in and out of existence? Or are they merely a mathematical bookkeeping device for quantum mechanics? Scientific American. https://www.scientificamerican.com/article/are-virtual-particles-rea/.

Kawabe, Rokuo. 1991. Yukawa's first two talks. *Progress of Theoretical Physics Supplement* 105: 241–261.

Kelly, Cynthia C. 2006. *Oppenheimer and the Manhattan Project. Insights into J. Robert Oppenheimer, "Father of the Atomic Bomb"*. Singapore: World Scientific Publishing.

Kemmer, Nicholas. 1935. Über die elektromagnetische Masse des Diracelektrons. *Annalen der Physik* 414: 674–712.

Kemmer, Nicholas. 1937. Über die Lichtstreuung an elektrischen Feldern nach der Theorie des Positrons. *Helvetica Physica Acta* 10: 112–122.

Kemmer, Nicholas, and Guido Ludwig. 1937. Über die Lichtstreuung an elektrischen Feldern nah der Theorie des Positrons II. *Helvetica Physica Acta* 10: 182–184.

Kemmer, Nicholas, and Victor Weisskopf. 1936. Deviations from the Maxwell equations resulting from the theory of the positron. *Nature* 137: 659.

Kevles, Daniel. 1979. *The physicists. The history of a scientific community in modern America, 1st Vintage*. Book Ed. New York: Random House.

Kikuchi, Seishi. 1930. Über die Fortpflanzung von Lichtwellen in der Heisenberg-Paulischen Formulierung der Quantenelektrodynamik. *Zeitschrift für Physik* 66: 558–571.

Kindi, Vasso. 2012. Concepts as vessels and concepts as use. In *Scientific concepts and investigative practice*, ed. by Uljana Feest and Friedrich Steinle, 23–46. Berlin, Boston: De Gruyter.

Klein, Martin J. 1964. Einstein and the wave-particle duality. *The Informant Philosopher* 3: 1–49.

Klein, Martin J. 1970. The first phase of Bohr-Einstein dialogue. *Historical Studies in the Physical Sciences* 2: 1–39.

Klein, Oskar. 1927. Elektrodynamik und Wellenmechanik vom Standpunkt des Korrespondenzprinzips. *Zeitschrift für Physik* 41: 407–442.

Klein, Oskar. 1929. Die Reflexion von Elektronen an einem Potentialsprung nach der relativistischen Dynamik von Dirac. *Zeitschrift für Physik* 53: 157–165.

Klein, Oskar, and Yoshio Nishina. 1929. Über die Streuung von Strahlung durch freie Elektronen nach der neuen relativistischen Quantendynamik von Dirac. *Zeitschrift für Physik* 52: 853–868.

Ko, Insok. 1997. *Korrespondenz und Komplementarität: ihre Rolle für Theoriendynamik und wissenschafltichen Fortschritt*. Doctoral thesis, Universität Konstanz, Konstanz.

Kockel, Bernhard. 1937. Über einige Mehrfachprozesse zwischen Elektronen, Positronen und Lichtquanten. *Zeitschrift für Physik* 107: 153–176.

Kockel, Bernhard. 1949. Prozesse zwischen leichten Teilchen nach der Diracschen Theorie (Mit 1 Abbildung). *Annalen der Physik* 439: 279–302.

Kojevnikov, Alexei. 1993. *Paul Dirac and Igor Tamm correspondence part I: 1928–1933*. München: Max-Planck-Institut für Physik.

Kojevnikov, Alexei. 2002. Dirac's quantum electrodynamics. In *Einstein studies in Russia*, ed. by Yuri Balashov and Vladimir Vizgin, 229–259. Boston, Basel: Birkhäuser.

Kojevnikov, Alexei. 2004. *Stalin's great science. The times and adventures of Soviet physics*. London: Imperial College Press.

Kojevnikov, Alexei. 2020. *The Copenhagen network. The birth of quantum mechanics from a postdoctoral perspective*. Cham: Springer Nature Switzerland.

Kojevnikov, Alexei, and Olga Novik. 1989. Analysis of informational ties dynamics in early quantum mechanics (1925–1927). *Acta Historiae rerum naturalium necnon technicarum* 20: 115–159.

Konno, Hiroyuki. 1993. Kramers' negative dispersion, the virtual oscillator model, and the correspondence principle. *Centaurus* 36: 117–166.

Kopfermann, Hans, and Rudolf Ladenburg. 1928. Experimenteller Nachweis der "negativenâŁž Dispersion. *Zeitschrift für Physikalische Chemie* 139A: 375–385.

Kossel, Walther. 1916. Bemerkungen zum Seriencharakter der Röntgenspektren. *Verhandlungen der Deutschen Physikalischen Gesellschaft* 18: 339–359.

Kragh, Helge. 1981. The genesis of Dirac's relativistic theory of electrons. *Archive for History of Exact Sciences* 24: 31–67.

Kragh, Helge. 1990. *Dirac. A scientific biography*. Cambridge: Cambridge University Press.

Kragh, Helge. 1992. Relativistic collisions: The work of Christian Møller in the early 1930s. *Archive for History of Exact Sciences* 43: 299–328.

Kragh, Helge. 1999. *Quantum Generations. A history of physics in the twentieth century*. Princeton, New Jersey: Princeton University Press.

Kragh, Helge. 2013. Paul Dirac and The Principles of Quantum Mechanics. In *Research and pedagogy: A history of quantum physics through its textbooks*, ed. by Massimiliano Badino and Jaume Navarro, 245–260. Berlin: Edition Open Access.

Kragh, Helge. 2017. Is the universe expanding? Fritz Zwicky and early tired-light hypotheses. *Journal of Astronomical History and Heritage* 20: 2–12.

Kramers, Hendrik Anthony. 1919. *Intensities of spectral lines*. Kopenhagen: Andr. Fred, Host. and Son, Kgl. Hof-Boghandel.

Kramers, Hendrik Anthony. 1924a. The law of dispersion and Bohr's theory of spectra. *Nature* 113: 673–674.

Kramers, Hendrik Anthony. 1924b. The quantum theory of dispersion. *Nature* 114: 310–311.

Kramers, Hendrik Anthony, and Werner Heisenberg. 1925. Über die Streuung von Strahlung durch Atome. *Zeitschrift für Physik* 31: 681–708.

Kramers, Hendrik Anthony, and Helge Holst. 1925. *Das Atom und die Bohrsche Theorie seines Baues. Gemeinverständlich dargestellt.* Translated by F. Arndt. Berlin: Verlag von Julius Springer.

Kromos-Buchwald, Diana, József Illy, Ze'ev Rosenkranz, and Tilman Sauer, eds. 2012. *The collected papers of Albert Einstein. Volume 13: The Berlin years: Writings and correspondence, January 1922–March 1923*. Princeton, New Jersey: Princeton University Press.

Kromos-Buchwald, Diana, József Illy, Ze'ev Rosenkranz, Tilman Sauer, and Osik Moses, eds. 2015. *The collected papers of Albert Einstein. Volume 14: The Berlin years: Writings and correspondence, April 1923–May 1925*. Princeton, New Jersey: Princeton University Press.

Kuhlmann, Meinard. 2018. Quantum field theory. The Stanford Encyclopedia of Philosophy (Winter 2018 Edition). https://plato.stanford.edu/archives/win2018/entries/quantum-field-theory/.

Lacki, Jan. 2009. Stueckelberg's covariant perturbation theory. In *E.C.G. Stueckelberg. An unconventional figure of twentieth century physics. Selected scientific papers with commentaries*, ed. by Henri Ruegg, Gérard Wanders, and Jan Lacki, 25–51. Basel, Boston, Berlin: Birkhäuser.

Lacki, Jan, Henri Ruegg, and Valentine Telegdi. 1999. The road to Stueckelberg's covariant perturbation theory as illustrated by successive treatments of Compton scattering. *Studies in History and Philosophy of Science Part B: Studies in History and Philosophy of Modern Physics* 30: 457–518.

Ladenburg, Rudolf. 1921. Die quantentheoretische Deutung der Zahl der Dispersionselektronen. *Zeitschrift für Physik* 4: 451–468.

Ladenburg, Rudolf, and Fritz Reiche. 1923. Absorption, Zerstreuung und Dispersion in der Bohrschen Atomtheorie. *Naturwissenschaften* 11: 584–598.

Landau, Lev, and Rudolf Peierls. 1931. Erweiterung des Unbestimmtheitsprinzips für die relativistische Quantentheorie. *Zeitschrift für Physik* 69: 56–69.

Landsberg, Grigori, and Leonid Mandelstam. 1928. Eine neue Erscheinung bei der Lichtzerstreuung. *Naturwissenschaften* 16: 557–558.

Langer, R. M. 1929. Incoherent scattering. *Nature* 123: 345.

Lauritsen, Charles, and Julius Robert Oppenheimer. 1934. On the scattering of the Th C gamma-rays. *Physical Review* 46: 80–81.

Loison, Laurent. 2016. Forms of presentism in the history of science. Rethinking the project of historical epistemology. *Studies in History and Philosophy of Science Part A* 60: 29–37.

Long, Derek A. 1988. Early history of the Raman effect. *International Reviews in Physical Chemistry* 7: 317–349.

MacKinnon, Edward. 1977. Heisenberg, models, and the rise of matrix mechanics. *Historical Studies in the Physical Sciences* 8: 137–188.

Majorana, Ettore. 1933. Über die Kerntheorie. *Zeitschrift für Physik* 82: 137–145.

Mandl, Franz, and Graham Shaw. 1984. *Quantum field theory*. Chichester, New York, Brisbane, Toronto, Singapore: Wiley.

Manneback, Charles. 1929. Die Intensität der sekundären Streustrahlung ("Raman-LinienâŁž). *Die Naturwissenschaften* 17: 364–365.

Martinez, Jean-Philippe. 2024. Virtuality in modern physics in the 1920s and 1930s: Meaning(s) of an emerging notion. *Perspectives on Science* 32: 350–371.

Masters, Barry R. 2013. The origins of Maria Göppert's dissertation on two-photon quantum transitions at Göttingen's institutes of physics 1920–1933. In *Traditions and transformations in the history of quantum physics. HQ-3: Third International Conference on the History of Quantum Physics, Berlin, June 28- July 2, 2010*, ed. by Jürgen Renn, Shaul Katzir, and Christoph Lehner, 209–230. Berlin: Edition Open Access.

McLennan, John C., and J. H. McLeod. 1929. The Raman effect with liquid oxygen, nitrogen, and hydrogen. *Nature* 123: 160.

Mehra, Jagdish, ed. 1973. *The physicist's conception of nature*. Dodrecht: D. Reidel Publishing Company.

Mehra, Jagdish. 1994. *The beat of a different drum. The life and science of Richard Feynman*. New York: Oxford University Press.

Mehra, Jagdish, and Helmut Rechenberg. 1982. *The historical development of quantum theory. Volume 2. The discovery of quantum mechanics 1925*. New York: Springer.

Mehra, Jagdish. and Helmut Rechenberg. 2000. *The historical development of quantum theory. Vol. 6: The completion of quantum mechanics 1926–1931. Part 1: The probability interpretation and the statistical transformation, the physical interpretation, and the empirical and mathematical foundations of quantum mechanics 1926–1932*. New York: Springer.

Meitner, Lise. 1934. Die Streuung harter Gamma-Strahlen. *Naturwissenschaften* 22: 174.

Meitner, Lise, and Heinrich Kösters. 1933. Über die Streuung kurzwelliger Gamma-Strahlen. *Zeitschrift für Physik* 84: 137–144.

Mercier, André. 1939. Sur l'énergie propre de l'électron. *Helvetica Physica Acta* 12: 55–75.

Mercier, André, and Torsten Gustafson. 1938. Remarque sur l'énergie propre de l'électron. *Comptes Rendus hebdomadaires de l'académie des sciences* 206: 1217–1219.

Meynell, Letitia. 2008. Why Feynman diagrams represent. *International Studies in the Philosophy of Science* 22: 39–59.

Meynell, Letitia. 2018. Picturing Feynman diagrams and the epistemology of understanding. *Perspectives on Science* 26: 459–481.

Miller, Arthur I. 1984. *Imagery in scientific thought. Creating 20th-century physics*. New York: Springer Science+Business Media.

Miller, Arthur I. 1994a. *Early quantum electrodynamics: a source book*. Cambridge: Cambridge University Press.

Miller, Arthur I. 1994b. Frame-setting essay. In *Early quantum electrodynamics: a source book*, ed. by Arthur I. Miller, 3–118. Cambridge: Cambridge University Press.

Mills, Robert. 1993. Tutorial on infinities in QED. In *Renormalization. From Lorentz to Landau (and beyond)*, ed. by Laurie Mark Brown, 57–86. New York: Springer.

Milonni, Peter W. 1994. *The quantum vacuum. An introduction to quantum electrodynamics*. San Diego: Academic Press.

Monaldi, Daniela. 2005. Life of μ: The observation of the spontaneous decay of mesotrons and its consequences, 1938–1947. *Annals of Science* 62: 419–455.

Monti, Dalida. 1995. Dirac's holes model: From proton to positron. *Nuncius* 10: 99–130.

Morgan, Mary S., and Margaret Morrison, eds. 1999. *Models as Mediators. Perspectives on natural and social science*. Cambridge: Cambridge University Press.

Morgan, Mary S., and M. Norton Wise. 2017. Narrative science and narrative knowing. Introduction to special issue on narrative science. *Studies in History and Philosophy of Science Part A* 62: 1–5.

Mott, Nevill Francis. 1982. Walter Heinrich Heitler, 2 January 1904–15 November 1981. *Biographical Memoires of the Fellows of the Royal Society* 28: 140–151.

Moyer, Donald F. 1981a. Evaluations of Dirac's electron, 1928–1932. *American Journal of Physics* 49: 1055–1062.

Moyer, Donald F. 1981b. Origins of Dirac's electron, 1925–1928. *American Journal of Physics* 49: 944–950.

Moyer, Donald F. 1981c. Vindications of Dirac's electron, 1932–1934. *American Journal of Physics* 49: 1120–1125.

Mukherji, Visvapriya. 1974. A history of the meson theory of nuclear forces from 1935 to 1952. *Archive for History of Exact Sciences* 13: 27–102.

Mulligan, Joseph. 2001. Karl Ferdinand Herzfeld. 1892–1978. *Biographical Memoirs of the National Academy of Sciences* 80: 1–24.

Nakane, Michiyo. 2019. Yoshikatsu Sugiura's contribution to the development of quantum physics in Japan. *Berichte zur Wissenschaftsgeschichte* 42: 338–356.

Navarro, Jaume. 2015. A peripheral centre. Early quantum physics at Cambridge. In *Sciences in the universities of Europe, nineteenth and twentieth centuries. Academic landscapes*, ed. by Ana Simoes, Maria Paula Diogo, and Kostas Gavroglu, 327–344. Dodrecht: Springer.

Nedelsky, Leo, and Julius Robert Oppenheimer. 1933. The production of positives by nuclear gamma-rays. *Physical Review* 44: 948–949.

Neumaier, Arnold. 2016. The physics of virtual partiles. Physics Forums Insights. https://www.physicsforums.com/insights/physics-virtual-particles/.

Nordsieck, Arnold. 1934. Neutron collisions and the beta-ray theory of Fermi. *Physical Review* 46: 234–235.

Oldenberg, Otto. 1928. Über ein Zusammenwirken von Stoß und Strahlung. *Zeitschrift für Physik* 51: 605–612.

Oldenberg, Otto. 1930. Über das Zusammenwirken von zwei Energieanteilen in einem Elementarvorgang. *Naturwissenschaften* 18: 789–791.

Oldofredi, Andrea, and Michael Esfeld. 2019. Observability, unobservability and the Copenhagen interpretation in Dirac's methodology of physics. *Quanta* 8: 68–87.

Oppenheimer, Julius Robert. 1930a. Note on the theory of the interaction of field and matter. *Physical Review* 35: 461–477.

Oppenheimer, Julius Robert. 1930b. On the theory of electrons and protons. *Physical Review* 35: 562–563.

Oppenheimer, Julius Robert. 1930c. Two notes on the probability of radiative transitions. *Physical Review* 35: 939–947.

Oppenheimer, Julius Robert. 1935. Note on charge and field fluctuations. *Physical Review* 47: 144–145.

Oppenheimer, Julius Robert, and Milton S. Plesset. 1933. On the production of the positive electron. *Physical Review* 44: 53–55.

Oseen, Carl Wilhelm. 1915. Das Bohrsche Atommodell und die Maxwellschen Gleichungen. *Physikalische Zeitschrift* 16: 395–405.

Pais, Abraham. 1986. *Inward bound. Of matter and forces in the physical world.* New York: Oxford University Press.

Pais, Abraham. 1987. Playing with equations, the Dirac way. In *Reminiscences about a Great Physicist: Paul Adrien Maurice Dirac*, ed. by Behram N. Kursunoglu and Eugene Wigner, 93–116. Cambridge: Cambridge University Press.

Pashby, Thomas. 2012. Dirac's prediction of the positron: A case study for the current realism debate. *Perspectives on Science* 20: 440–475.

Passon, Oliver. 2019. On the interpretation of Feynman diagrams, or, did the LHC experiments observe $H \rightarrow \gamma\gamma$? *European Journal for Philosophy of Science* 9: 20.

Pauli, Wolfgang. 1923. Über das thermische Gleichgewicht zwischen Strahlung und freien Elektronen. *Zeitschrift für Physik* 18: 272–286.

Pauli, Wolfgang. 1925a. Über den Einfluß der Geschwindigkeitsabhängigkeit der Elektronenmasse auf den Zeemaneffekt. *Zeitschrift für Physik* 31: 373–385.

Pauli, Wolfgang. 1925b. Ueber die Intensitäten der im elektrischen Feld erscheinenden Kombinationslinien. *Det Kgl. Danske Videnskabernes Selskab. Mathematisk-fysiske Meddelelser* 7: 1–20.

Pauli, Wolfgang. 1926. Quantentheorie. In *Quanten. Handbuch der Physik*, 1st ed., ed. by Hans Geiger and Karl Scheel, 1–278. Berlin: Springer.

Pauli, Wolfgang. 1933. Die allgemeinen Prinzipien der Wellenmechanik. In *Quantentheorie. Handbuch der Physik* 2nd ed., ed. by Hans Geiger and Karl Scheel, 83–272. Berlin: Springer.

Pauli, Wolfgang. 1936. On the quantization of the scalar relativistic wave equation. In *The theory of the positron and related topics: report of a seminar at the Institute for Advanced Study, 1935–36*, ed. by Wolfgang Pauli, 1–27. The Institute for Advanced Study: Princeton.

Pauli, Wolfgang, ed. 1936. *The theory of the positron and related topics: report of a seminar at the Institute for Advanced Study, 1935–36*. The Institute for Advanced Study: Princeton.

Pauli, Wolfgang. 1979. *Wissenschaftlicher Briefwechsel mit Bohr, Einstein, Heisenberg u.a. Band I: 1920–1929*, ed. by Armin Hermann, Karl von Meyenn and Victor Weisskopf. New York, Heidelberg, Berlin: Springer.

Pauli, Wolfgang. 1985. *Wissenschaftlicher Briefwechsel mit Bohr, Einstein, Heisenberg u.a. Band II: 1930–1939*, ed. by Karl von Meyenn. Berlin, Heidelberg, New York, Tokyo: Springer.

Pauli, Wolfgang. 1993. *Wissenschaftlicher Briefwechsel mit Bohr, Einstein, Heisenberg u.a. Band III: 1940-1949*, ed. by Karl von Meyenn. Berlin, Heidelberg: Springer.

Pauli, Wolfgang, and Markus Fierz. 1938. Zur Theorie der Emission langwelliger Lichtquanten. *Il Nuovo Cimento (1924–1942)* 15: 167–188.

Pauli, Wolfgang, and Victor Weisskopf. 1934. Über die Quantisierung der skalaren relativistischen Wellengleichungen. *Helvetica Physica Acta* 7: 709–731.

Pawlak, Alexander, and Karl-Martin Hartmann. 2004. Physik sakral. Interview mit Karl-Martin Hartmann. *Physik Journal* 3: 36–37.

Peierls, Rudolf. 1934. The vacuum in Dirac's theory of the positive electron. *Proceedings of the Royal Society of London A: Mathematical, Physical and Engineering Sciences* 146: 420–441.

Peierls, Rudolf. 1939. The meson. *Reports on Progress in Physics* 6: 78–94.

Peirce, Charles Sanders. 1902. Virtual. In *Dictionary of philosophy and psychology. Vol. 2*, ed. by James Mark Baldwin, 763–764. London: Macmillan and Co.

Peskin, Michael E., and Daniel V. Schroeder. 1995. *An introduction to Quantum Field Theory*. Boulder: Westview Press.

Pessoa JR, Osvaldo. 2022. The measurement problem. In *The Oxford handbook of quantum interpretations*, ed. by Olival Freire JR, Guido Bacciagaluppi, Olivier Darrigol, Christian Joas, Alexei Kojevnikov, and Osvaldo Pessoa JR, 281–302. Oxford: Oxford University Press.

Placzek, Georg. 1929. Zur Theorie des Ramaneffekts. *Zeitschrift für Physik* 58: 585–594.

Placzek, Georg. 1959. The Rayleigh and Raman scattering (Translation from Handbuch der Radiologie, VI, 2, 1934, p. 209–374, by Ann Werbin). Lawrence Radiation Laboratory, University of California.

Pringsheim, Peter. 1928. Der Ramaneffekt, ein neuer von C. V. Raman entdeckter Strahlungseffekt. *Naturwissenschaften* 16: 597–606.

R. P. 1936. The Quantum Theory of Radiation. *Nature* 138: 483–484.

Raman, Chandrasekhara Venkata. 1928. A new radiation. *Indian Journal of Physics* 2: 387–398.

Raman, Chandrasekhara Venkata. 1929. Investigations of the scattering of light. *Nature* 123: 50.

Raman, Chandrasekhara Venkata. 1929. Part II.-The Raman effect. Investigation of molecular structure by light scattering. *Transactions of the Faraday Society* 25: 781–792.

Raman, Chandrasekhara Venkata. 1930. The molecular scattering of light. Nobel Lecture, December 11, 1930. https://www.nobelprize.org/prizes/physics/1930/raman/lecture/.

Raman, Chandrasekhara Venkata, and Kariamanikkam Srinivasa Krishnan. 1928. A new class of spectra due to secondary radiation. *Part I. Indian Journal of Physics* 2: 399–419.

Raman, Chandrasekhara Venkata, and Kariamanikkam Srinivasa Krishnan. 1928. A new type of secondary radiation. *Nature* 121: 501–502.

Raman, Chandrasekhara Venkata, and Kariamanikkam Srinivasa Krishnan. 1928. The optical analogue of the Compton effect. *Nature* 121: 711.

Ramanathan, Kalpathi Ramakrishna. 1923. Electromagnetic theory of the scattering of light in fluids. *Proceedings of the Indian Association for the Cultivation of Science* 8: 181–198.

Rasetti, Franco. 1929. On the Raman effect in diatomic gases. *Proceedings of the National Academy of Sciences of the United States of America* 15: 234–237.

Rasetti, Franco. 1929. On the Raman effect in diatomic gases. II. *Proceedings of the National Academy of Sciences* 15: 515–519.

Rasetti, Franco. 1929. Raman effect in gases. *Nature* 123: 205.

Rasetti, Franco. 1929. Selection rules in the Raman effect. *Nature* 123: 757–759.

Rechenberg, Helmut. 2010. *Werner Heisenberg. Die Sprache der Atome. Leben und Wirken–Eine wissenschaftliche Biographie. Die "Fröhliche Wissenschaft" (Jugend bis Nobelpreis)*. Berlin, Heidelberg: Springer.

Redhead, Michael. 1988. A philosopher looks at quantum field theory. In *Philosophical foundations of quantum field theory*, ed. by Harvey R. Brown and Rom Harré, 9–24. New York: Oxford University Press.

Reich, Karin. 2011. Der erste Professor für Theoretische Physik an der Universität Hamburg: Wilhlem Lenz. In *Mathematics meets physics. A contribution to their interaction in the 19th and the first half of the 20th century*, ed. by Karl-Heinz Schlote and Martina Schneider, 89–143. Frankfurt am Main: Wissenschaftlicher Verlag Harri Deutsch GmbH.

Rheinberger, Hans-Jörg. 2002. *Experimentalsysteme und epistemische Dinge. Eine Geschichte der Proteinsynthese im Reagenzglas*, 2nd ed. Göttingen: Wallstein Verlag.

Rheinberger, Hans-Jörg. 2013. *Historische Epistemologie zur Einführung*, 3rd ed. Hamburg: Junius Verlag.

Röhl, Johannes. 2005. Zur Wirklichkeit virtueller Prozesses in der Teilchenphysik. In *Kreativität. XX. Deutscher Kongress für Philosophie 26.–30. September in Berlin. Sektionsbeiträge*, Vol. 1, 467–477. Berlin: Universitätsverlag der TU Berlin.

Roqué, Xavier. 1992. Møller scattering: a neglected application of early quantum electrodynamics. *Archive for History of Exact Sciences* 44: 197–264.

Roqué, Xavier. 1997. The manufacture of the positron. *Studies in History and Philosophy of Science Part B: Studies in History and Philosophy of Modern Physics* 28: 73–129.

Rorvig, Mordechai. 2020. OK, WTF are "virtual particles" and do they actually exist? *Motherboard. Tech by Vice.* https://www.vice.com/en_us/article/3az8g3/ok-wtf-are-virtual-particles-and-do-they-actually-exist.

Rosenfeld, Léon. 1929. Über die longitudinalen Eigenlösungen der Heisenberg-Paulischen elektromagnetischen Gleichungen. *Zeitschrift für Physik* 58: 540–555.

Rosenfeld, Léon. 1930. Über die Gravitationswirkung des Lichtes. *Zeitschrift für Physik* 65: 589–599.

Rosenfeld, Léon. 1931. Zur Kritik der Diracschen Strahlungstheorie. *Zeitschrift für Physik* 70: 454–462.

Rosenfeld, Léon. 1932. La théorie quantique des champs. *Annales de l'institut Henri Poincaré* 2: 25–91.

Rosenfeld, Léon. 1932. Über eine mögliche Fassung des Diracschen Programms zur Quantenelektrodynamik und deren formalen Zusammenhang mit der Heisenberg-Paulischen Theorie. *Zeitschrift für Physik* 76: 729–734.

Rosenfeld, Léon. 1979. Quantum theory in 1929: Recollections from the first Copenhagen Conference (initially published 1929). In *Selected Papers of Léon Rosenfeld*, ed. by Robert S. Cohen and John J. Stachel, 302–312. Dodrecht: Springer.

Rosenfeld, Léon, and Jacques Solomon. 1931. Sur la théorie quantique du rayonnement. *Le journal de physique et le radium* 2: 139–147.

Rovenchak, Andrij. 2013. Bibliography of the department for theoretical physics, University of Lviv, in 1914–1939. *Journal of Physical Studies* 17: 3002(13p.).

Rueger, Alexander. 1992. Attitudes towards infinities: Responses to anomalies in quantum electrodynamics, 1927–1947. *Historical Studies in the Physical Sciences* 22: 309–337.

Sachs, Robert G. 1979. Maria Goeppert Mayer. 1906–1972. *Biographical Memoirs of the National Academy of Sciences* 50: 311–328.

Sakurai, Jun John. 1967. *Advanced quantum mechanics*. Reading, Massachusetts: Addison-Wesley Publishing.

Schiemann, Gregor. 2008. *Werner Heisenberg*. München: C.H. Beck.

Schirrmacher, Arne. 2019. *Establishing quantum physics in Göttingen. David Hilbert, Max Born, and Peter Debye in context, 1900–1926*. Cham: Springer Nature Switzerland.

Schrödinger, Erwin. 1926. Quantisierung als Eigenwertproblem. *Erste Mitteilung. Annalen der Physik* 384: 361–376.

Schrödinger, Erwin. 1926. Quantisierung als Eigenwertproblem. *Vierte Mitteilung. Annalen der Physik* 386: 109–139.

Schrödinger, Erwin. 1931. Zur Quantendynamik des Elektrons. Sitzungsberichte der Preussischen Akademie der Wissenschaften. *Physikalisch-Mathematische Klasse* 3: 63–72.

Schweber, Silvan S. 1986. The empiricist temper regnant: Theoretical physics in the United States 1920–1950. *Historical Studies in the Physical and Biological Sciences* 17: 55–98.

Schweber, Silvan S. 1994. *QED and the men who made it: Dyson, Feynman, Schwinger, and Tomonaga*. Princeton, New Jersey: Princeton University Press.

Schweber, Silvan S. 2002. Enrico Fermi and quantum electrodynamics, 1929–1932. *Physics Today* 55: 31–36.

Schweber, Silvan S. 2012. *Nuclear forces. The making of the physicist Hans Bethe*. Cambridge, Mass; London, England: Harvard University Press.

Segrè, Emilio. 1970. *Enrico Fermi. Physicist*. Chicago, London: University of Chicago Press.

Serber, Robert. 1935. Linear modifications in the Maxwell field equations. *Physical Review* 48: 49–54.

Serber, Robert. 1936. A note on positron theory and proper energies. *Physical Review* 49: 545–550.

Serwer, Daniel. 1977. Unmechanischer Zwang: Pauli, Heisenberg, and the rejection of the mechanical atom, 1923–1925. *Historical Studies in the Physical Sciences* 8: 189–256.

Seth, Suman. 2010. *Crafting the quantum: Arnold Sommerfeld and the practice of theory, 1890–1926*. Cambridge, Massachusetts: MIT Press.

Shrader-Frechette, Kristin. 1977. Atomism in crisis: An analysis of the current high energy paradigm. *Philosophy of Science* 44: 409–440.

Singh, Rajinder. 2002. C. V. Raman and the discovery of the Raman effect. *Physics in Perspective* 4: 399–420.

Singh, Rajinder. 2004. *Nobel Laureate C. V. Raman's work on light scattering. Historical contribution to a scientific biography*. Berlin: Logos Verlag.

Singh, Rajinder, and Falk Riess. 2001. The 1930 Nobel Prize for physics: A close decision? *Notes and Records of the Royal Society of London* 55: 267–283.

Slater, John Clarke. 1925. A quantum theory of optical phenomena. *Physical Review* 25: 395–428.

Slater, John Clarke. 1927. Action of radiation and perturbations on atoms. *Proceedings of the National Academy of Sciences of the United States of America* 13: 104–111.

Slater, John Clarke. 1927. Radiation and absorption on Schrödinger's theory. *Proceedings of the National Academy of Sciences of the United States of America* 13: 7–12.

Smekal, Adolf. 1923a. Bericht über Versuch einer allgemeinen, einheitlichen Anwendung der Quantentheorie und einer Quantentheorie der Dispersion. *Die Naturwissenschaften* 11: 411–412.

Smekal, Adolf. 1923b. Zur Quantentheorie der Dispersion. *Die Naturwissenschaften* 11: 873–875.

Smekal, Adolf. 1925. Über "metastationäreâŁž Atom- und Molekülzustände. *Zeitschrift für Physik* 34: 81–93.

Smekal, Adolf. 1925. Zur Quantentheorie der Streuung und Dispersion. *Zeitschrift für Physik* 32: 241–244.

Smekal, Adolf. 1926. *Allgemeine Grundlagen der Quantenstatistik und Quantentheorie*. Leipzig, Berlin: B. G. Teubner.

Smekal, Adolf. 1926. Zur Quantenstatistik der Hohlraumstrahlung und ihrer Wechselwirkungen mit der Materie. *Zeitschrift für Physik* 37: 319–341.

Smekal, Adolf. 1928. Über die Inversprozesse spontaner strahlungsloser Quantenvorgänge (strahlungslose Zweierstoßbindung). *Annalen der Physik* 392: 959–964.

Smekal, Adolf. 1928. Zur Quantentheorie der Streuung und Dispersion. *Naturwissenschaften* 16: 612–613.

Smith, Alice Kimball, and Charles Weiner, eds. 1981. *Robert Oppenheimer. Letters and recollections*. Cambridge, Massachusetts: Harvard Universtiy Press.

Sokolov, Arseny. 1941. Die Streuung der Mesonen unter Berücksichtigung der Dämpfung. *Journal of Physics (Akademija Nauk SSSR)* 5: 231–237.

Sommerfeld, Arnold. 1924. Grundlagen der Quantentheorie und des Bohrschen Atommodells. *Die Naturwissenschaften* 12: 1047–1049.

Sondheimer, E. H. 1999. Sir Alan Herries Wilson. *Biographical Memoires of the Fellows of the Royal Society* 45: 549–562.

Sopka, Katherine Russell. 1988. *Quantum physics in America. The years through 1935*. New York: American Institute of Physics.

Steinle, Friedrich. 2009. Scientific change and empirical concepts. *Centaurus* 51: 305–313.

Steinle, Friedrich. 2024. How to conceive of virtual entities: Peirce's proposal. *Perspectives on Science* 32: 269–277.

Stöltzner, Michael. 2017. Feynman diagrams as models. *The Mathematical Intelligencer* 39: 46–54.

Stöltzner, Michael. 2017. The variety of explanations in the Higgs sector. *Synthese* 194: 433–460.

Stöltzner, Michael. 2018. Feynman diagrams: Modeling between physics and mathematics. *Perspectives on Science* 26: 482–500.

Stolzenburg, Klaus, ed. 1984. *The Emergence of Quantum Mechanics (Mainly 1924–1926). Niels Bohr: Collected Works,* Vol. 5. Amsterdam, New York, Oxford, Tokyo: North-Holland.

Stoner, E. C. 1926. VII. X-ray term values, absorption limits, and critical potentials. *The London, Edinburgh, and Dublin Philosophical Magazine and Journal of Science* 2: 97–113.

Strassler, Matt. 2011. Virtual particles: What are they? Blog: Of Particular Significance. https://profmattstrassler.com/articles-and-posts/particle-physics-basics/virtual-particles-what-are-they/.

Stueckelberg, Ernst Carl, and Gerlach. 1934. Relativistisch invariante Störungstheorie des Diracschen Elektrons I. Teil: Streustrahlung und Bremsstrahlung. *Annalen der Physik* 413: 367–389.

Stuewer, Roger Harry. 1975. *The Compton effect: Turning point in physics*. New York: Science History Pbulications.

Stuewer, Roger Harry. 2018. *The age of innocence. Nuclear physics between the First and Second World Wars*. Oxford: Oxford University Press.

Suárez, Mauricio. 2004. An inferential conception of scientific representation. *Philosophy of Science* 71: 767–779.

Suárez, Mauricio, ed. 2009. *Fictions in science. Philosophical essays on modelling and idealization*. New York: Taylor and Francis.

Tamm, Igor. 1930. Eine Bemerkung zur Diracschen Theorie der Lichtzerstreuung und Dispersion. *Zeitschrift für Physik* 65: 705–708.

Tamm, Igor. 1930. Über die Wechselwirkung der freien Elektronen mit der Strahlung nach der Diracschen Theorie des Elektrons und nach der Quantenelektrodynamik. *Zeitschrift für Physik* 62: 545–568.

Tamm, Igor. 1934. Exchange forces between neutrons and protons, and Fermi's Theory. *Nature* 133: 981.

Teller, Paul. 1995. *An interpretive introduction to quantum field theory*. Princeton, New Jersey: Princeton University Press.

The GW Astrophysics Group. 2022. Washington Conferences on Theoretical Physics. https://home.gwu.edu/~kargaltsev/HEA/washington-conferences.html.

Uehling, Edwin Albrecht. 1935. Polarization effects in positron theory. *Physical Review* 48: 55–63.

Vaihinger, Hans. 1911. *Die Philosophie des Als Ob. System der theoretische, praktischen und religiösen Fiktionen der Menschheit auf Grund eines idealisitschen Positivismus; mit einem Anhang über Kant und Nietzsche.* Berlin: Reuther und Reichard.

van der Waerden, Bartel, and Leendert, eds. 1967. *Sources of quantum mechanics.* Amsterdam: North-Holland.

van Vleck, John, and Hasbrouck. 1929. On the vibrational selection principles in the Raman effect. *Proceedings of the National Academy of Sciences* 15: 754–764.

Venkataraman, Ganeshan. 1995. *Raman and his effect.* Hyderguda: Universities Press (India) Private Limited.

von Meyenn, Karl, ed. 2011. *Eine Entdeckung von ganz außerordentlicher Tragweite. Schrödingers Briefwechsel zur Wellenmechanik und zum Katzenparadoxon,* Vol. 1. Berlin: Springer.

von Neumann, John. 1932. *Mathematische Grundlagen der Quantenmechanik.* Berlin: Verlag von Julius Springer.

von Weizsäcker, Carl, and Friedrich. 1934. Ausstrahlung bei Stößen sehr schneller Elektronen. *Zeitschrift für Physik* 88: 612–625.

von Weizsäcker, Carl, and Friedrich. 1936. Über die Spinabhängigkeit der Kernkräfte. *Zeitschrift für Physik* 102: 572–602.

von Weizsäcker, Carl, and Friedrich. 1937. *Die Atomkerne.* Leipzig: Akademische Verlagsgesellschaft.

Walker, Mark. 1992. Physics and propaganda: Werner Heisenberg's foreign lectures under National Socialism. *Historical Studies in the Physical and Biological Sciences* 22: 339–389.

Waller, Ivar. 1927. CIII. On the scattering of radiation from atoms. *The London, Edinburgh, and Dublin Philosophical Magazine and Journal of Science* 4: 1228–1237.

Waller, Ivar. 1928. Über eine verallgemeinerte Streuungsformel. *Zeitschrift für Physik* 51: 213–231.

Waller, Ivar. 1929. Die Streuung kurzwelliger Strahlung durch Atome nach der Diracschen Strahlungstheorie. *Zeitschrift für Physik* 58: 75–94.

Waller, Ivar. 1930. Bemerkungen über die Rolle der Eigenenergie des Elektrons in der Quantentheorie der Strahlung. *Zeitschrift für Physik* 62: 673–676.

Waller, Ivar. 1930. Die Streuung von Strahlung durch gebundene und freie Elektronen nach der Diracschen relativistischen Mechanik. *Zeitschrift für Physik* 61: 837–851.

Waller, Ivar. 1934. Über die Rückwirkung der Strahlung bei der Streuung durch freie Elektronen. *Zeitschrift für Physik* 88: 436–448.

Weinberg, Steven. 1977. The search for unity: Notes for a history of quantum field theory. *Daedalus* 106: 17–35.

Weingard, Robert. 1982. Do virtual particles exist. *PSA: Proceedings of the Biennial meeting of the philosophy of science association*, 235–242.

Weingard, Robert. 1988. Virtual particles and the interpretation of quantum field theory. In *Philosophical foundations of quantum field theory*, ed. by Harvey R. Brown and Rom Harré, 43–58. New York: Oxford University Press.

Weisskopf, Victor. 1931. Zur Theorie der Resonanzfluoreszenz. *Annalen der Physik* 401: 23–66.

Weisskopf, Victor. 1933. Die Streuung des Lichts an angeregten Atomen. *Zeitschrift für Physik* 85: 451–481.

Weisskopf, Victor. 1934. Über die Selbstenergie des Elektrons. *Zeitschrift für Physik* 89: 27–39.

Weisskopf, Victor. 1935. Probleme der neueren Quantentheorie des Elektrons (Schluß). *Die Naturwissenschaften* 23: 669–674.

Weisskopf, Victor. 1936. Über die Elektrodynamik des Vakuums auf Grund der Quantentheorie des Elektrons. *Det Kgl. Danske Videnskabernes Selskab. Mathematisk-fysiske Meddelelser* 14: 1–40.

Weisskopf, Victor. 1939. On the self-energy and the electromagnetic field of the electron. *Physical Review* 56: 72–85.

Weisskopf, Victor. 1949. Recent developments in the theory of the electron. *Reviews of Modern Physics* 21: 305–315.

Weisskopf, Victor. 1983. Growing up with field theory: the development of quantum electrodynamics. In *The birth of particle physics. Based on a Fermilab Symposium*, ed. by Laurie Mark Brown and Lillian Hoddeson, 56–81. Cambridge: Cambridge University Press.

Weisskopf, Victor. 1991. *The joy of insight. The passions of a physicist*. New York: Basic Books.

Weisskopf, Victor, and Eugene Wigner. 1930. Berechnung der natürlichen Linienbreite auf Grund der Diracschen Lichttheorie. *Zeitschrift für Physik* 63: 54–73.

Weisskopf, Victor, and Eugene Wigner. 1930. Über die natürliche Linienbreite in der Strahlung des harmonischen Oszillators. *Zeitschrift für Physik* 65: 18–29.

Welton, Theodore Allen. 1948. Some observable effects of the quantum-mechanical fluctuations of the electromagnetic field. *Physical Review* 74: 1157–1167.

Wentzel, Gregor. 1924. Zur Quantenoptik. *Zeitschrift für Physik* 22: 193–199.

Wentzel, Gregor. 1925. Die Theorien des Compton-Effektes. *I. Physikalische Zeitschrift* 26: 436–454.

Wentzel, Gregor. 1927. Über strahlungslose Quantensprünge. *Zeitschrift für Physik* 43: 524–530.

Wentzel, Gregor. 1933. Über die Eigenkräfte der Elementarteilchen. *I. Zeitschrift für Physik* 86: 479–494.

Wentzel, Gregor. 1933. Über die Eigenkräfte der Elementarteilchen. *II. Zeitschrift für Physik* 86: 635–645.

Wentzel, Gregor. 1933c. Wellenmechanik der Stoß- und Strahlungsvorgänge. In *Quantentheorie. Handbuch der Physik*, 2nd ed., ed. by Hans Geiger and Karl Scheel, 695–784. Berlin: Springer.

Wentzel, Gregor. 1934. Über die Eigenkräfte der Elementarteilchen. *III. Zeitschrift für Physik* 87: 726–733.

Wentzel, Gregor. 1934. Zur Frage der Äquivalenz von Lichtquanten und Korpuskelpaaren. *Zeitschrift für Physik* 92: 337–358.

Wentzel, Gregor. 1935. Zur Frage der Stabilität des Protons und des Neutrons. *Die Naturwissenschaften* 23: 35–36.

Wentzel, Gregor. 1937. Zur Frage der Beta-Wechselwirkung. *Helvetica Physica Acta* 10: 107–111.

Wentzel, Gregor. 1937. Zur Theorie der Beta-Umwandlung und der Kernkräfte. *I. Zeitschrift für Physik* 104: 34–47.

Wentzel, Gregor. 1937. Zur Theorie der Beta-Umwandlung und der Kernkräfte. *II. Zeitschrift für Physik* 105: 738–746.

Wentzel, Gregor. 1938. Schwere Elektronen und Theorien der Kernvorgänge. *Die Naturwissenschaften* 26: 273–279.

Wentzel, Gregor. 1943. *Einführung in die Quantentheorie der Wellenfelder*. Wien: Franz Deuticke.

Weyl, Hermann. 1929. Elektron und Gravitation. *I. Zeitschrift für Physik* 56: 330–352.

Weyl, Hermann. 1931. *Gruppentheorie und Quantenmechanik*, 2nd ed. Leipzig: Verlag von S. Hirzel.

Wheeler, John Archibald, and Richard Phillips Feynman. 1949. Classical electrodynamics in terms of direct interparticle action. *Reviews of Modern Physics* 21: 425–433.

Wick, Gian-Carlo. 1935. Teoria dei raggi Beta e momento magneitco del protone. Rendiconti Accademia Nazionale dei Lincei. *Classe di Scienze Fisiche, Matematiche e Naturali* 21: 170–173.

Wick, Gian-Carlo. 1938. Range of nuclear forces in Yukawa's theory. *Nature* 142: 993–994.

Wigner, Eugene. 1933. On the mass defect of helium. *Physical Review* 43: 252–257.

Williams, Evan James. 1933. Applications of the method of impact parameter in collisions. *Proceedings of the Royal Society of London A: Mathematical, Physical and Engineering Sciences* 139: 163–186.

Williams, Evan James. 1935. Correlation of certain collision problems with radiation theory. Det Kgl. *Danske Videnskabernes Selskab. Mathematisk-fysiske Meddelelser.* 13: 3–50.

Wilson, Alan Herries. 1941. The quantum theory of radiation damping. *Proceedings of the Cambridge Philosophical Society* 37: 301–316.

Wood, Robert Williams. 1929. The Raman effect with hydrochloric acid gas: the 'Missing Line'. *Nature* 123: 279.

Wood, Robert Williams, and Gerhard Heinrich Dieke. 1930. The Raman effect in HCL gas. *Physical Review* 35: 1355–1359.

Wright, Aaron Sidney. 2016. A beautiful sea: P. A. M. Dirac's epistemology and ontology of the vacuum. *Annals of Science* 73: 225–256.

Wright, Aaron Sidney. 2024. *More than nothing. A history of the vacuum in theoretical physics 1925–1980*. New York: Oxford University Press.

Wüthrich, Adrian. 2010. *The genesis of Feynman diagrams*. Dodrecht, Heidelberg, London, New York: Springer.

Wüthrich, Adrian. 2012. Interpreting Feynman diagrams as visual models. *Spontaneous Generations: A Journal for the History and Philosophy of Science* 6: 172–181.

Wüthrich, Adrian. 2018. The exigencies of war and the stink of a theoretical problem: Understanding the genesis of Feynman's quantum electrodynamics as mechanistic modelling at different levels. *Perspectives on Science* 26: 501–520.

Yazaki, Yuji. 2017. How the Klein-Nishina formula was derived: Based on the Sangokan Nishina Source Materials. *Proceedings of the Japan Academy. Series B. Physical and Biological Sciences* 93: 399–421.

Yukawa, Hideki. 1935. On the interaction of elementary particles. I. *Proceedings of the Physico-Mathematical Society of Japan (3rd Series)* 17: 48–57.

Yukawa, Hideki, and Shoichi Sakata. 1937. On the interaction of elementary particles II. *Proceedings of the Physico-Mathematical Society of Japan (3rd Series)* 19: 1084–1093.

Yukawa, Hideki, Shoichi Sakata, and Mitsuo Taketani. 1938. On the interaction of elementary particles. III. *Proceedings of the Physico-Mathematical Society of Japan (3rd Series)* 20: 319–340.

Index

M. Ehberger, *Representing the Unobservable*, Science Networks. Historical Studies 68,
https://doi.org/10.1007/978-3-032-09188-8

MIX
Papier aus verantwortungsvollen Quellen
Paper from responsible sources
FSC® C105338

If you have any concerns about our products,
you can contact us on
ProductSafety@springernature.com

In case Publisher is established outside the EU,
the EU authorized representative is:
Springer Nature Customer Service Center GmbH
Europaplatz 3, 69115 Heidelberg, Germany

Printed by Libri Plureos GmbH
in Hamburg, Germany